X-Ray Free-Electron Laser

Special Issue Editor
Kiyoshi Ueda

MDPI • Basel • Beijing • Wuhan • Barcelona • Belgrade

MDPI

Special Issue Editor
Kiyoshi Ueda
Tohoku University
Japan

Editorial Office
MDPI
St. Alban-Anlage 66
Basel, Switzerland

This edition is a reprint of the Special Issue published online in the open access journal *Applied Sciences* (ISSN 2076-3417) from 2017–2018 (available at: http://www.mdpi.com/journal/applsci/special_issues/x_ray_fel).

For citation purposes, cite each article independently as indicated on the article page online and as indicated below:

Lastname, F.M.; Lastname, F.M. Article title. *Journal Name* **Year**, *Article number*, page range.

First Edition 2018

ISBN 978-3-03842-879-4 (Pbk)
ISBN 978-3-03842-880-0 (PDF)

Table of Contents

About the Special Issue Editor vii

Preface to "X-Ray Free-Electron Laser" ix

Bart Faatz, Markus Braune, Olaf Hensler, Katja Honkavaara, Raimund Kammering,
    Marion Kuhlmann, Elke Ploenjes, Juliane Roensch-Schulenburg, Evgeny Schneidmiller,
    Siegfried Schreiber, et al.
    The FLASH Facility: Advanced Options for FLASH2 and Future Perspectives
    doi:10.3390/app7111114 1

R. W. Schoenlein, S. Boutet, M. P. Minitti and A.M. Dunne
    The Linac Coherent Light Source: Recent Developments and Future Plans
    doi:10.3390/app7080850 11

Makina Yabashi, Hitoshi Tanaka, Kensuke Tono and Tetsuya Ishikawa
    Status of the SACLA Facility
    doi:10.3390/app7060604 31

Luca Giannessi and Claudio Masciovecchio
    FERMI: Present and Future Challenges
    doi:10.3390/app7060640 41

In Soo Ko, Heung-Sik Kang, Hoon Heo, Changbum Kim, Gyujin Kim, Chang-Ki Min,
    Haeryong Yang, Soung Youl Baek, Hyo-Jin Choi, Geonyeong Mun, et al.
    Construction and Commissioning of PAL-XFEL Facility
    doi:10.3390/app7050479 57

Thomas Tschentscher, Christian Bressler, Jan Grünert, Anders Madsen, Adrian P. Mancuso,
    Michael Meyer, Andreas Scherz, Harald Sinn and Ulf Zastrau
    Photon Beam Transport and Scientific Instruments at the European XFEL
    doi:10.3390/app7060592 68

Christopher J. Milne, Thomas Schietinger, Masamitsu Aiba, Arturo Alarcon, Jürgen Alex,
    Alexander Anghel, Vladimir Arsov, Carl Beard, Paul Beaud, Simona Bettoni, et al.
    SwissFEL: The Swiss X-ray Free Electron Laser
    doi:10.3390/app7070720 103

Zhentang Zhao, Dong Wang, Qiang Gu, Lixin Yin, Ming Gu, Yongbin Leng and Bo Liu
    Status of the SXFEL Facility
    doi:10.3390/app7060607 160

Wolfram Helml, Ivanka Grguraš, Pavle N. Juranić, Stefan Düsterer, Tommaso Mazza,
    Andreas R. Maier, Nick Hartmann, Markus Ilchen, Gregor Hartmann, Luc Patthey, et al.
    Ultrashort Free-Electron Laser X-ray Pulses
    doi:10.3390/app7090915 173

Jakub Szlachetko, Maarten Nachtegaal, Daniel Grolimund, Gregor Knopp, Sergey Peredkov, Joanna Czapla–Masztafiak and Christopher J. Milne
A Dispersive Inelastic X-ray Scattering Spectrometer for Use at X-ray Free Electron Lasers
doi:10.3390/app7090899 219

Sergey Usenko, Andreas Przystawik, Leslie Lamberto Lazzarino, Markus Alexander Jakob, Florian Jacobs, Christoph Becker, Christian Haunhorst, Detlef Kip and Tim Laarmann
Split-And-Delay Unit for FEL Interferometry in the XUV Spectral Range
doi:10.3390/app7060544 229

Sandeep Kumar, Alexandra S. Landsman and Dong Eon Kim
Terawatt-Isolated Attosecond X-ray Pulse Using a Tapered X-ray Free Electron Laser
doi:10.3390/app7060614 241

Edwin Kukk, Koji Motomura, Hironobu Fukuzawa, Kiyonobu Nagaya and Kiyoshi Ueda
Molecular Dynamics of XFEL-Induced Photo-Dissociation, Revealed by Ion-Ion Coincidence Measurements
doi:10.3390/app7050531 252

Thomas J. A. Wolf, Fabian Holzmeier, Isabella Wagner, Nora Berrah, Christoph Bostedt, John Bozek, Phil Bucksbaum, Ryan Coffee, James Cryan, Joe Farrell, et al.
Observing Femtosecond Fragmentation Using Ultrafast X-ray-Induced Auger Spectra
doi:10.3390/app7070681 266

Li Fang, Hui Xiong, Edwin Kukk and Nora Berrah
X-ray Pump–Probe Investigation of Charge and Dissociation Dynamics in Methyl Iodine Molecule
doi:10.3390/app7050529 277

Carlo Callegari, Tsukasa Takanashi, Hironobu Fukuzawa, Koji Motomura, Denys Iablonskyi, Yoshiaki Kumagai, Subhendu Mondal, Tetsuya Tachibana, Kiyonobu Nagaya, Toshiyuki Nishiyama, et al.
Application of Matched-Filter Concepts to Unbiased Selection of Data in Pump-Probe Experiments with Free Electron Lasers
doi:10.3390/app7060621 288

Shingo Yamamoto and Iwao Matsuda
Measurement of the Resonant Magneto-Optical Kerr Effect Using a FreeElectron Laser
doi:10.3390/app7070662 297

Zhibin Sun, Jiadong Fan, Haoyuan Li and Huaidong Jiang
Current Status of Single Particle Imaging with X-ray Lasers
doi:10.3390/app8010132 320

Nastasia Mukharamova, Sergey Lazarev, Janne-Mieke Meijer, Matthieu Chollet, Andrej Singer, Ruslan P. Kurta, Dmitry Dzhigaev, Oleg Yu. Gorobtsov, Garth Williams, Diling Zhu, et al.
Probing Dynamics in Colloidal Crystals with Pump-Probe Experiments at LCLS: Methodology and Analysis
doi:10.3390/app7050519 348

Toshiaki Inada, Takayuki Yamazaki, Tomohiro Yamaji, Yudai Seino, Xing Fan, Shusei Kamioka, Toshio Namba, Shoji Asai
Probing Physics in Vacuum Using an X-ray Free-Electron Laser, a High-Power Laser, and a High-Field Magnet
doi:10.3390/app7070671 361

William Hanks, John T. Costello and Lampros A.A. Nikolopoulos
Two- and Three-Photon Partial Photoionization Cross Sections of Li^+, Ne^{8+} and Ar^{16+} under XUV Radiation
doi:10.3390/app7030294 . . . . **373**

Krzysztof Tyrała, Klaudia Wojtaszek, Marek Pajek, Yves Kayser, Christopher Milne, Jacinto Sá and Jakub Szlachetko
State-Population Narrowing Effect in Two-Photon Absorption for Intense Hard X-ray Pulses
doi:10.3390/app7070653 . . . . **387**

Keisuke Hatada and Andrea Di Cicco
Modeling Non-Equilibrium Dynamics and Saturable Absorption Induced by Free Electron Laser Radiation
doi:10.3390/app7080814 . . . . **395**

Adam Kirrander and Peter Weber
Fundamental Limits on Spatial Resolution in Ultrafast X-ray Diffraction
doi:10.3390/app7060534 . . . . **408**

Sung Soon Kim, Sandi Wibowo, and Dilano K. Saldin
Algorithm for Reconstruction of 3D Images of Nanorice Particles from Diffraction Patterns of Two Particles in Independent Random Orientations with an X-ray Laser
doi:10.3390/app7070646 . . . . **427**

Mats Larsson
Nobel Symposium on Free Electron Laser Research
doi:10.3390/app7040408 . . . . **440**

About the Special Issue Editor

Kiyoshi Ueda, Professor of Tohoku University, completed a graduate course at Kyoto University in 1982, receiving his PhD. Since 1982, he has been working at Tohoku University, as an assistant professor (1982–1990), as an associate professor (1990–2003), and as a full professor (2003–present). He was also a visiting scientist at University of Maryland in the United States (1985–1987), an invited senior scientist at Daresbury Laboratory in the United Kingdom (1992–1993), and an invited professor of Universite Paris Sud in France (1998). He has been working in the field of molecular science; his areas of specialization include ultrafast dynamic imaging, probing ultrafast electron and nuclear dynamics, and developing new methods for these studies. He has been the author or coauthor of about 500 papers, including more than ten review articles and chapters of books and giving more than 100 invited and plenary lectures in international conferences, workshops, symposiums, and seminars.

Preface to "X-Ray Free-Electron Laser"

During the last decades, the advent of the short-wavelength Free Electron Lasers (FELs) in the range from extreme ultraviolet (XUV) to hard x-rays has opened a new research avenue for investigations of ultrafast electronic and structural dynamics in any form of matter. FELs deliver coherent laser pulses, combining unprecedented power densities up to 1020 W/cm2 and extremely short pulse durations down to a few femtoseconds, offering important advantages over conventional short-wavelength light sources for many applications. Time-resolved spectroscopic and structural studies on the timescale of femtoseconds allow us to probe electrons and atoms in action. Indeed, FELs have been applied to the study of ultrafast charge transfer in a molecule and a molecular complex, chemical bond breaking and formation, and non-thermal phase transitions in solids. The intense, coherent, focused FEL pulse makes single-shot diffraction imaging of non-crystalized biomolecules and nanometer-size objects a reality. On the other hand, since the FEL pulses are entering a new regime of intensities, they are opening a new research field of studying interaction between intense short wavelength laser pulses and various forms of matter. The extremely intense FEL pulse strips so many electrons from an isolated atom, leads to a violent Coulomb explosion of an isolated molecule, and instantaneously transforms a nonometer-size object into a dense nano-plasma and a single crystal into completely new, disordered matter. Furthermore, rapidly developing FEL technologies make not only fully coherent FEL pulses available routinely but also pulse shaping and phase-controlling of multi-color harmonic pulses a reality, opening other novel research areas of short-wavelength, non-linear, four-wave mixing spectroscopy and attosecond coherent control.

The aim of the present special issue is to provide an overview of recent developments of XFELs and science in this area, and to foresee the future. For this purpose, this issue features reports on the current status and future plans of all eight XFEL facilities in the world. Namely, Faatz et al. describe the FLASH (the Free Electron LASer in Hamburg) in Germany, focusing on advanced options for FLASH2 and future perspectives [1]; Schoenlein et al. describe recent developments and future plans of LCLS (the Linac Coherent Light Source) in the United Sates [2]; and Yabashi et al. describe the status and future plans of SACLA (the Spring-8 Angstrom Compact free electron LAser) in Japan [3]. Giannessi and Masciovecchio discuss present and future challenges at FERMI (the Free Electron laser Radiation for Multidisciplinary Investigations) in Italy, focusing on phase-coherent multicolor pulse generations as a unique feature of FERMI [4]. Besides these four facilities that have been in operation for the last decade, three new facilities started operations for users in the last couple of years. This issue includes the first reports of these three facilities. Ko et al. describe the construction and commissioning of PAL-XFEL (the Pohang Accelerator Laboratory X-ray Free Electron Laser) in Korea [5], Tschentscher et al. describe photon beam transport and scientific instruments at the European XFEL (European X-ray Free Electron Laser) [6], and Milne et al describes SwissFEL (Swiss x-ray Free Electron Laser) [7]. In addition, Zhao et al. report on the status of the Shanghai SXFEL (Soft X-ray Free Electron Laser) project in China [8].

The present issue also includes XFEL-pulse characterizations and instrumentations at the XFELs. Helml et al. review progress on ultrashort pulse characterization at XFELs [9], while Inbushi et al. communicate on measurements of the X-ray spectra of XFEL (SACLA) with a wide-range high-resolution single-shot spectrometer, aiming at measurements of the temporal pulse duration [10]. Szlchetko et al. describe a dispersive, inelastic x-ray scattering spectrometer at XFELs [11], while Usenko et al. describe the split-and-delay unit for FEL interferometry in the XUV spectral range [12].

Kumar et al. discuss a plan for terawatt-isolated, attosecond X-ray pulse generation using a tapered XFEL (PAL-XFEL) [13].

Sciences at XFELs are also included in this issue. Kukk et al. review molecular dynamics of XFEL-induced photo-dissociation, placing an emphasis on the data analysis of ion-ion coincidence measurements [14]; Wolf et al. discuss observing the femtosecond fragmentation of the thymine molecule using ultrashort X-ray-induced Auger spectra [15], while Fang et al. describe the X-ray pump-probe investigation of the charge and dissociation dynamics in the methyl iodine molecule [16]. Callegari et al. describe the application of matched-filter concepts to an unbiased selection of data in pump-probe experiments with FELs [17]. Yamamoto and Matsuda review measurements of the resonant magneto-optical Kerr effect using FEL [18], Sun et al. review the current status of single particle imaging with XFELs [19], while Mukharanova et al. discuss methodology and analysis for probing dynamics in colloidal crystals with pump-probe experiments at LCLS [20]. Inada et al. review probing physics in vacuum using XFEL, a high-power laser, and a high-field magnet [21].

Theoretical developments related to science with XFELs are also included in this issue. Hanks et al. report two- and three-photon photoionization cross sections of Li+, Ne8+, and Ar16+ under XUV radiation [22]; Tyarla et al discuss the state-population narrowing effect in two-photon absorption for intense XFEL pulses [23], while Hatada and Di Cicco describe modeling non-equilibrium dynamics and saturable absorption induced by XFEL radiation [24]. Kirrander and Weber discuss fundamental limits on spatial resolution in ultrafast diffraction with XFELs [25], while Kim et al. describe the algorithm for the reconstruction of 3D images of nanorice particles from the diffraction pattern of two particles in independent random orientations with XFEL [26].

Last but not least, this issue includes the meeting report by Larsson on a historic Nobel Symposium on Free Electron Laser Research, which was held in Stigtuna outside Stockholm in June 2015 [27]. There, the inventor of FEL, John Madey, delivered a keynote address. He passed away about one year later, on 5 July 2016.

I believe this special issue will be helpful for providing an overview of the current status of XFELs and sciences at XFELs and foreseeing the future of the fields.

References

1. "The FLASH Fcacility: Advanced Options for FLASH2 and Future Perspectives" B. Faatz et al.
2. "The Linac Coherent Light Source: Recent Developments and Future plans" R.W. Schoenlein et al.
3. "Status of the SACLA Facility" M. Yabashi et al.
4. "FERMI; Present and Future Challenges" L. Giannessi and C. Masciovecchio
5. "Construction and Commissioning of PAL-XFEL Facility" I.S. Ko et al.
6. "Photon Beam Transport and Scientific Instruments at the European XFEL" T. Tschenscher et al.
7. "SwissFEL: The Swiss X-ray Free Electron Laser" C.J. Milne et al.
8. "Status of the SXFEL Facility" Z. Zhao et al.
9. "Ultrashort Free-Electron Laser X-ray pulses" W. Helml et al.
10. "Measurement of the X-ray Spectrum of a Free Electron Laser with a Wide-Range High-Resolution Single-Shot Spectrometer" Y. Inubushi et al.
11. "A Dispersive Inelastic X-ray Scattering Spectrometer for Use at X-ray Free Electron Lasers" J. Szlachetko et al.
12. "Split-And-Delay Unit for FEL Interferometry in the XUV Spectral Range" S. Usenko et al.
13. "Terawatt-Isolated Attosecond X-ray Pulse Using a Tapered X-ray Free Electron Laser" S. Kumar et al.

14. "Molecular Dynamics of XFEL-Induced Photo-Dissociation, Revealed by Ion-Ion Coincidence Measurements" E. Kukk et al.

15. "Observing Femtosecond Fragmentation Using Ultrafast X-ray-Induced Auger Spectra" T.J.A. Wolf et al.

16. "X-ray Pump-Probe Investigation of Charge and Dissociation Dynamics in Methyl Iodine Molecule" L. Fang et al.

17. "Application of Matched-Filter Concepts to Unbiased Selection of Data in Pump-Probe Experiments with Free Electron Lasers" C. Callegari et al.

18. "Measurement of the Resonant Magneto-Optical Kerr Effect Using a Free Electron Laser" S. Yamakoto and I. Matsuda

19. "Current Status of Single Particle Imaging with X-ray Lasers" Z. Sun et al.

20. "Probing Dynamics in Colloidal Crystals with Pump-Probe Experiments at LCLS: Methodology and Analysis" N. Mukharamova et al.

21. "Probing Physics in Vacuum Using an X-ray Free-Electron Laser, a High-Power Laser, and a High Field Magnet" T. Inada et al.

22. "Two- and Three-Photon Partial Photoionization Cross Sections of Li+, Ne8+ and Ar16+ under XUV Radiation" W. Hanks et al.

23. "State-Population Narrowing Effect in Two-Photon Absorption for Intense Hard X-ray Pulses" K. Tyala et al.

24. "Modeling Non-Equilibrium Dynamics and Saturable Absorption Induced by Free Electron Laser Radiation" K. Hatada and A. D. Cicco

25. "Fundamental Limits on Spatial Resolution in Ultrafast X-ray Diffraction" A. Kirrander and P. M. Weber

26. "Algorithm for Reconstruction of 3D Images of Nanorice Particles from Diffraction Patterns of Two Particles in Independent Random Orientations with an X-ray Laser" S.S. Kim et al.

27. "Nobel Symposium on Free Electron Laser Research" Mats Larsson

Kiyoshi Ueda
Special Issue Editor

Article

The FLASH Facility: Advanced Options for FLASH2 and Future Perspectives

Bart Faatz [1], Markus Braune [1], Olaf Hensler [1], Katja Honkavaara [1], Raimund Kammering [1], Marion Kuhlmann [1], Elke Ploenjes [1], Juliane Roensch-Schulenburg [1], Evgeny Schneidmiller [1], Siegfried Schreiber [1], Kai Tiedtke [1], Markus Tischer [1], Rolf Treusch [1], Mathias Vogt [1], Wilfried Wurth [1,2,*], Mikhail Yurkov [1] and Johann Zemella [1]

[1] Deutsches Elektronen-Synchrotron (DESY), 22607 Hamburg, Germany; Bart.faatz@desy.de (B.F.); Markus.braune@desy.de (M.B.); Olaf.hensler@desy.de (O.F.); katja.honkavaara@desy.de (K.H.); raimund.kammering@desy.de (R.K.); marion.kuhlmann@desy.de (M.K.); elke.ploenjes@desy.de (E.P.); juliane.roensch@desy.de (J.R.-S.); evgeny.schneidmiller@desy.de (E.S.); siegfried.schreiber@desy.de (S.S.); kai.tiedtke@desy.de (K.T.); markus.tischer@desy.de (M.T.); rolf.treusch@desy.de (R.T.); mathias.vogt@desy.de (M.V.); mikhail.yurkov@desy.de (M.Y.); johann.zemella@desy.de (J.Z.)
[2] Physics Department and Center for Free-Electron Laser Science, University of Hamburg, 22761 Hamburg, Germany
* Correspondence: wilfried.wurth@desy.de

Academic Editor: Kiyoshi Ueda
Received: 10 September 2017; Accepted: 16 October 2017; Published: 28 October 2017

Abstract: Since 2016, the two free-electron laser (FEL) lines FLASH1 and FLASH2 have been run simultaneously for users at DESY in Hamburg. With the installation of variable gap undulators in the new FLASH2 FEL line, many new possibilities have opened up in terms of photon parameters for experiments. What has been tested so far is post-saturation tapering, reverse tapering, harmonic lasing, harmonic lasing self-seeding and two-color lasing. At the moment, we are working on concepts to enhance the capabilities of the FLASH facility even further. A major part of the upgrade plans, known as FLASH2020, will involve the exchange of the fixed gap undulators in FLASH1 and the implementation of a new flexible undulator scheme aimed at providing coherent radiation for multi-color experiments over a broad wavelength range. The recent achievements in FLASH2 and the current status of plans for the further development of the facility are presented.

Keywords: free-electron lasers, variable gap undulators, tapering, harmonic lasing, frequency doubling

1. Introduction

FLASH began operation for experiments in the extended ultraviolet (XUV) and soft X-ray regime in summer 2005 as the world's first short-wavelength free-electron laser (FEL) facility [1–3]. Due to its superconducting accelerator technology, FLASH is currently the only high-repetition rate XUV and soft X-ray FEL which can deliver up to 8000 photon pulses per second for experiments, while normal conducting FELs typically run at rates between 10 and 120 pulses per second. Since 2016, after the installation and commissioning of a second undulator line, it has been possible for two experiments to receive a beam simultaneously [4,5]. Both FEL lines FLASH1 and FLASH2 (see Figure 1 for a layout of the facility) are currently run in self-amplified stimulated emission (SASE) mode. The superconducting linac of FLASH is operated in a so-called burst mode and can deliver up to 800 electron bunches in a train with a bunch-to-bunch separation of 1 µs and a 10 Hz repetition rate of the bunchtrains. Using two independent photocathode lasers for FLASH1 and FLASH2, the number of bunches from a bunchtrain as well as the intra-train bunch separation going to either of the two FEL lines can be chosen freely, taking into account that 20 to 50 µs are needed to switch bunches between FLASH1 and FLASH2 [4]. The independent photocathode lasers also ensure that the bunch charge can be adjusted individually for the two FEL lines. In combination with fast radio frequency (RF) changes in the time window needed for switching, this enables different compression schemes and hence different pulse durations for user experiments in FLASH1 and FLASH2 [4]. The most important parameters for FLASH2 are shown in the Table 1. While the original FLASH1 FEL line is equipped with fixed gap undulators, which requires a change in the electron beam energy to change the photon energy, the new FLASH2 FEL line has variable gap undulators, which allow for scanning of the photon energy. Furthermore, the possibility for tuning the undulators in FLASH2 has opened up the opportunity to implement and test a variety of novel lasing schemes. First results and future plans will be discussed below.

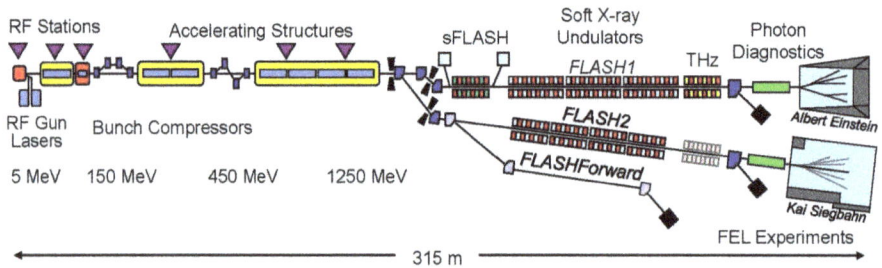

Figure 1. Schematic layout of the FLASH facility. The electron gun is on the left, and the experimental halls are on the right. Behind the last accelerating module the beam is switched between FLASH1, which is the original undulator line, and FLASH2, which has been in operation since 2015. Also shown is sFLASH, the seeding R&D setup in FLASH1 [6] and FLASHForward [7], which is a plasma wakefield experiment in the FLASH3 beamline under construction planned to start operation at the end of 2017. FEL: free-electron laser.

Table 1. Parameters for FLASH2. SASE: self-amplified stimulated emission.

Electron Beam	Value
Energy range	0.45–1.2 GeV
Peak current	$\leq$2.5 kA
Bunch charge	0.02–1 nC
Normalized emittance	1.4 mm mrad
Energy spread	0.5 MeV
Average β-function	6 m
Rep. rate	10 Hz
Bunch separation	1–25 µs
Undulator	**Value**
Period	31.4 mm
K_{rms}	0.5–2
Segment length	2.5 m
Number of segments	12
Photon Beam SASE	**Value**
Wavelength range (fundamental)	4–90 nm
Average single pulse energy	up to 1 mJ
Pulse duration (FWHM)	10–200 fs
Spectral width (FWHM)	$\approx$0.5–2%
Peak brilliance	10^{28}–10^{31} B

Appl. Sci. **2017**, *7*, 1114

2. Fast Wavelength Scans

The advantage of variable gap undulators in terms of enhanced flexibility regarding wavelength tunability is demonstrated in Figures 2 and 3 . While the setup of FLASH1 for SASE for a new user experiment at a specific wavelength takes up to several hours because of the required electron beam energy change (with FLASH1 undulators being fixed-gap), setup of FLASH2 can normally be done within one hour or less, depending on the wavelength requested compared to the one at FLASH1. As a consequence, the time it takes for the initial setup of FLASH2 which is done combined with FLASH1, is almost completely determined by the FLASH1 setup time. After that, any change in wavelength and pulse pattern for FLASH2 can be done in a matter of minutes. The longest wavelength that can be reached is produced when the undulators are closed and is given by approximately three times the FLASH1 wavelength for the specific electron beam energy. The minimum FLASH2 wavelength in normal SASE operation depends on the electron beam energy and beam quality, but it is always less than or equal to the FLASH1 wavelength. Therefore, factor of 3 wavelength tunability can be offered at any time. For lower electron beam energies, one can even go significantly beyond that, with up to factor of 4 tunability at lower energies.

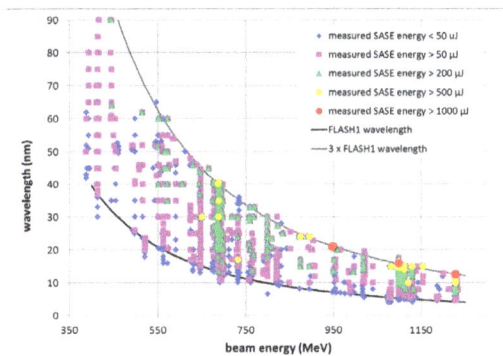

Figure 2. Wavelength tunability of FLASH2 for specific electron beam energies. The different symbols and colors refer to the achieved photon pulse energies. The upper and lower limits are given by three times the FLASH1 wavelength as a maximum (for a closed undulator gap) and the FLASH1 wavelength as a minimum.

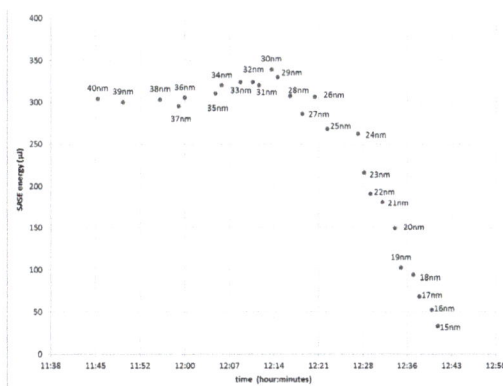

Figure 3. Fast wavelength scan at FLASH2 performed while FLASH1 delivered the beam to users at 13.5 nm (electron beam energy 699 MeV). This scan was performed with standard undulator optics (see below).

3

Figure 3 shows a fast wavelength scan. In principle, the only action needed to decrease the wavelength is opening of the undulators. In practice, minor orbit corrections of the electron beam are necessary. The online, non-invasive measurement of crucial photon parameters including wavelength, intensity, and beam position need only a few seconds to average before these values are available again. Hence, setting a new wavelength takes only minutes as long as the wavelength change is moderate. In particular, scanning across a photoabsorption resonance is almost as easy as at a storage ring.

For large wavelength changes, the setup is still fast, but needs further adjustments concerning electron beam optics. For electron beam energies, at which FLASH is operated, the undulator focusing can be rather strong when the undulators are closed. In extreme cases, as shown in simulations in Figure 4, the focusing becomes so strong that the beam size along the undulator grows and leads to losses unless the focusing is adjusted. Even if these losses would not trigger the machine protection system and consequently switch off the beam, the growing beam size would pose a problem. In this case, the last undulators would no longer contribute to the FEL amplification process, resulting in lower pulse energy. Furthermore, the source point would be no longer in the last undulator, forcing the experiment to either move the instrument or adapt the focusing, assuming that either is possible. Neither solution is straightforward and they can only be performed once the saturation source point is remeasured. At FLASH2, we therefore now adjust the focusing automatically, as shown in the simulations in the right picture of Figure 4. Theoretically, this would also require a rematching of the electron beam at the undulator entrance, which can be calculated in special and staightforward cases, but not easily in general. With more exotic schemes, such as two-color lasing or any fast switching schemes, that will become more important in the future, because a mismatch can in general no longer be avoided. However, as can also be seen in Figure 4, for now even without rematching, the result of the simple adjustment procedure used is more than sufficient.

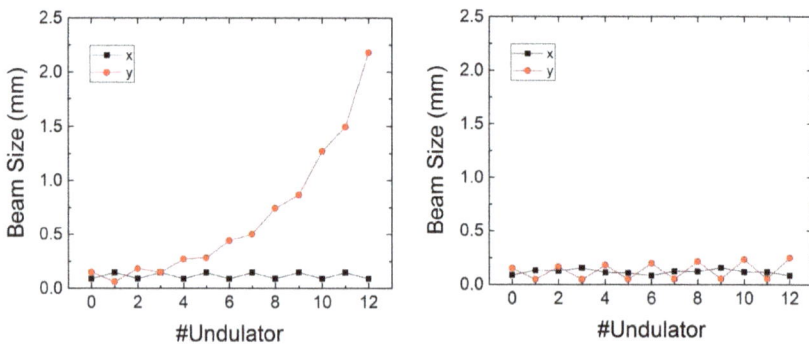

Figure 4. Simulated beam size (in mm) along the undulator without automatic adjustment (**left**) and including adjusted focusing to compensate for the additional undulator focusing (**right**) for a beam energy of 380 MeV and an undulator $K_{rms} = 2$.

The results in Figure 4 are from simulations as mentioned. However, the effect of the automatic optics adjustment can be clearly seen experimentally, as shown in Figure 5, for the same beam energy of 380 MeV as in the simulations. The figure shows the photon beam on a YAG (Yttrium aluminium garnet)-screen for different wavelengths from 50 to 150 nm in a single scan, with automatic optics adjustment switched on, but no other parameters touched during this scan (The interference pattern visible on the screen is caused by a mesh, which is inserted in the photon beam upstream of one of the photon detectors. Furthermore, the YAG screen already shows some beam-induced damage, which makes the beam quality seem poorer than it actually is). In contrast, without adjusting the optics, the spot would look identical at $\lambda = 50$ nm, where the undulator focusing is not yet important, but at $\lambda = 85$ nm, the beam would have already become extremely large and the radiation power would have dropped. For wavelengths longer than λ = 85 nm, the beam losses in the undulator would have exceeded the alarm threshold of the machine protection system and as a result, the beam would have been switched off. Including adjusted optics, one can continue to close the gaps down to 9 mm, corresponding to λ = 150 nm without any problem. It is also clear from Figure 5 that further improvement is still needed. Because no beam-based alignment was performed prior to this experiment, a small movement of the center of the beam is visible.

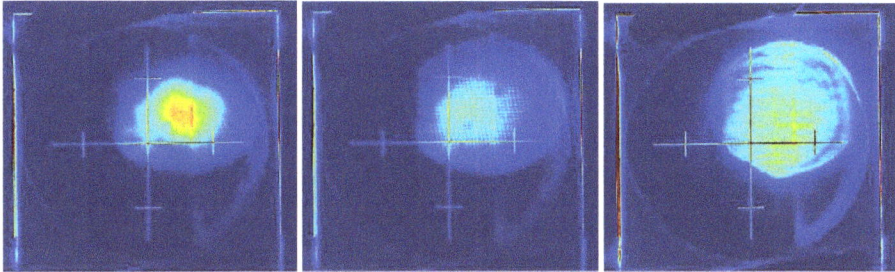

Figure 5. Photon spotsize for 50 (**left**), 85 (**middle**) and 150 nm (**right**) wavelength, 15 m downstream of the undulator with automatic optics adjustment in the undulator switched on. Without adjustment, the undulators cannot be closed to produce wavelengths longer than 85 nm because of beam loss.

However, even given the need for further beam optics automation, first user experiments where the photon energy has been scanned across a resonance within minutes have already demonstrated the great advantages of tunable undulators for experiments at FLASH.

3. New Operation Modes

The variable gap undulators not only allow fast wavelength scans but also enable novel operation modes, such as advanced tapering schemes, frequency doubling, two-color operation, and harmonic lasing self-seeding. With optimized undulator tapering, photon pulse energies up to 1 mJ have been demonstrated at FLASH2 [8]. A particularly interesting option in this respect is reverse tapering [9,10]. In combination with a harmonic afterburner for circular polarization currently under design this should in the future allow experiments with variable polarization at photon energies beyond the water window at FLASH2. Tuning the FLASH2 undulators individually it is also possible to push the photon energy range of FLASH beyond the current limit of 300 eV in the fundamental.

Setting the first part of the undulator section to twice the final wavelength in a frequency doubling scheme it has been shown that the photon energy range of FLASH2 can be extended up to 400 eV with stable pulse energies of a few µJ, significantly higher than what has been achieved when the full undulator section is set to the final wavelength at the same electron energy [11]. Another interesting option is harmonic lasing self-seeding (HLSS) which had been proposed a while ago [12] as a method to reach higher photon energies with increased brightness. With FLASH2, HLSS was recently demonstrated for the first time experimentally and the theoretical predictions were confirmed [13]. In the following some results for the different schemes will be presented.

Harmonic lasing self-seeding (HLSS), while proposed some time ago, could never be tested at FLASH1 with its fixed gap undulators. HLSS requires the setting of the first part of the undulator section to a sub-harmonic ($h\lambda$) of the final wavelength which is schematically shown in Figure 6. As can be seen in Figure 7, saturation is reached earlier with HLSS than with conventional SASE. This is clear from the higher power and the reduction in fluctuations when saturation is reached. More importantly, the bandwidth is also reduced and therefore the brightness is higher, as shown in Figure 8. Due to the limitations of FLASH2, which was originally not built with this concept in mind, one can only go to the second or third harmonic of the wavelength selected for the first part of the undulator, but in principle, higher harmonics could be considered when an FEL line is specifically designed for efficient use of the HLSS scheme.

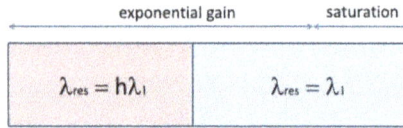

Figure 6. Conceptual scheme of a harmonic lasing self-seeded (HLSS) FEL.

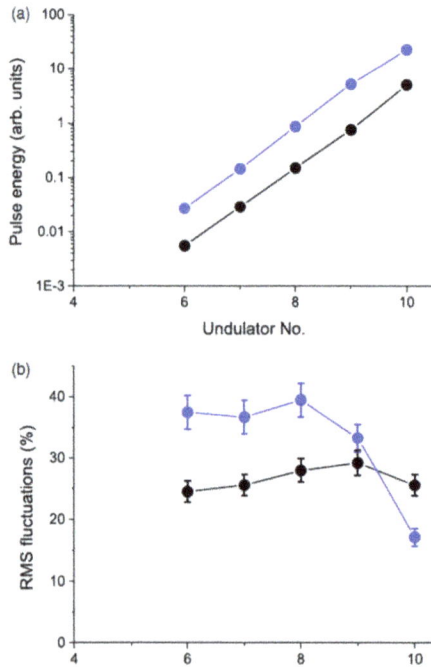

Figure 7. Growth of pulse energy (**a**) and fluctuations (**b**) along the undulator for SASE (black) and HLSS (blue). In the HLSS experiment the first four undulators were set to 33 nm and the next six undulators to 11 nm, while in the SASE case all ten undulators modules were set to 11 nm.

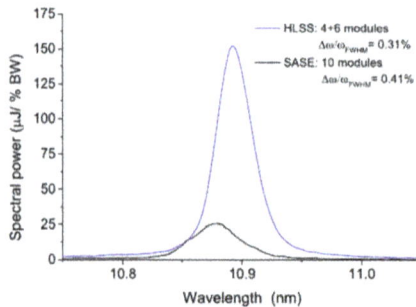

Figure 8. Spectral bandwidth for SASE and HLSS in the same experiment.

The wavelength limits can be pushed at FLASH2 by tuning the undulator sections individually as shown in the frequency doubler scheme in Figure 9. Given the maximum electron beam energy of FLASH of 1.25 GeV, radiation in the water window at 4 nm can only be reached with the present FLASH2 undulator design and normal SASE operation with all undulators set to the same wavelength using the complete undulator length. Going to shorter wavelengths, the SASE intensity drops fast, because saturation is no longer reached. In addition, the SASE fluctuations increase for the same reason. Compared to this, the radiation intensity is clearly higher than it would be without a frequency doubler scheme, as can be seen in Figure 10. In the experiment, first the fundamental at frequency ω is tuned for maximum SASE gain in the uniform undulator. Analysis of the gain curve and fluctuations of the radiation pulse energy allows for determination of the optimum length of the ω-section. Then, the remaining sections are tuned to the frequency 2ω, and after adjustment of the phase shifters and electron beam orbit, the frequency doubler starts to generate radiation at the second harmonic.

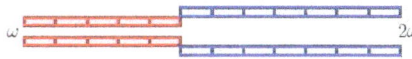

Figure 9. Scheme of frequency doubler operation.

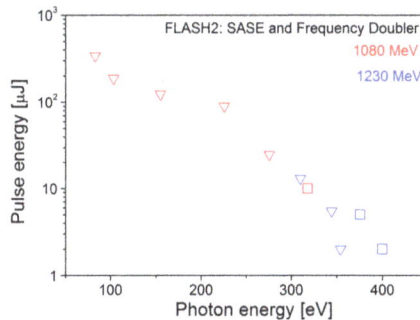

Figure 10. Setting the first part of the undulator to twice the final wavelength (squares), the final wavelength that can be reached with reasonable pulse energy is much shorter and has more stable intensity than with setting all undulators to the same resonant wavelength (triangles).

The same scheme can also be used for two-color operation of the FEL line. In Figure 11 the spectral power obtained is shown for specific settings of the two undulator sections. The experiments show that two-color operation is possible with similar pulse energies of roughly 10 µJ. Moreover, the relative pulse energies of the two colors (ω vs. 2ω) can be tuned in wide limits.

For an extreme case of frequency multiplication, using a short afterburner at for example the third harmonic, the power output is rather small because of the large energy spread generated in the main undulator section. In addition, if the afterburner generates circular polarized light at either the fundamental or odd harmonic, there is the problem that the linearly polarized light from the main undulator section produces a radiation pulse with roughly the same intensity. A proposal to improve this situation is found in [9]. In this scheme, the radiation of the main undulator is suppressed by using an inverse taper. This scheme keeps the bunching to a large extent, but suppresses the radiation and therefore also the energy spread induced during the amplification process. The first experimental demonstration is given in [10]. In the experiment shown in Figure 12, the first 10 undulators of FLASH2 were set up with reversed taper, whereas the last two undulators where used as an afterburner, showing that even though the first 10 undulators did not produce a high radiation intensity, they did produce bunching. In this demonstration experiment, the last two undulators were set to the same wavelength, resulting in an increase in pulse energy exceeding two orders of magnitude.

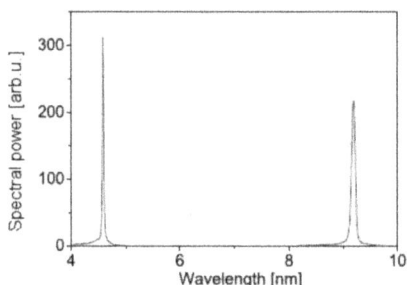

Figure 11. Radiation spectra generated in a frequency doubler experiment at FLASH2 tuned to equal pulse energies for the two wavelength taken with a grating spectrometer [14] in the experimental hall. In blue, the spectral line for the second harmonic at 4.5 nm, and in red, the first harmonic at 9 nm are shown. The electron beam energy in this experiment was 1080 MeV and the bunch charge 300 pC.

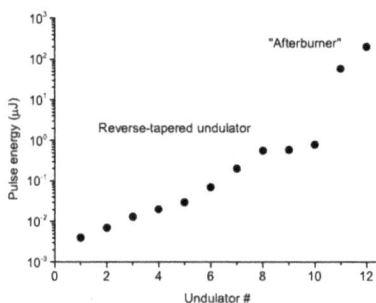

Figure 12. The figure shows the photon pulse energy measured after the respective undulator in FLASH2. Undulators 1 to 10 were set to reverse tapering while the final two act as an afterburner, leading to an increase in pulse energy by a factor of 200.

4. Future Upgrades

The experimental hall of FLASH2 can accommodate up to six beamlines and experimental stations. Since spring 2016, the beamlines FL24 and FL26 have been open for users. FL24 provides an open port for user-supplied experiments and has been equipped with KB (Kirkpatrick-Baez) focusing optics with bendable mirrors in order to adapt focus size and focal length to user demands. At FL26, the permanent end station REMI, a reaction microscope from the Max-Planck Institute for Nuclear Physics in Heidelberg, has been installed for advanced AMO (atomic, molecular, and optical) physics and molecular femtochemistry experiments [15]. As one of the next beamlines, a new time-compensating monochromator will be installed in the FLASH2 experimental hall. The design for this beamline has recently been finalized after intense discussions with the user community. In addition, it is planned to install a THz undulator at FLASH2 and to integrate a THz streaking station based on a single cycle source in the FLASH2 photon diagnostics section for online pulse duration monitoring.

The FLASH2 FEL control system is not yet completely finished. Regarding the control system, the undulator server, which controls the undulator gaps, phase shifters, and aircoils to correct gap-dependent kicks, now includes the automatic focusing. This means that with one click, one can change the wavelength, keeping the phases, undulator gaps, and focusing at an optimum. However, the undulator length and the starting point for and degree of tapering will also be determined in the near future automatically [16].

The new opportunities with the variable gap undulators in the FLASH2 line outlined above will significantly enhance the FLASH performance for users in the coming years. In the period from 2018 to 2020 we plan to refurbish two accelerator modules in the linac and to install a variable polarization afterburner in FLASH2. Furthermore, a new flexible injector laser for FLASH will be developed which can provide flexible electron bunch patterns at the full repetition rate for simultaneous operation of FLASH1 and FLASH2. In FLASH2, an X-band-deflecting cavity will be installed behind the undulators for advanced diagnostics of SASE pulses. The electron beam diagnostics will be upgraded with particular emphasis on low-charge operation required for the shortest SASE pulses. DESY is also currently in a preparation phase for a long-term upgrade plan of FLASH known as "FLASH2020". A major part of FLASH2020 will be the exchange of the fixed gap undulators in FLASH1 and the implementation of a new flexible undulator scheme aimed at providing coherent radiation for multi-color experiments over a broad wavelength range.

Author Contributions: All authors contributed equally to the work presented in the manuscript

Conflicts of Interest: The authors declare no conflict of interest.

References

1. Ackermann, W.; Asova, G.; Ayvazyan, V.; Azima, A.; Baboi, N.; Bähr, J.; Balandin, V.; Beutner, B.; Brandt, A.; Bolzmann, A.; et al. Operation of a free-electron laser from the extreme ultraviolet to the water window. *Nat. Photonics* **2007**, *1*, 336.
2. Honkavaara, K.; DESY. Status of the FLASH FEL user facility at DESY. In Proceedings of the FEL2017, Santa Fe, NM, USA, 15 October 2017.
3. Schreiber, S.; Faatz, B. The free-electron laser FLASH. *High Power Laser Sci. Eng.* **2015**, *3*, e20.
4. Faatz, B.; Plönjes, E.; Ackermann, S.; Agababyan, A.; Asgekar, V.; Ayvazyan, V.; Baark, S.; Baboi, N.; Balandin, V.; von Bargen, N.; et al. Simultaneous operation of two soft x-ray free-electron lasers driven by one linear accelerator. *New J. Phys.* **2016**, *18*, 062002.
5. Rönsch-Schulenburg, J.; Faatz, B.; Honkavaara, K.; Kuhlmann, M.; Schreiber, S.; Treusch, R.; Vogt, M. Experience with Multi-Beam and Multi-Beamline FEL-Operation. *J. Phys. Conf. Ser.* **2017**, *874*, 012023.
6. Böedewadt, J.; Aßmann, R.; Ekanayak, N.; Faat, B.; Hartl, I.; Kazem, M.M.; Laarmann, T.; Lechne, C.; Przystaw, A.; DESY; et al. Experience in Operating sFLASH with High-Gain Harmonic Generation. In Proceedings of the IPAC2017, Copenhagen, Denmark, 14–19 May 2017; p. 2596.
7. Aschikhin, A.; Behrens, C.; Bohlen, S.; Dale, J.; Delbos, N.; di Lucchio, L.; Elsen, E.; Erbe, J.-H.; Felber, M.; Foster, B.; et al. The FLASHForward Facility at DESY. *Nucl. Instrum. Method A* **2016**, *806*, 175.
8. Schneidmiller, E.A.; Yurkov, M.V. Optimum Undulator Tapering of SASE FEL: From the Theory to Experiment. In Proceedings of the 8th International Particle Accelerator Conference (IPAC 2017), Copenhagen, Denmark, 14–19 May 2017; p. 2639.
9. Schneidmiller, E.A.; Yurkov, M.V. Obtaining high degree of circular polarization at x-ray free electron lasers via a reverse undulator tape. *Phys. Rev. ST Accel. Beams* **2013**, *16*, 110702.
10. Schneidmiller, E.A.; Yurkov, M.V. Background-Free Harmonic Production in XFELs via a Reverse Undulator Taper. In Proceedings of the 8th International Particle Accelerator Conference (IPAC 2017), Copenhagen, Denmark, 14–19 May 2017; p. 2618.
11. Kuhlmann, M.; Schneidmiller, E.A.; Yurkov, M.V. Frequency Doubler and Two-Color Mode of Operation at the Free-Electron Laser FLASH2. In Proceedings of the X-ray Free-Electron Lasers: Advances in Source Development and Instrumentation, Prague, Czech Republic, 24–27 April 2017; p. 1023735.
12. Schneidmiller, E.A.; Yurkov, M.V. Harmonic lasing in x-ray free electron lasers. *Phys. Rev. ST Accel. Beams* **2012**, *15*, 080702.
13. Schneidmiller, E.A.; Faatz, B.; Kuhlmann, M.; Roensch-Schulenburg, J.; Schreiber, S.; Tischer, M.; Yurkov, M.V. First operation of a harmonic lasing self-seeded free electron laser. *Phys. Rev. ST Accel. Beams* **2017**, *20*, 020705.

Appl. Sci. **2017**, *7*, 1114

14. Tanikawa, T.; Hage, A.; Kuhlmann, M.; Gonschior, J.; Grunewald, S.; Plönjes, E.; Düsterer, S.; Brenner, G.; Dziarzhytski, S.; Braune, M.; et al. First observation of SASE radiation using the compact wide-spectral-range XUV spectrometer at FLASH2. *Nucl. Instrum. Method A* **2016**, *830*, 170.
15. Plönjes, E.; Faatz, B.; Kuhlmann, M.; Treusch, R. FLASH2: Operation, beamlines, and photon diagnostics. *AIP Conf. Proc.* **2016**, *1741*, 020008.
16. Faatz, B. et al. Automation of FLASH2 operation, in preparation.

applied
sciences

MDPI

Review

The Linac Coherent Light Source: Recent Developments and Future Plans

R. W. Schoenlein *, S. Boutet, M. P. Minitti and A.M. Dunne

SLAC National Accelerator Laboratory, Linac Coherent Light Source, 2575 Sand Hill Rd,
Menlo Par, CA 94025, USA; sboutet@slac.stanford.edu (S.B.); minitti@slac.stanford.edu (M.P.M.);
mdunne@slac.stanford.edu (A.M.D.)
* Correspondence: rwschoen@slac.stanford.edu; Tel.: +1-650-926-5155

Received: 31 July 2017; Accepted: 10 August 2017; Published: 18 August 2017

Abstract: The development of X-ray free-electron lasers (XFELs) has launched a new era in X-ray science by providing ultrafast coherent X-ray pulses with a peak brightness that is approximately one billion times higher than previous X-ray sources. The Linac Coherent Light Source (LCLS) facility at the SLAC National Accelerator Laboratory, the world's first hard X-ray FEL, has already demonstrated a tremendous scientific impact across broad areas of science. Here, a few of the more recent representative highlights from LCLS are presented in the areas of atomic, molecular, and optical science; chemistry; condensed matter physics; matter in extreme conditions; and biology. This paper also outlines the near term upgrade (LCLS-II) and motivating science opportunities for ultrafast X-rays in the 0.25–5 keV range at repetition rates up to 1 MHz. Future plans to extend the X-ray energy reach to beyond 13 keV (<1 Å) at high repetition rate (LCLS-II-HE) are envisioned, motivated by compelling new science of structural dynamics at the atomic scale.

Keywords: ultrafast; X-ray; XFEL; X-ray free-electron laser

1. Introduction

A new era in X-ray science has been launched by the development of X-ray free-electron lasers (XFELs) which provide ultrafast coherent X-ray pulses with a peak brightness that is approximately one billion times higher than previous X-ray sources. The FLASH facility at DESY in Hamburg (the first extreme ultraviolet XFEL) [1], the LCLS facility at SLAC National Accelerator Laboratory (the first hard X-ray XFEL) and the SACLA facility in Japan [2] represent the first generation of XFEL sources, and they have already demonstrated the tremendous scientific potential and impact across broad areas of science.

A comprehensive overview of the scientific impact of the first five years of operation of LCLS was published in Reviews of Modern Physics in 2016 [3]. This paper will highlight some of the more recent accomplishments from LCLS in the areas of atomic, molecular, and optical science; chemistry; condensed matter physics; matter in extreme conditions; and biology. This paper also presents an outline of the upgrade plans for the facility along with the motivating science opportunities. In particular, LCLS is now in the middle of an upgrade project (LCLS-II) which will provide ultrafast X-rays in the 0.25 keV–5 keV range at repetition rates up to 1 MHz with two independent XFELs based on adjustable-gap undulators: 0.25 keV–1.25 keV soft X-ray undulator (SXU) and 1 keV–5 keV hard X-ray undulator (HXU) [4]. LCLS-II is based on a new continuous-wave radio-frequency superconducting accelerator (CW-SCRF) operating at 4 GeV, and first-light is projected for 2020. A second phase upgrade is envisioned to extend the X-ray energy reach to beyond 13 keV (<1 Å) at a high repetition rate by doubling the CW-SCRF linac energy to 8 GeV. This is motivated by compelling new science of structural dynamics at the atomic scale.

2. LCLS Recent Representative Highlights

Following are some representative recent highlights that illustrate the breadth and depth of LCLS science. These pioneering results are expected to open new areas of science enabled by the unique capabilities of LCLS.

2.1. Gase-Phase Atomic, Molecular, and Optical Science

Ultrafast X-ray pulses from LCLS have revealed for the first time the atomic motions associated with an ultrafast chemical reaction triggered by light, as shown in Figure 1 [5]. The ring opening reaction of 1,3-cyclohexadiene (CHD) to form the linear 1,3,5-hexatriene molecule is a prototypical example of a class of organic electrocyclic reactions [6] that are relevant for a wide range of synthetic chemical processes, photochemical switches, and natural biochemical production. Time-resolved hard X-ray scattering studies at LCLS mapped the structural dynamics by providing snapshots roughly every 25 fs (over the 200 fs lifetime) following initiation of the reaction via ultrafast UV excitation. This time-resolved observation of an evolving photochemical reaction paves the way for a wide range of X-ray studies examining gas phase chemistry and the structural dynamics associated with chemical reactions.

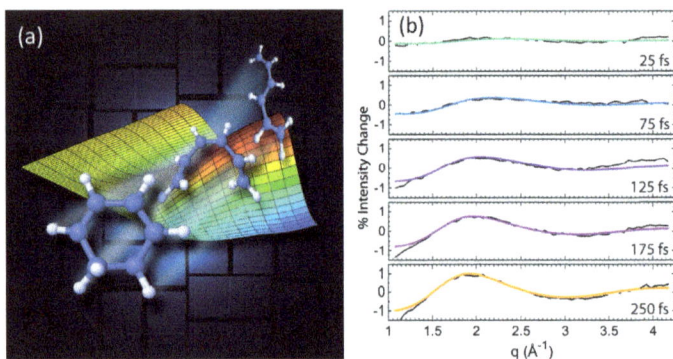

Figure 1. (a) Illustration of femtosecond electrocyclic ring opening of 1,3-cyclohexadiene. (b) Transient X-ray scattering data at a sequence of time delays following initiation of the reaction by an ultrafast UV laser pulse. Reprinted figure with permission from [5]. Copyright The American Physical Society, 2017.

2.2. Condensed-Phase Chemistry

Transition-metal complexes catalyze many important reactions, and their performance is coupled to charge and spin density changes at the metal site caused by electronic excitation and/or ligand loss from the metal center. LCLS results (Figure 2) show that femtosecond X-ray spectroscopy and quantum chemistry theory can provide an unprecedented molecular-level insight into the dynamics of the model transition-metal complex $Fe(CO)_5$, revealing that light-induced dissociation creates a previously unreported excited singlet species and its subsequent reactions [7]. Time-resolved resonant inelastic X-ray scattering (RIXS) spectroscopy is a powerful tool for mapping the evolution of frontier-orbitals with element-specificity and is expected to be applicable to a wide range of molecular dynamics (and a major area of science for LCLS-II). RIXS probing of electronic structural dynamics complements X-ray scattering approaches that probe the atomic structural changes associated with ultrafast chemical reactions.

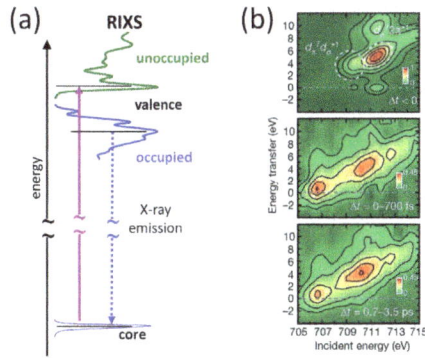

Figure 2. (a) Resonant inelastic X-ray scattering (RIXS) is an element-specific probe of occupied and unoccupied valence states in molecules. (b) Measured Fe L_3-RIXS intensity maps for $Fe(CO)_5$ ground-state (top); and difference intensities for the time intervals 0–700 fs (middle) and 0.7–3.5 ps (bottom). Reprinted with permission from [7]. Copyright Nature Publishing Group, 2015.

2.3. Materials Physics

Ultrafast X-ray scattering studies at LCLS resolved for the first time the mechanism responsible for the incipient ferroelectric behavior in the bulk thermoelectric PbTe (Figure 3), and demonstrate the critical importance of electron-phonon interactions [8]. In PbTe and related materials, the ferroelectric instability is associated with thermoelectricity, phase-change behavior, and superconductivity. However, the origin of the instability has long been controversial. Recent studies have focused on the role of anharmonic phonon–phonon interactions in these materials, while largely overlooking the role of electron-phonon coupling. LCLS studies, using the novel approach of Fourier-transform inelastic X-ray scattering (FT-IXS) [9,10], show that ultrafast infrared excitation transiently stabilizes the paraelectric phase, coupling the transverse optical and acoustic phonons propagating along the bonding direction. An important conclusion is that near band-gap electrons preferentially interact with the soft-phonons to induce ferroelectric instability. These results further reconcile the band and bond pictures of ferroelectricity, which has broad implications for broken-symmetry states in materials with strong electron-phonon interactions.

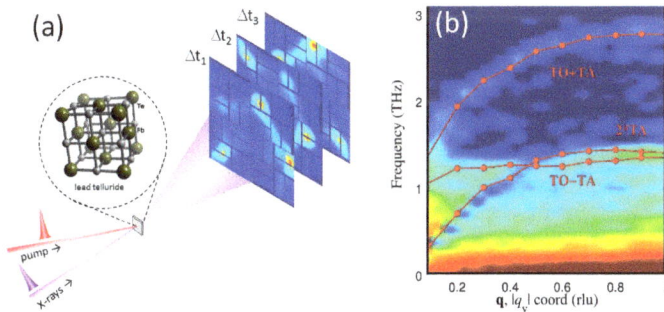

Figure 3. (a) Set-up for time-resolved inelastic X-ray scattering for probing nonequilibrium lattice dynamics. (b) Ultrafast two-phonon spectrum of PbTe obtained through Fourier transform of the femtosecond X-ray scattering signal. The results show a combination of modes (squeezed state) indicative of photo-induced stabilization of the paraelectric state [8].

2.4. Quantum Materials

LCLS provides new insight into quantum materials by enabling transient X-ray scattering studies in the presence of pulsed high magnetic fields. In cuprates and related materials exhibiting unconventional (high-Tc) superconductivity, the universal existence of charge density wave (CDW) correlations raises profound questions regarding the role of CDW phenomena in the emergence of high-Tc superconductivity. Uncovering the evolution of CDW upon suppression of superconductivity by an external magnetic field provides valuable insight into these issues. Studies at LCLS (Figure 4) directly revealed the structure of the long-sought field-induced CDW phase in the high-Tc cuprate $YBa_2Cu_3O_{6.67}$ via time-resolved X-ray scattering in the presence of a transient high magnetic field (28 Tesla) [11]. An unexpected three-dimensionally ordered CDW emerges at low temperatures with magnetic fields above 15 Tesla. This is a distinctly different ordering pattern than that observed previously at zero-field CDW. This discovery of the field-induced CDW provides long-sought information to bridge the gap in cuprate phenomenology, which is critical to uncover the mechanism of high-Tc superconductivity.

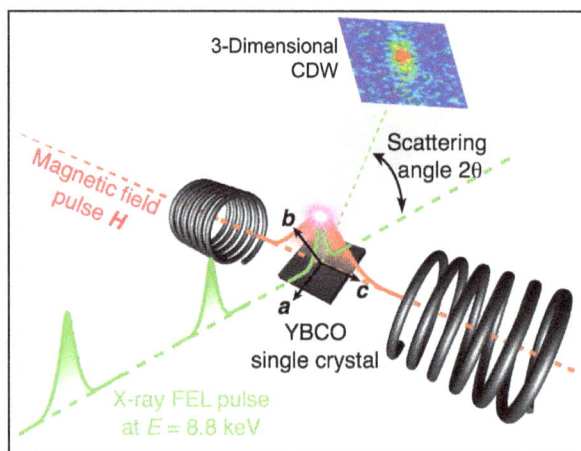

Figure 4. The msec pulsed magnetic field and femtosecond X-ray free-electron laser (FEL) pulses are synchronized to obtain a diffraction pattern from the $YBa_2Cu_3O_{6.67}$ YBCO single crystal at the maximum magnetic field. The 3-dimensional charge density wave (CDW) pattern (inset) was captured with an applied 28 Tesla magnetic field. Reprinted with permission from [11]. Copyright AAAS, 2015.

2.5. Matter in Extreme Conditions

A wealth of new and complex crystal structures emerge from elements exposed to high pressure, an important example of which is the incommensurate composite structure comprised of interpenetrating host and guest components (HG structure) [12]. LCLS shock compression studies (shown in Figure 5) demonstrate for the first time that that these complex crystal structures can develop in less than a few nanoseconds [13]. Time-resolved X-ray scattering studies of scandium under shock compression map the evolving crystal structure along the Hugoniot up to a pressure of 82 GPa. The complex HG crystal structure forms in less than a few nanoseconds, with the guest atoms disordered inside the channels. The onset of melting in scandium is directly observed at the highest compression. This observation of the rapid formation of a complex crystal structure provides an important benchmark of the time scale for atomic rearrangement that is expected to be relevant for a wide range of materials.

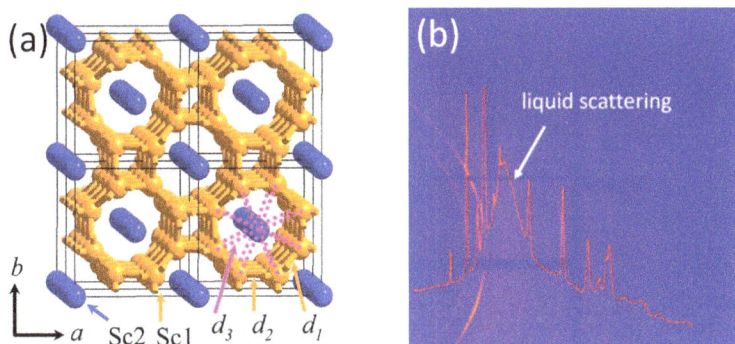

Figure 5. (a) Incommensurate host-guest crystal structure of scandium phase II at 23 GPa. Reprinted with permission from [14]. Copyright The American Physical Society, 2005. **(b)** Linac Coherent Light Source (LCLS) transient X-ray scattering snapshot of shock-compressed scandium showing liquid scattering (broad diffuse peak) along with uncompressed material ahead of the shockwave (sharp peaks) [13].

2.6. Structural Biology

Serial femtosecond crystallography (SFX) at XFEL sources [1,15–17] has revolutionized the ability to determine macromolecular structures that are inaccessible by synchrotron sources. Furthermore, time-resolved SFX maps the conformational dynamics that determine biological function, and enables studies at room temperature (near physiological conditions).

Recent LCLS time-resolved SFX studies of riboswitches, structural elements of messenger RNA (mRNA), capture the dynamic structural response to the binding of a ligand for the first time, as shown in Figure 6 [18]. This ligand-triggered conformational reaction in mRNA mediates gene expression. By using ultra-small riboswitch crystals, the diffusion of a ligand can be timed to initiate a reaction just prior to X-ray diffraction, thereby capturing the transient structure. Four transient structures identified support a reaction mechanism model with at least four states, and illustrate the structural basis for signal transmission. These results further demonstrate the potential of "mix-and-inject" time-resolved serial crystallography to study important interactions between biological macromolecules and ligands.

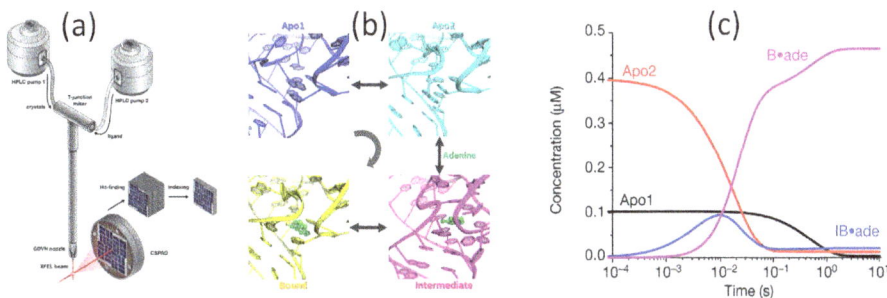

Figure 6. (a) Apparatus for "mix-and-inject" time-resolved serial crystallography at LCLS. **(b)** Structural snapshots of the binding pocket in the riboswitch viewed from the same angle (figure from Y.-X. Wang of NCI). **(c)** Evolution of species concentrations over time. Reprinted with permission from [18]. Copyright Nature Publishing Group, 2017.

Room temperature time-resolved SFX studies at LCLS have substantially advanced our understanding of the mechanism underlying sunlight-driven oxidation of water by photosystem II (PS II) [19]. The four-electron redox chemistry of water oxidation is accomplished by the Mn_4CaO_5 cluster in the oxygen-evolving complex (OEC) within PS II. A grand science challenge is to elucidate the transient structures of the OEC in the different chemical transition states, particularly the transient binding of two waters molecules at the catalytic site. LCLS SFX studies have captured the transient structures of the OEC in the dark S_1 and illuminated S_3 state at high resolution (2.25 to 3.0 Å), in situ and at room temperature. Substrate water and ammonia (water-analog) binding to the Mn_4Ca catalyst reveal new details about the mechanism of water oxidation.

2.7. XFEL Physics

The development of powerful new XFEL capabilities has been a hallmark of LCLS since its inception. Prominent examples include ultrashort X-ray pulse generation [20,21], self-seeding in both the hard X-ray [22] and soft X-ray ranges [23], and novel approaches for characterizing the X-ray pulse duration [24]. These developments are driven primarily by scientific need and potential impact, and are facilitated by close interaction between X-ray scientists and XFEL physicists at LCLS. At the same time, new XFEL capabilities trigger the development of creative new experimental approaches that enhance the scientific impact of the facility.

Recently, LCLS has developed a novel fresh-slice technique for multicolor XFEL pulse production (Figure 7), wherein different temporal slices of an electron bunch lase to saturation in separate undulator sections [25]. This method combines electron bunch tailoring from a passive wakefield device with trajectory control to provide multicolor pulses. The fresh-slice scheme outperforms existing techniques at soft X-ray wavelengths. It produces femtosecond pulses with a power of tens of gigawatts and flexible color separation. The pulse delay can be varied from temporal overlap to almost one picosecond. We have further demonstrated the first three-color XFEL and variably polarized two-color pulses.

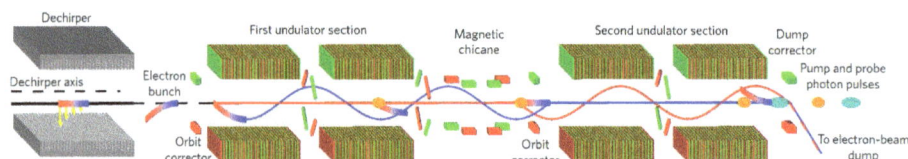

Figure 7. Fresh-slice multi-pulse scheme. The electron bunch propagates off-axis in the dechirper to create a strong transverse head–tail kick. The subsequent oscillating orbit (in combination with fixed-magnet corrections) is exploited so that the tail of the bunch (orange) lases in the first undulator section (at energy E_1) and the head of the bunch (blue) lases in the second undulator section (at energy E_2). The current LCLS layout allows for up to three pulses with controlled photon energies and pulse delays. Reprinted with permission from [25]. Copyright Nature Publishing Group, 2016.

3. LCLS-II Science Opportunities

The representative recent highlights illustrated above are just a glimpse of the remarkable scientific impact achieved by LCLS and related facilities that comprise the first generation of X-ray free-electron lasers. Similar facilities are just beginning operation or are under constructions around the world, including PAL-FEL in the Republic of Korea and Swiss-FEL in Switzerland [26]. However, despite the enormous peak brightness, the average X-ray brightness from this initial generation of XFEL facilities is quite modest, owing to the low repetition rate associated with pulsed-RF accelerator technology. This restricts their impact in many important areas of science that require both high average brightness and ultrafast time resolution.

A new generation of XFELs is now under development that will overcome this limitation by exploiting superconducting RF accelerator technology (SCRF) to provide ultrafast X-ray pulses at high repetition rate. This development is driven by important new science opportunities that have been identified and advanced over the past decade through scientific workshops around the world. The European-XFEL in Germany is the first of this new generation, spanning the hard and soft X-ray ranges and delivering 10 Hz bursts of 2700 pulses at ~4.5 MHz, representing an average repetition rate of up to 27 kHz. Most recently, a series of science workshops held at SLAC National Accelerator Laboratory in 2015 focused on the new science opportunities [27] that will be enabled by the LCLS upgrade project (LCLS-II), based on a novel continuous wave SCRF (CW-SCRF) accelerator technology. As shown in Figure 8, LCLS-II will provide ultrafast X-rays in the 0.25 keV–5 keV range at repetition rates up to 1 MHz with two independent XFELs based on adjustable-gap undulators: soft X-ray undulator (SXU) spanning the range from 0.25 to 1.6 keV, and the hard X-ray undulator (HXU) spanning the range from 1 keV to 5 keV [4].

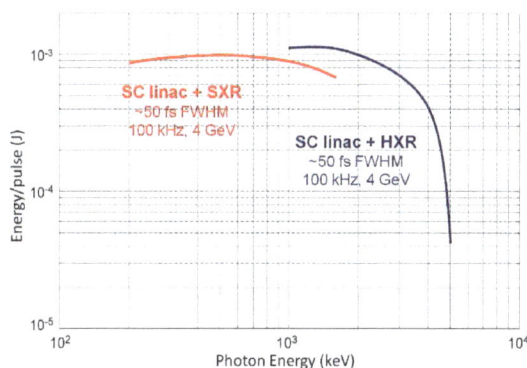

Figure 8. Projected LCLS-II X-ray pulse energies for the soft X-ray undulator (SXU) (red) and the hard X-ray undulator (HXU) (blue) undulators for 100 pC per bunch (from ref. [4]). The X-ray pulse energy is expected to be constant up to ~300 kHz, and will scale inversely with repetition rate (i.e., constant average X-ray power) for repetition rates above ~300 kHz.

Here we highlight some of the important new science opportunities enabled by such a facility in the areas of (1) fundamental charge and energy flow in molecular complexes; (2) photo-catalysis and coordination chemistry; (3) quantum materials; and (4) coherent imaging at the nanoscale. These examples represent just a few of the many science opportunities where a high repetition rate is particularly enabling, and is not intended to be comprehensive of the broad range of science that is driving the development of such facilities.

3.1. Energy and Charge Dynamics in Atoms and Molecules

The fundamental processes of charge migration, redistribution and localization are at the heart of complex processes such as photosynthesis, catalysis, and bond formation/dissolution that govern all chemical reactions. Such charge dynamics are closely coupled with atomic motion, and substantial evidence points to the importance of the concurrent evolution of electronic and nuclear wave functions (i.e., beyond the Born–Oppenheimer approximation [28]) in many molecular systems. Recent evidence also suggests that quantum coherence in chemical and biological complexes may play a more important role than previously appreciated [29]. Our understanding of these processes at the quantum level is limited, even for simple molecules. We have not been able to directly observe these processes to date, and they are beyond the description of conventional chemistry models.

Ultrafast soft X-rays at a high repetition rate from advanced XFELs will enable new experimental methods that will directly map valence charge distributions and reaction dynamics in the molecular frame. One powerful approach that will be qualitatively advanced by the capabilities of LCLS-II is the dynamic molecular reaction microscope. As illustrated in Figure 9, the molecular reaction microscope [30–32] employs sophisticated coincidence measurements of photoelectrons (scattered from a molecular structure at the moment of photo absorption) and ion fragments (of the dissociating molecule) to enable a reconstruction of the molecule at a fixed orientation in space. The unique combination of ultrafast X-rays and high repetition rate at LCLS-II will advance these techniques to the time domain to follow molecular dynamics in the excited-state on fundamental time scales. Here specific molecular dynamics are initiated via tailored transient excitations, such as charge transfer, vibrational excitation, the creation of a valence hole via ionization, or the creation of non-equilibrium Rydberg wavepackets.

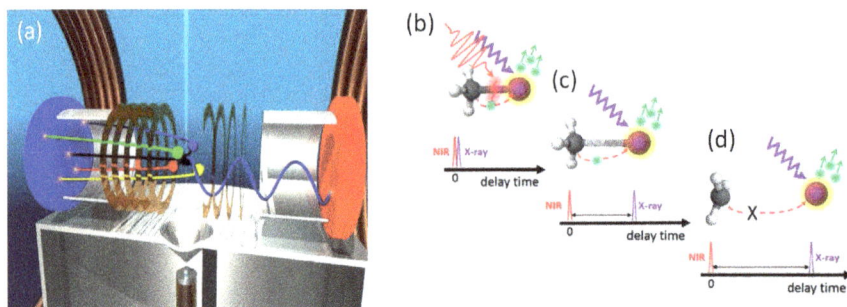

Figure 9. (**a**) Illustration of a molecular reaction microscope (also known as COLd Target Recoil Ion Momentum Spectroscopy, COLTRIMS). Only one molecule is in the X-ray beam on each pulse (i.e., less than one ionization event per pulse). The electron and ion momenta are fully characterized in coincidence via position-sensitive time-of-flight detectors (graphic courtesy of R. Dörner, Goethe U. Frankfurt). (**b–d**) Schematic of charge transfer in iodomethane during dissociation. Reprinted with permission from [33]. Copyright AAAS, 2014. Dynamic reaction microscope studies at high repetition rate would enable a full reconstruction of the charge-sharing dynamics in iodomethane and similar complexes in the molecular frame.

Recent experiments highlight the promising opportunities for dynamic reaction microscope studies at high repetition rate XFELs [34–37]. For example, LCLS experiments at 120 Hz by Erk et al. on charge-transfer processes in gas-phase iodomethane [33] identified three dissociative channels based on the time-dependent kinetic energy distributions of the charged fragments (Figure 9 right). Measurements at high repetition rate will enable the complete spatial reconstruction of the excited-state charge transfer and subsequent dissociation of iodomethane at each time step for a particular molecular orientation. The more than 1000-fold increase in coincidence rates (and information content) from LCLS-II will transform this into a powerful approach for visualizing a broad range of ultrafast molecular dynamics from dissociation of simple diatomic molecules, to charge-transfer processes, to isomerization and ring-opening reactions [5], to non-Born–Oppenheimer relaxation processes [38], to quantum symmetry breaking events from which chirality emerges [39].

3.2. Photochemistry and Catalysis

The directed design of photo-catalytic systems for chemical transformation and solar energy conversion (particularly systems that are efficient, chemically selective, robust, and based on earth-abundant elements) remains a major scientific and technological challenge. Meeting this challenge requires a much deeper understanding of the fundamental processes of photo-chemistry that influence the performance of photo-catalysts; namely, stable charge separation, transport, and localization. In molecular systems, these events are mediated by internal conversion, intersystem crossing, and conformational changes on the ultrafast time scale. Understanding such processes in molecular systems is hindered by the limited ability of conventional experimental or theoretical approaches to directly observed or calculate these charge dynamics. For example, ultrafast optical spectroscopy can capture charge dynamics (and reveal quantum coherences), but is limited to probing electronic states delocalized over multiple atomic sites (or overlapping vibrational spectra), and thus lack element specificity and struggle to identify nuclear degrees of freedom coupled to excited-state dynamics.

Since charge separation, charge transport, and catalysis are local phenomena, the resolving power and element specificity of X-rays can provide insight that is unavailable from non-local probes. Ultrafast X-ray can thus disentangle the coupled motion of electrons and nuclear dynamics, making them uniquely powerful for studying chemical dynamics. High repetition rate X-rays from LCLS-II will enable a powerful suite of spectroscopy tools for understanding the physics and chemistry of photo-catalysts in operating environments. One important example where LCLS-II capabilities will provide a qualitative advance beyond what is presently possible at LCLS is resonant inelastic X-ray scattering (RIXS). As illustrated in Figure 2 RIXS measures the energy distribution of both occupied and unoccupied molecular orbitals, via resonant transitions from a specific atomic core level, thus providing sensitivity to the local chemistry with high resolution.

As highlighted in Section 2.2, recent demonstration experiments at LCLS have applied femtosecond time-resolved RIXS to investigate the charge transfer and ligand exchange dynamics of $Fe(CO)_5$ in solution [7]. By comparing 2D RIXS maps with quantum chemical calculations, the dynamics of the frontier orbitals and their interactions are revealed for the first time with element specificity. These studies demonstrate the potential of time-resolved RIXS to capture short-lived reaction intermediates and correlate the underlying orbital symmetry with spin multiplicity and reactivity. However, the low repetition rate (and corresponding low average spectral flux: ph/s/meV) of LCLS presently limits the application of time-resolved RIXS to studies of gross structural changes in model molecular systems at high concentrations (e.g., ~1 M in the case of the $Fe(CO)_5$ studies [7]). With the more than 1000-fold increase in average brightness provided by LCLS-II, time-resolved RIXS with high spectral resolution will enable complete time-sequenced measurements at high fidelity. The resulting detailed mapping of frontier orbital energies and subtle conformational changes will drive a major advance in our understanding of charge separation and transfer in complex functioning systems where the active sites are often in dilute concentrations.

3.3. Quantum Materials

"Quantum materials" are materials where charged particles behave collectively in ways we are unable to predict from the conventional single-electron band models that effectively describe simple metals and semiconductors. For quantum materials, reductionist approaches that consider only individual atoms, electrons and their orbitals are inadequate. Rather, hallmarks of quantum materials are competing or entwined order, phase separation, and heterogeneity (e.g., fluctuating nanoscale texture of charge, spin and orbitals) that result from strong coupling between constituent particles (charge, spin, orbitals, and phonons). These quantum interactions give rise to important "emergent" macroscopic properties such as high-temperature superconductivity, colossal magnetoresistivity, and topologically protected phases.

One fundamental model for understanding the charged collective modes of an interacting electron system is the two-particle dynamic structure factor, $S_e(q,\omega) \sim \chi(q,\omega)$ that describes inter-particle correlations as a function of momentum-transfer (q) and energy (ω). Although the study of quasiparticles in quantum materials is now well advanced, we still lack effective experimental methods to directly probe $S_e(q,\omega)$ in relevant materials. Because of the subtle balance among competing interactions in quantum materials, the important ground states are determined by collective modes in the 1 to 100 meV energy range as illustrated in Figure 10. In this region, modern X-ray sources and X-ray emission spectrometers struggle to achieve the combination of photon flux and energy resolution required for incisive measurements. This capability gap for measuring the essential observable of an interacting electron system substantially limits our understanding of quantum materials. High repetition rate XFELs will bridge this capability gap and offer transformative capabilities; for both characterizing ground-state collective modes (energy and momentum dependence throughout the Brillouin zone), and for following their response to tailored excitations.

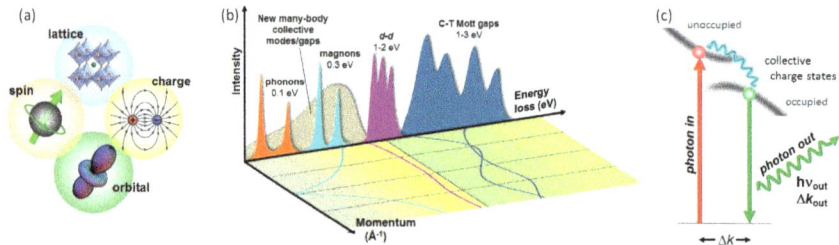

Figure 10. (**a**) Illustration of coupling between charge, spin, orbitals, and lattice. (**b**) Collective excitations that can be characterized by RIXS—including excitations within d-orbital manifolds (*d-d*) and charge-transfer excitations (C-T). Higher resolution is essential to reveal collective excitations at energy scales comparable to that of superconducting gap and pseudogap ~kBT (<25 meV) (image adapted from ref. [40]). (**c**) Resonant inelastic X-ray scattering (RIXS) probe of collective charge states.

Momentum-resolved resonant inelastic X-ray scattering (qRIXS) records the energy and momentum-transfer of scattered X-rays and is a powerful tool to map the energy-momentum dispersion of collective excitations [41]. This approach has been applied to characterize the energy-momentum dispersion of a range of collective excitations (such as magnons [42], paramagnons [43], triplons [44], two-spinons [45], phonons [46], orbitons [45] etc.) at a routinely available energy resolution of ~100 meV (resolving power R ~10,000). State-of-the art RIXS instruments (e.g., ESRF ID32 [47], NSLS-II 2-ID SIX [48]) provide R > 30,000 to investigate collective excitations at energy scales comparable to the superconducting gap and pseudogap in unconventional high-Tc superconductors (<50 meV). In spite of these advances, the inherent lack of longitudinal coherence (and therefore limited spectral flux, ph/s/meV) of synchrotron X-ray sources limits the ultimate scientific impact of qRIXS. High repetition rate seeded XFELs are anticipated to drive a qualitative advance in applications of qRIXS to quantum materials by providing nearly a 1000-fold increase in available X-ray spectral flux for studies at unprecedented resolution and sensitivity.

Beyond conventional qRIXS methods, *time-domain* approaches have tremendous potential to provide an important new perspective on collective excitations by selectively perturbing (or suppressing) one of the interacting degrees of freedom; e.g., by transient charge, spin, or vibrational excitation. Of particular interest is the stimulation of materials directly on the low-energy scales characteristic of the collective excitations (phonons, plasmons, magnons etc.). Ultrafast pulses spanning the visible-to-THz regimes are effective stimuli of coherent collective excitations and can transiently disrupt intertwined degrees of freedom. For example, recent studies have shown that broadband THz pulses can selectively couple to electronic order, and thereby transiently decouple charge and lattice modes [49]. Such approaches can also trigger phase transitions and create new phases that are inaccessible in thermal equilibrium [50–53], thus pointing the way toward control of quantum materials. For example, tailored ultrafast vibrational excitation has been shown to drive insulator-to-metal phase transitions in colossal magnetoresistive manganites [54], photo-induced superconductivity has been reported [50], and enhanced superconductivity is claimed to result from transiently-driven nonlinear lattice dynamics in YBCO [55] and K-doped C60 [56].

An unambiguous interpretation and characterization of these novel photo-induced phenomena is still lacking, but time- and momentum-resolved RIXS can provide much clearer insight. For example, time-resolved RIXS at the Cu L-edge can map the evolution of magnetic excitations and phonons in time, energy, and momentum to provide a more complete microscopic picture about the transient photo-enhanced superconducting phase. The role of charge order in high-Tc superconductivity remains the subject of ongoing debate, and following the evolution of coexisting charge-stripe order through the transient phase will be incredibly informative. Time-resolved RIXS is applicable to a wide range of problems in quantum materials, such as the recently discovered branch of collective modes near the zone center in $Nd_{2-x}Ce_xCuO_4$ and their putative connection to magnetic fluctuations [57].

3.4. Coherent X-ray Imaging at the Nanoscale—Heterogeneity and Dynamics

The ability to image individual non-identical particles in three dimensions at the atomic scale via coherent X-ray scattering remains one of the driving scientific visions for XFELs. This is based on the concept of "diffraction before destruction" whereby the interpretable scattering patterns are generated from individual X-ray pulses (of sufficient intensity and short duration) before X-ray damage effects degrade the achievable resolution [58]. Initial demonstration experiments and applications of single particle imaging methods at LCLS include single-shot coherent diffraction images of viruses [59], bacteriophages [60], organelles [61], and cyanobacteria [62].

Active ongoing research is refining the optimum conditions for single particle imaging. Biological samples are typically comprised of low-Z elements with low X-ray scattering cross-sections, thus yielding scattering snapshots with low signal-to-background ratios. Studies have suggested that the optimum photon energy for single particle imaging is in the tender X-ray range between 2 keV and 6 keV, which may represent the best compromise between scattering cross-section and resolution [63]. The assembly of complete data sets via single particle imaging is hampered by the low number of snapshots (at current low XFEL repetition rates), and further complicated by sample heterogeneity. In order to advance single particle imagine methods, LCLS has launched the single particle imaging (SPI) initiative to develop a roadmap towards the goal of imaging at 3 Å resolution [64]. The initiative consists of over 100 scientists from 20 international institutions and covers all aspects of SPI, from ultrafast X-ray induced damage processes to sample delivery and algorithm development. Recently, the SPI initiative reported scattering data from rice dwarf virus particles out to 3.0 Å (with scattering signals significantly above background) [65], and have pushed the state-of-the-art for 3D image reconstruction to below 10 nm [66].

LCLS-II presents some important opportunities for single particle imaging—particularly the possibility to image the dynamics of macromolecules and assemblies in near-native environments at room temperature. The high repetition rate of LCLS-II in the near-optimum tender X-ray range can potentially generate 10^8 to 10^{10} scattering snapshots per day, thereby driving a qualitative advance in our ability to map the conformational energy landscapes traversed by biological nanomachines. Cryogenic electron microscopy (cryo-EM) results demonstrate the potential to extract three-dimensional structure [67], conformational movies, and energy landscapes from ultralow-signal snapshots of biological complexes cryo-trapped in random orientations at statistically determined points in their work cycle [68]. In the low-signal regime, the number of available snapshots ultimately determines the information content of a conformational movie and the detail with which an energy landscape can be mapped.

Complementing single-particle imaging, fluctuation X-ray scattering (fSAXS) has emerged as a method bridging SPI and crystallography and is a potentially powerful approach for understanding protein interactions in native environments. fSAXS is a multi-particle scattering approach (based on a limited ensemble of particles) and is enabled by the combination of ultrafast X-ray pulses and high repetition rate [69–71]. It potentially provides ~100 times more information than conventional SAXS—sufficient for 3-D reconstruction. As an example, the dynamic fluctuations and conformational response of enzymes to active substrates or small molecules in physiological environments is central to their biological function, and current structural biology methods provide only limited insight. The repetition rate of LCLS-II combined with advanced microfluidic mixing liquid jets may open entirely new opportunities for investigating enzyme dynamics and their response to rapidly introduced substrates. While rapid mixing enhances the population of transient intermediate states, they nevertheless constitute "rare events," with their observation probability (and distinguishability from other states) determined by the reaction kinetics. High repetition rate XFELs will be essential to capture such rare events and characterize a distribution of transient intermediate structures.

4. Future Developments and Science Opportunities—LCLS-II-HE

The extension of high repetition rate XFELs to the hard X-ray regime is motivated by the scientific need for ultrafast atomic resolution at a high average power. LCLS-II-HE represents a significant next step in the ongoing revolution in X-ray lasers and is a natural extension to LCLS-II, based on known CW-SCRF accelerator technology and using existing LCLS infrastructure. LCLS-II-HE will extend operation of the high-repetition-rate LCLS-II beam into the critically important "hard X-ray" regime that has been used in more than 75% of LCLS experiments to date, providing a major advance in performance to the broadest cross-section of the user community. The energy reach of LCLS-II-HE (stretching from 5 keV to at least 13 keV and potentially up to 20 keV) will enable the study of atomic-scale dynamics with the penetrating power and pulse structure needed for in situ and *operando* studies of real-world materials, functioning assemblies, and biological systems. The projected performance of LCLS-II-HE in comparison to other X-ray sources is shown in Figure 11.

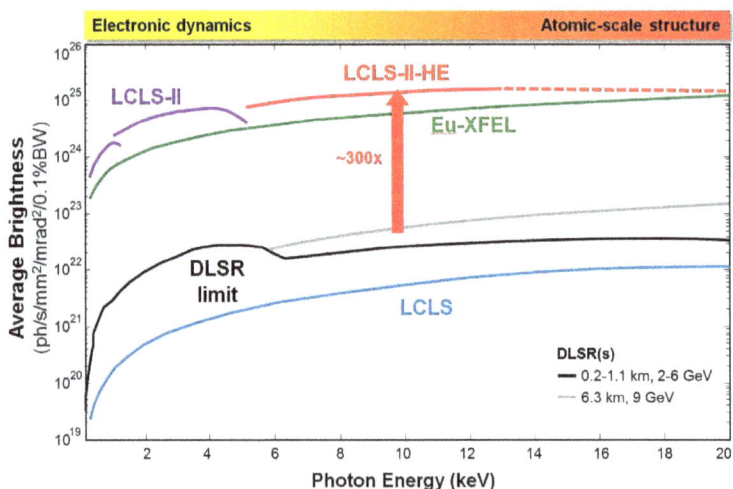

Figure 11. The performance of LCLS-II-HE will allow access to the 'hard X-ray' regime, providing atomic resolution capability, with an average brightness roughly 300 times the ultimate capability of a diffraction-limited storage ring (DLSR) [72]. Self-seeding will further increase the average brightness of the X-ray free-electron laser (XFEL) facilities by an additional factor of 20 to 50.

4.1. LCLS-II-HE Overview and Technical Capabilities

4.1.1. The Envisioned LCLS-II-HE Upgrade Will:

- Deliver two to three orders of magnitude of increase in average spectral brightness in the hard X-ray range beyond any proposed or envisioned diffraction-limited storage ring (DLSR) and exceed the anticipated performance of the European-XFEL.
- Provide temporal coherence for high-resolution spectroscopy near the Fourier transform limit with more than a 300-fold increase in average spectral flux (ph/s/meV) for high-resolution studies beyond any proposed or envisioned DLSR.

- Generate ultrafast hard X-ray pulses in a uniform (or programmable) time structure at a repetition rate of up to 1 MHz.
- Double the electron beam energy of the CW-SCRF linac to 8 GeV, thereby creating three independent accelerators within a single facility: (1) a new 8 GeV superconducting linac; (2) a separately tunable 3.6 GeV by-pass line for the LCLS-II instruments; and (3) the existing 15 GeV Cu-linac.

4.1.2. LCLS-II-HE Technical Capabilities

- **Access to the energy regime above 5 keV** for the analysis of key chemical elements and for atomic resolution. This regime encompasses Earth-abundant elements that are expected to comprise future photocatalysts for electricity and fuel production; it also accesses elements with strong spin-orbit coupling that are of significant interest for future quantum materials; and it reaches the biologically important selenium K-edge for phasing in protein crystallography.
- **High-repetition-rate, ultrafast hard X-rays** from LCLS-II-HE will reveal coupled atomic and electronic dynamics in unprecedented detail. Advanced X-ray techniques will simultaneously measure electronic structure and subtle nuclear displacements at the atomic scale, on fundamental timescales (femtosecond and longer), and in operating environments that require the penetrating capabilities of hard X-rays and the sensitivity provided by high repetition rate.
- **Temporal resolution:** LCLS-II-HE will deliver coherent X-rays on the fastest timescales, opening up experimental opportunities that were previously unattainable due to low signal-to-noise from LCLS (at 120 Hz) and that are simply not possible on non-laser sources. The performance of LCLS has progressed from initial pulse durations of 300 fs down to 5 fs, coupled to the capability for double pulses with independent control of energy, bandwidth, and timing. Ongoing development programs offer the potential for 0.5 fs pulses.
- **Temporal coherence:** Control over the XFEL bandwidth will be a major advance for high-resolution inelastic X-ray scattering and spectroscopy in the hard X-ray range (RIXS and IXS). The present scientific impact of RIXS and IXS is substantially limited by the available spectral flux (ph/s/meV) from temporally incoherent synchrotron sources. In the hard X-ray regime, LCLS-II-HE will provide more than a 300-fold increase in average spectral flux compared to synchrotron sources, opening new areas of science and exploiting high energy resolution and dynamics near the Fourier transform limit.
- **Spatial Coherence:** The high average coherent power of LCLS-II-HE in the hard X-ray range, with programmable pulses at high repetition rate, will enable studies of spontaneous ground-state fluctuations and heterogeneity at the atomic scale from μs (or longer) down to fundamental femtosecond timescales using powerful time-domain approaches such as X-ray photon correlation spectroscopy (XPCS). LCLS-II-HE capabilities will further provide a qualitative advance for understanding non-equilibrium dynamics and fluctuations via time-domain inelastic X-ray scattering (FT-IXS) and X-ray Fourier-transform spectroscopy approaches using Bragg crystal interferometers.
- **Structural dynamics and complete time sequences:** LCLS achieved early success in the determination of high-resolution structures of biological systems and nanoscale matter before the onset of damage. X-ray scattering with ultrashort pulses represents a step-change in the field of protein crystallography. An important scientific challenge is to understand function as determined by structural dynamics, at the atomic scale (requiring ~1 Å resolution) and under operating conditions or in physiologically relevant environments (e.g., aqueous, room temperature). The potential of dynamic pump-probe structure studies has been demonstrated in model systems, but the much higher repetition rates of LCLS-II-HE are needed in order to extract complete time sequences from biologically relevant complexes. Here, small differential scattering signals that originate from dilute concentrations of active sites and low photolysis levels are essential in order to provide interpretable results.

- **Heterogeneous sample ensembles and rare events:** The high repetition rate and uniform time structure of LCLS-II-HE provide a transformational capability to collect 10^8-10^{10} scattering patterns (or spectra) per day with sample replacement between pulses. By exploiting revolutionary advances in data science (e.g., Bayesian analysis, pattern recognition, manifold maps, or machine-learning algorithms) it will be possible to characterize heterogeneous ensembles of particles or identify and extract new information about rare transient events from comprehensive data sets.

4.2. LCLS-I-HE Science Opportunities

LCLS-II-HE will enable precision measurements of structural dynamics on atomic spatial scales and fundamental timescales—providing detailed insight into the behavior of complex matter in real-world heterogeneous samples on fundamental scales of energy, time, and length. The solutions to many important challenges facing humanity, such as developing alternative sources of energy, mitigating environmental and climate problems, and delivering precision medical tools, depend on an improved understanding and control of matter.

We highlight seven broad classes of science for which LCLS-II-HE will uniquely address critical knowledge gaps.

4.2.1. Coupled Dynamics of Energy and Charge in Atoms and Molecules

Flows of energy and charge in molecules are the fundamental processes that drive chemical reactions and store or release energy. They are central to energy processes ranging from combustion to natural and man-made molecular systems that convert sunlight into fuels. Understanding and controlling these processes remains a fundamental science challenge, in large part because the movement of charge is closely coupled to subtle structural changes of the molecule, and conventional chemistry models are inadequate to fully describe this. Sharper experimental tools are needed to probe these processes—simultaneously at the atomic level and on natural (femtosecond) time scales. LCLS-II-HE will image dynamics at the atomic scale via hard X-ray scattering and coherent diffractive imaging (CDI) to reveal the coupled behavior of electrons and atoms with unprecedented clarity. The combination of hard X-rays with high peak power and high average power will enable new nonlinear spectroscopies that promise important new insights into reactive chemical flows in complex chemical environments such as combustion.

4.2.2. Catalysis, Photocatalysis, Environmental & Coordination Chemistry

A deeper understanding of the fundamental processes in catalysis, photocatalysis, and interfacial chemistry is essential for the directed design of new systems for chemical transformations, energy storage, and solar energy conversion that are efficient, chemically selective, robust, and based on Earth-abundant elements. LCLS-II-HE will reveal the critical (and often rare) transient events in these multistep processes, from light harvesting to charge separation, migration, and accumulation at catalytically active sites. Time-resolved, high-sensitivity, element-specific scattering and spectroscopy enabled by LCLS-II-HE will provide the first direct view of atomic-scale chemical dynamics at interfaces. The penetrating capability of hard X-rays will probe operating catalytic systems across multiple time and length scales. The unique LCLS-II-HE capability for simultaneous delivery of hard and soft X-ray pulses opens the possibility to follow chemical dynamics (via spectroscopy) concurrent with structural dynamics (substrate scattering) during heterogeneous catalysis. Time-resolved hard X-ray spectroscopy with high fidelity, enabled by LCLS-II-HE, will reveal the fine details of functioning biological catalysts (enzymes) and inform the design of artificial catalysts and networks with targeted functionality.

4.2.3. Imaging Biological Function and Dynamics

The combination of high spatial and time resolution with a high repetition rate will make LCLS-II-HE a revolutionary machine for many biological science fields. At high repetition rates, serial femtosecond crystallography (SFX) will advance from successful demonstration experiments to address some of the most pressing challenges in structural biology for which only very limited sample volumes are available (e.g., human proteins); or only very small crystal sizes can be achieved (<1 μm); or where current structural information is significantly compromised by damage from conventional X-ray methods (e.g., redox effects in metalloproteins). In all of these cases, high throughput and near-physiological conditions of room temperature crystallography will be qualitative advances. X-ray energies spanning the Se K-edge (12.6 keV) will further enable de novo phasing and anomalous scattering. Time-resolved SFX and solution SAXS will advance from present few-time snapshots of model systems at high photolysis levels to full time sequences of molecular dynamics that are most relevant for biology. Hard X-rays and high repetition rates will further enable advanced crystallography methods that exploit diffuse scattering from imperfect crystals, as well as advanced solution scattering and single particle imaging methods to map sample heterogeneity and conformational dynamics in native environments.

4.2.4. Materials Heterogeneity, Fluctuations, and Dynamics

Heterogeneity and fluctuations of atoms and charge-carriers—spanning the range from the atomic scale to the mesoscale—underlie the performance and energy efficiency of functional materials and hierarchical devices. Conventional models of ideal materials often break down when trying to describe the properties that arise from these complex, non-equilibrium conditions. Yet, there exists an untapped potential to enhance material performance and create new functionality if we can achieve a much deeper insight into these statistical atomic-scale dynamics. Important examples include structural dynamics associated with ion transport in materials for energy storage devices and fuel cells; nanostructured materials for manipulating nonequilibrium thermal transport; two-dimensional materials and heterostructures with exotic properties that are strongly influenced by electron-phonon coupling, light-matter interactions, and subtle external stimuli; and perovskite photovoltaics where dynamic structural fluctuations influence power conversion efficiency. LCLS-II-HE will open an entirely new regime for time-domain coherent X-ray scattering of both statistical (e.g., XPCS) and triggered (pump-probe) dynamics with high average coherent power and penetrating capability for sensitive real-time, in situ probes of atomic-scale structure. This novel class of measurements will lead to new understanding of materials, and, ultimately, device performance, and will couple directly to both theory efforts and next-generation materials design initiatives.

4.2.5. Quantum Materials and Emergent Properties

There is an urgent technological need to understand and ultimately control the exotic quantum-based properties of new materials—ranging from superconductivity to ferroelectricity to magnetism. These properties emerge from the correlated interactions of the constituent matter components of charge, spin, and phonons, and are not well described by conventional band models that underpin present semiconductor technologies. A comprehensive description of the ground-state collective modes that appear at modest energies, 1 meV–100 meV, where modern X-ray sources and spectrometers lack the required combination of photon flux and energy resolution, is critical to understanding quantum materials. High-resolution hard X-ray scattering and spectroscopy at close to the Fourier limit will provide important new insights into the collective modes in 5*d* transition metal oxides—where entirely new phenomena are now being discovered, owing to the combination of strong spin-orbit coupling and strong charge correlation. The ability to apply transient fields and forces (optical, THz, magnetic, pressure) with the time-structure of LCLS-II-HE will be a powerful approach for teasing apart intertwined ordering, and will be a step toward materials control that exploits coherent light-matter interaction. Deeper insight into the coupled electronic and atomic structure in quantum materials will be achieved via simultaneous atomic-resolution scattering and bulk-sensitive photoemission enabled by LCLS-II-HE hard X-rays and high repetition rate.

4.2.6. Materials in Extreme Environments

LCLS-II-HE studies of extreme materials will be important for fusion and fission material applications and could lead to important insights into planetary physics and geoscience. The unique combination of capabilities from LCLS-II-HE will enable the high-resolution spectroscopic and structural characterization of matter in extreme states that is far beyond what is achievable today. High peak brightness combined with high repetition rates and high X-ray energies are required to (i) penetrate dynamically heated dense targets and diamond anvil cells (DAC); (ii) achieve high signal-to-noise data above the self-emission bremsstrahlung background; (iii) probe large momentum transfers on atomic scales to reveal structure and material phases; and (iv) measure inelastic X-ray scattering with sufficient energy resolution and sensitivity to determine the physical properties of materials.

4.2.7. Nonlinear X-ray Matter Interactions

A few seminal experiments on the first generation of X-ray free-electron lasers, LCLS and SACLA, have demonstrated new fundamental nonlinear hard X-ray-matter interactions, including phase-matched sum frequency generation, second harmonic generation, and two-photon Compton scattering. While nonlinear X-ray optics is still in the discovery-based science phase, advances in our understanding of these fundamental interactions will lead to powerful new tools for atomic and molecular physics, chemistry, materials science, and biology via measurement of valence charge density at atomic resolution and on the attosecond-to-femtosecond timescale of electron motion. The combination of high repetition rate and high peak intensity pulses from LCLS-II-HE will enable high-sensitivity measurements that exploit subtle nonlinear effects. This will transform the nonlinear X-ray optics field from demonstration experiments to real measurements that utilize the nonlinear interactions of "photon-in, photon-out" to simultaneously access transient spectroscopic and structural information from real materials.

5. Conclusions

Recent accomplishments from LCLS in the areas of atomic, molecular, and optical science; chemistry; condensed matter physics; matter in extreme conditions; and biology are just a few of the many examples illustrating the ongoing rapid development of XFEL science. At the same time, important new science opportunities are driving the development of a new generation of XFELs that will exploit continuous-wave superconducting accelerator technology to provide ultrafast X-ray pulses at a high repetition rate (~MHz) in a uniform or programmable time structure. The LCLS upgrade project (LCLS-II) will provide ultrafast X-rays in the 0.25–5.0 keV range at repetition rates up to 1 MHz with two independent XFELs based on adjustable-gap undulators: 0.25–1.25 keV soft X-ray undulator (SXU) and 1–5 keV hard X-ray undulator (HXU), with first-light projected for 2020 [4]. In this paper, we have highlighted a few of the important new science opportunities enabled by such a facility in the areas of (1) fundamental charge and energy flow in molecular complexes; (2) photo-catalysis and coordination chemistry; (3) quantum materials; and (4) coherent imaging at the nanoscale. A second phase upgrade is envisioned to extend the X-ray energy reach to beyond 13 keV (<1 Å) at high repetition rate by doubling the CW-SCRF linac energy to 8 GeV. This is motivated by compelling new science of structural dynamics at the atomic scale.

Acknowledgments: The Linac Coherent Light Source (LCLS) at the SLAC National Accelerator Laboratory is an Office of Science User Facility operated for the U.S. Department of Energy Office of Science by Stanford University. This work was: supported by the U.S. Department of Energy, Office of Science, Basic Energy Sciences under Contract No. DEAC02-76SF00515.

Conflicts of Interest: The authors declare no conflict of interest.

References

1. Ayvazyan, V.; Baboi, N.; Bähr, J.; Balandin, V.; Beutner, B.; Brandt, A.; Bohnet, I.; Bolzmann, A.; Brinkmann, R.; Brovko, O.I.; et al. First operation of a free-electron laser generating gw power radiation at 32 nm wavelength. *Eur. Phys. J. D* **2006**, *37*, 297–303. [CrossRef]

2. Ishikawa, T.; Aoyagi, H.; Asaka, T.; Asano, Y.; Azumi, N.; Bizen, T.; Ego, H.; Fukami, K.; Fukui, T.; Furukawa, Y.; et al. A compact X-ray free-electron laser emitting in the sub-angstrom region. *Nat. Photonics* **2012**, *6*, 540–544. [CrossRef]

3. Bostedt, C.; Boutet, S.; Fritz, D.M.; Huang, Z.; Lee, H.J.; Lemke, H.T.; Robert, A.; Schlotter, W.F.; Turner, J.J.; Williams, G.J. Linac coherent light source: The first five years. *Rev. Mod. Phys.* **2016**, *88*, 015007. [CrossRef]

4. SLAC National Accelerator Laboratory. *Linac Coherent Light Source II (LCLS-II) Project Final Design Report—LCLSII-1.1-dr-0251-r0*; SLAC National Accelerator Laboratory: Menlo Park, CA, USA, 2015.

5. Minitti, M.P.; Budarz, J.M.; Kirrander, A.; Robinson, J.S.; Ratner, D.; Lane, T.J.; Zhu, D.; Glownia, J.M.; Kozina, M.; Lemke, H.T.; et al. Imaging molecular motion: Femtosecond X-ray scattering of an electrocyclic chemical reaction. *Phys. Rev. Lett.* **2015**, *114*, 255501. [CrossRef] [PubMed]

6. Woodward, R.B.; Hoffmann, R. The conservation of orbital symmetry. *Angew. Chem. Int.* **1969**, *8*, 781–853. [CrossRef]

7. Wernet, P.; Kunnus, K.; Josefsson, I.; Rajkovic, I.; Quevedo, W.; Beye, M.; Schreck, S.; Grubel, S.; Scholz, M.; Nordlund, D.; et al. Orbital-specific mapping of the ligand exchange dynamics of $Fe(CO)_5$ in solution. *Nature* **2015**, *520*, 78–81. [CrossRef] [PubMed]

8. Jiang, M.P.; Trigo, M.; Savić, I.; Fahy, S.; Murray, É.D.; Bray, C.; Clark, J.; Henighan, T.; Kozina, M.; Chollet, M.; et al. The origin of incipient ferroelectricity in lead telluride. *Nat. Commun.* **2016**, *7*, 12291. [CrossRef] [PubMed]

9. Trigo, M.; Fuchs, M.; Chen, J.; Jiang, M.P.; Cammarata, M.; Fahy, S.; Fritz, D.M.; Gaffney, K.; Ghimire, S.; Higginbotham, A.; et al. Fourier-transform inelastic X-ray scattering from time- and momentum-dependent phonon-phonon correlations. *Nat. Phys.* **2013**, *9*, 790–794. [CrossRef]

10. Zhu, D.; Robert, A.; Henighan, T.; Lemke, H.T.; Chollet, M.; Glownia, J.M.; Reis, D.A.; Trigo, M. Phonon spectroscopy with sub-mev resolution by femtosecond X-ray diffuse scattering. *Phys. Rev. B* **2015**, *92*, 054303. [CrossRef]

11. Gerber, S.; Jang, H.; Nojiri, H.; Matsuzawa, S.; Yasumura, H.; Bonn, D.A.; Liang, R.; Hardy, W.N.; Islam, Z.; Mehta, A.; et al. Three-dimensional charge density wave order in $YBa_2Cu_3O_{6.67}$ at high magnetic fields. *Science* **2015**, *350*, 949–952. [CrossRef] [PubMed]

12. McMahon, M.; Nelmes, R. Incommensurate crystal structures in the elements at high pressure. *Z. Kristallogr.—Cryst. Mater.* **2004**, *219*, 742. [CrossRef]

13. Briggs, R.; Gorman, M.G.; Coleman, A.L.; McWilliams, R.S.; McBride, E.E.; McGonegle, D.; Wark, J.S.; Peacock, L.; Rothman, S.; Macleod, S.G.; et al. Ultrafast X-ray diffraction studies of the phase transitions and equation of state of scandium shock compressed to 82 gpa. *Phys. Rev. Lett.* **2017**, *118*, 025501. [CrossRef] [PubMed]

14. Fujihisa, H.; Akahama, Y.; Kawamura, H.; Gotoh, Y.; Yamawaki, H.; Sakashita, M.; Takeya, S.; Honda, K. Incommensurate composite crystal structure of scandium-ii. *Phys. Rev. B* **2005**, *72*, 132103. [CrossRef]

15. Martin-Garcia, J.M.; Conrad, C.E.; Coe, J.; Roy-Chowdhury, S.; Fromme, P. Serial femtosecond crystallography: A revolution in structural biology. *Arch. Biochem. Biophys.* **2016**, *602*, 32–47. [CrossRef] [PubMed]

16. Levantino, M.; Yorke, B.A.; Monteiro, D.C.F.; Cammarata, M.; Pearson, A.R. Using synchrotrons and xfels for time-resolved X-ray crystallography and solution scattering experiments on biomolecules. *Curr. Opin. Struct. Biol.* **2015**, *35*, 41–48. [CrossRef] [PubMed]

17. Schlichting, I. Serial femtosecond crystallography: The first five years. *IUCrJ* **2015**, *2*, 246–255. [CrossRef] [PubMed]

18. Stagno, J.R.; Liu, Y.; Bhandari, Y.R.; Conrad, C.E.; Panja, S.; Swain, M.; Fan, L.; Nelson, G.; Li, C.; Wendel, D.R.; et al. Structures of riboswitch rna reaction states by mix-and-inject xfel serial crystallography. *Nature* **2017**, *541*, 242–246. [CrossRef] [PubMed]

19. Young, I.D.; Ibrahim, M.; Chatterjee, R.; Gul, S.; Fuller, F.D.; Koroidov, S.; Brewster, A.S.; Tran, R.; Alonso-Mori, R.; Kroll, T.; et al. Structure of photosystem ii and substrate binding at room temperature. *Nature* **2016**, *540*, 453–457. [CrossRef] [PubMed]

20. Emma, P.; Bane, K.; Cornacchia, M.; Huang, Z.; Schlarb, H.; Stupakov, G.; Walz, D. Femtosecond and subfemtosecond X-ray pulses from a self-amplified spontaneous-emission-based free-electron laser. *Phys. Rev. Lett.* **2004**, *92*, 074801. [CrossRef] [PubMed]

21. Ding, Y.; Brachmann, A.; Decker, F.J.; Dowell, D.; Emma, P.; Frisch, J.; Gilevich, S.; Hays, G.; Hering, P.; Huang, Z.; et al. Measurements and simulations of ultralow emittance and ultrashort electron beams in the linac coherent light source. *Phys. Rev. Lett.* **2009**, *102*, 254801. [CrossRef] [PubMed]

22. Amann, J.; Berg, W.; Blank, V.; Decker, F.-J.; Ding, Y.; Emma, P.; Feng, Y.; Frisch, J.; Fritz, D.; Hastings, J.; et al. Demonstration of self-seeding in a hard-X-ray free-electron laser. *Nat. Photonics* **2012**, *6*, 693–698. [CrossRef]

23. Ratner, D.; Abela, R.; Amann, J.; Behrens, C.; Bohler, D.; Bouchard, G.; Bostedt, C.; Boyes, M.; Chow, K.; Cocco, D.; et al. Experimental demonstration of a soft X-ray self-seeded free-electron laser. *Phys. Rev. Lett.* **2015**, *114*, 054801. [CrossRef] [PubMed]

24. Behrens, C.; Decker, F.J.; Ding, Y.; Dolgashev, V.A.; Frisch, J.; Huang, Z.; Krejcik, P.; Loos, H.; Lutman, A.; Maxwell, T.J.; et al. Few-femtosecond time-resolved measurements of X-ray free-electron lasers. *Nat. Commun.* **2014**, *5*, 3762. [CrossRef] [PubMed]

25. Lutman, A.A.; Maxwell, T.J.; MacArthur, J.P.; Guetg, M.W.; Berrah, N.; Coffee, R.N.; Ding, Y.; Huang, Z.; Marinelli, A.; Moeller, S.; et al. Fresh-slice multicolour X-ray free-electron lasers. *Nat. Photonics* **2016**, *10*, 745–750. [CrossRef]

26. Schlichting, I.; White, W.E.; Yabashi, M. Journal of synchrotron radiation: Special issue on X-ray free-electron lasers. *J. Synchrotron Radiat.* **2016**, *22*, 471–866. [CrossRef] [PubMed]

27. SLAC National Accelerator Laboratory. *Menlo Park New Science Opportunities Enabled by LCLS-II X-ray Lasers*; SLAC-R-1053; SLAC National Accelerator Laboratory: Menlo Park, CA, USA, 2015.

28. Worth, G.A.; Cederbaum, L.S. Beyond born-oppenheimer: Molecular dynamics through a conical intersection. *Annu. Rev. Phys. Chem.* **2004**, *55*, 127–158. [CrossRef] [PubMed]

29. Scholes, G.D.; Fleming, G.R.; Chen, L.X.; Aspuru-Guzik, A.; Buchleitner, A.; Coker, D.F.; Engel, G.S.; van Grondelle, R.; Ishizaki, A.; Jonas, D.M.; et al. Using coherence to enhance function in chemical and biophysical systems. *Nature* **2017**, *543*, 647–656. [CrossRef] [PubMed]

30. Dörner, R.; Mergel, V.; Jagutzki, O.; Spielberger, L.; Ullrich, J.; Moshammer, R.; Schmidt-Böcking, H. Cold target recoil ion momentum spectroscopy: A 'momentum microscope' to view atomic collision dynamics. *Phys. Rep.* **2000**, *330*, 95–192. [CrossRef]

31. Landers, A.; Weber, T.; Ali, I.; Cassimi, A.; Hattass, M.; Jagutzki, O.; Nauert, A.; Osipov, T.; Staudte, A.; Prior, M.H.; et al. Photoelectron diffraction mapping: Molecules illuminated from within. *Phys. Rev. Lett.* **2001**, *87*, 013002. [CrossRef] [PubMed]

32. Ullrich, J.; Moshammer, R.; Dorn, A.; Dörner, R.; Schmidt, L.P.H.; Schmidt-Böcking, H. Recoil-ion and electron momentum spectroscopy: Reaction-microscopes. *Rep. Prog. Phys.* **2003**, *66*, 1463. [CrossRef]

33. Erk, B.; Boll, R.; Trippel, S.; Anielski, D.; Foucar, L.; Rudek, B.; Epp, S.W.; Coffee, R.; Carron, S.; Schorb, S.; et al. Imaging charge transfer in iodomethane upon X-ray photoabsorption. *Science* **2014**, *345*, 288–291. [CrossRef] [PubMed]

34. Rudenko, A.; Rolles, D. Time-resolved studies with FELs. *J. Electron Spectrosc. Relat. Phenom.* **2015**, *204*, 228–236. [CrossRef]

35. Zeller, S.; Kunitski, M.; Voigtsberger, J.; Kalinin, A.; Schottelius, A.; Schober, C.; Waitz, M.; Sann, H.; Hartung, A.; Bauer, T.; et al. Imaging the He_2 quantum halo state using a free electron laser. *Proc. Natl. Acad. Sci. USA* **2016**, *113*, 14651–14655. [CrossRef] [PubMed]

36. Schnorr, K.; Senftleben, A.; Kurka, M.; Rudenko, A.; Foucar, L.; Schmid, G.; Broska, A.; Pfeifer, T.; Meyer, K.; Anielski, D.; et al. Time-resolved measurement of interatomic coulombic decay in Ne_2. *Phys. Rev. Lett.* **2013**, *111*, 093402. [CrossRef] [PubMed]

37. Schnorr, K.; Senftleben, A.; Kurka, M.; Rudenko, A.; Schmid, G.; Pfeifer, T.; Meyer, K.; Kübel, M.; Kling, M.F.; Jiang, Y.H.; et al. Electron rearrangement dynamics in dissociating $i2^{n+}$ molecules accessed by extreme ultraviolet pump-probe experiments. *Phys. Rev. Lett.* **2014**, *113*, 073001. [CrossRef] [PubMed]

38. McFarland, B.K.; Farrell, J.P.; Miyabe, S.; Tarantelli, F.; Aguilar, A.; Berrah, N.; Bostedt, C.; Bozek, J.D.; Bucksbaum, P.H.; Castagna, J.C.; et al. Ultrafast X-ray auger probing of photoexcited molecular dynamics. *Nat. Commun.* **2014**, *5*, 4235. [CrossRef] [PubMed]

39. Pitzer, M.; Kunitski, M.; Johnson, A.S.; Jahnke, T.; Sann, H.; Sturm, F.; Schmidt, L.P.H.; Schmidt-Bocking, H.; Dorner, R.; Stohner, J.; et al. Direct determination of absolute molecular stereochemistry in gas phase by coulomb explosion imaging. *Science* **2013**, *341*, 1096–1100. [CrossRef] [PubMed]

40. Zhu, Y.; Durr, H. The future of electron microscopy. *Phys. Today* **2015**, *68*, 32. [CrossRef]

41. Ament, L.J.P.; van Veenendaal, M.; Devereaux, T.P.; Hill, J.P.; van den Brink, J. Resonant inelastic X-ray scattering studies of elementary excitations. *Rev. Mod. Phys.* **2011**, *83*, 705–767. [CrossRef]

42. Guarise, M.; Dalla Piazza, B.; Moretti Sala, M.; Ghiringhelli, G.; Braicovich, L.; Berger, H.; Hancock, J.N.; van der Marel, D.; Schmitt, T.; Strocov, V.N.; et al. Measurement of magnetic excitations in the two-dimensional antiferromagnetic $sr_2cuo_2cl_2$ insulator using resonant X-ray scattering: Evidence for extended interactions. *Phys. Rev. Lett.* **2010**, *105*, 157006. [CrossRef] [PubMed]

43. Le Tacon, M.; Ghiringhelli, G.; Chaloupka, J.; Sala, M.M.; Hinkov, V.; Haverkort, M.W.; Minola, M.; Bakr, M.; Zhou, K.J.; Blanco-Canosa, S.; et al. Intense paramagnon excitations in a large family of high-temperature superconductors. *Nat. Phys.* **2011**, *7*, 725–730. [CrossRef]

44. Schlappa, J.; Schmitt, T.; Vernay, F.; Strocov, V.N.; Ilakovac, V.; Thielemann, B.; Rønnow, H.M.; Vanishri, S.; Piazzalunga, A.; Wang, X.; et al. Collective magnetic excitations in the spin ladder $Sr_{14}Cu_{24}O_{41}$ measured using high-resolution resonant inelastic X-ray scattering. *Phys. Rev. Lett.* **2009**, *103*, 047401. [CrossRef] [PubMed]

45. Schlappa, J.; Wohlfeld, K.; Zhou, K.J.; Mourigal, M.; Haverkort, M.W.; Strocov, V.N.; Hozoi, L.; Monney, C.; Nishimoto, S.; Singh, S.; et al. Spin-orbital separation in the quasi-one-dimensional mott insulator sr_2cuo_3. *Nature* **2012**, *485*, 82–85. [CrossRef] [PubMed]

46. Lee, W.S.; Johnston, S.; Moritz, B.; Lee, J.; Yi, M.; Zhou, K.J.; Schmitt, T.; Patthey, L.; Strocov, V.; Kudo, K.; et al. Role of lattice coupling in establishing electronic and magnetic properties in quasi-one-dimensional cuprates. *Phys. Rev. Lett.* **2013**, *110*, 265502. [CrossRef] [PubMed]

47. ESRF ID-32 Soft X-ray Spectroscopy Beamline. Available online: http://www.esrf.eu/ID32 (accessed on 14 August 2017).

48. NSLS-II Soft Inelastic X-ray Scattering (SIX) Beamline (2-ID). Available online: https://www.bnl.gov/ps/beamlines/beamline.php?b=SIX (accessed on 14 August 2017).

49. Porer, M.; Leierseder, U.; Ménard, J.M.; Dachraoui, H.; Mouchliadis, L.; Perakis, I.E.; Heinzmann, U.; Demsar, J.; Rossnagel, K.; Huber, R. Non-thermal separation of electronic and structural orders in a persisting charge density wave. *Nat. Mater.* **2014**, *13*, 857–861. [CrossRef] [PubMed]

50. Fausti, D.; Tobey, R.I.; Dean, N.; Kaiser, S.; Dienst, A.; Hoffmann, M.C.; Pyon, S.; Takayama, T.; Takagi, H.; Cavalleri, A. Light-induced superconductivity in a stripe-ordered cuprate. *Science* **2011**, *331*, 189–191. [CrossRef] [PubMed]

51. Schmitt, F.; Kirchmann, P.S.; Bovensiepen, U.; Moore, R.G.; Rettig, L.; Krenz, M.; Chu, J.-H.; Ru, N.; Perfetti, L.; Lu, D.H.; et al. Transient electronic structure and melting of a charge density wave in TbTe$_3$. *Science* **2008**, *321*, 1649–1652. [CrossRef] [PubMed]

52. Hinton, J.P.; Koralek, J.D.; Lu, Y.M.; Vishwanath, A.; Orenstein, J.; Bonn, D.A.; Hardy, W.N.; Liang, R.X. New collective mode in $yba_2cu_3o_{6+x}$ observed by time-domain reflectometry. *Phys. Rev. B* **2013**, *88*, 060508. [CrossRef]

53. Stojchevska, L.; Vaskivskyi, I.; Mertelj, T.; Kusar, P.; Svetin, D.; Brazovskii, S.; Mihailovic, D. Ultrafast switching to a stable hidden quantum state in an electronic crystal. *Science* **2014**, *344*, 177–180. [CrossRef] [PubMed]

54. Rini, M.; Tobey, R.; Dean, N.; Itatani, J.; Tomioka, Y.; Tokura, Y.; Schoenlein, R.W.; Cavalleri, A. Control of the electronic phase of a manganite by mode-selective vibrational excitation. *Nature* **2007**, *449*, 72–74. [CrossRef] [PubMed]

55. Mankowsky, R.; Subedi, A.; Forst, M.; Mariager, S.O.; Chollet, M.; Lemke, H.T.; Robinson, J.S.; Glownia, J.M.; Minitti, M.P.; Frano, A.; et al. Nonlinear lattice dynamics as a basis for enhanced superconductivity in $YBa_2Cu_3O_{6.5}$. *Nature* **2014**, *516*, 71–73. [CrossRef] [PubMed]

56. Mitrano, M.; Cantaluppi, A.; Nicoletti, D.; Kaiser, S.; Perucchi, A.; Lupi, S.; Di Pietro, P.; Pontiroli, D.; Riccò, M.; Clark, S.R.; et al. Possible light-induced superconductivity in K_3C_{60} at high temperature. *Nature* **2016**, *530*, 461–464. [CrossRef] [PubMed]

57. Lee, W.S.; Lee, J.J.; Nowadnick, E.A.; Gerber, S.; Tabis, W.; Huang, S.W.; Strocov, V.N.; Motoyama, E.M.; Yu, G.; Moritz, B.; et al. Asymmetry of collective excitations in electron- and hole-doped cuprate superconductors. *Nat. Phys.* **2014**, *10*, 883–889. [CrossRef]

58. Neutze, R.; Wouts, R.; Spoel, D.v.d.; Weckert, E.; Hajdu, J. Potential for biomolecular imaging with femtosecond X-ray pulses. *Nature* **2000**, *406*, 752–757. [CrossRef] [PubMed]

59. Seibert, M.M.; Ekeberg, T.; Maia, F.R.N.C.; Svenda, M.; Andreasson, J.; Jonsson, O.; Odic, D.; Iwan, B.; Rocker, A.; Westphal, D.; et al. Single mimivirus particles intercepted and imaged with an X-ray laser. *Nature* **2011**, *470*, 78–81. [CrossRef] [PubMed]

60. Kassemeyer, S.; Steinbrener, J.; Lomb, L.; Hartmann, E.; Aquila, A.; Barty, A.; Martin, A.V.; Hampton, C.Y.; Bajt, S.A.; Barthelmess, M.; et al. Femtosecond free-electron laser X-ray diffraction data sets for algorithm development. *Opt. Express* **2012**, *20*, 4149–4158. [CrossRef] [PubMed]

61. Hantke, M.F.; Hasse, D.; Maia, F.R.N.C.; Ekeberg, T.; John, K.; Svenda, M.; Loh, N.D.; Martin, A.V.; Timneanu, N.; Larsson, D.S.D.; et al. High-throughput imaging of heterogeneous cell organelles with an X-ray laser. *Nat. Photonics* **2014**, *8*, 943–949. [CrossRef]

62. Van der Schot, G.; Svenda, M.; Maia, F.R.N.C.; Hantke, M.; DePonte, D.P.; Seibert, M.M.; Aquila, A.; Schulz, J.; Kirian, R.; Liang, M.; et al. Imaging single cells in a beam of live cyanobacteria with an X-ray laser. *Nat. Commun.* **2015**, *6*, 5704. [CrossRef] [PubMed]

63. Bergh, M.; Huldt, G.; Timneanu, N.; Maia, F.R.N.C.; Hajdu, J. Feasibility of imaging living cells at subnanometer resolutions by ultrafast X-ray diffraction. *Q. Rev. Biophys.* **2008**, *41*, 181–204. [CrossRef] [PubMed]

64. Aquila, A.; Barty, A.; Bostedt, C.; Boutet, S.; Carini, G.; dePonte, D.; Drell, P.; Doniach, S.; Downing, K.H.; Earnest, T.; et al. The linac coherent light source single particle imaging road map. *Struct. Dyn.* **2015**, *2*, 041701. [CrossRef] [PubMed]

65. Munke, A.; Andreasson, J.; Aquila, A.; Awel, S.; Ayyer, K.; Barty, A.; Bean, R.J.; Berntsen, P.; Bielecki, J.; Boutet, S.; et al. Coherent diffraction of single rice dwarf virus particles using hard X-rays at the linac coherent light source. *Sci. Data* **2016**, *3*, 160064. [CrossRef] [PubMed]

66. Hosseinizadeh, A.; Mashayekhi, G.; Copperman, J.; Schwander, P.; Dashti, A.; Sepehr, R.; Fung, R.; Schmidt, M.; Yoon, C.H.; Hogue, B.G.; et al. Conformational landscape of a virus by single-particle X-ray scattering. *Nat. Methods* **2017**. [CrossRef] [PubMed]

67. Frank, J. *Three-Dimensional Electron Microscopy of Macromolecular Assemblies*, 2nd ed.; Oxford University Press: New York, NY, USA, 2006.

68. Dashti, A.; Schwander, P.; Langlois, R.; Fung, R.; Li, W.; Hosseinizadeh, A.; Liao, H.Y.; Pallesen, J.; Sharma, G.; Stupina, V.A.; et al. Trajectories of the ribosome as a brownian nanomachine. *Proc. Natl. Acad. Sci. USA* **2014**, *111*, 17492–17497. [CrossRef] [PubMed]

69. Kam, Z. Determination of macromolecular structure in solution by spatial correlation of scattering fluctuations. *Macromolecules* **1977**, *10*, 927–934. [CrossRef]

70. Kam, Z.; Koch, M.H.; Bordas, J. Fluctuation X-ray scattering from biological particles in frozen solution by using synchrotron radiation. *Proc. Natl. Acad. Sci. USA* **1981**, *78*, 3559–3562. [CrossRef] [PubMed]

71. Saldin, D.K.; Shneerson, V.L.; Howells, M.R.; Marchesini, S.; Chapman, H.N.; Bogan, M.; Shapiro, D.; Kirian, R.A.; Weierstall, U.; Schmidt, K.E.; et al. Structure of a single particle from scattering by many particles randomly oriented about an axis: Toward structure solution without crystallization. *New J. Phys.* **2010**, *12*, 14. [CrossRef]

72. Borland, M.; Advanced Photon Source (APS), Argonne National Laboratory; Steier, C.; ALS. Personal communication: DLSR contributions, 2016.

applied
sciences

MDPI

Review

Status of the SACLA Facility

Makina Yabashi [1,*], **Hitoshi Tanaka** [1], **Kensuke Tono** [2] **and Tetsuya Ishikawa** [1]

1   RIKEN SPring-8 Center, 1-1-1 Kouto, Sayo-cho, Sayo-gun, Hyogo 679-5148, Japan; tanaka@spring8.or.jp (H.T.); ishikawa@spring8.or.jp (T.I.)

2   Japan Synchrotron Radiation Research Institute, 1-1-1 Kouto, Sayo-cho, Sayo-gun, Hyogo 679-5198, Japan; tono@spring8.or.jp

*   Correspondence: yabashi@spring8.or.jp; Tel.: +81-791-58-0802

Academic Editor: Kiyoshi Ueda

Received: 12 May 2017; Accepted: 8 June 2017; Published: 10 June 2017

Abstract: This article reports the current status of SACLA, SPring-8 Angstrom Compact free electron LAser, which has been producing stable X-ray Free Electron Laser (XFEL) light since 2012. A unique injector system and a short-period in-vacuum undulator enable the generation of ultra-short coherent X-ray pulses with a wavelength shorter than 0.1 nm. Continuous development of accelerator technologies has steadily improved XFEL performance, not only for normal operations but also for fast switching operation of the two beamlines. After upgrading the broadband spontaneous-radiation beamline to produce soft X-ray FEL with a dedicated electron beam driver, it is now possible to operate three FEL beamlines simultaneously. Beamline/end-station instruments and data acquisition/analyzation systems have also been upgraded to allow advanced experiments. These efforts have led to the production of novel results and will offer exciting new opportunities for users from many fields of science.

Keywords: X-ray free electron laser; SACLA; linac; undulator; X-ray optics; photon diagnostics; damage-free analysis; ultrafast science

---

1. Introduction

SACLA, SPring-8 Angstrom Compact free electron LAser, is an X-ray Free Electron Laser (XFEL) facility at SPring-8, Japan. It was inaugurated in March 2012 [1], becoming the second XFEL facility in the world, following the 2009 inauguration of LCLS, the Linac Coherent Light Source, at SLAC in the US [2]. The LCLS and the European XFEL [3] at DESY in Germany, which were first proposed in the 1990s, utilize high-energy linacs with beam energies around 15 GeV and lengths of a few kilometers to produce short wavelength XFEL radiation. In contrast, SACLA was designed as the first compact XFEL facility to produce brilliant and stable XFEL radiation with substantially lower costs for construction and operations. To enable this, we employed unique accelerator technologies: a low emittance injector with a thermionic electron gun (e-gun) and a velocity bunching system; a high-gradient normal-conducting C-band linac; and a short-period in-vacuum undulator [4]. Combining these devices with state-of-the-art X-ray optics [5,6], we have steadily generated XFEL light for users more than 4000 h in FY2016, allowing researchers to produce a number of important scientific results in the fields of biology [7–12], chemistry [13–16], materials science [17–20], high-energy density science [21], and non-linear X-ray optics [22–26]. The success of SACLA has promoted the development of similar compact XFEL facilities, such as SwissFEL at Paul Scherrer Institut in Switzerland [27].

In parallel to conducting user operations, we have continued to upgrade the facility. One of the most critical demands from users is to increase beam time, which recently led us to construct new beamlines and to develop an innovative scheme to switch XFEL over multiple beamlines in a pulse-by-pulse manner [28]. We have also developed various new beamline/end-station instruments and data acquisition/analysis systems to advance experimental capabilities [29–37].

Over the years, we have published several articles about the SACLA facility. In [1], we reported the concept, design, and initial performance of SACLA. Tono et al. also reported the initial beamline design and performance [5]. In [6], we provided updated information on beam performance, beamline instruments, and early scientific highlights. In this article, we report the current status of SACLA, including recent scientific achievements and new capabilities enabled in early 2017. In Section 2, we summarize the typical performance of a hard X-ray FEL beamline, BL3, with the updated design. Section 3 describes recent scientific highlights and the advanced technologies that enable these achievements. In Section 4, we review the construction and operation of two new beamlines, BL2 and BL1. Finally, Section 5 presents our plans for the future.

2. Typical Performance

Table 1 shows basic radiation characteristics for the first XFEL beamline, BL3. The unique injector system combines an e-gun using a thermionic cathode and a multi-stage bunch compression system, instead of an RF-photo cathode system. This configuration can produce high peak power with an X-ray pulse duration shorter than 10 fs under normal operating conditions. Such a short pulse duration assures analysis with a "diffraction-before-destruction" scheme and potentially facilitate ultrafast experiments.

Table 1. Typical characteristics for the X-ray Free Electron Laser (XFEL) light of beamline 3 (BL3).

Parameter	Typical Value
Pulse energy	~0.5 mJ at 10 keV
Pulse duration	<10 fs
Peak power	>50 GW
Photon energy (Wavelength)	4.0–20 keV [1] (0.062–0.31 nm)
Bandwidth	0.5% (FWHM [2])
Repetition rate	60 Hz maximum

[1] 4.0–15 keV in daily operation. [2] Full width at half maximum.

The beamline provides hard X-ray FEL with a high photon energy, greater than ~15 keV, even with a moderate beam energy of 8 GeV. This distinct capability is possible due to the in-vacuum short-period undulator, a core device adopted for our compact XFEL machine. Furthermore, shorter wavelength X-rays can be produced while fixing the electron beam energy by decreasing the magnetic field by opening a gap between the undulator magnets. This variable-gap design enables the production of two-color XFEL pulses with a wavelength separation above 30% using a simple split undulator technique, where the undulator gaps between the upstream and downstream sections are changed [38].

In the autumn of 2016, the first user experiment to use the maximum repetition rate of 60 Hz took place. The standard repetition rate in daily operations has increased steadily and will be 60 Hz during the latter half of 2017. Figure 1 shows a typical trend graph of pulse energy for user experiments. The constant output with a small fluctuation of 16% in root mean square (rms) demonstrates the high stability of operations, another important feature of SACLA operations. The mean fault interval in 2016 was more than 1 h at a repetition rate of 30 Hz.

Figure 2 shows the current schema for BL3. In the original design, the beamline had only basic functions, such as beam transport, monochromatization, focusing, and monitoring of fundamental beam parameters [5]. The beam-transport optical system basically consists of two sets of double plane mirrors and a double crystal monochromator (DCM). One of the mirror sets or DCM is selected to deliver a beam that is pink or monochromatic, respectively. The double mirror systems with different glancing angles (2 and 4 mrad) reflect X-rays below cutoff energies and attenuates the higher order harmonics above them. The DCM with Si (111) crystals delivers a monochromatic beam with a bandwidth of ~10^{-4} ($\Delta E/E$). The standard focusing optical device is a Kirkpatrick–Baez mirror pair, which is stationed at an experimental hutch to provide a 1 μm X-ray spot [30]. In the upgraded beamline, new functions have been added to support broader and more advanced applications. The following major upgrades have been applied:

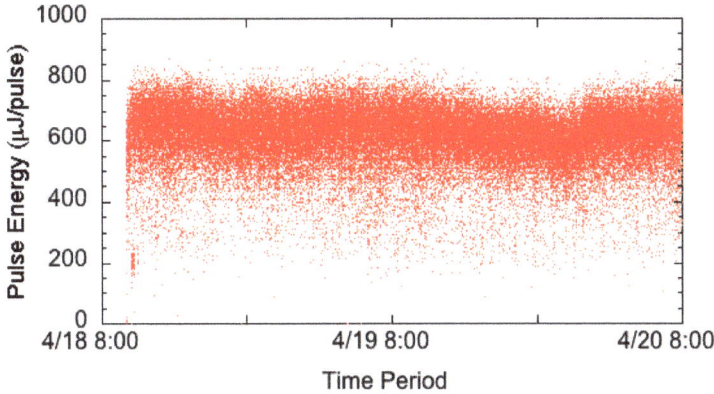

Figure 1. Trend of XFEL (X-ray Free Electron Laser) pulse energy at BL3 over 48 h on 18–20 April 2016. The average and peak pulse energies were 0.61 mJ and 0.87 mJ, respectively, at a photon energy of 10 keV. The rms pulse-energy fluctuation was 16%.

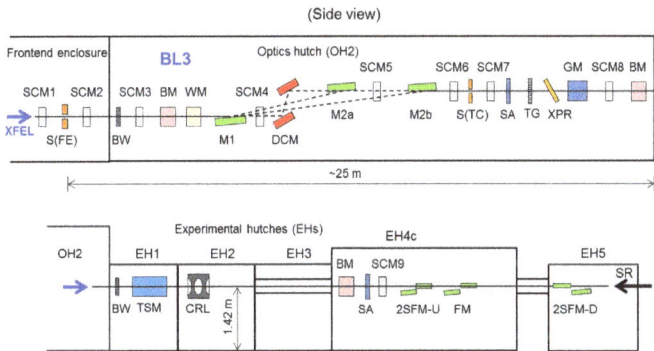

Figure 2. Major optical and diagnostic systems of SACLA BL3. SCM: screen monitor; S(FE): frontend slit; BW: beryllium window; BM: beam intensity and position monitor; WM: wavelength monitor; M1, M2a, M2b: plane mirrors; DCM: double crystal monochromator; S(TC): transport-channel slit; SA: solid attenuator; TG: transmission grating; XPR: X-ray phase retarder; GM: gas intensity monitor; TSM: timing and spectrum monitor; CRL: compound refractive lens; 2SFM-U(-D): two-stage focusing mirrors on the upstream (downstream) side (50 nm spot size); FM: focusing mirrors (1 μm spot size); SR: synchrotron radiation from SPring-8.

Beam-transport and 1 μm focusing mirrors were replaced with longer ones to increase acceptable beam sizes, especially in the lower photon energy ranges with a larger beam size [39]. Transmission ratios from the source to the sample position were improved from 30% to 55% at 5.5 keV, from 46% to 61% at 7.0 keV, and from 59% to 65% at 10 keV.

- Transmission gratings working as beam splitters were installed in an optics hutch (OH2). The gratings are used for simultaneous diagnostics of timing [36] and/or spectrum [29] at the first experimental hutch EH1 with user experiments performed at another experimental hutch [37].
- A diamond phase retarder was stationed in OH2 for the polarization control of XFEL light [40].

- Compound refractive lenses made of beryllium were installed in EH2.
- A two-stage focusing system was deployed at EH4c and EH5 to produce a 50 nm X-ray spot [32].

A more detailed description of the new components/functions will be provided in a separate article [41]. These state-of-the-art X-ray optics and diagnostics enable advanced experiments, such as X-ray nonlinear optics under ultrahigh intensity over 10^{20} W/cm^2, achieved with the two-stage focusing system.

3. Recent Scientific Highlights and New Instruments

To enable a deeper understanding of the photosynthetic process so that we can design artificial photosynthesis, it is important to determine the structures and functions of photosystem II (PSII), a key catalytic protein complex for photosynthesis. Shen and team achieved a milestone by determining the "radiation-damage-free" structure of PSII in the S_1 state at a resolution of 1.95 Å with SACLA. Their results show differences at the sub-angstrom level, compared to those obtained with a quasi-CW X-ray source of synchrotron radiation [10]. Furthermore, they determined the structure of an intermediate S_3 state with two-flash illumination at room temperature at a resolution of 2.35 Å with a time-resolved (TR) serial femtosecond crystallography (SFX) method. This finding suggests the insertion of a new oxygen atom close to an existing oxygen atom in the molecule [11]. These results provide a critical basis for understanding the mechanisms underlying oxygen evolution in photosynthesis.

The TR-SFX method was also used to determine conformational changes in bacteriorhodopsin (bR), a light-driven proton pump, and a model membrane transport protein, at 13 time points in a scale ranging from nanoseconds to milliseconds following photo activation [12]. The resulting molecular movie elucidated a fundamental mechanism for directional proton transport in bR.

These achievements were supported by an experimental platform named DAPHNIS developed by the SACLA team [34]. This instrument is composed of a small He chamber for sample injection and a short-work-distance MPCCD detector with octal sensors that is separated from the chamber [31]. This design offers great flexibility to efficiently meet various demands from users, including introduction of new types of sample injectors, such as grease-matrix and droplet injectors [9,42], and extensions to pump-probe experiments.

The pump-probe scheme was applied for X-ray analysis based on wide-angle scattering (WAXS) of solution. Ihee and Adachi et al. observed a formation process for a gold complex $[Au(CN)_2]_3$ in solution after excitation of an ultrafast laser pulse with a wavelength of 267 nm [14]. The time resolution in this experiment was sub-picosecond, limited mainly by arrival timing jitter between the XFEL and optical laser pulses.

To improve the time resolution, we developed an arrival timing monitor by probing the ultrafast change in optical transmittance induced by intense XFEL light with a spatial decoding technique [36,37]. A unique feature of our optical design is the utilization of an X-ray elliptical mirror for increasing X-ray intensity to form a line-focused profile, which suppresses the X-ray pulse energy that is required to be as small as several microjoules at ~10 keV. Furthermore, we developed a beam branching system for enabling timing diagnostics to be determined simultaneously with experiments. The system is based on a transmission grating that creates two branches dedicated for timing and spectral diagnostics in addition to the 0th order branch used for the main experiments [43]. We evaluated the accuracy of the monitor by constructing a similar setup in the main branch and found that the error was as small as 7 fs in rms. We compared our system with another timing monitor based on the THz streaking method constructed by the PSI group, which assured a relative accuracy of 16.7 fs [44]. This system is now routinely used for pump-probe experiments and contributes to improving time resolution down to a few tens of fs.

The two-color generation configuration based on the split-undulator technique can be combined with the two-stage tight X-ray focusing system, enabling researchers to investigate the non-linear interactions between intense X-ray fields and matter [32,38]. The generation of a Cu Kα laser marked a significant achievement. The Cu target was excited with ultra-intense 9-keV X-rays to produce K-shell vacancies and to form the population inversion condition, leading to the generation of amplified spontaneous emission on Cu Kα lines [26]. Furthermore, while operating SACLA in the two-color mode, we observed the efficient amplification of 8 keV X-rays induced as a seeding.

In this two-color mode, the temporal separation of two XFEL pulses could be tuned with a sub-fs resolution by using a small chicane of the electron beam in the middle of the undulator line. In the summer of 2016, the maximum separation of 40 fs for the 8 GeV electron beam was extended to ~300 fs by increasing the maximum current for the chicane magnets. Based on this scheme, an X-ray pump and X-ray probe experiment was performed to investigate the fundamental damage processes in a diamond induced by intense 6.1 keV X-rays [45]. It was found that the diffraction signal of the 5.9 keV probe pulse decreased after 20 fs following pump pulse irradiation with an intensity of 10^{19} W/cm^2 due to the X-ray–induced atomic displacement. This finding offers a valuable opportunity for experimental evaluation of the "diffraction-before-destruction" scheme.

We developed a hard X-ray split-and-delay optical (SDO) system based on the Bragg diffraction in crystal optics for generating two split pulses with a variable temporal separation [46–48]. To achieve both high stability and operational flexibility, the SDO system was designed to include both variable-delay and fixed-delay branches. As key optical elements, we fabricated high-quality thin crystals and channel-cut crystals by applying the plasma chemical vaporization machining technique. The SDO system using Si(220) crystals covered a photon energy range of 6.5–11.5 keV and a delay time range from a negative value to >45 ps over the photon energy range (up to 220 ps at 6.5 keV). We developed a simple alignment method for achieving a spatial overlap between the split pulses. This SDO system was tested at BL29XU of SPring-8 in combination with a focusing system. We achieved an excellent overlap with an accuracy of 30 nm for ~200 nm focused beams in both the horizontal and vertical directions. This result marks a milestone towards the realization of time-resolved studies using multiple X-ray pulses with a time range from femtosecond to sub-nanosecond scales at XFEL facilities.

4. New Beamlines

Figure 3 shows the schematic view of the current SACLA facility with three beamlines. In 2015, we constructed a second hard X-ray FEL beamline, BL2, and tested pulse-to-pulse switching operations between BL2 and BL3, the first hard XFEL beamline, by using a fast kicker magnet in the upstream location of the undulator lines [28]. At that time, we found that the quality of an electron beam was degraded at a dog-leg transport with a deflection angle of ±3 degrees to the BL2 undulators. The highly compressed electron beam with a peak current greater than 10 kA produced unwanted coherent synchrotron radiation (CSR), which significantly enlarged the electron beam emittance through large energy modulation in the dipole magnets. To resolve this issue, we redesigned the optics of the dog-leg transport to include a combination of two double-bend achromat structures while maintaining high symmetry. Furthermore, we developed a pulsed high power supply to increase the deflection angle of the switching magnet. In February 2017, we tested this system and found that high pulse energies of several hundred microjoules were simultaneously generated at the two beamlines by employing the highly compressed electron beam. Later in 2017, we plan to offer simultaneous operations of these beamlines for users.

In 2015, we also upgraded the broadband spontaneous radiation beamline BL1 to a soft X-ray FEL beamline. A key machine used for this project was the SPring-8 Compact SASE Source (SCSS) test accelerator [4]. This machine was built in 2005 with a beam energy of 250 MeV, originally for performing proof-of-principle tests for the compact XFEL scheme. Following the first lasing at 49 nm in 2006, SCSS was also used for R&D on FEL utilization and experiments with intense FEL radiation in the extreme ultraviolet (EUV) region. In 2013, SCSS was decommissioned after the successful inauguration of SACLA. However, we found space in the upstream position of the BL1 undulator to accommodate this machine as a compact electron beam driver dedicated to BL1. In the summer of 2015, we relocated the accelerator components of SCSS while increasing the beam energy to 450 MeV. This upgraded machine is called SCSS+. In October 2015, we achieved first lasing at a photon energy of 37 eV, and started user operations the following July. In the summer of 2016, we further added two accelerator units to increase the beam energy to 800 MeV. We observed a high pulse energy above 100 µJ at a photon energy around 100 eV before the winter shutdown of 2016. The typical energy resolution is ~2%. Since we use an in-vacuum variable-gap undulator for BL1, i.e., the same as that used for BL2 and BL3, we can cover a higher photon energy up to ~150 eV by opening the undulator gap while maintaining pulse energy above ~10 µJ. It should be emphasized that SCSS+ and SACLA are operated independently, so we are able to increase the available user time of soft X-ray FEL without limiting the availability of hard X-ray FEL at BL2 and BL3.

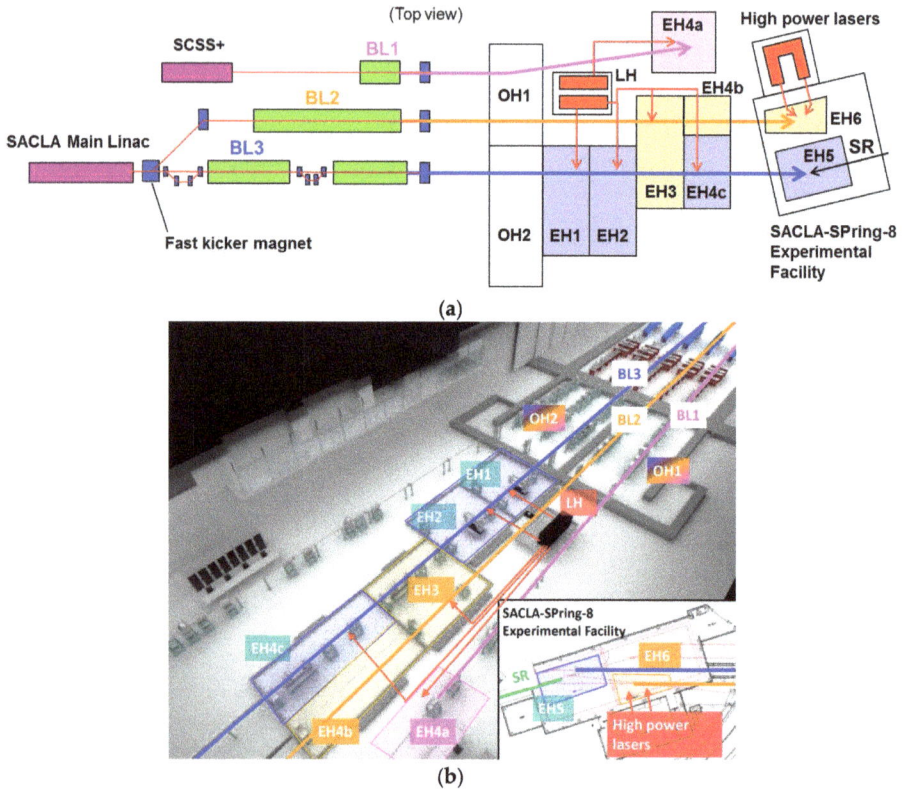

(a)

(b)

Figure 3. (a) Schematic top view of the SACLA facility. Purple bars in (a) indicate linacs and green bard indicate undulators. OH: optics hutch; EH: experimental hutch; LH: laser hutch; SR: synchrotron radiation from SPring-8. (b) Schematic bird's eye view of the SACLA experimental hall.

5. Plans for the Future

We continue to upgrade SACLA not only to enhance the performance and availability of XFEL but also to establish a technological foundation for the future SPring-8-II, our upgrade of the existing SPring-8 facility. The unique capability to operate multiple FEL beamlines allows us to offer exciting research opportunities with new experimental schemes. For example, BL1 and BL2/BL3 can be synchronized to provide FEL pulses with a finely controlled time interval because both the SCSS+ and SACLA linacs use a common trigger source. This synchronized operation will enable pump and probe experiments with both soft and hard X-ray FELs.

Appl. Sci. **2017**, *7*, 604

Pulse-by-pulse switching is presently realized in a scheme that distributes laser pulses equally over multiple beamlines. In this scheme, an equivalent number of pulses is delivered to each beamline. However, depending on the combination of experiments at BL2 and BL3, one beamline may require more pulses than the other. A flexible control system enabling an arbitrary pulse distribution, e.g., 40 pulses/s at BL2 and 20 at BL3, is under development to improve the utilization efficiency.

On the other hand, we plan to use the SACLA linac as an injector for SPring-8-II. Although the dynamic aperture of SPring-8-II is much narrower than the current one, the small emittance beams from the SACLA linac can maintain excellent injection efficiency with high stability. Another advantage of this scheme is that we can save massive amounts of electricity by stopping operations of the existing SPring-8 injector composed of the 8 GeV booster synchrotron and 1 GeV linac. However, this scenario of sharing the SACLA linac between SACLA and SPring-8-II requires two major developments: one is to change the bunch length in a pulse-by-pulse manner to keep the small beam emittance at the injection point of the ring; the other is to enable "on-demand beam injection" that accepts injection requests from the ring for the top-up operation. The first test of beam injection from the SACLA linac to the current storage ring is scheduled for FY2018.

Acknowledgments: The authors are grateful to all of the SACLA/SPring-8 staff who have participated in the construction and operation of SACLA.

Author Contributions: All authors wrote the paper on behalf of the SACLA team.

Conflicts of Interest: The authors declare no conflict of interest.

References

1. Ishikawa, T.; Aoyagi, H.; Asaka, T.; Asano, Y.; Azumi, N.; Bizen, T.; Ego, H.; Fukami, K.; Fukui, T.; Furukawa, Y.; et al. A compact X-ray free-electron laser emitting in the sub-ångstrom region. *Nat. Photonics* **2012**, *6*, 540–544. [CrossRef]

2. Emma, P.; Akre, R.; Arthur, J.; Bionta, R.; Bostedt, C.; Bozek, J.; Brachmann, A.; Bucksbaum, P.; Coffee, R.; Decker, F.-J.; et al. First lasing and operation of an ångstrom-wavelength free-electron laser. *Nat. Photonics* **2010**, *4*, 641–647. [CrossRef]

3. Altarelli, M. The European X-ray Free-Electron Laser Facility in Hamburg. *Nucl. Instrum. Methods B* **2011**, *269*, 2845–2849. [CrossRef]

4. Shintake, T.; Tanaka, H.; Hara, T.; Tanaka, T.; Togawa, K.; Yabashi, M.; Otake, Y.; Asano, Y.; Bizen, T.; Fukui, T.; et al. A compact free-electron laser for generating coherent radiation in the extreme ultraviolet region. *Nat. Photonics* **2008**, *2*, 555–559. [CrossRef]

5. Tono, K.; Togashi, T.; Inubushi, Y.; Sato, T.; Katayama, T.; Ogawa, K.; Ohashi, H.; Kimura, H.; Takahashi, S.; Takeshita, K.; et al. Beamline, experimental stations and photon beam diagnostics for the hard X-ray free electron laser of SACLA. *New J. Phys.* **2013**, *15*, 083035. [CrossRef]

6. Yabashi, M.; Tanaka, H.; Ishikawa, T. Overview of the SACLA facility. *J. Synchrotron Rad.* **2015**, *22*, 477–484. [CrossRef] [PubMed]

7. Kimura, T.; Joti, Y.; Shibuya, A.; Song, C.; Kim, S.; Tono, K.; Yabashi, M.; Tamakoshi, M.; Moriya, T.; Oshima, T.; et al. Imaging live cell in micro-liquid enclosure by X-ray laser diffraction. *Nat. Commun.* **2014**, *5*, 3052. [CrossRef] [PubMed]

8. Hirata, K.; Shinzawa-Itoh, K.; Yano, N.; Takemura, S.; Kato, K.; Hatanaka, M.; Muramoto, K.; Kawahara, T.; Tsukihara, T.; Yamashita, E.; et al. Determination of damage-free crystal structure of an X-ray–sensitive protein using an XFEL. *Nat. Methods* **2014**, *11*, 734–736. [CrossRef] [PubMed]

9. Sugahara, M.; Mizohata, E.; Nango, E.; Suzuki, M.; Tanaka, T.; Masuda, T.; Tanaka, R.; Shimamura, T.; Tanaka, Y.; Suno, C.; et al. Grease matrix as a versatile carrier of proteins for serial crystallography. *Nat. Methods* **2015**, *12*, 61–63. [CrossRef] [PubMed]

10. Suga, M.; Akita, F.; Hirata, K.; Ueno, G.; Murakami, H.; Nakajima, Y.; Shimizu, T.; Yamashita, K.; Yamamoto, M.; Ago, H.; Shen, J.-R. Native structure of photosystem II at 1.95 Å resolution viewed by femtosecond X-ray pulses. *Nature* **2015**, *517*, 99–103. [CrossRef] [PubMed]

11. Suga, M.; Akita, F.; Sugahara, M.; Kubo, M.; Nakajima, Y.; Nakane, T.; Yamashita, K.; Umena, Y.; Nakabayashi, M.; Yamane, T.; et al. Light-induced structural changes and the site of O=O bond formation in PSII caught by XFEL. *Nature* **2017**, *543*, 131–135. [CrossRef] [PubMed]

12. Nango, E.; Royant, A.; Kubo, M.; Nakane, T.; Wickstrand, C.; Kimura, T.; Tanaka, T.; Tono, K.; Song, C.; Tanaka, R.; et al. A three-dimensional movie of structural changes in bacteriorhodopsin. *Science* **2016**, *354*, 1552–1557. [CrossRef] [PubMed]

13. Obara, Y.; Katayama, T.; Ogi, Y.; Suzuki, T.; Kurahashi, N.; Karashima, S.; Chiba, Y.; Isokawa, Y.; Togashi, T.; Inubushi, Y.; et al. Femtosecond time-resolved X-ray absorption spectroscopy of liquid using a hard X-ray free electron laser in a dual-beam dispersive detection method. *Opt. Express* **2014**, *22*, 1105–1113. [CrossRef] [PubMed]

14. Kim, K.H.; Kim, J.G.; Nozawa, S.; Sato, T.; Oang, K.Y.; Kim, T.W.; Ki, H.; Jo, J.; Park, S.; Song, C.; et al. Direct observation of bond formation in solution with femtosecond X-ray scattering. *Nature* **2015**, *518*, 385–389. [CrossRef] [PubMed]

15. Uemura, Y.; Kido, D.; Wakisaka, Y.; Uehara, H.; Ohba, T.; Niwa, Y.; Nozawa, S.; Sato, T.; Ichiyanagi, K.; Fukaya, R.; et al. Dynamics of photoelectrons and structural changes of tungsten trioxide observed by femtosecond transient XAFS. *Angew. Chem. Int. Ed.* **2016**, *55*, 1364–1367. [CrossRef] [PubMed]

16. Canton, S.E.; Kjær, K.S.; Vankó, G.; van Driel, T.B.; Adachi, S.; Bordage, A.; Bressler, C.; Chabera, P.; Christensen, M.; Dohn, A.O.; et al. Visualizing the non-equilibrium dynamics of photoinduced intramolecular electron transfer with femtosecond X-ray pulses. *Nat. Commun.* **2015**, *6*, 6359. [CrossRef] [PubMed]

17. Takahashi, Y.; Suzuki, A.; Zettsu, N.; Oroguchi, T.; Takayama, Y.; Sekiguchi, Y.; Kobayashi, A.; Yamamoto, M.; Nakasako, M. Coherent diffraction imaging analysis of shape-controlled nanoparticles with focused hard X-ray free-electron laser pulses. *Nano Lett.* **2013**, *13*, 6028–6032. [CrossRef] [PubMed]

18. Mitrofanov, K.V.; Fons, P.; Makino, K.; Terashima, R.; Shimada, T.; Kolobov, A.V.; Tominaga, J.; Bragaglia, V.; Giussani, A.; Calarco, R.; et al. Sub-nanometre resolution of atomic motion during electronic excitation in phase-change materials. *Sci. Rep.* **2016**, *6*, 20633. [CrossRef] [PubMed]

19. Dean, M.P.M.; Cao, Y.; Liu, X.; Wall, S.; Zhu, D.; Mankowsky, R.; Thampy, V.; Chen, X.M.; Vale, J.G.; Casa, D.; et al. Ultrafast energy- and momentum-resolved dynamics of magnetic correlations in the photo-doped Mott insulator Sr_2IrO_4. *Nat. Mater.* **2016**, *15*, 601–605. [CrossRef] [PubMed]

20. Matsubara, E.; Okada, S.; Ichitsubo, T.; Kawaguchi, T.; Hirata, A.; Guan, P.F.; Tokuda, K.; Tanimura, K.; Matsunaga, T.; Chen, M.W.; Yamada, N. Initial atomic motion immediately following femtosecond-laser excitation in phase-change materials. *Phys. Rev. Lett.* **2016**, *117*, 135501. [CrossRef] [PubMed]

21. Hartley, N.J.; Ozaki, N.; Matsuoka, T.; Albertazzi, B.; Faenov, A.; Fujimoto, Y.; Habara, H.; Harmand, M.; Inubushi, Y.; Katayama, T.; et al. Ultrafast observation of lattice dynamics in laser-irradiated gold foils. *Appl. Phys. Lett.* **2017**, *110*, 071905. [CrossRef]

22. Fukuzawa, H.; Son, S.-K.; Motomura, K.; Mondal, S.; Nagaya, K.; Wada, S.; Liu, X.-J.; Feifel, R.; Tachibana, T.; Ito, Y.; et al. Deep inner-shell multiphoton ionization by intense X-ray free-electron laser pulses. *Phys. Rev. Lett.* **2013**, *110*, 173005. [CrossRef] [PubMed]

23. Tamasaku, K.; Shigemasa, E.; Inubushi, Y.; Katayama, T.; Sawada, K.; Yumoto, H.; Ohashi, H.; Mimura, H.; Yabashi, M.; Yamauchi, K.; Ishikawa, T. X-ray two-photon absorption competing against single and sequential multiphoton processes. *Nat. Photonics* **2014**, *8*, 313–316. [CrossRef]

24. Shwartz, S.; Fuchs, M.; Hastings, J.B.; Inubushi, Y.; Ishikawa, T.; Katayama, T.; Reis, D.A.; Sato, T.; Tono, K.; Yabashi, M.; et al. X-ray second harmonic generation. *Phys. Rev. Lett.* **2014**, *112*, 163901. [CrossRef] [PubMed]

25. Yoneda, H.; Inubushi, Y.; Yabashi, M.; Katayama, T.; Ishikawa, T.; Ohashi, H.; Yumoto, H.; Yamauchi, K.; Mimura, H.; Kitamura, H. Saturable absorption of intense hard X-rays in iron. *Nat. Commun.* **2014**, *5*, 5080. [CrossRef] [PubMed]

26. Yoneda, H.; Inubushi, Y.; Nagamine, K.; Michine, Y.; Ohashi, H.; Yumoto, H.; Yamauchi, K.; Mimura, H.; Kitamura, H.; Katayama, T.; et al. Atomic inner-shell laser at 1.5-ångström wavelength pumped by an X-ray free-electron laser. *Nature* **2015**, *524*, 446–449. [CrossRef] [PubMed]

27. Ganter, R. (Ed.) *SwissFEL Conceptual Design Report*; PSI Report 10-04; PSI: Villigen, Switzerland, 2010.

28. Hara, T.; Fukami, K.; Inagaki, T.; Kawaguchi, H.; Kinjo, R.; Kondo, C.; Otake, Y.; Tajiri, Y.; Takebe, H.; Togawa, K.; et al. Pulse-by-pulse multi-beam-line operation for X-ray free-electron lasers. *Phys. Rev. ST AB* **2016**, *19*, 020703. [CrossRef]

29. Inubushi, Y.; Tono, K.; Togashi, T.; Sato, T.; Hatsui, T.; Kameshima, T.; Togawa, K.; Hara, T.; Tanaka, T.; Tanaka, H.; et al. Determination of the pulse duration of an X-ray free electron laser using highly resolved single-shot spectra. *Phys. Rev. Lett.* **2012**, *109*, 144801. [CrossRef] [PubMed]

30. Yumoto, H.; Mimura, H.; Koyama, T.; Matsuyama, S.; Tono, K.; Togashi, T.; Inubushi, Y.; Sato, T.; Tanaka, T.; Kimura, T.; et al. Focusing of X-ray free-electron laser pulses with reflective optics. *Nat. Photonics* **2013**, *7*, 43–47. [CrossRef]

31. Kameshima, T.; Ono, S.; Kudo, T.; Ozaki, K.; Kirihara, Y.; Kobayashi, K.; Inubushi, Y.; Yabashi, M.; Horigome, T.; Holland, A.; et al. Development of an X-ray pixel detector with multi-port charge-coupled device for X-ray free-electron laser experiments. *Rev. Sci. Instrum.* **2014**, *85*, 033110. [CrossRef] [PubMed]

32. Mimura, H.; Yumoto, H.; Matsuyama, S.; Koyama, T.; Tono, K.; Inubushi, Y.; Togashi, T.; Sato, T.; Kim, J.; Fukui, R.; et al. Generation of 10^{20} Wcm^{-2} hard X-ray laser pulses with two-stage reflective focusing system. *Nat. Commun.* **2014**, *5*, 3539. [CrossRef] [PubMed]

33. Song, C.; Tono, K.; Park, J.; Ebisu, T.; Kim, S.; Shimada, H.; Kim, S.; Gallagher-Jones, M.; Nam, D.; Sato, T.; et al. Multiple application X-ray imaging chamber for single-shot diffraction experiments with femtosecond X-ray laser pulses. *J. Appl. Cryst.* **2014**, *47*, 188–197. [CrossRef]

34. Tono, K.; Nango, E.; Sugahara, M.; Song, C.; Park, J.; Tanaka, T.; Tanaka, R.; Joti, Y.; Kameshima, T.; Ono, S.; et al. Diverse application platform for hard X-ray diffraction in SACLA (DAPHNIS): Application to serial protein crystallography using an X-ray free-electron laser. *J. Synchrotron Rad.* **2015**, *22*, 532–537. [CrossRef] [PubMed]

35. Joti, Y.; Kameshima, T.; Yamaga, M.; Sugimoto, T.; Okada, K.; Abe, T.; Furukawa, Y.; Ohata, T.; Tanaka, R.; Hatsui, T.; et al. Data acquisition system for X-ray free-electron laser experiments at SACLA. *J. Synchrotron Rad.* **2015**, *22*, 571–576. [CrossRef] [PubMed]

36. Sato, T.; Togashi, T.; Ogawa, K.; Katayama, T.; Inubushi, Y.; Tono, K.; Yabashi, M. Highly efficient arrival timing diagnostics for femtosecond X-ray and optical laser pulses. *Appl. Phys. Exp.* **2015**, *8*, 012702. [CrossRef]

37. Katayama, T.; Owada, S.; Togashi, T.; Ogawa, K.; Karvinen, P.; Vartiainen, I.; Eronen, A.; David, C.; Sato, T.; Nakajima, K.; et al. A beam branching method for timing and spectral characterization of hard X-ray free electron lasers. *Struct. Dyn.* **2016**, *3*, 034301. [CrossRef] [PubMed]

38. Hara, T.; Inubushi, Y.; Katayama, T.; Sato, T.; Tanaka, H.; Tanaka, T.; Togashi, T.; Togawa, K.; Tono, K.; Yabashi, M.; Ishikawa, T. Two-colour hard X-ray free-electron laser with wide tenability. *Nat. Commun.* **2013**, *4*, 2919. [CrossRef] [PubMed]

39. Koyama, T.; Yumoto, H.; Miura, T.; Tono, K.; Togashi, T.; Inubushi, Y.; Katayama, T.; Kim, J.; Matsuyama, S.; Yabashi, M.; et al. Damage threshold of coating materials on X-ray mirror for X-ray free electron laser. *Rev. Sci. Instrum.* **2016**, *87*, 051801. [CrossRef] [PubMed]

40. Suzuki, M.; Inubushi, Y.; Yabashi, M.; Ishikawa, T. Polarization control of an X-ray free-electron laser with a diamond phase retarder. *J. Synchrotron Rad.* **2014**, *21*, 466–472. [CrossRef] [PubMed]

41. Tono, K.; Togashi, T.; Inubushi, Y.; Katayama, T.; Owada, S.; Yabuuchi, T.; Kon, A.; Inoue, I.; Osaka, T.; Yumoto, H.; et al. Overview of optics, photon diagnostics and experimental instruments at SACLA: Development, operation and scientific applications. *Proc. SPIE* **2017**. to be published.

42. Mafuné, F.; Miyajima, K.; Tono, K.; Takeda, Y.; Kohno, J.; Miyauchi, N.; Kobayashi, J.; Joti, Y.; Nango, E.; Iwata, S.; Yabashi, M. Microcrystal delivery by pulsed liquid droplet for serial femtosecond crystallography. *Acta Cryst. D* **2016**, *72*, 520–523. [CrossRef] [PubMed]

43. David, C.; Nöhammer, B.; Ziegler, E. Wavelength tunable diffractive transmission lens for hard X rays. *Appl. Phys. Lett.* **2001**, *79*, 1088–1090. [CrossRef]

44. Gorgisyan, I.; Ischebeck, R.; Erny, C.; Dax, A.; Patthey, L.; Pradervand, C.; Sala, L.; Milne, C.; Lemke, H.T.; Hauri, C.P. THz streak camera method for synchronous arrival time measurement of two-color hard X-ray FEL pulses. *Opt. Express* **2017**, *25*, 2080–2091. [CrossRef]

45. Inoue, I.; Inubushi, Y.; Sato, T.; Tono, K.; Katayama, T.; Kameshima, T.; Ogawa, K.; Togashi, T.; Owada, S.; Amemiya, Y.; et al. Observation of femtosecond X-ray interactions with matter using an X-ray-X-ray pump-probe scheme. *Proc. Natl. Acad. Sci. USA* **2016**, *113*, 1492–1497. [CrossRef] [PubMed]

46. Osaka, T.; Yabashi, M.; Sano, Y.; Tono, K.; Inubushi, Y.; Sato, T.; Matsuyama, S.; Ishikawa, T.; Yamauchi, K. A Bragg beam splitter for hard X-ray free-electron lasers. *Opt. Express* **2013**, *21*, 2823–2831. [CrossRef] [PubMed]

47. Osaka, T.; Hirano, T.; Sano, Y.; Inubushi, Y.; Matsuyama, S.; Tono, K.; Ishikawa, T.; Yamauchi, K.; Yabashi, M. Wavelength-tunable split-and-delay optical system for hard X-ray free-electron lasers. *Opt. Express* **2016**, *24*, 9187–9201. [CrossRef] [PubMed]
48. Hirano, T.; Osaka, T.; Sano, Y.; Inubushi, Y.; Matsuyama, S.; Tono, K.; Ishikawa, T.; Yabashi, M.; Yamauchi, K. Development of speckle-free channel-cut crystal optics using plasma chemical vaporization machining for coherent X-ray applications. *Rev. Sci. Instrum.* **2016**, *87*, 063118. [CrossRef] [PubMed]

applied
sciences

MDPI

Article

FERMI: Present and Future Challenges

Luca Giannessi [1,2,*] and Claudio Masciovecchio [2]

[1]   ENEA C.R. Frascati, Via E. Fermi 45, Frascati, 00044 Rome, Italy
[2]   Elettra Sincrotrone Trieste, I-34149 Basovizza, Trieste, Italy; claudio.masciovecchio@elettra.eu
*   Correspondence: luca.giannessi@elettra.eu; Tel.: +39-04-0375-8043

Academic Editor: Kiyoshi Ueda
Received: 12 May 2017; Accepted: 12 June 2017; Published: 21 June 2017

Abstract: We present an overview of the FERMI (acronym of Free Electron laser Radiation for Multidisciplinary Investigations) seeded free electron laser (FEL) facility located at the Elettra laboratory in Trieste. FERMI is now in user operation with both the FEL lines FEL-1 and FEL-2, covering the wavelength range between 100 nm and 4 nm. The seeding scheme adopted for photon pulse production makes FERMI unique worldwide and allows the extension of table top laser experiments in the extreme ultraviolet/soft X-ray region. In this paper, we discuss how advances in the performance of the FELs, with respect to coherent control and multi-colour pulse production, may push the development of original experimental strategies to study non-equilibrium behaviour of matter at the attosecond-nanometer time-length scales. This will have a tremendous impact as an experimental tool to investigate a large array of phenomena ranging from nano-dynamics in complex materials to phenomena that are at the heart of the conversion of light into other forms of energy.

Keywords: free electron laser; non-linear optics; four wave mixing; coherent control

1. Introduction

The newest light sources, extreme ultraviolet (XUV) and X-ray free electron lasers (FELs), are extending laboratory laser experiments to shorter wavelengths, adding element and chemical state specificity by exciting and probing electronic transitions from core levels. The high pulse energies available ensure that spectroscopies belonging to table top laser, such as, for instance, non-linear optics, can cross the frontier into this new wavelength range. Since the first short wavelength FEL FLASH began operation in Hamburg [1], development has been very rapid. The Linac Coherent Light Source (LCLS) was specified to produce 200 fs pulses, but soon produced few fs pulses [2]. Schemes have been developed to create multi pulse and polychromatic radiation, and FEL light has been used to pump atomic lasers in the X-ray region [3,4]. The XUV/soft X-ray laser FERMI is the first fully coherent FEL facility, as the light possesses the full longitudinal coherence lacking in a pure SASE (Self-Amplified Stimulated Emission) configuration [5]. Coherence is achieved by initiating the FEL process with an external seed. The coherence properties of the seed are indeed transferred to the electron modulation, leading to coherent emission at the undulator resonance and at its harmonics. A similar configuration is adopted in the new facility in China; the Dalian Coherent Light Source (DCLS, Dalian, China) covers the vacuum-ultraviolet (VUV) photon energy range [6]. Advanced schemes of self seeding, both in the soft and hard X-rays, were adopted at LCLS [7] to improve coherence of a SASE FEL. With the development of coherent FEL radiation sources, a new era of X-ray spectroscopy commenced, which may have a comparable impact to that of lasers in optics and spectroscopy [8]. In non-linear ultrafast time-resolved techniques, state-specific information is often provided through multiphoton resonances with combinations of sequential photons. Theoretically, combinations of multiple X-ray photons resonant with core transitions can also characterize different excitation processes due to specific sequences of light-matter interaction. Thus, particular sub-processes can be enhanced by matching the pulse frequencies to transitions between molecular Eigen-states. This provides high selectivity and flexibility due to momentum and energy conservation of the interacting photons with the material. The different non-linear processes can typically be ordered by the number of involved photons: sum-frequency generation, two-photon absorption and stimulated emission with two photons, followed by other non-linear X-ray phenomena, time resolved transient gratings (XUV-TG) or four wave mixing (XUV-FWM) spectroscopy (see [8] and references therein). Fluorescent decays in the soft X-ray region allow unique access to the structure of the occupied valence states, while keeping the element selectivity and chemical state specificity of soft X-ray spectroscopies [9]. Unfortunately, the probability for a fluorescent decay in the soft X-ray range is below 1%. With a typical spectrometer acceptance of less than 10^{-5} of the full solid angle into which the fluorescence is emitted, soft X-ray emission spectroscopy (XES) or resonant inelastic X-ray scattering (RIXS) are experiments where single photons have to be counted for hours in order to obtain useful information. Using stimulated emission spectroscopy combined to high intensity, small bandwidth, tunable soft X-ray beams of different colors, a stimulated beam containing the same information as a fluorescence spectrum can be formed—mitigating the low acceptance angle of the spectrometers, and suppressing the dominating Auger decays that also create electronic damage to the sample. With such techniques, two to three orders of magnitude in the signal levels can be gained through the suppression of Auger processes, while the beam directed into the spectrometer promises to gain about five orders of magnitude in detected signal levels—clearly having the potential to revolutionize how we can study matter with XES or RIXS based spectroscopies [10].

Beyond the understanding of molecular processes, one also wishes to manipulate them. In the quantum mechanical world this can be done by the established methods of coherent control [11,12]. When this approach is developed for FELs, the new frontiers opened up include control of inner valence and core levels, and consequently element sensitivity. Optical coherent control is temporally limited to pulses that are several times the duration of an optical cycle, which is several femtoseconds. In the XUV to X-ray range, the periods are from hundreds to a few attoseconds. Thus, extremely fast processes can be controlled by tuning the relative phases and intensities of pulses delivered to the sample [13].

We provide an overview of the status and perspectives of the FERMI facility as an experimental tool to investigate a large array of phenomena ranging from nano-dynamics in complex materials to phenomena that are at the heart of conversion of light into other forms of energy.

2. FERMI Overview

FERMI is located at the Elettra laboratory in Trieste. The FEL facility covers the VUV to soft X-ray photon energy range with two FELs, FEL-1 and FEL-2 both based on the High Gain Harmonic Generation seeded mode (HGHG) [14]. The HGHG scheme consists in preparing the electron beam phase space in a first undulator, called modulator in Figure 1, where the interaction with an external laser, the seed, induces a controlled and periodic modulation in the electron beam longitudinal energy distribution.

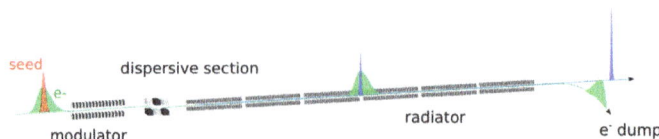

Figure 1. Sketch of the High Gain Harmonic Generation (HGHG) free electron laser (FEL) configuration. Seed laser pulse and electron beam are superimposed in a first undulator indicated as modulator. The FEL interaction induces an electron energy modulation with the periodicity of the seed wavelength. The dispersive section converts this energy modulation into a density modulation containing higher order Fourier components of the original modulation. One of these components is finally amplified in the radiator.

The beam propagates through a "dispersive section", a magnetic device introducing a strong correlation between the electron energy and the path length. This dispersive section converts the energy modulation into a density modulation, which is characterized by higher order harmonic components and retains the phase properties of the seed. The "density" modulated beam is then injected into a long FEL amplifier, similar to the one adopted in SASE FELs. The amplification process is initially enhanced by the presence of the modulation. The modulation depth is calibrated by tuning the seed intensity, in order to reach FEL saturation and efficient energy extraction at the end of the amplifier.

The HGHG cascade scheme is implemented in FERMI FEL-1, to generate fully coherent radiation pulses in the VUV spectral range [15]. The seed signal, continuously tunable, typically in the range 230–260 nm, is obtained from a sequence of nonlinear harmonic generation and mixing conversion processes from an optical parametric amplifier. The radiation resulting from conversion in the FEL up to the 13th harmonic is routinely delivered to user experiments [16].

The amplitude of the energy modulation necessary to initiate the HGHG process grows with the order of the harmonic conversion. The induced energy dispersion has a detrimental effect on the high gain amplification in the final radiator, for this reason the design of FEL-1 relied on harmonic conversions up to harmonic 13. During the last years of operations, we have demonstrated the ability to operate the FEL at even higher harmonic orders with reduced performances, e.g., up to the 20th harmonic, but substantially higher orders can be reached with a double stage HGHG cascade, where the harmonic conversion is repeated twice. The double conversion is done with the fresh bunch injection technique [17,18] on FERMI FEL-2 [19], shown in Figure 2. The FEL is composed by a first stage, analogous to FEL-1 (see Figure 1), followed by a delay line, a magnetic chicane slowing down the electron beam with respect to the light pulse generated in the first stage. The light pulse from the first stage is shifted to a longitudinal portion of the beam unperturbed by the seed in the first stage. In this way the light from the first stage functions as a short wavelength seed for the second stage. This scheme was implemented for the first time on FERMI FEL-2 and was used to demonstrate the seeded FEL coherent emission in the soft-X rays, up to harmonic orders of 65, and more [19].

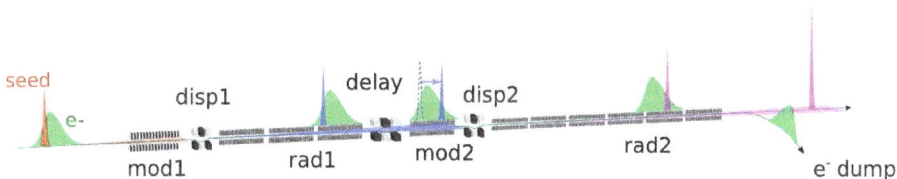

Figure 2. Schematic layout of FERMI (acronym of Free Electron laser Radiation for Multidisciplinary Investigations) FEL-2, implementing the HGHG double stage cascade in the fresh-bunch high-gain harmonic generation configuration. The first stage (mod1-disp1-rad1) is analogous to the high-gain harmonic generation scheme of FEL-1 shown in Figure 1. The second stage (mod2-disp2-rad2) is based on the same concept, but the seed is the radiation produced in the first stage. The two stages are separated by a delay line (delay) which lengthens the electron path with respect to the radiation path allowing to shift the seed over a "fresh" portion of the electron beam.

3. FEL Performances

3.1. Longitudinal Coherence and Pulse Duration

FEL-1 and FEL-2 may be operated even without the presence of an external seed, i.e., in SASE mode. The emission may be indeed enhanced by tuning the dispersive section of FEL-1 [20] or the multiple dispersion sections of FEL-2 [21], to increase the electron bunching at entrance of the final amplifier. In this configuration, the light pulse resembles the longitudinal structure of the beam current, with a correlation length much shorter than the electron bunch length. The consequence is a longer pulse, of several hundred of fs, with broad spectrum and spiky spectral features, characteristic of SASE FELs. This mode of operation can be used in those experiments requiring longer pulses and broadband emission, for alignments, or in the eventuality the seed laser system is not available. In Figure 3 we show the seeded FEL spectrum measured at harmonic order 8 compared to an optimized Optical-Klystron-SASE spectrum (OK-SASE spectrum). In in the example shown in the figure, in seeded mode the relative linewidth drops by about an order of magnitude. Typical relative linewidths of 2 to 5×10^{-4} (rms) are routinely achieved on both FEL-1 and FEL-2. Less evident from the images in Figure 3, the presence of the seed also affects the pulse duration.

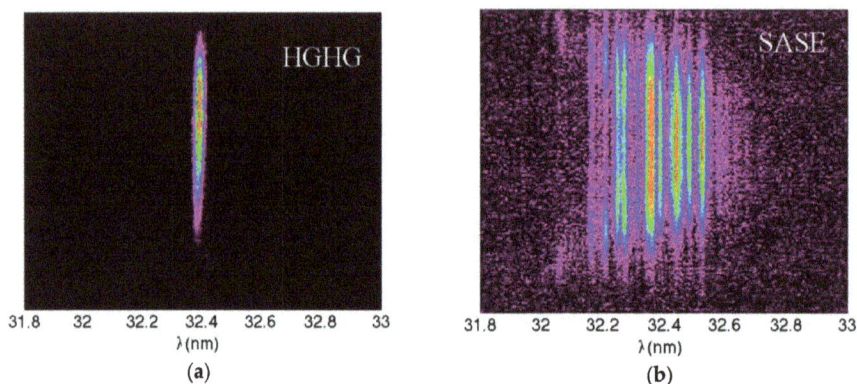

Figure 3. Spectrum in seeded mode (**a**) and in Self-Amplified Stimulated Emission (SASE) mode; (**b**) Wavelengths are dispersed along the horizontal axis. The vertical axis represents the vertical position at the spectrometer detector. The distribution on the vertical axis gives the projection of the beam spatial distribution on the vertical plane.

In seeded mode the pulse duration depends primarily on the duration of the seed, scaled according to the harmonic order n by a coefficient $7/6n^{1/3}$ [22]. Typical pulse duration with a seed duration of 120–150 fs are 50–70 fs (fwhm) on FEL-1. FEL-2 seeded with a short seed laser (70 fs) is expected to deliver FEL pulses of about 20 fs, depending on the laser setup and harmonic conversion order. We may therefore compare the longitudinal coherence of the source in the two modes of operation, in terms of distance from the Fourier transform limit for a Gaussian pulse (FTL),

$$\Delta t \Delta \lambda = 0.44 \lambda^2 / c$$

where Δt and $\Delta \lambda$ correspond to the full width half maximum of the temporal and spectral distributions of the optical pulse. If we consider the FERMI SASE conditions of operation in Figure 3b where the spectral width is about 0.3 nm, and where the optical pulse duration should be comparable to the electron bunch length, of the order of 0.5–0.7 ps, a rough estimate would indicate the pulse about 120–200 × FTL. This value, with the input seed and with the proper tuning of the FEL parameters, drops to ≈1.2–1.5FTL [23]. Specific tuning of the e-beam and the seed laser parameters may be set to bring the FEL even closer to the ideal Fourier limit [24]. The example in Figure 3 corresponds to harmonic order 8 of FEL-1. A narrow, single mode spectral line, such as the one shown in Figure 3 corresponding to harmonic 8 of FEL-1 seeded at 260 nm, is available in the entire spectral window of photon energies emitted by the FERMI FELs (20–60 eV on FEL-1 and 60–330 eV on FEL-2). As a second example in Figure 4, the spectrum of the second stage of FEL-2 is shown. The high end of the nominal photon energy range is reached upshifting the seed frequency by harmonic order 13 in the first stage and harmonic 5 in the second stage, for a total frequency upshift of 65.

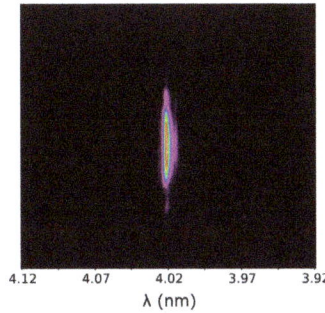

Figure 4. Spectrum of FERMI FEL-2 at harmonic 65 of the input seed. As in Figure 3, wavelengths are dispersed along the horizontal axis. As in Figure 3 the vertical axis represents the vertical position at the spectrometer detector.

3.2. Energy Per Pulse & Temporal Jitter

This dramatic increase of the longitudinal coherence with respect to a SASE source is only the first evident advantage of starting the amplification process from an external seed. The seed introduces indeed a number of additional handles to control amplification and the resulting output pulse properties. The central emission frequency at the lowest order of approximation is determined by the harmonic of the seed laser. The central emission frequency jitter is therefore normally lower than the linewidth by up to an order of magnitude. A SASE source is characterized by an intrinsic pulse energy fluctuation associated with the startup from the electron shot noise. From this point of view a seeded FEL is, conversely, a completely deterministic system.

The fluctuations of the pulse energy are determined by fluctuations of the external parameters driving the amplifier, as the energy of the electron beam, the timing between electrons and seed, and the intensity of the seed itself. The energy jitter may be lower than 5%, as in the example shown in Figure 5, where the histogram of the energy per pulse from FEL-1 in optimized conditions at 51 nm is displayed. This value may be compared to the typical energy jitter of the input seed laser in the UV, which is about 1%. A typical figure for the energy jitter is 10–15% for FEL-1 and 25–35% for FEL-2 with the double cascade. In terms of temporal synchronization, the seed timing largely determines the arrival time of the FEL photon pulse. At FERMI the optical synchronization between the seed and the lasers used in pump an probe experiments, which originates from the same oscillator, is associated to a jitter between the FEL light and the other sources of about 5 fs [25]. Recent measurements have shown even lower values, below 3 fs [22].

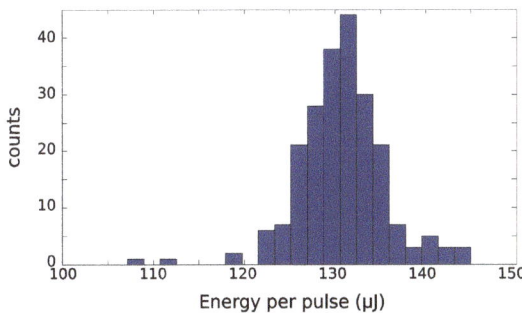

Figure 5. Histogram of the energy per pulse measured from FEL-1 at harmonic 5 (51 nm). Acquisition of 220 consecutive shots, average 131 μJ, standard deviation 4.8 μJ.

The typical energy per pulse depends on the wavelength of operation and in the requirements on the spectral linewidth. Highest energy per pulse may be achieved in saturated conditions, where the pulse shape may be affected by non-linear processes in the longitudinal electron dynamics. On FEL-1 an energy per pulse larger than 100 μJ is available in the entire range of operation (20–100 nm), in conditions of single mode, narrow linewidth spectral line. On FEL-2 this is possible in the low end of the spectral energy range, at wavelengths above 8–10 nm. The average energy per pulse decreases while increasing the photon energy, down to about 10 μJ at 4 nm. The energy per pulse may be lower than the one achievable in an optimized equivalent SASE FEL where most of the charge in the bunch participates to the FEL action. The seeded portion of the electron beam is typically of the order of 100 fs, which can be 10–20% or less of the entire bunch. Compressing the entire electron bunch down to the seed duration in an ultrashort multi-kA spike of current would ensure a higher efficiency, but would also introduce a much higher sensitivity to longitudinal jitter between the seed pulse and the current, and the compression process would be strongly affected by self-field effects, such as space charge, coherent synchrotron radiation and space charge fields [26], which alter the smooth distribution of the electron longitudinal phase space necessary to preserve the quality of the pulse during amplification. Moreover, an excess of peak current would increase the gain to a value leading the electron shot noise background to compete with the seed at saturation, and this would remove most of the advantages of seeding the amplifier. An alternative to extend the seeded region of the electron current is to use a temporally stretched, chirped seed pulse, thus extracting a larger amount of energy from the electrons. The chirp properties of the output pulse are strictly correlated to those of the seed. The FEL pulse can be therefore optically compressed to increase the peak power. The scheme is equivalent to the chirped pulse amplification scheme (CPA), commonly adopted in ultrashort pulse-solid state lasers [27]. Here the issue is not that of reducing the peak power in an active medium, but to allow the use of an extended electron beam longitudinal region, increasing the active bunch charge, without compressing the electrons to the ultimate limit. Chirped pulse amplification (CPA) in FEL amplifiers was proposed originally in [28] and further studied in the framework of a SASE amplifier in [29]. The scheme was later demonstrated at FERMI in [30].

3.3. Multiple Pulses

The intensity of the seed determines the depth level of saturation reached in the amplifier. This parameter can be tuned to control the intensity and to some extent the longitudinal shape of the laser pulse. An excess of seed can indeed be used for the generation of virtually jitter-free twin pulses delayed in time [31], which can also be separated in frequency by a proper frequency chirp of the input seed [32,33]. There are several other methods for generating multiple color—multiple pulses at FERMI. From FEL-2 for example, two color pulses almost superimposed in time, are naturally available as the emission from the first stage and the second stage. In normal operation, the light from the first stage has to be removed or attenuated by gas attenuators or filters, but in some specific experiments the VUV light from the first stage can be used in combination with XUV-Soft-X ray emitted by the second stage. This setup was used in the study of the ultrafast dynamic of melting in Si monitored by tracking the L(2,3)-edge shift [34]. Multiple pulses of different colors can be also generated on FEL-1. In Figure 6 we sketch three methods which were demonstrated in the past. In Figure 6a a second color is obtained by simply tuning part of the radiator at one harmonic and part of it at a second, different harmonic. The temporal jitter between the two components was estimated in an experiment of coherent control where we manipulated the relative phase of two colors (63.0 nm and 31.5 nm) to control the asymmetry of the photoelectron angular distribution of ionized neon with a temporal resolution of 3 as [13]. Temporally separated pulses can be generated in the setup of Figure 6b. A double seed can be indeed injected with a temporal delay between the two seed pulses, with delays comprised between 200 and 600–700 fs, to ensure overlapping with the electron beam current. Even in this case the temporal jitter between the two pulses can be sufficiently low to preserve a phase locking between the two pulses [35]. The two seed pulses can be also separated in frequency to generate two distinguishable colors, with the condition that they are both included in the gain bandwidth of the FEL amplifier (0.7–0.8%) [36]. A larger frequency separation between the two pulses can be achieved (Figure 6c) if the amplifier is tuned to different harmonics of the seed. In this case two seed pulses are still separated in frequency by less than the bandwidth of the modulator (3%) and both generate energy, and then density modulation at the entrance of the amplifier. The amplifier is then separated in two parts, one resonant with the modulation resulting from the first seed pulse and the other resonant with the one from the second. The two amplified pulses have to be separated in frequency to be generated and amplified separately in these two undulator parts and the amplification can happen on different harmonics of the seed [37].

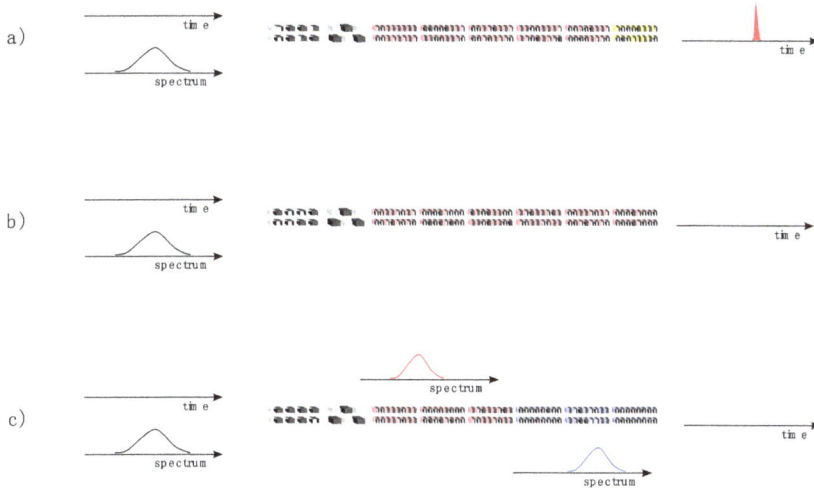

Figure 6. Various method for the generation of multiple pulses adopted on FERMI FEL-1. (**a**) Two colors can be generated by tuning the final amplifier to different harmonics of the seed. The output pulse is composed by the superposition of the two harmonic components; (**b**) A double seed can be injected with a temporal delay between the two seed pulses comprised between 200 and 600–700 fs. The two seed pulses can be separated in frequency to generate two distinguishable colors, by ensuring they are both included in the gain bandwidth of the FEL amplifier (0.7–0.8%); (**c**) A larger frequency separation between the two pulses is also possible if the amplifier is tuned to different harmonics of the seed.

4. XUV Wave Mixing and Coherent Control

In this section, we discuss the advantages offered by a fully coherent FEL source in the extension of experimental techniques as non-linear optics at shorter wavelength. An interesting aspect of the development of non-linear optics is that the main concepts at the basis of this field were already available for J. C. Maxwell (if in 1861 he had considered a power series expansion in the Maxwell equations) and H. A. Lorentz (if in 1878 he had introduced anharmonic terms in the oscillator model of the atom). The point is that neither Maxwell nor Lorentz had the experimental tools for inspiring their research in this direction. Until a few years ago non-linear light-matter interactions in the extreme ultraviolet (XUV) and X-ray range were basically ignored for the same reason. However nowadays FELs have allowed undertaking relevant steps towards the exploitation of XUV/X-ray non-linear optics. For instance, FELs have been employed to demonstrate: stimulated X-ray emission [3], amplified spontaneous XUV emission [10], X-ray-optical sum-frequency generation [38], X-ray two-photon absorption [39], X-ray second harmonic generation [40]; all these achievements were gained in the last few years, reflecting rapid growth in the field. A common motivation for many of the aforementioned investigations is the outlook represented by XUV/X-ray wave mixing experiments.

4.1. Four Wave Mixing

Among wave-mixing processes, four wave mixing (FWM) processes have a special place, since they are at the basis of most experimental methods, besides being the lowest order non-linear processes that are not vanishing by reason of symmetry. FWM arise from the 3rd-order non-linear interactions of three coherent electromagnetic fields ($E_{1,2,3}$) that might have different frequencies ($\omega_{1,2,3}$), wavevectors ($k_{1,2,3}$), polarizations, bandwidth, time delays, etc. (see Figure 7).

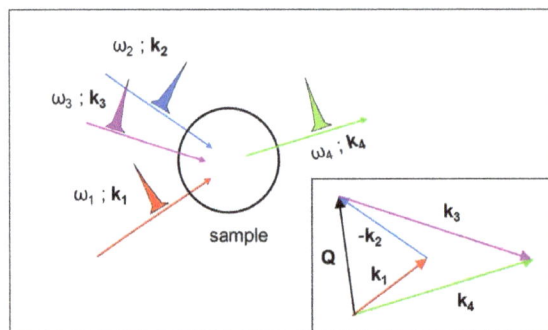

Figure 7. Sketch of the four wave mixing (FWM) experimental scheme. After the full stop: The non-linear interaction of the three photon pulses (i.e., coherent electromagnetic fields $E_{1,2,3}$) that may have different frequencies and wavevectors (1,2,3) is represented in the figure. The pulses may have as well different polarization, bandwidth and time delays. In the inset is depicted the phase matching condition.

Such interacting fields drive the radiation of a fourth (signal) field, whose photon parameters (frequency, ω_4, wavevector, k_4, polarization, etc.) may differ from those of the input fields. The possibility to experimentally control the input field parameters and to determine the output field parameters turns into the capability to stimulate and detect different FWM processes, which often carry out distinct and complementary information on the sample under study. Such a high degree of selectivity and richness of information makes FWM an extremely versatile and informative tool. For instance, through a FWM process termed impulsive stimulated scattering (ISS) it is possible to determine in real time and with wavevector selectivity the dynamics of both ultrafast (sub-picosecond) molecular vibrations or slow (millisecond) thermal and structural relaxations, depending on the bandwidth and time delays of the interacting fields [41]. Information on the dynamics of higher energy (i.e., faster) sample excitations can be achieved, e.g., by coherent Raman scattering (CRS) and coherent multi-dimensional spectroscopy; methods also able to provide insights into the correlations between different dynamical variables, such as those related to vibrational and electronic modes [42]. FWM approaches are also bringing relevant technological advances, ranging from sub-wavelength microscopy [43] to wavefunction tomography during chemical reactions [44]. The hard-limit of optical FWM, as any other optical spectroscopy, is inherently related to the long wavelength (some 100's of nm or more) and low photon energy (a few eV's or less) of the optical radiation itself. This prevents to probe matter at atomic and molecular length scales (a few nm or less) as well as to study excitations with energies larger than a few eV's, a limitation that basically confines optical CRS to the study of vibrational dynamics and that of low-energy electronic excitations. Furthermore, optical photons cannot exploit core-level electronic resonances, typically located in the 10's eV to 10's keV range, adding elemental selectivity to the FWM approach. The latter point combined with the possibility to detect high energy excitation (e.g., valence band excitons in the 1–10 eV energy range) on a Q-range comparable with the inverse molecular size ($\sim$nm^{-1}) are the main advantages expected from the XUV/X-ray analogue of CRS (XCRS). By tuning the frequencies of the input beams involved in the

XCRS process to core resonance of distinct atoms it would be possible to determine where a given electronic wave-packet is created and where it is probed, as well as to follow in real time the dynamics of such charge-transfer between the selected atoms. This unique capability arises from the multi-wave nature of XCRS, which permits to overcome the basic limitation of any linear X-ray method, where the light-matter interaction occurs in correspondence of a single atomic site and prevents the detection of real-time dynamics between distinct atoms. Moreover, XCRS is not limited to the study of valence band excitons, since the same concept in principle applies for all kind of excitations, related to any dynamical variable coupled to the field (in some circumstances also to those uncoupled in the linear regime), with energies lower than that of core-resonances. Such class basically includes all kind of modes related to nuclear (phonons, structural relaxations, heat diffusion, etc.), electronic (excitons, plasmons, etc.) and "mixed" (polarons, polaritons, etc.) degrees of freedom. The XCRS approach may hence allow to study, e.g., charge and energy transfer processes among different atoms in molecules, the delocalization and correlations of electronic excitations as well as structural fluctuations, nuclear motions and relaxation processes, other than detecting the dynamics of elementary excitations such as, e.g., phonons, plasmons and polarons. Among the worldwide existing FEL facilities, FERMI possesses unique characteristics for FWM experiments in the XUV/soft X-ray domain. Firstly, most of the photon parameters of the FEL radiation emitted (the longitudinal coherence properties among them) are related to those of the seed laser, which is fully controllable in all relevant parameters, so that it is possible to control the FEL output by simply acting on the seed. Moreover, it is possible to use two (or more) independently controllable seed laser pulses, which result into the radiation of two (or more) FEL-pulses with controllable photon energy, polarization, time delay, etc.; a further development of such scheme includes the simultaneous use of multiple resonances in the FEL amplifier, which largely extends the separation in the photon energy (i.e., up to several eV's) of the multiple FEL pulses [37]. Finally, the layout of FERMI could be adapted to generate a few fs to sub-fs FEL-pulses with the aforementioned benefits. In a nutshell, today FERMI is the FEL source most similar to a "multi-color" conventional laser and, therefore, it could be used to attempt the development of XUV/soft X-ray FWM. A significant endorsement of the latter statement is represented by the recent demonstration of a time-resolved FEL-stimulated FWM response [45] (see Figure 8). In this experiment two time-space coincident ultrafast (time duration ~70 fs) FEL pulses (photon energy ω_{XUV} ~45 eV) were crossed at the sample (amorphous SiO_2) position.

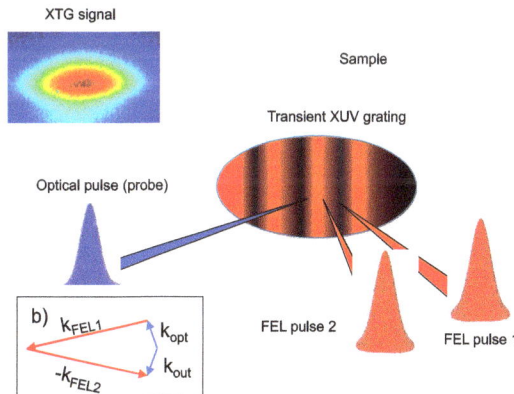

Figure 8. FEL-based FWM experiment stimulated by transient extreme ultraviolet (XUV) gratings (inset (b) reports the phase matching geometry: k_{FEL1}, k_{FEL2}, k_{opt} and k_{out} are the wavevectors of the two FEL pulses, the optical pulse and the FWM signal, respectively). The signal (XTG) has been registered on a charged coupled detector (CCD).

These beams were expected to generate an XUV grating able to scatter off, through FWM processes, a third coherent beam, provided that the latter is send into the sample in phase matching conditions. Indeed, using a third "phase matched" optical pulse (time duration ~ 100 fs, photon energy ω_{opt} ~ 3.1 eV) in the interaction region allowed the observation of an optical beam emerging from the sample along the expected "phase matched" direction (see Figure 8 XTG signal).

4.2. Perspective for the XUV/X-ray FWM

Further steps can be undertaken to make XUV/X-ray FWM experiments feasible on routine basis. One of the most useful would be to replace the optical pulse with a XUV/X-ray. This will permit to probe, via ISS-type FWM methods, low energy modes (e.g., acoustic modes) in a wavevector range (~0.1–1 nm^{-1}) hardly accessible by both optical methods and linear X-ray spectroscopy [46]. Such a wavevector range matches the characteristic length-scales of heterogeneities in the local structures of several classes of materials (e.g., block copolymer, relaxor ferroelectrics, glasses etc.), the incommensurate dimensions of many crystalline phases showing super-lattices of different natures, as well as the characteristic dimensions and periodicities of many nanostructures. Among these applications we mention the study of acoustic dynamics in glasses, which is largely believed to be the origin of the anomalous thermal properties of glasses, still a lively debated issue [47]. A debate that essentially comes from the impossibility to fully access such a ~0.1–1 nm^{-1} wavevector range and from the stark disagreement often observed between the extrapolations of the trends found at lower and larger wavevectors. Also, couplings between acoustic modes and local vibrations in nm-sized elastic domains, inherently connected to the amorphous local structure, may also have a relevant role; such a small length-scale is potentially in the range of XUV/X-ray FWM. Another relevant step to extend the range of applications of FEL-based FWM is to exploit the two-color emission scheme in order to demonstrate FEL-based CRS-type experiments. This would be a major step forward towards the realization of XCRS with atomic selectivity (see Figure 9), which in many respects may be considered as a main target application of XUV/X-ray FWM.

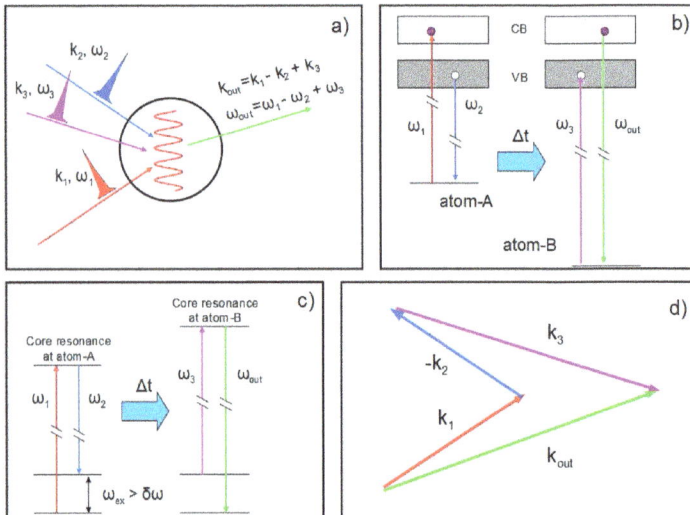

Figure 9. (**a**) Sketch of the FEL-based X-ray analogue of coherent Raman scattering (XCRS) experiment; (**b**,**c**) are the excitation processes and level scheme for a XUV/X-ray CRS experiment involving core transitions in both the excitation and probing process; (**d**) Phase matching diagram.

Indeed, the potential of XCRS to follow charge and energy flows between constituent atoms in materials would be of the greatest importance to address some fundamental scientific issues, such as: (a) the study of intramolecular relaxation dynamics in metal complexes, which are the doorway to photo-induced charge separation (the high selectivity of XCRS would be exploited to understand and disentangle processes, such as intramolecular vibrational redistribution, internal conversion and intersystem crossing, occurring upon photo-excitation; those are processes of fundamental interest in devising efficient molecular systems for applications as diverse as solar energy conversion, biology or data storage [48]); and (b) the dynamics of charge injection and transport in photocatalytic reactions taking place in metal oxides nanoparticles, such as TiO_2, that are key materials for renewable energy; for instance, the key event occurring in devices for solar energy conversion is the generation of a charge-separated state through ultrafast electron injection from an excited metal-complex, adsorbed on a nanoporous metal oxide substrate, to the conduction band of the substrate [49]. In this context the atomic selectivity of XCRS may allow us to understand whether such ultrafast electronic excited states have an O or Ti character.

4.3. Coherent Control

In order to manipulate molecular processes, coherent control could be the technique of choice. In this kind of experiment Phase coherent light of one or more colours interacts with a target. The outcome of the interaction is determined by the phase and amplitude of the light. Ion yield, direction of emission of ions and electrons, and so on, can be controlled. In the XUV to X-ray range, the times of the duration of an optical cycle periods are from hundreds to a few attoseconds thus increasing the temporal resolution of orders of magnitude when comparing to optical lasers. Recently FERMI has been employed to carry out the first XUV coherent control experiment where the first and the second harmonics were ionizing Neon on the $2p^5 4s$ resonance [13]. While the second harmonic was producing a single photon ionization process, the first was exiting two photon ionization processes (see Figure 10a).

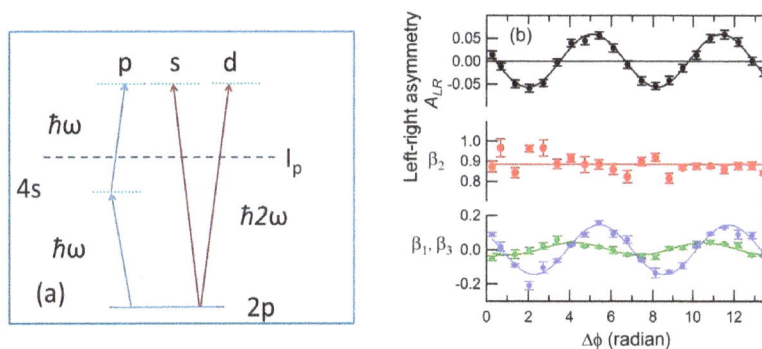

Figure 10. (a) Scheme of the experiment. Left: a $2p$ electron is excited to $4s$ by one photon and then emitted as a p-wave by a second photon. Right: a $2p$ electron process is emitted as an $(s + d)$-wave by a one-photon process. (b) Asymmetry parameter A_{LR} as a function of $\Delta\phi$ (black curve), and β_1 (green), β_3 (blue) and β_2 (red) parameters as a function of $\Delta\phi$. Experimental data are shown as markers with error bars. The lines are sinusoidal or straight line fits for β_1, β_3, and β_2 respectively.

The possibility offered by FERMI to change the phase among the two harmonics (see Figure 10b) allowed the first experimental demonstration that coherent control experiments can be carried out in the XUV X-ray region.

Much effort is being invested to develop ever shorter pulses at FELs. The arrival of phase control of multi-color pulses now means that they can be designed to produce trains of very short pulses, since a train of attosecond pulses can be constructed from a coherent superposition of coherent femtosecond pulses with commensurate wavelengths. This ability to tailor pulses may eventually lead to "design" single pulses of attosecond duration.

5. Discussion on Future Perspectives

A number of critical developments would boost the impact of the FERMI FELs in nonlinear multi-wave spectroscopies and the other applications of the seeded FERMI FEL sources. The Elettra-Sincrotrone Trieste scientific advisory committee, has recently outlined the importance of extending the spectral range of FEL-2 toward higher photon energies, that would allow to reach the K-edges of C (284 eV), N (410 eV) and O (543 eV). Higher order harmonics could be exploited to extend the photon delivery to the L-edges of the 3d transition metals relevant for magnetism catalysis and solar energy production. The required increase in electron beam energy would be modest and within the capabilities of the present infrastructure. On the other side, the generation of shorter pulses (sub-10 fs) would allow the investigation of the faster electron dynamics at the level of typical core hole lifetimes (4–8 fs for O, N and C). We already mentioned the chirped pulse amplification as a means for the generation of ultrashort pulses. Other options are under investigation to reduce the pulse duration while maintaining the synchronization and longitudinal coherence properties of the seeded source. On one side a proper shaping of the beam properties could limit the longitudinal region where lasing takes place [50], and the FERMI linac laser heater is an additional handle to control the properties of the FEL light [51,52]. On the other side, nonlinear dynamics at saturation may lead to a longitudinal self-focusing regime [53–55] which is well suited for reaching the sub-10 fs regime in the FERMI configuration. A second direction for the future development of FERMI in multidimensional spectroscopy and the control of the light properties in advanced pump and probe configurations is in a further extension of the flexibility for the generation of multiple colour, multiple pulses. While the generation of multiple pulses on FEL-1, with the constraints imposed by the double lasing condition, is routinely achieved, it is all but straightforward in the higher photon energy range, on the double stage configuration of FEL-2. The same spectral range could be reached on FEL-2 with an Echo Enabled harmonic generation configuration (EEHG) [56–58] that would allow on FEL-2 a similar flexibility as the one demonstrated on FEL-1. For this purpose, an experiment on FEL-2 is under study [59] and is planned for 2018. The requirement for different pump and probe pulses even calls for frequency tripling for the few-fs FEL emission scheme. In this respect, the possibility of combining FEL1 and FEL2 on the same experiment, with a variable controlled delay, could lead to a further extension in multicolour multiple pulses generation. The coherence properties of FEL-1 and FEL-2 could be combined while maintaining higher pulse energy and the full FELs flexibility in the choice of the radiation colour for the two pulses [60].

6. Conclusions

Free electron laser radiation, through its unique combination of parameters, namely ultrashort pulses with high brightness and coherence, has led to revolutionary steps forward in time resolved X-ray methodologies. A decade ago these developments were just dreams for the researchers who have recognized that a broad class of phenomena can be investigated by ultrafast X-rays [61]. The field of non-linear XUV/X-ray optics is one of the youngest within fields opened by FELs and we expect that it will advance fast, paralleling the ongoing development of FEL technology, and will contribute to gaining a deeper comprehension of fast dynamics and learning how to manipulate materials. In this framework, the FERMI facility would play a major role, thanks to its unique characteristics related to the pioneering laser-seeding scheme.

Acknowledgments: The authors gratefully acknowledge the FERMI teams for the outstanding work done in running the facility and for kindly providing the reference data and pictures on the FEL operation reported in this document.

Author Contributions: L.G. and C.M. equally contributed to the preparation of text and figures of this manuscript.

Conflicts of Interest: The authors declare no conflict of interest.

References

1. Tiedtke, K.; Azima, A.; von Bargen, N.; Bittner, L.; Bonfigt, S.; Düsterer, S.; Faatz, B.; Frühling, U.; Gensch, M.; Gerth, C.; et al. The soft X-ray free-electron laser FLASH at DESY: Beamlines, diagnostics and end-stations. *New J. Phys.* **2009**, *11*, 023029. [CrossRef]
2. Emma, P.; Akre, R.; Arthur, J.; Bionta, R.; Bostedt, C.; Bozek, J.; Brachmann, A.; Bucksbaum, P.; Coffee, R.; Decker, F.J.; et al. First lasing and operation of an angstrom-wavelength free-electron laser. *Nat. Photonics* **2010**, *4*, 641–647. [CrossRef]
3. Rohringer, N.; Ryan, D.; London, R.A.; Purvis, M.; Albert, F.; Dunn, J.; Bozek, J.D.; Bostedt, C.; Graf, A.; Hill, R.; et al. Atomic inner-shell X-ray laser at 1.46 nanometres pumped by an X-ray free-electron laser. *Nature* **2012**, *481*, 488–491. [CrossRef] [PubMed]
4. Yoneda, H.; Inubushi, Y.; Nagamine, K.; Michine, Y.; Ohashi, H.; Yumoto, H.; Yamauchi, K.; Mimura, H.; Kitamura, H.; Katayama, T.; et al. Atomic inner-shell laser at 1.5-ångström wavelength pumped by an X-ray free-electron laser. *Nature* **2015**, *524*, 446–449. [CrossRef] [PubMed]
5. Allaria, E.; Badano, L.; Bassanese, S.; Capotondi, F.; Castronovo, D.; Cinquegrana, P.; Danailov, M.B.; D'Auria, G.; Demidovich, A.; De Monte, R.; et al. The FERMI free-electron lasers. *J. Synchrotron Radiat.* **2015**, *22*, 485–491. [CrossRef] [PubMed]
6. Nannan, Z. Dalian Coherent Light Source Produces Laser Output for the First Time. Available online: http://english.cas.cn/newsroom/research_news/201610/t20161013_168621.shtml (accessed on 12 May 2017).
7. Amann, J.; Berg, W.; Blank, V.; Decker, F.-J.; Ding, Y.; Emma, P.; Feng, Y.; Frisch, J.; Fritz, D.; Hastings, J.; et al. Demonstration of self-seeding in a hard-X-ray free-electron laser. *Nat. Photonics* **2012**, *6*, 693–698. [CrossRef]
8. Bencivenga, F.; Capotondi, F.; Kiskinova, M.; Masciovecchio, C. Coherent and transient states studied with X-ray FELs: Present and future prospects. *Adv. Phys.* **2015**, *63*, 327–404. [CrossRef]
9. Nordgren, J.; Bray, G.; Cramm, S.; Nyholm, R.; Rubensson, J.E.; Wassdahl, N. Soft X-ray Emission Spectroscopy Using Monochromatized Synchrotron Radiation (Invited). *Rev. Sci. Inst.* **1989**, *60*, 1690. [CrossRef]
10. Beye, M.; Schreck, S.; Sorgenfrei, F.; Trabant, C.; Pontius, N.; Schüßler-Langeheine, C.; Wurth, W.; Föhlisch, A. Stimulated X-ray Emission for Materials Science. *Nature* **2013**, *501*, 191. [CrossRef] [PubMed]
11. Brif, C.; Chakrabarti, R.; Rabitz, H. Control of quantum phenomena: Past, present and future. *New J. Phys.* **2010**, *12*, 075008. [CrossRef]
12. Ehlotzky, F. Atomic phenomena in bichromatic laser fields. *Phys. Rep.* **2001**, *345*, 175–264. [CrossRef]
13. Prince, K.C.; Allaria, E.; Callegari, C.; Cucini, R.; de Ninno, G.; Di Mitri, S.; Diviacco, B.; Ferrari, E.; Finetti, P.; Gauthier, D.; et al. Coherent control with a short-wavelength Free Electron Laser. *Nat. Photonics* **2016**, *10*, 176–179. [CrossRef]
14. Yu, L.H. Generation of intense UV radiation by subharmonically seeded single-pass free-electron lasers. *Phys. Rev. A* **1991**, *44*, 5178–5193. [CrossRef] [PubMed]
15. Allaria, E.; Appio, R.; Badano, L.; Barletta, W.A.; Bassanese, S.; Biedron, S.G.; Borga, A.; Busetto, E.; Castronovo, D.; Cinquegrana, P.; et al. Highly coherent and stable pulses from the FERMI seeded free-electron laser in the extreme ultraviolet. *Nat. Photonics* **2012**, *6*, 699–704. [CrossRef]
16. Allaria, E.; Battistoni, A.; Bencivenga, F.; Borghes, R.; Callegari, C.; Capotondi, F.; Castronovo, D.; Cinquegrana, P.; Cocco, D.; Coreno, M.; et al. Tunability experiments at the FERMI@Elettra free-electron laser. *New J. Phys.* **2012**, *14*, 113009.
17. Ben-Zvi, I.; Yang, K.M.; Yu, L.H. The "fresh-bunch" technique in FELS. *Nucl. Instrum. Methods Phys. Res. A* **1992**, *318*, 726–729. [CrossRef]
18. Yu, L.H.; Ben-Zvi, I. High-gain harmonic generation of soft X-rays with the "fresh bunch" technique. *Nucl. Instrum. Methods Phys. Res. A* **1997**, *393*, 96–99. [CrossRef]
19. Allaria, E.; Castronovo, D.; Cinquegrana, P.; Craievich, P.; Dal Forno, M.; Danailov, M.B.; D'Auria, G.; Demidovich, A.; de Ninno, G.; Di Mitri, S.; et al. Two-stage seeded soft-X-ray free-electron laser. *Nat. Photonics* **2013**, *7*, 913–917. [CrossRef]

20. Penco, G.; Allaria, E.; de Ninno, G.; Ferrari, E.; Giannessi, L. Experimental Demonstration of Enhanced Self-Amplified Spontaneous Emission by an Optical Klystron. *Phys. Rev. Lett.* **2015**, *114*, 013901. [CrossRef] [PubMed]

21. Penco, G.; Allaria, E.; de Ninno, G.; Ferrari, E.; Giannessi, L.; Roussel, E.; Spampinati, S. Optical Klystron Enhancement to Self-Amplified Spontaneous Emission at FERMI. *Photonics* **2017**, *4*, 15. [CrossRef]

22. Finetti, P.; Höppner, H.; Allaria, E.; Callegari, C.; Capotondi, F.; Cinquegrana, P.; Coreno, M.; Cucini, R.; Danailov, M.B.; Demidovich, A.; et al. Pulse duration of seeded free-electron lasers. *Phys. Rev. X* **2017**, *7*, 021043. [CrossRef]

23. De Ninno, G.; Gauthier, D.; Mahieu, B.; Rebernik, P.; Allaria, E.; Cinquegrana, P.; Danailov, M.B.; Demidovich, A.; Ferrari, E.; Giannessi, L.; et al. Single-shot spectro-temporal characterization of XUV pulses from a seeded free-electron laser. *Nat. Commun.* **2015**, *6*, 8075. [CrossRef] [PubMed]

24. Gauthier, D.; Ribič, P.R.; De Ninno, G.; Allaria, E.; Cinquegrana, P.; Danailov, M.B.; Demidovich, A.; Ferrari, E.; Giannessi, L.; Mahieu, N.; et al. Spectro-temporal shaping of free-electron laser pulses. *Phys. Rev. Lett.* **2015**, *115*, 114801. [CrossRef] [PubMed]

25. Danailov, M.B.; Bencivenga, F.; Capotondi, F.; Casolari, F.; Cinquegrana, P.; Demidovich, A.; Giangrisostomi, E.; Kiskinova, M.P.; Kurdi, G.; Manfredda, M.; et al. Towards jitter-free pump-probe measurements at seeded free electron laser facilities. *Opt. Express* **2014**, *22*, 12869–12879. [CrossRef] [PubMed]

26. Giannessi, L. Simulation codes for high brightness electron beam free-electron laser experiments. *Phys. Rev. Spec. Top. Accel. Beams* **2003**, *11*, 114802. [CrossRef]

27. Strickland, D.; Mourou, G. Compression of amplified chirped optical pulses. *Opt. Commun.* **1985**, *56*, 219. [CrossRef]

28. Yu, L.H.; Johnson, D.; Li, D.; Umstadter, D. Femtosecond free-electron laser by chirped pulse amplification. *Phys. Rev. E* **1994**, *49*, 4480–4486.

29. Frassetto, F.; Giannessi, L.; Poletto, L. Compression of XUV FEL pulses in the few-femtosecond regime. *Nucl. Instrum. Methods Phys. Res. Sect. A* **2008**, *593*, 14–16. [CrossRef]

30. Gauthier, D.; Allaria, E.; Coreno, M.; Cudin, I.; Dacasa, H.; Danailov, M.B.; Demidovich, A.; Di Mitri, S.; Diviacco, B.; Ferrari, E.; et al. Chirped pulse amplification in an extreme-ultraviolet free-electron laser. *Nat. Commun.* **2016**, *7*, 13688. [CrossRef] [PubMed]

31. Labat, M.; Joly, N.; Bielawski, S.; Szwaj, C.; Bruni, C.; Couprie, M.E. Pulse splitting in short wavelength seeded free electron lasers. *Phys. Rev. Lett.* **2009**, *103*, 264801. [CrossRef] [PubMed]

32. De Ninno, G.; Mahieu, B.; Allaria, E.; Giannessi, L.; Spampinati, S. Chirped Seeded Free-Electron Lasers: Self-Standing Light Sources for Two-Color Pump-Probe Experiments. *Phys. Rev. Lett.* **2013**, *110*, 064801. [CrossRef] [PubMed]

33. Mahieu, B.; Allaria, E.; Castronovo, D.; Danailov, M.B.; Demidovich, A.; De Ninno, G.; Di Mitri, S.; Fawley, W.M.; Ferrari, E.; Frohlich, L.; et al. Two-colour generation in a chirped seeded Free-Electron Laser. *Opt. Express* **2013**, *21*, 022728. [CrossRef] [PubMed]

34. Principi, E.; Danailov, M.B.; Cucini, R.; D'Amico, F.; Pelizzo, M.G.; Mewes, L.-H.; Gessini, A.; Bencivenga, F.; Battistoni, A.; Giangrisostomi, E.; et al. Ultrafast dynamic of melting in Si monitored by tracking the L(2,3)-edge shift. In preparation.

35. Gauthier, D.; Ribič, P.R.; De Ninno, G.; Allaria, E.; Cinquegrana, P.; Danailov, M.B.; Demidovich, A.; Ferrari, E.; Giannessi, L. Generation of Phase-Locked Pulses from a Seeded Free-Electron Laser. *Phys. Rev. Lett.* **2016**, *116*, 024801. [CrossRef] [PubMed]

36. Allaria, E.; Bencivenga, F.; Borghes, R.; Capotondi, F.; Castronovo, D.; Charalambous, P.; Cinquegrana, P.; Danailov, M.B.; De Ninno, G.; Demidovich, A.; et al. Two-colour pump-probe experiments with a twin-pulse-seed extreme ultraviolet free-electron laser. *Nat. Commun.* **2013**, *4*, 2476. [CrossRef] [PubMed]

37. Ferrari, E.; Spezzani, C.; Fortuna, F.; Delaunay, R.; Vidal, F.; Nikolov, I.; Cinquegrana, P.; Diviacco, B.; Gauthier, D.; Penco, G.; et al. Widely tunable two-colour seeded free-electron laser source for resonant-pump resonant-probe magnetic scattering. *Nat. Commun.* **2016**, *7*, 10343. [CrossRef] [PubMed]

38. Glover, T.E.; Fritz, D.M.; Cammarata, M.; Allison, T.K.; Sinisa Coh; Feldkamp, J.M.; Lemke, H.; Zhu, D.; Feng, Y.; Coffee, R.N.; et al. X-ray and optical wave mixing. *Nature* **2012**, *488*, 603–608. [CrossRef] [PubMed]

39. Tamasaku, K.; Shigemasa, E.; Inubushi, Y.; Katayama, T.; Sawada, K.; Yumoto, H.; Ohashi, H.; Mimura, H.; Yabashi, M.; Yamauchi, K.; et al. X-ray two-photon absorption competing against single and sequential multiphoton processes. *Nat. Photonics* **2014**, *8*, 313–316. [CrossRef]

40. Shwartz, S.; Fuchs, M.; Hastings, J.B.; Inubushi, Y.; Ishikawa, T.; Katayama, T.; Reis, D.A.; Sato, T.; Tono, K.; Yabashi, M.; et al. X-ray Second Harmonic Generation. *Phys. Rev. Lett.* **2014**, *112*, 163901. [CrossRef] [PubMed]

41. Dhar, L.; Rogers, J.A.; Nelson, K.A. Time-Resolved Vibrational Spectroscopy in the Impulsive Limit. *Chem. Rev.* **1994**, *94*, 157–193. [CrossRef]

42. Hochstrasser, R.M. Two-dimensional spectroscopy at infrared and optical frequencies. *Proc. Natl. Acad. Sci. USA* **2007**, *104*, 14190–14196. [CrossRef] [PubMed]

43. Lewis, A.; Lieberman, K. Near-field optical imaging with a non-evanescently excited high-brightness light source of sub-wavelength dimensions. *Nature* **1991**, *354*, 214–216. [CrossRef]

44. Avisar, D.; Tannor, D.J. Complete Reconstruction of the Wave Function of a Reacting Molecule by Four-Wave Mixing Spectroscopy. *Phys. Rev. Lett.* **2011**, *106*, 170405. [CrossRef] [PubMed]

45. Bencivenga, F.; Cucini, R.; Capotondi, F.; Battistoni, A.; Mincigrucci, R.; Giangrisostomi, E.; Gessini, A.; Manfredda, M.; Nikolov, I.P.; Pedersoli, E.; et al. Four wave mixing experiments with extreme ultraviolet transient gratings. *Nature* **2015**, *520*, 205–208. [CrossRef] [PubMed]

46. Bencivenga, F.; Masciovecchio, C. FEL-based transient grating spectroscopy to investigate nanoscale dynamics. *Nuc. Instrum. Methods Phys. Res. A* **2009**, *606*, 785–789. [CrossRef]

47. Schirmacher, W.; Ruocco, G.; Scopigno, T. Acoustic Attenuation in Glasses and its Relation with the Boson Peak. *Phys. Rev. Lett.* **2007**, *98*, 025501. [CrossRef] [PubMed]

48. Bressler, Ch.; Milne, C.; Pham, V.-T.; ElNahhas, A.; van der Veen, R.M.; Gawelda, W.; Johnson, S.; Beaud, P.; Grolimund, D.; Kaiser, M.; et al. Femtosecond XANES Study of the Light-Induced Spin Crossover Dynamics in an Iron(II) Complex. *Science* **2009**, *323*, 489–492. [CrossRef] [PubMed]

49. Engel, G.S.; Calhoun, T.R.; Read, E.L.; Ahn, T.-K.; Mancal, T.; Cheng, Y.-C.; Blankenship, R.E.; Fleming, G.R. Evidence for wavelike energy transfer through quantum coherence in photosynthetic systems. *Nature* **2007**, *446*, 782–786. [CrossRef] [PubMed]

50. Marinelli, A.; Coffee, R.; Vetter, S.; Hering, P.; West, G.N.; Gilevich, S.; Lutman, A.A.; Li, S.; Maxwell, T.; Galayda, J.; et al. Optical Shaping of X-ray Free-Electron Lasers. *Phys. Rev. Lett.* **2016**, *116*, 254801. [CrossRef] [PubMed]

51. Roussel, E.; Ferrari, E.; Allaria, E.; Penco, G.; Di Mitri, S.; Veronese, M.; Danailov, M.; Gauthier, D.; Giannessi, L. Multicolor High-Gain Free-Electron Laser Driven by Seeded Microbunching Instability. *Phys. Rev. Lett.* **2015**, *115*, 214801. [CrossRef] [PubMed]

52. Grattoni, V.; Roussel, E.; Allaria, E.; Di Mitri, S.; Giannessi, L.; Ferrari, E.; Sigalotti, P.; Penco, G.; Veronese, M.; Badano, L.; et al. Control of the Seeded Fel Pulse Duration Using Laser Heater Pulse Shaping. Presented at the IPAC'17, Copenhagen, Denmark, 14–19 May 2017. Paper Number WEPAB034.

53. Giannessi, L.; Musumeci, P.; Spampinati, S. Nonlinear pulse evolution in seeded free-electron laser amplifiers and in free-electron laser cascades. *J. Appl. Phys.* **2005**, *98*, 043110. [CrossRef]

54. Watanabe, T.; Wang, X.J.; Murphy, J.B.; Rose, J.; Shen, Y.; Tsang, T.; Giannessi, L.; Musumeci, P.; Reiche, S. Experimental Characterization of Superradiance in a Single-Pass High-Gain Laser-Seeded Free-Electron Laser Amplifier. *Phys. Rev. Lett.* **2007**, *98*, 034802. [CrossRef] [PubMed]

55. Giannessi, L.; Artioli, M.; Bellaveglia, M.; Briquez, F.; Chiadroni, E.; Cianchi, A.; Couprie, M.E.; Dattoli, G.; Di Palma, E.; Di Pirro, G.; et al. High-order-harmonic generation and superradiance in a seeded free-electron laser. *Phys. Rev. Lett.* **2012**, *108*, 164801. [CrossRef] [PubMed]

56. Stupakov, G. Using the Beam-Echo Effect for Generation of Short-Wavelength Radiation. *Phys. Rev. Lett.* **2009**, *102*, 74801. [CrossRef] [PubMed]

57. Xiang, D.; Stupakov, G. Echo-enabled harmonic generation free electron laser. *Phys. Rev. Spec. Top. Accel. Beams* **2009**, *12*, 30702. [CrossRef]

58. Hemsing, E.; Dunning, M.; Garcia, B.; Hast, C.; Raubenheimer, T.; Stupakov, G.; Xiang, D.; et al. Echo-enabled harmonics up to the 75th order from precisely tailored electron beams. *Nat. Photonics* **2016**, *10*, 512–515. [CrossRef]

59. Rebernik Ribic, P.; Roussel, E.; Penn, G.; De Ninno, G.; Giannessi, L.; Penco, G.; Allaria, E. Echo-Enabled Harmonic Generation Studies for the FERMI Free Electron Laser. *Photonics* **2017**, *4*, 19. [CrossRef]

60. Penco, G.; Allaria, E.; Bassanese, S.; Cinquegrana, P.; Cleva, S.; Danailov, M.B.; Demidovich, A.A.; Ferianis, M.; Gaio, G.; Gauthier, D.; et al. Two-Bunch Operation at the FERMI FEL Facility. Presented at the IPAC'17, Copenhagen, Denmark, 14–19 May 2017. Paper Number WEPAB037.

61. Bennett, K.; Zhang, Y.; Kowalewski, M.; Hua, W.; Mukamel, S. Multidimensional resonant nonlinear spectroscopy with coherent broadband X-ray pulses. *Phys. Scr.* **2016**, 014002. [CrossRef]

applied sciences

MDPI

Article

Construction and Commissioning of PAL-XFEL Facility

In Soo Ko *, Heung-Sik Kang, Hoon Heo, Changbum Kim, Gyujin Kim, Chang-Ki Min,
Haeryong Yang, Soung Youl Baek, Hyo-Jin Choi, Geonyeong Mun, Byoung Ryul Park,
Young Jin Suh, Dong Cheol Shin, Jinyul Hu, Juho Hong, Seonghoon Jung, Sang-Hee Kim,
KwangHoon Kim, Donghyun Na, Soung Soo Park, Yong Jung Park, Young Gyu Jung,
Seong Hun Jeong, Hong Gi Lee, Sangbong Lee, Sojeong Lee, Bonggi Oh, Hyung Suck Suh,
Jang-Hui Han, Min Ho Kim, Nam-Suk Jung, Young-Chan Kim, Mong-Soo Lee, Bong-Ho Lee,
Chi-Won Sung, Ik-Su Mok, Jung-Moo Yang, Yong Woon Parc, Woul-Woo Lee, Chae-Soon Lee,
Hocheol Shin, Ji Hwa Kim, Yongsam Kim, Jae Hyuk Lee, Sang-Youn Park, Jangwoo Kim,
Jaeku Park, Intae Eom, Seungyu Rah, Sunam Kim, Ki Hyun Nam, Jaehyun Park, Jaehun Park,
Sangsoo Kim, Soonnam Kwon, Ran An, Sang Han Park, Kyung Sook Kim, Hyojung Hyun,
Seung Nam Kim, Seonghan Kim, Chung-Jong Yu, Bong-Soo Kim, Tai-Hee Kang,
Kwang-Woo Kim, Seung-Hwan Kim, Hee-Seock Lee, Heung-Soo Lee, Ki-Hyeon Park,
Tae-Yeong Koo, Dong-Eon Kim and Ki Bong Lee

Pohang Accelerator Laboratory, POSTECH, Pohang 37673, Korea; hskang@postech.ac.kr (H.-S.K.);
heohoon@postech.ac.kr (Hoo.H.); chbkim@postech.ac.kr (C.K.); ilyoukim@postech.ac.kr (G.K.);
minck@postech.ac.kr (C.-K.M.); highlong@postech.ac.kr (H.Y.); sybeak@postech.ac.kr (S.Y.B.);
choihyo@postech.ac.kr (H.-J.C.); gymun@postech.ac.kr (G.M.); brp@postech.ac.kr (B.R.P.);
yjseo@postech.ac.kr (Y.J.S.); dcshin@postech.ac.kr (D.C.S.); hjy@postech.ac.kr (J.H.);
npwinner@postech.ac.kr (J.H.); optichoon@postech.ac.kr (S.J.); ksangh@postech.ac.kr (S.-H.K.);
kkhoon@postech.ac.kr (K.K.); dhna3154@postech.ac.kr (D.N.); sspark@postech.ac.kr (S.S.P.);
parkyj@postech.ac.kr (Y.J.P.); jyg@postech.ac.kr (Y.G.J.); jsh@postech.ac.kr (S.H.J.); lhg@postech.ac.kr (H.G.L.);
sblee77@postech.ac.kr (San.L.); sojung8681@postech.ac.kr (Soj.L.); jjambbob@postech.ac.kr (B.O.);
suhhs@postech.ac.kr (H.S.S.); janghui_han@postech.ac.kr (J.-H.H.); minho@postech.ac.kr (M.H.K.);
nsjung@postech.ac.kr (N.-S.J.); kimyc@postech.ac.kr (Y.-C.K.); mslee@postech.ac.kr (M.-S.L.);
lbo4444@postech.ac.kr (B.-H.L.); scw@postech.ac.kr (C.-W.S.); mokalis@postech.ac.kr (I.-S.M.);
genstano7@postech.ac.kr (J.-M.Y.); young1@postech.ac.kr (Y.W.P.); lww@postech.ac.kr (W.-W.L.);
leech@postech.ac.kr (C.-S.L.); striter@postech.ac.kr (H.S.); jihkim@postech.ac.kr (J.H.K.);
yongsam_kim@postech.ac.kr (Y.K.); jaehyuk.lee@postech.ac.kr (J.H.L.); klide@postech.ac.kr (S.-Y.P.);
jkpal@postech.ac.kr (J.K.); pjaeku@postech.ac.kr (Jaeku.P.); neplus@postech.ac.kr (I.E.);
syrah@postech.ac.kr (S.R.); ksn7605@postech.ac.kr (Sun.K.); structure@postech.ac.kr (K.H.N.);
fermi13@postech.ac.kr (Jaehyun.P.); jaehunpa@postech.ac.kr (Jaehun.P.); sangsookim@postech.ac.kr (San.K.);
snkwon@postech.ac.kr (S.K.); anran@postech.ac.kr (R.A.); sh0912@postech.ac.kr (S.H.P.);
kyungkim@postech.ac.kr (K.S.K.); hjhyun@postech.ac.kr (Hyo.H.); ksn@postech.ac.kr (S.N.K.);
kimsh80@postech.ac.kr (Seo.K.); cjyu@postech.ac.kr (C.-J.Y.); kbs007@postech.ac.kr (B.-S.K.);
thkang@postech.ac.kr (T.-H.K.); xraykim@postech.ac.kr (K.-W.K.); yunsori@postech.ac.kr (S.-H.K.);
lee@postech.ac.kr (Hee.-S.L.); lhs@postech.ac.kr (Heung.-S.L.); pkh@postech.ac.kr (K.-H.P.);
ktypmk@postech.ac.kr (T.-Y.K.); dekim@postech.ac.kr (D.-E.K.); kibong@postech.ac.kr (K.B.L.)
* Correspondence: isko@postech.ac.kr; Tel.: +82-54-279-1003

Academic Editor: Kiyoshi Ueda
Received: 23 March 2017; Accepted: 26 April 2017; Published: 17 May 2017

Abstract: The construction of Pohang Accelerator Laboratory X-ray Free-Electron Laser (PAL-XFEL),
a 0.1-nm hard X-ray free-electron laser (FEL) facility based on a 10-GeV S-band linear accelerator
(LINAC), is achieved in Pohang, Korea by the end of 2016. The construction of the 1.11 km-long
building was completed by the end of 2014, and the installation of the 10-GeV LINAC and undulators
started in January 2015. The installation of the 10-GeV LINAC, together with the undulators and
beamlines, was completed by the end of 2015. The commissioning began in April 2016, and the

first lasing of the hard X-ray FEL line was achieved on 14 June 2016. The progress of the PAL-XFEL construction and its commission are reported here.

Keywords: FEL; free electron laser; PAL; PAL-XFEL; construction; commissioning; LINAC; beamline

1. Introduction

The Pohang Accelerator Laboratory X-ray Free-Electron Laser (PAL-XFEL) project was started in 2011 for the generation of X-ray FEL radiation in a range of 0.1 to 10 nm for users. The Korean government launched the project on 1 April 2011 with a budget of 400 billion Won (~400 million USD). The facility has the capacity for five undulator lines in total; three hard X-ray (HX) undulator lines and two soft X-ray (SX) undulator lines. However, the budget was limited to two undulator lines; one for a hard, and the other for a soft X-ray line. Since the PAL is the host institution carrying out the project, the project budget was able to avoid significant costs; for example, there was no need to purchase the land for the building site and a significant portion of the necessary infrastructures, such as power transmission lines and substations, were already in place. A total of 75 members were involved during the project. Among them, there were 35 newly hired members and 39 experienced members from an existing Pohang Light Source-II (PLS-II) team. A yearly budget is shown in Table 1.

Table 1. Budget of PAL-XFEL Project.

Year	2011	2012	2013	2014	2015	Total
Budget in billion Won	20	45	105	120	113.8	403.8

PAL-XFEL includes a 10-GeV S-band (2856 MHz) normal-conducting LINAC, which is about 700 m long. The LINAC consists of a photocathode RF gun, 176 S-band accelerating structures with 50 klystrons and matching modulators, one X-band RF system for linearization, and three bunch compressors in the HX line and one more for the SX line [1]. We also chose out-vacuum/variable-gap undulators for the easy change of beam parameters, and the fast development and manufacturing of undulators. Beyond the 10-GeV LINAC, a 250-m long hard X-ray undulator hall follows. An experimental hall, which is 60 m long and 16 m wide, is located at the end of the facility. The total length of the building is 1110 m, and the entire floor is 36,764 m². The building can withstand a maximum wind load of 63 m/s (US building code) and a seismic intensity of 0.19-g [2]. The facility suffered no damage from the earthquake of a 5.8 magnitude on 12 September 2016, nor from the typhoon Chava on 5 October 2016. The facility is shown in Figure 1.

Figure 1. Overview of Pohang Accelerator Laboratory. The Pohang Light Source (PLS) is shown in the middle (circular building) and PAL-XFEL is shown above the PLS.

2. LINAC

The PAL-XFEL LINAC is divided into four acceleration sections (L1, L2, L3, and L4), three bunch compressors (BC1, BC2, BC3), and a dogleg transport line to the undulators, as shown in Figure 2. The L1 section consists of two RF stations, where both are comprised of one klystron and two S-band structures, while L2 has 10, L3 has four, and L4 has 27 RF stations where each station has one klystron, four accelerating structures, and one energy doubler. A laser heater to mitigate micro-bunching instability is placed right after the injector, and an X-band cavity for linearization is placed right before the BC1. The major parameters of PAL-XFEL are summarized in Table 2.

Table 2. Major parameters of PAL-XFEL.

LINAC	
FEL radiation wavelength	0.1 nm (Hard X-ray)/1 nm (Soft X-ray)
Electron energy	10 GeV
Slice emittance	0.5 mm-mrad
Beam charge	0.2 nC
Peak current at undulator	3.0 kA
Pulse repetition rate	60 Hz
Electron source	Photo-cathode RF-gun
LINAC structure	S-band normal conducting
Undulator	
Type	out-vacuum, variable gap
Length	5 m
Undulator period	26 mm (HX)/35 mm (SX)
Undulator min. Gap	8.3 mm (HX)/9.0 mm (SX)
K value	1.9727 (HX)/3.3209 (SX)
Peak B (in Tesla)	0.8124 (HX)/1.0159 (SX)
Vacuum chamber dimension	13.4×6.7 mm^2

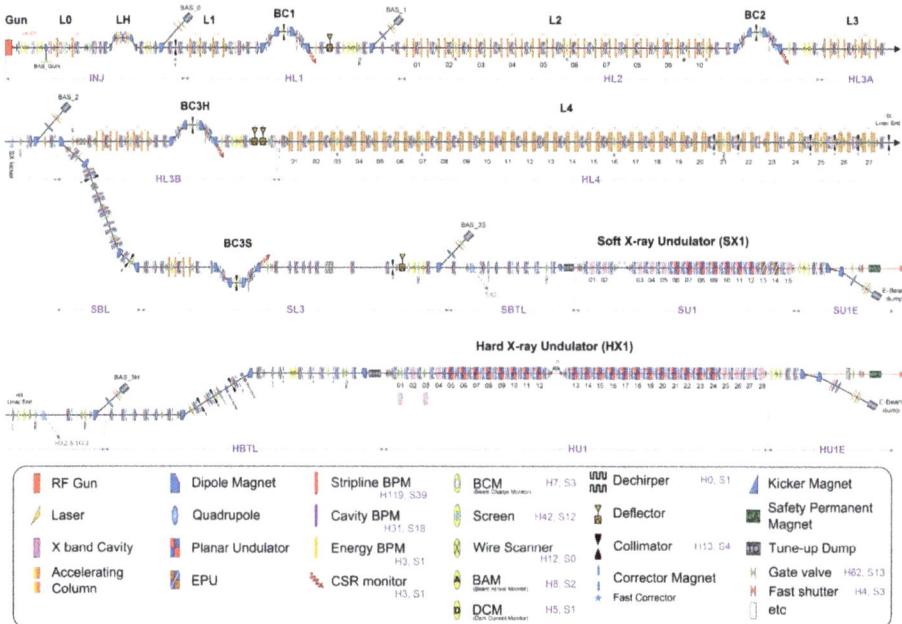

Figure 2. Schematic layout of PAL-XFEL.

The total length of the LINAC tunnel is about 710 m. There are 176 S-band accelerating structures and 42 energy doublers. The major high power devices of the 10-GeV linear accelerator are the modulators, the klystrons, the energy doublers (ED), and the accelerating structures (AS). An energy doubler increases the peak power of the RF pulse by reducing the RF pulse length to increase the energy gain at the accelerating structure. The energy doubler was designed by PAL and fabricated by a domestic company. There are 50 modulators for the S-band klystrons, and there is one modulator for the X-band klystron that is used for linearizing the electron beam. The LINAC requires 46 S-band klystrons to obtain an electron beam energy of 10 GeV. One S-band klystron is dedicated for the RF gun, and three RF stations are designated for deflectors to measure the electron bunch length.

The klystron requires an RF drive signal at the level of a few hundred watts. A low-level radio frequency (LLRF) and a solid-state amplifier (SSA) are necessary to supply the drive signal to a klystron. To achieve beam energy stability of below 0.02% and an arrival time jitter of 20-fs for PAL-XFEL, the LINAC RF parameters should be as stable as 0.03 degrees for the RF phase and 0.02% for the RF amplitude for S-band RF systems, and 0.1 degree/0.04% for the X-band linearizer RF system. The pulse-to-pulse klystron RF stability is determined by the klystron beam voltage driven by a modulator. Therefore, the klystron modulator beam voltage should be as stable as 50 ppm for the 0.03-degree S-band RF and 0.1-degree X-band RF [3].

An LLRF system consists of an SSA, a phase and amplitude detector (PAD), and a phase and amplitude control (PAC) unit. The function of the PAC is to control the phase and amplitude of the RF drive signal to a klystron, to provide a pulsed RF signal, and to reverse the RF phase of the klystron drive signal in the middle of the pulse by turning on the Phase Shift Key (PSK) 180-degree phase shifter). The RF pulse length of the drive signal is 4 μs and the PSK is on after 3.17 μs from the starting time of the pulse. The PAL-XFEL LINAC tunnel and the klystron gallery are shown in Figures 3 and 4, respectively.

Figure 3. LINAC tunnel of PAL-XFEL.

Figure 4. Klystron gallery of PAL-XFEL.

3. Undulator

The PAL-XFEL undulator system consists of 20 planar undulators for the hard X-ray line (HX) and seven planar undulators for the soft X-ray undulator line (SX) [4]. The HX covers a wavelength of $\lambda = 0.1{\sim}0.6$ nm using a 4 to 10-GeV electron beam. They are all out-vacuum undulators with variable gaps. The SX covers a wavelength of $\lambda = 1.0{-}4.5$ nm using a 3.15-GeV electron beam. The HX undulators have a 26-mm undulator period. The gap is controlled remotely within 1 μm repeatability and the minimum gap is 8.3 mm. The height is controlled remotely, too. The SX undulators have a 35-mm undulator period and a minimum gap of 9.0 mm. Both hard and soft X-ray undulators are planar type, and have the same structures except for the magnets. A self-seeding section is prepared in HX undulator line. Two elliptically polarizing undulators are planned to be installed at the SX beamline in coming years. The installed HX undulators are shown in Figure 5.

Figure 5. Hard X-ray undulator tunnel of PAL-XFEL.

4. Diagnostic System

For the operation of PAL-XFEL, electron beam parameters, such as beam positions, energy, charge, transverse beam size, bunch length, and arrival time, should be measured and monitored. A total of 209 beam position monitors (BPMs) are used for the electron beam position measurement [5]. Forty-nine of them are cavity type BPMs, which can measure the beam position with sub-micrometer resolution in the undulator beamline. Ten bunch charge monitors are installed for the bunch charge measurement from the gun as well as beam loss monitoring though the accelerator. The beam profile is measured with 54 screen monitors with YAG and/or Optical Transition Radiation (OTR) screens. Six spectrometer dipole systems are located at the gun section, laser heater, BC1, soft X-ray branch, BC3S and hard X-ray LINAC end. At both beam dumps at the ends of the hard and soft X-ray beamlines, the beam energy can also be measured with the screens. Three S-band deflector systems after BC1, BC3H and BC3S are used to measure the bunch longitudinal phase space. Table 3 summarizes the major components of beam diagnostics and their functions.

Table 3. Major components of beam diagnostics and their functions.

Parameter	Instruments	Number
Position Beam Energy	Stripline Beam Position Monitor	160
	Cavity BPM	49
Beam Charge	Turbo Integrated Current Transformer (ICT)	10
Beam Size	Screen Monitor	54
	Wire Scanner	9
Bunch Length	Coherent Radiation Monitor	4
	Transverse Cavity	3
Arrival Time	Arrival Time Monitor	10
Beam Loss	Beam Loss Monitor	26

5. Beamlines

There is one undulator line for the HX application and another for the SX application. However, there are several end-stations for each undulator line to support various requests from users. For hard X-ray application, there are two end-stations called HEH1 for the pump-probe experiment and HEH2 for the imaging [6]. These two stations are located in tandem. In order to provide HX FEL photons to HEH2, the entire HEH1 stage is able to move in the transverse direction by 1 m. When HX/SX FEL photons emerge from their corresponding undulator system, they pass through various components located in the undulator hall (UH) and the optics hall (OH). Both halls are isolated with proper concrete shielding.

For the HX case, photons are then allowed to enter the experimental hall (EH) where HEH1 and HEH2 are located. Mirrors and a double crystal monochromator (DCM) are located in UH/OH as well as various collimators and safety shutter for radiation safety. In HEH1, beam position monitors (BPM) and profile intensity monitors (PIM) are installed to measure the position and the intensity of the HX FEL beam. An optical laser/X-ray correlator (OXC) is also installed to measure the offset of photon arrival times in fs accuracy. Two Be compound refractive lenses (Be-CRL) are installed along with three slits to define and optimize the XFEL beam. In the final stage, there is a hexapod diffractometer and 4-circle goniometer to adjust sample's location. Finally, there is a robotic arm to control its position of the detector. The HEH1 beamline is shown in Figure 6.

Figure 6. Inside view of HEH1.

HEH2 is designed based on forward scattering geometry, and will be used for the coherent X-ray imaging (CXI) or the serial femtosecond crystallography (SFX) [7]. The XFEL beam is focused to about 2 μm by the K-B mirror. A wire scanning method was used to measure the focusing beam profiles. A tungsten wire with a diameter of 200 μm was placed at the focal point. A photo diode detector placed behind the wire was also used to measure the beam intensity during the wire scanning. There are also several diagnostic devices such as Pop-in, Quadrupole BPM (QBPM), and photo-diodes. The HEH2 beamline is shown in Figure 7.

Figure 7. Inside view of HEH2.

For the SX beamline, there are also two end-stations: one for coherent diffraction imaging (CDI) or X-ray emission/absorption spectroscopy (XES/XAS), and another for SX resonant scattering.

6. FEL Commissioning

After the Injector Test Facility (ITF) had stopped its operation by the end of September 2015, the photocathode RF gun and the two S-band accelerating structures at the ITF were moved to the main PAL-XFEL LINAC. The installation of 51 klystron modulators in the LINAC gallery was finished as of 30 November 2015. Twenty (20) HX undulators for the HX line were installed as of December 2015 in the 250-m long HX undulator tunnel. Since the PAL-XFEL LINAC has all new RF components, an RF aging or conditioning period is required. This RF conditioning had started in November 2015, and continued until the Korean Nuclear Safety and Security Commission (NSSC) issued the operation permission on 12 April 2016. The actual beam commissioning was started on 14 April 2016. Since the ITF parts have already been conditioned during their use from 2012 to 2015, the electron beam emitted from the photocathode gun quickly arrived at the first beam analyzing station (BAS0) on the same day. By 25 April, the 10-GeV electron beam reached BAS3, and the commissioning of the LINAC was completed [8].

Even though the 10-GeV beam is available, we have decided to reduce the electron beam energy to 4-GeV in order to send the electron beam to the main dump through the 6.7-mm gap undulator chambers. Also, the gaps of all undulators were fully opened. These two actions were intentionally chosen to minimize radiation damages to the permanent magnets of the undulator system. After the 4-GeV electron beam reached the main dump, an effort to lase the photon beam was carried out with several feedback algorithms including the beam-based alignment (BBA) technique. Finally, we lased the photons at 0.5 nm on 06:00 14 April 2016. The third harmonic spectrum of 6.6 keV was measured with a single shot spectrometer located in the HEH1, as shown in Figure 8 [9]. Its width was 30 eV or 0.45%. Figure 9 shows the snapshot of the 0.2-nm lasing. A bunch length of 12.7-fs was measured by S-band Transverse Deflecting Cavity (TCAV) with a peak current of 3.7-kA. Major achievements and unexpected interruptions during the commissioning are summarized in Table 4.

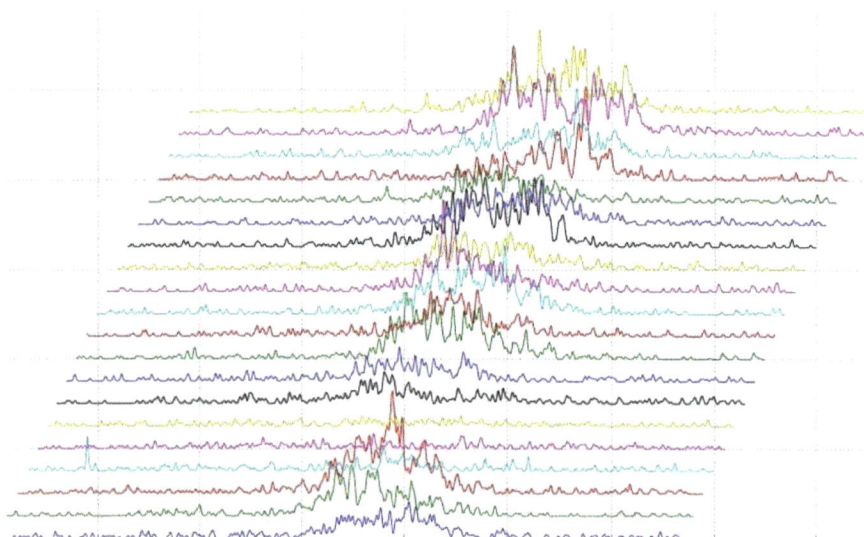

Figure 8. The third harmonic spectrum of 6.6 keV was measured with a single shot spectrometer located in the HEH1.

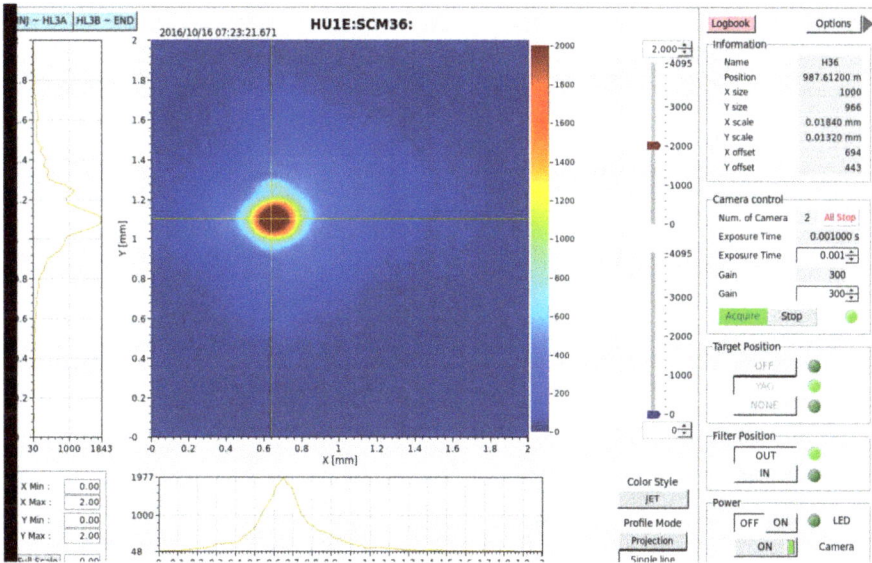

Figure 9. Snapshot of the 0.2-nm lasing.

Table 4. Major achievements during the commissioning.

Date	Energy (GeV)	Remarks
4/12 (2016)		Permission issued by National Nuclear Safety and Security Commission (NSSC)
4/14	0.152	E-Gun and BAS0
4/18	0.355	BAS1
4/19	0.355	Before BC2 (No acceleration by L2)
4/20	2.545	BAS2
4/21	3.15	BAS3 (No acceleration after BAS2)
04/25 (5:30 p.m.)	10	BAS3
5/19	10	Tuneup dump
6/2	10	Passing the HX undulator line, beam at the main dump
06/14 (6:00 a.m.)	4	First lasing (0.5-nm) observed at SCM36
06/21 (3:00 a.m.)	4	Photon beam at Digital Current Monitor (DCM) in Optical Hutch
July~August	–	Summer maintenance
8/16	4	Commissioning resumed
8/30	4	Recovered 0.5-nm lasing
9/9	4	HX Beamline commissioning started
09/12 (8:30 p.m.)	4	Earthquake stopped commissioning
9/29	–	Dedication ceremony
10/3	4	Commissioning resumed
10/8	5.2	0.35-nm lasing
10/16	6.7	0.2-nm lasing
11/27	8.04	0.144-nm lasing and saturation
12/2	8.04	First experiments
2/1 (2017)	3	1.5-nm SX lasing and saturation
3/16	9.78	0.1-nm HX lasing

7. Summary

The PAL-XFEL project has been successfully constructed and commissioned by the end of 2016. It has now provided XFEL photons of 0.1 nm as designed. The Pohang Accelerator Laboratory has already issued a call for proposals to potential (domestic and international) users on February 6, 2017. The successful user will use PAL-XFEL's first light in June 2017.

Acknowledgments: In Soo Ko thanks to all members of the PAL-XFEL project (2011–2016) supported by Ministry of Science, ICT and Future Planning, Korea for their endeavor and dedications.

Author Contributions: Contributions of authors are following: H.-S.K. Accelerator system design & commissioning; Hoo.H. Linac RF system design & commissioning; C.K. Diagnostics system design & commissioning; G.K. Diagnostics system design & commissioning; C.-K.M. Laser system design & commissioning; H.Y. Accelerator system design & commissioning; S.Y.B. Control system design & construction; H.-J.C. Control System design & construction; G.M. Control system design & construction; B.R.P. Machine interlock system design & construction; Y.J.S. Control system design & construction; D.C.S. Control system design & construction; J.Hu. LLRF system design & construction; J.Ho. RF-Gun design & construction; S.J. Laser System design & construction; S.-H.K. Klystron modulator system design & construction; K.K. Linac RF system design & commissioning; D.N. Vacuum system design & construction; S.S.P. Klystron modulator system design & construction; Y.J.P. Linac RF system design & construction; Y.G.J. Undulator system design & construction; S.H.J. Magnet power supply system design & construction; H.G.L. Undulator system design & construction; San.L. Undulator system design & construction; Soi.L. BPM system design & construction; B.O. Diagnostic system design & construction; H.S.S. Magnet system design & construction; J.-H.H. Injector system design and commissioning; M.H.K. Radiation safety system design and construction; N.-S.J. Radiation safety system design and construction; Y.-C.K. Building design and construction; M.-S.L. Building design and construction; B.-H.L. Utility system design and construction; C.-W.S. Utility system design and construction; I.-S.M. Building design and construction; J.-M.Y. Building design and construction; Y.W.P. Beam Dynamics; W.-W.L. Undulator system design & construction; C.-S.L. Beamline interlock system design & construction; H.S. Data center design & construction; J.H.K. Network system design & construction; Y.K. Beamline commissioning; J.H.L. Beamline commissioning; S.-Y.P. Control & DAQ system; J.K. X-ray optics commissioning; Jaeku P. Control & DAQ system; I.E. Optical laser system design, construction & commissioning; S.R. X-ray optics; Sun.K. Beamline design & commissioning; K.H.N. Serial femtosecond crystallography instrumentation design & commissioning; Jaehyun P. Serial femtosecond crystallography instrumentation design & commissioning; Jaehun P. Optical laser system commissioning; San.K. Coherent X-ray imaging instrumentation design & commissioning; Soo.K. Soft X-ray beamline design & commissioning; R.A. Soft X-ray beamline design & commissioning; S.H.P. Soft X-ray beamline design & commissioning; K.S.K. X-ray detector; Hyo.H. X-ray detector; S.N.K. Vacuum & beamline construction; Seo.K. Mechanical design & beamline construction; C.-J.Y. Beamline planning & management; B.-S.K. Beamline planning & management; T.-H.K. Beamline planning & management; K.-W.K. Beamline planning & management; S.H.K. Building design and construction; Hee-S.L. Radiation safety system design and construction; Heung.-S.L. RF system design and commissioning; K.-H.P. Magnet power supply system design & commissioning; T.-Y.K. Beamline commissioning; D.-E.K. Undulator system design & commissioning; K.B.L. Beamline commissioning.

Conflicts of Interest: The authors declare no conflict of interest.

References

1. Nam, S.H.; Kang, H.S.; Ko, I.S.; Cho, M. Upgrade of Pohang Light Source (PLS-II) and Challenge to PAL-XFEL. *Synchrotron Radiat. News* **2013**, *26*, 24–31. [CrossRef]
2. Ko, I.S. Status of PAL-XFEL Construction. *Bull. AAPPS* **2016**, *26*, 25–31.
3. Lee, H.S.; Park, S.S.; Kim, S.H.; Park, Y.J.; Heo, H.; Heo, I.; Kim, K.H.; Kang, H.S.; Kim, K.W.; Ko, I.S.; et al. PAL-XFEL Linac RF System. In Proceedings of the 7th International Particle Accelerator Conference, Busan, Korea, 8–13 May 2016; Petit-Jean-Genaz, C., Ed.; JACOW: Geneva, Switzerland, 2016; pp. 3192–3194.
4. Kim, D.E.; Jung, Y.G.; Lee, W.W.; Kang, H.S.; Ko, I.S.; Lee, H.G.; Lee, S.B.; Oh, B.G.; Suh, H.S.; Park, K.H.; et al. Development of PAL-XFEL Undulator System. In Proceedings of the 7th International Particle Accelerator Conference, Busan, Korea, 8–13 May 2016; Petit-Jean-Genaz, C., Ed.; JACOW: Geneva, Switzerland, 2016; pp. 4044–4046.
5. Kim, C.; Lee, S.; Kim, G.; Oh, B.; Yang, H.; Hong, J.; Choi, H.J.; Mun, G.; Baek, S.; Shin, D.; et al. Diagnostic System of the PAL-XFEL. In Proceedings of the 7th International Particle Accelerator Conference, Busan, Korea, 8–13 May 2016; Petit-Jean-Genaz, C., Ed.; JACOW: Geneva, Switzerland, 2016; pp. 2091–2094.
6. Park, J.; Eom, I.; Kang, T.-H.; Rah, S.; Nam, K.-H.; Park, J.; Kim, S.; Kwon, S.; Park, S.H.; Kim, K.S.; et al. Design of a hard X-ray beamline and end-station for pump and probe experiments at Pohang Accelerator Laboratory X-ray Free Electron Laser Facility. *Nucl. Instrum. Methods Phys. A* **2016**, *810*, 74–79. [CrossRef]
7. Park, J.; Kim, S.; Nam, K.-H.; Kim, B.S.; Ko, I.S. Current Status of the CXI Beamline at the PAL-XFEL. *J. Korean Phys. Soc.* **2016**, *69*, 1089–1093. [CrossRef]

8.   Han, J.H. Beam Commissioning of PAL-XFEL. In Proceedings of the seventh International Particle Accelerator Conference, Busan, Korea, 8–13 May 2016; Petit-Jean-Genaz, C., Ed.; JACOW: Geneva, Switzerland, 2016; pp. 6–10.

9.   Zhu, D.; Cammarata, M.; Feldkamp, J.M.; Fritz, D.M.; Hastings, J.B.; Lee, S.; Lemke, H.T.; Robert, A.; Turner, J.L.; Feng, Y. A single-shot transmissive spectrometer for hard X-ray free electron laser. *Appl. Phys. Lett.* **2012**, *101*, 034103. [CrossRef]

applied sciences

MDPI

Article

Photon Beam Transport and Scientific Instruments at the European XFEL

Thomas Tschentscher *, Christian Bressler, Jan Grünert, Anders Madsen, Adrian P. Mancuso, Michael Meyer, Andreas Scherz, Harald Sinn and Ulf Zastrau

European XFEL, Holzkoppel 4, 22869 Schenefeld, Germany; christian.bressler@xfel.eu (C.B.); jan.gruenert@xfel.eu (J.G.); anders.madsen@xfel.eu (A.M.); adrian.mancuso@xfel.eu (A.P.M.); michael.meyer@xfel.eu (M.M.); andreas.scherz@xfel.eu (A.S.); harald.sinn@xfel.eu (H.S.); ulf.zastrau@xfel.eu (U.Z.)

* Correspondence: thomas.tschentscher@xfel.eu; Tel.: +49-(0)40-8998-3904

Academic Editor: Kiyoshi Ueda
Received: 1 May 2017; Accepted: 1 June 2017; Published: 9 June 2017

Featured Application: This article describes the layout of the European XFEL, a soft and hard X-ray free-electron laser user facility starting operation in 2017. Emphasis is put on the photon beam systems, scientific applications and the instrumentation of the scientific instruments of the European XFEL.

Abstract: European XFEL is a free-electron laser (FEL) user facility providing soft and hard X-ray FEL radiation to initially six scientific instruments. Starting user operation in fall 2017 European XFEL will provide new research opportunities to users from science domains as diverse as physics, chemistry, geo- and planetary sciences, materials sciences or biology. The unique feature of European XFEL is the provision of high average brilliance in the soft and hard X-ray regime, combined with the pulse properties of FEL radiation of extreme peak intensities, femtosecond pulse duration and high degree of coherence. The high average brilliance is achieved through acceleration of up to 27,000 electron bunches per second by the super-conducting electron accelerator. Enabling the usage of this high average brilliance in user experiments is one of the major instrumentation drivers for European XFEL. The radiation generated by three FEL sources is distributed via long beam transport systems to the experiment hall where the scientific instruments are located side-by-side. The X-ray beam transport systems have been optimized to maintain the unique features of the FEL radiation which will be monitored using build-in photon diagnostics. The six scientific instruments are optimized for specific applications using soft or hard X-ray techniques and include integrated lasers, dedicated sample environment, large area high frame rate detector(s) and computing systems capable of processing large quantities of data.

Keywords: free-electron lasers; average brilliance; peak brilliance; photon diagnostics; X-ray optics; femtosecond time resolution; coherent X-ray diffraction imaging; ultrafast diffraction; ultrafast absorption and emission spectroscopy; non-linear X-ray processes

1. Introduction

During the last decades, the development of X-ray light sources based on low emittance electron accelerators has enabled spectacular increases in the average and peak brilliances. Brilliance corresponds to the number of photons per phase space element of the emitted X-rays and is the parameter best describing the performance of these sources. Electron accelerators optimized for free-electron lasers (FEL) use low emittance injectors to create electron bunches, linear accelerators and electron beam optics to minimize the emittance growth during acceleration and transport, and bunch compression to generate ultrashort bunches. The resulting low emittance and high peak current of the electron bunches are the key performance parameters for these facilities. In undulator sections much longer than for synchrotron radiation sources, the electron bunches are transported with high precision and collimation to enable the self-amplified spontaneous emission (SASE) process leading to FEL gain and occurring in a single-pass of the electron bunch [1,2]. In the SASE process, the electron bunch typically undergoes a degradation of its properties and cannot be reused for another FEL source. It is instead dumped at the end of the beam transport. FELs therefore, in general, are single-user machines making their operation costly and the access to them much more limited than storage ring sources where electron bunches are circulated and are reused by several insertion devices. Since X-ray FEL radiation is generated by an intrinsically coherent SASE process it provides huge pulse energies of 10 mJ and possibly beyond, and pulse durations as short as single femtoseconds. These properties correspond to peak brilliances eight to nine orders of magnitude higher than obtained by storage ring sources. Exploiting this brilliance in FEL experiments allows embarking on yet impossible X-ray experiments that will lead to interesting and valuable scientific and technological applications.

Two technologies have been pursued to construct electron accelerators for FEL applications: warm, normal conducting machines and cold, super-conducting accelerators. The latter allowing acceleration of electron bunches at a much higher repetition rate, thereby boosting the average brilliance and enabling to distribute electron bunches to many FEL sources. The first short-wavelength FEL user facility starting user operation was FLASH at Deutsches-Elektronen-Synchrotron (DESY) in Hamburg (Germany), which uses super-conducting accelerator technology [3] and provides FEL radiation in the XUV and soft X-ray spectral region up to the water window [4]. In the following years, several normal conducting accelerator-based FELs started operation at SLAC National Accelerator Laboratory [5], ELETTRA [6], SPring-8 [7], and the Pohang Accelerator Laboratory (PAL) [8]. The SwissFEL facility at the Paul-Scherrer-Institute (PSI) [9] is nearing completion. The European XFEL [10,11] employs the same super-conducting accelerator technology [12] used for FLASH and is currently under commissioning for first user experiments in 2017. European XFEL enables the acceleration of up to 27,000 electron bunches per second with an electron energy of up to 17.5 GeV (compare Table 1) which are distributed to several FEL sources. Starting from 2019 European XFEL will operate a regular user program with initially six instruments continuously receiving X-ray beams. At SLAC currently a 4 GeV super-conducting accelerator is under construction for LCLS-II, based on the same technology used for FLASH and European XFEL plus enabling continuous wave (cw) acceleration [13].

Table 1. Comparison of accelerator parameters of hard X-ray FEL facilities.

Parameter	LCLS	SACLA	PAL-XFEL	SwissFEL	European XFEL
Technology	Warm	Warm	Warm	Warm	Super-conducting
Accelerator frequency	2.856 GHz	5.7 GHz	2.856 GHz	6 GHz	1.3 GHz
Maximum energy	15 GeV	8 GeV	10 GeV	5.8 GeV	17.5 GeV
Bunch charge	0.2 nC	0.2 nC	0.2 nC	0.2 nC	1.0 nC
Repetition rate	120 Hz	60 Hz	60 Hz	100 Hz	27,000 bunches/s [1]
Maximum power [2]	300 W	100 W	100 W	100 W	500 kW
User operation	2009	2012	2017	2018[2]	2017
Reference	[5]	[7]	[8]	[9]	[14,15]

[1] Bunches are generated and distributed in 10 Hz bursts of 2700 bunches each; [2] Using nominal operation parameters for energy, bunch charge and repetition rate.

The article is organized as follows: Following a description of the overall European XFEL facility, we shall first describe the photon beam transport and photon diagnostics systems, before introducing the individual science instruments. Finally, an outlook to future developments of the facility is provided.

2. Overview European XFEL

X-ray FEL radiation is characterized by its ultrashort pulse duration, high pulse energies and a high degree of coherence. Scientific applications of soft and hard X-ray FEL radiation make use of these properties, in particular in the investigation of ultrafast processes in atoms, ions, simple and very complex molecules, clusters or condensed matter. The high pulse energies allow the collection of meaningful data sets from single pulses, thereby enabling the study of non-reversible processes. Coherence properties are exploited in imaging techniques that aim to obtain atomic spatial resolution for weakly scattering systems, in part combined with a corresponding temporal resolution [16]. Finally, the very high X-ray pulse energies, which combined with ultrashort pulse durations correspond to very high peak powers of up to several tens of GW, promise to enable access to new information of excited solids through non-linear X-ray scattering [17].

There are several classes of these experiments requiring not only a high peak brilliance, but also high average brilliance. Here European XFEL has a clear advantage compared to other X-ray FEL facilities. Examples comprise studies of ultra-dilute systems, very small cross-section processes, non-linear X-ray processes, or particle-particle/particle-X-ray coincidence spectroscopy experiments. Furthermore it is possible to use subsequent X-ray pulses to probe equilibrium dynamics at frequencies up to 4.5 MHz, and beyond [18].

To enable the use of the highest repetition rates in a pulse-resolved (non-integrating) manner has been the biggest instrumentation challenge for European XFEL. Such a mode of operation requires that diagnostic and X-ray detection systems operate at repetition rates of up to 4.5 MHz. In addition, optical lasers used to excite samples in a well-controlled manner and sample injections systems need to accommodate these high event rates. The development of this non-standard high repetition rate instrumentation is one key expertise of European XFEL and its partners.

2.1. Layout of the European XFEL Facility

The European XFEL facility consists principally of three sections. The first section includes the superconducting low emittance 17.5 GeV electron accelerator and the distribution of electron bunches to two beam lines comprising the FEL undulator sources. The electron beam transport is designed to accommodate up to five FEL sources. Each FEL source has a dedicated photon beam transport section to transport, steer, focus, and diagnose the X-ray FEL beams prior to their entry to the experiment hall. Mirrors in the photon beam transports will direct the X-ray FEL beam to one of the scientific instruments located at the respective FEL source. The third section is the experiment hall in which the scientific instruments are located and where the experiment program is run. In its first installment, only three FEL sources are constructed each leading to two scientific instruments. Figure 1 provides an overview of the European XFEL facility. Completion of the facility with five FEL sources and up to fifteen scientific instruments is expected to take place in the years following start of user operation.

The layout of European XFEL is governed by a few basic conditions. First, the goal to reach FEL radiation exceeding 20 keV with high pulse energies and outstanding coherence properties has been driving the definition of the maximum electron energy to be 17.5 GeV. Using an acceleration gradient of 23.6 MeV/m this alone results in a length of nearly 1000 m for electron acceleration. A second condition was to expand the highly collimated FEL beam to a size of order 1 mm, requiring with divergences of order of 1 μrad for hard X-rays another approximately 1000 m free-space transport. This is accompanied by another requirement of achieving lateral separations on the order of 17 m for the beam lines of the different FEL sources when arriving at the experiment hall.

The facility is located in the western part of the Metropolitan area of Hamburg, reaching from the DESY campus in Hamburg Bahrenfeld to the town of Schenefeld, Schleswig-Holstein. Located in a partly inhabited area, the facility had to be built in underground tunnels, fully immersed in the ground water in this location. Access to tunnels is enabled by shaft buildings at the start and end of each tunnel, and access to the experiment hall is provided from the office and laboratory building build on top of this hall.

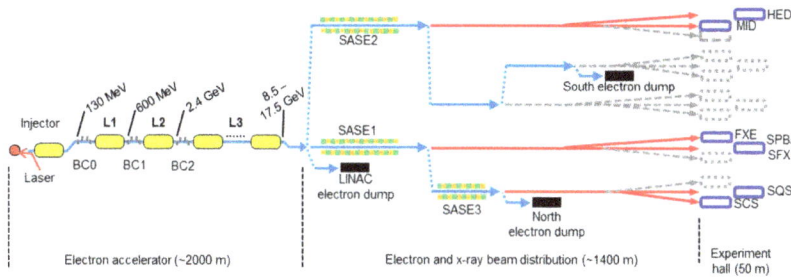

Figure 1. Overall layout of the European XFEL facility. The electron accelerator leads into two electron beamlines with up to five FEL sources. Each of these has dedicated X-ray transport sections leading towards the experiment hall, where the up to fifteen scientific instruments can be installed. For abbreviations see text.

2.2. The Super-Conducting Electron Accelerator

The European XFEL accelerator has the task of providing electron bunches for the FEL process. It consists of a photo-injector, the main linac and the different beam line sections. In the 43 m long photo-injector section [19,20], electron bunches are generated by means of the photoelectric effect from a CsTe cathode. The photocathode is located inside a normal-conducting cavity to immediately accelerate the electrons to 6 MeV before injection into the first super-conducting accelerator module. This module is directly followed by a super-conducting 3.9 GHz acceleration module needed to linearize the longitudinal phase of accelerated electrons. At 130 MeV the electrons enter a diagnostic section enabling the measurement of the phase space properties of individual electron bunches in the bunch train and even of slices of these bunches. At the end of the injector an electron beam dump allows standalone operation of the injector over its full parameter range, such that commissioning and further development can be performed independently of the operation of the main linac. Electron extraction from the cathode is achieved by frequency-quadrupled 257 nm laser pulses supplied by a dedicated Nd:YLF photo-injector laser. The laser is synchronized to the accelerator radio-frequency and delivers a time pattern corresponding to the burst mode repetition rate of electron bunch delivery. The photo-injector performance determines the smallest obtainable emittance of the entire accelerator and its design has been optimized in this regard. Space charge driven emittance growth is the most important limiting effect and is minimized by relatively long laser pulses (up to 20 ps) and very high acceleration gradients of up to 50–60 MV/m at the cathode. Furthermore, the spatial and temporal profile of the laser pulse ideally has a top-hat-like shape when hitting the cathode. The operation with electron bunch charges from 0.02 to 1.0 nC at different emittances and enabling different bunch durations is foreseen. Initial commissioning of the injector was concluded in summer 2016 [21].

The main linac accelerates the electrons to a final energy of up to 17.5 GeV by means of 96 accelerator modules operated at 2.2 K, built by an international collaboration for European XFEL based on the TESLA design [14]. Each module is 12 m long, weighs eight tons and comprises eight nine-cell Nb cavities. A total of 768 couplers provide the radio-frequency (RF) fields generated by 24 RF stations. The accelerator is operated in a 10 Hz pulsed RF mode (see Figure 2) for a maximum beam power of 500 kW, exceeding by far the power that can be reached by any normal conducting machine. The design gradient is 23.6 MV/m and the final installation considers the actual performance of each accelerator module by tuning the RF distribution system. The injector and the three linac sections L1, L2 and L3 are separated by three electron bunch compressors BC0, BC1 and BC2. These are used to compress the electron bunches in steps from their initial approximately 20 ps duration to as short as a few fs, hence reaching the design peak current of 5 kA, depending on the bunch charge. The electron energy at the end of L2 is 2.4 GeV and will be kept constant during operation in order to optimize the performance of the accelerator. Dedicated diagnostics sections for the measurement of integrated and slice bunch parameters are located after compressors BC1 and BC3. Cool-down of the linac was started end of 2016 and commissioning with electron beam commenced in 2017 [22,23].

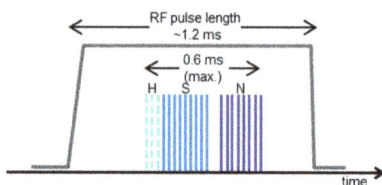

Figure 2. Time pattern of the electron bunch train in the linac. The RF field is pulsed with 10 Hz and has a flat top region of ~1.2 ms clearly exceeding the duration of the electron bunch train of 600 μs. The bunch train can be separated into portions with different function. The header H is typically used for fast intra-bunch feedback. The next portion S is dedicated for the South branch with SASE2. Following a short gap to switch the flat top kicker magnet the last portion N will be send to the North branch with SASE1 and SASE3. The smallest separation of electron bunches is 222 ns in standard operation, corresponding to 4.514 MHz and up to 2700 bunches per train. Operation at bunch separations of 886 ns (1.128 MHz) and 10 μs (0.1 MHz) is possible, too.

The last section comprises a total of approximately 3 km of electron beam transport systems and starts with a collimation section removing halo and electrons at non-matching energies, i.e., dark current, from the beam. Downstream of this section electron bunches are distributed to either one of the two beam lines with the FEL undulators, denoted the North and South electron branches, or to a dump beam line. Distribution between the two branch lines is performed by a precise flat top kicker magnet with fast falling edge operating at 10 Hz to switch once during each bunch train. Likewise a dedicated portion of the 600 μs bunch train is first kicked to the South branch line. After a switching time of approximately 20 μs electrons continue without kicking into the North branch line. An additional fast kicker can operate at up to 4.5 MHz and is used to deflect single bunches to the dump beam line. It is used to, e.g., generate the time window needed for switching between south and north branch and furthermore enables a free choice of the bunch pattern delivered to the two FEL beam lines while the accelerator is operated at constant loading. At the end of each electron beam line a solid state dump is capable of absorbing up to 300 kW of beam power. In case the accelerator is operated at full beam power, electrons will be distributed over more than one dump thereby limiting the absorbed power. In the beam transport section of the accelerator many important electron diagnostics systems are located. They measure beam position with μm accuracy, arrival time of bunches relative to a precise laser synchronization system with down to a few fs accuracy, and electron energy in dispersive sections to a level better than 10^{-4}. The long pulse trains allow using an initial portion of the train for intra-train feedback scheme, thereby enabling a higher stability and performance of the electron beam delivery in the remaining portion of the train delivered to the two branch lines.

2.3. The FEL Undulator Sources

The initially three FEL undulator sources denoted SASE1, SASE2 and SASE3 will provide FEL radiation ranging from the carbon K-edge to very hard X-rays for user experiments. SASE1 and SASE2 serve the hard X-ray regime from approximately 3 to 25 keV in the first harmonic. SASE3 produces soft X-rays from approximately 250 eV to 3 keV. These ranges are achieved by a combination of electron energy set points and gap tuning (see Table 2). The lengths of the FEL undulators have been determined after simulation of the saturation length for the highest photon energy at the largest considered electron beam emittance. All FELs are therefore much longer than the saturation length in the middle of the tuning range, around 10 keV (SASE1 and SASE2) or up to 1 keV (SASE3), which allows for special modes of operation, e.g., the implementation of self-seeding. In addition, each FEL has additional space before and after the device for optional extensions, e.g., for laser-driven beam manipulation or so-called afterburners.

Table 2. FEL undulator source parameters [24].

Parameter	SASE1/SASE2	SASE3
Period length (mm)	40	68
Maximum B-field (T; @10 mm)	1.11	1.68
Number of poles per segment	248	146
Number of segments	35	21
Total system length (m)	205	121
Gap range (mm)	10–20	10–25
K-parameter range	1.65–3.9	4–9
Photon energy range (keV; @8.5 GeV)	1.99 [1]–7.2	0.243–1.08
Photon energy range (keV; @12 GeV)	3.97–14.5	0.485–2.16
Photon energy range (keV; @17.5 GeV)	8.44–30.8 [1]	1.031–4.6 [1]

[1] Beam transport does not allow reaching these photon energies.

All FELs are segmented into 5 m long planar undulators and 1.1 m long intersections. Undulators are equipped with permanent hybrid NdFeB magnet technology for a minimum magnetic gap of 10 mm allowing the use of aluminum vacuum chambers with an inner opening of 8.8 mm for the electron beam. Out of vacuum magnetic structures were chosen to minimize radiation damage of the magnets, but also to reduce resistive wall wake fields. The intersections carry a quadrupole for electron beam focusing, a phase shifter for matching the radiation field and the micro-bunched electron beam, an electron beam position monitor, and vacuum devices. All parts for the total 91 undulator segments have been produced and assembled by industry. Magnetic tuning was performed by the European XFEL undulator group using the pole-height tuning technique [25].

The FEL source properties depend on a large number of parameters, not only of the FEL undulators, but also the electron beam properties, e.g., peak current, emittance, or energy spread. Table 3 shows simulation results for the saturation point for a selection of photon energies and electron beam parameters. The full set of properties can be found in refs. [26,27]. In practice, FELs are often operated well beyond saturation thereby boosting the emitted pulse energies but also sacrificing some of the other properties.

Table 3. FEL radiation properties at saturation for selected photon energies and electron parameters optimized for specific bunch charge working points [26].

X-ray Beam Energy	243 eV	790 eV	1680 eV	6.2 keV	13.3 keV	30.8 keV [1]
FEL source	SASE3	SASE3	SASE3	SASE1/2	SASE1/2	SASE1/2
Electron energy (GeV)	8.5	12	17.5	12	17.5	17.5
Bunch charge 0.02 nC						
Pulse energy (mJ)	0.14	0.14	0.17	0.05	0.06	0.03
Peak brilliance [2]	5.0×10^{31}	2.3×10^{32}	6.9×10^{32}	1.5×10^{33}	4.4×10^{33}	6.4×10^{33}
Average intensity [3]	9.6×10^{16}	3.0×10^{16}	1.7×10^{16}	1.5×10^{15}	7.7×10^{14}	1.5×10^{14}
Saturation length (m)	29	38	47	47	59	101
Bunch charge 0.25 nC						
Pulse energy (mJ)	2.04	2.06	2.34	0.73	0.65	0.235
Peak brilliance [2]	5.8×10^{31}	2.6×10^{32}	7.9×10^{32}	1.6×10^{33}	4.1×10^{33}	4.1×10^{33}
Average intensity [3]	1.4×10^{18}	4.4×10^{17}	2.3×10^{17}	2.0×10^{16}	8.3×10^{15}	1.3×10^{15}
Saturation length (m)	31	42	52	54	73	161
Bunch charge 1.0 nC						
Pulse energy (mJ)	8.51	8.36	9.25	2.29	1.68	-
Peak brilliance [2]	5.9×10^{31}	2.6×10^{32}	7.8×10^{33}	1.3×10^{33}	2.4×10^{33}	-
Average intensity [3]	5.9×10^{18}	1.8×10^{18}	9.3×10^{17}	6.2×10^{16}	2.1×10^{16}	-
Saturation length (m)	35	48	60	68	105	252

[1] Radiation parameters simulated for electron beam and FEL undulator. However, the photon beam transport in its present configuration does not allow propagating the photons to the instruments; [2] in units of photons/s/mm²/mrad²/0.1% BW; [3] in units of photons/s assuming 27,000 pulses/s.

2.4. The Experiment Hall and Ancillary Instrumentation

The experiment hall has a size of 50 m along the beam direction and 90 m across to install five beam line areas. The tunnels housing the X-ray beam transports enter with a separation of approximately 17 m. For each of the five beam line areas installation of up to three scientific instruments is considered with an X-ray beam separation of 1.4 m at the entrance to the hall. Each beam line area includes dedicated enclosures for X-ray optics, X-ray experiment, controlling the experiment, the pump-probe laser system, instrument laser hutches, and in some cases also preparatory labs. Control and data acquisition electronics are generally placed in separate rack rooms located on-top of the beam lines for fire protection purposes. Here also most of the air-conditioning systems are located, used to stabilize dedicated temperature zones to ± 0.1 °C while special care has been taken to avoid vibrations. The hall is connected via stairs and elevators to the laboratory and office floors in the building above. In the ground floor in total ~2500 m^2 is available for laboratories, comprising rooms for sample preparation and characterization, chemistry and biochemistry laboratories. Furthermore, cleanrooms for optics, detector and vacuum part assembly and testing and several laser labs for research and development are found here.

2.4.1. Large Area Detectors for European XFEL

Already in 2006 European XFEL launched a significant program for the development of large area detectors for FEL experiments, since it was clear that the requirements to detectors for FEL experiments in general and European XFEL specifically could not be fulfilled by existing devices [28]. General requirements include an integrating operation mode, enabling the detection of several X-ray photons per pixel and per pulse, very low noise, enabling to detect single X-ray photons, and a high dynamic range, enabling to count 10^4 or more X-ray photons in a single pixel. Specific requirements for European XFEL include the need for frame rates of up to 4.5 MHz, in order to be compatible with X-ray pulse delivery within the pulse train structure (compare Figure 2), high throughput, to collect as many images as possible per unit time, and radiation hardness. More recently, the possibility of vetoing specific events was added as a requirement. Three large projects were selected and are pursued together with external partners. Laboratory infrastructure, in particular for detector calibration and characterization, has been designed and is operated by the European XFEL detector group [29]. In addition, a few smaller projects consisted in modifying existing cameras, mostly designed for 10 Hz operation, and in upgrading the Gotthard one-dimensional strip detector [30] to 4.5 MHz repetition rate. In the following the three large area detector projects are described briefly.

The Adaptive Gain Integrated Pixel Detector (AGIPD) is developed by a consortium led by DESY [31]. The main features of this detector are 200 × 200 µm^2 pixels, dynamic gain switching with 3 stages, a dynamic range of ~10^4 at 12 keV, single photon detection (6σ) above 7 keV, in-vacuum operation, the capability of storing up to 352 images within the 600 µs pulse train, and to read out these data in-between pulse trains. As sensor material 500 µm thick Si is used. Two 1 Mpixel AGIPD devices constitute the primary 2D detectors at the SPB/SFX and MID instruments. A 4 Mpixel device is under development as part of the SFX User Consortium, as is a 1 Mpixel device with GaAs sensor for the HIBEF User Consortium.

The Large Pixel Detector (LPD) is developed by a consortium led by STFC [32]. The main features of this detector are 500 × 500 µm^2 pixels, three amplifiers with different gain per pixel, a dynamic range of up to ~10^5 at 12 keV, single photon detection (3σ) above ~12 keV, the capability of storing up to 512 images within the 600 µs pulse train, and to read out these data in-between pulse trains. As sensor material, 500 µm thick Si is used. A 1 Mpixel LPD device will be employed at the FXE instrument, primarily for liquid scattering experiments.

The DepFET Sensor with Signal Compression (DSSC) detector is developed by a consortium initially led by the Max-Planck-Society (MPG) [33]. The main features of this detector are hexagonal 236 µm diameter pixels, a non-linear gain, a dynamic range of ~6×10^3 at 1 keV, single photon detection (5σ) above 0.7 keV, in-vacuum operation, the capability of storing up to 800 images within the 600 µs pulse train, and to transfer up to 640 frames in-between pulse trains. The highest frame rate of the DSSC detector therefore is 6.4 kHz. As sensor material 300 µm thick Si is used. One 1 Mpixel DSSC device is considered as the primary 2D detector for the SCS and SQS instruments. In a first phase, a simplified detector with 1 Mpixel Si drift diodes with reduced performance will be available for experiments.

2.4.2. Optical Lasers for European XFEL

The usage of synchronized optical laser pulses is foreseen at all scientific instruments and opens various possibilities for time-resolved pump-probe studies and laser-controlled manipulation of electronic relaxation and excitation processes. A dedicated development has been initiated to meet the requirements of delivering 800 nm radiation, 10–100 fs pulse duration, 0.1–4.5 MHz selectable pulse delivery and 0.1–1 mJ pulse energy. After the successful completion of a first design phase [34] the implementation of three pump-probe burst-mode optical (PP) lasers systems was started, each serving one beam line area. The final amplification of the laser pulses employs the Non-collinear Optical Parametric Amplifier (NOPA) scheme. Three NOPA stages allow providing highest pulse energies at reduced repetition rate, 3.25 mJ for 0.1 MHz, and reduced pulse energies at highest repetition rate, 0.08 mJ at 4.5 MHz. A Pockels cell and polarizer before the NOPA amplifiers enable picking of arbitrary pump pulse sequences from the amplified burst at frequencies up to 4.5 MHz. An additional output delivers 1030 nm pulses with energies up to 40 mJ at 0.1 MHz and duration of 400 ps or 800 fs (compressed). Full performance of the system has been demonstrated recently [35]. In order to synchronize the delivery of optical laser and X-ray pulses, RF and optical lasers a laser-based synchronization system [36] is employed with the goal of reaching an accuracy better than 20 fs rms [37].

The PP lasers are placed in dedicated laser rooms at each beam line. Laser beams are transported to dedicated instrument laser hutches (ILH) adjacent to the X-ray experiment areas. The separation from the X-ray hutch provides the possibility to work on these systems without disturbing the X-ray program. In the ILHs, e.g., delay stages, frequency conversion optics, and laser diagnostics are placed. Laser pulses are transported in a time stretched mode and final compression occurs close to the experiment location. Particular care needs to be taken with respect to the dispersion management of the optical laser pulses in order to achieve the shortest pulse duration.

2.5. The User Program

The European XFEL is conceived as a user facility with the main emphasis of providing excellent conditions for FEL research with soft and hard X-rays. To reach this goal an accelerator operation of approximately 5600 h annually is foreseen to provide 4000 h of user operation, 800 h for accelerator and another 800 h for X-ray systems maintenance, research and development. Operation will be continuous for several weeks with interruptions during two workdays, mainly for setup changes, maintenance, and tuning for the next experiments. With initially two scientific instruments per FEL source each of them schedules ~2000 h per year for users. In regular operation this should allow for >200 user experiments annually, thereby significantly increasing the European and worldwide accessibility of FEL experiments. User experiments will be selected by peer-review using scientific excellence as criterion given that technical feasibility and safety requirements are fulfilled. User groups will be supported by the instrument staff and the scientific support groups in preparing and executing the experiment and analyzing the data. Due to the complexity of FEL experiments, often requiring expertise in X-rays, optical lasers, sample delivery, detectors and data analysis, it is the goal to provide these systems to the users, thereby facilitating the use of European XFEL and lowering the entry level to FEL experiments. During experiments scientific staff of European XFEL will continuously support user groups and ensure that the various sub-systems are functional.

2.6. European XFEL Governance and Organization

The operation of the overall European XFEL facility is entrusted to the European XFEL GmbH, based on an intergovernmental agreement between the participating countries Denmark, France, Germany, Hungary, Italy, Poland, Russia, Slovakia, Spain, Sweden, Switzerland, and the United Kingdom. The largest contributors are Germany with 58% and Russia with 27% of the total construction cost. Each participating country determines a legal entity to hold their shares and to represent the country in the Council, the superior governance board of European XFEL. The construction costs of the facility can be sub-divided into the three major areas: civil construction, accelerator complex, and X-ray systems with shares of roughly 30:50:20. About 50% of the construction cost was provided through in-kind contributions by the participating countries. The annual costs for operation are shared amongst the participating countries initially according to the participation in the construction period. Starting in 2023, 50% of the annual operation costs will be distributed according to real usage of the facility by research groups from the participating countries, calculated using a three-year average.

The construction, commissioning and operation of the superconducting accelerator and its ancillary systems depend on the expertise residing at DESY, a major accelerator and photon science laboratory located in Hamburg, Germany. During the construction phase DESY led the international Accelerator Consortium that designed, built, and commissioned the accelerator. European XFEL staff has been responsible for the X-ray systems and ancillary instrumentation, including the undulators. For the operation phase European XFEL and DESY have concluded an agreement according to which DESY provides the personnel and expertise to operate and further develop the accelerator. European XFEL takes the responsibility for the X-ray systems and the user program of the facility.

3. X-ray Photon Beam Transports

The X-ray optical systems that transport X-ray photons from the undulators to the experiment hall are located in long underground tunnels. From the source point of FEL radiation located within the last segments of the FEL undulators to the scientific instruments in the experiment hall these beam transport paths are up to 1 km long [38,39]. The key optical elements of each transport system are three mirrors: Mirrors "1" and "2" create a horizontal offset of the X-ray FEL beam. This offset prevents unwanted background radiation, consisting of Bremsstrahlung and high-energy spontaneous radiation also produced in the long FEL undulator, to be transported into the experiment areas. The spontaneous radiation has a critical energy of typically 200 keV and is not reflected by the offset mirrors, but is rather absorbed by the first mirror or transmitted and then stopped by a massive tungsten beam stop. Only the desired X-ray FEL photons in the energy ranges of 3–25 keV (SASE1 and SASE2) and 0.25–3 keV (SASE3) can pass the offset mirror chicane. Mirror "3" (distribution mirror) can be optionally inserted to reflect the photons to the HED, FXE, or SCS instruments, while the undeflected beam passes to the MID, SPB, and SQS instruments, respectively (compare Figure 3). In the case of the FXE, MID and HED instruments, multi-bounce crystal monochromators (optional) are integrated in the beam transport. For SASE3 a grating monochromator has been integrated that can be used by the SCS and SQS instruments at this FEL source. The beam transport layout includes for each of the systems the possibility of integrating a third beamline to a third instrument. Such an extension is currently in preparation at SASE3.

The distance from the source point to the first mirror is between ~245 and 290 m, which is enough to expand the beam to a size filling the about 1 m long mirrors under grazing incidence angle and thereby reducing damage and heat load effects. In order to handle the demanding power densities, a surface coating with boron carbide is applied on the single crystalline silicon mirrors. A liquid indium gallium eutectic film, which is in contact with water-cooled copper blades, is used to remove excess heat from all mirrors of the beam transport system. Because the X-ray FEL radiation is close to the diffraction limit, its divergence is roughly proportional to the photon wavelength. To utilize the full length of the offset mirrors for all photon energies, their reflection angle can be varied from 1.1–3.6 mrad for the hard X-ray beam transports and from 6–20 mrad for SASE3. The reflection angles define the energy cut-off and, thereby, the transport of higher harmonic radiation, too. The distribution mirror operates at a fixed angle, which is defined by the distance from the mirror to the experiment hall and the lateral distance (1.4 m) between the shutters of instruments operating at the same SASE beamline. To avoid over-illumination of the distribution mirror, the second offset mirror is bendable and can slightly focus the beam towards the distribution mirror. Alternatively, Be Compound Refracting Lenses (CRL) positioned upstream of the offset mirrors of the SASE1 and SASE2 beam transports can be used to collimate the beam or to produce a similar confocal beam situation with an intermediate focus behind the distribution mirror.

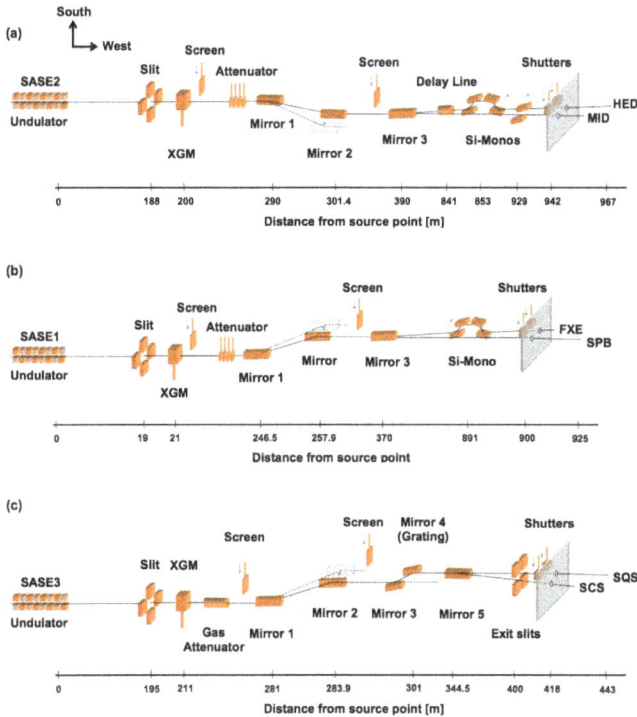

Figure 3. Optical layout of the three photon beam transports with some of the most important elements in the sequence towards the instruments located in the experiment hall. The most southern beam line is: SASE2 (**a**); then SASE1 (**b**); and SASE3 (**c**).

The most crucial requirement to the X-ray mirrors is preservation of the almost perfect wave front created by the lasing process [40]. Source properties and source-to-mirror distances at the European XFEL lead to requirements of about 2 nm peak-to-valley shape error for all mirrors of the beam transport systems, corresponding to roughly 50 nrad rms in slope error. The mirrors were manufactured in Japan and Germany by deterministic polishing techniques, where material is iteratively removed on atomic length scales according to a very precise metrology map of the mirror before each polishing step.

One important constraint of X-ray beam optics at an X-ray FEL is the so-called single-pulse damage. Because a large number of photons (corresponding to pulse energies of the order of mJ per pulse) arrive within 10–100 fs, thermal transport does not remove any heat during the pulse, even for excellent heat conductors like copper or diamond. For focused beam conditions (typically smaller than 50 μm diameter), most materials will vaporize on an ultrafast time scale due to the absorption of energy from a single X-ray FEL pulse within the X-ray penetration depth. More resistant materials are the ones with low atomic number where the absorption per atom is lower. Most components directly exposed to X-ray FEL radiation, or at least their beam facing surfaces, are therefore made of boron-carbide or diamond, for example slits, beam stops, shutters, collimators and attenuator plates. Exceptions are the cryogenically-cooled silicon monochromators in the hard X-ray beam transports SASE1 and SASE2, but also beam position screens and a few other components. For these components, it is required to carefully monitor the impinging beam size to avoid single-pulse damage effects.

Another big challenge is the total power of a train of X-ray pulses which could reach values as high as several kW depending on pulse energy and number of pulses within the train of 600 μs duration. An automatic protection system will trigger a reduction of the maximal number of pulses if equipment is inserted that does not withstand full pulse trains. In addition, long X-ray pulse trains, when missteered, could easily damage the stainless steel pipes of the vacuum system. To prevent this, photon beam loss monitors have been implemented at strategic places along the beam transport. Up to four diamond plates can be adjusted around the beam trajectory. In case of unwanted beam motions (e.g., due to mechanical drifts of the mirror mounts) the X-ray beam would hit a diamond and produce optical light fluorescence. This light is captured by a photomultiplier and triggers via the machine protection system a rapid interruption of the beam (within the same pulse train).

4. Photon Diagnostics Systems

X-ray photon diagnostics is required for monitoring the photon pulse parameters generated by the European XFEL [41–44]. The diagnostics systems provide essential information to the machine for setup, operation and optimization of the accelerator, undulator and X-ray optics, especially during commissioning. Diagnostics is also mandatory for normalization and interpretation of the experimental data. Several beam properties will be measured by so-called online methods, that is, for each photon pulse and with minimal distortion of the pulse. Examples are the pulse energy and beam position, but also spectral content and information about temporal properties can be collected through these systems. Pulse-to-pulse capability is challenging because of the 4.5 MHz repetition rate, but it is particularly important to be able to normalize data for fluctuations of photon pulse parameters due to the SASE process or due to electron or X-ray beam instabilities. In addition, for setup and specific measurements several invasive photon diagnostic systems are installed which stop the X-ray pulses, or at least severely modify the pulse properties.

In this section, we only describe photon diagnostic devices that will be employed in the photon transport sections inside the tunnels. Further systems are integrated in the scientific instruments in the experiment hall. These include, in particular, the temporal diagnostic systems [45,46] employed to monitor the X-ray pulse arrival, pulse duration, and, ideally, the temporal shape as shown previously [47–50].

4.1. Online Photon Diagnostic Systems

These systems can be separated into residual gas systems, naturally interfering only minimally with the X-ray beam, and systems using very thin solid films or crystals, thereby only absorbing a minor fraction of the FEL pulse. This latter method is only applicable to hard X-ray radiation as otherwise the absorption is too strong.

For residual gas diagnostic systems photoionization of rare gases (Xe, Ne, Ar or Kr) or nitrogen is applied making these devices indestructible and highly transparent [51]. This non-invasive diagnostic method is best suited for high peak energies and high average flux since there is no issue with damage or heating due to the absorbed X-ray pulse energy. At European XFEL these systems are employed in the beam transports to measure pulse energy, beam position and polarization of the X-ray pulse. Residual gas monitors can operate continuously up to very high pulse repetition rates, limited by the flight time of ions and electrons used for the measurement of pulse properties, and work even for hard X-rays if a sufficient sensitivity is able to compensate for the reduced cross-sections. As of today no reports about distortion of coherence and wavefront properties due to residual gas monitors have been reported, however for highest repetition rates and elevated gas pressures depletion may occur [52].

Online solid-state systems employ either thin foils to scatter a fraction of the X-ray beam, using the detection of this scattered fraction to measure the pulse energy and, in a special configuration, beam position [53], or thin curved crystals to disperse the incident spectrum on a position sensitive detector [54]. In both cases, only a small fraction of the X-ray beam is absorbed or scattered, however, these systems face limitations when it comes to very high pulse energies. In particular, heat transport limitations of thin films restrict their high repetition rate applications.

4.1.1. X-ray Gas Monitors

The X-ray gas monitors (XGM) are pulse energy (photon number and flux) and position monitors that resolve individual photon pulses at MHz rates (temporal resolution better than 100 ns) [51]. Due to a gain of up to 10^6, individual X-ray pulses with femtosecond durations containing 10^7 up to 10^{15} photons can be measured with better than 10% absolute accuracy, and with better than 1% relative (pulse-to-pulse) accuracy for pulses with more than 10^{10} photons. The beam position is monitored in both transverse directions with an accuracy on the order of ± 10 μm within a range of ± 1 mm. There is an XGM installed in the direct beam of each FEL, upstream of the double mirror systems, monitoring the source properties. Three more XGMs are placed closely upstream of the scientific instruments SPB/SFX, SCS, and HED, to monitor the pulse properties actually delivered to the experiments after passing several X-ray optics elements in the tunnels.

4.1.2. Photoelectron Spectrometer

The photo-electron spectrometer (PES) measures the spectrum and polarization of the photon pulse based on an angular resolved time-of-flight measurement of photo-electrons [55,56]. This device is integrated initially only in the SASE3 beam transport, because for soft X-rays one cannot employ crystal-based schemes to measure the spectrum, and instead the energy distribution of XFEL-generated photo-electrons can be used to deduce the center and width of the photon energy spectrum. In addition, it is planned to employ variable polarization schemes at the SASE3 FEL source, hence requiring measuring and monitoring of the polarization state. The PES has a spectral resolving power of $\Delta E/E \leq 10^{-4}$ and the polarization direction and degree can be measured with an accuracy of 1%.

4.1.3. The HIREX Spectrometer

The HIgh REsolution hard X-ray single-pulse diagnostic spectrometer (HiREX) spectrometer is an online device, based on a diamond diffraction grating used in transmission to split off a small fraction (0.1%) of the photon beam, a bent crystal as a dispersive element, and a MHz-repetition rate strip detector. The grating and crystal chambers are separated by 10 m distance. Gratings with pitches of 150 nm and 200 nm were installed. While beam transmission depends on the photon energy, typically 95% transmission is achieved. Five percent is then spread into all diffraction orders. The first order diffracted beam from the grating is sent to a bent crystal for energy dispersion under Bragg condition [54]. The 10 μm thick bent silicon Si crystals have (110) or (111) orientations and are mounted with fixed bending radii of 75 mm, 100 mm or 150 mm. Two detectors are available for data acquisition: an optical camera for full transverse 2D imaging at low repetition rate, and a modified Gotthard-II 1D strip detector for fast data acquisition at 4.5 MHz.

4.2. Invasive Photon Diagnostics Systems

The invasive diagnostics is either used for initial commissioning with spontaneous radiation, for FEL commissioning, or for setup purposes prior to or during measurements.

4.2.1. MCP Based Detector

When all undulator segments are inserted to establish the SASE condition, this detector measures intensities from the initial signs of lasing up to saturation [57]. Two horizontal manipulators insert either 15 mm diameter MCP discs for integral intensity monitoring with 1% rel. accuracy over a large pulse energy range (1 nJ–10 mJ), a photodiode (Hamamatsu, 10×10 mm^2, 300 µm thick), or a larger MCP-intensified phosphor screen providing an intensified beam image with 30 µm resolution via an optical camera setup.

4.2.2. Undulator Commissioning Spectrometer

This spectrometer analyses spontaneous radiation from one or few undulator segments to measure their individual undulator parameter K [58,59]. These measurements are necessary for an independent measurement and setting of all undulator segments with $\Delta K/K < 10^{-4}$ and to further adjust the individual phase shifters in-between undulators. The filter chambers of the systems at SASE1, SASE2 and SASE3 contain five filter foils of Al, Mo, Cu, Ni and Al with a diameter of 30 mm and varying thicknesses for attenuation and also for spectroscopy by scanning across their K-edges. The monochromator itself, called K-mono, contains two Si channel-cut crystals which can be used in two- or four-bounce geometry. The Bragg angle range is 7° to 55° to cover an energy range from 2.5 keV to 16 keV with Si (111) (7.5 to 48 keV with Si (333)). The resolution is $\Delta E/E = 2 \times 10^{-4}$ for Si (111) (10^{-5} for Si (333)). The crystals are retracted in horizontal direction from the beam. Detection is realized by a photodiode or the highly sensitive SR-imager (see below).

4.2.3. Imagers

There are almost 30 imaging units distributed over the photon tunnels which serve different purposes and therefore have different resolutions, fields of view, and geometries [60]. All of them contain one or more scintillators, mostly Ce:YAG, sometimes additionally polycrystalline diamond, and all but one type have stationary optics with sCMOS GigE cameras and fixed focus lenses.

- Transmissive imagers (1 per FEL) are closest to the source and have the thinnest scintillators to allow transmitting the beam for recording another image of the same photon pulse at a downstream imager. By this method beam pointing and beam offset data can be obtained simultaneously.
- The SR imagers (1 per FEL) are optimized for highest photon sensitivity to detect spontaneous radiation from single undulator segments when applied in conjunction with the K-mono in undulator commissioning. Their optical resolution is 25 µm (FWHM) and field of view (FOV) 26.6×15 mm^2 using YAG:Ce and ceramic Gd2O2S:Pr scintillators.
- The FEL imagers (1 per FEL) are optimized for detailed spatial characterization of the FEL beam to measure the transverse intensity profile with beam position, size and shape. Their optical resolution is 28 µm (FWHM) and FOV is 16×22 mm^2. These imagers have redundancy scintillators of several different materials.
- Pop-in monitors (15 in total) are the basic imagers for beam finding and alignment. These monitors are placed downstream of major optical elements like mirrors and monochromators. Their horizontal FOV is large as to cover the variable beam offset without scintillator or optics movements. Various geometries are employed. Most devices put the scintillator at 45° to the XFEL beam, but some have the scintillator at normal incidence and an additional optical mirror. Optical resolutions range from 35 to 83 µm (FWHM) and FOVs from 22.7×40 up to 150×30 mm^2.
- Exit slit imagers are installed on the two exit slits of the SASE3 monochromator for beam alignment, but more importantly to deliver single-pulse soft X-ray spectra with a resolution of $\Delta E/E \geq 10^{-5}$.

5. The SPB/SFX Instrument

5.1. Scientific Scope and X-ray Techniques

The Single Particles, Biomolecules and Serial Crystallography (SPB/SFX) scientific instrument's [61] primary goal is to enable three-dimensional imaging, or three-dimensional structure determination, of micrometer-scale and smaller objects. A particular focus is placed on biological objects—including viruses, biomolecules, and protein crystals—though the instrument will also be capable of investigating non-biological samples using similar techniques. This structure determination is not limited to static structures—three-dimensional time-resolved structures are within scope too. One of the main driving factors for such studies is to ultimately enable rational drug design through understanding the structure, and hence the function, of arbitrary biomolecules. Studies in structural biology with X-rays have a long history and have exploited ever-brighter X-ray sources as they have been developed [62].

X-ray FELs, as the most recent phase in X-ray source development, offer yet additional benefits to structural biology with X-rays. In particular, they offer the possibility to investigate radiation damage sensitive samples (such as proteins with important metal centers), samples that scatter only weakly (small crystals or non-crystalline specimens), time-resolved processes that are irreversible, as well as other cases that inherently require many incident X-ray photons in a single pulse [63,64]. Unprecedented possibilities are opening to observe weakly scattering samples, such as small crystals of proteins or perhaps even non-crystalline bio-samples such as viruses, which are largely unable to be seen at synchrotron or lab based X-ray sources. Nevertheless, techniques that are relatively simple at conventional sources, such as tomography, are not viable at an X-ray FEL where samples are typically destroyed by the act of illumination in a single projection. This reality means that many frames of data from different projections of a crystal or (reproducible) particle must be combined to form a complete three-dimensional diffraction volume that can be interpreted later as structure [65,66]. These methods require as many as tens of thousands or hundreds of thousands of "good" hits for a single structure [67,68]—and many more should one wish to look at a series of structures resolved in time for example.

5.2. Requirements

These experiments require X-ray instrumentation in a traditional forward scattering geometry to collect diffraction at angles up to those commensurate with atomic resolution. Crystallography requires photon energies up to about 16 keV—beyond the Selenium edge—to aid in anomalous diffraction measurements for structure determination. On the low energy side, single particle imaging, which deals with typically very low diffraction signals, requires as low a photon energy that permits the desired resolution for the system under study. Furthermore, mitigation of radiation damage requires optimization of the beam power, that is, to carefully trade highest pulse energy versus shortest pulse duration.

X-ray FEL serial crystallography and imaging experiments are primarily performed in a mode that is destructive to the sample. The goal is to illuminate the specimen with as many X-ray photons per pulse as possible, to maximize the scattered signal from each particle. To do so, one must have an optical system that is highly transmissive, as well as focusing to a spot size that is comparable to the size of the sample(s) under investigation. The "small" crystals used in serial crystallography at XFELs tend to be around 1 μm diameter in size, with some variation larger or smaller. Relevant biological single particles range from biomolecules some tens of nanometers across to large viruses up to 500 nm in diameter. To accommodate this wide range of sample sizes, the SPB/SFX instrument plans to deliver two different focal spots—a 1 μm-scale focus and a 100 nm-scale focus. Coherent diffraction imaging of individual particles further requires precise knowledge of the wavefront of the incident X-ray pulse leading to stringent requirements on the selection of the X-ray optical components and their performance.

Of particular importance for the experiments to be performed at SPB/SFX will be the performance of the large area detectors. The two primary requirements are a very high dynamic range and single photon sensitivity. An ideal 2D detector for serial crystallography should have a high dynamic range much higher than the four [31] or five [32] orders of magnitude presently achievable, as the intensities of individual Bragg peaks can vary enormously and these intensities must be determined very accurately for successful phasing and structure determination (though nevertheless a detector with 10^4 or smaller dynamic range can be successfully used). Single photon sensitivity is important for detecting weaker scattering, such as from single non-crystalline particles or weak Bragg peaks. The detectors should be compatible with detection at the 4.5 MHz intra-train repetition rate to ensure collection of as much images as possible within a meaningful time frame during which particles can be injected. This requirement leads to the further need for stringent data reduction techniques to avoid a data deluge. Finally the detector(s)' mechanical design must be compatible with the instrument. This means a pixel size that is not too large (≤ 200 µm), a number of pixels commensurate with the number of resolution elements desired in any given structure (i.e., ≥ 1 MPixel for ~200 linear resolution elements) and an operation that is ideally compatible with the sample environment (for the upstream interaction region at SPB/SFX this means in vacuum operation). The detector is required to operate in vacuum and be placed as close as 129 mm from the upstream interaction region, resulting in a better than 2 Å geometrical resolution limit for 9 keV photon energy. It can also be placed downstream as far as 6 m from the interaction region, allowing for appropriate sampling of diffraction data from samples as large as almost 1 µm at the lowest energies.

5.3. SPB/SFX Instrumentation and Capabilities

The SPB/SFX instrument is a 3 to 16 keV, forward scattering instrument [61] with a 1 µm-scale and a 100 nm-scale focus in the upstream interaction region [69,70], and optics to refocus the upstream focal point to a second interaction region further downstream (about 12 m) in the experiment hutch. This refocused beam allows for a second, in series, experiment to be performed simultaneously with a measurement in the upstream interaction region (see Figure 4). The Serial Femtosecond Crystallography (SFX) User Consortium provides the vast majority of the instrumentation for the downstream interaction region including, but not limited to, a 4 Mpixel detector (AGIPD) and an alternative detector (Jungfrau), the refocusing CRLs, sample delivery technologies (largely liquid jet delivery in various forms and fixed target systems) as well as various diagnostics and sundry apparatus. The 1 µm-scale and 100 nm-scale focal spots for the upstream interaction region of the SPB/SFX instrument are to be produced by mirror optics due to their high transmission and potential for making very neat and well confined focal spots. The mirrors are all designed with 950 mm clear aperture, with working angles of 4 mrad and 3.5 mrad, respectively.

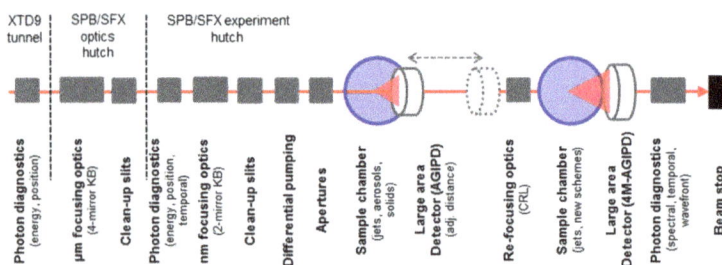

Figure 4. Overview schematic of the SPB/SFX scientific instrument at the SASE1 FEL beamline. The sketch indicates major instrumentation items installed in the photon beam transport tunnel, optics and experiment hutch. Not shown is the PP-laser instrumentation.

The 100 nm-scale design is a traditional Kirkpatrick–Baez (KB) design. The 1 μm-scale mirrors are four-bounce—with a plan horizontal mirror followed by a focusing ellipse in the horizontal and then a focusing vertical mirror with a plane vertical mirror. This four bounce design mitigates vibrational issues and a large displacement from the direct beam over the long (~24 m) mirror to interaction region distance. For early user operation in 2017, the mirrors will not yet be installed. Instead, Beryllium compound refractive lenses (CRLs) will be used to produce an approximately 2.5 μm spot in the interaction region. After mirror installation, the CRL unit will be moved to the refocusing position and new lenses installed to refocus the upstream spot to ~3–5 μm in the second interaction region downstream.

In addition to focusing elements, a variety of beam conditioning apparatus (slits, apertures and attenuator) will be installed to aperture the beam upstream of the optics (the so-called power slits), clean up tails and streaks from the optics immediately downstream of them (the so-called cleanup slits) and apertures near the focal plane that further clean up the beam (termed apertures and will likely be sacrificial). This beam conditioning is essential for single particle imaging where a very neat, clean and well-understood beam is necessary for the successful observation and interpretation of the weak diffraction data collected.

The primary 2D detector at SPB/SFX is a 1 Mpixel AGIPD detector [31]. It will be mounted in a vacuum chamber directly attached to the upstream sample chamber. Using a longitudinal translation, it can be placed as close as 129 mm from the interaction region, resulting in a better than 0.2 nm geometrical resolution limit for 9 keV photon energy. It can also be placed downstream as far as 6 m from the interaction region, allowing for appropriate sampling of diffraction data from samples as large as almost 1 μm at the lowest energies. The detector mechanics consists of four panels mounted on x-y-translations to adjust the central hole for letting the X-ray beam pass.

The destructive nature of these experiments and the high repetition rate of the European XFEL necessitate rapid delivery (and replenishment) of sample at the interaction region. Furthermore, for biological systems the samples must be appropriately hydrated and handled to ensure an intact and representative sample is brought to the XFEL beam. Three primary sample delivery mechanisms exist for the delivery and replenishment of samples: liquid jet injectors, aerosol injectors and fixed target stages, all of which will be deployed at the SPB/SFX instrument.

6. The FXE Instrument

6.1. Scientific Scope and X-ray Techniques

The Femtosecond X-ray Experiment (FXE) scientific instrument has a primary scientific focus in the field of photo-induced chemical dynamics in liquid environments [71]. The interplay between nuclear, electronic, and spin degrees of freedom during the course of an ongoing reaction will be monitored using a suite of X-ray techniques, thereby offering new observables in the femtosecond time domain to deliver this information. The FXE instrument will permit structural studies on the 25 fs time scale and below, with ultrafast X-ray Absorption Near Edge Structures (XANES), Extended X-ray Absorption Fine Structure (EXAFS), Resonant Inelastic X-ray Scattering (RIXS), non-resonant X-ray Emission Spectroscopy (XES), and Wide Angle X-ray Scattering (WAXS) from liquids being key techniques to unravel new details about the very first steps in these reacting systems. One fundamental goal is to eventually record a complete molecular movie, observing not only the structural rearrangements occurring in the system but also of the underlying electronic structure changes. Together with ultrafast optical spectroscopy techniques it will become possible to understand the ensuing photo-physical behavior.

One particular interesting area of research concerns catalytic activity and solar energy conversion schemes, which occur in several transition metal compounds. Such compounds are key ingredients in certain proteins, and are often at the very beginning of light-driven biological functions [72]. They are also studied in chemistry due to their rich magnetic switching behavior [73], their charge-transfer properties in light-harvesting applications [74], or for their ability to form highly reactive intermediate species, which enhance further reaction steps, e.g., towards more efficient catalytic behavior [75]. These compounds are believed to exhibit correlated electron dynamics in a regime in which the Born-Oppenheimer rule is not valid. The direct observation of elementary steps towards, e.g., spin transition dynamics has so far been impossible which is expected to change due to the possibility of studying new observables in X-ray FEL experiments [76,77].

6.2. Requirements

The different X-ray techniques offered at FXE have quite different requirements to the FEL source. XES requires merely the photon energy to be well above the absorption edge of the selected element. The same condition applies to WAXS while the bandwidth of SASE radiation is perfectly suited for diffuse scattering measurements [77]. Therefore both techniques can be applied simultaneously. More demanding X-ray beam properties exist for XANES, EXAFS or RIXS techniques, requiring smaller bandwidth of the incident beam (typical $\Delta E/E \sim 10^{-4}$), and scanning the photon energy over a certain range at the selected absorption edge. This scanning requires tuning of the undulator gap together with a primary monochromator.

All X-ray techniques need to be used in concert with an incident laser beam, whose femtosecond pulses are synchronized to the X-ray source, and with sufficient intensity to trigger the desired reactions. The experiments also require appropriate handling of the probed samples. For (bio)chemical systems in liquid solutions, the sample should be removed after each pump-probe event, to permit the next measurement to be recorded on a fresh sample, which has not been exposed to neither the optical laser nor X-ray beams before. To preserve the femtosecond time resolution, the time spread through the sample via the group velocity mismatch between optical and X-ray pulses needs to be minimized. In general, this requires a jet thickness below 10–20 um.

6.3. FXE Instrumentation and Capabilities

X-ray FEL radiation from the SASE1 FEL is collimated by means of Be compound refractive lenses (CRL) 900 m upstream from the experiment hall in XTD2 in order to maintain a beam size in the 1–2 mm (FWHM) diameter range (for all X-ray energies in the 5–20 keV range), when the beam enters the primary four-crystal monochromator, diamond beam splitter grating and eventually the FXE experiment hutch (compare Figure 5).

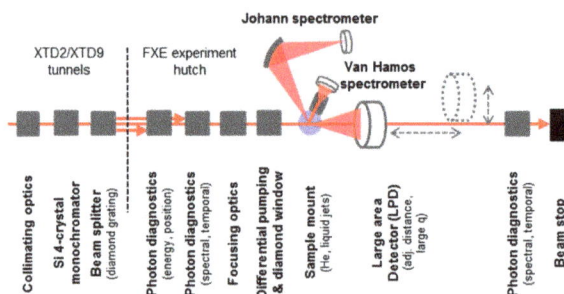

Figure 5. Overview schematic of the FXE scientific instrument at the SASE1 FEL beamline. The sketch indicates major instrumentation items installed in the photon beam transport tunnel and experiment hutch. Not shown is the PP-laser instrumentation.

The primary Si four-crystal monochromator ($\Delta E/E = 10^{-5}$–10^{-4}) maintains the same beam axis for the X-ray beam onto the sample as the pink beam (thus without monochromator), this way we ensure that the laser beam always strikes the X-ray illuminated volume, and that the X-rays always take the same path through the entire optics branch including the long stretch downstream to the beam stop. This arrangement eliminates the need to geometrically adjust the beam(s) for varying conditions, as demanded by the specific experiment. Only the timing changes considerably between pink and monochromatic pulses entering the sample, especially the monochromatic beam has different arrival times (with respect to the exciting optical laser pulses), which can be tabulated for each energy.

At the end of the tunnel section a diamond grating generates side maxima (diffraction orders) of the main X-ray beam, which are used for X-ray beam diagnostics inside the experiment hutch (similar to what has been described in Ref. [78]). The side beams can be used to measure the incident spectrum of the X-ray beam via a curved crystal spectrometer, and the actual arrival time of the X-ray pulse with respect to the optical laser pulse. Together with beam shaping slits and an intensity position monitor the conditions of the beam entering the sample are thus well characterized.

With a second stack of Be CRL lenses the X-ray spot size on sample can be freely tailored to values in the 2–200 um range (FWHM). The X-rays enter the sample area via a diamond window separating the ultrahigh vacuum optics branch from the sample environment under ambient conditions (He atmosphere at room temperature). A liquid flat sheet jet with adjustable thickness in the 2–200 um range provides a defined surface for optical excitation and X-ray probing. Two secondary spectrometers are available for XES experiments, and each can also be rotated around the sample from forward to nearly backward scattering angles: a Johann spectrometer with up to 5 spherically bent crystals collects single emission wavelengths with a resolution of $\Delta E/E = 10^{-4}$. This spectrometer has a large solid angle and spectra are obtained by scanning both the crystal rotation with the collecting detector on a Rowland circle. Alternatively, a 16 element von Hamos type spectrometer collects the entire XES spectra at a resolution of $\Delta E/E \sim 10^{-3}$ without moving elements, thus enabling single-pulse experiments.

The forward WAXS scattering pattern is collected using the LPD detector [32] having moveable quadrants for a central hole for the X-ray beam with adjustable size in the 1–10 mm range. A post-diagnostics bench can then record the beam properties (spectrum, intensity, and timing) of the transmitted beam, before it finally strikes the copper beam stop.

7. The SQS Instrument

7.1. Scientific Scope and X-ray Techniques

The SQS (Small Quantum Systems) scientific instrument is dedicated to investigations of fundamental processes of light-matter interaction in the soft X-ray wavelength regime. In particular, studies of non-linear phenomena, such as multiple ionization and multi-photon processes, time-resolved experiments following dynamical processes on the femtosecond timescale, and investigations using coherent scattering techniques are targeted [79]. Principal research targets are isolated species in the gas phase, such as atoms, molecules, ions, clusters, nanoparticles and large bio-molecules. The use of soft X-ray photons enables controlled excitations of specific electronic subshells in atomic and site- or element specific excitation in molecular targets. One of the main goals of the SQS instrument is the complete characterization of the ionization and fragmentation process, at least for smaller systems, by analyzing all products created in the interaction of the target with the FEL pulses.

Experiments at SQS typically use X-ray pulses of highest intensity to drive the probed system to highly excited states or initiate non-linear X-ray processes. The additional use of synchronized optical laser pulses will be applied to controlled manipulations of the electronic states and nuclear movement. Probing of the X-ray FEL interaction with the sample system will be performed either by direct coherent X-ray scattering to obtain structural information or by spectroscopic techniques. A focus is put on a variety of particle spectroscopy techniques, such as energy- and angle-resolved electron and ion spectroscopy allowing the determination of kinetic energies and momenta of the charged particles, and additional options for XUV and soft X-ray spectroscopy. In particular, the very open and flexible arrangement of the spectrometers will enable the application of various coincidence techniques, such as electron–electron, electron–ion and photon–electron/ion coincidences, which all require and therefore take full advantage of the high repetition rate available at the European XFEL.

7.2. Requirements

Located at the SASE3 FEL the photon energy of the radiation will range from about 250 eV up to 3000 eV, i.e., covering the energy range of ionization thresholds for numerous relevant atoms such as the K-edges of carbon, nitrogen, oxygen as well as of phosphor and sulfur, the L-edges of the 3d transition and rare earth metals and K- and L-edges of various ions. Pulse durations as short as 2 fs, available in the 0.02 nC low-charge mode, enable in combination with the synchronized optical laser time-resolved studies in the few-femtosecond time domain. Furthermore, pulse energies of up to 10 mJ are produced at 1 nC high-charge mode. This high pulse energy corresponds to 2×10^{14} photons per pulse and is the main requirement for the study of non-linear processes, since intensities of more than 10^{18} W/cm^2 can be reached by focusing, e.g., the 10 mJ/100 fs FEL beam to a diameter of about 1 µm.

Ultra-high vacuum conditions in the experimental area are required for coincidence techniques in order to minimize signals caused by ionization of residual gas. For this reason most of the experiments will operate at background pressures of about 10^{-10} mbar or less and supersonic molecular jets and specially designed quantum-state-, size-, and isomer-selected beams of polar molecules and clusters (COMO for "Controlled Molecules") will be used for sample delivery. These vacuum conditions will be reduced for experiments on larger targets requiring the use of the large DSSC imaging detector and of dedicated cluster or nanoparticle beam devices.

7.3. SQS Instrumentation and Capabilities

The optical layout of the beam transport system enables experiments using the direct beam from the variable gap SASE3 undulator or the reduced bandwidth radiation ($\Delta E/E \leq 10^{-4}$) from the soft X-ray monochromator. A Kirkpatrick–Baez adaptive mirror system assures a tight focusing of the beam down to spot sizes as small as about 1 micron (see Figure 6). The bendable high-polished mirrors allow adjustments of the focal spot size and displacement of the focus to three different interaction regions separated by 39 and 200 cm, respectively. The FEL radiation properties, such as pulse energy, pulse duration, arrival time, spectral distribution and focal spot size, are monitored with the help of several diagnostic devices installed downstream and upstream of the interaction regions inside the dedicated and enclosed experiment area.

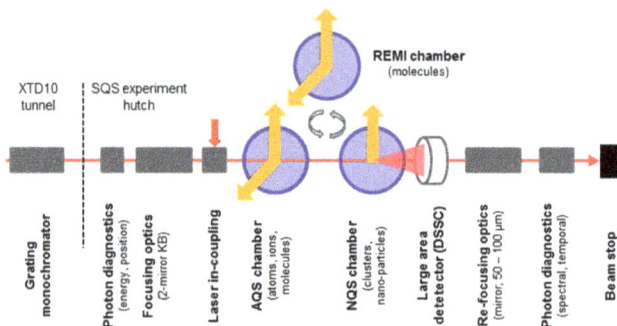

Figure 6. Schematic outline of the SQS scientific instrument at the SASE3 FEL. Shown are major instrumentation items installed in the photon beam transport tunnel and experiment hutch comprising the beam transport, focusing and diagnostic devices as well as the three interchangeable experimental vacuum chambers AQS, NQS and SQS-REMI.

The general concept of the instrument is based on a two-chamber system thus separating applications on "Atomic-like Quantum Systems" (AQS), such as free atoms, atomic ions, and small molecules, and on "Nano-size Quantum Systems" (NQS), such as clusters, nanoparticles and large biomolecules, all typically larger objects [79]. The AQS chamber will be equipped with a set of spectrometers enabling the analysis of electrons, ions and photons with high-energy resolution and the determination of the angular distribution of the particles. Six electron time-of-flight (TOF) analyzers can be used for angle-resolved high kinetic energy resolution experiments at distinct angles in the dipole and in the non-dipole planes. A velocity-map-imaging (VMI) spectrometer provides the full information about the angular distribution of the emitted electrons and ions, and is designed for electrons up to about 1000 eV kinetic energy. The single pulse analysis of very dilute samples or of processes characterized by extremely low cross sections is possible by means of a magnetic bottle electron spectrometer, which collects electrons over the full solid angle. Finally, a specially designed 1D-imaging XUV spectrometer is dedicated to the analysis of fluorescence emission at high spectral resolution. The use of Wolter optics and a 2D-imaging detector is enabling a spatial resolution of about 10 μm along the beam propagation direction and thereby a temporal resolution of about 30 fs in crossed beam experiments.

The NQS chamber will have as particular feature the option to use the DSSC detector [33] in forward diffraction geometry. Due to the high scattering cross sections in the soft X-ray regime, single pulse imaging of larger molecules and particles becomes possible and will be applied to structural analysis at reduced spatial resolution. The DSSC detector is also used in combination with various particle spectrometers (TOF, VMI) to characterize size and shape of clusters and nanoparticles in parallel to the determination of kinetic energies, fragmentation patterns and emission angles of ions and electrons produced in the interaction volume.

In addition, a third, specially designed ultra-high vacuum chamber will host a reaction microscope (SQS-REMI) for the complete characterization of molecular fragmentation processes by the application of electron–ion coincidence techniques [80], taking full advantage of the high repetition rate (until 27,000 pulses per second) available at the European XFEL. Using large area position sensitive delay-line detectors and a well-defined arrangement of magnetic and electric fields to extract and guide the electrons and ions, the kinetic energies of all fragments as well as their relative emission angles can be determined in a single molecule ionization event.

Specially designed in- and out-coupling units for the optical laser are available to provide the optical radiation to all three interaction points in collinear geometry. In general, great emphasis is placed on a flexible design and arrangement of the experimental chambers and the various spectrometers, which will enable users to make optimal use of all the specific characteristics of the European XFEL, in particular of its uniquely high repetition rate. Furthermore, several extensions of the FEL beam parameters (e.g., variable polarization or two-color operation), beam delivery capabilities (beam splitter and delay device) and instrument layout are already decided or under investigation.

8. The SCS Instrument

8.1. Scientific Scope and X-ray Techniques

The Spectroscopy and Coherent Scattering (SCS) scientific instrument is located at the SASE3 FEL source and aims at time-resolved experiments to unravel the electronic, spin and structural properties of materials in their fundamental space-time dimensions. Scientific objectives include, but are not limited to the understanding and control of complex materials [81–83], the investigation of ultrafast magnetization processes on the nanoscale [84,85], the real-time observation of chemical reactions at surfaces and in liquids [86,87], and the exploration of nonlinear X-ray spectroscopic techniques that are cornerstones at optical wavelengths [17,88].

The SCS instrument operates in the soft to tender X-ray regime (250 eV–3000 eV) covering a wide range of core level resonances: K-edges of most 2p and 3p elements (starting from carbon), $L_{2,3}$-edges of 3d and 4d elements (transition metals) and $M_{4,5}$ edges of 4f elements (lanthanides). Time-resolved resonant spectroscopy offers element-, site-, orbital-, and spin-selective probing of complex material dynamics that is either directly related to or indirectly coupled to the valence electrons. Physical properties such as oxidation state, magnetism, local symmetries and ordering as well as elementary excitations can be investigated using X-ray Absorption Spectroscopy (XAS) [86,89], X-ray Resonant Diffraction (XRD) [81–83] and Resonant Inelastic X-ray Scattering (RIXS) [87]. A particular aim of the SCS instrument is to combine these powerful spectroscopic techniques with X-ray diffraction and microscopy methods, which provide nanometer spatial- and femtosecond time resolutions. Such experiments open up a route to follow the dynamics in complex systems on their relevant length and time scales. The SCS instrument further implements Coherent Diffraction Imaging (CDI) techniques, i.e., X-ray holography [90,91]. A time series of reconstructed CDI images can elucidate excited state dynamics in real space.

8.2. Requirements

The monochromatic-beam operations described in Ref. [92] are key to the success of the SCS instrument. The SASE3 soft X-ray monochromator is equipped with two gratings and a flat mirror that allows for monochromatic beam operation at high ($\Delta E/E = 2.5 \times 10^{-5}$) and medium energy resolutions ($\Delta E/E = 1 \times 10^{-4}$) as well as non-monochromatized beam operations without changing the beam transport to the sample. A tunable grating illumination concept is therefore implemented to provide a minimum spectral bandwidth-time duration product [92]. In this way, RIXS experiments with high energy-resolution and lower time resolution as well as ultrafast dynamics studies at reduced energy resolutions (e.g., femtosecond surface chemistry and magnetism) can be performed at the same experiment station of SCS. New developments for the FEL source, such as full polarization control and undulator gap scanning techniques, will be implemented at SCS. These are nowadays standard capabilities at synchrotron facilities for X-ray spectroscopy investigations.

The majority of the experiments will require X-rays to impinge on fixed solid targets that cannot easily be replenished between X-ray pulses, in contrast to liquid jet or particle injection schemes. This sets particular constraints on the optical and X-ray pulse energies in pump-probe experiments when sample damage or degradation by the radiation and heating have to be mitigated to a level that sufficient data acquisition is possible between the sample exchanges. Heat dissipation schemes have to be developed in order to reach the ultimate 4.5 MHz burst mode operation, where the sample relaxation time and the heat dissipation in the probed area must be shorter than 220 ns.

While time-resolved spectroscopy experiments on fixed targets require high average photon flux, CDI and nonlinear X-ray-matter–interaction experiments need the highest pulse energies. In single shot imaging experiments the number of incoming photons determines the attainable resolution and therefore, depending on the damage threshold, requires the experiment to be carried out in a "diffraction before destruction" mode [93]. In this case, a new sample has to be repositioned in the beam between the X-ray bursts at 10 Hz repetition rate and a pulse selection mode is necessary.

8.3. SCS Instrumentation and Capabilities

One of the major goals of the technical design was to implement a diverse platform for spectroscopy and coherent scattering techniques that is realized in a modular instrumentation of experiment stations and detectors. The mirror benders of the SCS Kirckpatrick–Baez refocusing optics deliver the beam to two X-ray interaction regions separated by 2 m (see Figure 7). This allows not only a small beam focus of 1–2 μm for CDI experiments but also a variable beam diameter of up to 500 μm. In this way, time-resolved spectroscopic studies can be carried out making the best use of the high-average photon flux without further beam attenuation for avoiding sample damage.

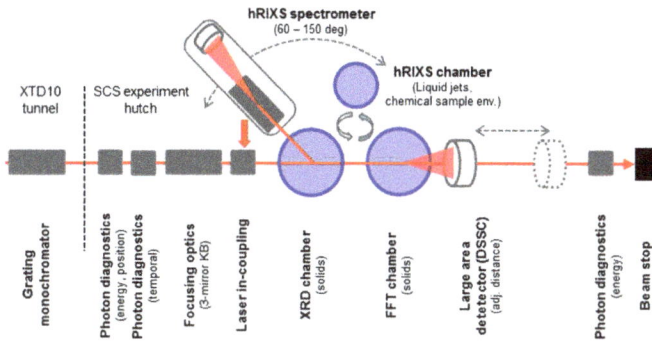

Figure 7. Schematic outline of the SCS scientific instrument at the SASE3 FEL. Shown are major instrumentation items installed in the photon beam transport tunnel and experiment hutch comprising the beam transport, focusing and diagnostic devices as well as the Heisenberg RIXS spectrometer contributed by the hRIXS user consortium.

The SCS instrument comprises two distinct experimental setups, the Forward-scattering Fixed-Target (FFT) chamber and the XRD chamber. Both chambers have the same mechanics on their base that locks to three fixation points on the floor, one set per interaction region. This allows for faster exchange of experiment stations and reproducible repositioning of the chambers.

The FFT chamber is optimized for forward-scattering geometries such as XAS in transmission, small-angle X-ray scattering (SAXS) and CDI experiments. Besides optical and THz beam delivery the sample environment encompasses static magnetic fields up to 0.5 T and a fast sample scanner that fits 50×50 mm^2 sample arrays. The diffraction signal from the samples is collected downstream on the primary area detector, DSSC. The detector is mounted on a girder with a 5 m long translation stage. The closest sample-detector distance is 350 mm, corresponding to spatial frequencies near the wavelength limit of a few nm at soft X-ray energies. Objects of up to 3–5 μm in diameter can be reconstructed using CDI at a sample-detector distance of 5 m and photon energies below 1.5 keV. The missing low-q data passes through a hole in the center of the detector and is recorded downstream as an integrated sample transmission signal. Since the monochromatic beam intensity jitter is large, data collected from low intensity pulses can be vetoed using the DSSC detector [33]. In this way, the signal to noise level of the data can be improved and data acquisition time is optimized.

The XRD chamber enables a range of time-resolved spectroscopy and scattering methods for which a variable scattering angle is needed. The most relevant techniques are time-resolved XRD and RIXS as well as nonlinear X-ray studies (stimulated emission and scattering). The XRD setup is equipped with a diffractometer where a diode array can be rotated by nearly ±180° in the horizontal scattering plane. The sample motion system provides six degrees of freedom and enables temperature-dependent studies between room temperature and cryogenic temperatures (liquid He cryostat). A detector flange with 90° continuous rotation can interface with large detectors and spectrometers. The Heisenberg-RIXS (hRIXS) User Consortium is contributing a high-resolution spectrometer ($\Delta E/E = 0.25 - 1 \times 10^{-4}$) that facilitates state-of-the-art RIXS experiments with unprecedented time-resolution at the SCS instrument. The 5 m long hRIXS spectrometer can rotate around the sample position, hovering 50 μm above a high-planarity floor (250 μm peak-to-valley over 37 m^2) on air pads.

Both experiment chambers are designed for solid targets and operate in the 10^{-9} mbar pressure regime depending on the detector vacuum. The chambers are equipped with a sample transfer system for exchanging samples under vacuum conditions. Optical laser delivery can be either collinear to the X-ray beam or arranged in off-axis geometry. THz generation and focusing takes place close to the interaction region and temporal diagnostics at the sample interaction point is realized. The hRIXS

User Consortium will contribute an additional experiment station that provides a chemical sample environment including liquid jet systems of different geometries and couples to the hRIXS spectrometer.

9. The MID Instrument

9.1. Scientific Scope and X-ray Techniques

The Materials Imaging and Dynamics (MID) instrument of the European XFEL facility, located at the SASE2 beamline, will provide unique capabilities in ultrafast imaging and dynamics of materials, with particular focus on the application of coherent X-ray scattering and diffraction techniques. Coherent diffractive imaging (CDI) [94,95] and X-ray photon correlation spectroscopy (XPCS) [18,96–98] experiments are at the heart of the activities planned. In addition, high resolution time-resolved scattering [99,100], nano-beam scattering/imaging [101,102] and novel correlation techniques [103] are foreseen at MID taking advantage of the unique time structure and high peak intensity of the European XFEL beam. The instrument can operate in small-angle (SAXS) and wide-angle (WAXS) X-ray scattering configurations with a movable large area detector. A large field-of-view configuration where the detector covers a maximum of reciprocal space is also possible. The instrument is optimized for windowless operation over a wide range of photon energies, 5–25 keV, and possibly higher in the future depending on the development of novel lasing schemes using the SASE2 FEL.

9.2. Requirements

The MID design has been guided by several goals. Firstly, the aim is to preserve the high average and peak brilliance provided by the source and make use of as many photons as possible in the experiments. At the same time, optimum conditions for beam tailoring must be ensured concerning focusing, energy selection, and spectral purity in a setup providing high beam stability (position, intensity) and fast and efficient data collection with the highest possible resolution. A versatile setup was required to enable the breadth of experiments that will take place at MID. The experimental setup is hence windowless (optional), multi-purpose and also contains beam diagnostics tools, both for the X-ray beam and the optical pump laser. MID strives to provide the best possible conditions for materials science experiments using hard X-ray FEL radiation, for instance in the studies of nanostructured materials, phase transitions and metastable states, liquid dynamics, and low-temperature physics and magnetism.

9.3. MID Instrumentation and Capabilities

The MID instrument is mainly installed in two safety hutches, an optics hutch (OH) and an experiment hutch (EH), but several essential components are also placed inside the SASE2 photon beam transport tunnel (see Figure 8). The OH contains a Si(220) monochromator to reduce the bandwidth of the SASE radiation to $\Delta E/E \sim 6.1 \times 10^{-5}$ if required. An additional Si(111) mono ($\Delta E/E \sim 1.4 \times 10^{-4}$) installed in the SASE2 tunnel can be used to pre-monochromatize or separately. Alternatively, the SASE beam can be applied directly ($\Delta E/E \sim 1 \times 10^{-3}$) or in self-seeded mode [104,105] ($\Delta E/E \sim 1 \times 10^{-5}$) once self-seeding becomes available at SASE2. Together with undulator tapering this will allow achieving more than 10^{12} ph/pulse and a record high spectral peak brightness of more than 10^{14} ph/s/meV at 9 keV [106]. The OH contains beam attenuators and slits for further beam tailoring as well as an imager system to provide in-situ visualization of the beam size, shape, and intensity. A split-and-delay line (SDL) [107,108] will also be installed in the OH and will give the possibility of modifying the time-structure of the beam. Normally, the European XFEL delivers ultrashort (~1–100 fs) pulses of photons every 220 ns (4.5 MHz), but with the SDL under construction for MID it is possible to reduce this spacing to any value from ~10 fs to 800 ps [109]. This enables particular experiments requiring such an X-ray pulse pattern, e.g., speckle visibility techniques [110,111] for ultrafast dynamics or X-ray pump X-ray probe, possibly in combination with an optical fs laser pump [112]. The latter gives the additional option of performing X-ray probe–Optical pump–X-ray probe measurements where the two pulses from the SDL are not only delayed in time but also are hitting the sample at different angles of incidence. This provides a unique possibility to distinguish the two X-ray diffraction patterns hitting the detector and a spatial encoding of ultrafast dynamics can hence be obtained to yield a time-resolution much better than the 4.5 MHz detector speed [112].

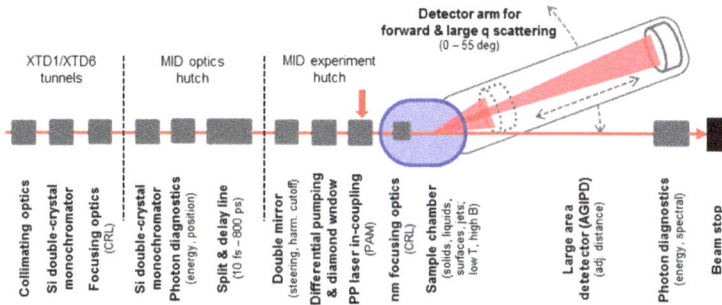

Figure 8. Schematic outline of the MID scientific instrument at the SASE2 FEL. Shown are major instrumentation items installed in the photon beam transport tunnel, optics hutch and experiment hutch comprising the beam transport, focusing and diagnostic devices.

In the EH, the beam first passes through a double mirror system (if inserted) that allows reflecting the X-ray beam downwards for grazing incidence liquid surface scattering. Another mirror reflecting upwards provides the aforementioned option of different incidence angles for the two split beams from the SDL. Downstream of the mirror system the ultra-high vacuum (~10^{-9} mbar) section of the instrument terminates and it is necessary to operate at a lower vacuum level or even at ambient conditions due to the presence of outgassing substances, sample environments, and electronics. This transition is ensured either by insertion of a beam transparent diamond window, or by use of the differential pumping section positioned immediately downstream of the mirror. A large multi-purpose sample chamber (MPC) hosts local optics for nano-focusing, a hexapod sample manipulation stage, as well as different sample environments, e.g., providing low-temperatures via He cryo-cooling, pulsed high magnetic fields, fast sample scanning, sample injection by liquid jets, aerosol injection, etc. To ensure a maximum of stability the stages carrying the nano-focusing setup and the sample hexapod are decoupled from the vacuum pipes and chamber walls and connected directly, via vacuum feedthroughs, to a several ton heavy granite block below the MPC. A focal spot down to 50×50 nm^2 or smaller is enabled by the nano-focusing system [112]. Assuming full transmission of the FEL pulses this could enable peak intensities of beyond 10^{21} W/cm^2 [39,113,114] allowing to explore non-linear X-ray interactions with matter, e.g., two- or multi-photon processes in scattering and absorption [115–117]. The PP laser beam is delivered to EH via a transfer pipe to a laser table next to the MPC allowing additional tailoring of the beam before it is directed towards the sample position. Temporal and spatial overlaps of the optical laser beam and the X-rays can be controlled through imaging and timing diagnostics [49] and tuned by adjusting optical components in the laser beam path located in the ILH.

The radiation scattered from the sample is measured using the AGIPD detector. In SAXS configuration the distance from sample to detector can be varied from ~200 to 8000 mm. This provides an angular detection resolution between 1 mrad and 25 µrad and a field-of-view between 1 rad and 25 mrad. With the direct beam in the center of the detector at 10 keV, it translates into a q-resolution and q-range of 5.0×10^{-3} and 2.3 Å^{-1} for 200 mm, and 1.3×10^{-4} and 6.3×10^{-2} Å^{-1} for 8000 mm, respectively. Special configurations with even shorter sample-detector distance and exploitation of the full energy range of the instrument (5–25 keV) allow tuning these values. In the SAXS case, a hole in the center of the detector (adjustable by movable quadrants) permits unhindered passage of the direct beam, i.e., without destroying the sensor. The exit port of AGIPD is connected to a diagnostics end-station where intensity, size and spectrum of the transmitted beam can be quantified with high resolution. In particular, a semi-transparent bent diamond spectral analyzer has been developed allowing to quantify the SASE spectrum down to a resolution of ~0.1 eV [118]. This spectrometer will operate in parallel with AGIPD and the spectral information together with scattering data enable a better data analysis as well as easy tuning of the self-seeded mode. A similar transparent diamond spectrometer can be inserted upstream of the MPC to measure the spectrum before interaction with the sample [118]. In this manner absorption spectroscopy [119] can be combined with, e.g., pump-probe and coherent scattering techniques providing unique new possibilities of investigating interactions of ultra-bright fs X-ray pulses with matter.

The MID instrument also features the option of measuring in a horizontal WAXS geometry (scattering angle up to ~55° with the sample-detector distance varying between 2000 and 8000 mm. This will enable high resolution detection at large q (beyond 10 Å^{-1}) investigating (coherent) diffraction originating from, e.g., structural, charge, or magnetic ordering in combination with the pulsed magnetic field or the fs pump laser to access ultrafast dynamics processes.

10. The HED Instrument

10.1. Scientific Scope and X-ray Techniques

The High-Energy Density (HED) instrument aims at the investigation of matter at extreme states of temperature, pressure, density, and/or electromagnetic fields using hard X-ray FEL radiation. For this goal the HED instrument will provide a unique combination of the drivers to create extreme states in the laboratory and hard X-ray laser pulses [120]. HED offers a wide range of time-resolved X-ray techniques reaching from diffraction, by imaging to different spectroscopy techniques for measuring various geometric and electronic structural properties. Research areas at HED include the investigation of properties of matter in solar and extra-solar planets, where high pressures of several 100 GPa at moderate temperatures (<10,000 K) are expected, and of properties of matter in the presence of both strong electric and magnetic fields. High-temperature superconductivity will be studied using pulsed magnetic fields generated in coils with field strengths up to 60 T. Extreme electromagnetic fields also occur during and after the interaction of short-pulse high-intensity lasers with solids and liquids, forming a dense plasma and accelerating electrons to up to several MeV kinetic energy. These induce very intense, transient magnetic fields, which could shed light on properties of matter at temperatures of several kT (~11,000 K).

10.2. Requirements

The use of a large variety of X-ray techniques creates a broad band of requirements to FEL operation and properties. Most important are the need for a small bandwidth, typically smaller than 10^{-4} in order to perform inelastic scattering experiments with sufficient resolution and throughput. Furthermore, as many experiments study or use low cross-section processes, high pulse energies are very important. This becomes particularly relevant for experiments at the highest photon energies above 20 keV. For experiments using high energy drivers to create extreme states and operating at reduced repetition rates of 1 Hz, or even far below, it is conceivable to switch beam to other stations at this FEL. Such an operation mode, however, requires that the experiments use the same, or at least very similar, X-ray properties.

10.3. HED Instrumentation and Capabilities

The HED instrument is installed at the SASE2 beamline and features an optics hutch (OH) and an experiment hutch (EH). In addition, an X-ray monochromator, focusing devices, a split and delay line optics and a pulse picker are placed inside the preceding tunnel section (see Figure 9). The four-bounce Si-(111) monochromator can reduce the SASE bandwidth to $\Delta E/E \sim 10^{-4}$ at 5–25 keV, while a high-resolution Si-(533) monochromator will allow for a 5×10^{-6} bandwidth at 7.5 keV. Focusing of 5–25 keV X-rays to foci of 1–200 μm at the sample position is established by several sets of Be compound refractive lenses (CRLs), located in the tunnel section (2×) and in OH. A fourth lens set close to the sample position will allow for sub-micron foci. A multilayer-based split-and-delay line has been designed, was constructed by the University of Münster (Germany) and is currently installed at HED [121]. This device allows splitting the X-ray pulse into two with a tunable intensity ratio and to separate them with a maximum delay of 2 ps (at 20 keV) and 23 ps (at 5 keV). A pulse picker will allow selecting X-ray pulses for 10 Hz, 1 Hz or pulse-on-demand operation, thereby synchronizing X-ray and optical laser delivery to the sample.

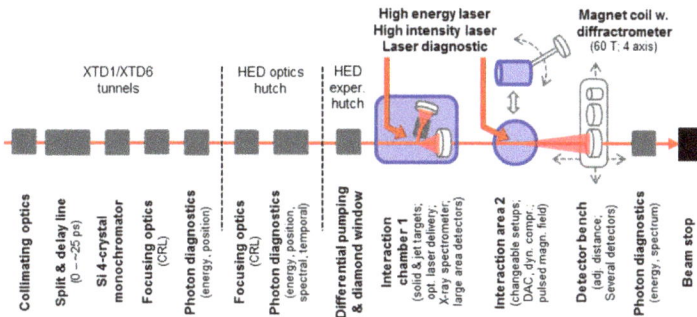

Figure 9. Schematic outline of the HED scientific instrument at the SASE2 FEL beamline. Shown are major instrumentation items installed in the photon beam transport tunnel, optics hutch and experiment hutch comprising the beam transport, focusing and diagnostic devices as well as the large optical lasers, second interaction area instrumentation and detectors contributed by the HiBEF User Consortium.

Power slits in OH can tailor the wings of the beam monitored by a beam-imaging unit. Using diamond gratings in first order a fraction of the incident beam can be steered to a single-pulse spectrometer using a bent Si crystal to monitor the incident X-ray spectrum. X-ray beam position and intensity are monitored by two intensity-position monitors, using backscattering from thin foils. Alternatively, real-time intensity monitoring is possible with a scintillator-coupled fast-frame CCD which picks up the other 1st order diffraction from the diamond grating. The quality of the photon beam can be further improved by cleanup slits for both high and low photon energies, located close to the interaction chamber in EH.

The 9×11 m^2 experiment hutch is enclosed by a heavy concrete wall of thicknesses between 0.5 and 1.0 m to establish radiation shielding for high energetic electrons generated by the relativistic laser–matter interaction processes when focusing the multi-100 TW laser on the sample. In EH two interaction areas IA1 and IA2 have been defined. In IA1, a large vacuum interaction chamber (IC1) with inner dimensions $2.6 \times 1.7 \times 1.5$ m^3 (LWH) accommodates several configurations for diffraction, imaging or low/high resolution spectroscopy and inelastic X-ray scattering. The IC1 vacuum of ~10^{-4} mbar is separated from the X-ray optics by differential pumping stage or, above 10 keV, by a diamond window. At IA2 various setups can be interchanged. A second interaction chamber (IC2) with 1 m diameter is dedicated to dynamic diamond anvil cell (DAC) experiments and high-precision dynamic laser compression experiments in a standardized configuration. Alternatively, a goniometer with a pulsed magnetic coil and a cryogenic sample environment shall be placed here. While in IA1 all X-ray and laser beams are available, IA2 has access to the X-ray FEL and the nanosecond laser beams only.

X-ray detectors inside IC1 need to be vacuum-compatible with compact dimensions, low weight, modular assembly, and >10 Hz repetition rate. HED plans to have several detectors installed. Two EPIX100 modules [122] offer a 35×38 mm^2 chip with 50 μm pixel pitch and 10^2 dynamic range at

8 keV. These detectors will be coupled, e.g., to crystal spectrometers. Three EPIX10k modules [123] have identical chip size, 100 μm pixel pitch, but offer 10^4 dynamic range by gain switching. With the same dynamic range, four Jungfrau modules offer 40×80 mm^2 chips each with 75 μm pixel pitch and 10^4 dynamic range at 12 keV [124]. The latter two gain-switching detectors are ideally suited to record dedicated parts of an X-ray diffraction pattern. For both, IA1 and IA2, a detector bench at the end of EH will offer a possibility to place large area detectors, e.g., for imaging or SAXS type experiments. This bench allows adjusting the distance from IA1 and IA2 to the detector. On this bench, the HIBEF consortium plans to integrate an AGIPD 1M detector [31], a Perkin-Elmer 4343CT flat-panel large-area detector, and high-resolution CCD cameras for X-ray phase contrast imaging and ptychography applications.

Several drivers to generate extreme states of matter will be available at HED, e.g., two high energy optical lasers, diamond anvil cells, and pulsed magnetic fields, contributed and operated by the international HIBEF user consortium. The all-diode pumped high energy (HE) nanosecond DiPOLE-100X laser is developed by STFC CLF (UK) [125]. It delivers up to 80 J at 515 nm wavelength with pulse durations of 2–15 ns with a maximum repetition rate of 10 Hz. This laser will be primarily used for shock compression experiments and its pulses can be temporally shaped to enable isentropic ramp compression techniques. The multi-100 TW Ti:Sapphire (HI) laser system, currently under construction by Amplitude (France), will deliver 4–10 J of 800 nm light in ultrashort pulses of less than 25 fs at a repetition rate of 10 Hz. The pulses of this laser can be focused to a few μm^2 spot by means of an off-axis parabola, reaching on-target intensities of the order of 10^{20} W/cm^2. This laser will primarily be used for relativistic laser–matter interaction experiments. In addition, the standard PP laser of the European XFEL will be available. All three lasers have to be precisely timed with respect to the X-ray pulses and are synchronized to the master oscillator. The timing jitter between the PP laser and the incident X-rays is monitored by photon-arrival diagnostics with a precision on the order of a few femtoseconds. Timing between the HI laser and the X-rays is realized indirectly using the characterized PP laser in an optical-optical balanced cross-correlator. Timing between the HE laser and the X-rays is less demanding and achieved via fast photo diodes that detect both X-rays and optical light with a resolution of few 10 ps. Matter in magnetic fields of up to 60 T can be studied in a solenoid coil. The timescale of the field build-up of 0.6 ms is perfectly adapted to the length of a 4.5 MHz pulse train of the facility.

11. Future Developments

Being a brand new facility and observing much progress in the field of FEL sources, FEL instrumentation, and novel types of scientific experiments, a rich variety of further developments is expected to be implemented during the coming years. Developments going beyond the baseline scope of European XFEL have already started using external funding. Most notable is the construction and implementation of self-seeding for the hard X-ray FEL sources [104,105]. This is on-going for the SASE2 FEL and under preparation for SASE1. The FEL radiation performances of SASE3 would benefit enormously by the provision of variable polarization that can be switched between linear and circular with full flexibility [56]. The installation of an SASE3 afterburner is therefore under preparation. This afterburner consists of several 2 m long APPLE-type undulators, which will be added to the main SASE3 undulator. Furthermore, the installation of a chicane in the SASE3 undulator will enable operation at two widely separated photon energies for time-resolved X-ray–X-ray pump-probe investigations. Another development concerns the construction of additional scientific instruments. Using funds from user consortia, the additional end-station at SPB/SFX for serial femtosecond crystallography and a third beam transport system, vacuum port and experiment hutch at SASE3 are pursued. A completely different area is that of further developing the PP laser towards providing much longer wavelengths. Pumping solids in the THz regime has many scientific applications and is vigorously requested by part of the user community. The feasibility and possible implementation of laser- and accelerator-based techniques to produce intense, ultrashort duration, and monochromatic THz pulses is currently studied.

Naturally, the completion of the remaining two yet unoccupied FEL sources and the construction of further scientific instruments are expected to become major activities of European XFEL once the regular operation of the facility is achieved successfully. In a more distant future, a modification of the superconducting accelerator to include a cw mode of operation is very interesting for scientific applications, as is also indicated by the LCLS-II project [13]. Such an upgrade first requires developing an additional low emittance injector operating in cw mode. Since the electron energy will be significantly smaller than with the present pulsed RF system, this upgrade also requires a modified concept for the FEL sources. One possibility would be to direct the electrons to a second switchyard with novel FEL undulators specifically designed for the smaller electron energies and providing at the same time space for a second experiment hall hosting additional scientific instruments.

Acknowledgments: All contributors, both internal and external, to European XFEL are acknowledged for their outstanding contributions to build this facility. Additional instrumentation for the scientific instruments and the laboratory infrastructure provided through external contributions, in particular by the HIBEF, hRIXS, COMO, SFX and XBI User Consortia, and through BMBF Verbundforschung projects 05K10PM2, 05K13RF4, 05K16BC1, 05K16PE1 and 05K13PE2 is most gratefully acknowledged.

Author Contributions: Thomas Tschentscher conceived the article, Christian Bressler, Jan Grünert, Anders Madsen, Adrian P. Mancuso, Michael Meyer, Andreas Scherz, Harald Sinn and Ulf Zastrau contributed sections to this paper.

Conflicts of Interest: The authors declare no conflict of interest.

References

1. Pellegrini, C.; Marinelli, A.; Reiche, S. The physics of X-ray free-electron lasers. *Rev. Mod. Phys.* **2016**, *88*, 015006. [CrossRef]
2. Saldin, E.L.; Schneidmiller, E.V.; Yurkov, M.V. *The Physics of Free-Electron Lasers*, 1st ed.; Springer: Berlin, Germany, 1999.
3. Ayvazyan, V.; Baboi, N.; Bähr, J.; Balandin, V.; Beutner, B.; Brandt, A.; Bohnet, I.; Bolzmann, A.; Brinkmann, R.; Brovko, O.I.; et al. First operation of a free-electron laser generating GW power radiation at 32 nm wavelength. *Eur. Phys. J. D* **2006**, *37*, 297–303. [CrossRef]
4. Ackermann, W.; Asova, G.; Ayvazyan, V.; Azima, A.; Baboi, N.; Bähr, J.; Balandin, V.; Beutner, B.; Brandt, A.; Bolzmann, A.; et al. Operation of a free-electron laser from the extreme ultraviolet to the water window. *Nat. Photonics* **2007**, *1*, 336–342. [CrossRef]
5. Emma, P.; Akre, R.; Arthur, J.; Bionta, R.; Bostedt, C.; Bozek, J.; Brachmann, A.; Bucksbaum, P.; Coffee, R.; Decker, F.-J.; et al. First lasing and operation of an Angstrom-wavelength free-electron laser. *Nat. Photonics* **2010**, *4*, 641–647. [CrossRef]
6. Allaria, E.; Appio, R.; Badano, L.; Barletta, L.W.; Bassanese, S.; Biedron, S.G.; Borga, A.; Busetto, E.; Castronovo, D.; Cinquegrana, P.; et al. Highly coherent and stable pulses from the FERMI seeded free-electron laser in the extreme ultraviolet. *Nat. Photonics* **2012**, *6*, 699–704. [CrossRef]
7. Ishikawa, T.; Aoyagi, H.; Asaka, T.; Asano, Y.; Azumi, Y.; Bizen, T.; Ego, H.; Fukami, K.; Fukui, T.; Furukawa, Y.; et al. A compact X-ray free-electron laser emitting in the sub-Ångström region. *Nat. Photonics* **2012**, *6*, 540–544. [CrossRef]
8. Ko, I.S.; Kang, H.-S.; Heo, H.; Kim, C.; Kim, G.; Min, C.-K.; Yang, H.; Baek, S.Y.; Choi, H.-J.; Mun, G.; et al. Construction and Commissioning of PAL-XFEL Facility. *Appl. Sci.* **2017**, *7*, 479. [CrossRef]
9. Patterson, B.D.; Abela, R.; Braun, H.-H.; Flechsig, U.; Ganter, R.; Kim, Y.; Kirk, E.; Oppelt, A.; Pedrozzi, M.; Reiche, S.; et al. Coherent science at the SwissFEL X-ray laser. *New J. Phys.* **2010**, *12*, 035012. [CrossRef]
10. Altarelli, M.; Brinkmann, R.; Chergui, M.; Decking, W.; Dobson, B.; Düsterer, S.; Grübel, G.; Graeff, W.; Graafsma, H.; Hajdu, J.; et al. (Eds.) *XFEL: The European X-ray Free-Electron Laser—Technical Design Report*; DESY 2006-097; DESY: Hamburg, Germany, 2006. [CrossRef]
11. Altarelli, A. The European X-ray free-electron laser facility in Hamburg. *Nucl. Instrum. Methods Phys. Res. B* **2011**, *269*, 2845–2849. [CrossRef]

12. Flöttmann, K.; Rossbach, J.; Schmüser, P.; Walker, N.; Weise, H.; Brinkmann, R. (Eds.) *TESLA Technical Design Report—Part II: The Accelerator*; DESY 2001-011; DESY: Hamburg, Germany, 2001.

13. Galayda, J.N. The New LCLS-II project: Status and Challenges. In Proceedings of the LINAC 2014 Conference, Geneva, Switzerland, 31 August–5 September 2014.

14. Brinkmann, R.; Faatz, B.; Flöttmann, K.; Rossbach, J.; Schneider, J.R.; Schulte-Schrepping, H.; Trines, D.; Tschentscher, Th.; Weise, H. (Eds.) *Design Report: First Stage of the TESLA XFEL Laboratory*; DESY 2002-167; DESY: Hamburg, Germany, 2002.

15. Decking, W.; Limberg, T. *European XFEL Post-TDR Description*; XFEL.EU TN-2013-004; European XFEL: Hamburg, Germany, 2013.

16. Vartanyants, I.A.; Robinson, I.K.; McNulty, I.; David, C.; Wochner, P.; Tschentscher, Th. Coherent X-ray scattering and lensless imaging at the European XFEL Facility. *J. Synchrotron Radiat.* **2007**, *14*, 453–470. [CrossRef] [PubMed]

17. Kimberg, V.; Sanchez-Gonzalez, A.; Mercadier, L.; Weninger, C.; Lutman, A.; Ratner, D.; Coffee, R.; Bucher, M.; Mucke, M.; Agaker, M.; et al. Stimulated X-ray Raman scattering—A critical assessment of the building block of nonlinear X-ray spectroscopy. *Faraday Discuss.* **2016**, *194*, 305–324. [CrossRef] [PubMed]

18. Grübel, G.; Stephenson, G.B.; Gutt, C.; Sinn, H.; Tschentscher, Th. XPCS at the European X-ray free electron laser facility. *Nucl. Instrum. Methods Phys. Res. B* **2007**, *262*, 357–367. [CrossRef]

19. Krasilnikov, M.; Stephan, F.; Asova, G.; Grabosch, H.-J.; Groß, M.; Hakobyan, L.; Isaev, I.; Ivanisenko, Y.; Jachmann, L.; Khojoyan, M.; et al. Experimentally minimized beam emittance from an L-band photoinjector. *Phys. Rev. ST Accel. Beams* **2012**, *15*, 100701. [CrossRef]

20. Rimjaem, S.; Stephan, F.; Krasilnikov, M.; Ackermann, W.; Asova, G.; Bähr, J.; Gjonaj, E.; Grabosch, H.J.; Hakobyan, L.; Hänel, M.; et al. Optimizations of transverse projected emittance at the photo-injector test facility at DESY, location Zeuthen. *Nucl. Instrum. Methods Phys. Res. A* **2012**, *671*, 62–75. [CrossRef]

21. Brinker, F. Commissioning of the European XFEL Injector. In Proceedings of the 7th International Particle Accelerator Conference (IPAC'16), Busan, Korea, 8–13 May 2016.

22. Decking, W.; Weise, H. Commissioning of the European XFEL. In Proceedings of the International Particle Accelerator Conference 2017, Copenhagen, Denmark, 14–19 May 2017.

23. Nölle, D. Commissioning of the European XFEL Facility. In Proceedings of the SPIE, X-ray Free-Electron Lasers: Advances in Source Development and Instrumentation, Prague, Czech Republic, 24–27 April 2017; Volume 10237.

24. Li, Y.; Abeghyan, S.; Berndgen, K.; Baha-Shanjani, M.; Deron, G.; Englisch, U.; Karabekyan, S.; Ketenoglu, B.; Knoll, M.; Wolff-Fabris, F.; et al. Magnetic Measurement Techniques for the Large-Scale Production of Undulator Segments for the European XFEL. *Synchrotron Radiat. News* **2015**, *28*, 23–28. [CrossRef]

25. Pflüger, J.; Lu, H.; Teichmann, T. Field fine tuning by pole height adjustment for the undulator of the TTF-FEL. *Nucl. Instrum. Methods Phys. Res. A* **1999**, *429*, 386–391. [CrossRef]

26. Schneidmiller, E.A.; Yurkov, M.V. An Overview of the Radiation Properties of the European XFEL. In Proceedings of the FEL 2014 Conference, Basel, Switzerland, 5–29 August 2014.

27. Schneidmiller, E.A.; Yurkov, M.V. *Photon Beam Properties at the European XFEL*, December 2010 Revision ed; Preprint DESY 11-152; DESY: Hamburg, Germany, 2011.

28. Kuster, M.; Boukhelef, D.; Donato, M.; Dambietz, J.-S.; Hauf, S.; Maia, L.; Raab, N.; Szuba, J.; Turcato, M.; Wrona, K.; et al. Detectors and Calibration Concept for the European XFEL. *Synchrotron Radiat. News* **2014**, *27*, 35–38. [CrossRef]

29. Raab, N.; Ballak, K.-E.; Dietze, T.; Ekmedzic, M.; Hauf, S.; Januschek, F.; Kaukher, A.; Kuster, M.; Lang, P.M.; Münnich, A.; et al. Status of the laboratory infrastructure for detector calibration and characterization at the European XFEL. *J. Instrum.* **2016**, *11*, C12051. [CrossRef]

30. Mozzanica, A.; Bergamaschi, A.; Dinapoli, R.; Graafsma, H.; Greiffenberg, D.; Henrich, B.; Johnson, I.; Lohmann, M.; Valeria, R.; Schmitt, B. The GOTTHARD charge integrating readout detector: Design and characterization. *J. Instrum.* **2012**, *7*, C01019. [CrossRef]

31. Henrich, B.; Becker, J.; Dinapoli, R.; Goettlicher, P.; Graafsma, H.; Hirsemann, H.; Klanner, R.; Krueger, H.; Mazzocco, R.; Mozzanica, A.; et al. The adaptive gain integrating pixel detector AGIPD a detector for the European XFEL. *Nucl. Instrum. Methods Phys. Res. A* **2011**, *633*, S11–S14. [CrossRef]

32. Koch, A.; Hart, M.; Nicholls, T.; Angelsen, C.; Coughlan, J.; French, M.; Hauf, S.; Kuster, M.; Sztuk-Dambietz, J.; Turcato, M.; et al. Performance of an LPD prototype detector at MHz frame rates under Synchrotron and FEL radiation. *J. Instrum.* **2013**, *8*, C11001. [CrossRef]

33. Porro, M.; Andricek, L.; Bombelle, L.; De Vita, G.; Fiorini, C.; Fischer, P.; Hansen, K.; Lechner, P.; Lutz, G.; Strüder, L.; et al. Expected performance of the DEPFET sensor with signal compression: A large format X-ray imager with mega-frame readout capability for the European XFEL. *Nucl. Instrum. Methods Phys. Res. A* **2010**, *624*, 509–519. [CrossRef]

34. Pergament, M.; Kellert, M.; Kruse, K.; Wang, J.; Palmer, G.; Wissmann, L.; Wegner, U.; Lederer, M.J. High power burst-mode optical parametric amplifier with arbitrary pulse selection. *Opt. Express* **2014**, *22*, 22202–22210. [CrossRef] [PubMed]

35. Pergament, M.; Palmer, G.; Kellert, M.; Kruse, K.; Wang, J.; Wissmann, L.; Wegner, U.; Emons, M.; Kane, D.; Priebe, G.; et al. Versatile optical laser system for experiments at the European X-ray Free-Electron Laser Facility. *Opt. Express* **2016**, *24*, 29349–29359. [CrossRef] [PubMed]

36. Sydlo, C.; Czwalinna, M.; Felber, M.; Gerth, C.; Jabłoński, S.; Müller, J.; Schlarb, H.; Zummack, F. Femtosecond Timing Distribution at the European XFEL. In Proceedings of the 7th International Particle Accelerator Conference (IPAC'16), Busan, Korea, 8–13 May 2016.

37. Schulz, S.; Grguras, I.; Behrens, C.; Bromberger, H.; Costello, J.T.; Czwalinna, M.K.; Felber, M.; Hoffmann, M.C.; Ilchen, M.; Liu, H.Y.; et al. Femtosecond all-optical synchronization of an X-ray free-electron laser. *Nat. Commun.* **2015**, *6*, 5938. [CrossRef] [PubMed]

38. Sinn, H.; Gaudin, J.; Samoylova, L.; Trapp, A; Galasso, G. *Conceptual Design Report: X-ray Optics and Beam Transport*; XFEL.EU TR-2011-002; European XFEL: Hamburg, Germany, 2011. [CrossRef]

39. Sinn, H.; Dommach, M.; Dong, X.; La Civita, D.; Samoylova, L.; Villanueva, R.; Yang, F. *Technical Design Report: X-ray Optics and Beam Transport*; XFEL.EU TR-2012-006; European XFEL: Hamburg, Germany, 2012. [CrossRef]

40. Geloni, G.; Saldin, E.; Samoylova, L.; Schneidmiller, E.; Sinn, H.; Tschentscher, Th.; Yurkov, M. Coherence properties of the European XFEL. *New J. Phys.* **2010**, *12*, 035021. [CrossRef]

41. Grünert, J. *Conceptual Design Report: Framework for X-ray Photon Diagnostics at the European XFEL*; XFEL.EU TR-2012-003; European XFEL: Hamburg, Germany, 2012. [CrossRef]

42. Grünert, J.; Buck, J.; Ozkan, C.; Freund, W.; Molodtsov, S. X-ray Photon Diagnostics Devices for the European XFEL. In Proceedings of the SPIE, X-ray Free-Electron Lasers: Beam Diagnostics, Beamline Instrumentation, and Applications, San Diego, CA, USA, 12 August 2012. [CrossRef]

43. Grünert, J.; Buck, J.; Freund, W.; Ozkan, C.; Molodtsov, S. Development status of the X-ray beam diagnostics devices for the commissioning and user operation of the European XFEL. *J. Phys. Conf. Ser.* **2013**, *425*, 072004. [CrossRef]

44. Gruenert, J.; Koch, A.; Kujala, N.; Freund, V.; Planas, M.; Dietrich, F.; Buck, J.; Liu, J.; Sinn, H.; Dommach, M. Photon Diagnostics and Photon Beamlines Installations at the European XFEL. In Proceedings of the 37th International Free-Electron Laser Conference 2015, Daejeon, Korea, 23–28 August 2015.

45. Liu, J. *Requirements and Concept for the Characterization of Photon Beam Temporal Properties at the SQS Scientific Instrument of the European XFEL Facility*; XFEL.EU TN-2015-002-01; European XFEL: Hamburg, Germany, 2015.

46. Liu, J.; Dietrich, F.; Grünert, J. *Technical Design Report: Photon Arrival Time Monitor (PAM) at the European XFEL*; XFEL.EU TR-2017-002; European XFEL: Schenefeld, Germany, 2017.

47. Maltezopoulos, Th.; Cunovic, S.; Wieland, M.; Beye, M.; Azima, A.; Redlin, H.; Krikunova, M.; Kalms, R.; Frühling, U.; Budzyn, F.; et al. Single-shot timing measurement of extreme-ultraviolet free-electron laser pulses. *New J. Phys.* **2008**, *10*, 033026. [CrossRef]

48. Grguras, I.; Maier, A.R.; Behrens, C.; Mazza, T.; Kelly, T.J.; Radcliffe, P.; Düsterer, S.; Kazansky, A.K.; Kabachnik, N.M.; Tschentscher, Th.; et al. Ultrafast X-ray pulse characterization at free-electron lasers. *Nat. Photonics* **2012**, *6*, 852–857. [CrossRef]

49. Harmand, M.; Coffee, R.; Bionta, M.R.; Chollet, M.; French, D.; Zhu, D.; Fritz, D.M.; Lemke, H.T.; Medvedev, N.; Ziaja, B.; et al. Achieving few-femtosecond time-sorting at hard X-ray free-electron lasers. *Nat. Photonics* **2013**, *7*, 215–218. [CrossRef]

50. Hartmann, N.; Helml, W.; Galler, A.; Bionta, M.R.; Grünert, J.; Molodtsov, S.L.; Ferguson, K.R.; Schorb, S.; Swiggers, M.L.; Carron, S.; et al. Sub-femtosecond precision measurement of relative X-ray arrival time for free-electron lasers. *Nat. Photonics* **2014**, *8*, 706–709. [CrossRef]

51. Tiedtke, K.; Feldhaus, J.; Hahn, U.; Jastrow, U.; Nunez, T.; Tschentscher, Th.; Bobashev, S.V.; Sorokin, A.A.; Hastings, J.B.; Möller, S.; et al. Gas-detector for X-ray lasers. *J. Appl. Phys.* **2008**, *103*, 094511. [CrossRef]

52. Feng, Y.; Raubenheimer, T.O. Duty-Cycle Dependence of the Filamentation Effect in Gas Devices for High Repetition Rate Pulsed X-ray FEL's. In Proceedings of the SPIE, X-ray Free-Electron Lasers: Advances in Source Development and Instrumentation, Prague, Czech Republic, 24–27 April 2017; Volume 10237.

53. Tono, K.; Kudo, T.; Yabashi, M.; Tachibana, T.; Feng, Y.; Fritz, D.; Hastings, J.; Ishikawa, T. Single-shot beam-position monitor for X-ray free electron laser. *Rev. Sci. Instrum.* **2011**, *82*, 023108. [CrossRef] [PubMed]

54. Zhu, D.; Cammarata, M.; Feldkamp, J.M.; Fritz, D.M.; Hastings, J.B.; Lee, S.; Lemke, H.T.; Robert, A.; Turner, J.L.; Feng, Y. A single-shot transmissive spectrometer for hard X-ray free electron lasers. *Appl. Phys. Lett.* **2012**, *101*, 034103. [CrossRef]

55. Buck, J. *Conceptual Design Report: Online Time-of-Flight Photoemission Spectrometer for X-ray Photon Diagnostics*; XFEL.EU TR-2012-002; European XFEL: Hamburg, Germany, 2012. [CrossRef]

56. Lutman, A.A.; MacArthur, J.P.; Ilchen, M.; Lindahl, A.O.; Buck, J.; Coffee, R.N.; Dakovski, G.L.; Dammann, L.; Ding, Y.; Dürr, H.A.; et al. Polarization control in an X-ray free-electron laser. *Nat. Photonics* **2016**, *10*, 468–472. [CrossRef]

57. Syresin, E.; Brovko, O.; Kapishin, M.; Shabunov, A.; Yurkov, M.; Freund, W.; Gruenert, J.; Sinn, H. Development of MCP Based Photon Detectors for the European XFEL. In Proceedings of the International Particle Accelerator Conference 2011, San Sebastian, Spain, 4–9 September 2011.

58. Freund, W. *The Undulator Commissioning Spectrometer for the European XFEL, XFEL.EU Technical Report*; XFEL.EU TN-2014-001-01; European XFEL: Hamburg, Germany, 2013.

59. Ozkan, C.; Freund, W.; Rehanek, J.; Buck, J.; Zizak, I.; Gruenert, J.; Schaefers, F.; Erko, A.; Molodtsov, S. Initial Evaluation of the European XFEL Undulator Commissioning Spectrometer with a Single Channel-Cut Crystal. In Proceedings of the SPIE, X-ray Free-Electron Lasers: Beam Diagnostics, Beamline Instrumentation, and Applications, San Diego, CA, USA, 12 August 2012; Volume 8504. [CrossRef]

60. Ozkan, C. *Conceptual Design Report: Imaging Stations for Invasive Photon Diagnostics*; XFEL.EU TR-2012-004; European XFEL: Hamburg, Germany, 2012.

61. Mancuso, A.P.; Aquila, A.L.; Borchers, G.; Giewekemeyer, K.; Reimers, N. *Technical Design Report: Scientific Instrument Single Particles, Clusters, and Biomolecules (SPB)*; XFEL.EU TR-2013-004; European XFEL: Hamburg, Germany, 2013. [CrossRef]

62. Garman, E.F. Developments in X-ray Crystallographic Structure Determination of Biological Macromolecules. *Science* **2014**, *343*, 1102–1108. [CrossRef] [PubMed]

63. Spence, J.C.H.; Weierstall, U.; Chapman, H.N. X-ray lasers for structural and dynamic biology. *Rep. Prog. Phys.* **2012**, *75*, 102601. [CrossRef] [PubMed]

64. Schlichting, I. Serial femtosecond crystallography: The first five years. *IUCrJ* **2015**, *2*, 246–255. [CrossRef] [PubMed]

65. White, T.A.; Kirian, R.A.; Martin, A.V.; Aquila, A.; Nass, K.; Barty, A.; Chapman, H.N. CrystFEL: A software suite for snapshot serial crystallography. *J. Appl. Crystallogr.* **2012**, *45*, 335–341. [CrossRef]

66. Loh, N.-T.D.; Elser, V. Reconstruction algorithm for single-particle diffraction imaging experiments. *Phys. Rev. E* **2009**, *80*, 026705. [CrossRef] [PubMed]

67. Barends, T.R.M.; Foucar, L.; Botha, S.; Doak, R.B.; Shoeman, R.L.; Nass, K.; Koglin, J.E.; Williams, G.J.; Boutet, S.; Messerschmidt, M.; et al. De novo protein crystal structure determination from X-ray free-electron laser data. *Nature* **2015**, *505*, 244–247. [CrossRef] [PubMed]

68. Schwander, P.; Fung, R.; Phillips, G.N.; Ourmazd, A. Mapping the conformations of biological assemblies. *New J. Phys.* **2010**, *12*, 035007. [CrossRef]

69. Aquila, A.; Sobierajski, R.; Ozkan, C.; Hajkova, V.; Burian, T.; Chalupský, J.; Juha, L.; Störmer, M.; Bajt, S.; Klepka, M.T.; et al. Fluence thresholds for grazing incidence hard X-ray mirrors. *Appl. Phys. Lett.* **2015**, *106*, 241905. [CrossRef]

70. Bean, R.J.; Aquila, A.; Samoylova, L.; Mancuso, A.P. Design of the mirror optical systems for coherent diffractive imaging at the SPB/SFX instrument of the European XFEL. *J. Opt.* **2016**, *18*, 1–10. [CrossRef]

71. Bressler, Ch.; Galler, A.; Gawelda, W. *Technical Design Report: Scientific Instrument FXE*; XFEL.EU TR-2012-008; European XFEL: Hamburg, Germany, 2012. [CrossRef]

72. Cho, H.S.; Dashdorj, N.D.; Schotte, F.; Graber, T.; Henning, R.; Anfinrud, P. Protein structural dynamics in solution unveiled via 100-ps time-resolved X-ray scattering. *Proc. Natl. Acad. Sci. USA* **2010**, *107*, 7281–7286. [CrossRef] [PubMed]

73. Brefuel, N.; Watanabe, H.; Toupet, L.; Come, J.; Matsumoto, N.; Collet, E.; Tanaka, K.; Tuchagues, J.P. Concerted Spin Crossover and Symmetry Breaking Yield Three Thermally and One Light-Induced Crystallographic Phases of a Molecular Material. *Angew. Chem. Int. Ed.* **2013**, *48*, 9304–9307. [CrossRef] [PubMed]

74. Smolentsev, G.; Sundström, V. Time-resolved X-ray absorption spectroscopy for the study of molecular systems relevant for artificial photosynthesis. *Coord. Chem. Rev.* **2015**, *304*, 117–132. [CrossRef]

75. Torres-Alacan, J.; Lindner, J.; Vöhringer, P. Probing the Primary Photochemical Processes of Octahedral Iron(V) Formation with Femtosecond Mid-Infrared Spectroscopy. *Chem. Phys. Chem.* **2015**, *16*, 2289–2293. [CrossRef] [PubMed]

76. Zhang, W.; Alonso-Mori, R.; Bergmann, U.; Bressler, C.; Chollet, M.; Galler, A.; Gawelda, W.; Hadt, R.G.; Hartsock, R.W.; Kroll, T.; et al. Tracking excited-state charge and spin dynamics in iron coordination complexes. *Nature* **2014**, *509*, 345–348. [CrossRef] [PubMed]

77. Haldrup, K.; Gawelda, W.; Abela, R.; Alonso-Mori, R.; Bergmann, U.; Bordage, A.; Cammarata, M.; Canton, S.; Dohn, A.O.; van Driel, T.B. Observing Solvatiuon Dynamics with Simultaneous Femtosecond X-ray Emission Spectroscopy and X-ray Scattering. *J. Phys. Chem. B* **2016**, *120*, 1158–1168. [CrossRef] [PubMed]

78. Katayama, T.; Owada, S.; Togashi, T.; Ogawa, K.; Karvinen, P.; Vartiainen, I.; Eronen, A.; David, C.; Sato, T.; Nakajima, K.; et al. A beam branching method for timing and spectral characterization of hard X-ray free electron lasers. *Struct. Dyn.* **2016**, *3*, 034301. [CrossRef] [PubMed]

79. Mazza, T.; Zhang, H.; Meyer, M. *Technical Design Report: Scientific Instrument SQS*; XFEL.EU TR-2012-007; European XFEL: Hamburg, Germany, 2012. [CrossRef]

80. Ullrich, J.; Moshammer, R.; Dorn, A.; Dörner, R.; Schmidt, H.L.P.; Schmidt-Böcking, H. Recoil-ion and electron momentum spectroscopy reaction-microscopes. *Rep. Prog. Phys.* **2003**, *66*, 1463–1545. [CrossRef]

81. De Jong, S.; Kukreja, R.; Trabant, C.; Pontius, N.; Chang, C.F.; Kachel, T.; Beye, M.; Sorgenfrei, F.; Back, C.H.; Bräuer, B.; et al. Speed limit of the insulator–metal transition in magnetite. *Nat. Mater.* **2013**, *12*, 882–886. [CrossRef] [PubMed]

82. Kubacka, T.; Johnson, J.A.; Hoffmann, M.C.; Vicario, C.; de Jong, S.; Beaud, P.; Grübel, S.; Huang, S.-W.; Huber, L.; Patthey, L.; et al. Large-Amplitude Spin Dynamics Driven by a THz Pulse in Resonance with an Electromagnon. *Science* **2014**, *343*, 1333–1336. [CrossRef] [PubMed]

83. Först, M.; Caviglia, A.D.; Scherwitzl, R.; Mankowsky, R.; Zubko, P.; Khanna, V.; Bromberger, H.; Wilkins, S.B.; Chuang, Y.-D.; Lee, W.S.; et al. Spatially resolved ultrafast magnetic dynamics initiated at a complex oxide heterointerface. *Nat. Mater.* **2015**, *14*, 883–888. [CrossRef] [PubMed]

84. Graves, C.E.; Reid, A.H.; Wang, T.; Wu, B.; de Jong, S.; Vahaplar, K.; Radu, I.; Bernstein, D.P.; Messerschmidt, M.; Müller, L.; et al. Nanoscale spin reversal by non-local angular momentum transfer following ultrafast laser excitation in ferrimagnetic GdFeCo. *Nat. Mater.* **2013**, *12*, 293–298. [CrossRef] [PubMed]

85. Vodungbo, B.; Tudu, B.; Perron, J.; Delaunay, R.; Müller, L.; Berntsen, M.H.; Grübel, G.; Malinowski, G.; Weier, C.; Gautier, J.; et al. Indirect excitation of ultrafast demagnetization. *Sci. Rep.* **2016**, *6*, 18970. [CrossRef] [PubMed]

86. Öström, H.; Öberg, H.; Xin, H.; LaRue, J.; Beye, M.; Dell'Angela, M.; Gladh, J.; Ng, M.L.; Sellberg, J.A.; Kaya, S.; et al. Probing the transition state region in catalytic CO oxidation on Ru. *Science* **2015**, *347*, 978–982. [CrossRef] [PubMed]

87. Wernet, Ph.; Kunnus, K.; Josefsson, I.; Rajkovic, I.; Quevedo, W.; Beye, M.; Schreck, S.; Grubel, S.; Scholz, M.; Nordlund, D.; et al. Orbital-specific mapping of the ligand exchange dynamics of Fe(CO)5 in solution. *Nature* **2015**, *520*, 78–81. [CrossRef] [PubMed]

88. Beye, M.; Schreck, S.; Sorgenfrei, F.; Trabant, C.; Pontius, N.; Schüßler-Langeheine, C.; Wurth, W.; Fohlisch, A. Stimulated X-ray emission for materials science. *Nature* **2013**, *501*, 191–194. [CrossRef] [PubMed]

89. Higley, D.J.; Hirsch, K.; Dakovski, G.L.; Jal, E.; Yuan, E.; Liu, T.; Lutman, A.A.; MacArthur, J.P.; Arenholz, E.; Chen, Z.; et al. Femtosecond X-ray magnetic circular dichroism absorption spectroscopy at an X-ray free electron laser. *Rev. Sci. Instrum.* **2016**, *87*, 033110. [CrossRef] [PubMed]
90. Eisebitt, S.; Luning, J.; Schlotter, W.F.; Lorgen, M.; Hellwig, O.; Eberhardt, W.; Stohr, J. Lensless imaging of magnetic nanostructures by X-ray spectro-holography. *Nature* **2004**, *432*, 885–888. [CrossRef] [PubMed]
91. Liu, T.-M.; Wang, T.; Reid, A.H.; Savoini, M.; Wu, X.; Koene, B.; Granitzka, P.; Graves, C.E.; Higley, D.J.; Chen, Z.; et al. Nanoscale Confinement of All-Optical Magnetic Switching in TbFeCo—Competition with Nanoscale Heterogeneity. *Nano Lett.* **2015**, *15*, 6862–6868. [CrossRef] [PubMed]
92. Scherz, A.; Krupin, O.; Buck, J.; Gerasimova, N.; Palmer, G.; Poolton, N.; Samoylova, L. *Conceptual Design Report: Scientific Instrument Spectroscopy and Coherent Scattering (SCS)*; XFEL.EU TR-2013-006; European XFEL: Hamburg, Germany, 2013. [CrossRef]
93. Wang, T.; Zhu, D.; Wu, B.; Graves, C.; Schaffert, S.; Rander, T.; Müller, L.; Vodungbo, B.; Baumier, C.; Bernstein, D.P.; et al. Femtosecond Single-Shot Imaging of Nanoscale Ferromagnetic Order in Co/Pd Multilayers Using Resonant X-ray Holography. *Phys. Rev. Lett.* **2012**, *108*, 267403. [CrossRef] [PubMed]
94. Miao, J.; Ishikawa, T.; Robinson, I.K.; Murnane, M.M. Beyond crystallography: Diffractive imaging using coherent X-ray light sources. *Science* **2015**, *348*, 530–535. [CrossRef] [PubMed]
95. Clark, J.; Beitra, L.; Xiong, G.; Higginbotham, A.; Fritz, D.M.; Lemke, H.T.; Zhu, D.; Chollet, M.; Williams, G.J.; Messerschmidt, M.; et al. Ultrafast Three-Dimensional Imaging of Lattice Dynamics in Individual Gold Nanocrystals. *Science* **2013**, *341*, 56–59. [CrossRef] [PubMed]
96. Madsen, A.; Fluerasu, A.; Ruta, B. Structural Dynamics of Materials Probed by X-ray Photon Correlation Spectroscopy. In *Synchrotron Light Sources and Free-Electron Lasers. Accelerator Physics, Instrumentation and Science Applications*, 1st ed.; Jaeschke, E., Khan, S., Schneider, J.R., Hastings, J.B., Eds.; Springer: Berlin, Germany, 2016; pp. 1617–1641.
97. Carnis, J.; Cha, W.; Wingert, J.; Kang, J.; Jiang, Z.; Song, S; Sikorski, M.; Robert, A.; Gutt, C.; Chen, S.-W.; et al. Demonstration of Feasibility of X-ray Free Electron Laser Studies of Dynamics of Nanoparticles in Entangled Polymer Melts. *Sci. Rep.* **2014**, *4*, 6017. [CrossRef] [PubMed]
98. Lehmkühler, F.; Kwasniewski, P.; Roseker, W.; Fischer, B.; Schroer, M.A.; Tono, K.; Katayama, T.; Sprung, M.; Sikorski, M.; Song, S.; et al. Sequential Single Shot X-ray Photon Correlation Spectroscopy at the SACLA Free Electron Laser. *Sci. Rep.* **2015**, *5*, 17193. [CrossRef] [PubMed]
99. Singer, A.; Patel, S.K.K.; Kukreja, R.; Uhlir, V.; Wingert, J.; Festersen, S.; Zhu, D.; Glownia, J.M.; Lemke, H.T.; Nelson, S.; et al. Photoinduced Enhancement of the Charge Density Wave Amplitude. *Phys. Rev. Lett.* **2016**, *117*, 056401. [CrossRef] [PubMed]
100. Trigo, M.; Fuchs, M.; Chen, J.; Jiang, M.P.; Cammarata, M.; Fahy, S.; Fritz, D.M.; Gaffney, K.; Ghimire, S.; Higginbotham, A.; et al. Fourier-transform inelastic X-ray scattering from time- and momentum-dependent phonon–phonon correlations. *Nat. Phys.* **2013**, *9*, 790. [CrossRef]
101. Liang, M.; Williams, G.J.; Messerschmidt, M.; Seibert, M.M.; Montanez, P.A.; Hayes, M.; Milathianaki, D.; Aquila, A.; Hunter, M.S.; Koglin, J.S.; et al. The Coherent X-ray Imaging instrument at the Linac Coherent Light Source. *J. Synchrotron Radiat.* **2015**, *22*, 514. [CrossRef] [PubMed]
102. Nagler, B.; Schropp, A.; Galtier, E.C.; Arnold, B.; Brown, S.B.; Fry, A.; Gleason, A.; Granados, E.; Hashim, A.; Hastings, J.B.; et al. The phase-contrast imaging instrument at the matter in extreme conditions endstation at LCLS. *Rev. Sci. Instrum.* **2016**, *87*, 103701. [CrossRef] [PubMed]
103. Wochner, P.; Gutt, C.; Autenrieth, T.; Demmer, T.; Bugaev, V.; Ortiz, A.D.; Duri, A.; Zontone, F.; Grübel, G.; Dosch, H. X-ray cross correlation analysis uncovers hidden local symmetries in disordered matter. *Proc. Natl. Acad. Sci. USA* **2009**, *106*, 11511. [CrossRef] [PubMed]
104. Geloni, G.; Kocharyan, V.; Saldin, E. A novel self-seeding scheme for hard X-ray FELs. *J. Mod. Opt.* **2011**, *58*, 1391. [CrossRef]
105. Amann, J.; Berg, W.; Blank, V.; Decker, F.-J.; Ding, Y.; Emma, P.; Feng, Y.; Frisch, J.; Fritz, D.; Hastings, J.; et al. Demonstration of self-seeding in a hard-X-ray free-electron laser. *Nat. Photonics* **2012**, *6*, 693. [CrossRef]

106. Chubar, O.; Geloni, G.; Kocharyan, V.; Madsen, A.; Saldin, E.; Serkez, S.; Shvyd'ko, Y.; Sutter, J. Ultra-high-resolution inelastic X-ray scattering at high-repetition-rate self-seeded X-ray free-electron lasers. *J. Synchrotron Radiat.* **2016**, *23*, 410. [CrossRef] [PubMed]

107. Roseker, W.; Franz, H.; Schulte-Schrepping, H.; Ehnes, A.; Leupold, O.; Zontone, F.; Lee, S.; Robert, A.; Grübel, G. Development of a hard X-ray delay line for X-ray photon correlation spectroscopy and jitter-free pump–probe experiments at X-ray free-electron laser sources. *J. Synchrotron Radiat.* **2011**, *18*, 481. [CrossRef] [PubMed]

108. Osaka, T.; Hirano, T.; Sano, Y.; Inubushi, Y.; Matsuyama, S.; Tono, K.; Ishikawa, T.; Yamauchi, K.; Yabashi, M. Wavelength-tunable split-and-delay optical system for hard X-ray free-electron lasers. *Opt. Express* **2016**, *24*, 9187. [CrossRef] [PubMed]

109. Lu, W.; Noll, T.; Roth, T.; Agapov, I.; Geloni, G.; Holler, M.; Hallmann, J.; Ansaldi, G.; Eisebitt, S.; Madsen, A. Design and throughput simulations of a hard X-ray split and delay line for the MID station at the European XFEL. *AIP Conf. Proc.* **2016**, *1741*, 030010. [CrossRef]

110. Gutt, C.; Stadler, L.-M.; Duri, A.; Autenrieth, T.; Leupold, O.; Chushkin, Y.; Grübel, G. Measuring temporal speckle correlations at ultrafast X-ray sources. *Opt. Express* **2009**, *17*, 55. [CrossRef] [PubMed]

111. Bandyopadhyay, R.; Gittings, A.S.; Suh, S.S.; Dixon, P.K.; Durian, D.J. Speckle-visibility spectroscopy: A tool to study time-varying dynamics. *Rev. Sci. Instrum.* **2005**, *76*, 093110. [CrossRef]

112. Van Thor, J.J.; Madsen, A. A split-beam probe-pump-probe scheme for femtosecond time resolved protein X-ray crystallography. *Struct. Dyn.* **2015**, *2*, 014102. [CrossRef] [PubMed]

113. Madsen, A.; Hallmann, J.; Roth, T.; Ansaldi, G. *Technical Design Report: Scientific Instrument MID*; XFEL.EU TR-2013-005; European XFEL: Hamburg, Germany, 2013. [CrossRef]

114. Roth, T.; Helfen, T.; Hallmann, J.; Samoylova, L.; Kwaśniewski, P.; Lengeler, B.; Madsen, A. X-ray Laminography and SAXS on Beryllium Grades and Lenses and Wavefront Propagation through Imperfect Compound Refractive Lenses. In Proceedings of the SPIE: Advances in X-ray/EUV Optics and Components IX, San Diego, CA, USA, 17 August 2014. [CrossRef]

115. Shwartz, S.; Coffee, R.N.; Feldkamp, J.M.; Feng, Y.; Hastings, J.B.; Yin, G.Y.; Harris, S.E. X-ray Parametric Down-Conversion in the Langevin Regime. *Phys. Rev. Lett.* **2012**, *109*, 013602. [CrossRef] [PubMed]

116. Temasaku, K.; Shigemasa, E.; Inubushi, Y.; Katayama, T.; Sawada, K.; Yumoto, H.; Ohashi, H.; Mimura, H.; Yabashi, M.; Yamauchi, K.; et al. X-ray two-photon absorption competing against single and sequential multiphoton processes. *Nat. Photonics* **2014**, *8*, 313. [CrossRef]

117. Stöhr, J. Two-Photon X-ray Diffraction. *Phys. Rev. Lett.* **2017**, *118*, 024801. [CrossRef] [PubMed]

118. Boesenberg, U.; Samoylova, L.; Roth, T.; Zhu, D.; Terentyev, S.; Vannoni, M.; Feng, Y.; van Driel, T.B.; Song, S.; Blank, V.; et al. X-ray spectrometer based on a bent diamond crystal for high repetition rate free-electron laser applications. *Opt. Express* **2017**, *25*, 2852. [CrossRef]

119. Obara, Y.; Katayama, T.; Ogi, Y.; Suzuki, T.; Kurahashi, N.; Karashima, S.; Chiba, Y.; Isokawa, Y.; Togashi, T.; Inubushi, Y.; et al. Femtosecond time-resolved X-ray absorption spectroscopy of liquid using a hard X-ray free electron laser in a dual-beam dispersive detection method. *Opt. Express* **2014**, *22*, 1105. [CrossRef] [PubMed]

120. Nakatsutsumi, M.; Appel, K.; Priebe, G.; Thorpe, I.; Pelka, A.; Muller, B.; Tschentscher, Th. *Technical Design Report: Scientific Instrument High Energy Density Physics (HED)*; XFEL.EU TR-2014-001; European XFEL: Hamburg, Germany, 2014. [CrossRef]

121. Roling, S.; Zacharias, H.; Samoylova, L.; Sinn, H.; Tschentscher, Th.; Chubar, O.; Buzmakov, A.; Schneidmiller, E.; Yurkov, M.V.; Siewert, F.; et al. Time-dependent wave front propagation simulation of a hard X-ray split-and-delay unit: Towards a measurement of the temporal coherence properties of X-ray free electron lasers. *Phys. Rev. STAB* **2014**, *17*, 110705. [CrossRef]

122. Carini, G.A.; Alonso-Mori, R.; Blaj, G.; Caragiulo, P.; Chollet, M.; Damiani, D.; Dragone, A.; Feng, Y.; Haller, G.; Hart, P.; et al. ePix100 camera: Use and applications at LCLS. *AIP Conf. Proc.* **2016**, *1741*, 040008. [CrossRef]

123. Nishimura, K.; Blaj, G.; Caragiulo, P.; Carini, G.A.; Dragone, A.; Haller, G.; Hart, P.; Hasi, J.; Herbst, R.; Herrmann, S.; et al. Design and performance of the ePix camera system. *AIP Conf. Proc.* **2016**, *1741*, 040047. [CrossRef]

124. Mozzanica, A.; Bergamaschi, A.; Cartier, S.; Dinapoli, R.; Greiffenberg, D.; Johnson, I.; Jungmann, J.; Maliakal, D.; Mezza, D.; Ruder, C.; et al. Prototype characterization of the JUNGFRAU pixel detector for SwissFEL. *J. Instrum.* **2014**, *9*, C05010. [CrossRef]

125. Banerjee, S.; Ertel, K.; Mason, P.D.; Phillips, P.J.; De Vido, M.; Smith, J.M.; Butcher, T.J.; Hernandez-Gomez, C.; Greenhalgh, R.J.S.; Collier, J.L. DiPOLE: A 10 J, 10 Hz cryogenic gas cooled multi-slab nanosecond Yb:YAG laser. *Opt. Express* **2015**, *23*, 19542–19551. [CrossRef] [PubMed]

applied
sciences

MDPI

Article

SwissFEL: The Swiss X-ray Free Electron Laser

Christopher J. Milne [1], Thomas Schietinger [1], Masamitsu Aiba [1], Arturo Alarcon [1], Jürgen Alex [1], Alexander Anghel [1], Vladimir Arsov [1], Carl Beard [1], Paul Beaud [1], Simona Bettoni [1], Markus Bopp [1], Helge Brands [1], Manuel Brönnimann [1], Ingo Brunnenkant [1], Marco Calvi [1], Alessandro Citterio [1], Paolo Craievich [1], Marta Csatari Divall [1], Mark Dällenbach [1], Michael D'Amico [1], Andreas Dax [1], Yunpei Deng [1], Alexander Dietrich [1], Roberto Dinapoli [1], Edwin Divall [1], Sladana Dordevic [1], Simon Ebner [1], Christian Erny [1], Hansrudolf Fitze [1], Uwe Flechsig [1], Rolf Follath [1], Franziska Frei [1], Florian Gärtner [1], Romain Ganter [1], Terence Garvey [1], Zheqiao Geng [1], Ishkhan Gorgisyan [1,†], Christopher Gough [1], Andreas Hauff [1], Christoph P. Hauri [1], Nicole Hiller [1], Tadej Humar [1], Stephan Hunziker [1], Gerhard Ingold [1], Rasmus Ischebeck [1], Markus Janousch [1], Pavle Juranić [1], Mario Jurcevic [1], Maik Kaiser [1], Babak Kalantari [1], Roger Kalt [1], Boris Keil [1], Christoph Kittel [1], Gregor Knopp [1], Waldemar Koprek [1], Henrik T. Lemke [1], Thomas Lippuner [1], Daniel Llorente Sancho [1], Florian Löhl [1], Carlos Lopez-Cuenca [1], Fabian Märki [1], Fabio Marcellini [1], Goran Marinkovic [1], Isabelle Martiel [1], Ralf Menzel [1], Aldo Mozzanica [1], Karol Nass [1], Gian Luca Orlandi [1], Cigdem Ozkan Loch [1], Ezequiel Panepucci [1], Martin Paraliev [1], Bruce Patterson [1,‡], Bill Pedrini [1], Marco Pedrozzi [1], Patrick Pollet [1], Claude Pradervand [1], Eduard Prat [1], Peter Radi [1], Jean-Yves Raguin [1], Sophie Redford [1], Jens Rehanek [1], Julien Réhault [1], Sven Reiche [1], Matthias Ringele [1], Jochen Rittmann [1,§], Leonid Rivkin [1,2], Albert Romann [1], Marie Ruat [1], Christian Ruder [1], Leonardo Sala [1], Lionel Schebacher [1], Thomas Schilcher [1], Volker Schlott [1], Thomas Schmidt [1], Bernd Schmitt [1], Xintian Shi [1], Markus Stadler [1,‖], Lukas Stingelin [1], Werner Sturzenegger [1], Jakub Szlachetko [1,‖], Dhanya Thattil [1], Daniel M. Treyer [1], Alexandre Trisorio [1], Wolfgang Tron [1], Seraphin Vetter [1], Carlo Vicario [1], Didier Voulot [1], Meitian Wang [1], Thierry Zamofing [1], Christof Zellweger [1], Riccardo Zennaro [1], Elke Zimoch [1], Rafael Abela [1,¶], Luc Patthey [1,*] and Hans-Heinrich Braun [1,*]

[1] Paul Scherrer Institute, 5232 Villigen-PSI, Switzerland; chris.milne@psi.ch (C.J.M.);
thomas.schietinger@psi.ch (T.S.); masamitsu.aiba@psi.ch (M.A.); Arturo.Alarcon@psi.ch (A.A.);
juergen.alex@psi.ch (J.A.); alexander.anghel@psi.ch (A.A.); vladimir.arsov@psi.ch (V.A.);
carl.beard@psi.ch (C.B.); paul.beaud@psi.ch (P.B.); simona.bettoni@psi.ch (S.B.);
markus.bopp@psi.ch (M.B.); helge.brands@psi.ch (H.B.); mbroenni@gmx.net (M.B.);
ingo.brunnenkant@psi.ch (I.B.); marco.calvi@psi.ch (M.C.); alessandro.citterio@psi.ch (A.C.);
paolo.craievich@psi.ch (P.C.); marta.divall@psi.ch (M.C.D.); mark.daellenbach@psi.ch (M.D.);
michael.damico@psi.ch (M.D.); andreas.dax@psi.ch (A.D.); Yunpei.Deng@psi.ch (Y.D.);
alexander.dietrich@psi.ch; (A.D.); roberto.dinapoli@psi.ch (R.D.); Edwin.Divall@psi.ch (E.D.);
sladana.dordevic@psi.ch (S.D.); Simon.Ebner@psi.ch (S.E.); Christian.Erny@psi.ch (C.E.);
hansruedi.fitze@psi.ch (H.F.); uwe.flechsig@psi.ch (U.F.); rolf.follath@psi.ch (R.F.);
franziska.frei@psi.ch (F.F.); f.gaertner@hispeed.ch (F.G.); romain.ganter@psi.ch (R.G.);
terence.garvey@psi.ch (T.G.); zheqiao.geng@psi.ch (Z.G.); ishkhan.gorgisyan@psi.ch (I.G.);
christopher.gough@psi.ch (C.G.); andreas.hauff@psi.ch (A.H.); christoph.hauri@psi.ch; (C.P.H.);
Nicole.Hiller@psi.ch (N.H.); Tadej.Humar@psi.ch (T.H.); stephan.hunziker@psi.ch (S.H.);
gerhard.ingold@psi.ch (G.I.); rasmus.ischebeck@psi.ch (R.I.); markus.janousch@psi.ch (M.J.);
Pavle.Juranic@psi.ch (P.J.); mario.jurcevic@psi.ch (M.J.); maik.kaiser@psi.ch (M.K.);
babak.kalantari@psi.ch (B.K.); roger.kalt@psi.ch (R.K.); boris.keil@psi.ch (B.K.); christoph.kittel@psi.ch (C.K.);
gregor.knopp@psi.ch (G.K.); waldemar.koprek@psi.ch (W.K.); Henrik.Lemke@psi.ch (H.T.L.);
Thomas.Lippuner@psi.ch (T.L.); daniel.llorente@psi.ch (D.L.S.); florian.loehl@psi.ch (F.L.);
carlos.lopez-cuenca@psi.ch (C.L.-C.); fabian.maerki@psi.ch (F.M.); fabio.marcellini@psi.ch (F.M.);
goran.marinkovic@psi.ch (G.M.); isabelle.martiel@psi.ch (I.M.); ralf.menzel@psi.ch (R.M.);
aldo.mozzanica@psi.ch (A.M.); karol.nass@psi.ch (K.N.); gianluca.orlandi@psi.ch (G.L.O.);
cigdem.ozkan@psi.ch (C.O.L.); ezequiel.panepucci@psi.ch (E.P.); martin.paraliev@psi.ch (M.P.);

bruce.patterson@empa.ch (B.P.); bill.pedrini@psi.ch (B.P.); marco.pedrozzi@psi.ch (M.P.);
patrick.pollet@psi.ch (P.P.); claude.pradervand@psi.ch (C.P.); eduard.prat@psi.ch (E.P.);
peter.radi@psi.ch (P.R.); jean-yves.raguin@psi.ch (J.-Y.R.); sophie.redford@psi.ch (S.R.);
Jens.Rehanek@psi.ch (J.R.); julien.rehault@gmail.com (J.R.); sven.reiche@psi.ch (S.R.);
matthias.ringele@psi.ch (M.R.); Jo.rittmann@gmail.com (J.R.); leonid.rivkin@psi.ch (L.R.);
Albert.Romann@psi.ch (A.R.); marie.ruat@psi.ch (M.R.); christian.ruder@psi.ch (C.R.);
Leonardo.Sala@psi.ch (L.S.); lionel.schebacher@yahoo.fr (L.S.); thomas.schilcher@psi.ch (T.S.);
volker.schlott@psi.ch (V.S.); thomas.schmidt@psi.ch (T.S.); bernd.schmitt@psi.ch (B.S.);
xintian.shi@psi.ch (X.S.); markus.stadler@psi.ch (M.S.); lukas.stingelin@psi.ch (L.S.);
werner.sturzenegger@psi.ch (W.S.); jakub.szlachetko@ujk.edu.pl (J.S.); dhanya.thattil@psi.ch (D.T.);
daniel.treyer@psi.ch (D.M.T.); alexandre.trisorio@psi.ch (A.T.); wolfgang.tron@psi.ch (W.T.);
seraphin.vetter@psi.ch (S.V.); Carlo.Vicario@psi.ch (C.V.); didier.voulot@psi.ch (D.V.);
meitian.wang@psi.ch (M.W.); thierry.zamofing@psi.ch (T.Z.); christof.zellweger@psi.ch (C.Z.);
riccardo.zennaro@psi.ch (R.Z.); elke.zimoch@psi.ch (E.Z.); rafael.abela@psi.ch (R.A.)
2   École Polytechnique Fédérale de Lausanne, 1015 Lausanne, Switzerland
*   Correspondence: luc.patthey@psi.ch (L.P.); hans.braun@psi.ch (H.-H.B.);
     Tel.: +41-56-310-4562 (L.P.); +41-56-310-3241 (H.-H.B.)
†   Current address: CERN, 1211 Geneva, Switzerland.
‡   Current address: Empa, 8600 Dübendorf, Switzerland.
§   Current address: Kistler AG, 8408 Winterthur, Switzerland.
‖   Current address: Institute of Physics, Jan Kochanowski University, 25406 Kielce, Poland.
¶   Current address: leadXpro AG, PARK innovAARE, 5234 Villigen, Switzerland.

Academic Editor: Kiyoshi Ueda
Received: 13 June 2017; Accepted: 30 June 2017; Published: 14 July 2017

Abstract: The SwissFEL X-ray Free Electron Laser (XFEL) facility started construction at the Paul Scherrer Institute (Villigen, Switzerland) in 2013 and will be ready to accept its first users in 2018 on the Aramis hard X-ray branch. In the following sections we will summarize the various aspects of the project, including the design of the soft and hard X-ray branches of the accelerator, the results of SwissFEL performance simulations, details of the photon beamlines and experimental stations, and our first commissioning results.

Keywords: X-ray free electron laser; linac; X-rays; undulator; SwissFEL; X-ray optics; X-ray photon diagnostics; ultrafast X-ray science; X-ray detector; JUNGFRAU; serial femtosecond crystallography

1. Introduction

X-ray free electron lasers (XFELs) represent a new generation of electron accelerators [1]. They produce bright bursts of X-rays at periodic intervals, where these pulses are both spatially coherent [2–4] and ultrashort in duration [5–9]. This combination of high brightness and ultrashort pulses produces extraordinary peak intensities, which has proven extremely attractive for certain fields of research in addition to creating entirely new fields which were not previously feasible, such as nonlinear X-ray signals [10–15]. Due to the extensive experience amongst researchers with storage ring X-ray techniques [16], many of the first experiments [17] applied well-established methods, such as X-ray spectroscopy and scattering, but in a time-resolved manner, taking advantage of the ultrashort pulse durations to measure dynamics in matter [18–21]. As experience with the facilities has increased in recent years, new techniques have been developed, including diffract-before-destruction methods, where the short X-ray pulse scatters from the sample before the atoms can move [22], providing the ability to measure room-temperature, radiation-damage-free structures. This ability has been applied to develop a technique called serial femtosecond crystallography (SFX), where a stream of tiny protein crystals is delivered into the focus of the XFEL, and, though the intense X-ray pulse destroys the crystal, its diffraction pattern is measured before the crystal is destroyed [23,24]. These facilities are still in their infancy, with the first hard X-ray FEL only in operation since 2009 [25], but even in the few short years since their arrival they have attracted significant interest from researchers around the world, with the result that several new XFEL projects are underway worldwide. Here we present an overview of an XFEL project located in Switzerland, which is expected to welcome its first users in 2018.

The SwissFEL XFEL facility is located at the Paul Scherrer Institute [26] (PSI) which is the Swiss national laboratory home for large-scale accelerator-based user facilities. It includes a 3rd-generation synchrotron light source (SLS), the Swiss Muon Source (SµS), the Swiss Spallation Neutron Source (SINQ), and a high-intensity proton accelerator (HIPA). The SwissFEL construction was preceded by an intense R&D period, with the goal to allow for a very compact and economical design, and to have several features which are unique amongst the XFEL facilities presently in operation or under commissioning worldwide. In this review we highlight SwissFEL's expected capabilities and how it fits into the worldwide XFEL community. This article is organized into several sections: Section 2 covers the beam dynamics, injector, linear accelerator, undulator, and beam diagnostics components of the accelerator; Section 3 describes the X-ray optics, photon diagnostics, and experimental laser systems of the photon beamlines; Section 4 provides details on the experimental stations and their instruments, with a subsection on the 2D X-ray detectors available at the facility; finally Section 5 summarizes the infrastructure common to all aspects of the project, including timing and synchronization, motion control and data acquisition.

2. Accelerator

The fundamental design concept behind SwissFEL is to construct an X-ray Free Electron Laser, capable of lasing at 1 Å, but with investment and operation costs substantially reduced in comparison with other facilities of similar scientific potential. In the following sections we will describe the accelerator components of SwissFEL, illustrating how the project has been able to achieve its goals without sacrificing performance. The accelerator layout is shown in Figure 1.

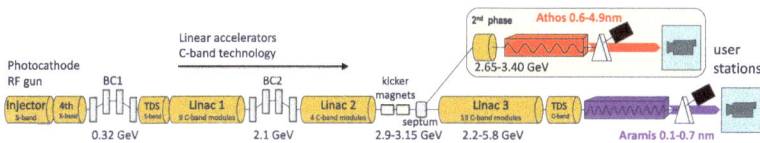

Figure 1. SwissFEL accelerator layout. It consists of an S-band injector, a C-band linear accelerator, and two undulator lines. The details of these components are described in the following sections.

2.1. Beam Dynamics and FEL Concept

The overall design goal of SwissFEL was to build a compact facility to produce FEL pulses down to 1 Å wavelength, with the lowest electron beam energy suitable to drive the FEL. This is constrained by the design of the undulator where a short undulator period (λ_u) reduces the required electron beam energy (γ) to fulfill the resonance condition of the FEL [27] with

$$\lambda = \frac{\lambda_u}{2\gamma^2} \left(1 + \frac{K^2}{2} \right) \tag{1}$$

and undulator strength $K = 0.93 \cdot B[T] \cdot \lambda_u[cm]$ in order to lase at a wavelength of $\lambda = 1$ Å. For a compact, in-vacuum undulator [28] we assumed that a period of 15 mm and an undulator K value of 1.2 are feasible. The FEL resonance condition then dictates that the maximally required beam energy to be provided by the linear accelerator is 5.8 GeV. With the nominal beam energy defined, the design value of the beam emittance (ϵ_n) can be estimated from the condition [29]

$$\frac{\epsilon_n}{\gamma} \approx \frac{\lambda}{4\pi} \tag{2}$$

to be met for all electrons to radiate into the fundamental mode of the FEL. This is, however, a rather soft limit, as larger values simply lead to a reduced coherence of the FEL output beam [3] and lower pulse energies. For the design we assumed a target emittance of 430 nm for the longitudinally central parts of the beam.

The second consideration concerns the wakefields within the undulator [30,31], which alter the electrons' local mean energy, thereby shifting them away from the resonance condition of the FEL. With an undulator gap of 4.4 mm for the nominal K of 1.2 the dominant wakefields are caused by the finite conductivity of the vacuum chamber. For the chosen material, copper, the characteristic length of the resistive wall wake potential is about 7.6 μm, resulting in a parabolic shape of the wakefields with a length of 20 μm. The mean energy loss can be compensated with a linear taper of the undulator, which recovers 90% of the FEL performance as calculated without wakefields or taper [32]. The SwissFEL machine parameters are listed in Table 1.

The basic operation mode requires a bunch with charge 200 pC, which is compressed in two stages to realize a peak current of 3 kA. The overall compression scheme is based on the stipulation that the wakefields of the C-band structures remove the induced energy chirp needed for the final compression [33]. This approach avoids wasting additional radio frequency (RF) power to remove the chirp actively by off-crest acceleration. To provide shorter but more efficient FEL pulses (in the sense that the number of emitted photons per electron is increased) SwissFEL can operate down to 10 pC bunch charge, starting with a shorter and smaller but brighter beam from the source [34]. Besides this flexibility of tuning to any bunch charge between 10 and 200 pC, SwissFEL offers two special modes. In the first a pulse similar to the one for the nominal 200 pC mode is generated, but with a large correlated energy chirp [35]. This is achieved by overcompressing the bunch in the last compression stage. In this case the wakefields in the main linac now add up to the reversed chirp, resulting in a peak-to-peak energy chirp of 1 to 1.5% at 5.8 GeV. The quadratic dependence of the photon energy on the electron energy Equation (1) yields a chirp in the photon pulse twice as large. The other special mode consists in the full compression of a 10 pC pulse to achieve sub-femtosecond FEL pulses. To avoid the transport of a high peak current through the main linac, which could degrade the beam quality due to the space-charge field [36], the full compression is achieved in the energy collimator right before the undulator, where a certain control over the energy-dependent path length of the electrons is given.

Table 1. SwissFEL hard X-ray Free Electron Laser (FEL) design parameters.

Electron Accelerator	
Beam energy	2.1–5.8 GeV
Energy spread (rms)	350 keV
Normalized emittance	430 nm
Current	3 kA
Undulator Parameters	
Period	15 mm
K value	1.2
Active length	48 m
Total length	60 m
Photon Parameters	
Wavelength	1–7 Å
Energy	1.77–12.4 keV
Pulse energy	0.01–1 mJ
Pulse length (rms)	0.2–20 fs
Bandwidth	0.04–3%

Appl. Sci. **2017**, *7*, 720

The SwissFEL accelerator is designed to drive a second beam line called Athos, providing FEL pulses in the soft X-ray regime between 5 and 0.65 nm wavelength. In the two-bunch operation mode, the machine simultaneously accelerates two electron bunches at 100 Hz, with 28 ns spacing between the two bunches. The Athos beam is extracted at around 3 GeV to allow for the independent tuning of both lines (see Figure 1). With an undulator periodicity of 38 mm it is not required to place the magnet array in vacuum. This in turn opens up the possibility of a more advanced design of the undulator, with better control and tunability of polarization, on-axis field and transverse gradient [37] as compared to the Aramis hard X-ray undulator, which is strictly planar. Another important distinction with respect to Aramis is the shorter length of the undulator modules and the inclusion of delaying chicanes between modules. The inter-undulator chicanes and the special undulator configurations give access to novel operation modes with improved control over power, pulse length, bandwidth and temporal coherence [38–42].

2.2. Injector

The generation and preservation of very high brightness electron beams able to drive an FEL requires particular care at the source and in the low-energy section of the machine. The electron source determines the best obtainable emittance at the FEL undulator line and hence the FEL performance. The SwissFEL injector consists of a 2.5-cells S-band (3 GHz) RF photoinjector gun followed by an S-band booster linac providing the necessary energy gain before the first compression stage (BC1). The PSI RF photoinjector gun [43] generates high brightness electron bunches with an energy of 7.1 MeV, an intrinsic emittance of 0.55 μm/mm and a peak current of 20 A. An IR Yb:CaF$_2$ laser system operating at 1040 nm with frequency multiplication to 260 nm drives the Cs$_2$Te coated copper photocathode installed in the backplane of the RF gun. A detailed description of the gun laser system is given in Section 2.2.1.

The ensuing booster linac consists of two sections. In booster 1 two S-band traveling-wave cavities [44] accelerate the electron beam on crest up to an energy of 150 MeV. After this first acceleration stage a set of five quadrupole magnets allows matching the optical functions through a laser heater chicane, whose purpose is the controlled enhancement of the uncorrelated energy spread of the beam to mitigate micro-bunching instabilities in the bunch compressors [45]. Another set of five quadrupoles follows the laser heater modulator undulator to control the matching into booster 2. The second booster section consists of two S-Band RF modules, each including one klystron amplifier and two accelerating cavities. In booster 2 the electrons are accelerated off crest, up to an energy of 345 MeV, to provide the necessary energy-time correlation needed for the longitudinal compression of the bunches. Enough space has been reserved to allow future energy upgrades with a third RF accelerating module. The focusing along booster 2 consists of three FODO cells with 11 m period. To suppress the second-order energy-time correlation two X-band RF cavities (4th harmonic of S-band) [46–48] running in decelerating mode precede the 13.5 m long compression chicane, which is typically set to yield compression factors between 10 and 15. The final nominal energy of the injector is 320 MeV.

The compact SwissFEL design hinges on the small beam emittance provided by the injector. Therefore a great deal of experimental effort, carried out mainly at the SwissFEL Injector Test Facility [49], has gone into the characterization and optimization of the emittance at the source [50–52], as well as its preservation under acceleration, transport and compression. The starting point of our injector optimization is the effective working point found by Ferrario et al. [53] during the redesign of the Linac Coherent Light Source (LCLS), further refined for the SwissFEL case by numerical optimization [54]. The main empirical tuning steps toward minimal emittance consist in achieving the optimal laser spot size on the cathode with homogeneous transverse and longitudinal pulse profiles, the adjustment of the relative phase between laser injection and gun RF, the optimal setting of the gun solenoid excitation current, the correction of coupling terms by means of small (regular and skew) quadrupole magnets integrated into the gun solenoid as well as further solenoid magnets [55], the centering of the orbit in the S-band booster structures and the correction of spurious dispersion downstream of the booster [56]. For compressed beams special care must be taken to keep adverse effects from coherent synchrotron radiation in the compression chicane under control, e.g., by ensuring a small transverse beam size in the last chicane dipole or by adopting a shallow bending angle for the compression. Our studies also revealed a strong sensitivity of the final slice emittance on the beam optics upstream of the bunch compressor. Therefore a small optics mismatch along the longitudinal position of the bunch turns out to be of great importance for the preservation of the emittance under compression [57].

The normalized slice emittance at the end of the injector with uncompressed beam is expected to be around 0.2 μm for a bunch charge of 200 pC, and nearly preserved under moderate compression (see Reference [49] for details).

2.2.1. Gun Laser

Compact and industrial-grade laser systems with high power stability and ultra-low timing jitter have become a key component in free electron lasers. At SwissFEL the drive laser for the electron gun consists of solid state Yb:CaF$_2$ chirped pulse amplifier. For electron production the stability of the drive laser plays a crucial role. While Ti:sapphire lasers are standard technology used in many FELs around the world, we considered an Ytterbium-doped gain medium. This laser system offers exceptional long-term amplitude stability, low intrinsic timing jitter, a compact design and very high up-time. The Yb-doped laser is pumped with a telecom-standard semiconductor diode emitting at 976 nm. Such pump diodes are favourable in view of long-term performance and low maintenance costs. The oscillator delivers sub-200 fs, transform-limited soliton pulses centered at 1041.3 nm with an amplitude stability of 0.19% rms over 18 hours. The measured free running jitter of 6.3 fs rms (integrated over 1 kHz to 1 MHz) is ultralow. The system can be actively synchronized with an RF reference signal with a locked timing jitter of 18 fs rms (10 Hz–1 MHz).

The oscillator seeds a commercial Yb:CaF$_2$ regenerative chirped pulse amplifier system. The amplifier is pumped with a single CW diode module delivering high power at 980 nm. This provides high reliability and a long lifetime of up to 20,000 h. After amplification to 2.4 mJ the stretched pulse is compressed to 700 fs FWHM by means of a transmission grating pulse compressor. Temporal drifts of the chirped-pulse amplification (CPA) system are compensated by employing a feedback loop which stabilizes the drift to <32 fs rms over 250 min. For UV generation two home-made nonlinear frequency conversion stages based on BBO crystals are employed which provide pulses of up to 600 μJ at 260 nm. To lower the electron beam emittance the Gaussian-like temporal pulse shape is transformed into a flattop-like pulse by a set of four birefringent α-BBO crystals. Three different sets of crystals provide the three pulse durations of 3.6, 6.7 and 10 ps for electron bunch production. For transverse beam shaping a variable circular aperture is used to produce a truncated Gaussian beam profile which is imaged over a 20 m distance onto the Cs$_2$Te cathode. At the cathode a pulse energy of approximately 100 nJ is used to produce the 200 pC charge.

A small part of the chirped amplified infrared laser beam (150 μJ) is split, compressed (40–50 ps) and directed to the laser heater. Overlapped with the electron bunch in time and space the scheme allows the energy spread of the electrons to be increased to avoid unwanted coherent radiation downstream of the linear accelerator.

2.3. Linac

The main acceleration of the electron beam is achieved in a C-band linac that increases the beam energy up to 5.8 GeV. The linac is divided into three segments: linac 1, linac 2 and linac 3. After linac 1, the electron bunches from the injector are further compressed in a second bunch compressor (BC2) at an energy of 2.1 GeV. At the end of linac 2, at an energy of 3.15 GeV, a switch-yard [58] is installed with which electron bunches can be sent either straight into linac 3 and subsequently the Aramis hard X-ray beam line, or into the Athos soft X-ray beam line, currently under construction. The accelerator scheme is shown in Figure 1. The switch-yard allows the parallel operation of both undulator lines at the full design repetition rate of 100 Hz. This is accomplished by generation of two electron pulses from the injector separated by 28 ns, both accelerated up to 3.15 GeV before separation. At the end of linac 3, two C-band transverse deflecting structures (TDS), provided by Mitsubishi Heavy Industries Mechatronics Systems, allow for measurements of the longitudinal charge profile with a resolution of a few femtoseconds.

The linac consists of a total of 26 C-band modules, where each module comprises four C-band structures that are mounted onto two granite girders (see [59] for a schematic and [60] for further information). The C-band structures [61] were stacked and brazed at PSI [62] from copper cells manufactured at VDL ETG (J-couplers) and VDL ETG Switzerland (regular cells) with micrometer precision using ultra-precision diamond milling and turning. This process renders further tuning steps of the structures unnecessary while still achieving excellent field flatness and phase advance errors (see [59] for an example). The achieved structure straightness is excellent: the maximum measured deviations from a straight line are typically below 20 μm, at the resolution limit of the applied laser tracker. In addition to the four structures, each linac module also comprises a barrel open cavity (BOC) RF pulse compressor [63], machined and brazed at PSI [64]. See Figure 2 for a photo of linac 2 installed at SwissFEL. Most parts of the waveguide distribution and the BOC pulse compression cavities are mounted on the granite support girder. This allowed for preassembly of most of the linac vacuum system in a cleanroom, while only the interconnects between girders and the waveguide run to the klystrons have to be done in the tunnel.

The produced structures have been sorted according to their resonance frequencies, as determined by RF measurements, with the goal of assigning four similar structures to the same module. This grouping is necessary since all structures of one module are cooled and temperature stabilized to the millikelvin level by a single cooling station. Another advantage of the structure sorting is that the RF power overhead required to compensate for the loss in energy gain by deviating slightly from their beam synchronous frequencies is less than 1%.

Figure 2. The C-band accelerating modules of linac 1.

Supplying four structures with a single RF power source represents a challenge for the waveguide network, furnished by Mitsubishi Heavy Industries Mechatronics Systems: the mechanical fit demands a mechanical tolerance of 200 µm between the structures, and the phases at the structure entrances must match within a few degrees. The first requirement was ensured by the manufacturer during the production process by measuring the dimensions of the individual waveguide components and providing accordingly machined correction pieces. To achieve the correct phase relations at the four structures, the complete horizontal waveguide network was assembled and moved away from the structures to measure the phase relations, which were then corrected through suitable deformations of the waveguides.

Linacs 1 and 2 are powered by Type-μ modulators manufactured by Ampegon, whereas linac 3 uses PSI C-band series modulators M1071 provided by ScandiNova. Both modulator types use IGBT for high voltage switching and very precise charging circuits. This allows the pulse-to-pulse voltage variation to be kept well below 20 ppm. All modulators drive Toshiba klystrons of type E37212. The E37212 klystron was specifically developed by Toshiba for an increased pulse length and average power rating in comparison with former C-band tubes. It delivers up to 50 MW with 3 µs pulselength at 100 Hz. The nominal operation point in SwissFEL is 40 MW, thus a 25% power margin is maintained. Furthermore the E37212 is operated with the collector water cooling circuit at 80 °C output temperature. This allows to recuperate the power lost in the collector for use in the PSI building heating network. The installation and commissioning of the main linac RF power stations [65] started in 2016 and will continue through 2017. During this process the beam energy is successively increased as more modulators become available.

2.4. Undulator Line

The Aramis undulator line of SwissFEL, shown in Figure 3, consists of 13 in-vacuum undulator modules with 4 m length and 265 periods of 15 mm each. High-performing NdFeB permanent magnets with diffused dysprosium (Hitachi metals; remanence $B_r = 1.25$ T, coercivity $H_{cJ} = 2400$ kA/m) and poles with a trapezoidal geometry made out of Vacoflux 50, Vacuumschmelze to focus the field on the beam axis, provide a peak field of $B = 1.3$ T (corresponding to a K value of 1.8) at a minimum gap of 3 mm. Thanks to a reduced pole tip width of 15 mm the magnetic forces can be limited to 25 kN (corresponding to the weight of 2.7 metric tons), under which a gap adjustment precision better than 1 µm is needed. The undulator module design was made by PSI in close collaboration with industry to include specific manufacturing know-how right from the beginning. The production of the modules was carried out by our industrial partners (Daetwyler Industries (Huntersville, NC, USA), RI, VDL, Comvat (Sennwald, Switzerland), Schaeffler Schweiz GmbH (Romanshorn, Switzerland), EPUCRET (Wangen, Germany), Agathon, Rollvis (Plan-les-Ouates, Switzerland)) but the optimization and characterization of the magnet structures were performed in the undulator laboratory inside the SwissFEL building at a rate of one undulator module per month.

An optimization of quality and cost in the early conceptual phase resulted in the following main design principles: a closed O-shaped support structure with cast mineral material, a wedge-based drive system, a common, modular design for in-vacuum, standard or APPLE II configurations, and a magnet keeper design enabling fast field optimization. Further constraints were the requirement for transport without crane and the target beam height of 1.2 m.

Figure 3. SwissFEL Aramis undulator line with a total length of 60 m (4 m each module and 0.75 m intersection).

The closed support structure was chosen because of the superior stiffness compared to a standard C-structure when accessibility for magnetic measurements and pre-installations of vacuum chambers are not an issue. The high stiffness is transferred to the I-beam by a small angle wedge system, a novel concept in undulator design. With our industrial partners we produced a full-size prototype in 2013. With a 70% support by bearings the height of the I-beam can be reduced, and the wedge works as a gear reduction. The drive system only consists of a servomotor and a satellite roller screw with a small pitch of 1 mm per turn. Two wedges move against each other, synchronized by the Beckhoff motion control system. The I-Beam is fixed in the longitudinal direction by a central guiding rod (Agathon). The backlash-free system allows gap changes with a reproducibility of 0.4 μm. The gap position is monitored by absolute linear encoders.

Twenty columns (Comvat) with integrated differential screws provide for the connection to the magnet arrays through the vacuum vessel produced by VDL. The top columns are shifted with respect to the lower ones, which reduces the critical variation of the gap (with an exponential field dependence) but comes at the price of a (less harmful) non-straightness of the magnetic field axis following a hyperbolic cosine dependence.

To meet the critical time schedule and to achieve optimum results in terms of magnetic field profile, a block keeper was designed and realized from extruded aluminum, which allows all magnet-pole pairs to be adjusted in height. An accuracy of better than 1 μm can be achieved with the 3° wedge shown in Figure 4. The pole height can be tuned within ±30 μm by means of a flexor system, enough for the shimming of all 13 undulators. The magnet arrays on the in-vacuum I-beams have been assembled aligned and pre-measured by RI.

Magnetic field measurements and local field corrections are carried out with an integrated measurement bench based on the SAFALI system developed for measurements of cryogenic undulators [66]: After defining two parallel axes by means of a pointing stabilized laser, the system uses two pinholes around the Hall probes and two position sensitive photodiodes to detect and correct for horizontal and vertical positon as well as pitch in a closed loop with 20 μm accuracy. A custom-made, low-noise Hall probe with a novel ceramic support was developed by SENIS to meet the stringent demands given by the small gap and the required accuracy [67].

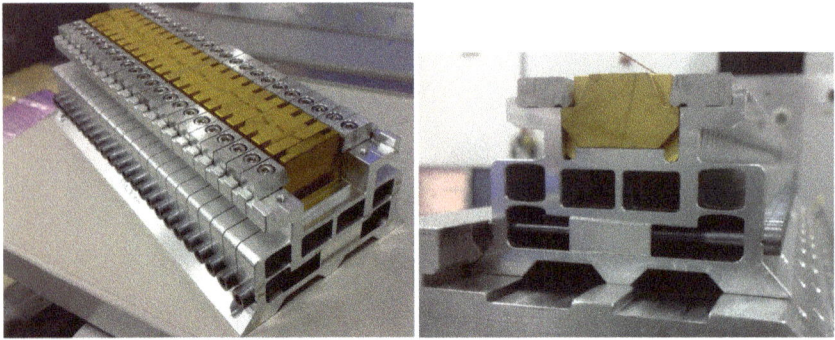

Figure 4. Magnet block keeper made of extruded aluminum. The height of the magnets can be adjusted by the wedge shown in the image to the right.

For the transport of the undulators inside the SwissFEL building an air-cushion vehicle (AEROFILM Systems) allows for smooth and precise positioning of the undulators onto the camshaft mover system. The 5-axis camshaft movers allow for adjustments of the module with respect to the reference axis. The straightening of the magnetic axis and the correction of long-range errors are accomplished with the adjustable columns. Local field corrections are automatically corrected by an integrated screwdriver robot; the fine adjustment of all magnets in a single module takes slightly more than one hour. All corrections, columns and local keeper are model based and result in a straightforward field optimization. For the installation of the vacuum vessel, however, the magnet array has to be disassembled. At the level of precision needed for the hard X-ray FEL the columns have to be readjusted afterwards using a similar measurement setup, but adapted to the limited space available inside the vessel. Field maps at various gaps provide the raw data needed for modelling the operation of the undulators [68].

Finally, dedicated small alignment quadrupole magnets, realized with permanent magnets, are located at the entrance and exit of each undulator module and adjusted to the magnetic axis. They can be moved in and out with pneumatic drives and are used in conjunction with the module's camshaft mover system for the beam-based alignment of the undulator modules during beam commissioning. This approach is inspired by commissioning work at the LCLS, where the quadrupole magnets in the intersections are used for the alignment of the undulator modules [69]. Since in our case no transfer via any mechanical fiducial system is involved, the achievable accuracy is expected to be better than 20 μm.

2.5. Accelerator Instrumentation

SwissFEL has a wide range of instruments that perform measurements on the electron beam properties, help set up different operation modes, and monitor the stable operation of the accelerator. Many of these diagnostics have been specifically developed to meet the stringent requirements of SwissFEL for beam quality and stability. A common design goal for the diagnostics is the relatively low bunch charge at SwissFEL, which ranges between 10 and 200 pC. In several cases, these requirements have led to the design of novel diagnostics or to iterative improvements on existing designs.

2.5.1. Bunch Charge

The total charge of the electron bunches influences several parameters of the accelerator, and its measurement is required to adjust the electron beam optics in the space-charge dominated region in the injector, the proper compensation of wake fields in the accelerating cavities, and finally the FEL process in the undulators. It is also a legal requirement to monitor and record the total charge accelerated by SwissFEL for radiation protection reasons.

The primary measurement of the electron bunch charge is provided by integrating current transformers (ICTs). There are two types of ICTs used at SwissFEL. The first is a conventional ICT, with BCM-IHR readout electronics for reading the total charge from the gun. The second is the Turbo-ICT, developed for the SwissFEL two-bunch operation, with BCM-RF-E readout electronics for signal processing of the bunch charges. The conventional ICT is calibrated for a charge range up to 800 pC [70], whereas the Turbo-ICT is calibrated on the Turbo-ICT/BCM-RF calibration test bench [71] for charges up to 300 pC.

The bunch charge is also monitored with beam position monitors, as described in the following section.

2.5.2. Orbit

The stabilization of the electron beam trajectory throughout SwissFEL is crucial for an optimal beam quality: In the accelerator the proper alignment with respect to the accelerating cavities and quadrupoles minimizes transverse wakefield effects and residual dispersion. In the undulator line the BPMs are essential to steer the electron beam on a straight line to keep the overlap between electron and photon beams.

Beam position monitors [72] (BPMs), based on dual-resonator cavity pickups, measure the beam position in a dipole cavity, while using the signal from a monopole cavity for charge normalization. Injector, linac and transfer line pickups operate at 3.3 GHz with a relatively low loaded quality factor (Q_L) of 40 to resolve the two bunches of the SwissFEL, which are separated by 28 ns when operating both Athos and Aramis simultaneously. The pickups in the undulator line, where only a single bunch is present, work at 4.9 GHz with a higher Q_L of about 1000 to achieve better position resolution. The low-Q pickup signals are converted directly to baseband, the high-Q signals to an intermediate frequency of 135 MHz. After sampling with fast 16-bit analog-to-digital converters (ADCs), the digitized signals are extensively post-processed in field-programmable gate arrays (FPGAs) to correct systematic measurement errors, yielding position readings with submicrometer resolution.

The signal arising from the monopole cavity is also used to monitor the bunch charge. It is processed and digitized with low noise, resulting in a charge-independent relative resolution well below 0.1% [72]. At very low charges an absolute resolution of a few fC is achieved.

BPMs at locations with suitable (moderately low) dispersion, e.g., in bunch compressor chicanes near the first and last bending magnets, are also used to measure the electron beam energy.

2.5.3. Emittance

Together with the short undulator period, the small normalized emittance of SwissFEL enables the generation of high-energy photons at a lower electron energy than at the first XFELs [25,73], thereby significantly reducing building size and overall facility cost. The generation of a low-emittance beam at the electron source, as well as the measurement of the emittance along the accelerator to ensure its preservation, are thus important elements for the successful operation of SwissFEL.

Scintillating transverse profile imagers are used to measure the beam profile at several locations along the beam line. SwissFEL uses a novel geometry to image the beam on scintillating crystals, which aims for the suppression of unwanted coherent optical transition radiation (COTR), but allows for a good resolution over a large field of view [74]. The emittance measurement is performed by scanning the electron beam optics while observing the beam profile (see, e.g., [56]).

As a complement to these viewscreens, wire scanners will be used in SwissFEL to measure the transverse electron beam properties. The wire scanners will perform a monitoring of the beam profile along the horizontal and vertical direction either with high spatial resolution—when the 5 μm tungsten wires are used for the scan—or with minimal perturbation of FEL operations—when the 12.5 μm aluminum wires are used instead [75].

2.5.4. Energy Spread

The operation of an FEL demands tight control over the electrons' energy spread: an excessive energy spread in the undulator impedes the FEL process, but a value that is too small may give rise to the coherent emission of synchrotron radiation in the bunch compression chicanes, potentially resulting in beam breakup.

In the bunch compressors the energy spread can be measured by inserting scintillating screens into the dispersed electron beam or by imaging the synchrotron radiation emitted in one of the dipole magnets. At SwissFEL SCMOS cameras (PCO Edge 5.5), equipped with a $f = 300$ mm lens of 107 mm diameter to avoid vignetting, are installed in both magnetic chicanes to image the synchrotron radiation light emanating from the third dipole of the chicane. As these synchrotron radiation detectors are completely non-destructive they will be used for the routine monitoring of the beam energy and energy spread as well as for the optimization of the bunch compression setup [76,77].

Furthermore, a precise measurement of the energy spread in the injector can be performed by observing the degradation of the coherent modulation of the beam following the laser heater. This measurement will be performed by monitoring coherent transition radiation at a photon energy which is an integer multiple of the laser heater photon energy [78] (see Section 2.2.1).

2.5.5. Time-Resolved Measurements

The electron bunch length and current profile are measured by directly streaking the electron beam with transverse deflecting RF cavities [79], installed at suitable locations in the accelerator beamline, and observing the streaked beam with regular transverse profile monitors [80,81]. The same method also allows the time-resolved measurement of emittance and energy spread along the bunch (so-called "slice" parameters).

Complementary to the transverse RF deflector, an effective streaking in both transverse directions all along the accelerator can be achieved by introducing dispersion to an energy-chirped beam [82].

2.5.6. Bunch Arrival Monitoring

The monitoring of the bunch arrival time is important for maintaining the longitudinal stability of the linac. The bunch arrival-time monitors (BAMs) [83] developed for SwissFEL provide non-destructive, shot-to-shot arrival time information relative to a highly stable pulsed optical reference (see Section 5.1) with resolution better than 5 fs and less than 10 fs drift per day [84,85]. The electron beam generates an S-shaped bipolar transient with a steep slope (15 ps peak-to-peak) in a pick-up with 40 GHz bandwidth [86]. This pickup-signal is probed by a single reference laser pulse and the arrival time is encoded in its amplitude. The bunch arrival-time monitors are used for feedback on machine parameters with impact on the arrival time, such as the accelerating cavity amplitudes and phases. By measuring the electron arrival time after the last undulator, the BAM allows event correlation for the experiments at the fs level.

2.5.7. Bunch Compression Monitoring

In addition to the destructive bunch length measurement with transverse deflecting cavities and screens, a non-destructive online monitoring of the compression process will be performed by a spectral analysis of the coherent radiation emitted at the edge of the bunch compressor dipoles.

At the first bunch compressor this radiation occurs in the THz spectral range and is monitored by two Schottky diodes, each equipped with a different high-pass filter to select specific wavelength regions. These detectors have been demonstrated to provide excellent sensitivity with good signal-to-noise ratio, therefore enabling very accurate measurements of accelerator changes [87].

Radiation emitted by the sub-picosecond bunches after the second bunch compressor is in the far infrared spectral region. At this location the bunch compression will be monitored by a spectrometer equipped with 32 mercury cadmium telluride (MCT) detectors.

In the sub-femtosecond pulse mode of SwissFEL (see Section 2.1) the final compression of the electron bunches takes place in the energy collimator chicane before the undulator line. Again a spectrometer will be used to analyze the coherent edge radiation, in this case reaching the near-infrared and visible regions of the spectrum.

2.5.8. Loss Monitoring

Electron beam losses along the accelerator give rise to radiation damage of accelerator components and may disturb accelerator performance by inducing measurement backgrounds. To protect the machine from excessive radiation we monitor such losses by measuring Cherenkov light emitted in fused silica fibers installed along the accelerator beamline. The loss location can be determined from the time of arrival of the light at the photodetectors.

2.6. First Commissioning Experience and Outlook

A detailed summary of the commissioning of elements of SwissFEL in the former SwissFEL test facility can be found in [49], here we will focus on our recent commissioning activities after installation in the facility, including the generation of our first FEL photons. The commissioning of the electron gun started in August 2016. Once the electron source was operating, electrons were further accelerated with two S-band RF stations to an energy of 145 MeV and transported to the injector beam dump for the first time in early September 2016. At that time only the first two S-band RF stations were available. Soon after the first injector transmission, however, the first of the linac C-band modules could be integrated into the acceleration process, thereby increasing the available beam energy to about 380 MeV. The first transmission through the entire accelerator including the undulator line was achieved at this energy in mid-November. The initial commissioning of essential accelerator instrumentation systems (charge and beam position monitors, screens) was also performed at this energy.

For the occasion of our official inauguration event in December 2016 an attempt was made to lase with the 380 MeV beam. To reach a sufficient charge density the beam was somewhat compressed in the first bunch compressor, while the accrued energy chirp was compensated for in the one available C-band module. After some further tuning, characteristic FEL radiation from the SASE process could be observed with a photodiode located after the undulator line. The photon wavelength derived from the beam energy and the undulator parameters was in the UV range (24 nm or 50 eV).

By mid-2017, all injector RF stations as well as two more linac RF stations had become operational, pushing the beam energy close to 1 GeV. This resulted in successful SASE soft X-ray operation at 4.1 nm (300 eV) in May of 2017 (see Figure 5). In the following months the beam energy will be further increased step-by-step in accordance with the availability of the remaining RF stations. First pilot experiments are foreseen at a beam energy of 3 GeV towards the end of 2017, with hard X-ray user operation in 2018.

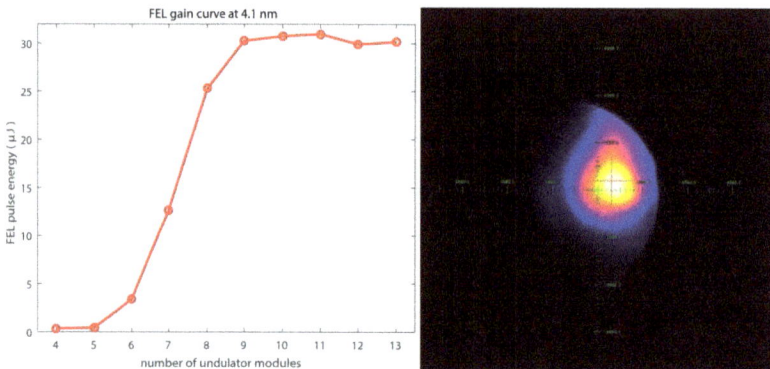

Figure 5. Measured SwissFEL pulse energy gain curve as a function of inserted undulator modules at an electron energy of 900 MeV (left). Saturation is reached after insertion of 9 modules. Pulse energy measured with the gas-based photon beam intensity and position monitor (see Section 3.3.1). An image of the photon beam at saturation, as recorded with a Ce:YAG scintillating screen, is shown on the right.

3. Photon Beamlines

The Aramis branch of SwissFEL is designed to produce photons from 1.77–12.4 keV (1–7 Å), at 100 Hz repetition rate, <50 fs pulse duration (FWHM), and around 10^{12} photons/pulse (see Section 2.1 for more details). In order to allow users to take full advantage of the unique properties of these photons, including the ultrashort pulse durations and the spatial coherence of the beam, careful design of X-ray optics and diagnostics is necessary. This is especially challenging at SwissFEL since the photon beamlines need to cover the "tender" X-ray photon energy range from 2–5 keV, which is unique amongst XFELs and requires careful beamline design. In addition to X-ray optics (Section 3.1) and diagnostics (Section 3.3), this section includes descriptions of the pulse picker (Section 3.2), to allow users to select the X-ray pulse frequency, and the experimental laser systems (Section 3.4), to allow users to perform pump-probe experiments. We will focus on describing the components in the Aramis-1 and Aramis-2 beamlines which will be the first two operational beamlines ready for user operation in 2018.

3.1. X-ray Optics

Free Electron Lasers are by design single user machines in the sense that every electron bunch serves only once as a source of light. This is in contrast to storage rings, where a finite number of bunches circulate for many hours, or even days, and repetitively deliver photons to many beamlines in parallel. From this it is evident that FELs can supply only one beamline at a time, meaning that beamtime is quite expensive and in high demand.

To make full use of the valuable beamtime it is mandatory to have several experimental setups in parallel with the ability to switch between them in a fast and reproducible way. The beamline design [88] has to provide the ability to alternate between several experimental stations, while giving full access to the unused experimental hutches. This is achieved with the help of pairs of offset mirrors (OM) in front of the beamline as shown in Figure 6. They are located as far upstream as possible to achieve a reasonable lateral separation at the end. The beam is subsequently directed towards two independent double crystal monochromators (DCM) and finally focused by sets of bendable Kirkpatrick-Baez mirrors [89] (KB) to the experiments. Two instruments are installed in line at experimental station Alvra on Aramis-1 (see Section 4.1) and will be operated by changing the focal distance of the KB system. A similar pair of KB mirrors are installed at experimental station Bernina on Aramis-2, but as instrument exchange is accomplished using a rail system perpendicular to the photon beam, all three planned instruments can take advantage of the minimum focus of the mirrors (see Section 4.2). A pair of harmonic rejection mirrors (HRM) after the DCM is foreseen in the Aramis-1 beamline from the beginning and as an optional extension for the Aramis-2 beamline. The Aramis-3 beamline is not yet specified and is under design concept phase in combination with the third experimental station, Cristallina. As one option, a simple pink beam beamline without monochromator and nanometer focusing at the end is considered and is shown in Figure 6.

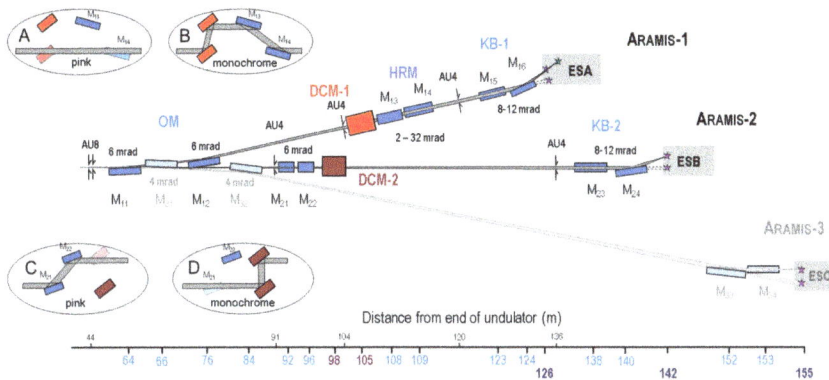

Figure 6. Optical layout of the hard X-ray beamlines of the Aramis undulator. The insets show the setup for pink and monochromatic operation of the beamlines Aramis-1 (A,B) and Aramis-2 (C,D), respectively.

A set of two horizontally deflecting offset mirrors direct the beam into the Aramis-1 beamline with a total deflection angle of 12 mrad. The central Aramis-2 beamline stays in the direction of the FEL beam and uses two vertically deflecting offset mirrors in a zigzag geometry. For Aramis-3 a second pair of horizontal offset mirrors with a total deflection angle of 8 mrad is considered as worst case scenario with respect to spatial restrictions along the beamlines and the end stations.

All offset mirrors are coated with low- and mid-Z materials to prevent single shot damage by the intense FEL beam [90–92]. However, this limits the deflection angle and two mirrors with small deflection angles instead of a single one with large deflection angle became necessary. The mirrors are coated with two bilayers leaving a blank area of uncoated silicon between them. The low-Z bilayer, 10 nm B_4C on top of 36 nm SiC, should withstand the FEL beam at all photon energies and reflect over the full energy range of the FEL. Its B_4C-layer is effective at lower photon energies and prevents the drop in reflectance at 1.8 keV due to the silicon K-edge absorption of the subjacent SiC-layer. At higher photon energies the B_4C layer becomes transparent and reflectance is supported by the SiC layer with a critical energy of 12.4 keV. The mid-Z bilayer, 15 nm B_4C on top of 20 nm Mo, extends the operation range up to the Mo K-edge at 20 keV but has a higher damage risk and may be used only at reduced fluence. In between both coatings a stripe of uncoated silicon serves as third reflecting area. It has the lowest critical energy of all coatings and may be used at low energies when high harmonic rejection becomes an issue. All mirrors are mounted in benders with two actuators, allowing bending radii from flat to ±10 km. By this, the offset mirrors can produce a line focus at the experiment stations.

Both beamlines include double crystal monochromators, each containing three pairs of crystals. Two pairs of silicon crystals, Si(111) and Si(311), are foreseen for standard and high resolution applications. An additional pair of InSb(111) crystals extends the wavelength range up to 7 Å (1.77 keV). The crystals are mounted on a common Bragg-rotation axis that sets the Bragg angle for both crystals from 5° to 80°. The translation of the second crystal perpendicular to its surface allows for a constant beam offset of 20 mm as well as for a variable beam offset in the harmonic rejection mode. The second crystals are long enough to omit the translation parallel to their surface and therefore the beam spot moves along the second crystal while the photon energy is scanned. As the average heat load on the crystals is quite low, a side cooling or even intrinsic cooling scheme is not required.

However, for temperature stabilization the first crystals are mounted on a common water-cooled copper block. The second crystals are temperature stabilized via copper braids connected to this block.

To allow for experimental flexibility both beamlines must be able to operate in both monochromatic and broad bandwidth (pink) beam modes with the same beam path downstream of the monochromator. The switching from monochromatic mode to pink beam mode is accomplished in two different ways as sketched in the insets of Figure 6. Aramis-1 has to retract the crystals as well as the HRMs, passing the beam untouched through both vessels, whereas Aramis-2 has to retract the crystals and instead insert the OMs into the beam path. Both methods differ in the number of optical elements that are in the beam at a time. Aramis-2 has always two optical elements in the beam, not counting the retractable refocusing KB mirrors. Aramis-1 utilizes two optical elements in pink beam and six optical elements in monochromatic mode. This rather large number of optical elements became necessary to improve the spectral purity in monochromatic mode. It is accomplished with the help of the two HRMs directly downstream of the DCM-1. Both HRMs are mounted in a fixed distance but the height of the first HRM as well as the beam offset in the DCM are variable. When setting a new deflection angle both mirrors are rotated. Due to the fixed separation of both mirrors, the height of the first HRM and the beam offset must be set to keep the beam with the new deflection angle in the center of the second (fixed height) HRM.

A pair of retractable KB mirrors is foreseen for refocusing in both beamlines. They combine achromaticity with a high throughput. The KB mirrors with 500 mm optical length and B_4C/Mo-coating are bendable and operate with two sets of deflecting angles. Above 4 keV a deflection angle of 8 mrad improves reflectance and below 10 keV a deflection angle of 12 mrad increases the acceptance. By this an acceptance of more than 5σ can be achieved down to 2.5 keV decreasing to 3.5σ at 1.8 keV. The surface quality of the mirrors must be extremely good to reflect the FEL beam without deteriorating the wavefront even at the shortest wavelengths. According to the Maréchal criterion, the maximum allowed rms-profile error σ in a beamline with N mirrors and grazing incidence angles θ must not be larger than $\lambda/(2 \times 14\sqrt{N}\sin\theta)$. This condition must be met over the illuminated length of the mirror. As the beam divergence is proportional to the photon wavelength, the central part of the mirrors must have the highest quality while its outer parts are only illuminated at longer wavelengths where the Maréchal criterion tolerates larger surface errors. Figure 7 shows the maximum allowed profile error as a function of the beam footprint, i.e., the length of central mirror part, for a KB mirror in the Aramis-2 beamline.

The beamline performance was evaluated with PHASE, a computer code for physical optics simulations. The results for the Aramis-1 beamline are summarized in [93]. Spot sizes below 1 μm and peak power densities of up to 10^{21} W/m^2 can be expected. Due to the larger distance to the focal spot from the KB mirrors, the Aramis-2 beamline has a slightly larger spot size with reduced power density.

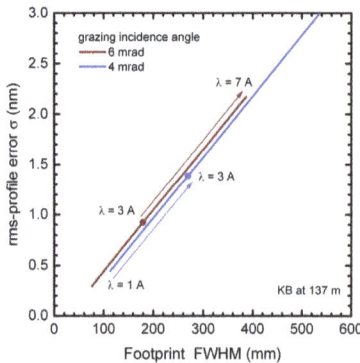

Figure 7. Required surface quality of the Kirkpatrick-Baez (KB) mirrors.

3.2. Pulse Picker

SwissFEL operates at a repetition rate of 100 Hz. For some experiments it is desirable to have only half or quarter of the base repetition rate or even any arbitrary pulse pattern. The SwissFEL accelerator is capable of producing any pulse pattern by controlling the gun laser but this is undesirable for two reasons: (1) The optimal stability and tuning of SwissFEL is achieved at 100 Hz repetition rate and (2) a non-standard repetition rate would affect other beamlines like the future Athos soft X-ray branch. Therefore an X-ray pulse picker has been designed and developed together with Dynamic Structures and Materials [94]. The pulse picker can operate at a continuous rate of 50 Hz to allow the selection of every other X-ray pulse. Furthermore it can generate any desired pulse pattern. The pulse picker blades, made of tungsten, are individually mounted which allows them to be exchanged in case of damage over time due to ablation by the mJ-level X-ray pulses.

The shutter mechanics are UHV compatible and are installed in a dedicated chamber, which is mounted on a translation stage based on the same design as for the photon backscattering monitor (PBPS, see Section 3.3). The mechanical mount also acts as a heat conductor to transfer the heat from the shutter to the chamber. The shutter is equipped with a Type-K thermocouple and measurements have shown that the temperature of the shutter does not exceed more than 40 °C over ambient temperature, without active cooling and at 50 Hz continuous operation.

3.3. Photon Diagnostics

Photon diagnostics for FELs is a new research field that has arisen as a response to the development and increased use of FELs. The SASE process used to generate the X-ray FEL pulses leads to changes in the characteristics of the beam on a shot-to-shot basis. The complicated structure of the machine and the many variables one has to oversee in the experiment are also prone to drifts that need to be measured and controlled. Data collected by researchers needs to be correlated with the intensity, position, spectral, and temporal properties of the FEL pulses to yield a clear picture of the effect that is being observed.

To facilitate the better operation of the machine, and to help the SwissFEL users better be able to use the full capabilities of the photon beam, PSI has implemented a full online photon diagnostics suite meant to measure every property that a researcher might need on a shot-to-shot basis. Wherever possible, the instruments were made for non-destructive measurements in the energy region the experiments are to take place. The devices installed at SwissFEL are presented here, and their capabilities are discussed.

3.3.1. Position and Intensity Diagnostics

The first photon diagnostics devices downstream of the undulators and the beam dump is the gas-based photon beam intensity and photon beam position monitor (PBIG and PBPG). This dual-purpose device, developed by the photon diagnostics group at DESY for hard X-ray FEL pulse characterization [95,96], uses a gas-filled ionization chamber and sets of split electrodes and multipliers to measure the position of the beam and its absolute and relative intensity on a shot-to-shot basis. The gas-based nature of the device leads to non-destructive diagnostics for photon energies between 25 and 12,000 eV, with the transverse position of every X-ray photon pulse position being measured to an accuracy of 10 μm, and the beam flux measured to a relative accuracy of about 1%, and an absolute value accuracy of about 10%. The gas-based detector is always on, but is located before any optics or mirrors in the beamline. The position and flux values measured before the optics section are not the same as the values at the experimental stations, since the intensity and position of the photon beam is affected by the beamline optics. However the position and flux values of the photon pulses at the experimental stations can be easily calculated when the data from the detector is combined with the X-ray transmission values of the beamline.

The next device downstream of the gas detector is the photon backscattering monitor (PBPS). The device, similar to those developed at LCLS [97] and SACLA [98], use thin CVD diamond disks with thicknesses of 30, 50 and 100 μm to scatter a portion of the incoming light onto four photodiodes that are placed out of the path of the beam in a backscattering geometry (see Figure 8). In addition to the device downstream of the gas detector, SwissFEL also has a PBPS placed at the entrance to the KB mirrors before each of the experimental stations, and has the option of being placed behind the experiment. This device measures the relative flux to about 1% accuracy and the absolute position of the beam to about 10 μm accuracy. The thickness of the disks means that the transmission of the incoming X-ray beam is somewhat compromised, with about 15% of the intensity lost at 5 keV. However, the transmission is higher than 90% at photon energies higher than 6 keV, and reaches 95% at energies above 8 keV. Operating the device at energies lower than 4 keV raises the risk of damaging the diamond crystals used for the backscattering due to the amount of energy that would be deposited in them from the incoming X-ray pulses. To allow the PBPS to operate a lower photon energies the disks can be switched out for thinner targets (down to 10 μm) if required.

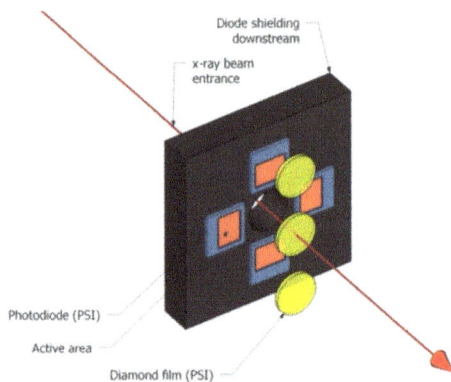

Figure 8. Schematic drawing of the photon backscattering monitor (PBPS) design.

The next device for profile and position measurement of the FEL beam is the photon profile monitor (PPRM). This is a Ce:YAG screen that scintillates when a photon pulse impacts it, with the image relayed to a 100 Hz camera via a mirror and a lens. This setup is destructive since no beam is transmitted through the device, though it can be used as an online diagnostic after the experimental stations where the experiment allows for this. Since this technique is common in both synchrotron [99,100] and FEL [101] facilities worldwide it does not need further discussion. The only unique features of the device are that it sports three different thicknesses of Ce:YAG crystals (30, 50, and 100 μm), and has an optical geometry developed at PSI [74] that has the mirror always being in the same spot relative to the camera, ensuring that any movement we see on the screen comes only from the motion of the beam itself and limiting the damage the mirror may suffer from the FEL beam.

Additional diagnostics are available for pre-SASE beam, mainly used for the setup and commissioning of SwissFEL. The photon diode intensity monitor (PDIM) is a simple Si PIN diode that can be inserted into the photon beam to measure the intensity of the spontaneous radiation before lasing is achieved to measure gain curves from different undulator configurations. The photon spontaneous radiation detector (PSRD) uses an MCP-and-screen setup to acquire profile images of the spontaneous radiation. It is meant to be used behind the Aramis-2 monochromator when the FEL is still emitting only spontaneous radiation to adjust the radiation to the right wavelength. The PDIM and PSRD are not meant for regular use with experiments, and are both destructive.

3.3.2. Temporal Diagnostics

The measurement of the arrival time of the X-ray pulse relative to a laser pulse is crucial for pump-probe experiments that are an important category of research performed at FELs. The monitoring of temporal lengths of X-ray pulses is similarly important for both experiments and for the optimization of the performance of the machine. PSI has installed several diagnostics to monitor and record both of these properties over a wide range of X-ray intensities and photon energies.

The first device is based on techniques developed at LCLS [102,103] and SACLA [104] that measure the arrival time of the FEL pulse relative to a pump laser using spectral encoding. The method works by using a chirped laser pulse that passes through a dielectric substance and with the change in transmission being measured. The wavelength corresponding to the change in transmission is directly linked to the arrival time of the FEL on the dielectric material due to the spectrally chirped nature of the laser beam. As the chirped laser pulse is derived from the same laser beam that is used for the experiment, the time-arrival information can be used to directly sort the experimental data. The photon spectral encoder (PSEN) is located a few meters before the experimental chamber and is projected to have an accuracy of measurement of the arrival time of 20 fs or better. It has a large number of thin membranes of different materials and thicknesses to allow for online measurement of the arrival time over a range of 1–1.5 ps without a significant loss of X-rays for the experiment. However, the device will require a minimum of 5–10% transmission losses to be able to measure the arrival times accurately.

The second device SwissFEL will have for temporal diagnostics is the photon arrival and length monitor (PALM), which uses THz streaking [7,105] to measure both the arrival time and pulse length of the incoming XFEL pulses relative to a THz beam generated from the same source as the experimental pump laser [106], as shown in Figure 9. The method has proven to work well with soft X-rays at FLASH [7] and with hard X-rays at SACLA [107], where the PALM demonstrated measured arrival time accuracies to a precision of about 5 fs [108]. The device measured pulse lengths with an HHG source down to 25 fs rms with accuracies typically between 4 and 10 fs [109]. The device is gas based, and is virtually non-destructive, with X-ray absorption typically being less than 0.1% of the incoming flux. However, unlike the PSEN with its large time window, the PALM has an acceptance window for arrival time measurement that is typically between 400 and 600 fs. If the X-ray/laser timing jitter is larger than this value the PALM requires additional timing information, for example from the PSEN, to be used. Future upgrades are planned to both increase the range of the PALM and to improve the time resolution of both the arrival time and pulse length measurements.

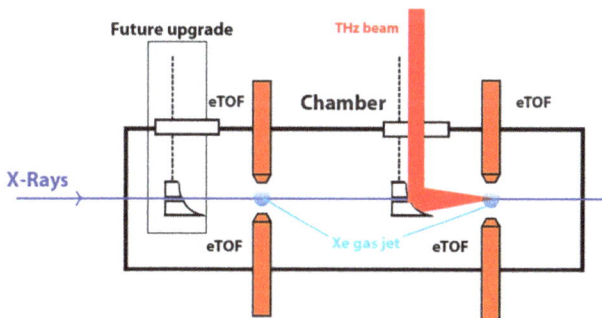

Figure 9. Schematic drawing of the photon arrival and length monitor (PALM) setup.

3.3.3. Spectral Diagnostics

As with the beam position, intensity, and temporal properties, the spectral content of the SASE beam changes on a shot-to-shot basis. Since both experimental data analysis and accelerator operation often benefit from high-resolution spectral information about the FEL beam, the SwissFEL photonics team designed a photon single-shot spectrometer (PSSS) [110]. The PSSS is meant to non-destructively measure the hard X-ray photon spectrum on a shot-to-shot basis.

The basic operating principle of the PSSS is simple: the FEL beam passes through a thin diamond transmission grating that splits off the first order light and illuminates a bent crystal spectrometer downstream. The spectrometer is meant to function at photon energies above 4 keV, covering a 0.5% bandwidth with a $\Delta E/E$ resolving power of 10^{-5} to 10^{-4}. The device has been successfully tested in its prototype phase at LCLS [111], and it will be fully operational at SwissFEL. Further concepts for spectral measurements at longer wavelengths are being developed with similar properties, and will be implemented as they are commissioned and fully developed.

3.4. Experimental Laser

One of the unique advantages of an FEL is the ultrashort pulse duration in the X-ray range. Therefore a large number of experiments are time resolved. The SwissFEL experimental laser facility [112] provides lasers for both pump-probe experiments, as well as for certain specialized diagnostic techniques which provide feedback on FEL operation (see Section 3.3.2). The critical parameters are availability, performance, and stability. The laser systems are based on a commercial 100 Hz Ti:Sapphire amplifier system from Coherent. An enhanced diagnostics system ensures the long-term stability of the laser. The laser output is combined with nonlinear conversion stages and allows access to a broad spectral range, including UV, visible, near-IR, mid-IR, and THz, as well as the generation of ultrashort (<10 fs) pulses. In addition to the femtosecond laser systems a tunable (UV to IR), ns pulse-duration laser system which will be fibre-coupled to the experimental stations is also planned to allow experiments to take advantage of the short XFEL probe pulses, for example using "diffract-before-destroy" techniques [22,113,114], but with more efficient optical excitation resulting from the longer duration excitation pulses. The accessible timescales from ns to ms with this laser system are especially relevant for dynamics in proteins [115–119].

The laser infrastructure is distributed over two floors. The two identical laser amplifiers are located in a dedicated pump laser room (LHx), on the floor above the experimental stations (Figure 10), while the nonlinear conversion stages are installed on optical tables inside the experimental hutches, close to the experiment. This layout allows the simultaneous and independent operation of both end stations, and, in case of a failure of one system, the beam from the remaining laser can be redirected to both experimental stations and assure uninterrupted operation.

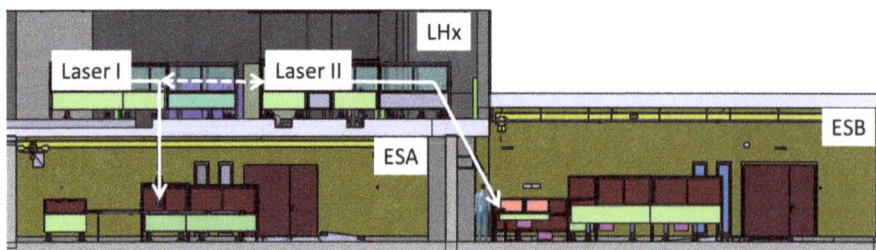

Figure 10. Laser I and Laser II are located in the laser lab LHx, on the floor above the two experimental stations Alvra (ESA) and Bernina (ESB).

The uncompressed amplifier output is sent through an evacuated transfer line from LHx to ESA and ESB (Figure 11). The pulse can be compressed to <30 fs and has a total pulse energy of 20 mJ. Arriving on the optical table inside the end station, the laser beam is split into two branches, one for diagnostics and one for the pump-probe experiment. Each branch is equipped with its individual optical compressor. Normally the available laser energy is split equally between the two branches. The diagnostics branch is used to operate the pulse arrival time and length monitor (PALM) [107,108], as well as the spectral encoding (PSEN) [102,120] for X-ray arrival time measurements (see Section 3.3.2). The major part of the diagnostics branch laser energy is used for the THz source to operate the PALM. It is based on the tilted pulse front scheme in $LiNbO_3$ [121] with a typical field strength of 100 kV/cm, centered at 0.5 THz [107]. The rest of the laser energy is used for chirped white light generation for the operation of the PSEN [102,122].

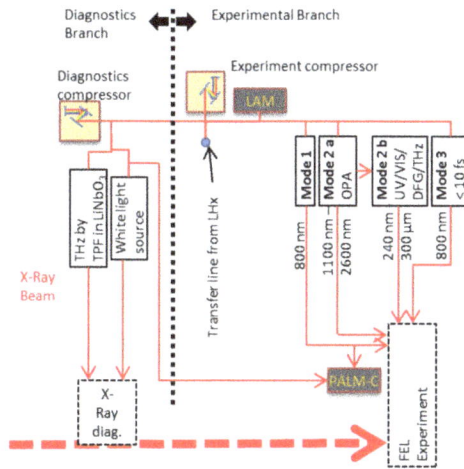

Figure 11. Available operation modes for X-ray diagnostics and FEL experiments. From [112].

For user operation the following modes are available: Mode 1 is the standard compressed output of the laser (10 mJ, <30 fs, 800 nm), which, when used in combination with an optical parametric amplifier (OPA), can generate wavelengths from 1100 to 2600 nm with up to 2 mJ pulse energy (Mode2a). Mode 2b combines the OPA output with subsequent conversion modules and extends the spectral range to the visible and UV (NirUVIS), to the IR (NDFG up to 15 µm) and to the THz range (1–10 THz, depending on the crystal used [123–125]). Mode 3 is the short-pulse option, delivering <10 fs pulses with >200 µJ. It is based on a hollow core fiber compressor [126]. For the first user experiments, Mode1, Mode2a and the UV/VIS module will be available at ESA and Mode1, Mode2a, DFG, THz, and <10 fs will be available for ESB. In the future all SwissFEL experimental stations will have access to all laser operation modes, with additional modes currently under development. The operation modes available for the user operation on the experimental branch are summarized in Table 2.

A laser diagnostics system will inform the user about the typical laser performance, such as energy, pulse duration, beam pointing, spot size, and spectrum. For the visible and near infrared range up to 1100 nm the corresponding data can be collected single shot and beam synchronous and allow a complete reconstruction of the experiment from the laser side.

Table 2. Wavelength-dependent laser performance using the optical parametric amplifier (OPA) (mode 2a and 2b), typically pumped with 30 fs, 8 mJ pulses.

Operation Mode (Module)	Wavelength Range	Output Energy	Output Pulse Duration
2b (NirUVis)	240–295 nm	>26 μJ at peak	<3 × pump pulse width
2b (NirUVis)	290–480 nm	>40 μJ at peak	1.2–2 × pump pulse width
2b (NirUVis)	475–533 nm	>466 μJ at peak	1–1.5 × pump pulse width
2b (NirUVis)	533–600 nm	>306 μJ at peak	1–1.5 × pump pulse width
2b (NirUVis)	600–1160 nm	>320 μJ at peak	1–1.5 × pump pulse width
2a	1160–2600 nm	>2000 μJ at peak	1.2–1.5 × pump pulse width ≤1550 nm <2 × pump pulse width >1550 nm
2b (NDFG)	2.6–9 μm	>22 μJ @ 4 μm	<3× pump pulse width
2b (NDFG)	9–15 μm	>10 μJ	n.a.

The two laser systems are locked to the optical reference timing distribution in several locations. In the final setup the two oscillators will be optically synchronized to the reference timing. For this stabilized optical links are installed between the timing hutch and the LHx, as well as the end stations (see Section 5.1). This will allow a timing jitter <10 fs rms between reference and oscillator and drifts of <10 fs over 24 h to be achieved. Nevertheless the optical path between the oscillator and the experiment is exposed to environmental changes, such as pressure, humidity and temperature. This will lead to drifts between FEL pulse and optical laser on the order of 100 fs and more within a few hours [127] at the experiment. The laser arrival time monitor (LAM) measures these drifts relative to the reference timing system and can be used for drift compensation and data binning [127,128]. For the same reasons drifts on the order of 100 fs over several hours are expected between the diagnostics branch and the experimental branch, though both are located within the same hutch environment. To compensate for this a reference pulse is guided through an evacuated beam pipe along the FEL beam from the diagnostics branch and the timing drift with respect to the experimental branch is measured by a balanced cross-correlator setup (PALM-C). This provides a correction value to the arrival time measured by PALM and increases the accuracy for long-term measurements.

4. Experimental Stations

The layout of the SwissFEL experimental hutches is shown in Figure 12. The Athos soft X-ray experimental area is contained within one large hutch with a floor space of 692 m^2. Space within this hutch is allocated for the optical laser. The Aramis hard X-ray experimental hutches have a combined floor space of 523 m^2 in three separate experimental hutches. The three experimental stations are called Alvra, Bernina, and Cristallina. Both Alvra and Bernina will be ready for users in 2018, with Cristallina to be installed in Phase II concurrently with the Athos installation. The optical laser room is located directly above the first experimental hutch and delivered to the experiment using vacuum transfer-lines (see Figure 10). Further details on the experimental lasers can be found in Section 3.4. In addition to the experimental areas there is lab space allocated in the SwissFEL building for both biological and chemical sample preparation, and a large work area for experimental testing and assembly. Users will also have access to the lab facilities at PSI, including the sample crystallization and characterization facilities at the Swiss Light Source for protein crystallography experiments [129]. In the following sections we will provide details on the two hard X-ray experimental stations, Alvra and Bernina, which will begin user operation in 2018.

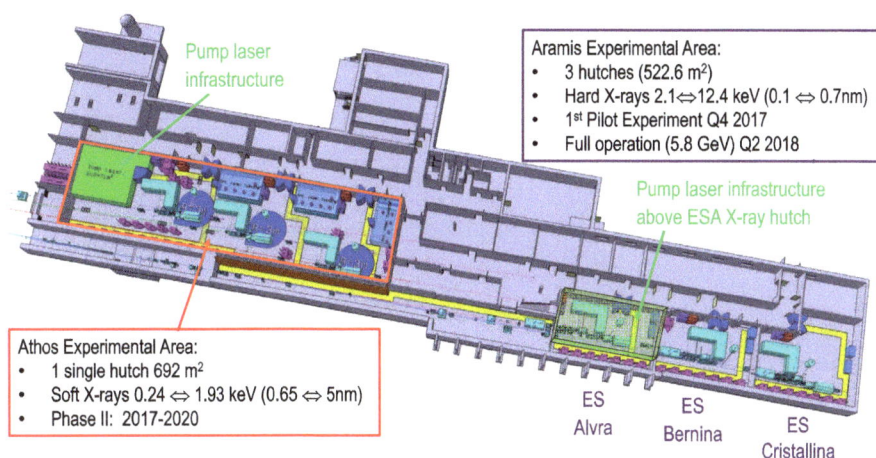

Figure 12. The experimental areas of SwissFEL. The accelerator is to the left of the above area, with the X-rays travelling from left to right.

4.1. Experimental Station Alvra

The layout of the Experimental Station Alvra (ESA) hutch is shown in Figure 13. Alvra is focussed primarily on two techniques: X-ray spectroscopy [18,130] and Serial Femtosecond Crystallography (SFX) [131,132]. X-ray absorption spectroscopy (XAS) involves measuring the X-ray transmission or X-ray fluorescence of a sample as a function of incident monochromatic X-ray energy [16]. These measurements can provide information on the local electronic and geometric structure around the absorbing atom [130] and can be applied to ordered or disordered samples in almost any form, including species in solution or solid state samples. SwissFEL will be particularly suited for these experimental techniques due to its variable-gap undulators [133,134] (see Section 2.4) which can easily scan the X-ray energy of the XFEL over a very wide range, allowing techniques such as X-ray absorption near-edge structure (XANES) and extended X-ray absorption fine structure (EXAFS) to be used [130]. SFX [19] is an XFEL technique that has been developed for protein crystallography where a stream of micrometer-sized crystals is delivered into the focussed X-ray beam using a range of different injector techniques [135–137], and the diffraction pattern from each crystal is recorded on a large two-dimensional pixel detector. Though the intense XFEL pulse destroys the crystal, because an ultrashort X-ray pulse (<50 fs) is used the diffraction occurs before the atoms have time to move from their lattice positions [22,138]. SFX can resolve room-temperature protein structures to better than 2 Å resolution on very small crystals [19,131], which expands the technique to include samples that are difficult to crystallize, such as membrane proteins [139] and 2D protein crystals [140,141]. A complementary technique with similar technical requirements to SFX that will also be possible at Alvra is wide-angle X-ray scattering (WAXS, also called X-ray diffuse scattering or XDS), where the sample under investigation is a liquid solution [142]. The result of the X-ray scattering measurement is powder-diffraction-like rings, containing information on the pair-distribution function of the sample [143,144]. This technique has proven particularly useful in measuring large-scale, light-activated functional protein motions [145–148]. ESA has the additional capability of performing X-ray emission spectroscopy (XES) which uses X-ray diffraction from an analyzer crystal to measure the scattered or fluorescence X-ray photons with high energy resolution [149]. ESA uses short focal length crystals (25 cm) in a dispersive von Hamos geometry [150] to measure a range of XES energies in a single measurement. This spectrometer can also be used for a variety of other scattering measurements include off-resonant techniques [151–154] and inelastic X-ray scattering (IXS) [155].

The ability to measure both X-ray scattering and spectroscopy simultaneously has proven to be a powerful combination for resolving structural and electronic dynamics in both molecules and proteins [115,156–158].

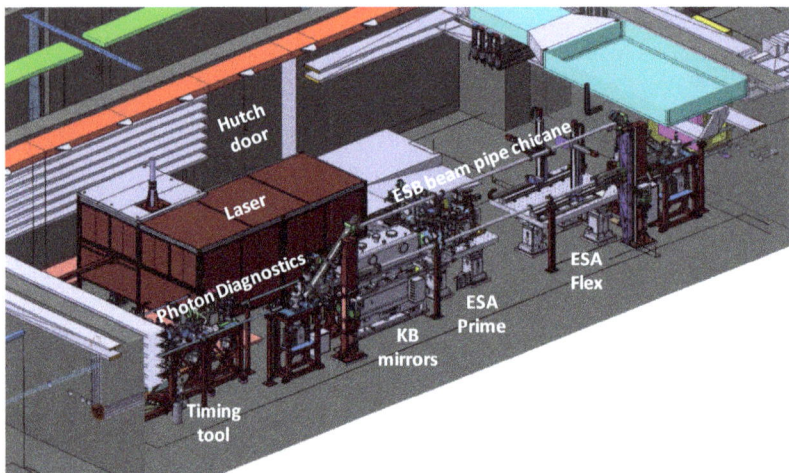

Figure 13. The Experimental Station Alvra X-ray hutch. The X-rays come from the Optics hutch, to the left of the schematic and move from left to right in the figure. The various components located in the hutch are labelled. The ESB beam pipe chicane is a motorized beam pipe that can be moved vertically to either allow the XFEL beam to be delivered to Bernina, or to allow easy access to the Alvra instruments. The same motorized elements will be used in a similar fashion for an ESC beam pipe chicane in the future.

4.1.1. X-ray Optics and Diagnostics

A detailed summary of the X-ray optics is given in Section 3.1 and of the X-ray diagnostics in Section 3.3, here we will focus on the aspects specific to Alvra. The layout of the Aramis-1 beamline is shown in Figure 14. The key components are two horizontal offset mirrors that work at incidence angles of 6 mrad, for a total deflection angle of 12 mrad, followed by a fixed-exit double-crystal monochromator which deflects the beam vertically, and two harmonic rejection mirrors which return the beam vertically to its original beam path. This allows both monochromatic and pink beam paths to be identical downstream of the optical elements (see insets A and B in Figure 6). The monochromator contains 3 crystal pairs, which include Si(111), Si(311), and InSb(111) to cover the full Aramis photon energy range with varying bandwidth and energy resolution. The final optical components are two KB mirrors, with a working distance of 1.5 m from the center of the last mirror. These achromatic focusing optics are capable of achieving a 1.5 μm focal spot (FWHM). The optics have been designed to be used over the full range of the Aramis X-ray energies: 1.77–12.4 keV.

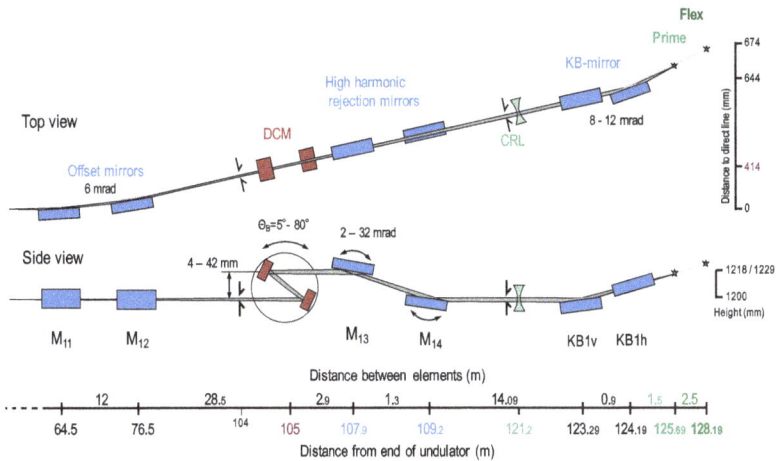

Figure 14. Schematic layout of the X-ray optics at the Aramis-1 beamline. Note the compound refractive lenses (CRLs) are a possible future upgrade.

One crucial aspect of the Aramis-1 beamline is to allow for measurements in the 2–5 keV energy range. At these photon energies the optics will all function flawlessly at the higher harmonic energies as well, providing no discrimination for these photons. The HRM optics allow the incidence beam angle to be tuned, allowing the harmonics to be greatly suppressed when the monochromator is used. Calculations for the HRMs indicate we can achieve a contrast of at least 10^{-3} for these higher photon energies, with up to 10^{-5} achievable under certain optics configurations [88]. Note that the expected harmonic contribution from the XFEL is ~1% for each subsequent odd harmonic, which could result in up to 10^9 photons per pulse in the 3rd harmonic if no suppression is used.

The photon diagnostics described in Section 3.3 have all either been tested with or designed for X-ray photon energies above 4 keV. This implies the development of these diagnostics components in the 2–4 keV range will need to be evaluated once SwissFEL is operational. Several of the elements are capable of lower energy measurements, including the gas-based photon beam intensity and position monitor (PBIG and PBIM). The solid photon backscattering monitors (PBPS) can be used with thinner scattering films, such as 200 nm of Si_3N_4, which will allow them to be used at lower photon energies.

4.1.2. ESA Prime and Flex

The techniques introduced in Section 4.1 will be applied at two instruments, which are located in line with the X-ray beam: ESA Prime and ESA Flex (see Figure 15). Due to the bendable KB mirrors the X-rays can be focussed at either instrument, with the minimum focus of 1.5 µm achieved at ESA Prime. Both instruments can be used with the optical laser for pump-probe experiments, with an anticipated time resolution of better than 50 fs [108].

ESA Prime is chamber that can be operated under vacuum, He, or neutral atmosphere and combines a large 2D 16 M JUNGFRAU scattering detector [159–161] and a dual-crystal von Hamos X-ray emission spectrometer [150,162,163]. This allows experiments to be performed using both scattering and emission techniques simultaneously, which has proven to be a powerful combination for molecular [156,157] and protein [115,158] samples. The chamber has the possibility of using different types of sample injectors, including several specifically for SFX sample delivery [135,164]. The expected achievable resolution of the crystallography measurements at 12.4 keV is better than 1.5 Å ($Q_{max} = 7$ Å^{-1}) and the X-ray spectrometer will be capable of measuring the full photon energy range of SwissFEL resonantly, with non-resonant measurements from 1–2 keV. This energy range is shown in Figure 16 with the elements labelled and the available analyzer crystals shown (see Table 3).

The Bragg angle range covered by the spectrometer is from 40° to 80°. Thanks to the dual-crystal design this spectrometer is capable of measuring multiple signals simultaneously, for example the Kα and Kβ X-ray emission from a 3*d* transition metal. The detector used for these measurements is a 4.5 M JUNGFRAU, which consists of 9 × 0.5 M modules in a linear geometry with an area of 4 × 72 cm. This allows the detector to cover all possible Bragg angles without requiring motion of the detector. The regions of interest containing the X-ray signals can be read out without reading out the entire 4.5 M detector. Due to the excellent noise characteristics of the detector, this spectrometer will be capable of single-photon sensitivity below 2 keV when operated in a high gain mode [159,165].

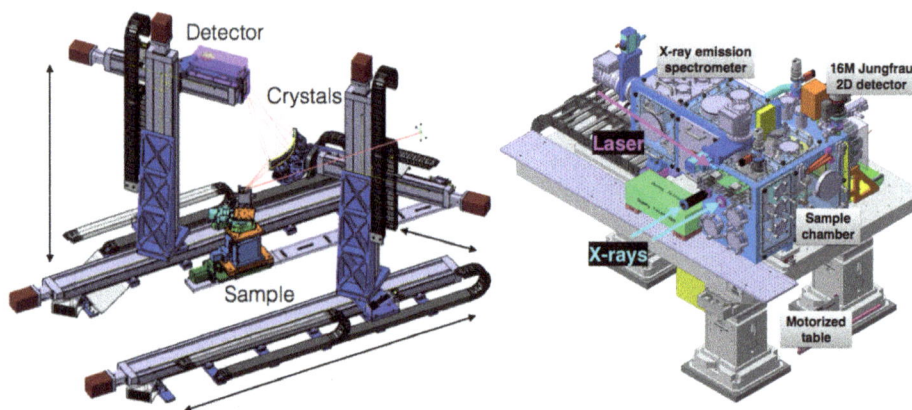

Figure 15. The instruments installed at Experimental Station Alvra (ESA). The ESA Flex instrument (**left**) is a flexible X-ray spectrometer that can be positioned according to the experimental requirements. The ESA Prime instrument (**right**) is a combined scattering and spectroscopy chamber designed to allow experiments in neutral, vacuum, or He atmosphere. ESA Prime is focussed on allowing experiments to be performed in the 2–5 keV tender X-ray regime.

The scattering or serial femtosecond crystallography experiments benefit greatly from the per-pixel dynamic gain switching of the JUNGFRAU detector, which has 3 levels of gain, resulting in a dynamic range of 10^4 photons at 12 keV. This gain switching occurs automatically, allowing the detector to handle transparently the broad range of incident intensities expected at SwissFEL. Further details on the JUNGFRAU detector can be found in Section 4.3. The detector is mounted on the back flange of the ESA Prime chamber, and has two holes to allow the beam to pass through the detector. This will allow the detector to be positioned in two scattering geometries: one where the beam is centered on the detector, to allow for fully symmetric scattering measurements, and one where the beam is close to one edge of the detector, which increases the scattering range of the detector, while simultaneously increasing the maximum accessible Bragg angle of the von Hamos spectrometer. The minimum sample-to-detector distance is fixed at 10 cm.

ESA Flex is a flexible instrument that allows users to build up the experiment as required. It is mounted on a motorized table, allowing user-supplied chambers to be installed for the measurement. ESA Flex also includes a configurable X-ray spectrometer that can be mounted in a variety of positions to measure a range of scattering angles, in both vertical and horizontal geometries. The 3-crystal spectrometer can be used from 40° to 85° Bragg angles, and uses a 1.5 M JUNGFRAU pixel detector. When used with large Bragg angles (>85°) and segmented X-ray crystals [150] this spectrometer is capable of 100 meV energy resolution. ESA Flex is shown on the left of Figure 15.

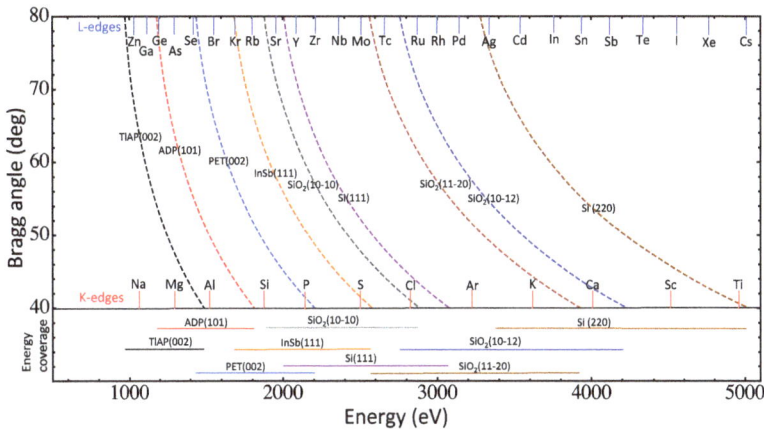

Figure 16. The tender X-ray energy range covered by the ESA Prime X-ray spectrometer. The dashed lines are the various crystals available for use with the spectrometer (see Table 3), covering the full range of energies at the various Bragg angles available in the spectrometer. The energies of various X-ray emission transitions are also marked to provide a sense as to what elements can be measured in this photon energy range.

The crystals available for both von Hamos spectrometers are listed in Table 3. As noted in the caption of Table 3 the crystals are either curved or segmented. In general the curved crystals provide excellent focusing capabilities (2 × X-ray spot size on the sample in the focusing direction), with some loss of energy resolution due to the crystal curvature, while the segmented crystals provide focal spots of 2 mm (2 × 1 mm segment size) but much improved energy resolution (essentially the Darwin width of the crystal reflection). The contributions of the various components in the spectrometer to the final energy resolution, including X-ray focal spot, detector pixel size, and geometry Bragg angle, can be found in reference [150].

Table 3. List of von Hamos geometry spectrometer crystals available for use at either ESA Prime or ESA Flex. Note that higher-order reflections have been omitted for clarity. Curved means smooth crystals, segmented mean diced along the focusing axis, but perfect flat crystals along the dispersive axis [150].

Crystal	Miller Indices	2d Spacing	Radius of Curvature	Type	Crystal Area
TlAP	002	12.95 Å	25 cm	Curved	5 × 10 cm
ADP	101	10.64 Å	25 cm	Curved	5 × 10 cm
PET	002	8.742 Å	25 cm	Curved	5 × 10 cm
InSb	111	7.4806 Å	25 cm	Curved	5 × 6 cm
SiO_2	$10\bar{1}0$	6.687 Å	25 cm	Curved	5 × 10 cm
Ge	111	6.532 Å	25 cm	1 mm segments	5 × 10 cm
Si	111	6.271 Å	7 and 25 cm	1 mm segments	5 × 10 cm
SiO_2	$11\bar{2}0$	4.912 Å	25 cm	Curved	5 × 10 cm
SiO_2	$10\bar{1}2$	4.564 Å	25 cm	Curved	5 × 10 cm
Ge	220	4.0 Å	25 cm	1 mm segments	5 × 10 cm
Si	220	3.840 Å	25 cm	1 mm segments	5 × 10 cm
Si	311	3.274 Å	25 cm	1 mm segments	5 × 10 cm
Ge	400	2.829 Å	25 cm	1 mm segments	5 × 10 cm
Si	400	2.714 Å	25 cm	1 mm segments	5 × 10 cm
Si	331	2.4916 Å	25 cm	1 mm segments	5 × 10 cm
Si	531	1.836 Å	25 cm	1 mm segments	5 × 10 cm

4.1.3. Optical Laser System

The femtosecond optical lasers available to the experimental stations have been described in detail in Section 3.4. The primary short-pulse excitation source anticipated for ESA is the 240–2600 nm output from the OPA (see Table 2), which covers the UV to IR range in which most molecular or biological samples absorb light. In addition to the femtosecond laser, ESA will also have access to a long pulse laser, which is tuneable from the UV to the IR. This laser will operate at 100 Hz, and can produce >1 mJ/pulse from 200 nm to 2 μm with pulses from 3–6 ns in duration. This laser can be used to more efficiently excite samples and measure on timescales ranging from ns out to ms. These experiments will take advantage of the fact that SwissFEL will have more photons available in a single pulse (>10^{11}) than a 3rd-generation storage ring in 1 ms ($\sim 10^9$), making them better at time-resolved experiments even on longer timescales, under certain conditions. The approach of using long-pulse excitation has proven particularly attractive for measurements on biological samples, where the relevant timescales can often be slow (μs to ms), but when performed at XFELs the measurements can take advantage of the room-temperature, damage-free conditions [115,117–119].

4.2. Experimental Station Bernina

The Experimental Station Bernina (ESB) is designed to pursue femtosecond (fs) time-resolved (tr) hard X-ray pump-probe (XPP) diffraction and scattering experiments in condensed matter systems [166]. Its main emphasis is on the electromagnetic response in correlated crystalline materials, but it is sufficiently flexible to host a variety of hard X-ray experiments that benefit from short pulses. Frequently in correlated systems many of the energy scales such as charge-transfer energy, magnetic exchange, Jahn-Teller splitting, hopping integral or Hubbard interaction, are of similar size ($\sim$0.15–4 eV). The ground states obtained by minimizing the total energy include ferro- and antiferromagnetism (FM, AFM), multiferroics, charge and spin density waves (CDW, SDW), high-T_c superconductors (SC) and topological insulators. Due to the fine balance of all the energy scales new electronic and magnetic phases emerge with competing ground states. Small changes of external conditions such as temperature, pressure, magnetic or electrical fields, and carrier doping, can cause switching of ground states and large changes in functional properties. A new avenue is to dynamically excite the system with photons employing pump-probe photon-in and photon-out techniques. The spin, charge, orbital and lattice degrees of freedom can be coupled via electron-phonon and spin-lattice interactions. One way to understand the competing interactions and to clarify cause and effect in such coupled systems is to follow their different response in the time domain by selective pumping and probing of specific modes using tailored pulses, properly matched in time structure, wavelength, polarization and intensity. This allows new type of pump-probe experiments: selective excitation and probing of low energy electronic, magnetic and structural dynamics; photo-induced (non-thermal) phase transitions away from equilibrium (such as new metastable states and light induced symmetry breaking); coupling, control and switching in multiferroic systems; correlations and fluctuations in non-equilibrium systems. In a pump-probe experiment the time response—which for coherent excitation is oscillatory followed by dephasing and relaxation—is measured by precisely varying the time distance between pump and probe. The pump pulses are derived from an optical fs laser system. The sample is probed by fs X-ray pulses in SASE mode employing time-resolved resonant and non-resonant X-ray diffraction (tr(R)XRD), diffuse scattering (trDS), and resonant inelastic X-ray scattering (trRIXS).

4.2.1. Aramis Beamline

Bernina is installed on the Aramis-2 beamline of SwissFEL [167]. The schematic of the optical layout is shown in Figure 17 and the engineering layout of the ESB hutch is shown in Figure 18. The beamline covers (2.7) 4.5–12.4 keV in the tender to hard X-ray range of the fundamental, which covers the absorption energies of the 3d K-edges, 4f L-edges and 5d L-edges. To cover the

4d L-edges, which occur at tender X-ray energies (2–5 keV), virtually windowless operation is foreseen where the X-ray been will be delivered entirely through vacuum up to the sample position. The SwissFEL Aramis modes of operation are described in Section 2.1. At 7 keV the flux and pulse length for the two operation modes are expected to be around 1.3×10^{12} ph/s/0.1% bw with 5 fs FWHM (10 pC, 100 Hz), and 2.1×10^{13} ph/s/0.1% bw with 50 fs FWHM (200 pC, 100 Hz). There is the plan to operate Aramis in self-seeding mode [168] in the future.

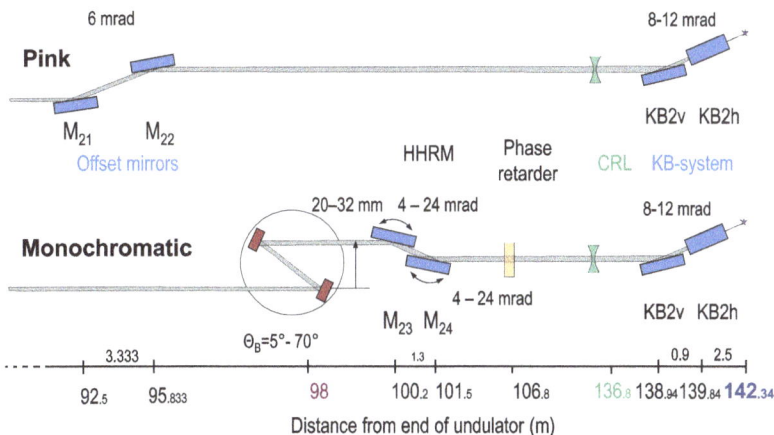

Figure 17. Schematic layout of the Aramis-2 X-ray optics for pink beam (**top**) and monochromatic beam (**bottom**).

Figure 18. The Experimental Station Bernina X-ray hutch. The various components located in the hutch are labelled.

4.2.2. X-ray Optics

A detailed summary of the X-ray optics is given in Section 3.1, here we will focus on the aspects specific to Bernina. The layout of the X-ray optics [88] is shown in Figure 17. For pink beam a pair of bendable plane elliptical mirrors (offset mirrors, coating SiC/B_4C, Si, Mo/B_4C) installed in the optics hutch (OH) deflect the beam vertically by 6 mrad. For monochromatic beam the offset mirrors are retracted. In this case the double crystal monochromator (DCM) is the first optical element in the beam. A motorized horizontal translation allows switching between Si(111), Si(311) and InSb(111) crystal pairs.

The DCM has a variable offset (20–32 mm) to allow for future operation in combination with harmonic rejection mirrors (HRM, coating B_4C/SiC) with variable deflection angles in a 4-bounce scheme.

Installed in the ESB hutch at working distance 2.6 m upstream of the sample position, a pair of bendable KB mirrors provide achromatic focusing with horizontal and vertical spot size of (1)2–100 μm FWHM for (ideal) real performance of the focusing optics. Upstream of the KB mirror pair free space is allocated to optionally install a movable in-vacuum Be compound refractive lens (CRL) transfocator for in-line focusing or to install an in-line electron time-of-flight (eTOF) polarization monitor to allow online monitoring of the incident polarization once a phase retarder is installed.

4.2.3. Phase Retarder

There is the option to install an in-vacuum double X-ray phase-retarder (XPR) setup in the OH hutch downstream of DCM-2 (see Figure 6). Circular and arbitrary linear polarization can be generated for X-rays with energy above 4 keV using diamond quarter-wave plates and half-wave plates, respectively. When operated in Bragg/Laue transmission geometry such an XPR can provide flexible polarization control with a high degree of polarization ($\geq$90%) [169,170].

4.2.4. Photon Diagnostics

Multiple X-ray diagnostics both upstream and downstream of the sample position are available including profile-intensity monitors, intensity-position monitors, and X-ray timing diagnostics (see Section 3.3 for further details on these diagnostics). The PALM and PSEN components (see Section 3.3.2) are installed in the ESB hutch upstream of the diamond/Si solid state attenuators. The attenuator stacks are available to tune the incident X-ray flux to the desired level. By taking advantage of all the available timing diagnostics (in order of distance from the experiment: BAM, LAM, PALM, PSEN) will allow pump-probe measurements with time resolution $\leq$50 fs FWHM. The BAM and LAM diagnostics are independent of X-ray intensity, allowing them to be used with all modes of the accelerator.

4.2.5. Optical Laser System

The femtosecond laser system available for experiments at Bernina is described in detail in Section 3.4. For pump-probe experiments several excitation options (UV/NIR/FIR) are available [112]. An optical parametric amplifier (OPA) with subsequent difference frequency generation covers the spectral range 1100–15,000 nm. This output can also be used to generate intense THz pulses in organic crystals with field strengths exceeding 1 MV/cm [123] and pulse energies up to 10 μJ in the frequency range 1–10 THz. Very short pulses (<10 fs) are available at 800 nm by pulse compression in a gas filled hollow core fibre [126]. To compensate for drifts in the amplifier system, a laser arrival time monitor (LAM) will be installed directly after the compressor. A variety of laser diagnostics will provide all relevant laser parameters for the user.

4.2.6. Laser In-Coupling

Downstream of the KB mirror pair the laser in-coupling (LIC) section is installed to facilitate pump-probe experiments under ambient conditions, nitrogen or helium atmosphere or in vacuum when HV/UHV conditions are required. Two vacuum chambers are mounted on a heavy load hexapod: the LIC diagnostic chamber, which is permanently installed, and the LIC mirror chamber which can be moved to clear space for in-air laser optics when experiments are performed at ambient conditions. The diagnostic chamber houses clean-up slits, an intensity-position monitor (PBPS) and a spectral encoding timing tool (PSEN). A motorized laser mirror assembly is mounted in the mirror chamber to allow for both collinear and non-collinear laser excitation geometries. The chambers are separated by valves. A diamond window of thickness 50 μm is mounted in the vacuum valve at the exit of the diagnostic chamber to pass the X-ray beam when experiments are performed in-air at atmospheric pressure.

4.2.7. Experimental Instruments

The layout of the experimental X-ray hutch is shown in Figure 18. The endstation design emphasizes rapid reconfiguration capability in terms of flexible sample environment, goniometers, spectrometers and detector geometries to support a wide variety of experiments. There are two instruments operated at a single focal position, the general purpose station (XPP-GPS) and the six-circle X-ray diffractometer (XPP-XRD). These stations are mounted on rails, which enables them to be easily moved in and out of the X-ray focus. Sample stages designed as single modules can be swapped and mounted at both stations. Heavy load capability allows the stations to support and precisely align a variety of custom built 'modules' such as spectrometers, analyzers and various sample environmental setups provided by an experimental team or user group. Sample modules would include in-air goniometer stages, HV/UHV vacuum chambers for temperatures below 20 K, cryostats for super-conducting high field magnets or cryogenically cooled pulsed magnet systems to study H-induced transient states [171], and a cold finger cryostat with a high pressure diamond anvil cell.

4.2.8. General Purpose Station

The XPP-GPS station is a general purpose station for XPP experiments in non-scanning mode. It consists of a non-magnetic heavy load sample goniometer with six degrees of freedom, a rotary table to mount spectrometers or detectors, and a large robot detector arm carrying a 16 M JUNGFRAU pixel detector. The JUNGFRAU pixel detector, which has been developed by the SLS detector group [172], is a charge integrating detector with pixel size $75 \times 75\ \mu m^2$. It has variable gain switching with dynamic range 10^4 (12 keV) that allows measurement of intense reflections and low intensity diffuse scattering simultaneously with a good signal-to-noise ratio (see Section 4.3). By swapping stages a variety of sample environmental modules can be accommodated. The large robot detector arm is mounted on a linear slide attached to the ceiling and can be retracted up to 3 m downstream ($\sim$3.7 m from the focal position) to enable coherent diffraction and small-angle X-ray scattering (SAXS) experiments. For these experiments a He flight tube between the sample and detector will be installed.

4.2.9. XRD and Six-Circle Diffractometer

The XPP-XRD station is dedicated to X-ray pump-probe resonant and non-resonant X-ray diffraction experiments. Whereas the change of regular Bragg peaks provides information about structural changes of the atomic lattice, incommensurate superlattice reflections are sensitive to symmetry changes driven by charge, orbital or spin dynamics. By tuning the X-ray energy to specific absorption edges, these dynamics can be directly probed by reflections which are allowed by the symmetry of the underlying unit cell. In the past at the FEMTO endstation at the SLS numerous trXRD experiments have been performed on the structural dynamics of semiconductors and semimetals, colossal magneto resistance (CMR) manganites, CDW systems, magnetic shape memory alloys, high-T_c SCs and ferroelectrics [173]. Recent trRXRD experiments were performed at the LCLS XFEL on a non-equilibrium phase transition in CMR manganites [174] and on THz-induced excitation of an electromagnon in a multiferroic, probed by soft X-ray resonant diffraction [175]. A first THz-pumped trXRD experiment in the hard X-ray range on a ferroelectric semiconductor has also been performed at FEMTO [176], demonstrating the feasibility of such experiments. The XPP-XRD station consists of a high precision '4S + 2D' six-circle Kappa-diffractometer with dual-detector arm carrying a 1.5 M 2D pixel detector (JUNGFRAU) and a polarization analyzer stage with beam collimation and point detector. Offset in scattering angle ($\sim$20°) allows rapid switching from measurements with a point detector to measurements with a 2D pixel detector. The design of the diffractometer allows the Kappa-goniometer to be replaced by a hexapod sample-stage or by an open χ-circle [170] with cryostat carrier. Operated in scanning mode both in energy and momentum (reciprocal space), tr(R)XRD experiments can be performed on samples under a variety of environmental conditions by replacing the in-air Kappa-goniometer by custom built sample environmental chambers.

4.2.10. trRXRD and Polarization Analyzer

To analyze the polarization properties of the diffracted beam the complete analyzer stage can be rotated around the scattered beam axis while the cross-slit system stays fixed. The analyzer crystal is mounted with angle $\theta_p = 45° \pm 5°$ towards the diffracted beam and the point detector is positioned at $2\,\theta_p$. Once phase retarders are installed to provide flexible linear and circular polarization of the incident X-ray beam (see Section 4.2.3), polarization scans will allow measuring the dependence of the diffracted intensity on the Stokes parameters to disentangle the different contributions to the resonant cross section, determining the magnetic moment or orbital orientation, and analyzing chiral spin structures.

4.2.11. trRIXS and Energy Analyzer

In 5d transition metal oxides such as iridates, the strong spin-orbit coupling—in addition to the complex interactions of spin, charge, orbital and lattice—mixes spin and electronic degrees of freedom. New classes of materials are predicted such as quantum spin liquids, topological insulators and new superconductors. High resolution hard X-ray spectrometers (resolving power $\simeq 2 \times 10^5$) to perform momentum resolved RIXS (qRIXS) experiments exist at synchrotrons [177,178] and a first fs time-resolved qRIXS experiment has recently been performed to directly determine the dynamics of magnetic correlations for the Mott insulator Sr_2IrO_4 [179]. At Bernina to study mixed electro-magnetic modes and spin correlation dynamics in Ir-perovskites, a dedicated compact qRIXS spectrometer is being designed that can be mounted both on the GPS goniometer and on the dual-arm of the six-circle diffractometer. It is operated in horizontal scattering geometry with the incident photon polarization in the scattering plane. Using the Si(444) reflection of the DCM Si(111) and a spherically-bent, diced Si(844) crystal located one meter away from the sample, the overall energy resolution at the Ir L_3-edge (11.216 keV) is <100 meV. The position-sensitive detector is a 0.5 M JUNGFRAU pixel detector or a position-sensitive silicon microstrip detector mounted in Rowland geometry with the diced spherical analyzer. Momentum-resolved trRIXS will greatly benefit from the enhanced spectral flux by self-seeding operation [168,180] of Aramis in the future.

4.2.12. ESB-MX Station

Until Experimental Station Cristallina (ESC) becomes operational, fixed target protein crystallography [140,141,181–185] will be implemented at Bernina [186]. A dedicated station (ESB-MX) is currently under construction. It will allow for serial femtosecond crystallography experiments at up to 100 Hz from 3D micro-crystals of size below 1 μm, as well as synchrotron-like data collection schemes from large crystals [187]. The crystals will be placed on a solid support, for example a chip with a silicon nitride window. The measurements can be performed in air or in helium atmosphere for photon energies from 5–12.4 keV. The samples will be either at room temperature or maintained under cryogenic conditions by a cold nitrogen or helium gas jet. The movable ESB-MX station can be installed and aligned within one day in the Bernina hutch.

The installation requires the GPS and XRD stations to be moved to the side along the rails. The main element of the station will be a sample diffractometer suitable for various data collection schemes which require a precise and fast tracking of the crystal positions. The station also includes an experimental chamber for the diffractometer and the cryo-jet, the room temperature and liquid nitrogen sample storage, as well as a robot for automated sample exchange. The diffraction images from the protein crystals will be collected with the 16 M pixel array JUNGFRAU detector positioned immediately downstream of the chamber with the large robot arm fixed to the ceiling. In this configuration, a resolution of better than 1 Å and 2.5 Å can be achieved with 12 keV and 5 keV radiation, respectively.

4.3. 2D X-ray Detectors

4.3.1. Overview of Current Detector Developments for XFELs

Several dedicated detection systems are under development for photon science at XFELs. Importantly, single photon counting detectors (like PILATUS [188], EIGER [189] or Medipix [190]) are not suitable for the high photon fluxes at XFELs, and the use of charge-integrating detection systems becomes crucial. Each of the following new detection systems for XFELs is tailored to meet the specific requirements of their respective light source. These detector systems comprise the Cornell-SLAC Pixel Array Detector (CSPAD) [191] and the new ePix family [192] for the SLAC Linac Coherent Light Source, the Adaptive Gain Integrating Pixel Detector (AGIPD) [193], the Large Pixel Detector (LPD) [194] and the DEPFET Sensor with Signal Compression (DSSC) [195] for the European X-ray Free Electron Laser and the Silicon-On-Insulator PHoton Imaging Array Sensor (SOPHIAS) [196] for the SACLA FEL.

None of these detection systems is capable (Table 4) of meeting the specific needs of the Aramis beam line and its user stations Alvra and Bernina. In particular, no detector system can provide at the same time single photon resolution at 2 keV photon energy and a dynamic range suitable for crystallography experiments. Therefore, a dedicated, state-of-the-art application-specific integrated circuit (ASIC) and detector system named JUNGFRAU (adJUstiNg Gain detector FoR the Aramis User station) has been developed in house [161].

Table 4. Comparison of the main characteristics of the current state-of-the-art pixel detection systems for XFEL hard X-ray light sources. The design characteristics which would limit the applicability of the detection system to SwissFEL are highlighted in yellow.

Detector System	Pixel Size μm × μm	Electronic Noise e$^-$	Single Photon Sensitivity @ 6 keV	Single Photon Sensitivity @ 2 keV	Dynamic Range Photons Per Pulse Per Pixel	Repetition Rate kHz
CSPAD	110 × 110	~330	Yes [†]	No	>2.5 × 10^3 (@ 8 keV) [‡]	0.12
ePix100	50 × 50	<60	Yes	Yes	100 (@ 8 keV)	~1
ePix100k	100 × 100	~120	Yes	No	10,000 (@ 8 keV)	~1
AGIPD	200 × 200	~265	Yes	No	>10^4 (@ 12 keV)	4500 burst
LPD	500 × 500	~1000	No	No	10^5 (@ 12 keV)	4500 burst
DSSC	Pitch 200 *	<50	Yes	Yes	>6 × 10^3 (@ 1 keV)	1000 burst
SOPHIAS	30 × 30	~150	Yes	No	~2000 (@ 12 keV)	0.06
JUNGFRAU	75 × 75	~65 G0 or ~50 HG0	Yes	Yes	>10^4 (@ 12 keV)	~2.4

[†] at >5 keV, for CSPAD in high gain; [‡] in low gain; * hexagonal pixels.

4.3.2. Detector Geometry

The development of the JUNGFRAU detector reuses, whenever possible, design blocks which have already been developed and successfully used for other PSI detectors in order to minimize the development time and to keep the project cost at an affordable level. In particular the geometrical dimension of the EIGER project [189] has been inherited by JUNGFRAU, so that a common silicon sensor can be used for the two projects.

Geometrically, the final JUNGFRAU chip comprises 256 × 256 pixels of 75 × 75 μm^2 each. Arrays of 2 × 4 chips (512 × 1024 pixels) are tiled to form modules of 38.4 × 76.8 mm^2 sensitive area, i.e., about 500 kPixel per module. The chips are coupled to 320 μm or 450 μm thick $p+$ on n silicon sensors (Hamamatsu Photonics, Hamamatsu, Japan). Multi-module systems of 1 MPixel (2 modules), 4 MPixel (8 modules, Figure 19), 8 MPixel (16 modules), 16 MPixel (32 modules) are envisioned and can be tiled in application specific geometries. The gap between two adjacent JUNGFRAU modules is about 500 μm (short) and 2.7 mm (long side) which corresponds to an insensitive area of about 7 pixels and 36 pixels respectively. The total dead area is 7% on a unit cell, or 5% on a typical 4 M detector.

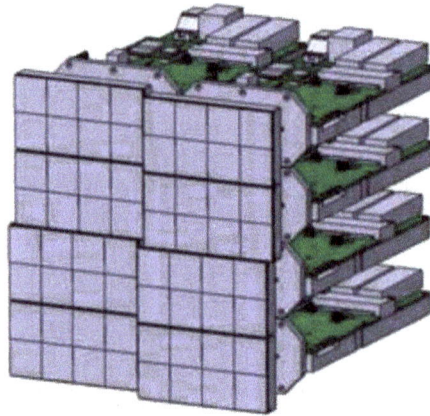

Figure 19. Mechanical drawing of a 4 M JUNGFRAU detector, showing a possible module arrangement around a small central hole.

4.3.3. Readout Chip Design

The JUNGFRAU readout chip (ROC) is designed in United Microelectronics Corporation (UMC) 110 nm CMOS technology with aluminum-only interconnects. JUNGFRAU is a charge-integrating detector rather than a single photon counter. Most hybrid pixel detectors for synchrotron radiation are based on the photon counting principle, which cannot meet the requirements of XFEL applications. In photon counting detectors, an event in the sensor causes a response in one or more ASIC pixels. Whenever such a response pulse reaches a user-set threshold level in a photon counting ASIC, the event is "counted". Photon counting is inaccurate when multiple photons arrive at the detector the same time, since two or more photons are counted as one. Charge integrating ASICs record a signal proportional to the number of photons interacting with the sensor, at least in the case of monochromatic light, independently of the time structure of the incoming beam.

4.3.4. Pixel Design

The design of the front-end block of the JUNGFRAU pixel is based on the automatic gain switching principle previously used in GOTTHARD [197] and AGIPD. A schematic view of the pixel circuit is shown in Figure 20. The pixel front end consists of several blocks:

- a preamplifier with three selectable gains: high (G0), medium (G1) and low (G2),
- an automatic gain switching block consisting of a comparator with tuneable threshold and switching control logic,
- a correlated double sampling (CDS) stage to remove the preamplifier low frequency and reset noise in high gain,
- a storage array for 16 images,
- a buffer needed to drive the column bus during the readout phase.

The preamplifier gain is variable: a fixed small size feedback capacitor is used for the high gain, while the insertion of two capacitors, ~10 and ~100 times bigger respectively, lowers the gain to a medium or low value. While the gain setting can be fixed with an external signal, in the normal mode of operation the control of the gain is automatically handled by the front-end circuit itself, pixel by pixel. For this purpose, the output of the preamplifier is monitored, during the integration, by a comparator. When the signal level crosses the threshold the gain switching logic is triggered. The threshold voltage is common for all the channels and is placed at the upper limit of the output range of the preamplifier. The digital logic, based on delay stages and latches, controls the insertion of the feedback capacitors and lowers the gain first to medium, then to low until the output of the preamplifier is brought back into the preamplifier output range.

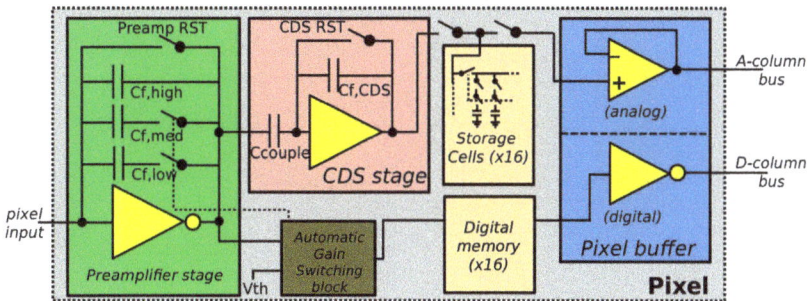

Figure 20. Block view of the pixel architecture. The bypass of the correlated double sampling (CDS) stage after switching is not shown.

In the idle state the preamplifier is kept in reset at the low gain mode, so that all the feedback capacitor are emptied. A few nanoseconds before the beginning of the measurement these capacitors are disconnected so that the gain is set to high. Then the reset switch is opened and the input charge starts to be integrated. The amplifier switches depending on the amount of input charge flowing into the readout channel. The output voltage and the gain (encoded in two digital bits) are sampled at the end of the integration time. Together, these allow the determination of the incoming charge with the help of a gain calibration curve.

A precharge scheme for the feedback capacitors has been introduced to achieve the requested dynamic range of 10^4 photons per pixel per pulse even if the maximum available feedback capacitance in a 75 µm pixel is rather small (<8 pF); during the preamplifier reset, the output side of the capacitor is not charged to the preamplifier reset level but is set to a tuneable voltage level which is chosen such that the preamplifier output is at the lower end of its range for signals just higher than the switching point. For low photon energy operation, the starting gain of the preamplifier can be increased, by a factor of ~2, to HG0, decreasing the noise but also the dynamic range before the first switching. In this case the gain switching sequence is HG0, G1 and G2. As in the GOTTHARD ROC, the CDS stage is bypassed in case of gain switching so that the (higher) noise of the high gain is not added to the (lower) noise of G1 and G2. 16 analogue storage cells are present in-pixel, each one with the corresponding 2 bit digital latches to store the gain information. In burst operation, up to 16 images can be stored on the ROC before readout.

4.3.5. Periphery and System Architecture

The 256 × 256 pixel matrix of the ROC is organized in four super-columns of 256 × 64 pixels, each of which is served by its own off-chip driver. This results in a readout time for the full ASIC of 256 × 64/(40 MHz) = 410 µs where 40 MHz is the design ADC clock speed. This readout time corresponds to a frame rate greater than 2 kHz. The fully differential off-chip driver is designed to directly drive the external ADCs, minimizing the number of components and the cost of the readout printed circuit board (PCB). The readout system is composed of a single PCB organized around an Altera Cyclone V Field Programmable Gate Array (FPGA). The FPGA generates the control signals for the readout ASICs and receives the data stream from the 14 bit 40 MHz ADCs digitizing the multiplexed analogue output from the chips. The data stream is reorganized such that contiguous images are formed and routed to a 10 Gigabit Ethernet (10 GbE) link for the data download to a receiving/storage server. This 10 GbE data link can sustain frame rates in the 1 kHz range, well in excess of the 100 Hz SwissFEL repetition rate. A CM-BF537 system-on-a-board (SOB) (BlueTechnix,

Vienna, Austria) is present on the module. The SOB features a 100 Mbit Ethernet connection and runs a server that implements the configuration interface for the detector (slow control).

4.3.6. First Characterization Results

The first full size JUNGFRAU modules have been extensively characterized with continuous X-ray sources and electrical or optical stimulation. The following performance has been measured:

- Noise in HG0 of 52 e.n.c. This allows single photon detection at energies <2 keV
- Noise in G0 lower than 70 e.n.c.
- Noise in G1 and G2 of 0.5 and 4 12 keV photons rms, well below Poisson statistical fluctuations
- Saturation level higher than 10,500 12 keV photons
- Linearity better than 1% rms.
- Radiation hard up to 10 MGy

A noise distribution for all the pixels of one module, measured using copper X-ray fluorescence radiation, is presented in Figure 21. A 1 M JUNGFRAU system has been tested at the LCLS X-ray Correlation Spectroscopy (XCS) instrument. Data was collected using Cu fluorescence, providing uniform illumination of the sensor, and with powder and single crystal diffraction samples. The data from LCLS confirmed the previous characterization, in terms of noise, dynamic range and spatial resolution. As an example, Figure 22 shows the measured number of photons hitting the sensor, averaged over a number of LCLS pulses, for a silver behenate powder sample and 9.3 keV illumination. The pixels in the inner rings recorded intensities covering most of the medium gain range (G1). The width and double structure of the rings is due to the geometry of the capillary.

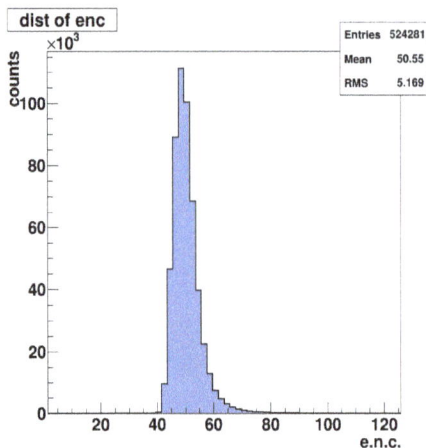

Figure 21. Distribution of the rms noise (in electrons) for all the pixels of a sample module. An average value of 50.5 e.n.c. is measured.

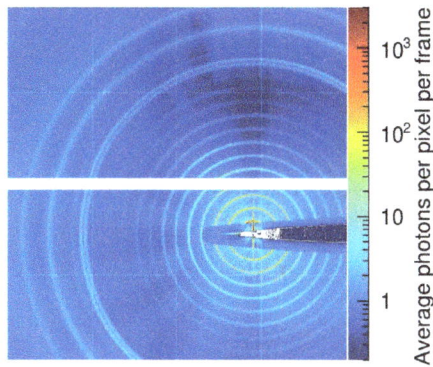

Figure 22. Image recorded with a 1 M JUNGFRAU system for a silver behenate powder sample at the LCLS XCS instrument. The image is averaged over 10,000 LCLS pulses.

5. Common Systems

The following sections describe hardware and software systems that are common to both the accelerator and photonics sections of the SwissFEL facility.

5.1. Timing and Synchronization

5.1.1. SwissFEL Timing Reference Generation and Distribution

The stringent stability requirements of SwissFEL sub-systems demand an ultra-stable timing reference distribution system with the highest flexibility, reliability and availability. In order to achieve all these goals and to optimize costs, a mixture of adapted technologies is applied to build the reference timing system. These are presented in the following sections.

The most critical clients require <10 fs jitter (rms) and down to 10 fs peak-to-peak temporal drift per day [49]. Therefore the key technology is the generation of both ultra-stable and low-noise reference signals in the RF as well as in the optical domain and the distribution of those signals as timing reference and precise clocks to the clients. Full remote control and monitoring of the components is necessary for smooth operation and debugging of the system and the facility as a whole.

The origin of the timing reference generation and distribution system resides in an environmentally controlled room at the electron gun end of the SwissFEL facility. The specified nominal temperature is 24 °C ± 0.1 °C and the relative humidity is controlled to within 42.5 %rH ± 2.5 %rH.

5.1.2. Optical Master Oscillator and Signal Generation

The optical master oscillator (OMO) Origami-15 from OneFive GmbH has two fiber-coupled outputs. One port with 2 mW optical power at 1 550 nm wavelength is used to produce the 142.8 MHz sinusoidal signal for the clock distribution. The other port, with 87 mW average output power, is used to generate 2.9988 GHz for S-band stations in the accelerator and laser synchronization as well as 2.856 GHz for C-band stations (see Figure 23). Both optical ports are connected to an ultra-low phase noise harmonic extraction unit (ULNHEU), where these signals originate from temperature-stabilized photo-detection, harmonic extraction and amplification. The output signals are linked to the corresponding distribution systems via temperature stable RF cables (Andrew Heliax tempered 1/4″, drifts −6 fs/m/K at 42.5 %rH around 24 °C for the relevant frequencies and about 0.2 fs/m/%rH at 24 °C). The absolute jitter of the microwave signal extracted from the free running Origami-15 and measured at the C-band frequency of 5.712 GHz including the noise of the optimized photo-receiver and the post amplifier is as low as 1.6 fs rms (1 kHz–10 MHz).

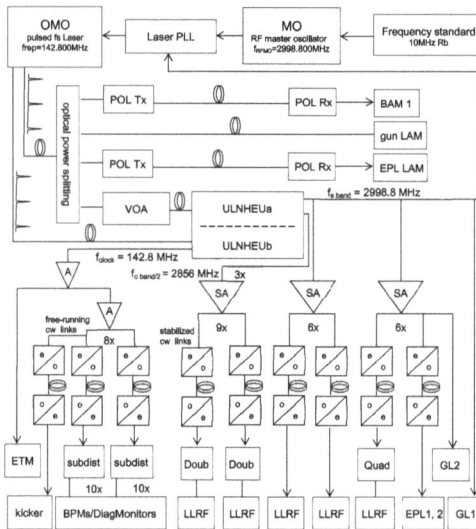

Figure 23. Schematic overview of the reference timing generation and distribution system in its current state. Only active elements are shown. OMO: Optical Master Oscillator, MO: Master Oscillator, PLL: Phase-locked loop, POL Tx/Rx: Pulsed Optical Link Transmitter/Receiver, BAM: Bunch Arrival Monitor, LAM: Laser Arrival Monitor, VOA: Variable Optical Attenuator (with power stabilization feedback), ULNHEU: Ultra-Low Noise Harmonic Extraction Unit, A: Amplifier, SA: Stabilized Amplifier, ETM: Event Timing Master, subdist: coaxial sub-distribution, BPM: Beam Position Monitor, DiagMonitors: various Diagnostic Monitors, Doub: Stabilized Frequency Doubler, LLRF: Low-Level RF, Quad: Stabilized Frequency Quadrupler, GL1: Gun Laser 1, GL2: Gun Laser 2, EPL1,2: Experiment Laser 1,2. Details are described in the text.

5.1.3. Pure Optical Pulse Distribution

The second port of the OMO is first split into four fiber-optic ports. One of these ports is fed through a variable optical attenuator (VOA) with a feedback loop for power stabilization down to 0.01 dB (Keysight N7764A). This helps reduce AM/PM conversion for subsequent components. That output is used as input to the harmonic extraction unit for RF signal generation. The three other fiber-optic connections can be used for clients who directly need the optical pulses (see Figure 23). Once ready, this optical splitter will be replaced by a stabilized optical amplifier including symmetric splitting for transmission in pulsed optical links.

5.1.4. Optical Master Oscillator Synchronization

For optimized overall phase noise performance, including the lower acoustic frequency regime, the OMO is synchronized to a master oscillator (MO, SMA 100A from Rohde & Schwarz GmbH) at 2.9988 GHz, which itself is locked to a 10 MHz Rubidium frequency standard (SRS FS725). The synchronization is implemented with a phase-locked loop (PLL), for which the master oscillator serves as reference input and the photo-detected filtered OMO signal coupled out from the harmonic extraction unit output is used as the comparator signal. Both are fed into a phase detector and the resulting error signal is sent to an amplifying piezo driver unit via a PID controller. The amplifier output drives a mirror mounted on a piezoelectric stage in order to keep the frequency output of the laser cavity extremely stable. Thus the integrated absolute jitter from 10 Hz to 1 MHz is as low as 14 fs rms. In the near term it is foreseen to develop an all-digital PLL with similar performance.

In addition to the piezo-electrical fine tuning, the OMO has a coarse tuning range of 6.4 kHz. Depending on the environmental conditions it can be necessary to bring the laser frequency into the fine tuning range by changing the frequency via the coarse tuning. This is done by regulating the temperature of the laser's base plate. This synchronization procedure is automated as well as the survey of all relevant signals such that the OMO can be kept synchronized for months. If necessary, the synchronization program corrects the laser temperature in order to keep the piezo-driving voltage in its range between 0 V and 150 V. Since the automated search for synchronization between the OMO and the MO via the laser temperature could be very time-consuming, an additional phase-frequency detector unit was implemented. This allows determining the sign of the phase difference between both signals and therefore simplifies the search.

5.1.5. Client Requirements

The clients of the ultra-stable timing reference generation and distribution system are to date: High power RF via LLRF, Bunch Arrival Monitor (BAM), Laser Arrival Monitor (LAM), Gun and Experimental Lasers to be synchronized to the reference, diagnostic monitors all along the machine, Event timing system, and switch-yard kicker.

The stability requirements concerning reference signal drift and integrated jitter depend very much on the various clients. Therefore distribution sub-systems based on adapted technologies and adequate performance were either developed or purchased with an emphasis on an optimized ratio of performance to investment in equipment. The types of links and the final performance goals are listed in Table 5. In the following sections the technologies of the different links and their realization/installation status is described in more detail.

Table 5. The clients of the Timing Reference Distribution, the types of links and their stability goals.

Client (#)	Reference Signal at Client	Distribution (Link Type)	Stability Goal jitter [1]/drift [2] ( ) [3]
Gun and Experiment Lasers (4)	142.8 MHz optical fs pulses	stabilized pulsed optical	few fs$_{rms}$/<10 fs$_{p-p}$ (<1 fs$_{rms}$/few fs$_{p-p}$[5])
BAM (4, later 6)	142.8 MHz optical fs pulses	stabilized pulsed optical	few fs$_{rms}$/<10 fs$_{p-p}$ (<1 fs$_{rms}$/few fs$_{p-p}$[5])
LAM (2)	142.8 MHz optical fs pulses	stabilized pulsed optical	few fs$_{rms}$/<10 fs$_{p-p}$ (<1 fs$_{rms}$/few fs$_{p-p}$[5])
S-band RF (6)	2998.8 MHz RF (21 × f$_{rep}$)	stabilized CW optical	<10 fs$_{rms}$/~30 fs$_{p-p}$ (~3 fs$_{rms}$/<20 fs$_{p-p}$)
C-band RF (27)	5712 MHz RF (40 × f$_{rep}$)	stabilized CW optical	<10 fs$_{rms}$/~40 fs$_{p-p}$ [4]
X-band RF (S-band front end) (1)	11,995.2 MHz RF (84 × f$_{rep}$)	stabilized CW optical + quadrupler	<10 fs$_{rms}$/~30 fs$_{p-p}$ (<3 fs$_{rms}$/<30 fs$_{p-p}$)
BPM (46)	142.8 MHz RF	VHF CW optical, coaxial	not critical
Event System (1)	142.8 MHz RF [6]	coaxial	not critical

[1] 10 Hz-10 MHz offset frequency range; [2] per 1/2 day – day; [3] potential of the technology; [4] up to 500 fs$_{p-p}$ (depends on station); [5] with polarization maintaining fibre as transmission medium; [6] eventually 1428 MHz.

5.1.6. Continuous Wave Fiber-Optic Links for RF Reference Distribution

The RF transmission to the LLRF stations is done with so-called "CW optical links" [198]. These are radio-over-fiber type links based on standard telecom single mode fibers as the transmission medium, which have been developed in collaboration between PSI and Instrumentation Technologies Ltd. (Solkan, Slovenia) The operating principle of the "Libera Sync 3" fiber-optic links is based on intensity modulation of CW laser diode light with additional group delay stabilization between the transmitter input signal and the photo-detected light reflected from the link end receiver.

The performance goal of these links was set to less than 40 fs drift peak-peak over 24 h (the typically measured drift is approximately 20 fs peak-peak for a 500 m link) and less than 10 fs rms integrated jitter in the frequency range from 10 Hz to 10 MHz (typically measured jitter is around 2.4 fs rms for a 500 m link and is limited by the measurement setup). Nearly 90% of the links which are needed for SwissFEL operation are currently at PSI (no spares in case of failure), all of which have been tested for drift and jitter performance and have gone through a burn-in phase of about 1000 h. To date, about 70% of the links are installed in SwissFEL and are in operation. For the operation of the links, communication between transmitter and receiver over ethernet is necessary. Because of initial problems with high network traffic and ethernet module failures of the link devices, the CW links have their own virtual LAN along the entire machine. Furthermore, for a full maneuverability during setup and operation, an EPICS based remote control interface was built, which can be used in parallel for all links.

For the injector, 6 slightly de-tuned European standard S-band frequencies and one European X-band frequency link are needed. The three linac sections need 9, 4 and 14 American standard C-band frequency links, correspondingly. Based on the de-tuned S-band (2.9988 GHz), the de-tuned X-band (11.9952 GHz) and the C-band (5.712 GHz) frequency, a reasonably high base frequency could still be found (142.8 MHz), which needs to be a common subharmonic of those frequencies and was chosen as the repetition rate of the OMO. All frequencies throughout the SwissFEL accelerator, which need to be synchronous in some way, are derived from this source of the timing reference distribution.

Concerning the transmission of the C-band frequency, it was decided to use the Libera Sync 3 links to transmit half of the actual C-band frequency (2.856 GHz), for which the S-band links could be used with only minor modifications. At the output of those links, an ultra-stable RF frequency doubler with amplitude/phase stabilized power amplifier will be used to provide the necessary frequency and amplitude level. The frequency doubler was developed at PSI and a prototype has been successfully tested.

For the X-band transmission, a similar philosophy is adapted: an S-band link together with an ultra-stable quadrupler unit built in-house is installed and in operation.

5.1.7. CW Fiber-Optic Links for Clock Distribution

For most diagnostic monitors along the machine, the kicker systems as well as the controls event timing distribution system, the drift and jitter requirements are less stringent. Therefore a distribution system was developed in collaboration with and purchased from SINTEC Microwave Systems GmbH, which is adapted for the needs of SwissFEL with respect to the number of clients, their location along the machine, the requested power levels and the jitter performance. Our clock signal at 142.8 MHz serves as input signal to a transmitter module, where the signal is amplified and split into 8 outputs. Each of these outputs is fed into an optical transmitter module where it modulates an optical carrier at 1310 nm wavelength. After transmission through the fiber connection, the optical carrier input signal is demodulated in a Receiver module, converted back to a 142.8 MHz electrical signal and amplified. Emanating from the 8 Receiver stations along the machine, a local coaxial T-shaped sub-distribution supplies several diagnostic monitors. The added jitter of the clock distribution in the offset range from 10 Hz to 10 MHz is 73 fs rms and lies well below the most demanding specified value among the clients of 200 fs (rms). This system is installed in SwissFEL, operational and remotely monitored with respect to RF power levels at the receiver stations, module temperatures, amplifier and laser diode currents.

5.1.8. Pulsed Optical Links for Reference Signal Distribution to Critical Clients

Clients like the BAM, the LAM, gun and experiment lasers, and X-band RF station require ultimate long-term stability below 10 fs peak-to-peak drift per day. In addition, they may need direct optical input signals. Therefore these clients will be fed with reference signals from stabilized pulsed optical links. This technology is commercially available and it is planned to purchase this system together with an optical power amplifier and splitter system. The evaluation process is ongoing.

5.1.9. Laser Arrival Monitor

The LAM is one example of a client that directly uses the pulses from the OMO. The Gun-LAM (next to the photocathode, including beam propagation through laser chain and transfer line) employs a novel scheme, where the current generated by the UV pulses impinging on a photodiode directly drives an electro-optic intensity modulator acting on the optical reference. Thus, arrival time is directly encoded onto the amplitude of the reference pulse train. A balanced photoreceiver in combination with a fast analog-to-digital converter (ADC) decodes this information shot-by-shot, limited by the ADC sampling rate. Expected temporal resolution is low tens of fs. A laser arrival time jitter of around 40 fs (rms) at the gun would lie within the tolerance budget to ensure that FEL beam fluctuations stay within intrinsic fluctuations at 200 pC as well as for 10 pC electron beam charge mode [199].

The experiment LAM is based on a spectrally resolved cross correlator [127]. The incoming IR pulses from the experiment laser and the reference pulses (stretched and chirped) mix in a nonlinear crystal and their relative timing results in wavelength change that is read out with a spectrometer. A resolution of 5 fs (rms) should be achievable for this application. The tolerable laser jitter at the Experimental Stations is 25 fs (rms).

For the LAM at the electron gun, which is still close to the source of the reference generation and distribution, the link will initially be realized with a few tens of meters of temperature stable fiber with low drift (3 fs/m/K and 0.4 fs/m/%rH). The LAM at the experimental stations will initially get a pulsed link prototype.

5.1.10. Bunch Arrival Monitor and Other Clients

The first BAM will initially get a pulsed link prototype. In addition, the second BAM will use a second redundant OMO, synchronized to the MO, as a pulsed source. With an additional RF phase shifter it will be possible to time delay the optical pulses as needed. It is likely that BAM 1 will then also be added to this system. The BAM is described in more detail in Section 2.5.6.

5.2. Motion Control

Motion control is an integral part of any large research facility. This motion ranges from moving several tons with micrometer accuracy (e.g., undulators) to moving samples with nanometer accuracy and almost everything in between. There are several hundred motion axes with diverse applications. In order to minimize the variety of hardware used and thus also the requirements from the Controls group to support all the different platforms, standards were set at an early stage. Although PSI has a motion control system for the SLS, which is controlling over 1000 axes and has proven to be very reliable, special requirements for SwissFEL made it necessary to evaluate a new system to cover all these requirements.

A wide range of requirements and specifications need to be met. These are among others:

- Complex coordinated motions,
- Large distances between motion controller and motor,
- Interface to the timing system and
- Low cost per axis.

In order to cover the widest possible range without having too many different systems, the motions were categorized into 3 different groups:

1. High performance motion axis
2. Simple motion axis
3. Piezo positioners for sub-μm motions

After an evaluation phase of one year, the DeltaTau PowerBrick IMS motion control system was chosen for the high performance motion axis. Based on the PowerPMAC motion controller, the PowerBrick IMS motion control system is custom built by DeltaTau UK. The PowerBrick is used at several research facilities around the world. The following sections describe the various requirements.

5.2.1. Complex Coordinated Motions

Complex motions become more important for motion control systems in research. For example, an X-ray mirror mounted on a tri-pod (mover with three vertical translation stages) can move vertically as well as being tilted in the horizontal plane. In order to define a motion around a fixed point in space, in this case the center of the mirror surface, coordinated motion of all three axes is required. This can be easily implemented with the PowerBrick. Another example is the energy change of a monochromator where several motions need to be coordinated for a smooth change of the energy. Furthermore, a linear scan of the energy requires non-linear motion of the motors. This can be achieved through coordinate transformation inside the PowerBrick.

5.2.2. Timing Interface

As SwissFEL is a pulsed source, the timing plays an important role in motion control. This can range from reading out a motor position, i.e., encoder, when an X-ray pulse arrives and "tagging" the value with the pulse ID, to controlling the exact positioning of a motor based on the arrival of the X-ray pulses. PSI uses a "Timing and Event System" from Micro Research Finland. This sends "event codes" over an optical fiber link to the "event receivers", which can then generate low jitter hardware trigger outputs and/or generate interrupts to a controlling CPU. It can also send real-time data to the event receivers which can be interpreted by the controlling CPU. The system must be able to synchronize and trigger motions on different remote motion controllers as well as to implement synchronous encoder readout during a measurement and tag the readings with real-time data from the timing system at a trigger rate of at least 100 Hz. An event receiver card is integrated into the PowerBrick and communicates through the PCIe bus directly with the motion control CPU. User programs and event receiver driver software run on the internal CPU of the PowerBrick.

5.2.3. Large Distances

SwissFEL spans well over 700 m from the electron gun to the last experimental station, Cristallina. Furthermore, for the accelerator, electronics need to be placed outside the tunnel due to high radiation levels. This results in the need for a distributed system as well as the need to drive motors over long cables. In contrast to the SLS, where motor cables are around 20 m, at SwissFEL they can be up to 50 m. As the motion controllers are decentralized and spread out over the entire length of the machine, remote control of all settings and parameters as well as status readback need to be possible. The PowerBrick allows for complete parametrization over the network. In fact, the system is built in such a way that if a driver unit would need to be replaced, a completely unconfirmed unit can be used simply by setting the host name, the PowerBrick motion controller becomes fully parameterized when booted.

5.2.4. Simple Motion Axis

The PowerBrick is a very powerful motion controller but beyond what is required for simple motion applications, such as moving a screen into the X-ray beam. For such simple motion axes the Schneider Electric MDrive Stepper Motor was selected. The MDrive come in a variety of sizes ranging from 36 mm to 85 mm flange size with holding torques from 13 Ncm to 770 Ncm. Another advantage is that the motion controller and driver are integrated into the motor. The controller can also read incremental encoders to form a positioning feedback. The Schneider Electric MDrive motors can be used wherever the radiation levels are limited, where no interface to the timing system is required and where no coordinated motion is required.

5.2.5. Piezo Positioners for Sub-μm Motions

There are several applications where small masses need to be moved but to very high precision, e.g., alignment of optical laser mirrors or positioning small samples in the micro-focused

X-ray beam. This is where piezoelectric positioners have an advantage. Making use of the piezoelectric effect, piezo positioners can be built to be compact and move with nm resolution. For SwissFEL we have chosen the piezo positioners from SmarAct in Germany. SmarAct offers a wide range of positioners with a small form factor for many different applications. Furthermore, SmarAct positioners have been used at the SLS for several years with very good results.

5.3. Data Acquisition

Within SwissFEL there are two categories of data: beam synchronous and asynchronous data. All beam synchronous data from any part of the machine can be correlated and can be associated to an individual FEL pulse/bunch. Asynchronous data neither can be correlated directly nor be related to an individual pulse/bunch. Regardless of the class of data, each reading from sensors, detectors, motors, etc. is available via so called channels. Each channel is identified by a channel name that is unique within the whole facility.

As there are two categories of data, the SwissFEL data acquisition system is split into two systems: a system dealing with beam synchronous data and a system dealing with asynchronous data. As the asynchronous system does not differ from the systems used at other facilities at PSI, it is therefore described only briefly in Section 5.3.1. The remaining sections will focus on the aspects related to the beam synchronous system.

Although SwissFEL is comprised of various sections (e.g., accelerator, beamlines, experimental stations, etc.) and include components that are operated and maintained by different groups, data from all parts of the facility needs to be collected and retrieved via the same systems and tools.

5.3.1. Asynchronous DAQ

The asynchronous data acquisition system is based on EPICS [200] and data can be retrieved and subscribed to via the Channel Access protocol. To retrieve data from applications and scripts, standard EPICS tools and libraries can be used.

For continuous recording and long-term archiving, a dedicated instance of the EPICS Archiver Appliance [201] is in place. Any asynchronous channel can be recorded and archived via this system. Data from the EPICS Archiver Appliance can be retrieved and viewed via the same APIs and tools as for the beam synchronous data which are described in the following sections.

5.3.2. Beam Synchronous DAQ

The beam synchronous data acquisition system is a new development centered around the idea of data streaming. Load and code complexity on data sources are reduced by immediately streaming raw data out as soon as it becomes available. For a given pulse, data from all configured channels of a source is collected, serialized and sent out in an atomic message. Along with the data of all channels, this message also contains the unique pulse-id of the machine pulse the data belongs to as well as other metadata. As the pulse-id is needed by all systems providing beam synchronous data simultaneously, all these systems need to be connected to the global SwissFEL real-time timing system.

Load balancing and fail-over is built into the system by design since it uses the ZeroMQ protocol [202] for data transport. Channels to be streamed out by the data sources can be configured dynamically without the need of restarting or interrupting the data collection.

To prevent data sources from being overwhelmed by client requests, the only client collecting data directly from sources is the Dispatching Layer. Beside data source isolation, this subsystem, as described in the next section, also provides transparent and synchronized access to all incoming data. This layer can also reduce the incoming data streams, e.g., by reducing the frequency, to meet client requirements and capabilities. As not all clients can cope with the incoming data rates and volumes the system also provides a buffering layer in order to provide data for retrieval at a later point in time, namely DataBuffer and ImageBuffer. Although these are two separate subsystems, data is accessible via a common API.

The subsystems of the beam synchronous data acquisition systems take care of all beam synchronous data, except the primary experimental station pixel detectors (e.g., JUNGFRAU). This data is collected and processed through a system derived from [203], but will be nevertheless synchronized with all the rest of beam synchronous data.

5.3.3. Dispatching Layer

The Dispatching Layer is a cluster of machines that take care of retrieving and providing data from all data sources except the primary experimental station 2D detectors. This layer shields all the data sources from other clients and acts as the single point of access to live data. Via this layer clients can subscribe to customized and synchronized data streams of channels from any part of the facility. Clients can request channels in various frequencies and combinations. This architecture allows clients to retrieve any combination of channels regardless of their origin as if they came from a single source, i.e., the data of a given machine pulse of all requested channels is in the same message alongside the pulse-id and other global metadata. The synchronization of data is close to real time and data emitted can be used for immediate online monitoring, data analysis and processing. Requests and calls to this layer are done via a central REST API. Data streams provided by the layer are based on ZeroMQ.

5.3.4. DataBuffer/ImageBuffer

All beam synchronous data is buffered and stored for a configurable amount of time. By doing so users are able to retrieve historical data if needed. The systems taking care of the buffering are DataBuffer and ImageBuffer. Both systems take care of and are optimized for different classes of data. DataBuffer takes care of small data, i.e., of all scalar and one dimensional arrays (waveform) data. The ImageBuffer system takes care of all beam synchronous images from cameras and detectors running at the facility except the JUNGFRAU detector data.

The DataBuffer system is currently a cluster of 12 machines and uses Cassandra [204] for persistent data. The ImageBuffer system is based on Spectrum Scale [205] and an in-house data format. Data from both systems is accessible to users via the DataAPI interface described in Section 5.3.5.

5.3.5. DataAPI

The DataAPI acts as the single point of access to all archived and buffered data from SwissFEL. It currently provides beam synchronous data from the DataBuffer, ImageBuffer as well as asynchronous data from the EPICS Archiver Appliance. It is a REST based API that is easily accessible from almost any programming language. Data can be retrieved based on channel names and time or pulse-id (for beam synchronous data) ranges. Besides the ability to retrieve raw data the API also provides the means to retrieve reduced data for the requested range, e.g., for later studies and documentation.

5.3.6. Data Web Frontend

The Data Web Frontend provides an easy-to-use and visual frontend to the DataAPI. It is a web based interface based on the latest web technologies, namely Web Components [206] and Plotly [207]. It is accessible from all computers within the Paul Scherrer Institute that have a modern web browser installed. The web interface provides advanced means to browse large amounts of data quickly and to extract the most important information, without retrieving and drawing the complete raw data. This is achieved through data reduction strategies provided by the DataAPI backend.

5.3.7. Experimental Data Container

Experimental data will be saved in the Experimental Data Container. HDF5 [208] has been chosen as file format due to its widespread adoption in the scientific community and the possibility to save data and metadata in the same file. Data will be retrieved directly from the Dispatching Layer: users will select a list of data channels to be written into the final file. Further data can be added at a later time from the Data/ImageBuffer.

6. Conclusions and Outlook

The SwissFEL project has made remarkable progress since it first broke ground in 2013, culminating in first SASE lasing in the soft X-ray domain in May 2017. The Aramis hard X-ray branch of SwissFEL will be operational for users in 2018, with its first call for proposals in the beginning of 2018. Installation of the Athos soft X-ray branch has begun, with expected user operation after 2020 on three photon beamlines. In addition to the components and systems we have described in this review the project has several near-term future upgrades planned, including hard X-ray self-seeding [168,180] and the addition of a third experimental station (Cristallina). Beyond the currently defined Aramis and Athos accelerator infrastructure there are plans to add a third undulator beamline. The building was designed to allow space for this installation in the accelerator tunnel and internal discussions have begun on what the parameters of this project might be. We look forward to joining the thriving hard X-ray free electron laser community [209–212] and welcoming researchers from around the world to our exciting new facility in the coming years.

Acknowledgments: The authors would like to extend their profound thanks to all of the facilities with whom we have collaborated on this project: ASTeC (Daresbury, UK), SPring-8 (JASRI/RIKEN, Hyogo, Japan), CERN (Geneva, Switzerland), SLAC (Stanford, CA, USA), Elettra Sincrotrone Trieste (Basovizza, Italy), DESY (Hamburg, Germany), and the European XFEL (Schenefeld, Germany). Furthermore the authors acknowledge the advice and support they received throughout the preparation and realization of SwissFEL by the members of the FLAC advisory committee, chaired by Jörg Rossbach (DESY and University of Hamburg), the ESA Review Panel, chaired by Wojciech Gawelda (European XFEL and Jan Kochanowski University), the ESB Review Panel, chaired by Steven Johnson (ETH Zürich), and the ESB-MX Review Panel, chaired by Matthias Frank (LLNL). We would also like to acknowledge the funding and support we have received from the Swiss Federal Government, the ETH Council, the government of the canton of Aargau, the Swiss National Science Foundation, the town of Würenlingen, and the Swiss Lottery Fund (Swisslos). Finally we would like to acknowledge the support and effort from all the groups at PSI who have contributed to this project.

Author Contributions: All authors have contributed in some way to the concept, design, installation and commissioning of different aspects of the SwissFEL project. All authors contributed to the paper either during the writing or editing phases.

Conflicts of Interest: The authors declare no conflict of interest.

Abbreviations

The following abbreviations are used in this manuscript:

ADC	Analog-to-digital converter
AGIPD	Adaptive Gain Integrating Pixel Detector
API	Application programming interface
ASIC	Application-specific integrated circuit
BAM	Electron bunch arrival-time monitor
BBO	beta barium borate
BC	Bunch compressor
BOC	Barrel open cavity
BPM	Beam position monitor
CDS	Correlated double sampling
COTR	Coherent optical transition radiation
CPA	Chirped-pulse amplification
CRL	compound refractive lenses
CSPAD	Cornell-SLAC Pixel Array Detector
CVD	Chemical vapour deposition
DCM	double-crystal monochromator
DAQ	Data acquisition
e.n.c.	equivalent noise charge in electrons
EPICS	Experimental Physics and Industrial Control System
ESA	Experimental Station Alvra
ESB	Experimental Station Bernina

ESC	Experimental Station Crystallina
eTOF	electron time-of-flight
FEL	Free Electron Laser
FPGA	Field-programmable gate arrays
FWHM	Full-width at half maximum
GOTTHARD	Gain Optimizing microsTrip sysTem witH Analog ReaDout
HDF	Hierarchical Data Format
HRM	Harmonic rejection mirrors
HV	High vacuum
IR	Infrared
ICT	Integrated current transformers
JUNGFRAU	adJUstiNg Gain detector FoR the Aramis User station
KB mirrors	Kirkpatrick-Baez mirrors
LAM	Laser arrival monitor
LCLS	Linac Coherent Light Source
Linac	Linear accelerator
LIC	Laser in-coupling
LPD	Large Pixel Detector
MCP	Microchannel plate
MCT	Mercury cadmium telluride
MO	Master oscillator
OM	Offset mirrors
OMO	Optical master oscillator
OPA	Optical parametric amplifier
PALM	Pulse arrival and length monitor
PCB	Printed circuit board
PBIG	Photon beam intensity monitor
PBPG	Photon beam position monitor
PBPS	Photon backscattering monitor
PDIM	Photon diode intensity monitor
PID	Proportional-integral-derivative controller
PLL	Phase-locked loop
PPRM	Photon profile monitor
PSEN	Photon spectral encoder
PSI	Paul Scherrer Institute
PSRD	Photon spontaneous radiation detector
PSSS	Photon single-shot spectrometer
REST	Representational state transfer
RF	Radio frequency
rms	root-mean-square
ROC	Readout chip
SACLA	SPring-8 Angstrom Compact free electron LAser
SASE	Self-amplified spontaneous emission
SOB	System-on-a-board
SOPHIAS	Silicon-On-Insulator PHoton Imaging Array Sensor
TDS	Transverse deflecting structure
UHV	Ultra-high vacuum
ULNHEU	Ultra-low phase noise harmonic extraction unit
UMC	United Microelectronics Corporation
VOA	Variable optical attenuator
XCS	X-ray Correlation Spectroscopy
XFEL	X-ray free electron laser
XPR	X-ray phase retarder
XRTD	X-ray timing diagnostic

References

1. Pellegrini, C.; Marinelli, A.; Reiche, S. The physics of X-ray free-electron lasers. *Rev. Mod. Phys.* **2016**, *88*, 015006.
2. Geloni, G.; Saldin, E.; Samoylova, L.; Schneidmiller, E.; Sinn, H.; Tschentscher, T.; Yurkov, M.V. Coherence properties of the European XFEL. *New J. Phys.* **2010**, *12*, 035021.
3. Saldin, E.L.; Schneidmiller, E.A.; Yurkov, M.V. Coherence properties of the radiation from X-ray free electron laser. *Opt. Commun.* **2008**, *281*, 1179–1188.

4. Saldin, E.L.; Schneidmiller, E.A.; Yurkov, M.V. Statistical and coherence properties of radiation from X-ray free-electron lasers. *New J. Phys.* **2010**, *12*, 035010.

5. Ding, Y.; Behrens, C.; Coffee, R.; Decker, F.J.; Emma, P.J.; Field, C.; Helml, W.; Huang, Z.; Krejcik, P.; Krzywinski, J.; et al. Generating femtosecond X-ray pulses using an emittance-spoiling foil in free-electron lasers. *Appl. Phys. Lett.* **2015**, *107*, 191104.

6. Duesterer, S.; Radcliffe, P.; Bostedt, C.; Bozek, J.; Cavalieri, A.L.; Coffee, R.; Costello, J.T.; Cubaynes, D.; Dimauro, L.F.; Ding, Y.; et al. Femtosecond X-ray pulse length characterization at the Linac Coherent Light Source free-electron laser. *New J. Phys.* **2011**, *13*, 093024.

7. Grguraš, I.; Maier, A.R.; Behrens, C.; Mazza, T.; Kelly, T.J.; Radcliffe, P.; Dusterer, S.; Kazansky, A.K.; Kabachnik, N.M.; Tschentscher, T.; et al. Ultrafast X-ray pulse characterization at free-electron lasers. *Nat. Photonics* **2012**, *6*, 852–857.

8. Helml, W.; Maier, A.R.; Schweinberger, W.; Grguraš, I.; Radcliffe, P.; Doumy, G.; Roedig, C.; Gagnon, J.; Messerschmidt, M.; Schorb, S.; et al. Measuring the temporal structure of few-femtosecond free-electron laser X-ray pulses directly in the time domain. *Nat. Photonics* **2014**, *8*, 950–957.

9. Inubushi, Y.; Tono, K.; Togashi, T.; Sato, T.; Hatsui, T.; Kameshima, T.; Togawa, K.; Hara, T.; Tanaka, T.; Tanaka, H.; et al. Determination of the Pulse Duration of an X-ray Free Electron Laser Using Highly Resolved Single-Shot Spectra. *Phys. Rev. Lett.* **2012**, *109*, 144801.

10. Fuchs, M.; Trigo, M.; Chen, J.; Ghimire, S.; Shwartz, S.; Kozina, M.; Jiang, M.; Henighan, T.; Bray, C.; Ndabashimiye, G.; et al. Anomalous nonlinear X-ray Compton scattering. *Nat. Phys.* **2015**, *11*, 964–970.

11. Shwartz, S.; Fuchs, M.; Hastings, J.B.; Inubushi, Y.; Ishikawa, T.; Katayama, T.; Reis, D.A.; Sato, T.; Tono, K.; Yabashi, M.; et al. X-ray Second Harmonic Generation. *Phys. Rev. Lett.* **2014**, *112*, 163901.

12. Tamasaku, K.; Nagasono, M.; Iwayama, H.; Shigemasa, E.; Inubushi, Y.; Tanaka, T.; Tono, K.; Togashi, T.; Sato, T.; Katayama, T.; et al. Double core-hole creation by sequential attosecond photoionization. *Phys. Rev. Lett.* **2013**, *111*, 043001.

13. Tamasaku, K.; Shigemasa, E.; Inubushi, Y.; Katayama, T.; Sawada, K.; Yumoto, H.; Ohashi, H.; Mimura, H.; Yabashi, M.; Yamauchi, K.; et al. X-ray two-photon absorption competing against single and sequential multiphoton processes. *Nat. Photonics* **2014**, *8*, 313–316.

14. Glover, T.E.; Fritz, D.M.; Cammarata, M.; Allison, T.K.; Coh, S.; Feldkamp, J.M.; Lemke, H.T.; Zhu, D.; Feng, Y.; Coffee, R.N.; et al. X-ray and optical wave mixing. *Nature* **2012**, *488*, 603–608.

15. Szlachetko, J.; Hoszowska, J.; Dousse, J.C.; Nachtegaal, M.; Błachucki, W.; Kayser, Y.; Sa, J.; Messerschmidt, M.; Boutet, S.; Williams, G.J.; et al. Establishing nonlinearity thresholds with ultraintense X-ray pulses. *Sci. Rep.* **2016**, *6*, 33292.

16. Wilmott, P. *An Introduction to Synchrotron Radiation: Techniques and Applications*; Wiley: Hoboken, NJ, USA, 2011.

17. Bostedt, C.; Boutet, S.; Fritz, D.M.; Huang, Z.; Lee, H.J.; Lemke, H.T.; Robert, A.; Schlotter, W.F.; Turner, J.J.; Williams, G.J. Linac Coherent Light Source: The first five years. *Rev. Mod. Phys.* **2016**, *88*, 015007.

18. Gawelda, W.; Szlachetko, J.; Milne, C.J. X-ray Spectroscopy at Free Electron Lasers. In *X-ray Absorption and X-ray Emission Spectroscopy*; John Wiley & Sons, Ltd.: Chichester, UK, 2016; pp. 637–669.

19. Spence, J.C.H.; Weierstall, U.; Chapman, H.N. X-ray lasers for structural and dynamic biology. *Rep. Prog. Phys.* **2012**, *75*, 102601.

20. Levantino, M.; Yorke, B.A.; Monteiro, D.C.; Cammarata, M.; Pearson, A.R. Using synchrotrons and XFELs for time-resolved X-ray crystallography and solution scattering experiments on biomolecules. *Curr. Opin. Struct. Biol.* **2015**, *35*, 41–48.

21. Feldhaus, J.; Krikunova, M.; Meyer, M.; Moeller, T.; Moshammer, R.; Rudenko, A.; Tschentscher, T.; Ullrich, J. AMO science at the FLASH and European XFEL free-electron laser facilities. *J. Phys. B Atom. Mol. Opt. Phys.* **2013**, *46*, 164002.

22. Barty, A.; Caleman, C.; Aquila, A.; Timneanu, N.; Lomb, L.; White, T.A.; Andreasson, J.; Arnlund, D.; Bajt, S.; Barends, T.R.M.; et al. Self-terminating diffraction gates femtosecond X-ray nanocrystallography measurements. *Nat. Photonics* **2011**, *6*, 35–40.

23. Schlichting, I. Serial femtosecond crystallography: The first five years. *IUCrJ* **2014**, *2*, 246–255.

24. Martin-Garcia, J.M.; Conrad, C.E.; Coe, J.; Roy-Chowdhury, S.; Fromme, P. Serial femtosecond crystallography: A revolution in structural biology. *Arch. Biochem. Biophys.* **2016**, *602*, 32–47.

25. Emma, P.J.; Akre, R.; Arthur, J.; Bionta, R.; Bostedt, C.; Bozek, J.; Brachmann, A.; Bucksbaum, P.; Coffee, R.; Decker, F.J.; et al. First lasing and operation of an ångstrom-wavelength free-electron laser. *Nat. Photonics* **2010**, *4*, 641–647.

26. The Paul Scherrer Institute. Available online: https://www.psi.ch/ (accessed on 3 July 2017).

27. Madey, J.M.J. Stimulated emission of bremsstrahlung in a periodic magnetic field. *J. Appl. Phys.* **1971**, *42*, 1906–1913.

28. Hara, M.; Tanaka, T.; Tanabe, T.; Maréchal, X.-M.; Okada, S.; Kitamura, H.; In-vacuum undulators of SPring-8. *J. Synchrotron Radiat.* **1998**, *5*, 403–405.

29. Kim, K.J. Brightness, coherence and propagation characteristics of synchrotron radiation. *Nucl. Instrum. Methods Phys. Res. A* **1986**, *246*, 71–76.

30. Bane, K.L.F. *The Short Range Resistive Wall Wakefields*; Technical Report, SLAC/AP-87; Stanford Linear Accelerator Center, Stanford University: Stanford, CA, USA, June 1991.

31. Reiche, S.; Emma, P.J.; Pellegrini, C. Pulse length control in an X-ray FEL by using wakefields. *Nucl. Instrum. Methods Phys. Res. A* **2003**, *507*, 426–430.

32. Prat, E.; Reiche, S. Update on FEL performance for SwissFEL. In Proceedings of the 36th International Free Electron Laser Conference (FEL 2014), Basel, Switzerland, 25–28 August 2014; pp. 140–143.

33. Beutner, B. Bunch compression layout and longitudinal operation modes for the SwissFEL Aramis Line. In Proceedings of the 34th International Free Electron Laser Conference (FEL 2012), Nara, Japan, 26–31 August 2012; pp. 297–300.

34. Rosenzweig, J.B.; Alesini, D.; Andonian, G.; Boscolo, M.; Dunning, M.; Faillace, L.; Ferrario, M.; Fukusawa, A.; Giannessi, L.; Hemsing, E.; et al. Generation of ultra-short, high brightness electron beams for single-spike SASE FEL operation. *Nucl. Instrum. Methods Phys. Res. A* **2008**, *593*, 39–44.

35. Saa Hernandez, A.; Prat, E.; Bettoni, S.; Beutner, B.; Reiche, S. Generation of large-bandwidth X-ray free-electron-laser pulses. *Phys. Rev. Accel. Beams* **2016**, *19*, 090702.

36. Wang, L.; Ding, Y.; Huang, Z. Optimization for single-spike X-ray FELS at LCLS with a low charge beam. In Proceedings of the 2nd International Particle Accelerator Conference (IPAC 2011), San Sebastián, Spain, 4–9 September 2011; pp. 3131–3133.

37. Calvi, M.; Camenzuli, C.; Prat, E.; Schmidt, T. Transverse gradient in Apple-type undulators. *J. Synchrotron Radiat.* **2017**, *24*, 600–608.

38. Prat, E.; Calvi, M.; Ganter, R.; Reiche, S.; Schietinger, T.; Schmidt, T. Undulator beamline optimization with integrated chicanes for X-ray free-electron-laser facilities. *J. Synchrotron Radiat.* **2016**, *23*, 861–868.

39. Prat, E.; Calvi, M.; Reiche, S. Generation of ultra-large-bandwidth X-ray free-electron-laser pulses with a transverse-gradient undulator. *J. Synchrotron Radiat.* **2016**, *23*, 874–879.

40. Reiche, S.; Prat, E. Two-color operation of a free-electron laser with a tilted beam. *J. Synchrotron Radiat.* **2016**, *23*, 869–873.

41. Prat, E.; Löhl, F.; Reiche, S. Efficient generation of short and high-power X-ray free-electron-laser pulses based on superradiance with a transversely tilted beam. *Phys. Rev. Spec. Top. Accel. Beams* **2015**, *18*, 100701.

42. Prat, E.; Reiche, S. Simple Method to Generate Terawatt-Attosecond X-ray Free-Electron-Laser Pulses. *Phys. Rev. Lett.* **2015**, *114*, 244801.

43. Raguin, J.-Y.; Bopp, M.; Citterio, A.; Scherer, A. The Swiss FEL RF gun: RF design and thermal analysis. In Proceedings of the 26th Linear Accelerator Conference (LINAC 2012), Tel Aviv, Israel, 9–14 September 2012; pp. 442–444.

44. Raguin, J.-Y. The Swiss FEL S-Band Accelerating Structure: RF Design. In Proceedings of the 26th Linear Accelerator Conference (LINAC 2012), Tel Aviv, Israel, 9–14 September 2012; pp. 498–500.

45. Pedrozzi, M.; Calvi, M.; Ischebeck, R.; Reiche, S.; Vicario, C.; Fell, B.D.; Thompson, N. The laser heater system of SwissFEL. In Proceedings of the 36th International Free Electron Laser Conference (FEL 2014), Basel, Switzerland, 25–28 August 2014; pp. 871–877.

46. Dehler, M.; Raguin, J.-Y.; Citterio, A.; Falone, A.; Wuensch, W.; Riddone, G.; Grudiev, A.; Zennaro, R. X-band rf structure with integrated alignment monitors. *Phys. Rev. Spec. Top. Accel. Beams* **2009**, *12*, 062001.

47. Dehler, M.; Atieh, S.; Gudkov, D.; Lebet, S.; Riddone, G.; Shi, J.; Citterio, A.; Zennaro, R.; Scherrer, P.; D'Auria, G.; et al. Fabrication of the CERN/PSI/ST X-band accelerating structures. In Proceedings of the 2nd International Particle Accelerator Conference (IPAC 2011), San Sebastián, Spain, 4–9 September 2011; pp. 86–88.

48. Dehler, M.; Zennaro, R.; Citterio, A.; Lebet, S.; Riddone, G.; Shi, J.; Samoshkin, A.; Gudkov, D.; D'Auria, G.; Serpico, C. A multi purpose x band accelerating structure. In Proceedings of the International Particle Accelerator Conference (IPAC 2012), New Orleans, LA, USA, 2–25 May 2012; pp. 70–72.

49. Schietinger, T.; Pedrozzi, M.; Aiba, M.; Arsov, V.; Bettoni, S.; Beutner, B.; Calvi, M.; Craievich, P.; Dehler, M.; Frei, F.; et al. Commissioning experience and beam physics measurements at the SwissFEL Injector Test Facility. *Phys. Rev. Accel. Beams* **2016**, *19*, 100702.

50. Divall, M.C.; Prat, E.; Bettoni, S.; Vicario, C.; Trisorio, A.; Schietinger, T.; Hauri, C.P. Intrinsic emittance reduction of copper cathodes by laser wavelength tuning in an rf photoinjector. *Phys. Rev. Spec. Top. Accel. Beams* **2015**, *18*, 033401.

51. Prat, E.; Bettoni, S.; Braun, H.-H.; Ganter, R.; Schietinger, T. Measurements of copper and cesium telluride cathodes in a radio-frequency photoinjector. *Phys. Rev. Spec. Top. Accel. Beams* **2015**, *18*, 043401.

52. Prat, E.; Bettoni, S.; Braun, H.-H.; Divall, M.C.; Schietinger, T. Measurements of intrinsic emittance dependence on rf field for copper photocathodes. *Phys. Rev. Spec. Top. Accel. Beams* **2015**, *18*, 063401.

53. Ferrario, M.; Clendenin, J.E.; Palmer, D.T.; Rosenzweig, J.B.; Serafini, L. HOMDYN Study for the LCLS rf Photo-Injector. In Proceedings of 2nd ICFA Advanced Accelerator Workshop on the Physics of High Brightness Beams, Los Angeles, CA, USA, 9–11 September 1999.

54. Bettoni, S.; Pedrozzi, M.; Reiche, S. Low emittance injector design for free electron lasers. *Phys. Rev. Spec. Top. Accel. Beams* **2015**, *18*, 123403.

55. Prat, E.; Aiba, M. Four-dimensional transverse beam matrix measurement using the multiple-quadrupole scan technique. *Phys. Rev. Spec. Top. Accel. Beams* **2014**, *17*, 052801.

56. Prat, E.; Aiba, M.; Bettoni, S.; Beutner, B.; Reiche, S.; Schietinger, T. Emittance measurements and minimization at the SwissFEL Injector Test Facility. *Phys. Rev. Spec. Top. Accel. Beams* **2014**, *17*, 104401.

57. Bettoni, S.; Aiba, M.; Beutner, B.; Pedrozzi, M.; Prat, E.; Reiche, S.; Schietinger, T. Preservation of low slice emittance in bunch compressors. *Phys. Rev. Accel. Beams* **2016**, *19*, 034402.

58. Paraliev, M.; Gough, C.; Dordevic, S.; Braun, H. High stability resonant kicker development for the SwissFEL switch yard. In Proceedings of the 36th International Free Electron Laser Conference (FEL 2014), Basel, Switzerland, 25–28 August 2014; pp. 103–106.

59. Loehl, F.; Alex, J.; Blumer, H.; Bopp, M.; Braun, H.; Citterio, A.; Ellenberger, U.; Fitze, H.; Joehri, H.; Kleeb, T.; et al. Status of the swissfel c-band linear accelerator. In Proceedings of the 35th International Free Electron Laser Conference (FEL 2013), New York, NY, USA, 26–29 August 2013; pp. 317–321.

60. Loehl, F.; Alex, J.; Blumer, H.; Bopp, M.; Braun, H.; Citterio, A.; Ellenberger, U.; Fitze, H.; Joehri, H.; Kleeb, T.; et al. Status of the SwissFEL C-band linac. In Proceedings of the 36th International Free Electron Laser Conference (FEL 2014), Basel, Switzerland, 25–28 August 2014; pp. 322–326.

61. Raguin, J.-Y.; Bopp, M. The Swiss FEL C-band accelerating structure: RF design and thermal analysis. In Proceedings of the 26th Linear Accelerator Conference (LINAC 2012), Tel Aviv, Israel, 9–14 September 2012; pp. 501–503.

62. Ellenberger, U.; Paly, L.; Blumer, H.; Zumbach, C.; Loehl, F.; Bopp, M.; Fitze, H. Status of the manufacturing process for the SwissFEL C-band accelerating structures. In Proceedings of the 35th International Free Electron Laser Conference (FEL 2013), New York, NY, USA, 26–29 August 2013; pp. 245–249.

63. Zennaro, R.; Bopp, M.; Citterio, A.; Reiser, R.; Stapf, T. C-band RF pulse compressor for SwissFEL. In Proceedings of the 4th International Particle Accelerator Conference (IPAC 2013), Shanghai, China, 12–17 May 2013; pp. 2827–2829.

64. Ellenberger, U.; Blumer, H.; Bopp, M.; Citterio, A.; Heusser, M.; Kleeb, M.; Paly, L.; Probst, M.; Stapf, T.; Zennaro, R. The SwissFEL C-band RF pulse compressor: Manufacturing and proof of precision by RF measurements. In Proceedings of the 36th International Free Electron Laser Conference (FEL 2014), Basel, Switzerland, 25–29 August 2014; pp. 859–863.

65. Loehl, F. on Behalf of the SwissFEL Team. Status of SwissFEL. In Proceedings of the 28th Linear Accelerator Conference (LINAC 2016), East Lansing, MI, USA, 25–30 September 2016; pp. 22–26.

66. Tanaka, T.; Tsusu, R.; Nakajima, T.; Seike, T.; Kitamura, H. In-situ undulator field measurement with the SAFALI system. In Proceedings of the 29th International Free Electron Laser Conference (FEL 2007), Novosibirsk, Russia, 26–31 August 2007; pp. 468–471.

67. Calvi, M.; Brügger, M.; Danner, S.; Imhof, A.; Jöhri, H.; Schmidt, T.; Scoular, C. SwissFEL U15 magnet assembly: First experimental results. In Proceedings of the 34th International Free Electron Laser Conference (FEL 2012), Nara, Japan, 26–31 August 2012; pp. 662–665.

68. Calvi, M.; Camenzuli, C.; Ganter, R.; Sammut, N.; Schmidt, T. Magnetic Measurement Optimisation and Modelling of the SwissFEL U15 In-Vacuum Undulators. *J. Synchrotron Radiat.* To be submitted.

69. Emma, P.J.; Carr, R.; Nuhn, H.D. Beam-based alignment for the LCLS FEL undulator. *Nucl. Instrum. Methods Phys. Res. A* **1999**, *429*, 407–413.

70. Available online: http://www.bergoz.com/ict-bcm-ihr?d=7 (accesed on 3 July 2017).

71. Stulle, F.; Bergoz, J. Turbo-ICT pico-Coulomb calibration to percent-level accuracy. In Proceedings of the 37th International Free Electron Laser Conference (FEL 2015), Daejeon, Korea, 23–28 August 2015; pp. 118–121.

72. Keil, B.; Baldinger, R.; Ditter, R.; Koprek, W.; Kramert, R.; Marcellini, F.; Marinkovic, G.; Roggli, M.; Rohrer, M.; Stadler, M.; et al. Design of the SwissFEL BPM system. In Proceedings of the 2nd International Beam Instrumentation Conference (IBIC 2013), Oxford, UK, 16–19 September 2013; pp. 427–430.

73. Ishikawa, T.; Aoyagi, H.; Asaka, T.; Asano, Y.; Azumi, N.; Bizen, T.; Ego, H.; Fukami, K.; Fukui, T.; Furukawa, Y.; et al. A compact X-ray free-electron laser emitting in the sub-ångström region. *Nat. Photonics* **2012**, *6*, 540–544.

74. Ischebeck, R.; Prat, E.; Thominet, V.; Loch, C.O. Transverse profile imager for ultrabright electron beams. *Phys. Rev. Spec. Top. Accel. Beams* **2015**, *18*, 082802.

75. Orlandi, G.L.; Heimgartner, P.; Ischebeck, R.; Loch, C.O.; Trovati, S.; Valitutti, P.; Schlott, V.; Ferianis, M.; Penco, G. Design and experimental tests of free electron laser wire scanners. *Phys. Rev. Accel. Beams* **2016**, *19*, 092802.

76. Orlandi, G.L.; Aiba, M.; Bettoni, S.; Beutner, B.; Brands, H.; Ischebeck, R.; Prat, E.; Peier, P.; Schietinger, T.; Schlott, V.; et al. Bunch-compressor transverse profile monitors of the SwissFEL Injector Test Facility. In Proceedings of the 1st International Beam Instrumentation Conference (IBIC 2012), Tsukuba, Japan, 1–4 October 2012; pp. 272–275.

77. Orlandi, G.L.; Aiba, M.; Baerenbold, F.; Bettoni, S.; Beutner, B.; Brands, H.; Craievich, P.; Frei, F.; Ischebeck, R.; Pedrozzi, M.; et al. Characterization of compressed bunches in the SwissFEL injector test facility. In Proceedings of the 2nd International Beam Instrumentation Conference (IBIC 2013), Oxford, UK, 16–19 September 2013; pp. 515–518.

78. Bettoni, S.; Reiche, S. High resolution method for uncorrelated energy spread measurement. Presented at Workshop on Physics and Applications of High Brightness Beams, Havana, Cuba, 28 March–1 April 2016. (unpublished)

79. Loew, G.A.; Altenmueller, O.H. Design and applications of R.F. deflecting structures at SLAC. In Proceedings of the 5th International Conference on High-Energy Accelerators, Frascati, Italy, 9–16 September 1965; Comitato Naz. Ener. Nucl.: Rome, Italy 1966.

80. Akre, R.; Bentson, L.; Emma, P.J.; Krejcik, P. A transverse RF deflecting structure for bunch length and phase space diagnostics. In Proceedings of the 19th Particle Accelerator Conference (PAC 2001), Chicago, IL, USA, 18–22 June 2001; pp. 2353–2355.

81. Akre, R.; Bentson, L.; Emma, P.J.; Krejcik, P. Bunch length measurements using a transverse RF deflecting structure in the SLAC linac. In Proceedings of the 8th European Particle Accelerator Conference (EPAC 2002), Paris, France, 3–7 June 2002; pp. 1882–1884.

82. Prat, E.; Aiba, M. General and efficient dispersion-based measurement of beam slice parameters. *Phys. Rev. Spec. Top. Accel. Beams* **2014**, *17*, 032801.

83. Löhl, F.; Arsov, V.; Felber, M.; Hacker, K.; Jalmuzna, W.; Lorbeer, B.; Ludwig, F.; Matthiesen, K.H.; Schlarb, H.; Schmidt, B.; et al. Electron bunch timing with femtosecond precision in a superconducting free-electron laser. *Phys. Rev. Lett.* **2010**, *104*, 144801.

84. Arsov, V.; Dehler, M.; Hunziker, S.; Kaiser, M.; Schlott, V. First results from the bunch arrival-time monitor at the SwissFEL test injector. In Proceedings of the 2nd International Beam Instrumentation Conference (IBIC 2013), Oxford, UK, 16–19 September 2013; pp. 8–11.

85. Arsov, V.; Aiba, M.; Dehler, M.; Frei, F.; Hunziker, S.; Kaiser, M.; Romann, A.; Schlott, V. Commissioning and results from the bunch arrival-time monitor downstream the bunch compressor at the SwissFEL injector test facility. In Proceedings of the 36th International Free Electron Laser Conference (FEL 2014), Basel, Switzerland, 25–29 August 2014; pp. 933–936.

86. Angelovski, A.; Kuntzsch, M.; Czwalinna, M.K.; Penirschke, A.; Hansli, M.; Sydlo, C.; Arsov, V.; Hunziker, S.; Schlarb, H.; Gensch, M.; et al. Evaluation of the cone-shaped pickup performance for low charge sub-10 fs arrival-time measurements at free electron laser facilities. *Phys. Rev. Spec. Top. Accel. Beams* **2015**, *18*, 012801.

87. Frei, F.; Gorgisyan, I.; Smit, B.; Orlandi, G.L.; Beutner, B.; Prat, E.; Ischebeck, R.; Schlott, V.; Peier, P. Development of electron bunch compression monitors for SwissFEL. In Proceedings of the 2nd International Beam Instrumentation Conference (IBIC 2013), Oxford, UK, 16–19 September 2013; pp. 769–771.

88. Follath, R.; Flechsig, U.; Milne, C.J.; Szlachetko, J.; Ingold, G.; Patterson, B.; Patthey, L.; Abela, R. Optical design of the ARAMIS-beamlines at SwissFEL. *AIP Conf. Proc.* **2016**, *1741*, 020009.

89. Kirkpatrick, P.; Baez, A.V. Formation of Optical Images by X-rays. *J. Opt. Soc. Am.* **1948**, *38*, 766–774.

90. Koyama, T.; Yumoto, H.; Tono, K.; Sato, T.; Togashi, T.; Inubushi, Y.; Katayama, T.; Kim, J.; Matsuyama, S.; Mimura, H.; et al. Damage threshold investigation using grazing incidence irradiation by hard X-ray free electron laser. In Proceedings of the SPIE 8848, Advances in X-ray/EUV Optics and Components VIII, San Diego, CA, USA, 18 October 2013; p. 88480T.

91. Aquila, A.; Ozkan, C.; Sobierajski, R.; Hajkova, V.; Burian, T.; Chalupsky, J.; Juha, L.; Störmer, M.; Ohashi, H.; Koyama, T.; et al. Results from single shot grazing incidence hard X-ray damage measurements conducted at the SACLA FEL. In Proceedings of the SPIE 8777, Damage to VUV, EUV, and X-ray Optics IV; and EUV and X-ray Optics: Synergy between Laboratory and Space III, Prague, Czech Republic, 3 May 2013; p. 87770H.

92. Aquila, A.; Sobierajski, R.; Ozkan, C.; Hájková, V.; Burian, T.; Chalupský, J.; Juha, L.; Störmer, M.; Bajt, S.; Klepka, M.T.; et al. Fluence thresholds for grazing incidence hard X-ray mirrors. *Appl. Phys. Lett.* **2015**, *106*, 241905.

93. Flechsig, U.; Bahrdt, J.; Follath, R.; Reiche, S. Physical optics simulations with PHASE for SwissFEL beamlines. *AIP Conf. Proc.* **2016**, *1741*, 040040.

94. Dynamic Structures & Materials, LLC, Franklin, TN, USA. Available online: http://www.dynamic-structures.com (accessed on 14 July 2017).

95. Kato, M.; Tanaka, T.; Kurosawa, T.; Saito, N.; Richter, M.; Sorokin, A.A.; Tiedtke, K.; Kudo, T.; Tono, K.; Yabashi, M.; et al. Pulse energy measurement at the hard X-ray laser in Japan. *Appl. Phys. Lett.* **2012**, *101*, 023503.

96. Tiedtke, K.; Sorokin, A.A.; Jastrow, U.; Juranic, P.; Kreis, S.; Gerken, N.; Richter, M.; Arp, U.; Feng, Y.; Nordlund, D.; et al. Absolute pulse energy measurements of soft X-rays at the Linac Coherent Light Source. *Opt. Express* **2014**, *22*, 21214–21226.

97. Feng, Y.; Feldkamp, J.M.; Fritz, D.M.; Cammarata, M.; Robert, A.; Caronna, C.; Lemke, H.T.; Zhu, D.; Lee, S.; Boutet, S.; et al. A single-shot intensity-position monitor for hard X-ray FEL sources. In Proceedings of the SPIE 8140, X-ray Lasers and Coherent X-ray Sources: Development and Applications IX, San Diego, CA, USA, 23–25 August 2011; p. 81400Q.

98. Tono, K.; Kudo, T.; Yabashi, M.; Tachibana, T.; Feng, Y.; Fritz, D.; Hastings, J.; Ishikawa, T. Single-shot beam-position monitor for X-ray free electron laser. *Rev. Sci. Instrum.* **2011**, *82*, 023108.

99. Martin, T.; Koch, A. Recent developments in X-ray imaging with micrometer spatial resolution. *J. Synchrotron Radiat.* **2006**, *13*, 180–194.

100. Naito, T. YAG:Ce screen monitor using a gated CCD camera. In Proceedings of the 3rd International Beam Instrumentation Conference (IBIC 2014), Monterey, CA, USA, 14–18 September 2014; pp. 426–429.

101. Sikorski, M.; Song, S.; Schropp, A.; Seiboth, F.; Feng, Y.; Alonso-Mori, R.; Chollet, M.; Lemke, H.T.; Sokaras, D.; Weng, T.C.; et al. Focus characterization at an X-ray free-electron laser by coherent scattering and speckle analysis. *J. Synchrotron Radiat.* **2015**, *22*, 599–605.

102. Bionta, M.R.; Lemke, H.T.; Cryan, J.P.; Glownia, J.M.; Bostedt, C.; Cammarata, M.; Castagna, J.C.; Ding, Y.; Fritz, D.M.; Fry, A.R.; et al. Spectral encoding of X-ray/optical relative delay. *Opt. Express* **2011**, *19*, 21855–21865.

103. Lemke, H.T.; Weaver, M.; Chollet, M.; Robinson, J.; Glownia, J.M.; Zhu, D.; Bionta, M.R.; Cammarata, M.; Harmand, M.; Coffee, R.N.; et al. Femtosecond optical/hard X-ray timing diagnostics at an FEL: Implementation and Performance. In Proceedings of the SPIE 8778, Advances in X-ray Free-Electron Lasers II: Instrumentation, Prague, Czech Republic, 8 May 2013; p. 87780S.

104. Katayama, T.; Owada, S.; Togashi, T.; Ogawa, K.; Karvinen, P.; Vartiainen, I.; Eronen, A.; David, C.; Sato, T.; Nakajima, K.; et al. A beam branching method for timing and spectral characterization of hard X-ray free-electron lasers. *Struct. Dyn.* **2016**, *3*, 034301.

105. Fruehling, U.; Wieland, M.; Gensch, M.; Gebert, T.; Schuette, B.; Krikunova, M.; Kalms, R.; Budzyn, F.; Grimm, O.; Rossbach, J.; et al. Single-shot terahertz-field-driven X-ray streak camera. *Nat. Photonics* **2009**, *3*, 523–528.

106. Juranić, P.N.; Stepanov, A.; Peier, P.; Hauri, C.P.; Ischebeck, R.; Schlott, V.; Radovic, M.; Erny, C.; Ardana-Lamas, F.; Monoszlai, B.; et al. A scheme for a shot-to-shot, femtosecond-resolved pulse length and arrival time measurement of free electron laser X-ray pulses that overcomes the time jitter problem between the FEL and the laser. *J. Instrum.* **2014**, *9*, P03006.

107. Juranić, P.N.; Stepanov, A.; Ischebeck, R.; Schlott, V.; Pradervand, C.; Patthey, L.; Radovic, M.; Gorgisyan, I.; Rivkin, L.; Hauri, C.P.; et al. High-precision X-ray FEL pulse arrival time measurements at SACLA by a THz streak camera with Xe clusters. *Opt. Express* **2014**, *22*, 30004–30012.

108. Gorgisyan, I.; Ischebeck, R.; Erny, C.; Dax, A.; Patthey, L.; Pradervand, C.; Sala, L.; Milne, C.J.; Lemke, H.T.; Hauri, C.P.; et al. THz streak camera method for synchronous arrival time measurement of two-color hard X-ray FEL pulses. *Opt. Express* **2017**, *25*, 2080–2091.

109. Ardana-Lamas, F.; Erny, C.; Stepanov, A.G.; Gorgisyan, I.; Juranić, P.N.; Hauri, C.P. Temporal characterization of individual harmonics of an attosecond pulse train by THz streaking. *Phys. Rev. A* **2016**, *93*, 043838.

110. Rehanek, J.; Makita, M.; Wiegand, P.; Heimgartner, P.; Pradervand, C.; Seniutinas, G.; Flechsig, U.; Thominet, V.; Schneider, C.W.; Fernandez, A.R.; et al. The hard X-ray Photon Single-Shot Spectrometer of SwissFEL—Initial characterization. *J. Instrum.* **2017**, *12*, P05024.

111. Makita, M.; Karvinen, P.; Zhu, D.; Juranić, P.N.; Grünert, J.; Cartier, S.; Jungmann-Smith, J.H.; Lemke, H.T.; Mozzanica, A.; Nelson, S.; et al. High-resolution single-shot spectral monitoring of hard X-ray free-electron laser radiation. *Optica* **2015**, *2*, 912–916.

112. Erny, C.; Hauri, C.P. The SwissFEL Experimental Laser facility. *J. Synchrotron Radiat.* **2016**, *23*, 1143–1150.

113. Neutze, R.; Wouts, R.; van der Spoel, D.; Weckert, E.; Hajdu, J. Potential for biomolecular imaging with femtosecond X-ray pulses. *Nature* **2000**, *406*, 752–757.

114. Nass, K.; Foucar, L.; Barends, T.R.M.; Hartmann, E.; Botha, S.; Shoeman, R.L.; Doak, R.B.; Alonso-Mori, R.; Aquila, A.; Bajt, S.; et al. Indications of radiation damage in ferredoxin microcrystals using high-intensity X-FEL beams. *J. Synchrotron Radiat.* **2015**, *22*, 225–238.

115. Kern, J.; Alonso-Mori, R.; Tran, R.; Hattne, J.; Gildea, R.J.; Echols, N.; Glockner, C.; Hellmich, J.; Laksmono, H.; Sierra, R.G.; et al. Simultaneous Femtosecond X-ray Spectroscopy and Diffraction of Photosystem II at Room Temperature. *Science* **2013**, *340*, 491–495.

116. Kern, J.; Tran, R.; Alonso-Mori, R.; Koroidov, S.; Echols, N.; Hattne, J.; Ibrahim, M.; Gul, S.; Laksmono, H.; Sierra, R.G.; et al. Taking snapshots of photosynthetic water oxidation using femtosecond X-ray diffraction and spectroscopy. *Nat. Commun.* **2014**, *5*, 4371.

117. Kupitz, C.; Basu, S.; Grotjohann, I.; Fromme, R.; Zatsepin, N.A.; Rendek, K.N.; Hunter, M.S.; Shoeman, R.L.; White, T.A.; Wang, D.; et al. Serial time-resolved crystallography of photosystem II using a femtosecond X-ray laser. *Nature* **2014**, *513*, 261–265.

118. Nango, E.; Royant, A.; Kubo, M.; Nakane, T.; Wickstrand, C.; Kimura, T.; Tanaka, T.; Tono, K.; Song, C.; Tanaka, R.; et al. A three-dimensional movie of structural changes in bacteriorhodopsin. *Science* **2016**, *354*, 1552–1557.

119. Suga, M.; Akita, F.; Sugahara, M.; Kubo, M.; Nakajima, Y.; Nakane, T.; Yamashita, K.; Umena, Y.; Nakabayashi, M.; Yamane, T.; et al. Light-induced structural changes and the site of O=O bond formation in PSII caught by XFEL. *Nature* **2017**, *543*, 131–135.

120. Bionta, M.R.; Hartmann, N.; Weaver, M.; French, D.; Nicholson, D.J.; Cryan, J.P.; Glownia, J.M.; Baker, K.; Bostedt, C.; Chollet, M.; et al. Spectral encoding method for measuring the relative arrival time between X-ray/optical pulses. *Rev. Sci. Instrum.* **2014**, *85*, 083116.

121. Hebling, J.; Almasi, G.; Kozma, I.Z.; Kuhl, J. Velocity matching by pulse front tilting for large-area THz-pulse generation. *Opt. Express* **2002**, *10*, 1161–1166.
122. Harmand, M.; Coffee, R.; Bionta, M.R.; Chollet, M.; French, D.; Zhu, D.; Fritz, D.M.; Lemke, H.T.; Medvedev, N.; Ziaja, B.; et al. Achieving few-femtosecond time-sorting at hard X-ray free-electron lasers. *Nat. Photonics* **2013**, *7*, 215–218.
123. Ruchert, C.; Vicario, C.; Hauri, C.P. Scaling submillimeter single-cycle transients toward megavolts per centimeter field strength via optical rectification in the organic crystal OH1. *Optics Lett.* **2012**, *37*, 899–901.
124. Ruchert, C.; Vicario, C.; Hauri, C.P. Spatiotemporal Focusing Dynamics of Intense Supercontinuum THz Pulses. *Phys. Rev. Lett.* **2013**, *110*, 123902.
125. Shalaby, M.; Hauri, C.P. Demonstration of a low-frequency three-dimensional terahertz bullet with extreme brightness. *Nat. Commun.* **2015**, *6*, 5976.
126. Nisoli, M.; DeSilvestri, S.; Svelto, O. Generation of high energy 10 fs pulses by a new pulse compression technique. *Appl. Phys. Lett.* **1996**, *68*, 2793–2795.
127. Divall, M.C.; Mutter, P.; Divall, E.J.; Hauri, C.P. Femtosecond resolution timing jitter correction on a TW scale Ti:sapphire laser system for FEL pump-probe experiments. *Opt. Express* **2015**, *23*, 29929–29939.
128. Divall, M.C.; Kaiser, M.; Hunziker, S.; Vicario, C.; Beutner, B.; Schietinger, T.; Lüthi, M.; Pedrozzi, M.; Hauri, C.P. Timing jitter studies of the SwissFEL Test Injector drive laser. *Nucl. Instrum. Methods Phys. Res. A* **2014**, *735*, 471–479.
129. The SLS Group for Macromolecular Crystallography. Available online: https://www.psi.ch/macromolecular-crystallography/ (accessed on 3 July 2017).
130. Milne, C.J.; Penfold, T.J.; Chergui, M. Recent experimental and theoretical developments in time-resolved X-ray spectroscopies. *Coord. Chem. Rev.* **2014**, *277*, 44–68.
131. Boutet, S.; Lomb, L.; Williams, G.J.; Barends, T.R.M.; Aquila, A.; Doak, R.B.; Weierstall, U.; DePonte, D.P.; Steinbrener, J.; Shoeman, R.L.; et al. High-Resolution Protein Structure Determination by Serial Femtosecond Crystallography. *Science* **2012**, *337*, 362–364.
132. Barends, T.R.M.; Foucar, L.; Ardevol, A.; Nass, K.; Aquila, A.; Botha, S.; Doak, R.B.; Falahati, K.; Hartmann, E.; Hilpert, M.; et al. Direct observation of ultrafast collective motions in CO myoglobin upon ligand dissociation. *Science* **2015**, *350*, 445–450.
133. Calvi, M.; Aiba, M.; Brügger, M.; Danner, S.; Schmidt, T.; Ganter, R.; Schietinger, T.; Ischebeck, R. General strategy for the commissioning of the Aramis undulators with a 3 GeV electron beam. In Proceedings of the 36th International Free Electron Laser Conference (FEL 2014), Basel, Switzerland, 25–29 August 2014; pp. 107–110.
134. Calvi, M.; Aiba, M.; Brügger, M.; Danner, S.; Ganter, R.; Ozkan, C.; Schmidt, T. Summary of the U15 prototype magnetic performance. In Proceedings of the 36th International Free Electron Laser Conference (FEL 2014), Basel, Switzerland, 25–29 August 2014; pp. 111–115.
135. Weierstall, U.; James, D.; Wang, C.; White, T.A.; Wang, D.; Liu, W.; Spence, J.C.H.; Doak, R.B.; Nelson, G.; Fromme, P.; et al. Lipidic cubic phase injector facilitates membrane protein serial femtosecond crystallography. *Nat. Commun.* **2014**, *5*, 3309.
136. Weierstall, U. Liquid sample delivery techniques for serial femtosecond crystallography. *Phil. Trans. R. Soc. B* **2014**, *369*, 20130337.
137. Oberthuer, D.; Knoska, J.; Wiedorn, M.O.; Beyerlein, K.R.; Bushnell, D.A.; Kovaleva, E.G.; Heymann, M.; Gumprecht, L.; Kirian, R.A.; Barty, A.; et al. Double-flow focused liquid injector for efficient serial femtosecond crystallography. *Sci. Rep.* **2017**, *7*, 44628.
138. Chapman, H.N.; Caleman, C.; Timneanu, N. Diffraction before destruction. *Phil. Trans. R. Soc. B* **2014**, *369*, 20130313.
139. Neutze, R.; Brändén, G.; Schertler, G.F.X. Membrane protein structural biology using X-ray free electron lasers. *Curr. Opin. Struct. Biol.* **2015**, *33*, 115–125.
140. Frank, M.; Carlson, D.B.; Hunter, M.S.; Williams, G.J.; Messerschmidt, M.; Zatsepin, N.A.; Barty, A.; Benner, W.H.; Chu, K.; Graf, A.T.; et al. Femtosecond X-ray diffraction from two-dimensional protein crystals. *IUCrJ* **2014**, *1*, 95–100.
141. Pedrini, B.; Tsai, C.J.; Capitani, G.; Padeste, C.; Hunter, M.S.; Zatsepin, N.A.; Barty, A.; Benner, W.H.; Boutet, S.; Feld, G.K.; et al. 7 A resolution in protein two-dimensional-crystal X-ray diffraction at Linac Coherent Light Source. *Phil. Trans. R. Soc. B* **2014**, *369*, 20130500.

142. Bratos, S.; Wulff, M. Time-resolved X-ray diffraction from liquids. *Adv. Chem. Phys.* **2008**, *137*, 1–29.
143. Haldrup, K.; Christensen, M.; Nielsen, M.M. Analysis of time-resolved X-ray scattering data from solution-state systems. *Acta Cryst. A* **2010**, *66*, 261–269.
144. Kim, J.; Kim, K.H.; Lee, J.H.; Ihee, H. Ultrafast X-ray diffraction in liquid, solution and gas: Present status and future prospects. *Acta Cryst. A* **2010**, *66*, 270–280.
145. Arnlund, D.; Johansson, L.C.; Wickstrand, C.; Barty, A.; Williams, G.J.; Malmerberg, E.; Davidsson, J.; Milathianaki, D.; Deponte, D.P.; Shoeman, R.L.; et al. Visualizing a protein quake with time-resolved X-ray scattering at a free-electron laser. *Nat. Methods* **2014**, *11*, 923–926.
146. Cammarata, M.; Levantino, M.; Schotte, F.; Anfinrud, P.A.; Ewald, F.; Choi, J.; Cupane, A.; Wulff, M.; Ihee, H. Tracking the structural dynamics of proteins in solution using time-resolved wide-angle X-ray scattering. *Nat. Methods* **2008**, *5*, 881–886.
147. Levantino, M.; Schirò, G.; Lemke, H.T.; Cottone, G.; Glownia, J.M.; Zhu, D.; Chollet, M.; Ihee, H.; Cupane, A.; Cammarata, M. Ultrafast myoglobin structural dynamics observed with an X-ray free-electron laser. *Nat. Commun.* **2015**, *6*, 6772.
148. Malmerberg, E.; Bovee-Geurts, P.H.M.; Katona, G.; Deupi, X.; Arnlund, D.; Wickstrand, C.; Johansson, L.C.; Westenhoff, S.; Nazarenko, E.; Schertler, G.F.X.; et al. Conformational activation of visual rhodopsin in native disc membranes. *Sci. Signal.* **2015**, *8*, ra26.
149. Bergmann, U.; Glatzel, P. X-ray emission spectroscopy. *Photosynth. Res.* **2009**, *102*, 255–266.
150. Szlachetko, J.; Nachtegaal, M.; de Boni, E.; Willimann, M.; Safonova, O.; Sa, J.; Smolentsev, G.; Szlachetko, M.; van Bokhoven, J.A.; Dousse, J.C.; et al. A von Hamos X-ray spectrometer based on a segmented-type diffraction crystal for single-shot X-ray emission spectroscopy and time-resolved resonant inelastic X-ray scattering studies. *Rev. Sci. Inst.* **2012**, *83*, 103105.
151. Kavčič, M.; Žitnik, M.; Bučar, K.; Mihelič, A.; Marolt, B.; Szlachetko, J.; Glatzel, P.; Kvashnina, K. Hard X-ray absorption spectroscopy for pulsed sources. *Phys. Rev. B* **2013**, *87*, 075106.
152. Szlachetko, J.; Nachtegaal, M.; Sa, J.; Dousse, J.C.; Hoszowska, J.; Kleymenov, E.; Janousch, M.; Safonova, O.V.; König, C.; van Bokhoven, J.A. High energy resolution off-resonant spectroscopy at sub-second time resolution: (Pt(acac)$_2$) decomposition. *Chem. Commun.* **2012**, *48*, 10898.
153. Szlachetko, J.; Milne, C.J.; Hoszowska, J.; Dousse, J.C.; Błachucki, W.; Sa, J.; Kayser, Y.; Messerschmidt, M.; Abela, R.; Boutet, S.; et al. Communication: The electronic structure of matter probed with a single femtosecond hard X-ray pulse. *Struct. Dyn.* **2014**, *1*, 021101.
154. Błachucki, W.; Szlachetko, J.; Hoszowska, J.; Dousse, J.C.; Kayser, Y.; Nachtegaal, M.; Sa, J. High Energy Resolution Off-Resonant Spectroscopy for X-ray Absorption Spectra Free of Self-Absorption Effects. *Phys. Rev. Lett.* **2014**, *112*, 173003.
155. Schülke, W. *Electron Dynamics by Inelastic X-ray Scattering*; Oxford Series on Synchrotron Radiation; Oxford University Press (OUP): Oxford, UK, 2007.
156. Canton, S.E.; Kjaer, K.S.; Vankó, G.; van Driel, T.B.; Adachi, S.I.; Bordage, A.; Bressler, C.; Chabera, P.; Christensen, M.; Dohn, A.O.; et al. Visualizing the non-equilibrium dynamics of photoinduced intramolecular electron transfer with femtosecond X-ray pulses. *Nat. Commun.* **2015**, *6*, 6359.
157. Haldrup, K.; Gawelda, W.; Abela, R.; Alonso-Mori, R.; Bergmann, U.; Bordage, A.; Cammarata, M.; Canton, S.E.; Dohn, A.O.; van Driel, T.B.; et al. Observing Solvation Dynamics with Simultaneous Femtosecond X-ray Emission Spectroscopy and X-ray Scattering. *J. Phys. Chem. B* **2016**, *120*, 1158–1168.
158. Kern, J.; Hattne, J.; Tran, R.; Alonso-Mori, R.; Laksmono, H.; Gul, S.; Sierra, R.G.; Rehanek, J.; Erko, A.; Mitzner, R.; et al. Methods development for diffraction and spectroscopy studies of metalloenzymes at X-ray free-electron lasers. *Phil. Trans. R. Soc. B* **2014**, *369*, 20130590.
159. Jungmann-Smith, J.H.; Bergamaschi, A.; Cartier, S.; Dinapoli, R.; Greiffenberg, D.; Johnson, I.; Maliakal, D.; Mezza, D.; Mozzanica, A.; Ruder, C.; et al. JUNGFRAU 0.2: Prototype characterization of a gain-switching, high dynamic range imaging system for photon science at SwissFEL and synchrotrons. *J. Instrum.* **2014**, *9*, P12013.
160. Jungmann-Smith, J.H.; Bergamaschi, A.; Brückner, M.; Cartier, S.; Dinapoli, R.; Greiffenberg, D.; Jaggi, A.; Maliakal, D.; Mayilyan, D.; Medjoubi, K.; et al. Radiation hardness assessment of the charge-integrating hybrid pixel detector JUNGFRAU 1.0 for photon science. *Rev. Sci. Inst.* **2015**, *86*, 123110.

161. Mozzanica, A.; Bergamaschi, A.; Cartier, S.; Dinapoli, R.; Greiffenberg, D.; Johnson, I.; Jungmann, J.; Maliakal, D.; Mezza, D.; Ruder, C.; et al. Prototype characterization of the JUNGFRAU pixel detector for SwissFEL. *J. Instrum.* **2014**, *9*, C05010.

162. Hoszowska, J.; Dousse, J.C.; Kern, J.; Rhême, C. High-resolution von Hamos crystal X-ray spectrometer. *Nucl. Instrum. Methods Phys. Res. A* **1996**, *376*, 129–138.

163. Dousse, J.; Hoszowska, J. Crystal Spectrometers. In *High-Resolution XAS/XES: Analyzing Electronic Structures of Catalysts*; CRC Press: Boca Raton, FL, USA, 2014; pp. 27–58.

164. Weierstall, U.; Spence, J.C.H.; Doak, R.B. Injector for scattering measurements on fully solvated biospecies. *Rev. Sci. Inst.* **2012**, *83*, 035108.

165. Redford, S.; Bergamaschi, A.; Brückner, M.; Cartier, S.; Dinapoli, R.; Ekinci, Y.; Fröjdh, E.; Greiffenberg, D.; Mayilyan, D.; Mezza, D.; et al. Calibration status and plans for the charge integrating JUNGFRAU pixel detector for SwissFEL. *J. Instrum.* **2016**, *11*, C11013.

166. Ingold, G.; Beaud, P. Available online: https://www.psi.ch/swissfel/internal-reports (accessed on 3 July 2017).

167. Ingold, G.; Rittmann, J.; Beaud, P.; Divall, M.; Erny, C.; Flechsig, U.; Follath, R.; Hauri, C.P.; Hunziker, S.; Juranic, P.; et al. SwissFEL instrument ESB femtosecond pump-probe diffraction and scattering. *AIP Conf. Proc.* **2016**, *1741*, 030039.

168. Amann, J.; Berg, W.; Blank, V.; Decker, F.J.; Ding, Y.; Emma, P.J.; Feng, Y.; Frisch, J.; Fritz, D.; Hastings, J.; et al. Demonstration of self-seeding in a hard-X-ray free-electron laser. *Nat. Photonics* **2012**, *6*, 693–698.

169. Suzuki, M.; Inubushi, Y.; Yabashi, M.; Ishikawa, T. Polarization control of an X-ray free-electron laser with a diamond phase retarder. *J. Synchrotron Radiat.* **2014**, *21*, 466–472.

170. Strempfer, J.; Francoual, S.; Reuther, D.; Shukla, D.K.; Skaugen, A.; Schulte-Schrepping, H.; Kracht, T.; Franz, H. Resonant scattering and diffraction beamline P09 at PETRA III. *J. Synchrotron Radiat.* **2013**, *20*, 541–549.

171. Gerber, S.; Jang, H.; Nojiri, H.; Matsuzawa, S.; Yasumura, H.; Bonn, D.A.; Liang, R.; Hardy, W.N.; Islam, Z.; Mehta, A.; et al. Three-dimensional charge density wave order in YBa$_2$Cu$_3$O$_{6.67}$ at high magnetic fields. *Science* **2015**, *350*, 949–952.

172. SLS Detectors Group. Available online: https://www.psi.ch/detectors (accessed on 3 July 2017).

173. FEMTO Group. Available online: https://www.psi.ch/femto/ (accessed on 3 July 2017).

174. Beaud, P.; Caviezel, A.; Mariager, S.O.; Rettig, L.; Ingold, G.; Dornes, C.; Huang, S.W.; Johnson, J.A.; Radovic, M.; Huber, T.; et al. A time-dependent order parameter for ultrafast photoinduced phase transitions. *Nat. Mater.* **2014**, *13*, 923–927.

175. Kubacka, T.; Johnson, J.A.; Hoffmann, M.C.; Vicario, C.; de Jong, S.; Beaud, P.; Grübel, S.; Huang, S.W.; Huber, L.; Patthey, L.; et al. Large-Amplitude Spin Dynamics Driven by a THz Pulse in Resonance with an Electromagnon. *Science* **2014**, *343*, 1333–1336.

176. Grübel, S.; Johnson, J.A.; Beaud, P.; Dornes, C.; Ferrer, A.; Haborets, V.; Huber, L.; Huber, T.; Kohutych, A.; Kubacka, T.; et al. Ultrafast X-ray diffraction of a ferroelectric soft mode driven by broadband terahertz pulses. *arXiv* **2016**, arXiv:1602.05435v1.

177. Sala, M.M.; Henriquet, C.; Simonelli, L.; Verbeni, R.; Monaco, G. High energy-resolution set-up for Ir L$_3$ edge RIXS experiments. *J. Electron Spectrosc. Relat. Phenom.* **2013**, *188*, 150–154.

178. Shvyd'ko, Y.V.; Hill, J.P.; Burns, C.A.; Coburn, D.S.; Brajuskovic, B.; Casa, D.; Goetze, K.; Gog, T.; Khachatryan, R.; Kim, J.H.; et al. MERIX-Next generation medium energy resolution inelastic X-ray scattering instrument at the APS. *J. Electron Spectrosc. Relat. Phenom.* **2013**, *188*, 140–149.

179. Dean, M.P.M.; Cao, Y.; Liu, X.; Wall, S.; Zhu, D.; Mankowsky, R.; Thampy, V.; Chen, X.M.; Vale, J.G.; Casa, D.; et al. Ultrafast energy- and momentum-resolved dynamics of magnetic correlations in the photo-doped Mott insulator Sr2IrO4. *Nat. Mater.* **2016**, *15*, 601–605.

180. Geloni, G.; Kocharyan, V.; Saldin, E. A novel self-seeding scheme for hard X-ray FELs. *J. Mod. Opt.* **2011**, *58*, 1391–1403.

181. Feld, G.K.; Heymann, M.; Benner, W.H.; Pardini, T.; Tsai, C.J.; Boutet, S.; Coleman, M.A.; Hunter, M.S.; Li, X.; Messerschmidt, M.; et al. Low-Z polymer sample supports for fixed-target serial femtosecond X-ray crystallography. *J. Appl. Cryst.* **2015**, *48*, 1072–1079.

182. Mueller, C.; Marx, A.; Epp, S.W.; Zhong, Y.; Kuo, A.; Balo, A.R.; Soman, J.; Schotte, F.; Lemke, H.T.; Owen, R.L.; et al. Fixed target matrix for femtosecond time-resolved and in situ serial micro-crystallography. *Struct. Dyn.* **2015**, *2*, 054302.

183. Roedig, P.; Vartiainen, I.; Duman, R.; Panneerselvam, S.; Stübe, N.; Lorbeer, O.; Warmer, M.; Sutton, G.; Stuart, D.I.; Weckert, E.; et al. A micro-patterned silicon chip as sample holder for macromolecular crystallography experiments with minimal background scattering. *Sci. Rep.* **2015**, *5*, 10451.

184. Roedig, P.; Duman, R.; Sanchez-Weatherby, J.; Vartiainen, I.; Burkhardt, A.; Warmer, M.; David, C.; Wagner, A.; Meents, A. Room-temperature macromolecular crystallography using a micro-patterned silicon chip with minimal background scattering. *J. Appl. Cryst.* **2016**, *49*, 968–975.

185. Opara, N.; Martiel, I.; Arnold, S.A.; Braun, T.; Stahlberg, H.; Makita, M.; David, C.; Padeste, C. Direct protein crystallization on ultrathin membranes for diffraction measurements at X-ray free-electron lasers. *J. Appl. Cryst.* **2017**, *50*, 909–918.

186. Pedrini, B.; Martiel, I. Available online: https://www.psi.ch/swissfel/internal-reports (accessed on 3 July 2017).

187. Hirata, K.; Shinzawa-Itoh, K.; Yano, N.; Takemura, S.; Kato, K.; Hatanaka, M.; Muramoto, K.; Kawahara, T.; Tsukihara, T.; Yamashita, E.; et al. Determination of damage-free crystal structure of an X-ray–sensitive protein using an XFEL. *Nat. Methods* **2014**, *11*, 734–736.

188. Broennimann, C.; Eikenberry, E.F.; Henrich, B.; Horisberger, R.; Huelsen, G.; Pohl, E.; Schmitt, B.; Schulze-Briese, C.; Suzuki, M.; Tomizaki, T.; et al. The PILATUS 1M detector. *J. Synchrotron Radiat.* **2006**, *13*, 120–130.

189. Dinapoli, R.; Bergamaschi, A.; Henrich, B.; Horisberger, R.; Johnson, I.; Mozzanica, A.; Schmid, E.; Schmitt, B.; Schreiber, A.; Shi, X.; et al. EIGER: Next generation single photon counting detector for X-ray applications. *Nucl. Instrum. Methods Phys. Res. A* **2011**, *650*, 79–83.

190. Campbell, M. 10 years of the Medipix2 Collaboration. *Nucl. Instrum. Methods Phys. Res. A* **2011**, *633*, S1–S10.

191. Hart, P.; Boutet, S.; Carini, G.; Dubrovin, M.; Duda, B.; Fritz, D.; Haller, G.; Herbst, R.; Herrmann, S.; Kenney, C.; et al. The CSPAD megapixel X-ray camera at LCLS. In Proceedings of the SPIE 8504, X-ray Free-Electron Lasers: Beam Diagnostics, Beamline Instrumentation, and Applications, San Diego, CA, USA, 17 October 2012; p. 85040C.

192. Dragone, A.; Caragiulo, P.; Markovic, B.; Herbst, R.; Nishimura, K.; Reese, B.; Herrmann, S.; Hart, P.; Blaj, G.; Segal, J.; et al. *ePix: A Class of Front-End ASICs for Second Generation LCLS Integrating Hybrid Pixel Detectors*; IEEE Nuclear Science Symposium Conference Record; Stanford Linear Accelerator Center: Menlo Park, CA, USA, 2013.

193. Göttlicher, P.; Graafsma, H.; Hirsemann, H.; Jack, S.; Nilsson, B.; Potdevin, G.; Sheviakov, I.; Tian, F.; Trunk, U.; Youngman, C.; et al. *The Adaptive Gain Integrating Pixel Detector (AGIPD): A Detector for the European XFEL. Development and Status*; IEEE Nuclear Science Symposium Conference Record; Deutsches Elektronen-Synchrotron: Hamburg, Germany, 2009; pp. 1817–1820.

194. Blue, A.; French, M.; Seller, P.; O'Shea, V. Edgeless sensor development for the LPD hybrid pixel detector at XFEL. *Nucl. Instrum. Methods Phys. Res. A* **2009**, *607*, 55–56.

195. Porro, M.; Andricek, L.; Aschauer, S.; Bayer, M.; Becker, J.; Bombelli, L.; Castoldi, A.; De Vita, G.; Diehl, I.; Erdinger, F.; et al. Development of the DEPFET sensor with signal compression: A large format X-ray imager with mega-frame readout capability for the European XFEL. *IEEE Trans. Nucl. Sci.* **2012**, *59*, 3339–3351.

196. Saji, C.; Ohata, T.; Kudo, T.; Sugimoto, T.; Tanaka, R.; Hatsui, T.; Yamaga, M. Evaluation of data-acquisition front ends for handling high-bandwidth data from X-ray 2D detectors: A feasibility study. *Nucl. Instrum. Methods Phys. Res. A* **2013**, *731*, 229–233.

197. Mozzanica, A.; Bergamaschi, A.; Dinapoli, R.; Graafsma, H.; Greiffenberg, D.; Henrich, B.; Johnson, I.; Lohmann, M.; Valeria, R.; Schmitt, B.; et al. The GOTTHARD charge integrating readout detector: Design and characterization. *J. Instrum.* **2012**, *7*, C01019.

198. Hunziker, S.; Arsov, V.; Buechi, F.; Kaiser, M.; Romann, A.; Schlott, V.; Orel, P.; Zorzut, S. Reference distribution and synchronization system for SwissFEL: Concept and first results. In Proceedings of the 3rd International Beam Instrumentation Conference (IBIC 2014), Monterey, CA, USA, 14–18 September 2014; pp. 29–33.

199. Beutner, B.; Reiche, S. Sensitivity and tolerance study for the SwissFEL. In Proceedings of the 32nd International Free Electron Laser Conference (FEL 2010), Malmö, Sweden, 23–27 August 2010; pp. 437–440.

200. Available online: http://www.aps.anl.gov/epics (accesed on 3 July 2017).

201. Available online: https://slacmshankar.github.io/epicsarchiver_docs/ (accesed on 3 July 2017).

202. Available online: http://zeromq.org (accesed on 3 July 2017).

203. Mokso, R.; Theidel, G.; Billich, H.; Schlepütz, C.; Schmid, E.; Celcer, T.; Mikuljan, G.; Marone, F.; Schlumpf, N.; Stampanoni, M. Gigabit Fast Readout System for Tomography. *J. Synchrotron Radiat.* In preparation.

204. Available online: http://cassandra.apache.org (accesed on 3 July 2017).

205. Available online: http://www.ibm.com/support/knowledgecenter/SSFKCN/gpfs_welcome.html (accesed on 3 July 2017).

206. Available online: https://www.webcomponents.org (accesed on 3 July 2017).

207. Available online: https://plot.ly (accesed on 3 July 2017).

208. Available online: https://support.hdfgroup.org/HDF5/ (accesed on 3 July 2017).

209. Available online: https://lcls.slac.stanford.edu/ (accesed on 3 July 2017).

210. Available online: http://xfel.riken.jp/ (accesed on 3 July 2017).

211. Available online: http://pal.postech.ac.kr/ (accesed on 3 July 2017).

212. Available online: http://www.xfel.eu/ (accesed on 3 July 2017).

applied sciences

MDPI

Project Report
Status of the SXFEL Facility

Zhentang Zhao *, Dong Wang, Qiang Gu, Lixin Yin, Ming Gu, Yongbin Leng and Bo Liu

Shanghai Institute of Applied Physics, Chinese Academy of Sciences, Shanghai 201800, China;
wangdong@sinap.ac.cn (D.W.); guqiang@sinap.ac.cn (Q.G.); yinlixin@sinap.ac.cn (L.Y.);
guming@sinap.ac.cn (M.G.); lengyongbin@sinap.ac.cn (Y.L.); liubo@sinap.ac.cn (B.L.)
* Correspondence: zhaozt@sinap.ac.cn; Tel.: +86-021-3393-3102

Academic Editor: Kiyoshi Ueda
Received: 24 April 2017; Accepted: 6 June 2017; Published: 12 June 2017

Abstract: The Shanghai soft X-ray Free-Electron Laser facility (SXFEL) is being developed in two steps; the SXFEL test facility (SXFEL-TF), and the SXFEL user facility (SXFEL-UF). The SXFEL-TF is a critical development step towards the construction a soft X-ray FEL user facility in China, and is under commissioning at the Shanghai Synchrotron Radiation Facility (SSRF) campus. The test facility is going to generate 8.8 nm FEL radiation using an 840 MeV electron linac passing through the two-stage cascaded HGHG-HGHG or EEHG-HGHG (high-gain harmonic generation, echo-enabled harmonic generation) scheme. The construction of the SXFEL-TF started at the end of 2014. Its accelerator tunnel and klystron gallery were ready for equipment installation in April 2016, and the installation of the SXFEL-TF linac and radiator undulators were completed by the end of 2016. In the meantime, the SXFEL-UF, with a designated wavelength in the water window region, began construction in November 2016. This was based on upgrading the linac energy to 1.5 GeV, and the building of a second undulator line and five experimental end-stations. Construction status and the future plans of the SXFEL are reported in this paper.

Keywords: X-ray FEL; SXFEL; cascaded HGHG; cascaded EEHG-HGHG; water window

1. Introduction

Free-electron lasers (FELs) hold the ability to generate extremely high intensity, ultra-short, and coherent radiation pulses, which can open up new frontiers of ultra-fast and ultra-small sciences at the atomic scale. In the X-ray region, most of the existing FEL facilities, such as FLASH [1], LCLS [2], SACLA [3], PAL [4], SwissFEL [5], and European XFEL [6], are based on the self-amplified spontaneous emission (SASE) principle [7,8]. While SASE FEL has the advantages of simple setup, technological maturity, and excellent transverse coherence, it typically has rather limited temporal coherence. In order to improve the temporal coherence of SASE, several seeding schemes, including external seeding [9–14] or self-seeding [15–18], have been developed in recent years. Among these schemes, high-gain harmonic generation (HGHG) [9] and echo-enabled harmonic generation (EEHG) [11,12] have been proven as promising candidates for generating nearly Fourier-transform limited pulses with better stabilities of central wavelength and intensity [19–23]. To further extend the output wavelength of an external seeding FEL down to the X-ray regime, cascading stages of HGHG FEL with the fresh bunch technique [10] have been demonstrated both with SDUV-FEL [21] and FERMI FEL [24]. The cascaded HGHG at FERMI has already been applied for FEL user experiments and has prominent advantages in temporal coherence and controllable longitudinal phase.

The Shanghai soft X-ray Free-Electron Laser Facility (SXFEL), as a phased project, is composed of the SXFEL test facility (SXFEL-TF), and the SXFEL user facility (SXFEL-UF). The main purpose of the SXFEL-TF is to promote FEL research in China, including exploring the possibility of the seeded X-ray FEL with two stages of cascaded HGHG-HGHG, or a new scheme based on an EEHG-HGHG cascade and performing research and development on X-ray FEL-related key technologies. After a series of discussions and comparisons organized by the Chinese Academy of Sciences, it was decided to establish this test facility in the campus of the Shanghai Synchrotron Radiation Facility (SSRF). The civil construction started at the end of 2014. The tunnel and technical buildings were ready in April 2016, and the installation was almost completed by the end of 2016. Currently, the test facility is under commissioning and is expected to be finished by the end of 2017. The upgrading of the test facility to the water window user facility, SXFEL-UF, has been undertaken by the collaboration between the Shanghai Institute of Applied Physics (SINAP) and Shanghai-Tech University. Shanghai-Tech University is in charge of developing science cases and experimental end-stations, and SINAP is responsible for the remaining parts of facility development, including upgrading the linac energy to 1.5 GeV, building a second undulator line, facility integration, and constructing the utility and SXFEL-UF buildings. The civil construction was started in November 2016, and the user facility is scheduled to be open to users in 2019.

2. The SXFEL Test Facility

2.1. Layout and Main Parameters

The SXFEL-TF consists of an 840 MeV electron linac and a two-stage cascaded seeding scheme-based undulator system, as shown in Figure 1. The initial proposal of the SXFEL-TF project in 2016 was to test the cascaded HGHG scheme [25]. In the following years, it was gradually optimized and more contents of the EEHG were added to the project when the construction started in 2014. A new cascaded EEHG-HGHG operation scheme was incorporated into the SXFEL-TF to further improve the ultra-high harmonic up-conversion efficiency.

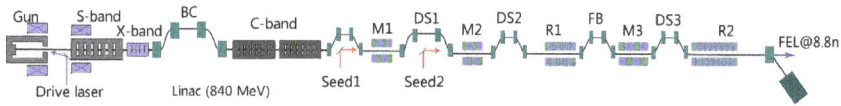

Figure 1. Schematic layout of the Shanghai soft X-ray Free-Electron Laser Test Facility (SXFEL-TF), including a photo-cathode injector, a main linac, and an undulator system (BC: bunch compressor, M: modulator, DS: dispersion section, R: radiator, FB: fresh bunch).

The main parameters of the SXFEL-TF are shown in Table 1. The wavelength of the seeding laser used in the first stage is 265 nm. The output radiation wavelength from the first stage is about 44 nm. In the second stage, an 8.8 nm soft X-ray FEL radiation pulse will be eventually produced based on the HGHG scheme. The total harmonic up-conversion number of the two stages is 30.

Table 1. Main parameters of the SXFEL-TF. (Reprint from reference [26]).

Linac	Values
Electron energy	840 MeV
Energy spread (rms)	$\leq$0.1%
Normalized emittance (rms)	$\leq$1.5 mm·mrad
Bunch length (FWHM)	$\leq$1.0 ps
Bunch charge	0.5 nC
Peak current at undulator	$\geq$500 A
Pulse repetition rate	10 Hz
Undulator	
Stage 1	
Seed laser wavelength	265 nm
FEL output wavelength	44 nm
Modulator undulator period	80 mm
Modulator undulator K value	5.81
Radiator undulator period	40 mm
Radiator undulator K value	2.22

Table 1. *Cont.*

Stage 2	
FEL output wavelength	8.8 nm
Modulator undulator period	40 mm
Modulator undulator K value	2.22
Radiator undulator period	23.5 mm
Radiator undulator K value	1.43

2.2. Injector

The SXFEL photo-injector, as shown in Figure 2, consists of an S-band photo-cathode radio frequency (RF) gun, emittance compensating solenoids, a drive laser, two S-band accelerating structures, and a laser heater. To generate a flat-top driving laser pulse, the pulse stacking technique is adopted in the temporal shaping system. Three diagnostic stations are employed in the injector. The 6D phase space of the electron bunch can be reconstructed by the combination of a transverse deflecting cavity and a 2-cell FODO lattice, each cell contains a focusing quadrupole (F), a space (O), a defocusing quadrupole (D) and a space (O).

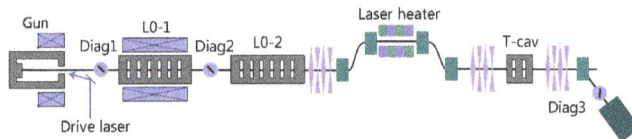

Figure 2. Layout of the SXFEL injector. (Reprint from reference [26]).

The beam parameters at the exit of the injector are shown in Table 2. The injector aims at achieving sub-μm level normalized emittance with the bunch charge of 0.5 nC, and the operation parameter errors of photo-cathode gun, drive laser, solenoids, and accelerating structures are well controlled to make sure the beam parameters lie within the specifications.

Table 2. Main electron beam parameters of the SXFEL injector. (Reprint from reference [26]).

Parameters	Value
Electron energy	130 MeV
Bunch charge	0.5 nC
Projected emittance (rms)	0.95 mm·mrad
Central slice emittance (rms)	0.65 mm·mrad
Bunch length (FWHM)	~10 ps
Projected energy spread (rms)	0.14%

2.3. Main Accelerator

The main accelerator of the SXFEL-TF is designed as a compact linac with high-gradient C-band RF accelerating structures. Initially, the main linac consists of 3 linac sections (L1 to L3) and two bunch compressors [25]. Later on, the two-stage bunch compression scheme was replaced by a single-stage bunch compressor with BC1 at a beam energy of approximately 200 MeV, based on the experience of FERMI. Currently, the main linac layout is shown in Figure 3, where L1 is the S-band accelerating section and L2 and L3 are C-band accelerating sections. To further boost the electron beam energy up to 1.5 GeV in the future, extra space between L2 and L3 has been reserved, in which a second bunch compressor and more C-band structures can be installed.

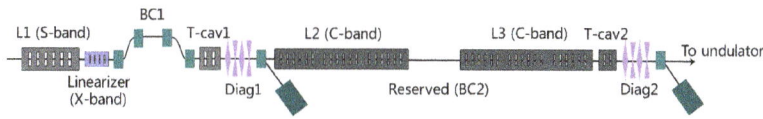

Figure 3. Layout of the main linac of the SXFEL-TF. (Reprint from reference [26]).

C-band accelerating structures have the advantage of compensating the energy spread at the phase around the crest due to the stronger longitudinal wakefield, and this, as a result, can reduce the beam energy jitter. In the meantime, simulation suggests that the transverse wakefield of the C-band structure will not obviously degrade the beam performance. Designed parameters for the main linac are shown in Table 3.

Table 3. Designed working parameters of the main linac. (Reprint from reference [26]).

Parameters	L1	LX	L2	L3
Effective accelerating gradient (MV/m)	15	25	32	32
Accelerating phase (deg)	−52.6	−180	14	14
Energy at the section exit (MeV)	184	165	389	840

In order to monitor the beam positions and optics, beam position monitors (BPMs) and beam profilers are used in the main linac. In the first bunch compressor (BC1), non-destructive bunch length and beam energy detectors are installed to give feedback on the beam energy, peak current, and arrival time with the Low-level RF system.

The triplets and FODO lattices are arranged for the main linac. Figure 4 shows the Twiss functions calculated by the ELEGANT program [27]. To minimize the horizontal beta function at the last bending magnet and hence mitigate the CSR effect, two doublets are used on each arm of the chicane. After L2, there is some extra space reserved for accommodating C-band accelerating structures for upgrading the linac energy in the future.

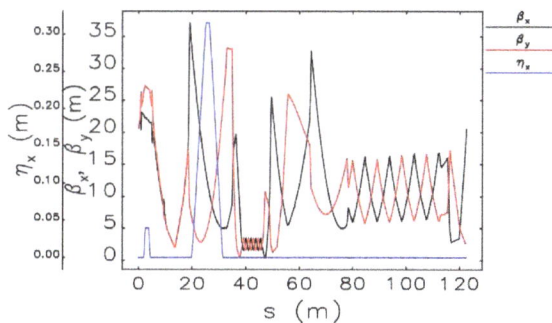

Figure 4. Twiss functions along the main linac.

2.4. Undulator Line

The main purpose of the SXFEL-TF is to test two-stage harmonic generation, such as the HGHG-HGHG or EEHG-HGHG cascading schemes. As shown in Figure 5, the undulator system consists of a seed laser system, three modulators, two radiator sections, and a couple of chicanes serving for laser injection, dispersive section, and fresh bunch delay purposes.

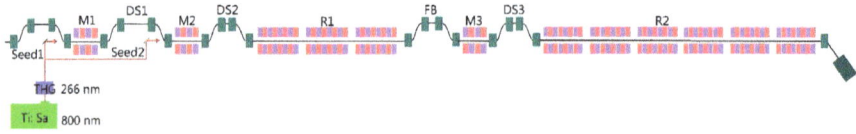

Figure 5. Layout of the undulator line.

The electron bunch at the exit of the linac is about 1 ps (FWHM) long. A small part is modulated in the modulators by a seed laser pulse with a pulse length of about 100 fs (FWHM), and the modulated part of the electron bunch generates coherent radiation in the first stage radiator. This radiation is shifted ahead to a fresh part of the electron bunch by the fresh bunch chicane and serves as the seed for the following stage. A short undulator of the same type as the first stage radiator is employed as the modulator and a small chicane is used as the dispersion section in the second stage. The harmonic up-conversion numbers of the two stages are six and five, respectively, which makes the final output wavelength ~8.8 nm.

2.5. Performance of the SXFEL-TF

The SXFEL-TF has been designed and constructed to be a flexible facility for testing various advanced FEL seeding concepts. A number of different operating modes have been considered for the SXFEL-TF. Here we only present some typical results for the two-stage HGHG cascade, the EEHG-HGHG cascade and single stage EEHG. More details of the simulations can be found in references [26,28].

Using the parameters listed in Table 1, Figure 6 shows the simulation results for the cascaded HGHG. The simulations were carried out by the time-dependent mode of GENESIS [29]. The seed power used for the first stage HGHG is about 200 MW. The whole electron beam was tracked though the two stages to obtain realistic simulation results. One can find in Figure 6 that a coherent soft X-ray radiation pulse at 8.8 nm with peak power exceeding 200 MW can be produced through the two-stage HGHG scheme.

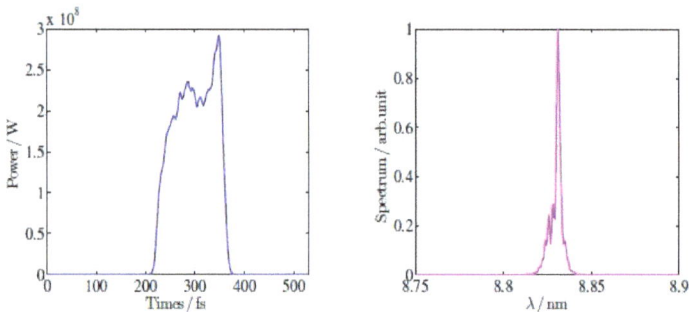

Figure 6. Simulated FEL performances of cascaded high-gain harmonic generation (HGHG) at the SXFEL-TF: final output pulse at 8.8 nm in time (**left**) and spectral (**right**) domain.

Besides the conventional cascaded HGHG, we also want to perform some proof-of-principle experiments for novel schemes. As shown in Figure 1, the first stage of the SXFEL-TF is a typical EEHG, which employs two modulator-chicane sections to introduce the echo effect into the electron beam. Therefore, the SXFEL provides a perfect platform for testing the cascaded EEHG-HGHG scheme. The simulation results for this scheme are shown in Figure 7.

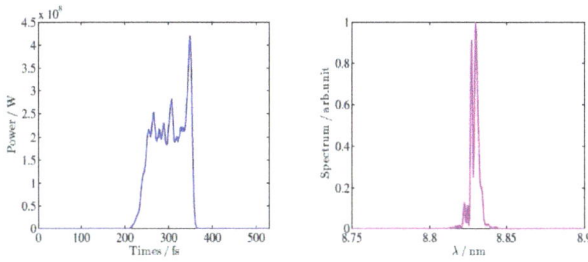

Figure 7. Simulated FEL performances of the cascaded echo-enabled harmonic generation (EEHG)-HGHG scheme at the SXFEL-TF: final output pulse at 8.8 nm in time (**left**) and spectral (**right**) domain.

Moreover, we also have the plan of operating a single-stage EEHG at ultra-high harmonics for directly generating the 8.8 nm radiation pulse [29]. The main simulation results for this case are shown in Figure 8. In the single stage EEHG, long seed laser pulses can be adopted to fully cover the electron bunch, which results in a much higher output pulse energy and much narrower output bandwidth.

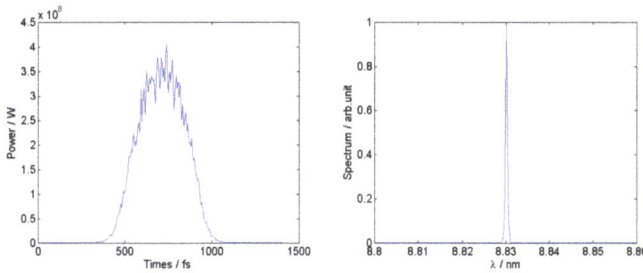

Figure 8. Simulated FEL performances of a single stage EEHG at the SXFEL-TF: final output pulse at 8.8 nm in time (**left**) and spectral (**right**) domain.

2.6. Construction, Installation and Commissioning of the SXFEL-TF

Construction of the SXFEL-TF started in December 2014. One year later, 547 pile foundations, a total area of about 7000 m^2 of civil construction, and a concrete tunnel were completed. The building was ready for machine installation in April 2016 and then installation started. Figures 9 and 10 show the status of the construction and installation at the SXFEL-TF site.

Installation of the linac, the undulator system, and the diagnostic beamline was almost completed by the end of 2016. The linac RF conditioning and beam commissioning started in late December 2016. The beam was then successfully accelerated to 700 MeV at the exit of the C-band linac and sent to the undulator line to check the installation and equipment's function. The electron beam went through the radiator undulators and the spontaneous undulator radiation at wavelength about 15 nm was characterized with the photodiode and X-ray charge-coupled-device camera at the diagnostic beamline on 31 December 2016. The RF conditioning of the C-band linac and the commissioning of the S-band injector have been performed at the same time. The normalized emittance of the injector electron beam at about 200 MeV and after the magnetic bunch compressor BC1 is 1.2 mm·mrad and 1.1 mm·mrad in the horizontal and vertical directions, respectively. Further optimization to achieve FEL lasing is ongoing.

Figure 9. Bird's eye view of the SXFEL-TF site.

Figure 10. The SXFEL linac (**left**) and undulator (**right**).

3. The SXFEL User Facility

3.1. Layout and Parameters

The SXFEL-TF will be upgraded to the soft X-ray user facility, SXFEL-UF, with the radiation wavelength extended to cover the water window region by boosting the electron beam energy to 1.5 GeV with more C-band accelerating structures. Two undulator lines, their associated beamlines, and five experimental end-stations are under construction for user experiments. The layout comparison between the SXFEL-TF and the SXFEL-UF is shown in Figure 11. Table 4 lists the main basic parameters of the SXFEL-UF.

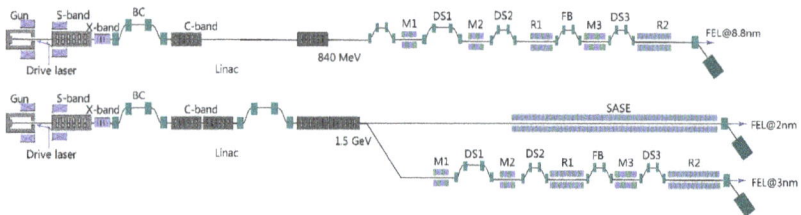

Figure 11. Schematic layouts: SXFEL-TF (**upper**) and Shanghai soft X-ray Free-Electron Laser User Facility (SXFEL-UF) (**lower**).

Table 4. Main parameters of the SXFEL-UF.

Linac	Values
Electron energy	1500 MeV
Energy spread (rms)	≤0.1%
Normalized emittance (rms)	≤1.5 mm·mrad
Bunch charge	0.5 nC
Peak current at undulator	≥700 A
Pulse repetition rate	50 Hz
Undulator	
Line 1	
FEL operation mode	SASE
FEL output wavelength	~2 nm
FEL output pulse peak power	≥100 MW
Line 2	
FEL operation mode	External seeding
FEL output wavelength	~3 nm
FEL output pulse peak power	≥100 MW

3.2. Energy Upgrade

The main linac of the SXFEL-UF will accelerate the electron beam from an energy of 130 MeV at the exit of the injector to 1.5 GeV at the end of the linac. In this process, the electron bunch length will be compressed from 10 ps to about 0.7 ps. As shown in Figure 12, an additional set of S-band 50 MW RF power source in section L1, four C-band accelerating units with eight RF accelerating structures, and a second bunch compressor section (BC2) are added to SXFEL-TF in its reserved space between L2 and L3 to constitute the SXFEL-UF main linac. In this upgrade, an X-band transverse deflecting cavity is placed at the end of this linac to obtain high resolution bunch length measurement at higher energy. The designed working parameters are shown in Table 5 and the simulated beam distributions are shown in Figure 13.

Table 5. Designed working parameters of the main linac for SXFEL-UF.

Linac Components	Eout (MeV)	σz-Out (mm)	σδ-Out (%)	E (MV/m)	Φrf\Θbend (Deg)	R56 (mm)
L0	130	0.86	0.14	-	-	-
L1	273	0.86	1.44	27	−29.2	-
X	256	0.86	1.51	19	180	-
BC1	-	0.13	-		3.968	−48
L2	640	0.13	0.42	38	4	-
BC2	-	0.07	-		2.217	−15
L3	1500	0.07	0.028	38	6	-

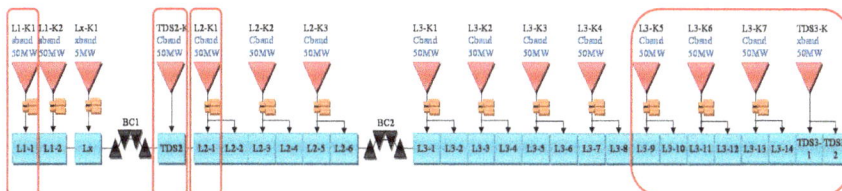

Figure 12. Layout of the main linac of the SXFEL-UF. (Reprint from reference [26]).

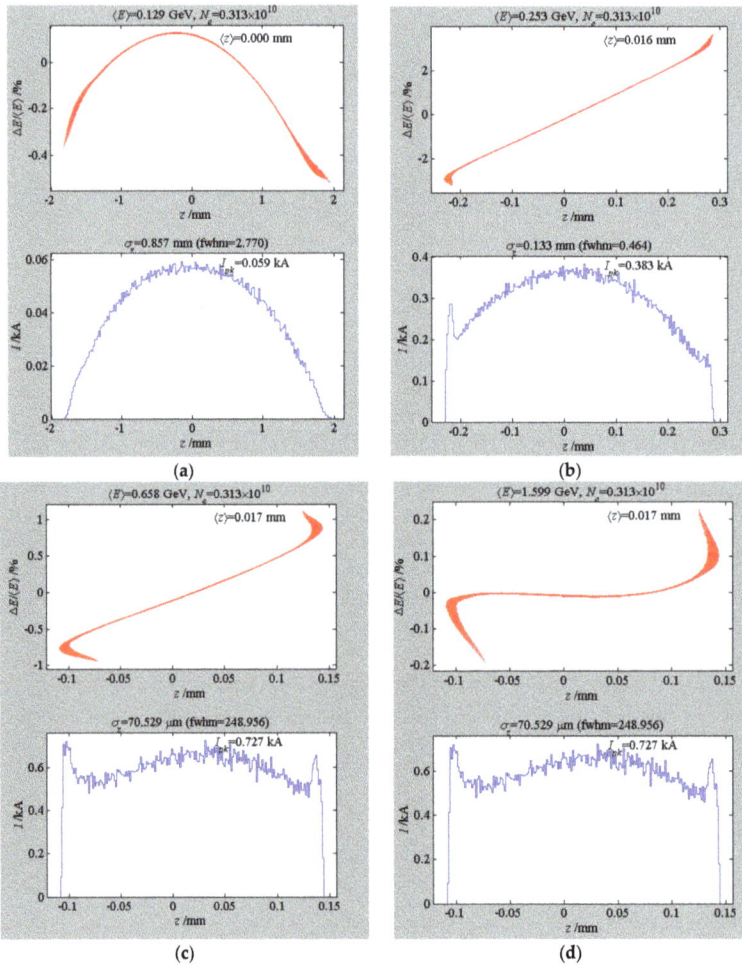

Figure 13. Simulation results of the electron beam distributions: energy distributions and current profiles at (**a**) injector exit; (**b**) BC1 exit; (**c**) BC2 exit and (**d**) L3 exit.

3.3. Undulator Lines and the FEL Performance

There are two undulator lines for the user facility. One is upgraded from the test facility and will be operated with either a cascaded HGHG or EEHG-HGHG mode, as shown in Figure 14a, and hereafter referred to as the seeded line, while the other will be a brand-new line operated in the SASE mode, as shown in Figure 14b.

With the beam energy boosted to 1.5 GeV and the peak current increased to 700 A, the output wavelength of the SXFEL-UF can cover the water window region. Accordingly, to make the FEL output of the seeded line saturate, the length of the original undulator line should be increased. Here, one radiator is added in the first stage and four planar undulators are added in the second stage. With such a configuration, a fully coherent saturated 3 nm FEL output could be obtained. To fulfill users' demands, two elliptical polarized undulators (EPU) will be added following the radiators of the second stage to form the so-called "afterburner" scheme and realize the full control of the soft X-ray FEL's polarization.

(a)

(b)

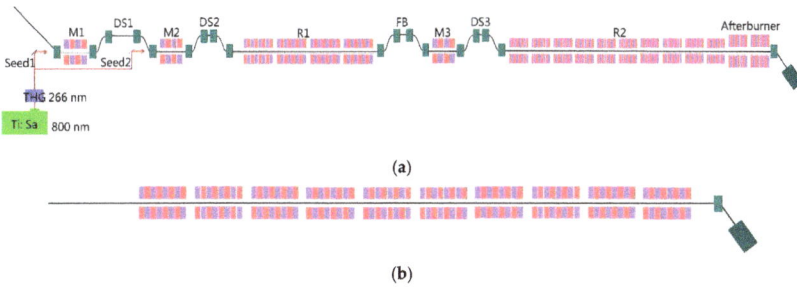

Figure 14. Schematic layout of the FEL undulator lines for the SXFEL-UF: (**a**) Seeded line of the SXFEL-UF (upgraded from SXFEL-TF); (**b**) new undulator line for the SXFEL-UF.

In the baseline design, the second undulator line is based on the SASE mode, and the undulator is the in-vacuum planar type. The undulator period is 16 mm and the working gap is around 3.7 mm, resulting in a K value of 1.8. For this undulator line, the final FEL output is around 8 nm at a beam energy of 840 MeV, and can be as low as 2 nm at a beam energy of 1.5 GeV, while the output peak power will be greater than 100 MW for both cases.

With the parameters shown in Table 4, the SXFEL-UF's performance with different undulator lines was simulated with the time-dependent mode of GENESIS. A 265 nm laser pulse with a longitudinal Gaussian profile, 200 MW peak power, and 100 fs (FWHM) pulse length is used as the seed laser for the seeded line. To obtain realistic simulation results, the whole electron beam was tracked through the undulator lines. The simulation results are illustrated in Figure 15 for the seeded line with the cascaded HGHG-HGHG mode, and in Figure 16 for the SASE line.

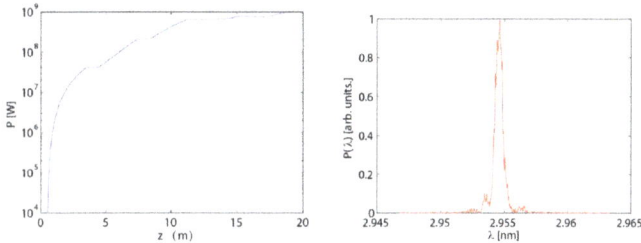

Figure 15. Simulated FEL performance of the seeded line at 3 nm output: gain curve and the output spectrum.

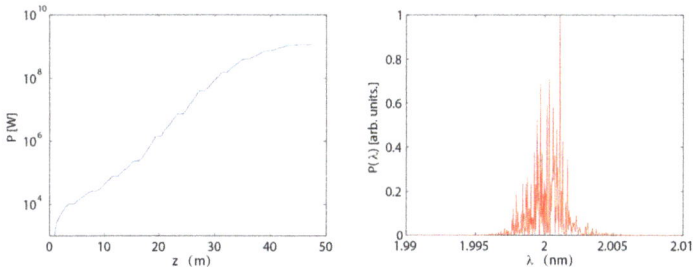

Figure 16. Simulated FEL performance of the SASE line at 2 nm output: gain curve and the output pulse in the time domain.

3.4. Beamlines and Experimental End-Stations

X-ray beamlines are important parts of the SXFEL user facility, precisely transmitting the FEL output into various end-stations. Online diagnostics of the FEL outputs will also be provided at the beamlines. In the initial phase of the user facility, there are two beamlines and five end-stations, including cell imaging, atomic molecular and optical physics (AMO), ultrafast physics, surface chemistry, etc. Figure 17 shows one of the SXFEL-UF beamlines together with two end-stations.

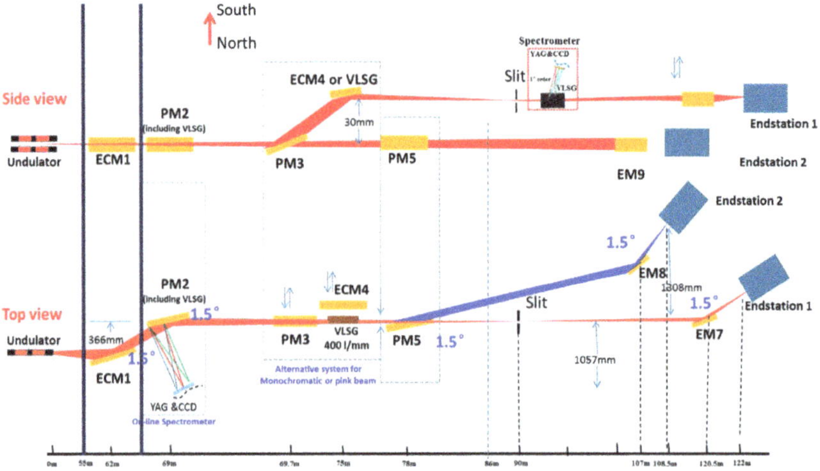

Figure 17. A typical beamline with two end-stations.

With its high light source performance properties and these well-designed instruments, SXFEL is dedicated to ultrafast X-ray science and fundamental research, especially on femtosecond chemistry, materials under extreme conditions, high-throughput bio-imaging, light-induced transient phenomenon and dynamic functions in real time, etc. Owing to its ultra-intense and ultrafast pulses, the facility will attract much attention from the science community and this will be the main driving force for the SXFEL.

To make use of the full coherence and femtosecond pulse duration of XFEL, technologies such as ultrafast scattering and imaging will be employed. Ultrafast scattering and imaging with unprecedented spatial resolution will permit the evolution of nano-materials, and will allow further investigation of the relationship between material synthesis and operating conditions. This will change our methods for the development of novel materials. Moreover, the capability of SXFEL will enable the unique characterization of structure and dynamic changes therein. Electron transfer, electronic fluctuation, transient phase transfer and hidden phase could be captured in a few femtoseconds. The ability to capture ultrafast processes will give a better understanding of motions at the sub-nanometer and femtosecond scales. Flash X-ray imaging, combined with intense lasers and mass spectroscopy, could be applied to bio-imaging and the investigation of materials under extreme conditions. To obtain the special properties of catalysis and photo-catalysis, methods such as laser-pump X-ray probing and two-color X-ray pump X-ray probing may achieve surprising results. Considering dynamic functions under natural states and in real time is the key to investigating any ultrafast problems. To this end, multiple pump-probes (Vis-/infrared-laser, THz, X-ray) with mixed particle injectors will be designed to capture high frame rates images. The understanding of any emergent phenomena is essential for designing new materials and chemical reactions. To cover this issue, SXFEL will reveal these critical events step by step, from scattering and diffraction to absorption. The data from different end-stations either specially-designed or based on general X-ray technologies at SXFEL will be combined. Multiple big data analyses would provide new ways to approach new materials and chemical reactions.

3.5. Construction and Project Schedule

The civil construction of the SXFEL-UF undulator tunnel and experimental hall started in November 2016. Procurement of the main components for the energy upgrade and new undulators is underway and on schedule. The civil construction will be finished in July 2018, and will be followed by the installation and commissioning of the apparatus. User experiments are expected to commence in early 2019.

4. Conclusions

The SXFEL (both the test facility and the user facility) projects are in good shape. The SXFEL-TF is under commissioning, aiming to commence lasing in the first half of 2017. The test facility will be quickly upgraded to a soft X-ray user facility by boosting the beam energy to 1.5 GeV, constructing two undulator lines and five experimental stations, and adding a new undulator tunnel and experimental hall. The SXFEL-UF started its civil construction in November 2016 and will provide high-brightness and ultrafast soft X-ray FEL beams covering the full water window to the user community in 2019.

Acknowledgments: The authors thank the whole FEL team in the Shanghai Institute of Applied Physics for their excellent support during the past decades, who made invaluable contributions and dedications to the design, construction, and commissioning of SXFEL. We also thank Zhi Liu and Huaidong Jiang for the science programs of the SXFEL user facility. Thanks also go to Senyu Chen, Li-Hua Yu, Zhirong Huang, Gennady Stupakov and Alexander Wu Chao for useful discussions and their continuous encouragement. This work was partially supported by the National Development and Reform Commission ([2013]2347) and the National Basic Research Program of China (2015CB859700).

Author Contributions: Zhentang Zhao, Dong Wang, Qiang Gu, Lixin Yin, Ming Gu, Yongbin Leng and Bo Liu led the design, construction and commissioning of the SXFEL. Qiang Gu and Bo Liu were in charge of the numerical simulations. All authors co-wrote the paper.

Conflicts of Interest: The authors declare no conflict of interest.

References

1. Ackermann, W.A.; Asova, G.; Ayvazyan, V.; Azima, A.; Baboi, N.; Bähr, J.; Balandin, V.; Beutner, B.; Brandt, A.; Bolzmann, A.; et al. Operation of a free-electron laser from the extreme ultraviolet to the water window. *Nat. Photonics* **2007**, *1*, 336–342. [CrossRef]
2. Emma, P.; Akre, R.; Arthur, J.; Bionta, R.; Bostedt, C.; Bozek, J.; Brachmann, A.; Bucksbaum, P.; Coffee, R.; Decker, F.-J.; et al. First lasing and operation of an ångstrom-wavelength free-electron laser. *Nat. Photonics* **2010**, *4*, 641–647. [CrossRef]
3. Ishikawa, T.; Aoyagi, H.; Asaka, T.; Asano, Y.; Azumi, N.; Bizen, T.; Ego, H.; Fukami, K.; Fukui, T.; Furukawa, Y.; et al. A compact X-ray free-electron laser emitting in the sub-angstrom region. *Nat. Photonics* **2012**, *6*, 540–544. [CrossRef]
4. Han, J.H.; Kang, H.S.; Ko, I.S. Status of the PAL-XFEL project. In Proceedings of the IPAC2012, New Orleans, LA, USA, 20–25 May 2012; pp. 1735–1737.
5. Ganter, R. *SwissFEL-Conceptual Design Report*; No. PSI-10-04; Paul Scherrer Institute (PSI): Villigen, Switzerland, 2010.
6. Massimo, A. *The European X-ray Free-Electron laser: Technical Design Report*; European XFEL Project Team: Hamburg, Germany, 2013.
7. Kondratenko, A.M.; Saldin, E.L. Generating of coherent radiation by a relativistic electron beam in an ondulator. *Part. Accel.* **1980**, *10*, 207–216.
8. Bonifacio, R.; Pellegrini, C.; Narducci, L.M. Collective instabilities and high-gain regime in a free electron laser. *Opt. Commun.* **1984**, *50*, 373–378. [CrossRef]
9. Yu, L.-H. Generation of intense uv radiation by subharmonically seeded single-pass free-electron lasers. *Phys. Rev. A* **1991**, *44*, 5178. [CrossRef] [PubMed]

10. Wu, J.H.; Yu, L.H. Coherent Hard X-ray Production by Cascading Stages of High Gain Harmonic Generation X-ray FEL. *Nucl. Instrum. Methods A* **2001**, *475*, 104–111. [CrossRef]

11. Stupakov, G. Using the Beam-Echo Effect for Generation of Short-Wavelength Radiation. *Phys. Rev. Lett.* **2009**, *102*, 074801. [CrossRef] [PubMed]

12. Xiang, D.; Stupakov, G. Enhanced tunable narrow-band THz emission from laser-modulated electron beams. *Phys. Rev. STAB* **2009**, *12*, 256–273. [CrossRef]

13. Deng, H.; Feng, C. Using off-resonance laser modulation for beam-energy-spread cooling in generation of short-wavelength radiation. *Phys. Rev. Lett.* **2013**, *111*, 084801. [CrossRef] [PubMed]

14. Feng, C.; Deng, H.; Wang, D.; Zhao, Z. Phase-merging enhanced harmonic generation free-electron laser. *New J. Phys.* **2014**, *16*, 043021. [CrossRef]

15. Feldhaus, J.; Saldin, E.L.; Schneider, J.R.; Schneidmiller, E.A.; Yurkov, M.V. Possible application of X-ray optical elements for reducing the spectral bandwidth of an X-ray SASE FEL. *Nucl. Instrum. Methods Phys. Res. Sect. A* **1997**, *393*, 162–166. [CrossRef]

16. Geloni, G.; Kocharyan, V.; Saldin, E. A novel self-seeding scheme for hard X-ray FELs. *J. Mod. Opt.* **2011**, *58*, 1391–1403. [CrossRef]

17. Amann, J.; Berg, W.; Blank, V.; Decker, F.J.; Ding, Y.; Emma, P.; Feng, Y.; Frisch, J.; Fritz, D.; Hastings, J.; et al. Demonstration of self-seeding in a hard-X-ray free-electron laser. *Nat. Photonics* **2012**, *6*, 693–698. [CrossRef]

18. Ratner, D.; Abela, R.; Amann, J.; Behrens, C.; Bohler, D.; Bouchard, G.; Bostedt, C.; Boyes, M.; Chow, K.; Cocco, D.; et al. Experimental demonstration of a soft x-ray self-seeded free-electron laser. *Phys. Rev. Lett.* **2015**, *114*, 054801. [CrossRef] [PubMed]

19. Yu, L.-H.; Babzien, M. High-gain harmonic-generation free-electron laser. *Science* **2000**, *289*, 932–935. [CrossRef] [PubMed]

20. Zhao, Z.T.; Wang, D.; Chen, J.H.; Chen, Z.H.; Deng, H.X.; Ding, J.G.; Feng, C.; Gu, Q.; Huang, M.M.; Lan, T.H.; et al. First lasing of an echo-enabled harmonic generation free-electron laser. *Nat. Photonics* **2012**, *6*, 360–363. [CrossRef]

21. Liu, B.; Li, W.B.; Chen, J.H.; Chen, Z.H.; Deng, H.X.; Ding, J.G.; Fan, Y.; Fang, G.P.; Feng, C.; Feng, L.; et al. Demonstration of a widely-tunable and fully-coherent high-gain-harmonic-generation free-electron laser. *Phys. Rev. Spec. Top. Accel. Beams* **2013**, *16*, 020704. [CrossRef]

22. Allaria, E.; Appio, R.; Badano, L.; Barletta, W.A.; Bassanese, S.; Biedron, S.G.; Borga, A.; Busetto, E.; Castronovo, D.; Cinquegrana, P.; et al. Highly coherent and stable pulses from the FERMI seeded free-electron laser in the extreme ultraviolet. *Nat. Photonics* **2012**, *6*, 699–704. [CrossRef]

23. Boedewadt, J.; Ackermann, S.; Aßmann, R.; Ekanayake, N.; Faatz, B.; Feng, G.; Hartl, I.; Ivanov, R.; Amstutz, P.; Azima, A.; et al. Recent results from FEL seeding at FLASH. In Proceedings of the IPAC2015, Richmond, VA, USA, 3–8 May 2015; pp. 1366–1369.

24. Allaria, E.; Castronovo, D.; Cinquegrana, P.; Craievich, P.; Dal Forno, M.; Danailov, M.B.; D'Auria, G.; Demidovich, A.; De Ninno, G.; Di Mitri, S.; et al. Two-stage seeded soft-X-ray free-electron laser. *Nat. Photonics* **2013**, *7*, 913–918. [CrossRef]

25. Zhao, Z.T.; Chen, S.Y.; Yu, L.H.; Tang, C.X.; Yin, L.X.; Wang, D.; Gu, Q. Shanghai soft X-ray free electron laser test facility. In Proceedings of the IPAC2011, San Sebastián, Spain, 4–9 September 2011; pp. 3011–3013.

26. Zhao, Z.T.; Wang, D.; Yin, L.X.; Gu, Q.; Fang, G.P.; Liu, B. The current status of the SXFEL project. *AAPPS Bull.* **2016**, *26*, 12–24.

27. Borland, M. *Elegant: A Flexible SDDS-Compliant Code for Accelerator Simulation*; Technical Report No. LS-287; Advanced Photon Source: Argonne, IL, USA, 2000.

28. Feng, C.; Huang, D.; Deng, H.; Chen, J.; Xiang, D.; Liu, B.; Wang, D.; Zhao, Z. A single stage EEHG at SXFEL for narrow-bandwidth soft X-ray generation. *Sci. Bull.* **2016**, *61*, 1202–1212. [CrossRef]

29. Reiche, S. GENESIS 1.3: A fully 3D time-dependent FEL simulation code. *Nucl. Instrum. Methods A* **1999**, *429*, 243–248. [CrossRef]

applied
sciences

MDPI

Review

Ultrashort Free-Electron Laser X-ray Pulses

Wolfram Helml [1,2], Ivanka Grguraš [3], Pavle N. Juranić [4], Stefan Düsterer [5], Tommaso Mazza [6], Andreas R. Maier [7], Nick Hartmann [8], Markus Ilchen [6], Gregor Hartmann [9], Luc Patthey [4], Carlo Callegari [10] , John T. Costello [11] , Michael Meyer [6], Ryan N. Coffee [12], Adrian L. Cavalieri [3] and Reinhard Kienberger [2,*]

[1] Department für Physik, Ludwig-Maximilians-Universität München, Am Coulombwall 1, 85748 Garching, Germany; Wolfram.Helml@lmu.de
[2] Physik-Department E11, Technische Universität München, James-Franck-Straße 1, 85748 Garching, Germany
[3] Center for Free-Electron Laser Science, Notkestraße 85, 22607 Hamburg, Germany; ivanka.grguras@mpsd.mpg.de (I.G.); adrian.cavalieri@mpsd.cfel.de (A.L.C.)
[4] Paul Scherrer Institut, 5232 Villigen, Switzerland; pavle.juranic@psi.ch (P.N.J.); luc.patthey@psi.ch (L.P.)
[5] Deutsches Elektronen-Synchrotron DESY, Notkestraße 85, 22607 Hamburg, Germany; stefan.duesterer@desy.de
[6] European XFEL GmbH, Holzkoppel 4, 22869 Schenefeld, Germany; tommaso.mazza@xfel.eu (T.M.); markus.ilchen@xfel.eu (M.I.); michael.meyer@xfel.eu (M.M.)
[7] Center for Free-Electron Laser Science & Department of Physics, University of Hamburg, 22761 Hamburg, Germany; andreas.maier@desy.de
[8] Coherent Inc., 5100 Patrick Henry Drive, Santa Clara, CA 95054, USA; hartmann.nick@gmail.com
[9] Institut für Physik und CINSaT, Universität Kassel, Heinrich-Plett-Straße 40, 34132 Kassel, Germany; gregor.hartmann@physik.uni-kassel.de
[10] Elettra—Sincrotrone Trieste, Strada Statale 14–km 163.5 in AREA Science Park, 34149 Basovizza, Trieste, Italy; carlo.callegari@elettra.eu
[11] National Center for Plasma Science and Technology and School of Physical Sciences, Dublin City University, Dublin 9, Ireland; john.costello@dcu.ie
[12] SLAC National Accelerator Laboratory, Linac Coherent Light Source, Menlo Park, CA 94025, USA; coffee@slac.stanford.edu
* Correspondence: reinhard.kienberger@tum.de; Tel.: +49-(0)89-289-12840

Received: 28 July 2017; Accepted: 30 August 2017; Published: 6 September 2017

Abstract: For the investigation of processes happening on the time scale of the motion of bound electrons, well-controlled X-ray pulses with durations in the few-femtosecond and even sub-femtosecond range are a necessary prerequisite. Novel free-electron lasers sources provide these ultrashort, high-brightness X-ray pulses, but their unique aspects open up concomitant challenges for their characterization on a suitable time scale. In this review paper we describe progress and results of recent work on ultrafast pulse characterization at soft and hard X-ray free-electron lasers. We report on different approaches to laser-assisted time-domain measurements, with specific focus on single-shot characterization of ultrashort X-ray pulses from self-amplified spontaneous emission-based and seeded free-electron lasers. The method relying on the sideband measurement of X-ray electron ionization in the presence of a dressing optical laser field is described first. When the X-ray pulse duration is shorter than half the oscillation period of the streaking field, few-femtosecond characterization becomes feasible via linear streaking spectroscopy. Finally, using terahertz fields alleviates the issue of arrival time jitter between streaking laser and X-ray pulse, but compromises the achievable temporal resolution. Possible solutions to these remaining challenges for single-shot, full time–energy characterization of X-ray free-electron laser pulses are proposed in the outlook at the end of the review.

Keywords: free-electron laser; ultrashort pulse characterization; ultrafast X-ray physics; laser-dressed electron spectroscopy

Appl. Sci. **2017**, *7*, 915

1. Introduction—Laser-Assisted Time-Domain Characterization at Free-Electron Lasers

Availability of ultrashort, high-brightness X-ray pulses

Exploring fundamental quantum building blocks of nature and their interactions on time scales of molecular, atomic and even electronic motion, has led to the development of scientific instruments with steadily enhanced spatial and temporal resolution. For modern day optical tools, that historical path reaches from the discovery of X-rays and their properties just before the turn to the 20th century [1], via synchrotron radiation detected as parasitic electron energy-loss effect in the middle of the last century [2], to the parasitic operation at electron synchrotrons [3]. Next came dedicated beamlines built as multiple synchrotron radiation stations around storage rings [4], and finally high-brightness X-ray beams through periodically alternating magnetic insertion devices [5] were achieved. Today, the most recent additions to the arsenal of X-ray physics stem from high-order harmonic generation (HHG) by intense few-cycle optical laser pulses [6] and from the realization of the concept of free-electron lasers (FELs) [7,8] at large-scale facilities around the world. These FEL sources provide ultrashort high-brightness pulses from the soft X-ray regime, like the free-electron laser in Hamburg (FLASH), operated by DESY in Germany [9] and the free-electron laser radiation for multidisciplinary investigations (FERMI) seeded FEL in Trieste, Italy [10], to the hard X-ray range, e.g., the Linac Coherent Light Source (LCLS) at the SLAC National Accelerator Laboratory in California, USA [11] and the SPring-8 angstrom compact free-electron laser (SACLA) in Japan [12], among others [13,14]. Future machines, using novel laser-based accelerator concepts promising much more compact X-ray sources [15–18] or utilizing superconducting linear accelerators for higher repetition rates in the MHz range (e.g., the European XFEL in Germany [19] or LCLS-II in the USA [20]), are expected to generate X-ray pulses with few- to sub-femtosecond durations as well, and are likewise crucially dependent on a precise X-ray pulse metrology.

Well-controlled X-ray pulses with durations shorter than few tens of femtoseconds that are currently produced at X-ray free-electron lasers (XFELs) are a necessary prerequisite for the investigation of ultrafast processes on the time scale of the motion of bound electrons, as have been successfully studied in seminal high harmonic generation-based attosecond experiments [21–31]. A number of schemes have been proposed and implemented over the last decade to shorten the pulse durations at XFELs to the few-femtosecond or even sub-femtosecond range, many of them based on the principle that the X-ray pulse duration is related to the driver electron bunch length. One approach, the so-called 'slotted spoiler', was invented at the LCLS and provides few-femtosecond to sub-femtosecond pulses [32,33]. The working principle of this scheme is illustrated in Figure 1 on the left. A thin foil with a narrow vertical slit is introduced into the electron beam path in one of the magnetic bunch compressor chicanes and spoils the beam quality (emittance) of the largest part of the spatially chirped electron bunch via Coulomb scattering. Only the small central slice of the bunch passes unhindered through the slit and will subsequently experience efficient FEL amplification in the undulator. Other proposed methods involve the modulation of the electron bunch with a high-power, few-cycle optical laser in a single-period wiggler before the FEL undulator to induce a pronounced energy modulation along the electron bunch and a sub-femtosecond spike in the electron peak current. Thus, only a small region of the bunch contributes efficiently to the X-ray amplification process and generates an attosecond spike of X-ray radiation, see for example [34–37].

Ultrafast temporal characterization for novel attosecond X-ray experiments

From the perspective of the scientific researcher, the special opportunities opened up by free-electron laser sources lie in their extreme spatial and temporal coherence, combined with the achievable ultrahigh intensity, broadly adjustable photon energy range and ultrashort pulse duration. Using these sources will ultimately enable novel ultrafast experiments with attosecond temporal resolution, like site-specific X-ray pump/X-ray probe measurements [39] and serial crystallography for studies of structural dynamics [40], or atomic-scale diffractive imaging as proposed by Yakovlev et al. [41] on energy-selected target states of molecules, surfaces and nanoparticles. Recent experiments with high-energy electrons [42,43] and X-rays [44] have already demonstrated the first steps towards this so-called molecular movie on the femtosecond scale. For an extensive overview of the experimental capabilities and scientific achievements during the last years at LCLS see Ref. [45].

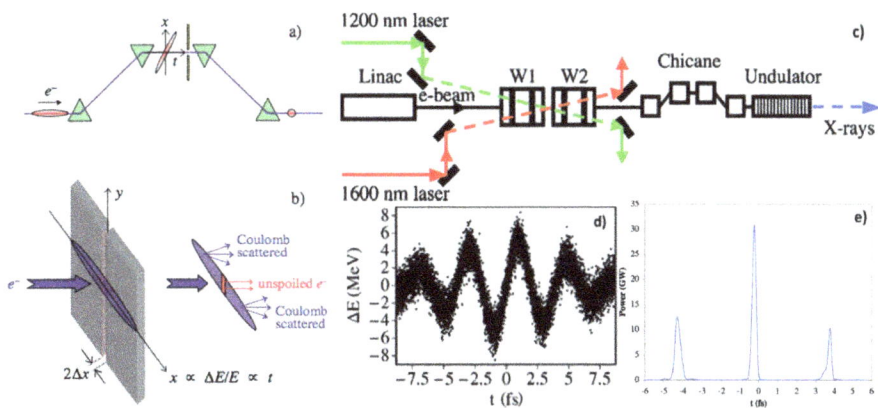

Figure 1. Sketch of the working principles for few-femtosecond and sub-femtosecond X-ray free-electron lasers (XFEL) pulse generation. (**a**) The upper part shows the spatially chirped electron bunch at the center of the compressor chicane with a marked tilt with respect to the direction of propagation. (**b**) The slotted foil is depicted, leaving only a narrow central region of the bunch pass unspoiled ((**a**,**b**) are reproduced with permission from [32], Copyright American Physical Society, 2004). (**c**) The upper panel shows a schematic of the components involved in attosecond X-ray pulse production. W1 and W2 are two wiggler sections that generate the modulation of the electron bunch by interaction with high-power, few-cycle optical lasers. (**d**) In panel, the calculated energy modulation of the electrons along the electron bunch produced in the interaction with a few-cycle, 1200 nm laser pulse in a wiggler magnet is shown. (**e**) The resulting attosecond X-ray power profile after 50 m of propagation in the undulator is plotted in panel ((**c**–**e**) are reprinted under the terms of the Creative Commons Attribution 3.0 License from [38], 2005).

To realize the goal of time-resolved X-ray measurement with attosecond resolution, one first has to consider the special complexities arising from the work with attosecond XFEL pulses, foremost the characterization of their temporal structure. As with any signal based on the self-amplified spontaneous emission (SASE) process [46], the temporal shape of each subsequent single XFEL pulse is implicitly different from the previous one, demanding a single-shot measurement technique for its investigation. However, until recently no direct experimental determination of the structure of these X-ray pulses in the time domain was accomplished. Therefore, in this paper we want to concentrate on the unique aspects and concomitant challenges that arise from the availability of X-ray free-electron laser pulses of femtosecond and even sub-femtosecond durations and their characterization on a suitable time scale.

Indirect measurement methods

SASE XFEL pulse durations have been indirectly inferred (see also the review in Ref. [47]) from measurements on the electron bunch length before the undulator [48] or from statistical analysis of the spectral coherence properties of the ultrashort X-ray pulses [49]. However, these characterization techniques cannot provide single-shot information on the temporal structure of the X-ray pulses and only statistically averaged quantities for possible pulse length predictions can be derived from these measurements. Furthermore, studies of fundamental ionization processes [50] and sideband measurements, also reported on in this review [51], have revealed a clear deviation of the actual X-ray pulse duration from the one inferred from electron bunch length measurements. A substantial improvement of the temporal resolution and the reliability for X-ray pulse characterization has been achieved by X-band radiofrequency transverse deflector (XTCAV) measurements on the spent electron bunch after propagation through the undulator [52]. Very recently, the concept of machine learning has successfully been used for the prediction of XFEL pulse properties, based on the online measurement of numerous accelerator specific parameters [53]. Nevertheless, these techniques all constitute indirect characterization methods and need direct time domain measurements for their exact calibration.

Direct measurement methods

The usual characterization techniques for ultrashort optical pulses are not readily transferrable to the X-ray regime. Autocorrelation setups have been developed for VUV wavelengths [54,55], but are hard to establish in the X-ray range due to the corresponding relatively small nonlinear interaction cross sections and phase-matching bandwidth for most materials, together with the lack of suitable beam splitting and combining optics and the ensuing stability issues. Other standard cross-correlation schemes like spectral phase interferometry for direct electric-field reconstruction (SPIDER) [56] are, on the other hand, hindered by the requirement for reliable phase modulators for X-ray pulses or are restricted to the implementation at seeded FELs [57].

Various methods with solid state targets have been used for FEL characterization, which are based on an FEL-induced transient change of the refractive index of material, as demonstrated in Ref. [58]. The most common scheme remains that of probing the cross-correlation between FEL and IR pulses, and single-shot streaking geometries have been implemented (see e.g., [59]). In general, their temporal resolution is inherently limited by the electron relaxation dynamics in condensed matter, and these measurements are usually destructive towards the X-ray beam. Thus, these concepts are not part of this review dealing specifically with ultrashort XFEL pulse characterization on the femtosecond and sub-femtosecond time scales.

For the sake of completeness, we should remind the reader of the particularly noteworthy four-wave mixing scheme pioneered by Masciovecchio and co-workers [60,61], in which the interaction of three photons with matter results in the generation of a fourth "signal" photon whose color and direction fulfils the phase-matching conditions; this method probes the third-order nonlinear susceptibility $\chi^{(3)}$, and although technically more complicated has an enormous advantage in terms of background suppression. Let us finally note that in the framework of high harmonic generation, a four-wave mixing scheme has been successfully used with a gaseous target [62].

In light of the reckoning above, the method of choice to directly measure the ultrashort XFEL pulse duration is the concept of time-resolved optical laser-dressed photoelectron (PE) spectroscopy in gaseous media, which is a well-established technique for temporal characterization of attosecond pulses in the XUV spectral region. This scheme has been theoretically described in [63,64] and successfully demonstrated in numerous time-resolved measurements of laser-based XUV sources [65–67]. This setup, shown exemplarily in Figure 2, provides a direct, non-invasive, single-shot, low-photon intensity, photon energy-independent and high-repetition rate-compatible pulse characterization method. The essential parts of the theory for the purpose of the temporal XFEL pulse characterization shall be briefly discussed in the following section.

As will be shown, a number of variants of this technique exist, each with their own individual merits for a specific range of X-ray pulse durations and required temporal resolution. While the method relying on the sideband measurement (Section 2) is relatively easy to implement with standard Ti:sapphire laser systems, its information about single X-ray shot characteristics is limited. On the other hand, longer laser wavelengths, only indirectly accessible via optical parametric amplification (OPA) [68], are preferable to avoid the breakdown of the streaking regime (cf. Section 3.1.1 of this paper). This is typically the case, when the X-ray pulse duration is longer than half the oscillation period of the streaking field, corresponding to only 1.3 fs for the gain peak wavelength of Ti:sapphire at 795 nm. Finally, using terahertz (THz) fields with a very long optical period (Sections 3.1.2 and 3.2) alleviates the problem of arrival time jitter between streaking laser and X-ray pulse, but compromises the achievable temporal resolution. Possible solutions to these remaining issues for single-shot, full time–energy characterization of XFEL pulses are proposed in the outlook at the end of the paper.

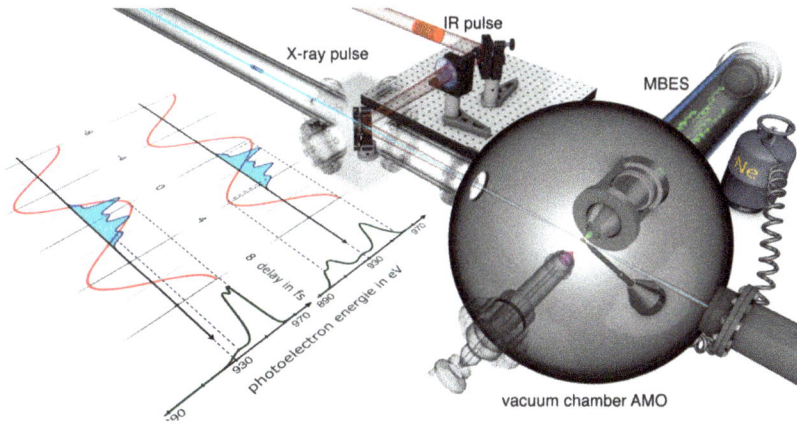

Figure 2. General setup for laser-dressed photoelectron spectroscopy. Typical experimental setup and measurement principle for a two-color ultrafast temporal characterization measurement. The picture shows the experimental setup in the atomic, molecular and optical science (AMO) hutch at the Linac Coherent Light Source (LCLS). The X-ray laser and the near-infrared (NIR) dressing laser are coupled into the vacuum chamber and are collinearly focused onto a neon gas target. The generated photoelectrons are then energy resolved with a magnetic bottle electron spectrometer. The inset on the left depicts two distinctive cases of temporal overlap of the XFEL with respect to the streaking field, one at the zero crossing and one at a maximum of the NIR vector potential. The respective photoelectron spectra are also shown. More details are given in the text in Section 3.1.1 and in Figure 7 (reproduced with permission from [69], Copyright Nature Publishing Group, 2014).

2. Sideband Method for X-ray Pulse Characterization

Temporal pulse characterization with photoelectrons

Two-color photoionization experiments using the combination of intense free-electron laser femtosecond pulses in the extreme ultraviolet (XUV) to X-ray wavelength range and optical or NIR pulses offer great potential to determine the temporal characteristics, i.e., pulse durations and the relative temporal arrival time, of the FEL pulses.

Although studies of two-color photoionization of atoms in weak resonant optical/NIR fields are well established at synchrotron radiation facilities [70], it is only with the availability of short wavelength FELs that experiments have been performed, in which electrons ejected by the XUV or X-ray field can also be effectively "dressed" by the intense NIR laser field. Overlapping the two intense ultrashort pulses temporally and spatially in a gaseous target leads to the appearance of additional structures in the photoelectron spectrum, which are the clear indication of the two-color process. Ionization by the XUV or soft X-ray pulse creates electrons with typically a few tens and up to few hundreds of eV kinetic energy. The additional interaction with the intense NIR fields induces absorption or emission of NIR photons resulting in extra lines in the photoelectron spectrum (Figure 3a) spaced by the photon energy of the NIR photons (1.55 eV for 800 nm) (e.g., [71]). These lines are called 'sidebands' and are ideal candidates for a first characterization of the FEL pulses. In addition, the quasi-monochromaticity of the FELs makes it possible to observe, for high dressing fields, absorption and emission of more than one NIR photon [72,73], avoiding interference effects present in similar experiments with broader-bandwidth XUV pulses of high-harmonic-generation sources [74]. In the following we will always use the term 'X-ray' for the FEL pulses, meant to cover the whole spectral range from XUV (>10 eV) over soft X-rays (>100 eV) to hard X-rays (>5 keV), to discern them from other short-wavelength sources as e.g., high harmonic-generated XUV pulses.

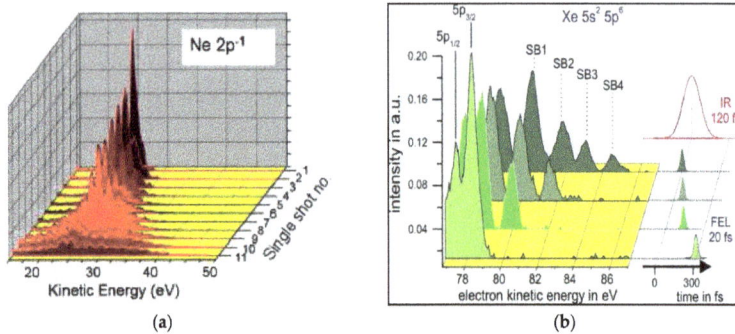

Figure 3. Single-shot sideband spectra. (**a**) Single-shot electron spectra in the region of the Ne 2p^{-1} photoemission line sorted with respect to different NIR intensities (reprinted with permission from [72], Copyright IOP Publishing, 2010). (**b**) Single-shot electron spectra in the region of the Xe 5p^{-1} photoemission line and the high-energy sidebands, taken at the nominal temporal overlap. Due to the relative arrival time jitter the actual timing varies by more than the pulse durations. The sidebands are used to get information about the actual degree of temporal (and spatial) overlap. In addition, a schematic of the temporal overlap between FEL and NIR pulses is shown (reprinted with permission from [75], Copyright AIP Publishing, 2007).

Besides using the sideband structures in the photoelectron spectra for the study of photoionization dynamics in strong external fields, the sideband intensity provides a clear indication of the temporal and spatial overlap between the FEL and the NIR laser beams. Since sidebands appear only when the two laser pulses overlap (see Figure 3b), their intensity exhibits a characteristic dependence on the relative time delay between the ionizing and the dressing pulses. For free-electron lasers based on the SASE process, such as FLASH, LCLS and SACLA, the X-ray and NIR pulses are generated separately by the FEL and the optical laser system. These two independent sources are only linked by a common externally working synchronization system [76]. Although the synchronization has been improved tremendously during the last years <30 fs root mean square (rms) [77], the pulse-to-pulse arrival time fluctuations exceed very often the pulse durations of the two pulses. The analysis of electron spectra for many individual pulses provides direct information about the temporal stability and can be used as inherent time marker for time-resolved pump/probe experiments.

The actual sideband amplitude depends on several parameters. First, good spatial overlap is crucial for the sideband generation. In addition, a meaningful analysis of sideband spectra is only possible when the X-ray focus is significantly smaller compared to the NIR focal spot size. Under such conditions the electrons are created only in the maximal field of the NIR pulse and are all exposed to (almost) the same NIR field strength. Nonetheless, pointing fluctuations of either of the two beams present a severe limitation for the practical use of the method.

Second, temporal overlap has to be assured, since the sidebands correspond to free–free transitions in the ionization continuum and both pulses have to interact with the electronic state at the same time. When the intensity conditions are such that a single sideband is generated, the interaction is determined by one X-ray and one NIR photon yielding a linear dependence on the relative overlap of the two pulses for the resulting sideband amplitude [78]. However, at high NIR intensities more than one sideband is present, implying the involvement of several NIR photons in the sideband formation process at the same time. This leads to a nonlinear response of the observed electron spectra, with a strong dependence on the intensity of the driving NIR laser. Thus, the sideband amplitudes are no longer linearly connected to the temporal overlap. Therefore, quantitative information about the X-ray pulse duration τ_{X-ray} can only be extracted by comparing the sideband features to a theoretical model.

The experimental line profiles can be fitted with a simplified analytical model based on the so-called soft-photon approximation (SPA), allowing us to simulate the sideband spectrum [79]. In general, the SPA gives excellent agreement with experimental data and can match the more elaborate non-perturbative methods such as the time-dependent Schrödinger equation (TDSE) when the electron kinetic energy is large compared to the photon energy of the dressing field [51,79–81]. The simulated sideband spectra are defined by four parameters: duration, τ_{NIR}, and intensity I_{NIR}, of the NIR laser pulse as well as duration (and shape) τ_{X-ray} of the FEL pulse and, finally, the relative temporal delay between the two pulses δ. The NIR pulse width and intensity can be determined with good accuracy before the experiment leaving τ_{X-ray} and δ as parameters to be determined during the measurement [51].

The sideband spectra can be analyzed in two ways. One approach comprises a statistical analysis of the single shot spectra while setting the relative delay to a fixed value, namely at nominal maximum overlap of X-ray and NIR pulses. Due to the inherent arrival time fluctuations the sideband amplitudes and the number of sidebands vary significantly from shot to shot. However, looking at the frequency distribution of the sideband amplitudes (histogram), information about the average temporal jitter and the X-ray pulse duration can be extracted [51,75]. Secondly, by changing the relative arrival time between the two pulses successively, a cross-correlation curve is acquired as shown in Figure 4 [82]. Again, the width of the correlation is determined by the dominant parameters of the X-ray and the NIR pulse durations as well as by the relative arrival time jitter.

In this scheme, photoelectrons generated via two-photon ionization in a single shot were measured at FLASH [9,83] using a magnetic bottle electron spectrometer (MBES) [84]. This efficient type of spectrometer allows the collection of all electrons emitted over the complete solid angle for small kinetic energies <100 eV [85], and still over approximately 0.8 sr corresponding to a 30° half-angle emission cone at higher electron kinetic energies around 0.8 keV [51]. Hence, the MBES setup can be applied to low-density targets and enables, in the case of the intense FEL beam, an analysis of the photoionization process on a shot-to-shot basis [75,84]. The best energy resolution, 1–2% of the electron kinetic energy, is achieved by decelerating the electrons with the aid of electrostatic retardation fields at the entrance of the flight tube.

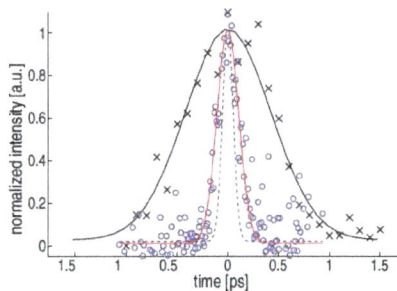

Figure 4. Sideband amplitudes. The amplitude of the first-order sideband measured in Xe is plotted as function of the set delay between the NIR laser and the free-electron laser in Hamburg (FLASH) free-electron laser (FEL) pulse. The uncorrected X-ray pump/NIR probe delay scan (x, fit as solid black line) is entirely dominated by the relative arrival time jitter and shows a width of 410 fs rms. A measurement technique based on electro-optical sampling [82] is able to provide an independent measurement of this jitter and can thus be used to correct the arrival time. This leads to an about 4 times narrower correlation width of 100 fs rms (o, fit as solid red line), which is already close to the simulated jitter-free case (dashed blue line). Normalized signal amplitudes are used throughout these plots (reproduced with permission from [82], Copyright AIP Publishing, 2009).

First experiments were mainly able to reveal the effect and show the NIR pulse duration, which was at least an order of magnitude larger than the other two parameters [78]. By reducing the NIR pulse duration from several picoseconds to about 120 fs it was possible to determine the relative arrival time jitter [84]. Eventually, by minimizing the timing jitter and NIR pulse duration even further, the final goal was reached to directly determine the X-ray pulse duration of a SASE XFEL [51].

For seeded FELs like FERMI, the relative arrival time jitter is strongly reduced in contrast to unseeded (SASE) FELs. Thus in this case the jitter is significantly less than the X-ray and NIR pulse durations, allowing for an even more precise measurement of the X-ray pulse duration [86]. Generally, the sideband method has been extensively used at FERMI to characterize the pulse length, and its results are in agreement with a complementary method and with theory [59]. A few distinctive aspects of its use at FERMI are related to the fact that this is a seeded source, resulting in predictable timing and coherence of the pulses. The measurements are intrinsically jitter-free, and concentrated on characterizing the pulse properties as a function of machine parameters. In particular, pulse coherence was exploited to implement a compression scheme [87], and the pulse duration was measured with the sideband method.

Sideband formation of Auger electrons

The typical bandwidth of a SASE FEL pulse is in the range of 0.1 to 1% of the nominal photon energy. For the photon energies at FLASH this results in a width less than the spacing of the sidebands (see Figure 3), such that they can be analyzed individually. On the other hand for shorter wavelengths in the X-ray regime as available at LCLS, the linewidth of the XFEL radiation can exceed the separation of the sidebands significantly (e.g., a linewidth of 7 eV at a photon energy of 1 keV was measured in Ref. [51]). In this case the emission of Auger electrons rather than photoelectrons can be measured instead, but still a de-convolution procedure and a comparison with simulated spectra has to be applied to extract the information about the X-ray pulse duration (Figure 5).

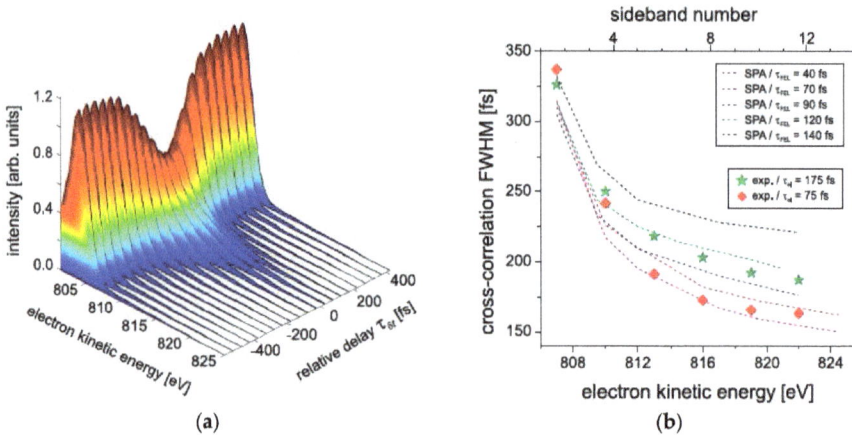

Figure 5. Auger sideband spectra. (**a**) NIR-dressed neon Auger electron energy spectra as a function of the relative delay time δ between the NIR ($\tau_{NIR} = 100$ fs) and the X-ray pulses ($\tau_{X-ray} = 75$ fs); (**b**) Comparison of experimentally obtained cross-correlation widths with those calculated within the soft-photon approximation (SPA) using the following parameters (in full width at half maximum (FWHM)): $\tau_{X-ray} = 40$ fs–140 fs, $\tau_{NIR} = 100$ fs and $\delta = 140$ fs, $I_{NIR} = 1.2 \times 10^{12}$ W cm^{-2} (reproduced with permission from [51], Copyright IOP Publishing, 2011).

Appl. Sci. **2017**, *7*, 915

X-ray pulse characterization including polarization

Sidebands were also used to determine other characteristics of the X-ray photon beam, namely its degree of polarization. For the first experiment using linearly polarized FEL and NIR pulses, it was demonstrated that the sideband intensity depends in a sensitive and characteristic way on the relative orientation between the two linear polarization vectors [88]. Since this dichroic effect is caused by a different ratio of the partial waves contributing to the outgoing electron (e.g., s- and d-waves in case of a two-photon ionization in He), which can be calculated very precisely by theory, it can in turn also be used to determine the degree of polarization of the FEL beam. This was demonstrated at FERMI, the first seeded FEL producing circularly polarized X-ray pulses [10]. By measuring the circular dichroism in the electron angular distribution of the first sideband (Figure 6), it was possible to determine the degree of circular polarization as well as the sign of the helicity. The measured value for the polarization at the experimental station (about 95%) [89] is in excellent agreement with estimates, taking into account an almost perfect polarization at the end of the undulator and the contribution from the optical elements of the beamline.

A similar method, i.e., the measurement of the circular dichroism in the sideband yields of electrons emitted from the inner shell of molecular oxygen, was also applied for the determination of the degree of circular polarization produced by the DELTA undulator at LCLS [90]. Here, the above discussed issue concerning the sideband separation in relation to the relatively large bandwidth of a SASE based XFEL was overcome by dressing the molecules with a frequency-doubled optical laser, ensuring a sufficient energy spacing of the individual sidebands.

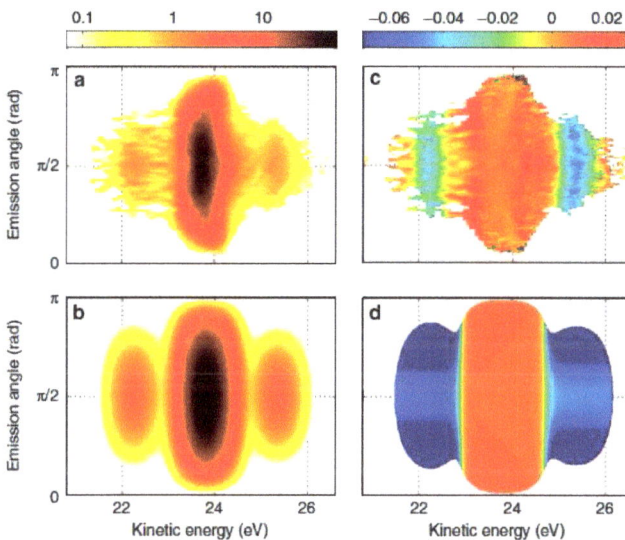

Figure 6. Angle-resolved spectra and dichroism. Experimental (top) and simulated (bottom) angle- and energy-resolved electron spectra (**a,b**) and circular dichroism (**c,d**) in He 1s photoionization at low NIR laser intensity (7.3×10^{11} Wcm^{-2}). The respective intensities are indicated by the color scales on top of the graphs (reproduced with permission from [89], Copyright Nature Publishing Group, 2014).

All these examples show that the measurements of sidebands, which are produced by the simultaneous interaction of the X-ray pulse and the NIR dressing field, provide detailed information about the X-ray pulse characteristics, especially on the pulse duration, the relative arrival time and its degree of polarization. This will certainly be used in future experiments at new, not yet well-characterized FEL facilities, but also the investigation of the dynamics of the two-color photoionization in itself is still of large interest. These measurements will permit a focus on particular aspects of the ionization process, e.g., the intensity dependence of the electron angular distribution or near-threshold phenomena [86,91].

3. Linear Streaking Measurements at Free-Electron Lasers

The method of sideband generation for the temporal characterization of X-ray pulses is valid in the regime where the X-ray pulse duration is longer than the cycle period of the applied dressing field [92]. For very short X-ray pulses, of few-femtosecond or even sub-femtosecond duration, a different approach is needed, one that relies on the ultrafast changes of the electric dressing field itself instead of its cycle-averaged intensity. In this section we will first discuss the characterization of soft X-ray pulses (Section 3.1), before describing the metrology of hard X-rays in 3.2.

3.1. Streaking Measurements of Soft X-ray Pulses

3.1.1. Infrared Streaking Experiments at Free-Electron Lasers

The presented diagnostic utilizes time-resolved photoelectron spectroscopy (also called laser-dressed photoionization), a technique that was originally developed for temporal characterization of attosecond XUV pulses from higher-order harmonic generation [66]. This technique is applicable over a broad range of photon energies produced at XFELs and can be applied to pulses ranging from several femtoseconds to hundreds of femtoseconds. The measurement works on a single-shot basis, requires only a small fraction of the available XFEL power and can be used in parallel with an actual experiment, thus allowing tagging of the XFEL temporal properties and post-experimental single-shot sorting of acquired results.

Photoelectron streaking

In a conventional streak camera, the temporal profile of the incident photon pulse is imprinted on an electron wavepacket through photoemission. The electronic wavepacket—or in the case of high X-ray intensities—the burst of photoelectrons is subsequently deflected transversely by a fast voltage ramp. In this way, the temporal profile of the incident light pulse is mapped to the spatial coordinate on the electron detection screen. Due to energy dispersion in the photocathode and limitations in the gradient of the voltage ramp, the temporal resolution that can be achieved in a conventional streak camera is limited to the picosecond level.

To reach few-femtosecond or even sub-femtosecond resolution, a similar concept is applied in the attosecond streak camera. Here, the electromagnetic field of a near-infrared laser pulse is used to project the temporal properties of the X-ray pulse onto a photoelectron wavepacket created through ionization of a noble gas.

In state-of-the art attosecond physics, a carrier–envelope phase-stable few-cycle laser pulse at a central wavelength of 750 nm is used to generate attosecond XUV pulses through high harmonic generation [6]. The resultant XUV pulses are intrinsically synchronized to the field of the driving laser pulse, assumed to be fully coherent and additionally assumed to be identical from shot to shot. Attosecond streaking spectroscopy was first applied to characterize these XUV pulses.

Figure 7 shows a sketch of a typical experimental setup and illustrates the basic principle of streaking. The X-rays eject electrons from gas atoms via the photoelectric effect, and set them free with a kinetic energy equal to the X-ray photon energy minus the specific electron binding energy of the atomic shell. These photoelectrons can be described as free electron wave packets for our experimental conditions. The temporal structure of the generated photoelectron wave packet mimics that of the incoming XFEL pulse and its width is directly related to the X-ray pulse duration [93].

For the streaking measurement, the XUV pulse is focused on a noble gas target and temporally and spatially overlapped with an intense laser pulse (the temporal overlap can be adjusted by varying the optical path lengths between the XUV and NIR beams). Photoelectrons emitted during the streaking pulse are accelerated by an amount that depends on the instantaneous electric field amplitude or vector potential $A(t)$ at the precise moment of ionization. For the polarization and detection oriented as in Figure 7, the shift in kinetic energy relative to the field-free case is given by:

$$\Delta W(t_0) \approx e\sqrt{\frac{2W_i}{m_e}} A(t_0), \tag{1}$$

where W_i is the initial kinetic energy of the photoelectron, and e and m_e are the electron charge and mass. For extended XUV pulses, the variation of the laser electric field during the photoemission results in different final momenta for electrons ejected at different times. If the photoemission is confined to within one half-cycle of the laser field (1.25 fs for a carrier wavelength of 750 nm), the initial kinetic energy distribution of the photoelectrons released by the XUV pulse from the noble gas target will be uniquely broadened and shifted by the overlapping NIR streaking field. In general, the temporal profile and spectral phase of the ionizing XUV pulse can be determined by analyzing the broadening of the detected photoelectrons.

Scanning the relative delay between the ionizing XUV pulse and NIR streaking pulse allows consecutive measurements to be accumulated and a time-resolved spectrogram to be constructed. By applying an appropriate retrieval algorithm [94], the streaking pulse and XUV pulse, with temporal profile and spectral phase can be retrieved from this multi-shot measurement.

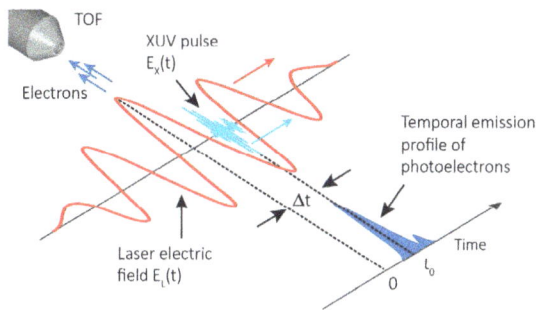

Figure 7. Photoelectron streaking schematic. A short extreme ultraviolet (XUV) pulse generates a photoelectron bunch by ionizing noble gas atoms in the presence of a strong, few-cycle phase-stable NIR laser pulse. The XUV pulse and a NIR streaking pulse are precisely synchronized to each other. The current profile of the generated photoelectrons is given by the temporal structure of the ionizing XUV pulse. After photoionization, the electrons are subject to the electric field $E_L(t)$ of the NIR pulse which introduces a change in the final momentum distribution of the photoelectrons, depending on the instant of release t_0 into the streaking laser field. The momentum change is detected with a time-of-flight (TOF) detector, which collects the electrons released along the direction of the streaking field polarization. The TOF is oriented such that it is axially aligned with the linear polarization of the laser and the XUV pulse. (Figure from [95]).

Experimental implementation

For the purpose of the streaking measurement at an XFEL an experiment was conducted (for further experimental details and theoretical background see [96]) at the atomic, molecular and optical science (AMO) instrumental end station at the LCLS [97]. It comprises a vacuum chamber within which an optical laser co-propagates with the XFEL and a magnetic bottle electron spectrometer [98], a high collection efficiency spectrometer that operates over a broad energy range capable of detecting the photoelectron spectra for every single shot.

The streaking laser was based on an optical parametric amplifier with the idler tuned to 2.4 µm central wavelength, horizontal polarization of the electric field, a pulse energy of ~30 µJ and a compressed pulse duration of ~50 fs full width at half maximum (FWHM). The output beam of the NIR laser was focused to a $1/e^2$ beam diameter of ~450 μm in the interaction region, yielding an optical intensity on target of roughly 4×10^{11} W/cm^2 or a corresponding peak vector potential of the streaking field of $A_{peak} \approx 0.18$ atomic units. As a target for photoexcitation neon gas was streaming from a nozzle at a background pressure of ~1×10^{-7} mbar in the experimental chamber, and the photoelectrons were detected with the MBES.

The LCLS was working at a repetition rate of 60 Hz in the so-called 'low-charge mode', with an electron bunch charge of 20 pC around the point of maximum compression at ~10 kA, producing XFEL pulses expected to be of sub-10 femtosecond duration [11]. In order to further minimize the XFEL pulse duration, the 'slotted spoiler' [32] was inserted into the electron beam at the second bunch compressor chicane along the SLAC Linac ahead of X-ray generation in the undulator. With an XFEL photon energy of $E_{X\text{-ray}} = 1790$ eV and a binding energy of $E_B = 870$ eV for the 1s electron in neon, the central energy of the photoelectrons is distributed around the mean value of 920 eV with a standard deviation of ~2 eV rms. The bandwidth of the photoline, corresponding to the energy spread of the X-ray pulse, was measured to be ~3.3 eV rms. The various measured parameters of the XFEL are summarized in Figure 8.

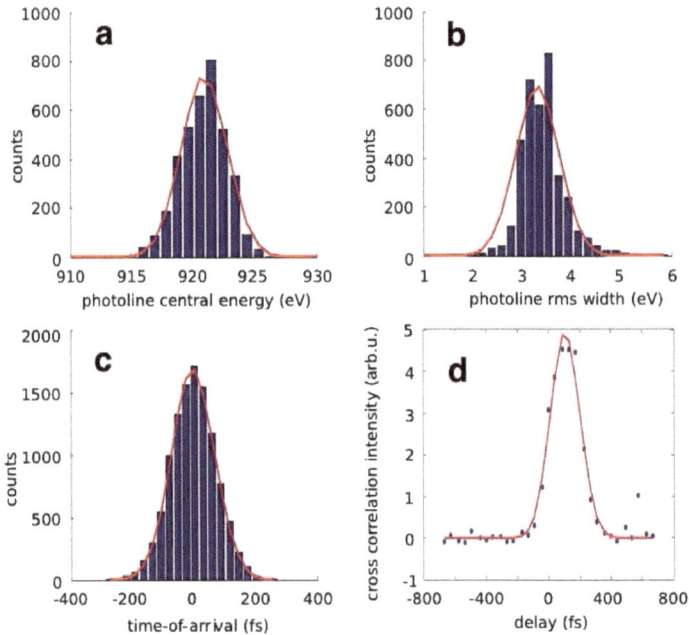

Figure 8. Measured XFEL parameters. (**a**) Single-shot distribution of central XFEL photon energy, measured with the magnetic bottle electron spectrometer (MBES); (**b**) Single-shot distribution of XFEL photon energy spread (rms); (**c**) Distribution of relative arrival time between NIR laser and XFEL, measured for a nominal delay set to zero (at full temporal overlap); (**d**) Cross-correlation intensity scan, by moving the NIR pulse in small steps through temporal overlap with the X-rays pulse and incorporating the arrival time jitter (Blue bars and points are the experimental data, and red full curves are fits to the data).

For the streaking experiment one has to carefully overlap the XFEL pulse with the NIR laser spatially and temporally. A fine-scan was performed by integrating the electron detection signal in an energy region close to the unstreaked photoline and changing the length of the path of the optical laser in small steps (compare Figure 9 for the measurement principle). The better the XFEL and the NIR laser are overlapped, the more electrons see the NIR field and the detected electron spectrum deviates from a pure photoline. Thus, one gets a time-dependent measure of the strength of the overlap. This signal can be used to find the maximum temporal overlap (see Figure 8d).

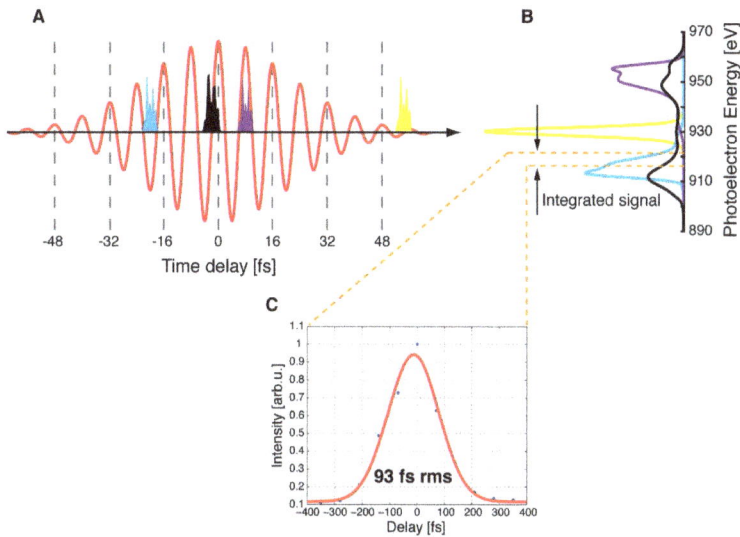

Figure 9. Determination of XFEL–NIR arrival time jitter. Panels (**A,B**) show the simulation of X-ray pulses (blue, black, purple and yellow) overlapping with the NIR streaking field (red) at different relative delays and the corresponding photoelectron spectra. Panel (**C**) depicts the measured cross-correlation obtained by integrating the signal over a small region of the spectrum at a point 4σ away from the unstreaked photoline for different delay settings. In the case of very poor overlap (the situation depicted as the yellow XFEL pulse in panel (**A**) there is nearly no streaking effect: the measured photoline stays narrow and the signal in the region of interest is small. As the X-ray pulse overlaps more and more with a vector potential of higher amplitude (the cases of the blue, purple and black pulse, in ascending order of streaking potential amplitude), the generated PE spectra become ever more broadened and the integrated signal rises accordingly. Therefore, the detected width of this cross-correlation can be interpreted as the convolution of the NIR pulse duration and the average arrival time jitter between the XFEL and the streaking field (panel (**C**)), which amounts to ~93 fs rms in our case (Figure 9). Figure from [96]).

Direct time-domain measurement of ultrashort X-ray pulses

In this experiment [69], the attosecond streaking technique was applied to X-ray FEL pulses for the first time. The XFEL and the streaking laser typically are two separate laser sources and need to be externally synchronized to keep them temporally overlapped. Nevertheless, even with synchronization a residual shot-to-shot arrival time jitter between the two lasers remains. Since the arrival time jitter between XFEL and NIR is of the same order as the NIR pulse duration, it is not possible to assign a definite delay and corresponding value of the vector potential with respect to the streaking field to each single XFEL pulse. With the presented scheme and following analysis one can overcome the synchronization issue and derive an upper limit of the X-ray pulse duration for each single shot in a manner that is decoupled from all machine parameters and can be used simultaneously with any ongoing measurement.

Before presenting the details of the evaluation, it is an important prerequisite for the following analysis to make an estimate of the X-ray pulse duration compared to the streaking laser period. Following the previous reasoning, for longer X-ray pulses (enclosing one or more cycles of the streaking field) the energy shift of the streaked photoelectron spectrum is negligible, thereby resulting in uniformly larger streaked spectral widths. This is evident when looking at the spectra presented in Figure 10a,d, which show 50 simulated (a) and measured (d) streaked spectra obtained with X-ray pulses stretched to around 25 fs (by changing the compression of the electron bunch) and a streaking field with a wavelength of 2 μm. For the X-ray pulses at optimized compression and inserting the spoiler [32], one can see in Figure 10b,e that when employing the same 2 μm streaking laser, the situation is globally unchanged, whereas when using a 2.4 μm wavelength NIR laser, one observes the apparition of a clear streaking regime [21,66] (Figure 10c,f). This is clear evidence that the XFEL pulses under these experimental conditions are at least shorter than 8 fs (one period at 2.4 μm wavelength) and one can apply the streaking concept for the X-ray pulse duration acquisition.

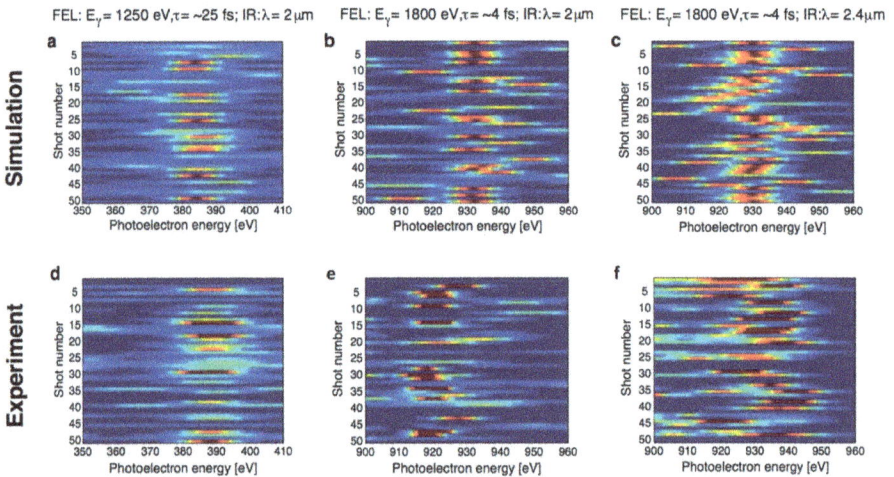

Figure 10. Streaked X-ray photoelectron spectra (top: Simulation, bottom: Experiment). Six color-coded intensity plots for photoelectron spectra of 50 consecutive shots, for two different X-ray pulse durations ($\tau_{X-ray} \approx 25$ fs and $\tau_{X-ray} \approx 4$ fs) streaked by optical lasers with two different central wavelengths ($\lambda = 2$ μm and $\lambda = 2.4$ μm). Panels (**a–c**) are simulations, while (**d–f**) are constructed from the corresponding measurements taken at the LCLS during our experiment. The long 25 fs X-ray pulses (panels (**a,d**)) exhibit nearly no spectral shift but rather uniform smearing of the photoelectron (PE) spectra, revealing that for these long pulses the measurement method breaks down. A similar behavior can also be observed for the short X-ray pulses (<5 fs) when streaked with the 2 μm NIR field, shown in the middle in panels (**b,e**): the ratio of the period of one optical cycle at this wavelength to the X-ray pulse duration is still not optimal, resulting in a high probability to find the spectral center near the unstreaked value. In contrast, for the 2.4 μm streaking field, right panels (**c,f**), the whole spectrum is hugely shifted away from the mean, as is expected when the PEs 'see' only part of the NIR vector potential, a clear demonstration of the onset of the streaking regime. Note that the overall shift of the central energy in panels (**e,f**) from 930 eV to 920 eV is due to an accordingly lower X-ray photon energy (1790 eV instead of 1800 eV) for this measurement and is not a streaking effect. (Figure from [96]).

Single-shot X-ray pulse duration measurement

As described above, one of the major challenges for these types of correlation measurements is the arrival time jitter of each single XFEL pulse with respect to the NIR laser (see Figure 9). Due to the statistical nature of the SASE process the actual pulse shape and thus the duration of consecutive XFEL pulses also varies from shot to shot, necessitating a single-shot measurement technique for its determination. That means that one has to adopt a different method to extract the X-ray pulse duration from the measured spectra than what is conventionally used in streaking experiments with precisely synchronized gas-phase high harmonic sources.

As the relative phase between the X-ray pulse and the streaking field is not known a priori, one has, in a first step, to determine the value of the streaking vector potential for each single XFEL pulse. This fact is critical for the analysis and restricts the characterization to the determination of upper limits for the X-ray pulse durations. The peak of each single-shot spectrum that is pushed farthest away from the unstreaked center is identified and is assumed to stem from an overlap of the X-ray pulse with a local maximum of the vector potential. This sets an experimental limit for the local electric field strength of the streaking laser for each specific X-ray shot and thus for the gradient of the vector potential, which will be underestimated on average.

In the second step of the evaluation process one uses this vector potential as a calibration factor to convert the measured spectral width of each streaked photoelectron wave packet into an upper limit for the duration of the corresponding X-ray pulse.

To illustrate the principle of evaluation, have a look at an example from the data, sketched in blue in Figure 11a. In this case a streaked spectrum ranging from 915 eV to 945 eV was measured, corresponding to a width of the streaked spectrum of δ_s = 30 eV and a maximal energy shift in relation to the unstreaked photoline at 920 eV of $\Delta\varepsilon$ = (945 − 920) eV = 25 eV. Following Ref. [99], one assumes the vector potential $A_{NIR}(t)$ of the streaking field to be approximately linear in time,

$$A_{NIR}(t) \approx -E_{peak} \times \left(t - t_{peak} \right), \tag{2}$$

with a slope given by the peak electric field $E_{peak} = E(t_{peak})$ of the half-cycle that the photoelectrons are overlapping with, which occurs at the moment t_{peak}. The period of the streaking laser was t_{period} = 8 fs, setting the time interval for the vector potential to rise from a zero-crossing to a maximum (a quarter period) to 8 fs/4 = 2 fs. This value corresponds to the above calculated maximum shift $\Delta\varepsilon$ = 25 eV in the energy domain. Now one can use the measured width of the streaked spectrum, δ_s = 30 eV, as a ruler for the pulse length, resulting in an upper limit for the X-ray pulse duration:

$$\tau_{X-ray} = \frac{\delta_S}{\Delta\varepsilon} \times \frac{t_{period}}{4}, \tag{3}$$

which is (30/25) × (8/4) = 2.4 fs in this case, represented by the blue X-ray pulse on the left in Figure 11a.

There is, in addition, a residual ambiguity in the analysis method that has to be considered. Figure 11a shows schematically a second XFEL pulse (green, in the middle) with a different duration that nevertheless leads to nearly the same measured width of the streaked spectrum. Since it was assumed that the X-ray pulse partly overlaps with a local maximum of the vector potential, one cannot discern streaking effects caused by the right or the left edge of this vector potential curve. To get a correct upper limit for the possible X-ray pulse duration, one therefore must double the above calculated value to 4.8 fs. All pulse lengths between 2.4 fs and 4.8 fs are compatible with the measurement.

Figure 11, panels (b)–(e), shows measured streaked photoelectron spectra generated by four different X-ray shots (solid blue curves) and their derived pulse duration upper limits (red bars on the right). From the maximum shifted peak in panel (b) a maximal energy shift $\Delta\varepsilon$ = | −32 | eV is inferred, i.e., the laser half-cycle of 4 fs is mapped onto an energy range of 2 × $\Delta\varepsilon$ = 64 eV. The complete spectrum spans from −32 eV to 27 eV, setting the overall width to δ_s = 59 eV and corresponding to a pulse duration of 3.7 fs. In a similar manner one derives upper limits for the other pulse durations of 3.4 fs in panel (c), 2.8 fs in panel (d) and 2.6 fs in panel (e), respectively. Considering the ambiguity of the energy-to-time mapping, all these shots are in principle also compatible with the doubled pulse lengths.

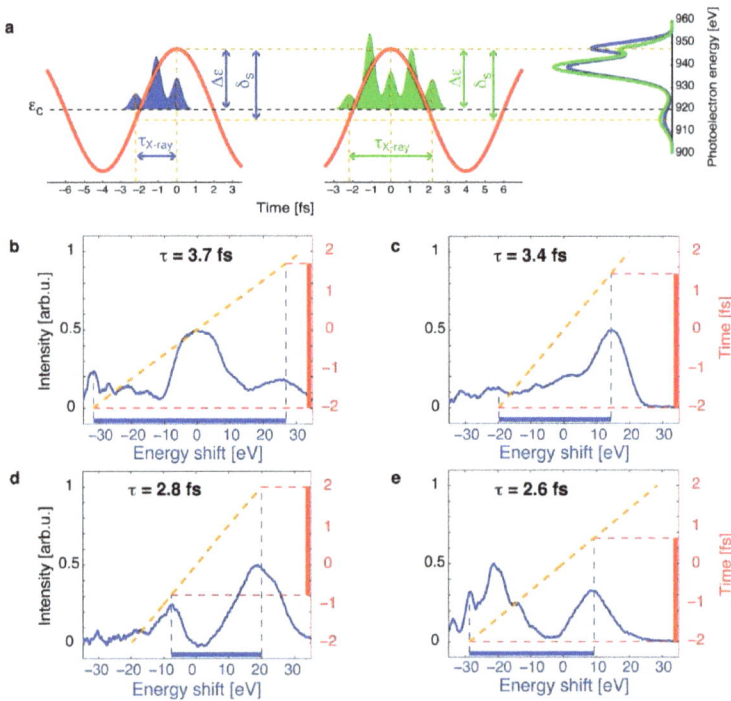

Figure 11. Streaking of few-femtosecond X-ray pulses. The top panel (**a**) shows the principle of the X-ray pulse duration evaluation: the maximal shifted peak of the measured PE spectrum is associated with an overlap of the X-ray pulse with a local maximum of the streaking field (blue pulse on the left). This energy shift $\Delta\varepsilon$ defines the gradient of the energy-to-time conversion and the overall width δ_s of the streaked spectrum is used for the determination of the XFEL pulse duration. The green X-ray pulse, twice as long and depicted in the middle, gives rise to a very similar PE spectrum, leading to a residual ambiguity of the XFEL pulse length measurement. Panels (**b–e**) show four different streaked photoelectron spectra as measured in the experiment (solid blue lines) at different values of the NIR vector potential and the respective linear streaking ramp (dashed orange lines) derived from the maximum shifted peak of the spectrum. Corresponding pulse duration upper limits are given and shown on the right vertical axes (red bars) (reproduced with permission from [69], Copyright Nature Publishing Group, 2014).

Owing to the statistical nature of the SASE generated XFEL pulses originating from noise, the X-ray pulse profiles differ greatly from shot to shot. For the first time, this variation in the intensity substructure of XFEL pulses can be shown directly in the time domain. In Figure 11b, an XFEL pulse consisting of only one single peak overlaps with the zero crossing of the vector potential and is broadened compared to the photoline. Different arrival times of the XFEL pulse relative to the streaking laser lead to an overall shift of the whole spectrum as depicted in Figure 11c,d, where the XFEL pulse overlaps with the negative or positive part of the vector potential and the resulting spectrum is shifted accordingly. An individual peak followed by a tail (c) can be distinguished from a pulse containing two well-separated spikes (d) or even three individual spikes as shown in Figure 11e.

If one repeats this analysis for each single shot and plots the number of shots versus the corresponding pulse durations, an estimation of the average FWHM X-ray duration can be built up as a histogram (Figure 12). The average pulse duration of the XFEL is best fitted by a Gaussian distribution with a mean value of ~4.5 fs and a standard deviation of 1.7 fs. Clearly, this is a very conservative upper limit and the actual pulse duration is likely to be considerably shorter for the vast majority of LCLS pulses in this mode of operation.

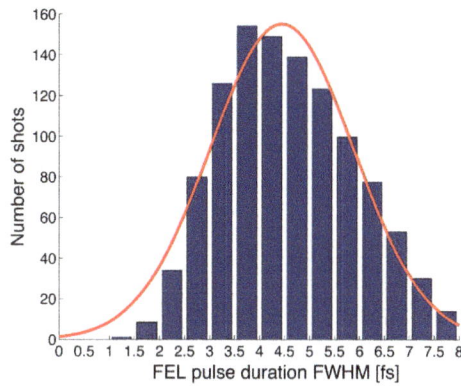

Figure 12. Average XFEL pulse duration. A histogram of the possible pulse duration values in agreement with the measurement (blue) for the described parameter range at LCLS. The red solid line shows a Gaussian fit to the distribution, indicating an upper limit for the average X-ray pulse duration of ~4.5 fs (reproduced with permission from [69], Copyright Nature Publishing Group, 2014).

Measurement of sub-femtosecond X-ray pulses

Due to pulse-to-pulse fluctuations there are some pulses that are considerably shorter than the average duration. XFEL theory predicts a temporal pulse substructure composed of a train of ultrashort spikes [100] with a typical width corresponding to the coherence time, characterized by the X-ray photon energy bandwidth and machine specific parameters. For the parameter range during the measurements at the LCLS, spike durations of about one femtosecond and even below are expected to occur (compare for example Ref. [48]). The average number of spikes per XFEL pulse can be estimated, but it will also be subject to shot-to-shot variations. Occasionally, an X-ray pulse consisting of just one single isolated spike will occur.

A closer analysis of the data confirms the existence of such ultrashort 'single-spike' pulses, constituting about 5% of the total number of shots. By sorting and collating the streaked spectra with just one peak and determining their centers, one can identify the energy shift $\Delta\varepsilon$ and the spectral width δ_s for each of these shots. Figure 13 shows the measurement principle for attosecond X-ray pulses: For the evaluation of these single-spike pulses one assumes that due to simple statistical considerations the predominant number of XFEL shots overlaps with a part of the ramp of the vector potential and hence again Equation (2) is valid in most cases. Then one can employ a reasoning similar to that described above to assess an upper limit for the pulse duration. In order to not overestimate the actual value of the vector potential for the conversion into the time domain, the minimum possible value of the vector potential compatible with each specific shot is used for the analysis, that is the value that corresponds to the observed shift of the central photoelectron energy for each of these single-spike pulses. This is a very conservative approach and will again result in the determination of an upper limit for the respective pulse duration.

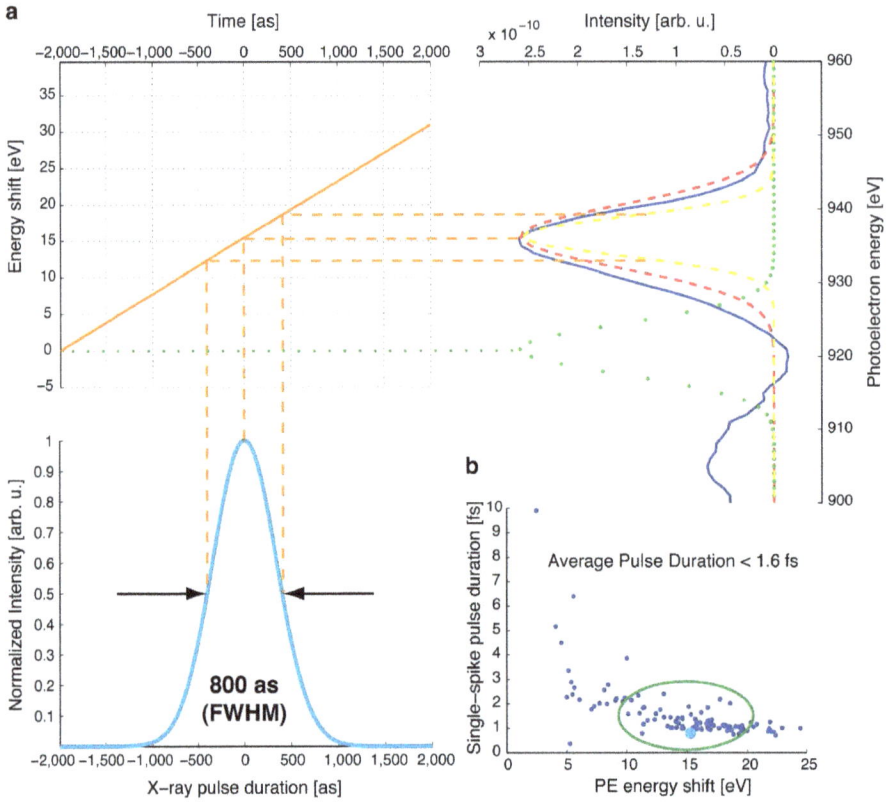

Figure 13. Measurement of an attosecond X-ray pulse. Panel (**a**) on the right-hand side shows a measured spectrum of a single XFEL intensity spike (solid dark blue line) and the corresponding Gaussian fit (dashed red line). The shift of the peak from the unstreaked central energy at 920 eV (green dotted line) defines the energy-to-time ramp (top left), with the help of which the width of the deconvoluted spectrum (yellow dashed-dotted line) can be converted into a pulse duration upper limit of 800 as, depicted as the light blue line on the bottom left. Panel (**b**) at the bottom right shows the derived upper limits for the pulse duration of all single-spike shots as a function of the spectral energy shift of the peak. The very short pulse described in panel (**a**) is marked as the bigger light blue dot. The outliers towards longer pulse durations are all associated with small energy shifts (≤5 eV), and therefore potentially stemming from an underestimation of the actual streaking vector potential by our method. Nevertheless, most of the measured single-spike X-ray pulse durations are in the range of 1.5 femtoseconds and below, highlighted by the green oval (reproduced with permission from [69], Copyright Nature Publishing Group, 2014).

For the shot marked as the bigger light blue dot in panel b in Figure 13, the energy shift amounts to $\Delta\varepsilon = 17.4$ eV. Following deconvolution of the initial unstreaked X-ray photon energy bandwidth, which was determined by separate measurements, from the measured signal and using $\Delta\varepsilon$ as a calibration factor the spectral bandwidth of this single peak was found to correspond to an X-ray pulse with a duration of maximum 800 attoseconds FWHM. In general, the derived upper limits for the duration of X-ray single-spike pulses are mainly in a range of 750 as to 1500 as (FWHM) (marked by the green oval in Figure 13b), which is in good agreement with the predictions based on equations for the SASE process under the conditions of the LCLS machine parameters [48]. However, in case the X-ray shot is actually overlapping with a maximum of the vector potential, this method breaks down; therefore, novel streaking concepts have to be developed for the unambiguous and complete characterization of X-ray FELs, as presented in Section 3.1.2 for femtosecond pulse structures, and in the outlook (Section 4) for the determination of the full time–energy information of SASE X-ray pulses with attosecond resolution.

3.1.2. Laser-Driven Terahertz Streaking for Complete Temporal Characterization

Single-cycle terahertz streaking at XFELs

In contrast to attosecond HHG sources, at SASE XFELs the temporal profile and arrival time of the X-ray pulse changes from shot to shot and cumulative measurements are typically not possible. Therefore, the pulse measurements should be performed on a single-shot basis. In addition, the duration of SASE FEL pulses can be as long as 100 femtoseconds. As a result, lower frequency streaking fields, in the THz regime, are required (1 THz corresponds to a half-cycle time of 500 fs) as the ionizing FEL pulse must be shorter than a half-cycle of the streaking field.

While phase-locked THz pulses do not typically exist at XFEL facilities, at FLASH it is possible to send the "spent" FEL-driving electron bunch through an additional dedicated undulator to generate multi-cycle, phase-stable THz pulses. In principle, this allows for direct application of attosecond techniques. Indeed, in previous work, this THz source was used as the streaking field for temporal characterization of the XUV FEL pulses [101]. However, this technique does not provide access to relative timing information for pump–probe experiments. Furthermore, the THz streaking field changes as the electron beam is varied, fluctuates or is tuned, influencing the measurement and effectively limiting the utility of this technique. In an extreme case, when the accelerator is tuned for the shortest FEL pulses using low-charge highly compressed bunches, the beam-based generated THz might not even be strong enough for streaking.

An independent laser-driven THz source for pulse characterization overcomes the limitations of streaking with accelerator-based THz sources, while maintaining the capability to characterize 100 fs long pulses. In addition, as laser-driven THz fields are locked to the external laser, the streaking measurement provides the capability to determine simultaneously the FEL pulse arrival time and its pulse profile on a time base that is synchronized to the pump–probe experiment environment.

Many laser-based THz sources rely on optical rectification of femtosecond Ti:sapphire laser pulses in a $LiNbO_3$ crystal [102]. With this technique, high-field phase-stable single-cycle THz pulses with pulse durations of ~2 ps can be generated. The THz vector potential, which is the relevant quantity in streaking spectroscopy, has a half-cycle of ~600 fs. The half-cycle duration, or streaking ramp gives the dynamic range of the measurement. Notably, it is significantly longer than the maximum expected FEL pulse duration and facility timing jitter. Therefore, once the temporal overlap between the X-ray pulse and the THz pulse has been established, all subsequent FEL pulses will overlap at some point on the single-valued streaking ramp allowing them to be characterized unambiguously.

The principle of a single-cycle THz streaking measurement is illustrated in Figure 14. Three characteristic measurement conditions are shown in the figure. In the first panel the FEL and THz pulses are not temporally overlapped, therefore, the photoelectron spectrum remains unperturbed and only the FEL pulse spectrum shifted by the ionizing potential of the target gas is measured. In the second panel the FEL pulse is overlapped with an extremum of the vector potential (it overlaps with the bottom of the streaking ramp), resulting in a maximally downshifted photoelectron spectrum with minimized spectral broadening. Consequently, the temporal structure of the pulse is not observed in the measured spectrum. In the last panel, the ionizing FEL pulse coincides with a zero crossing of the THz vector potential $A(t)$. Photoelectrons emitted by the leading and trailing edges of the FEL pulse are streaked with a momentum shift of opposite sign, resulting in a maximally broadened photoelectron around the field-free central energy. At this temporal overlap, the arrival time as well as the temporal profile and duration of the FEL pulse can be accessed with highest resolution.

If the streaked photoelectron spectrum is significantly broader than the initial bandwidth of the ionizing FEL pulse, the temporal profile of the FEL pulse can be recovered directly from the measured spectrum. In this case, also referred to as the strong streaking regime, the energy gradient that is imprinted onto the electron wavepacket by the streaking field is much larger than the initial energy variation caused by an FEL pulse with an initial energy chirp. As a result, the degree of spectral broadening is assumed to depend only on the duration of the FEL pulse and the gradient of the THz streaking field.

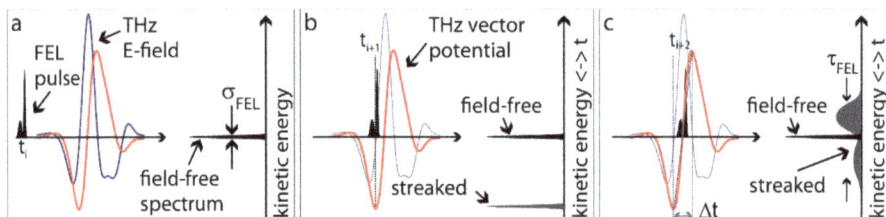

Figure 14. Schematic of FEL pulse characterization by THz streaking. Three distinct measurements are depicted in frames (**a**–**c**). FEL induced photoemission from a noble gas, which replicates the FEL temporal profile (black filled area), is broadened and shifted, or "streaked", depending on the instant of overlap with the THz pulse. The THz streaking pulse electric field is shown in blue, while the corresponding vector potential is drawn in red in all three panels. On the right vertical axis, the resulting kinetic energy of the photoelectrons is indicated: in black the field-free spectrum, defined by the SASE FEL bandwidth in the unstreaked case (**a**), and in gray spectra for the streaked cases, at overlap with an extremum (**b**) or at the zero-crossing (**c**) of the THz vector potential. In (**c**), overlap occurs near the zero-crossing of the THz vector potential where the arrival time and temporal profile can be accessed with the highest resolution (reproduced with permission from [103], Copyright Nature Publishing Group, 2012).

To make a calibrated transformation from the streaked photoelectron spectra to the time domain, the THz vector potential must be known. Due to timing jitter at FELs, the vector potential cannot typically be determined directly from the streaking spectrogram as is the case in attosecond spectroscopy. Rather, at XFELs the transformation map is determined in a two-step process. First, the electric field shape is evaluated independently by electro-optic-sampling. Second, the electric field is scaled by evaluating a distribution of streaking measurements over all possible time delays. Measurements that overlap with a maximum of the vector potential result in maximal kinetic energy shift, as depicted in Figure 14, panel b. The maximum kinetic energy shift can then be inverted to determine the absolute THz electric field strength:

$$\Delta W_{\max} \approx \frac{e\lambda_L}{\pi c \sqrt{2m_e}} \sqrt{W_i} E_0 \tag{4}$$

Here, λ_L is the streaking laser wavelength, E_0 is the maximum electric field, W_i is the initial electron energy and c is the velocity of light in vacuum. With the transformation map and under the assumptions listed above, the temporal profile of the FEL pulse can be retrieved from a single-shot streaked spectrum as illustrated in Figure 15.

THz streaking can be used to simultaneously determine the arrival time of the FEL pulse. But, it is important to note that the information provided by this diagnostic is far more complete than a simple arrival time. In fact, the FEL pulse profile is given on a time-base defined by the THz streak ramp that is locked to the experiment environment, if the same femtosecond laser pulse that is used the generate the THz streaking pulse is also used to pump dynamics in an independent experiment. With this complete information, it is possible to define a single point as the "arrival time". This point could be given by the center-of-mass of the FEL pulse, or the position of the peak intensity of the FEL pulse, or at a certain intensity threshold. The exact definition would depend on the dynamics being observed, but crucially, any definition can be chosen when the pulse is characterized on an experimentally relevant time base.

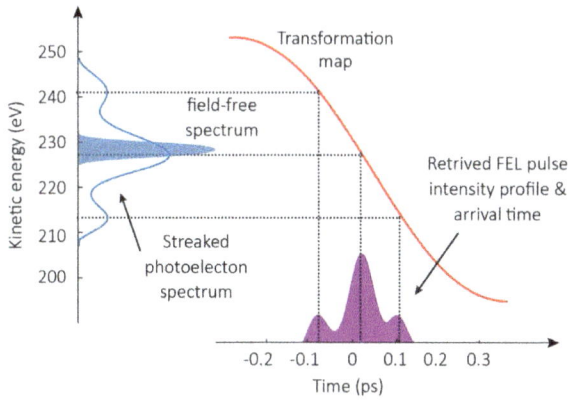

Figure 15. Mapping THz streaked spectrum to temporal profile. When the streaked photoelectron spectrum (blue line) is significantly broader than the initial spectral bandwidth of the FEL photon pulse (blue filled area), the temporal profile (violet filled area) of the ionizing FEL pulse can be easily retrieved. The red curve indicates the streaking vector potential used as a transformation map, to obtain temporal information from a measurement in the spectral domain. (Figure from [95]).

For FEL pulses with a broad initial bandwidth and large initial energy variation across the pulse, it may not be possible to neglect the initial chirp in these streaking measurements. Moreover, it may be crucial for experimental applications to characterize the spectral chirp in the FEL pulse.

For linearly chirped FEL pulses, at least two measurements of the streaked photoelectron spectrum at opposite slopes of the vector potential are required to infer the temporal duration and shape of the ionizing pulse [104]. At FELs, different experimental approaches can be taken to achieve this on a single shot basis. For example, two time-of-flight detectors, both mounted in the plane of the streaking field polarization but in opposite directions could be used to record two spectra with opposite slope. Alternatively, it has been proposed that different phases of the streaking pulse can be realized by taking advantage of the Gouy phase shift, which occurs when a Gaussian beam propagates through a focus [105].

If single-shot information cannot be obtained, information regarding the average chirp of the FEL pulses could be obtained by making two sets of measurements with THz pulses generated with two distinctly different phases. This can be realized easily for certain THz generation methods. In particular, for THz pulses generated using collinear rectification in ZnTe or organic crystals with higher electro-optic coefficients.

Measurement resolution

The temporal resolution in THz streaking of X-ray pulses depends primarily on the energy resolution of the photoelectron spectrometer and the THz streaking strength. In addition, the initial FEL pulse bandwidth and the evolution and Gouy phase advance of the THz pulse in the interaction region can also limit the temporal resolution. Moreover, shot-to-shot intensity fluctuation of the streaking pulse, as well as energy jitter in the FEL pulse can degrade the temporal resolution of the measurement.

Typically, the streaked photoelectron spectrum is recorded with a time-of-flight (TOF) spectrometer. The energy of the photoelectron is determined by measuring the time t required by the electron to travel from the interaction region to the detector placed at a known distance L. The kinetic energy of the electron is given by:

$$E_{\text{kin}} = \frac{1}{2}m_e\frac{L^2}{t^2} \tag{5}$$

From this simple kinematic equation, a measure of the energy resolution ΔE_{kin} can be written:

$$\Delta E_{\text{kin}} = \frac{2\sqrt{2}}{\sqrt{m_e}}\frac{\Delta t}{L}E_{\text{kin}}^{\frac{3}{2}}, \tag{6}$$

where Δt is the smallest interval that can be resolved by the electronic detection system.

The acceptance angle of the TOF spectrometer is defined by its geometry (length of the drift tube and detector diameter). In order to enhance the sensitivity, the TOF is equipped with an electrostatic lens at the entrance of the drift tube, which increases the effective acceptance angle for electrons within a specific kinetic energy range. Additionally, a post acceleration stage in the TOF ensures a flat response of the electron detector (multi-channel plate, MCP) regardless of the initial electrons' kinetic energy. Wide acceptance angle, and flat detector response is especially important for measurements at FELs, as the full spectrum of photoelectrons must be collected in a single shot. High efficiency through a larger acceptance angle is usually accompanied by a decrease of the energy resolution due to variations in the electron flight path for electrons of same energy, but with different initial momentum. Moreover, the need to collect single-shot spectra, means that many photoelectrons must be generated during each exposure which can lead to space charge and additional degradation of the energy resolution.

Beyond these fundamental factors, the relationship written above for ΔE_{kin} indicates that the energy resolution of the TOF is limited by the length of the drift tube and the temporal resolution of the detection electronics. The latter one includes the response time of the MCP electron detector and the resolution of the analog-to-digital (ADC) converter [106]. To improve the energy resolution, retardation potentials can be applied to increase the flight time of the electrons, before they enter the drift tube.

At FELs, the relative energy resolution of electron TOF spectrometers is usually not much better than ~1%. The dominant issue lies in the temporal resolution of the detection electronics for multiple photoelectron acquisition.

The issue is illustrated in Figure 16. In the first panel, the signal resulting from a single electron response is illustrated. The single electron response is amplified and a pulse with a characteristic duration determined by the capacitive properties of the MCP amplifier is produced. For a TOF operating in a single-electron mode, it is sufficient to determine the time-of-flight from the rising edge of this signal, which can be determined with high accuracy, depending mostly on the temporal resolution of the ADC that samples the pulse. In panel b, the MCP response for multiple distinct single-electron events are depicted. This situation is typical in the case of experiments at synchrotrons, where, on average, less than one electron per X-ray shot is detected. The accumulated arrival time distribution of the single photoelectron events gives the photoelectron spectrum. The time resolution is mostly determined by the rise time of the MCP signal and the resolution of the digitizer.

In contrast, in measurements at FELs, where the entire photoelectron spectrum must be recorded in one FEL shot, the sum of the individual electron signals is collected and the detector acts as a current amplifier, rather than a discriminator for counting single-electron events. The measured photoelectron spectrum is then a convolution of the actual signal and the width of the single-electron signal. Therefore, to increase the resolution of single-shot TOF measurements, the pulse width of a single-electron event must be reduced. To achieve this, MCP amplifiers with a smaller diameter can be used; however, this would reduce the collection efficiency.

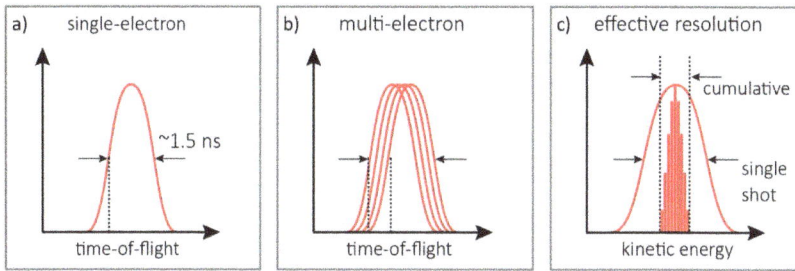

Figure 16. Time-of-flight (TOF) Energy resolution at FELs. In contrast to application at synchrotrons, the resolution in time-of-flight spectra collected at XFELs is dependent on the single electron response of the detector. Panel (**a**) depicts the single-electron response; panel (**b**) shows multiple electron hits; panel (**c**) shows the effective resolution, if the multiple electron hits are accumulated within a single-shot acquisition or a cumulative acquisition. At XFELs, the acquisition is single-shot and the observed signal is a convolution of the photoelectron bandwidth with the single-electron response. If the acquisition is cumulative, the observed signal is the convolution of the photoelectron bandwidth and the resolution with which the edge of the single-electron signal can be determined. (Figure from [95]).

Independent of TOF detector issues, the THz streaking strength determines the degree of broadening of the photoelectron spectra and thus directly influences the achievable temporal resolution. The FEL pulse can be retrieved with the highest resolution, if it is overlapped at the zero crossing of the streaking pulse vector potential. At this overlap, the maximum streaking strength is present and given by [101]:

$$s = \frac{\partial(\Delta W)}{\partial t} \tag{7}$$

The average streaking speed is given by the ratio of the peak-to-peak amplitude of the kinetic energy shift to the rise time of the streaking vector potential:

$$s_{\text{avg}} = \frac{\Delta W_{\text{max}-\text{min}}}{\Delta t_{\text{rise}}} \tag{8}$$

With these formulas, a measure of the temporal resolution of the streaking measurement can be written as:

$$\tau_{\text{res}} = \frac{\sigma_{\text{field}-\text{free}}}{s}, \tag{9}$$

where $\sigma_{\text{field}-\text{free}}$ is the width of the photoelectron spectrum without streaking field.

Clearly, the temporal resolution can be increased by applying steeper streaking fields. Such fields can be achieved by scaling the THz generation to higher peak intensities, or alternatively by using higher frequency THz pulses with shorter streaking ramps.

Experimental implementation

Proof-of-principle experiments on THz streaking were first performed at the free-electron laser FLASH in Hamburg. This facility operates in the XUV spectral range emitting pulses with ~100 femtosecond duration and below. Typical pulse energies are on the order of 100 µJ. The experimental setup is shown in Figure 17.

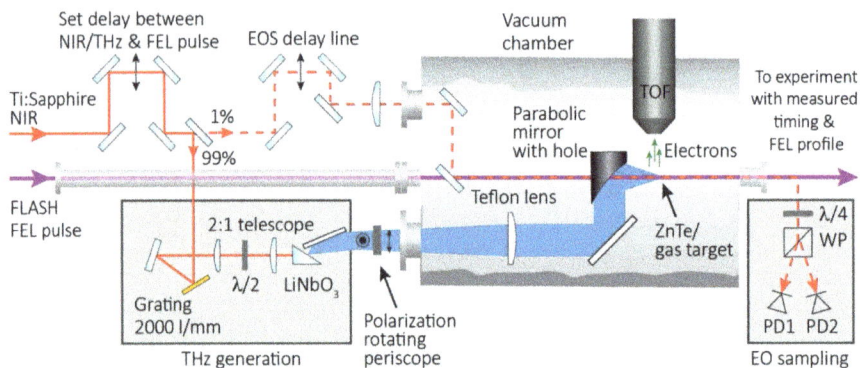

Figure 17. THz streaking apparatus. Layout of the THz streaking experimental setup used at FLASH: The NIR laser pulse is split into two parts. Most of the pulse energy is used for the tilted pulse-front THz generation in LiNbO₃, in setup geometry as shown. The remaining 1% pulse energy is used as a probe for electro-optic sampling (EOS) of the THz pulses in a 500 μm thick zinc telluride (ZnTe) crystal. The NIR probe pulse and the FEL pulse pass through a 2 mm hole in the parabolic mirror and are collinearly overlapped with the THz beam and focused into the interaction region, which is fixed by the position of the time-of-flight spectrometer. For the EO-sampling a ZnTe crystal is placed in the exact interaction region. The crystal is also used as an observation screen for monitoring the spatial overlap of the FEL and the NIR probe pulses. For the streaking measurement a gas nozzle providing neon or helium is placed in the interaction region. The FEL pulses eject a burst of photoelectrons from the atomic gas system and the kinetic energy of the photoelectrons is modified by the THz pulse. Recording the photoelectron spectrum with the TOF spectrometer consequently enables characterization of the FEL pulse (reproduced with permission from [103], Copyright Nature Publishing Group, 2012).

THz pulses were generated with user pump–probe laser, ~3 mJ, 60 fs pulses from a Ti:sapphire laser amplifier by tilted pulse front optical rectification in LiNbO₃. The peak on-target electric field strength was observed to be 165 kV/cm. The THz electric-field shape was characterized independently by electro-optic sampling, and the absolute field strength determined from a distribution of streaking measurements as discussed before. The THz electric field is shown in Figure 18, panel (a).

Single-shot streaking measurements on XUV pulses at ~260 eV photon energy were made in helium. The corresponding single-shot and averaged THz streaking spectrograms are shown in Figure 18, panels (b) and (c), respectively. To build the spectrogram, the delay between the THz and the soft X-ray pulses was scanned in 100 fs steps. Approximately 400 single-shot spectra were averaged at each time delay to generate the averaged spectrogram. For large delays, where the THz field is weak, the He 1s photoemission peak is nearly unaffected and located near its field-free energy. Close to temporal overlap the kinetic energy of the photoelectrons is shifted and broadened depending on the THz streaking field phase. In the averaged spectrogram the dynamic range given by the single valued THz streaking ramp is indicated with dashed white lines. As the streaking ramp of ~600 fs is significantly longer than the maximum expected FEL pulse duration and timing jitter, once the streaking THz pulse and the FEL are temporally overlapped, all single-shot acquisitions will occur on a uniquely defined position on the streaking ramp.

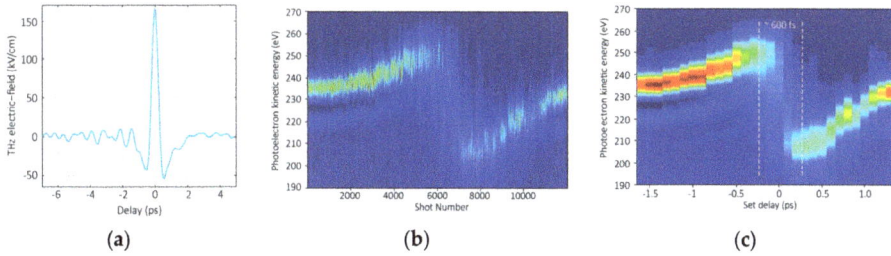

Figure 18. Proof-of-principle streaking measurement at FLASH. (**a**) The single-cycle THz electric field. (**b**) Single-shot spectrogram composed of 12,000 consecutive shots. (**c**) The averaged spectrogram, with approximately 400 shots binned at each 100-femtosecond time step. The duration of the THz streaking ramp is indicated by the dashed white lines in the right panel, giving the dynamic range of the measurement, which is ~600 fs. (Figure from [95]).

A characteristic single-shot measurement reproduced from reference [103] is shown in Figure 19. The streaked spectrum is shown in panel a. The retrieved temporal profile, with conversion map as a dashed line, is shown in panel b. The statistical error in the single-shot spectrum is calculated according to the number of electrons collected within the energy resolution window of the detector. As the photoelectron spectrum is heavily oversampled, boxcar integration is performed across the spectrum, resulting in a smooth error envelope that bounds the measured spectrum. The reconstructed FEL pulse exhibits a dominant central feature with a width of (53 ± 5) fs FWHM and has weaker satellite features at approximately 100 fs from the main peak.

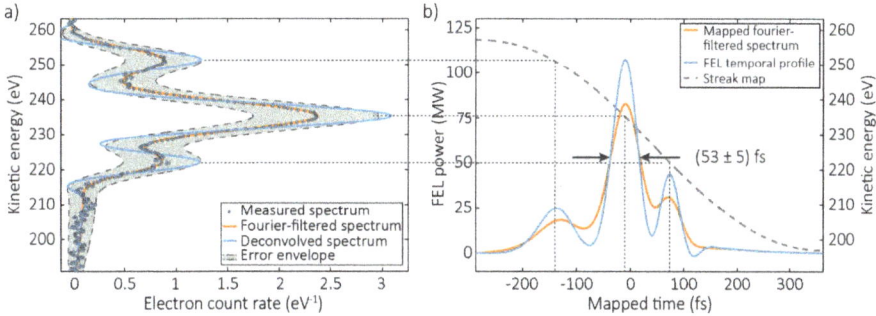

Figure 19. Single-shot FEL pulse profile. (**a**) Example of a measured streaked photoelectron spectrum. The shaded error envelope is calculated by boxcar integration, based on the number of electrons collected within the TOF energy resolution window of the time-of-flight spectrometer. The blue dots are raw data points in the measured streaked spectra. Orange curve shows the spectra after Fourier filtering to remove high-frequency noise. Blue curves are filtered, streaked spectra following deconvolution of the photoelectron spectrometer resolution. (**b**) Retrieved temporal profile for this FEL shot. Orange and blue curves are the mapped intensity structures corresponding to the Fourier-filtered spectrum and the deconvolved spectrum in panel (**a**), respectively. The energy-to-time conversion map is depicted as the dashed gray line in panel (**b**), with the corresponding energy axis on the right (reproduced with permission from [103], Copyright Nature Publishing Group, 2012).

The temporal profile shown in Figure 19 is given on a time base that is locked to the laser that generates the streaking THz pulse. This information can be condensed into a single arrival timestamp by calculating the center-of-mass of the retrieved FEL pulse profile. To evaluate the short-term timing jitter at FLASH ~450 consecutive FEL pulses, collected near time-zero were analyzed. The results are displayed in Figure 20. The shot-to-shot analysis is shown in panel (a), and the distribution of arrival times, which is the timing jitter, is shown in panel (b).

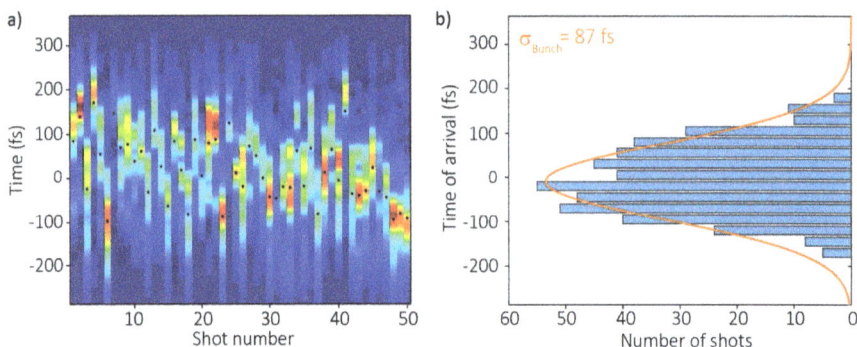

Figure 20. Arrival time and jitter. Panel (**a**), false color plot of 50 consecutive single-shot measurements recorded close to time zero. The center of mass of the individual photoelectron spectra (black dots) is used to determine the arrival time of each pulse. Panel (**b**) shows the distribution of arrival times collected over ~450 consecutive shots with a Gaussian fit. The corresponding width of the fit is ~87 fs rms, which is a measure of the short-term timing jitter between the pump–probe laser that generates the THz, and the FEL (reproduced with permission from [103], Copyright Nature Publishing Group, 2012).

The Gaussian fit to the distribution has a width 87 fs rms, which is a measure of the short-term timing jitter between the FEL and the pump–probe laser. This value is consistent with the expected performance of the electronic laser synchronization [83]. The accuracy of the relative arrival time information is mainly influenced by the stability of the mean FEL photon energy. During our measurements, it fluctuated with a level of 1 eV rms, which corresponds to a time-of-arrival uncertainty of ~6 fs rms.

THz generation in $LiNbO_3$ is attractive, as the rise time of the streaking field is ~600 fs, resulting in a dynamic range that allows for measurement of all pulses, with lengths that can exceed 100 fs in duration. But the streaking power is limited. Alternatively, THz streaking fields can be generated by collinear optical rectification of infrared laser pulses in the organic electro-optic salt 4-*N*,*N*-dimethylamino-4′-*N*′-methyl-stilbazolium tosylate (DAST). THz fields generated in the DAST crystal have steeper half-cycle rise-times of ~180 fs, which in principle can lead to higher resolution measurements.

At LCLS measurements on 1.1 keV soft X-rays were made using THz pulses generated in DAST. High field strengths approaching 1 MV/cm were observed in the interaction region. However, these measurements suffered from timing jitter due to the reduced dynamic range. The timing jitter between XFEL and THz was measured at ~200 fs rms. Therefore, an additional method for sorting the arrival time of the X-ray pulses with respect to the THz pulses was required to isolate shots that arrived on the single-valued THz streaking ramp [107].

Three streaking spectrograms collected as the relative delay between DAST-THz pulse and X-ray pulse was smoothly varied are shown in Figure 21. The panel on the left shows the single-shot measurements in the order as they were collected and the panel in the center shows the single-shot spectra re-ordered according to the additional timing information from the spectral encoding measurements. In the sorted spectrogram, the shots arriving on the single-valued streaking ramp are clearly identified, allowing for calibrated transformation from streaked kinetic energy to time. The panel on the right shows the measured spectrogram after rotating the DAST crystal by 180 degrees to reverse the sign of the streaking field. In principle, as in attosecond spectroscopy and discussed earlier, this provides access to the average spectral phase, or chirp, of the X-ray pulse in addition to the temporal profile.

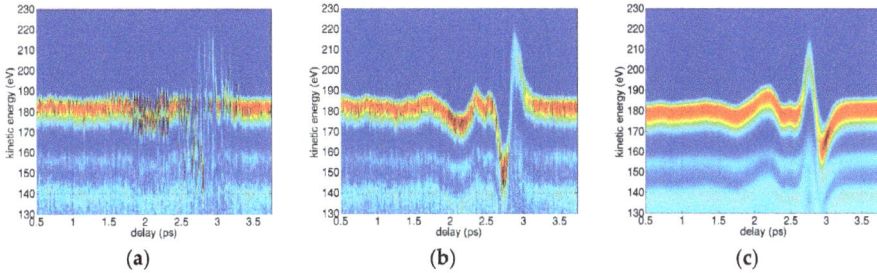

Figure 21. Streaking measurement made with THz fields generated in DAST at the LCLS. (a) In the panel, the single-shot measurements are displayed in the order that they were collected. (b) The single-shot measurements are sorted based on the X-ray pulse arrival time measured by spectral encoding. (c) The panel shows the sorted and binned spectrogram collected with the opposite sign THz field, providing access to the spectral phase.

The streaking power s (Equation (7)) in the DAST measurements was ~1.7 fs/eV. With 1% relative energy resolution in the measured photoelectron spectrum, these fields have the potential to be used to achieve ~5 fs resolution in the X-ray profile measurements.

For higher time resolution, to measure pulses compressed to several femtoseconds in duration and shorter to the sub-femtosecond, attosecond regime the wavelength of the streaking fields needs to be decreased further. In attosecond physics, ~10 attosecond resolution is routinely achieved using near infrared streaking fields. However, these fields provide dynamic range of only ~1.5 fs and are only suitable for attosecond pulse characterization.

3.2. Terahertz Streaking with Hard X-rays

Issues at high photon energies

THz streaking with hard X-rays should, theoretically, be easier to achieve than streaking with soft X-rays or HHG sources for a given ponderomotive potential $U_p = e^2 E_0^2 / 4 m_e \omega_L^2$, where $\omega_L = \frac{2\pi c}{\lambda_L}$ is the angular frequency of the streaking field. The energy shift term of Equation (4) predicting the maximum THz streaking effect, $\sqrt{8 W_i U_p}$, shows that the energy shift increases as the square root of the electron kinetic energy after photoionization W_i. Thus, the streaking effect is indeed stronger and more pronounced, but the difficulties in streaking with hard X-rays come from different sources.

Hard X-rays, especially those over 5 keV photon energy, tend to be fairly non-interactive with most materials. The photoionization cross sections for most elements and molecules drop by orders of magnitude when compared to soft X-rays [108], and make the observation of any kind of photoelectron spectrum on a single-shot basis difficult. Since most FELs that can generate high-energy X-ray radiation tend to have shot-to-shot temporal jitter, and THz streaking is used as a diagnostic to measure that jitter, this difficulty has to be overcome, and there must be enough signal generated in a single shot to measure the temporal properties of the FEL beam. Simply ramping up the gas pressure in the interaction region to get more signal is also not an option, since the TOF spectrometers used for THz streaking measurements have MCPs as their detectors, and they tend to begin malfunctioning if the background pressure of the vacuum chamber becomes too high. Last, but not least, the choice of observable photoionization lines is very limited for hard X-rays, and the large streaking effect heavily influences the settings on the TOF spectrometers themselves. Yet, the experience with the Photon Arrival and Length Monitor (PALM) at the beamtimes performed at the SACLA FEL [12] have conclusively shown that these challenges can be overcome.

Solutions and experience

The experiments performed with hard X-ray streaking at FELs have used photon energies as high as 10 keV [109,110], and had the challenge of performing the streaking measurements for single FEL pulses. The experiment was performed at SACLA, which works at a repetition rate of 30 Hz and pulse energies between 150 μJ and 300 μJ. The choice of a gaseous target with suitable electron binding energies at 10 keV photon energy is limited, and the closest easily reachable noble gas absorption edge is that of the xenon L shell. In this particular case, the Xe $2p_{3/2}$ photoionization line was the best suited for the measurements, as it had the largest cross section at this photon energy, and xenon was easily available for the experiment.

To ensure that the number of photoelectrons generated by the ionization process was as large as possible, while keeping the mean background pressure in the vacuum chamber low enough for safe MCP operation at the same time, the experimental setup used pulsed piezo valves [111] that were synchronized to the FEL trigger signal and allowed short 15 μs (FWHM) bursts of xenon gas into the chamber very close to the interaction region. A large 1200 L/s pump was placed directly below the gas jet to suck away the particles as quickly as possible and additional differential pumping was installed for the TOFs. The valve was backed by xenon gas at a pressure between 2 bar and 3 bar above atmosphere and was equipped with a conical nozzle with a 150 μm exit diameter, which forced the gas atoms to cluster upon entering the chamber. This arrangement both increased the local density of the atoms at the point of photoionization due to the formation of large, thousand-strong Xe clusters, and prevented a large broadening of the gas jet as it streamed from the nozzle into the vacuum [112]. Overall, the choice of xenon gas as the interaction medium together with the specific nozzle type ensured that there was sufficient signal for single-shot measurements even at 10 keV, and worked well even when the intensity of the X-rays was reduced by introducing a monochromator into the beam path. Moreover, a non-streaked TOF photoionization measurement was implemented upstream of the THz streaking region, since both the bandwidth and the photon energy of FEL pulses change from shot to shot. The availability of a calibration measurement for every shot makes the evaluation of the single-shot arrival time jitter and the temporal characteristics of FEL pulses much easier.

Because the kinetic energy of electrons set free at 10 keV X-ray photon energy amounts to about 5200 eV for the Xe $2p_{3/2}$ shell, the streaking shift in electron momentum according to Equation (1) is much larger than that experienced with soft X-rays, even when fairly weak THz fields are applied. In the experiments performed in [109,110], the electric field strength of the focused THz beam in the interaction region was only between 5 MV/m and 6 MV/m, but the maximum streaking energy shift observed was about 100 eV. To accommodate this large shift, the TOFs had to be designed to be able to detect electrons over a large energy window with constant resolution over the whole range and special care had to be taken to not accidentally cut off any signal by choosing the retardation potentials on the TOFs too high. For example, though the kinetic energies of streaked photoelectrons shifted by ±100 eV lie between 5100 eV and 5300 eV in this case, the retardation potentials were set to an upper limit of about 4850 V to ensure that no signal was lost and the spectral resolution was not compromised due to contributions from the full FEL bandwidth or dispersive focusing effects of the TOF.

In addition to these special components, the setup also included the standard components for a THz streak camera. A parabolic mirror with a centered hole inside the vacuum chamber focused the collimated THz beam collinearly with the FEL propagation direction onto the streaking interaction region. The schematic setup of the experiment from [110] is shown in Figure 22.

The experiment conducted in [110] compared the arrival time measurements of the spatial-encoding SACLA timing tool and the PALM, finding that the two agreed to within about 17 fs of each other at 9 keV photon energy. Earlier measurements presented in [109] showed a better arrival time agreement, which can be explained by the altered geometry for the PALM in [109]. The entrance opening to the TOF was twice as far away from the interaction region than ideal due to a manufacturing issue with the PALM chamber, which affected the number of electrons that were detected. This in turn caused the peaks to be less well-defined and reduced the accuracy of the measurement accordingly. The measurements are presented in Figure 23.

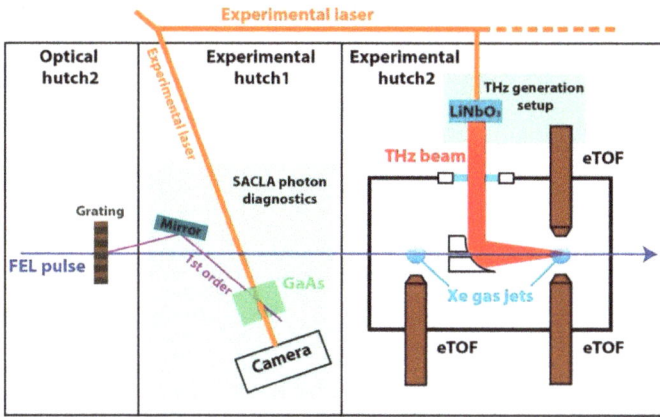

Figure 22. THz streak camera setup for experiments in [110] (reproduced with permission from [110], Copyright Optical Society, 2017).

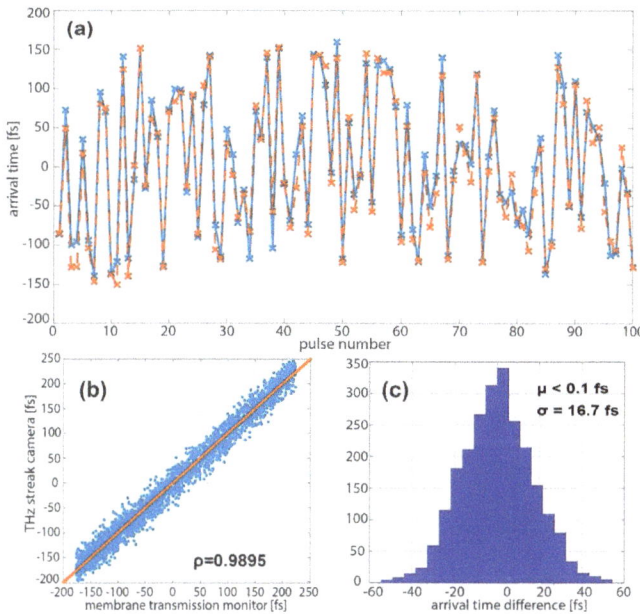

Figure 23. Comparison of the spatial-encoding timing tool with the Photon Arrival and Length Monitor (PALM) at SPring-8 angstrom compact free-electron laser (SACLA) (reproduced with permission from [110], Copyright Optical Society, 2017). (**a**) compares the SACLA timing tool (blue solid line) and the PALM (red dashed line) arrival time measurements for a set of consecutive pulses; (**b**) shows the correlation ρ between the two methods for several thousand shots; (**c**) shows the histogram of the deviations between the two methods from (**b**).

To demonstrate the flexibility of the PALM, two-color mode measurements at SACLA showed that the PALM could also measure the temporal separation of multiple FEL pulses. In this mode, the FEL was set to create two FEL pulses at photon energies of 9 keV and 8.8 keV, with the ability to shift their relative arrival times with respect to each other by manipulating the X-ray generating electron beam with the help of a magnetic chicane. The average rms jitter of the two pulses with respect to each other at a given delay setting was about 31 fs, of which about 17 fs are attributed to the uncertainty of the PALM measurements. These results are presented in Figure 24.

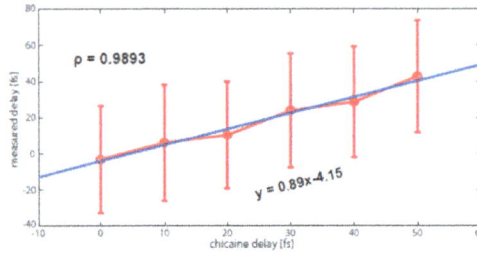

Figure 24. Chicane delay vs. two-color pulse delay measured with the PALM, and their correlation ρ (reproduced with permission from [110], Copyright Optical Society, 2017).

The experiment presented in [109] also checked the streaking effect at several photon energies ranging from 5 keV to 9 keV. The goal was to observe the streaking shift, and the accompanying streaking slope, get stronger or weaker depending on the X-ray photon energy. The experiment showed the expected relationship between the accuracy of the arrival time measurement and the kinetic energy of the electrons ionized by the FEL pulse, shown in Figure 25.

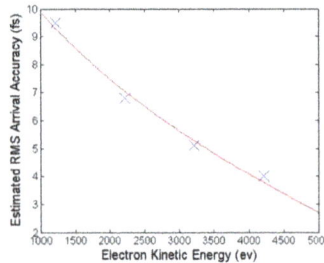

Figure 25. Arrival time accuracy as a function of electron kinetic energy, as shown in [109] (reproduced with permission from [109], Copyright Optical Society, 2014). The blue crosses are the data points, while the red curve is the square root function that best fits them.

In conclusion, the experiments on THz streaking of hard X-ray pulses worked as expected and the signal strengths on a shot-to-shot basis, as reported in [109], were more than sufficient for single-shot temporal measurements. Based on these experiments, there seems to be no reason why the technique could not be extended to even higher photon energies, and ideas for adapting the device to handle larger jitters between the FEL laser and the experimental laser are also being considered [113].

4. Outlook

4.1. Angular Streaking—Full Time–Energy SASE Pulse Retrieval

All intensity dependent experiments at free-electron lasers are highly affected by the effective pulse durations and SASE caused sub-spikes in the XFEL pulses. Mostly, such experiments rely on pulse energy measurements by gas detectors or other methods averaging over the intensity distribution of the XFEL pulses. However, the physical processes under investigation often depend on the peak intensity, which is hidden in the unresolved spike structure of the pulses. In addition to the strong impact on the broad field of non-linear physics, developments towards mode-locking [114–116] and the generation of X-ray pulses on the attosecond timescale [34–37] are currently in progress and will ultimately originate the scientific field of attosecond X-ray physics. This imminent breakthrough demands unprecedented time resolution for the temporal characterization of the structure of SASE XFEL pulses under various modes of operation. Such possibilities desperately need a direct single-shot method to diagnose not merely the X-ray pulse duration, but also the full temporal coherence properties of the XFEL pulses including the spectral chirp of the X-rays and the stochastic sub-spikes of the SASE pulses.

Another essential requirement for every ultrafast experiment that relies on a synchronized optical laser for pump or probe is precise information about the relative arrival time between the two lasers. Steady progress in this regard has been made [51,52,58,69,101,103,117–120], as was demonstrated throughout this review, and electron acceleration designs relying on superconducting linear accelerators [19,20] open up even better prospects for temporal synchronization [121]. Nevertheless, a complete and direct single-shot measurement in the time domain that reports both the exact XFEL pulse temporal structure and arrival time simultaneously and which is applicable over a broad range of X-ray pulse durations remains elusive.

A possible solution to all these challenges could be the method of two-color angle-resolved circular streaking spectroscopy of photoelectrons or Auger electrons, dubbed 'angular streaking' in the following for brevity. The measurement builds on the above described experiences with linear X-ray photoelectron streaking spectroscopy [69,103]. The need for a measurement technique capable of coping with varying pulse shapes and arrival times with respect to the lab time frame on a shot-to-shot basis is recognized and taken into account in this setup by employing a streaking laser with circular polarization in combination with a ring-like arrangement of 16 high-efficiency time-of-flight spectrometers [122]. This scheme has recently also been theoretically described [123,124] and evaluated for application at realistic XFEL parameters by a quasiclassical description supported with numerical calculations based on a quantum-mechanical approach (see Figure 26).

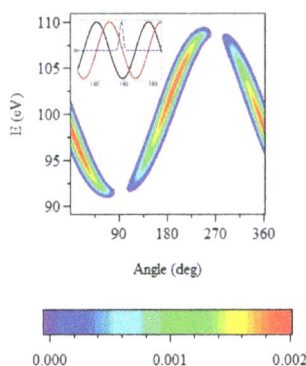

Figure 26. Simulation of circular streaking. Color-scaled streaking spectra as a function of the emission angle of photoelectrons produced by an ultrashort X-ray pulse (FWHM of the field is 2.8 fs) assisted by the THz field of 1.7×10^{10} W cm^{-2} intensity, 300 fs duration and 28.2 THz frequency. The electron kinetic energy (without the THz field) is 100 eV. In the inset, the X-ray and THz pulses are shown in arbitrary units. Time (in fs) is counted from the onset of the THz pulse. The black and red lines are, respectively, the *x*- and *y*-component of the THz field. (reproduced with permission from [124], Copyright IOP Publishing, 2017).

The vector potential of the streaking field rotates in the lab frame with a well-specified period, like the hand of a clock, and imprints the XFEL pulse temporal structure onto the angularly resolved distribution of streaked photoelectrons [63,125,126]. Varying the wavelength of the circular streaking field allows to set the period of the 'clock' and thus the temporal resolution of the measurement. While capable of measuring the shortest X-ray FEL pulses with a resolution of ~160 attoseconds at 800 nm, the resolution can easily be adapted to longer pulse durations via the adjustable streaking wavelength. To avoid ambiguities, the period of the streaking laser field has to be larger than the XFEL pulse duration.

At the same time, the relative strength of the streaking effect gives information about the arrival time shift from shot to shot. Moreover, this setup allows to investigate the shift of the photoelectron energy versus delay within a single X-ray pulse and thus measure its instantaneous frequency. For the first time, this would provide the direct measurement of residual X-ray pulse chirp from the different electron bunch compression settings, pointing the way toward full XFEL spectral phase control and chirped pulse amplification.

The proposed extended and angle-resolving method of streaking can be realized with an array of 16 independently working time-of-flight spectrometers which are aligned in the dipole plane of the incoming light (see Figure 27, left). As described previously, the circularly polarized streaking laser and the XFEL have to be overlapped spatially and temporally in the experimental chamber, where the X-ray pulse ionizes a diluted noble gas e.g., xenon. In the presence of the optical field the original angular distribution of photoelectrons is redistributed according to the rotational angle of the vector potential at the moment of their generation. Thus, photoelectrons generated from different parts within the XFEL pulse envelope are streaked at different angles, effectively mapping time to angle (Figure 27, right). We therefore do not require precise knowledge of the X-ray arrival time on the scale of the carrier phase of the streaking laser, as was the case in past streaking experiments, where temporal overlap with the linear ramp of the linearly polarized vector potential enabled the photoelectron energy shift-to-time mapping. Since the information of the XFEL pulse structure is given through the relative change in angular distribution within each single XFEL shot, we expect this technique to be more robust with regards to pulse energy instabilities and less demanding with regards to the optical pulse shape.

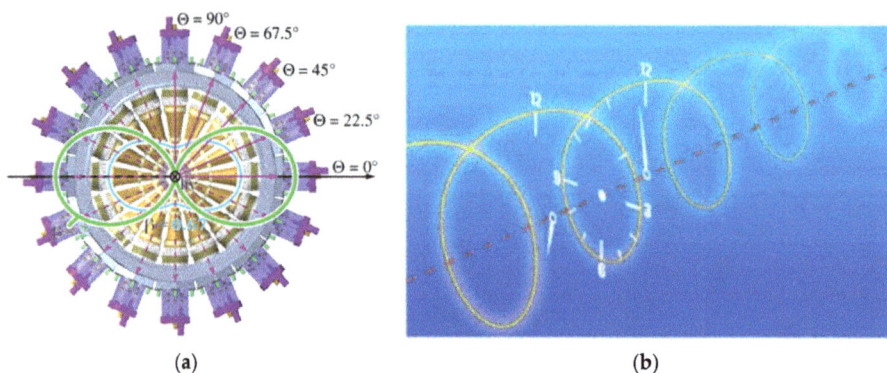

(a)   (b)

Figure 27. Conceptual design of angular streaking. (a) Sketch of the experimental chamber with 16 independent time-of-flight spectrometers (reprinted under the terms of the Creative Commons Attribution 3.0 License from [122]). (b) The vector potential of the optical streaking laser rotates similar to the hand of a watch. Photoelectrons resulting from different parts of the X-ray pulse are therefore streaked at different angles, allowing the reconstruction of the X-ray temporal profile (reproduced with permission from [127], Copyright Nature Publishing Group, 2011).

First demonstration measurements with this setup have successfully been conducted at the Linac Coherent Light Source in California, USA, and the results are being analyzed at the moment by some of the authors of this article. The first impression of the data is very encouraging, giving rise to the expectation of an adjustable online characterization method for complete shot-to-shot arrival time and time–energy X-ray FEL pulse reconstruction in the near future.

4.2. Tandem Phase-Shifted Geometry for Infrared Streaking and Attosecond Resolution

As discussed, to further improve the resolution of the linear THz streaking measurements to the sub-femtosecond level, streaking pulses with a shorter rise time in the IR spectral region must be used. These pulses, however, consist of multiple electric field cycles and are typically not carrier–envelope phase stabilized. As a result, it not possible to control the streaking field amplitude and phase due to the temporal jitter at XFELs. Crucially, without defined streaking parameters it would not be possible to perform a calibrated transformation from energy to time.

This problem can be addressed by recording two streaked photoelectron spectra for each X-ray pulse, using a phase-shifted streaking pulse in a tandem streaking geometry as illustrated in Figure 28. As usual, the instantaneous vector potential $A(t)$ is determined by analysing the shift in the centroid of the dressed photoemission. Qualitatively, due to the fact that the measurements are made with phase-shifted streaking fields, the value of the instantaneous vector potential of one measurement gives the gradient of the field in the other and vice versa, thus providing all of the information required to yield a calibrated measurement of the temporal profile of the photoemission and corresponding X-ray pulse. The instantaneous amplitude and phase of the streaking field can also be combined to determine the cycle-averaged amplitude of the IR streaking laser pulse, which gives the precise overlap between the X-ray pulse and envelope of the streaking laser pulse for use in pump/probe experiments.

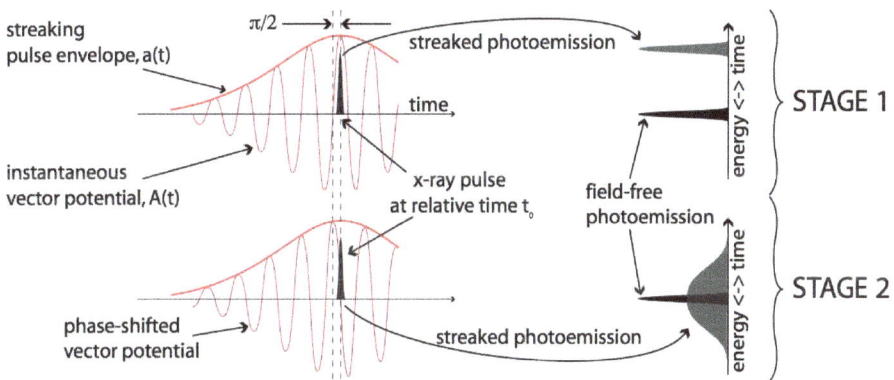

Figure 28. Principle of self-calibrated streaking measurements made with multi-cycle infrared fields. In the first stage the X-ray pulse overlaps with a maximum of the vector potential, resulting in a maximal energy shift of the streaked photoelectron spectrum by an amount proportional to the instantaneous vector potential. In the second stage the photoemission occurs at a zero-crossing of the vector potential, resulting in a maximally broadened photoelectron spectrum. The magnitude of the shift in kinetic energy, observed in the first stage measurement, gives the gradient of the vector potential in the second stage measurement. Combined, this information enables a calibrated transformation of the second stage measurement from energy to time.

In practice, the Gouy phase shift can be used to introduce a controlled and fixed phase shift between the two different stages of the streaking measurement required for self-calibration. To achieve a $\pi/2$ phase shift, streaking measurements would be performed at one Rayleigh length upstream and downstream of the laser focus. A precise $\pi/2$ phase shift between the measurements allows for full temporal characterization of the X-ray pulse, but the technique is applicable for any phase shift as long as it is constant.

For a $\pi/2$ phase shift the instantaneous streaking fields in each stage of the streaking measurement for an arbitrary overlap t_0 between streaking pulse (of arbitrary carrier–envelope offset φ_0) and the X-ray pulse are:

$$A_1(t_0) = a(t_0)\cos(\omega_L t_0 + \phi_0)$$

and

$$A_2(t_0) = a(t_0)\cos(\omega_L t_0 + \phi_0 + \pi/2) = a(t_0)\sin(\omega_L t_0 + \phi_0)$$

As usual, $A_1(t_0)$ and $A_2(t_0)$ would be found by inverting the expression for the energy shift in the streaked photoemission, which is to first order: $\Delta W(t_0) \approx -p_c A(t_0)$, compare also Equation (1). Here, p_c is the initial photoelectron momentum.

Conceptually, to transform broadened or streaked spectra from kinetic energy to time, the streaking strength or derivative of the instantaneous streaking field must also be known. From the expressions for the arbitrary streaking field at the $\pi/2$ phase-shifted streaking measurements, it is clear that the instantaneous vector potential in the second measurement is proportional to the derivative of the instantaneous vector potential in the first and vice versa, where the proportionality constant is ω_L the carrier frequency of the IR streaking field.

Using this crucial information, each streaked spectrum provides an independent measurement of the X-ray pulse profile. Furthermore, if the phase relationship between the streaking fields in the two measurements is $\pi/2$, the value of the cycle-averaged amplitude at the point of overlap is simply $a(t_0) = \sqrt{A_1(t_0)^2 + A_2(t_0)^2}$. Then, if the laser pulse envelope is well characterized, this can be inverted to provide the precise time of arrival of the X-ray pulse and the temporal characterization is complete.

Beyond pulse characterization, the streaking technique may also be used to study few-femtosecond electron dynamics at XFELs. For example, the interaction of matter with high-fluence X-ray pulses of durations of few femtoseconds triggers complex dynamics in the electronic system that are not yet fully understood. Photoelectron streaking spectroscopy provides the possibility to study these dynamics on a femtosecond time scale. As a first approach, Auger decay lifetime of simple gas-phase molecules could be measured directly in the temporal domain. Synchrotron sources have been used to measure the lifetime of core holes from the spectral width of the electrons emitted during the decay of the excited state. However, these measurements fail to provide details of the temporal evolution of the excited state, when multi-electron dynamics are involved. In contrast, at XFELs with photoelectron streaking spectroscopy, the temporal evolution of highly excited states and corresponding relaxation dynamics can also be investigated.

Acknowledgments: W.H. acknowledges financial support from a Marie Curie International Outgoing Fellowship. W.H. and R.K. acknowledge financial support by the BaCaTeC program. T.M., M.I. and M.M. acknowledge support by the Deutsche Forschungsgemeinschaft (DFG) under grant nos. SFB925/A1 and SFB 925/A3. M.I. acknowledges funding from the VW foundation within a Peter Paul Ewald-Fellowship. A.R.M. acknowledges funding through BMBF 05K16GU2. J.T.C. acknowledges support from Science Foundation Ireland under grant nos. 12/IA/1742 and 16/RI/3696. Part of this research was carried out at the Linac Coherent Light Source (LCLS) at the SLAC National Accelerator Laboratory. LCLS is an Office of Science User Facility operated for the U.S. Department of Energy (DOE) Office of Science by Stanford University.

Author Contributions: All authors contributed equally to the publication.

Conflicts of Interest: The authors declare no conflict of interest.

References

1. Roentgen, W.C. Ueber eine neue Art von Strahlen. *Sitzungsberichte der Würzburger Phys.-medic. Ges.* **1895**, 132.
2. Elder, F.R.; Gurewitsch, A.M.; Langmuir, R.V.; Pollock, H.C. Radiation from Electrons in a Synchrotron. *Phys. Rev.* **1947**, *71*, 829–830. [CrossRef]

3. Madden, R.P.; Codling, K. New Autoionizing Atomic Energy Levels in He, Ne, and Ar. *Phys. Rev. Lett.* **1963**, *10*, 516–518. [CrossRef]
4. Rowe, E.M.; Mills, F.E. TANTALUS I: A Dedicated Storage Ring Synchrotron Radiation Source. *Part. Accel.* **1973**, *4*, 211–227.
5. Halbach, K. Design of permanent multipole magnets with oriented rare earth cobalt material. *Nucl. Instrum. Methods* **1980**, *169*, 1–10. [CrossRef]
6. Christov, I.P.; Murnane, M.M.; Kapteyn, H.C. High-Harmonic Generation of Attosecond Pulses in the "Single-Cycle" Regime. *Phys. Rev. Lett.* **1997**, *78*, 1251–1254. [CrossRef]
7. Deacon, D.; Elias, L.; Madey, J.; Ramian, G.; Schwettman, H.; Smith, T. First Operation of a Free-Electron Laser. *Phys. Rev. Lett.* **1977**, *38*, 892–894. [CrossRef]
8. McNeil, B.W.J.; Thompson, N.R. X-ray free-electron lasers. *Nat. Photonics* **2010**, *4*, 814–821. [CrossRef]
9. Ackermann, W.; Asova, G.; Ayvazyan, V.; Azima, A.; Baboi, N.; Bähr, J.; Balandin, V.; Beutner, B.; Brandt, A.; Bolzmann, A.; et al. Operation of a free-electron laser from the extreme ultraviolet to the water window. *Nat. Photonics* **2007**, *1*, 336–342. [CrossRef]
10. Allaria, E.; Appio, R.; Badano, L.; Barletta, W.A.; Bassanese, S.; Biedron, S.G.; Borga, A.; Busetto, E.; Castronovo, D.; Cinquegrana, P.; et al. Highly coherent and stable pulses from the FERMI seeded free-electron laser in the extreme ultraviolet. *Nat. Photonics* **2012**, *6*, 699–704. [CrossRef]
11. Emma, P.; Akre, R.; Arthur, J.; Bionta, R.; Bostedt, C.; Bozek, J.D.; Brachmann, A.; Bucksbaum, P.H.; Coffee, R.N.; Decker, F.-J.; et al. First lasing and operation of an ångstrom-wavelength free-electron laser. *Nat. Photonics* **2010**, *4*, 641–647. [CrossRef]
12. Ishikawa, T.; Aoyagi, H.; Asaka, T.; Asano, Y.; Azumi, N.; Bizen, T.; Ego, H.; Fukami, K.; Fukui, T.; Furukawa, Y.; et al. A compact X-ray free-electron laser emitting in the sub-ångström region. *Nat. Photonics* **2012**, *6*, 540–544. [CrossRef]
13. Ganter, R. PSI—SwissFEL Conceptual Design Report. 5232 Villigen PSI, Switzerland. 2012. Available online: https://www.psi.ch/swissfel/CurrentSwissFELPublicationsEN/SwissFEL_CDR_V20_23.04.12_small.pdf (accessed on 1 September 2017).
14. Kang, H.-S.; Han, J.H.; Kim, C.; Kim, D.E.; Kim, S.H.; Park, K.-H.; Park, S.-J.; Kang, T.-H.; Ko, I.S. Current status of PAL-XFEL project. In Proceedings of the IPAC2013, Shanghai, China, 12–17 May 2013; pp. 2074–2076. Available online: https://accelconf.web.cern.ch/accelconf/IPAC2013/papers/weodb103.pdf (accessed on 1 September 2017).
15. Nakajima, K. Compact X-ray sources: Towards a table-top free-electron laser. *Nat. Phys.* **2008**, *4*, 92–93. [CrossRef]
16. Maier, A.R.; Meseck, A.; Reiche, S.; Schroeder, C.B.; Seggebrock, T.; Grüner, F. Demonstration Scheme for a Laser-Plasma-Driven Free-Electron Laser. *Phys. Rev. X* **2012**, *2*, 31019. [CrossRef]
17. Maier, A.R.; Kirchen, M.; Grüner, F. Brilliant Light Sources driven by Laser-Plasma Accelerators. In *Synchrotron Light Sources and Free-Electron Lasers*; Springer International Publishing: Cham, Switzerland, 2016; pp. 1–22.
18. Kärtner, F.X.; Ahr, F.; Calendron, A.-L.; Çankaya, H.; Carbajo, S.; Chang, G.; Cirmi, G.; Dörner, K.; Dorda, U.; Fallahi, A.; et al. AXSIS: Exploring the frontiers in attosecond X-ray science, imaging and spectroscopy. *Nucl. Instrum. Methods Phys. Res. Sect. A Accel. Spectrom. Detect. Assoc. Equip.* **2016**, *829*, 24–29. [CrossRef] [PubMed]
19. Altarelli, M.; Brinkmann, R.; Chergui, M.; Decking, W.; Dobson, B.; Düsterer, S.; Grübel, G.; Graeff, W.; Graafsma, H.; Hajdu, J.; et al. *The European X-ray Free-Electron Laser—Technical Design Report*; Notkestrasse: Hamburg, Germany, 2007. Available online: http://xfel.desy.de/localfsExplorer_read?currentPath=/afs/desy.de/group/xfel/wof/EPT/TDR/XFEL-TDR-final.pdf (accessed on 1 September 2017).
20. Galayda, J. The new LCLS-II project: Status and challenges. In Proceedings of the LINAC2014, Geneva, Switzerland, 31 August–5 September 2014; Available online: https://inspirehep.net/record/1363411 (accessed on 1 September 2017).
21. Drescher, M.; Hentschel, M.; Kienberger, R.; Uiberacker, M.; Yakovlev, V.S.; Scrinzi, A.; Westerwalbesloh, T.; Kleineberg, U.; Heinzmann, U.; Krausz, F. Time-resolved atomic inner-shell spectroscopy. *Nature* **2002**, *419*, 803–807. [CrossRef] [PubMed]

22. Uiberacker, M.; Uphues, T.; Schultze, M.; Verhoef, A.J.; Yakovlev, V.S.; Kling, M.F.; Rauschenberger, J.; Kabachnik, N.M.; Schröder, H.; Lezius, M.; et al. Attosecond real-time observation of electron tunnelling in atoms. *Nature* **2007**, *446*, 627–632. [CrossRef] [PubMed]

23. Cavalieri, A.L.; Müller, N.; Uphues, T.; Yakovlev, V.S.; Baltuška, A.; Horvath, B.; Schmidt, B.; Blümel, L.; Holzwarth, R.; Hendel, S.; et al. Attosecond spectroscopy in condensed matter. *Nature* **2007**, *449*, 1029–1032. [CrossRef] [PubMed]

24. Schultze, M.; Fieß, M.; Karpowicz, N.; Gagnon, J.; Korbman, M.; Hofstetter, M.; Neppl, S.; Cavalieri, A.L.; Komninos, Y.; Mercouris, T.; et al. Delay in photoemission. *Science* **2010**, *328*, 1658–1662. [CrossRef] [PubMed]

25. Klünder, K.; Dahlström, J.; Gisselbrecht, M.; Fordell, T.; Swoboda, M.; Guénot, D.; Johnsson, P.; Caillat, J.; Mauritsson, J.; Maquet, A.; et al. Probing Single-Photon Ionization on the Attosecond Time Scale. *Phys. Rev. Lett.* **2011**, *106*. [CrossRef] [PubMed]

26. Schultze, M.; Ramasesha, K.; Pemmaraju, C.D.; Sato, S.A.; Whitmore, D.; Gandman, A.; Prell, J.S.; Borja, L.J.; Prendergast, D.; Yabana, K.; et al. Attosecond band-gap dynamics in silicon. *Science* **2014**, *346*, 1348–1352. [CrossRef] [PubMed]

27. Neppl, S.; Ernstorfer, R.; Cavalieri, A.L.; Lemell, C.; Wachter, G.; Magerl, E.; Bothschafter, E.M.; Jobst, M.; Hofstetter, M.; Kleineberg, U.; et al. Direct observation of electron propagation and dielectric screening on the atomic length scale. *Nature* **2015**, *517*, 342–346. [CrossRef] [PubMed]

28. Sabbar, M.; Heuser, S.; Boge, R.; Lucchini, M.; Carette, T.; Lindroth, E.; Gallmann, L.; Cirelli, C.; Keller, U. Resonance Effects in Photoemission Time Delays. *Phys. Rev. Lett.* **2015**, *115*, 133001. [CrossRef] [PubMed]

29. Kraus, P.M.; Mignolet, B.; Baykusheva, D.; Rupenyan, A.; Horny, L.; Penka, E.F.; Grassi, G.; Tolstikhin, O.I.; Schneider, J.; Jensen, F.; et al. Measurement and laser control of attosecond charge migration in ionized iodoacetylene. *Science* **2015**, *350*, 790–795. [CrossRef] [PubMed]

30. Förg, B.; Schötz, J.; Süßmann, F.; Förster, M.; Krüger, M.; Ahn, B.; Okell, W.A.; Wintersperger, K.; Zherebtsov, S.; Guggenmos, A.; et al. Attosecond nanoscale near-field sampling. *Nat. Commun.* **2016**, *7*, 11717. [CrossRef] [PubMed]

31. Ossiander, M.; Siegrist, F.; Shirvanyan, V.; Pazourek, R.; Sommer, A.; Latka, T.; Guggenmos, A.; Nagele, S.; Feist, J.; Burgdörfer, J.; et al. Attosecond correlation dynamics. *Nat. Phys.* **2016**, *1*. [CrossRef]

32. Emma, P.; Bane, K.; Cornacchia, M.; Huang, Z.; Schlarb, H.; Stupakov, G.; Walz, D. Femtosecond and Subfemtosecond X-ray Pulses from a Self-Amplified Spontaneous-Emission–Based Free-Electron Laser. *Phys. Rev. Lett.* **2004**, *92*, 74801. [CrossRef] [PubMed]

33. Ding, Y.; Behrens, C.; Coffee, R.; Decker, F.-J.; Emma, P.; Field, C.; Helml, W.; Huang, Z.; Krejcik, P.; Krzywinski, J.; et al. Generating femtosecond X-ray pulses using an emittance-spoiling foil in free-electron lasers. *Appl. Phys. Lett.* **2015**, *107*, 191104. [CrossRef]

34. Zholents, A.; Fawley, W. Proposal for Intense Attosecond Radiation from an X-ray Free-Electron Laser. *Phys. Rev. Lett.* **2004**, *92*, 224801. [CrossRef] [PubMed]

35. Saldin, E.L.; Schneidmiller, E.A.; Yurkov, M.V. A new technique to generate 100 GW-level attosecond X-ray pulses from the X-ray SASE FELs. *Opt. Commun.* **2004**, *239*, 161–172. [CrossRef]

36. Prat, E.; Reiche, S. Simple Method to Generate Terawatt-Attosecond X-ray Free-Electron-Laser Pulses. *Phys. Rev. Lett.* **2015**, *114*, 244801. [CrossRef] [PubMed]

37. Kumar, S.; Parc, Y.W.; Landsman, A.S.; Kim, D.E. Temporally-coherent terawatt attosecond XFEL synchronized with a few cycle laser. *Sci. Rep.* **2016**, *6*, 37700. [CrossRef] [PubMed]

38. Zholents, A.A.; Penn, G. Obtaining attosecond X-ray pulses using a self-amplified spontaneous emission free electron laser. *Phys. Rev. Spec. Top. Accel. Beams* **2005**, *8*, 50704. [CrossRef]

39. Picón, A.; Lehmann, C.S.; Bostedt, C.; Rudenko, A.; Marinelli, A.; Osipov, T.; Rolles, D.; Berrah, N.; Bomme, C.; Bucher, M.; et al. Hetero-site-specific X-ray pump-probe spectroscopy for femtosecond intramolecular dynamics. *Nat. Commun.* **2016**, *7*, 11652. [CrossRef] [PubMed]

40. Pande, K.; Hutchison, C.D.M.; Groenhof, G.; Aquila, A.; Robinson, J.S.; Tenboer, J.; Basu, S.; Boutet, S.; DePonte, D.P.; Liang, M.; et al. Femtosecond structural dynamics drives the trans/cis isomerization in photoactive yellow protein. *Science* **2016**, *352*, 725–729. [CrossRef] [PubMed]

41. Yakovlev, V.S.; Stockman, M.I.; Krausz, F.; Baum, P. Atomic-scale diffractive imaging of sub-cycle electron dynamics in condensed matter. *Sci. Rep.* **2015**, *5*, 14581. [CrossRef] [PubMed]

42. Yang, J.; Guehr, M.; Vecchione, T.; Robinson, M.S.; Li, R.; Hartmann, N.; Shen, X.; Coffee, R.; Corbett, J.; Fry, A.; et al. Diffractive imaging of a rotational wavepacket in nitrogen molecules with femtosecond megaelectronvolt electron pulses. *Nat. Commun.* **2016**, *7*, 11232. [CrossRef] [PubMed]

43. Yang, J.; Guehr, M.; Shen, X.; Li, R.; Vecchione, T.; Coffee, R.; Corbett, J.; Fry, A.; Hartmann, N.; Hast, C.; et al. Diffractive Imaging of Coherent Nuclear Motion in Isolated Molecules. *Phys. Rev. Lett.* **2016**, *117*, 153002. [CrossRef] [PubMed]

44. Glownia, J.M.; Natan, A.; Cryan, J.P.; Hartsock, R.; Kozina, M.; Minitti, M.P.; Nelson, S.; Robinson, J.; Sato, T.; van Driel, T.; et al. Self-Referenced Coherent Diffraction X-ray Movie of Ångstrom- and Femtosecond-Scale Atomic Motion. *Phys. Rev. Lett.* **2016**, *117*, 153003. [CrossRef] [PubMed]

45. Bostedt, C.; Fritz, D.M.; Huang, Z.; Lee, H.J.; Lemke, H.; Schlotter, W.F.; Turner, J.J.; Williams, G.J. Linac Coherent Light Source: The First Five Years. *Rev. Mod. Phys.* **2015**, *88*. [CrossRef]

46. Milton, S.V.; Gluskin, E.; Arnold, N.D.; Benson, C.; Berg, W.; Biedron, S.G.; Borland, M.; Chae, Y.C.; Dejus, R.J.; Den Hartog, P.K.; et al. Exponential gain and saturation of a self-amplified spontaneous emission free-electron laser. *Science* **2001**, *292*, 2037–2041. [CrossRef] [PubMed]

47. Düsterer, S.; Rehders, M.; Al-Shemmary, A.; Behrens, C.; Brenner, G.; Brovko, O.; DellAngela, M.; Drescher, M.; Faatz, B.; Feldhaus, J.; et al. Development of experimental techniques for the characterization of ultrashort photon pulses of extreme ultraviolet free-electron lasers. *Phys. Rev. Spec. Top. Accel. Beams* **2014**, *17*, 120702. [CrossRef]

48. Ding, Y.; Brachmann, A.; Decker, F.-J.; Dowell, D.; Emma, P.; Frisch, J.; Gilevich, S.; Hays, G.; Hering, P.; Huang, Z.; et al. Measurements and Simulations of Ultralow Emittance and Ultrashort Electron Beams in the Linac Coherent Light Source. *Phys. Rev. Lett.* **2009**, *102*, 254801. [CrossRef] [PubMed]

49. Lutman, A.; Ding, Y.; Feng, Y.; Huang, Z.; Messerschmidt, M.; Wu, J.; Krzywinski, J. Femtosecond X-ray free electron laser pulse duration measurement from spectral correlation function. *Phys. Rev. Spec. Top. Accel. Beams* **2012**, *15*, 1–13. [CrossRef]

50. Young, L.; Kanter, E.P.; Krässig, B.; Li, Y.; March, A.M.; Pratt, S.T.; Santra, R.; Southworth, S.H.; Rohringer, N.; Dimauro, L.F.; et al. Femtosecond electronic response of atoms to ultra-intense X-rays. *Nature* **2010**, *466*, 56–61. [CrossRef] [PubMed]

51. Düsterer, S.; Radcliffe, P.; Bostedt, C.; Bozek, J.D.; Cavalieri, A.L.; Coffee, R.N.; Costello, J.T.; Cubaynes, D.; DiMauro, L.F.; Ding, Y.; et al. Femtosecond X-ray pulse length characterization at the Linac Coherent Light Source free-electron laser. *New J. Phys.* **2011**, *13*, 93024. [CrossRef]

52. Behrens, C.; Decker, F.-J.; Ding, Y.; Dolgashev, V.A.; Frisch, J.; Huang, Z.; Krejcik, P.; Loos, H.; Lutman, A.; Maxwell, T.J.; et al. Few-femtosecond time-resolved measurements of X-ray free-electron lasers. *Nat. Commun.* **2014**, *5*, 3762. [CrossRef] [PubMed]

53. Sanchez-Gonzalez, A.; Micaelli, P.; Olivier, C.; Barillot, T.R.; Ilchen, M.; Lutman, A.A.; Marinelli, A.; Maxwell, T.; Achner, A.; Agåker, M.; et al. Accurate prediction of X-ray pulse properties from a free-electron laser using machine learning. *Nat. Commun.* **2017**, *8*, 15461. [CrossRef] [PubMed]

54. Mitzner, R.; Sorokin, A.A.; Siemer, B.; Roling, S.; Rutkowski, M.; Zacharias, H.; Neeb, M.; Noll, T.; Siewert, F.; Eberhardt, W.; et al. Direct autocorrelation of soft-X-ray free-electron-laser pulses by time-resolved two-photon double ionization of He. *Phys. Rev. A* **2009**, *80*, 25402. [CrossRef]

55. Moshammer, R.; Pfeifer, T.; Rudenko, A.; Jiang, Y.H.; Foucar, L.; Kurka, M.; Kühnel, K.U.; Schröter, C.D.; Ullrich, J.; Herrwerth, O.; et al. Second-order autocorrelation of XUV FEL pulses via time resolved two-photon single ionization of He. *Opt. Express* **2011**, *19*, 21698. [CrossRef] [PubMed]

56. Iaconis, C.; Walmsley, I.A. Spectral phase interferometry for direct electric-field reconstruction of ultrashort optical pulses. *Opt. Lett.* **1998**, *23*, 792–794. [CrossRef] [PubMed]

57. Mahieu, B.; Gauthier, D.; De Ninno, G.; Dacasa, H.; Lozano, M.; Rousseau, J.-P.; Zeitoun, P.; Garzella, D.; Merdji, H. Spectral-phase interferometry for direct electric-field reconstruction applied to seeded extreme-ultraviolet free-electron lasers. *Opt. Express* **2015**, *23*, 17665–17674. [CrossRef] [PubMed]

58. Harmand, M.; Coffee, R.; Bionta, M.R.; Chollet, M.; French, D.; Zhu, D.; Fritz, D.M.; Lemke, H.T.; Medvedev, N.; Ziaja, B.; et al. Achieving few-femtosecond time-sorting at hard X-ray free-electron lasers. *Nat. Photonics* **2013**, *7*, 215–218. [CrossRef]

59. Finetti, P.; Höppner, H.; Allaria, E.; Callegari, C.; Capotondi, F.; Cinquegrana, P.; Coreno, M.; Cucini, R.; Danailov, M.B.; Demidovich, A.; et al. Pulse Duration of Seeded Free-Electron Lasers. *Phys. Rev. X* **2017**, *7*, 21043. [CrossRef]

60. Bencivenga, F.; Cucini, R.; Capotondi, F.; Battistoni, A.; Mincigrucci, R.; Giangrisostomi, E.; Gessini, A.; Manfredda, M.; Nikolov, I.P.; Pedersoli, E.; et al. Four-wave mixing experiments with extreme ultraviolet transient gratings. *Nature* **2015**, *520*, 205–208. [CrossRef] [PubMed]

61. Bencivenga, F.; Masciovecchio, C. Results and Perspectives for Short-Wavelength, Four-Wave-Mixing Experiments with Fully Coherent Free Electron Lasers. *Synchrotron Radiat. News* **2016**, *29*, 15–20. [CrossRef]

62. Mairesse, Y.; Zeidler, D.; Dudovich, N.; Spanner, M.; Levesque, J.; Villeneuve, D.M.; Corkum, P.B. High-order harmonic transient grating spectroscopy in a molecular jet. *Phys. Rev. Lett.* **2008**, *100*, 15–18. [CrossRef] [PubMed]

63. Itatani, J.; Quéré, F.; Yudin, G.; Ivanov, M.; Krausz, F.; Corkum, P. Attosecond Streak Camera. *Phys. Rev. Lett.* **2002**, *88*, 173903. [CrossRef] [PubMed]

64. Kitzler, M.; Milosevic, N.; Scrinzi, A.; Krausz, F.; Brabec, T. Quantum Theory of Attosecond XUV Pulse Measurement by Laser Dressed Photoionization. *Phys. Rev. Lett.* **2002**, *88*, 173904. [CrossRef] [PubMed]

65. Hentschel, M.; Kienberger, R.; Spielmann, C.; Reider, G.A.; Milosevic, N.; Brabec, T.; Corkum, P.B.; Heinzmann, U.; Drescher, M.; Krausz, F. Attosecond metrology. *Nature* **2001**, *414*, 509–513. [CrossRef] [PubMed]

66. Kienberger, R.; Goulielmakis, E.; Uiberacker, M.; Baltuska, A.; Yakovlev, V.; Bammer, F.; Scrinzi, A.; Westerwalbesloh, T.; Kleineberg, U.; Heinzmann, U.; et al. Atomic transient recorder. *Nature* **2004**, *427*, 817–821. [CrossRef] [PubMed]

67. Goulielmakis, E.; Uiberacker, M.; Kienberger, R.; Baltuška, A.; Yakovlev, V.S.; Scrinzi, A.; Westerwalbesloh, T.; Kleineberg, U.; Heinzmann, U.; Drescher, M.; et al. Direct measurement of light waves. *Science* **2004**, *305*, 1267–1269. [CrossRef] [PubMed]

68. Baumgartner, R.; Byer, R. Optical parametric amplification. *IEEE J. Quantum Electron.* **1979**, *15*, 432–444. [CrossRef]

69. Helml, W.; Maier, A.R.; Schweinberger, W.; Grguraš, I.; Radcliffe, P.; Doumy, G.; Roedig, C.; Gagnon, J.; Messerschmidt, M.; Schorb, S.; et al. Measuring the temporal structure of few-femtosecond free-electron laser X-ray pulses directly in the time domain. *Nat. Photonics* **2014**, *8*, 950–957. [CrossRef]

70. Wuilleumier, F.J.; Meyer, M. Pump–probe experiments in atoms involving laser and synchrotron radiation: An overview. *J. Phys. B At. Mol. Opt. Phys.* **2006**, *39*, R425–R477. [CrossRef]

71. Glover, T.; Schoenlein, R.; Chin, A.; Shank, C. Observation of Laser Assisted Photoelectric Effect and Femtosecond High Order Harmonic Radiation. *Phys. Rev. Lett.* **1996**, *76*, 2468–2471. [CrossRef] [PubMed]

72. Meyer, M.; Costello, J.T.; Düsterer, S.; Li, W.B.; Radcliffe, P. Two-colour experiments in the gas phase. *J. Phys. B At. Mol. Opt. Phys.* **2010**, *43*, 194006. [CrossRef]

73. Radcliffe, P.; Arbeiter, M.; Li, W.B.; Düsterer, S.; Redlin, H.; Hayden, P.; Hough, P.; Richardson, V.; Costello, J.T.; Fennel, T.; et al. Atomic photoionization in combined intense XUV free-electron and infrared laser fields. *New J. Phys.* **2012**, *14*, 43008. [CrossRef]

74. O'Keeffe, P.; López-Martens, R.; Mauritsson, J.; Johansson, A.; L'Huillier, A.; Véniard, V.; Taïeb, R.; Maquet, A.; Meyer, M. Polarization effects in two-photon nonresonant ionization of argon with extreme-ultraviolet and infrared femtosecond pulses. *Phys. Rev. A* **2004**, *69*, 51401. [CrossRef]

75. Radcliffe, P.; Düsterer, S.; Azima, A.; Redlin, H.; Feldhaus, J.; Dardis, J.; Kavanagh, K.; Luna, H.; Gutierrez, J.P.; Yeates, P.; et al. Single-shot characterization of independent femtosecond extreme ultraviolet free electron and infrared laser pulses. *Appl. Phys. Lett.* **2007**, *90*, 131108. [CrossRef]

76. Redlin, H.; Al-Shemmary, A.; Azima, A.; Stojanovic, N.; Tavella, F.; Will, I.; Düsterer, S. The FLASH pump-probe laser system: Setup, characterization and optical beamlines. *Nucl. Instrum. Methods Phys. Res. Sect. A Accel. Spectrom. Detect. Assoc. Equip.* **2011**, *635*, S88–S93. [CrossRef]

77. Schulz, S.; Grguraš, I.; Behrens, C.; Bromberger, H.; Costello, J.T.; Czwalinna, M.K.; Felber, M.; Hoffmann, M.C.; Ilchen, M.; Liu, H.Y.; et al. Femtosecond all-optical synchronization of an X-ray free-electron laser. *Nat. Commun.* **2015**, *6*, 5938. [CrossRef] [PubMed]

78. Meyer, M.; Cubaynes, D.; O'Keeffe, P.; Luna, H.; Yeates, P.; Kennedy, E.T.; Costello, J.T.; Orr, P.; Taïeb, R.; Maquet, A.; et al. Two-color photoionization in xuv free-electron and visible laser fields. *Phys. Rev. A* **2006**, *74*, 11401. [CrossRef]

79. Maquet, A.; Taïeb, R. Two-colour IR + XUV spectroscopies: the "soft-photon approximation". *J. Mod. Opt.* **2007**, *54*, 1847–1857. [CrossRef]

80. Meyer, M.; Cubaynes, D.; Glijer, D.; Dardis, J.; Hayden, P.; Hough, P.; Richardson, V.; Kennedy, E.T.; Costello, J.T.; Radcliffe, P.; et al. Polarization Control in Two-Color Above-Threshold Ionization of Atomic Helium. *Phys. Rev. Lett.* **2008**, *101*, 193002. [CrossRef] [PubMed]

81. Hayden, P.; Dardis, J.; Hough, P.; Richardson, V.; Kennedy, E.T.; Costello, J.T.; Düsterer, S.; Redlin, H.; Feldhaus, J.; Li, W.B.; et al. The Laser-assisted photoelectric effect of He, Ne, Ar and Xe in intense extreme ultraviolet and infrared laser fields. *J. Mod. Opt.* **2016**, *63*, 358–366. [CrossRef]

82. Azima, A.; Düsterer, S.; Radcliffe, P.; Redlin, H.; Stojanovic, N.; Li, W.; Schlarb, H.; Feldhaus, J.; Cubaynes, D.; Meyer, M.; et al. Time-resolved pump-probe experiments beyond the jitter limitations at FLASH. *Appl. Phys. Lett.* **2009**, *94*, 144102. [CrossRef]

83. Ayvazyan, V.; Baboi, N.; Bähr, J.; Balandin, V.; Beutner, B.; Brandt, A.; Bohnet, I.; Bolzmann, A.; Brinkmann, R.; Brovko, O.I.; et al. First operation of a free-electron laser generating GW power radiation at 32 nm wavelength. *Eur. Phys. J. D* **2006**, *37*, 297–303. [CrossRef]

84. Radcliffe, P.; Düsterer, S.; Azima, A.; Li, W.B.; Plönjes, E.; Redlin, H.; Feldhaus, J.; Nicolosi, P.; Poletto, L.; Dardis, J.; et al. An experiment for two-color photoionization using high intensity extreme-UV free electron and near-IR laser pulses. *Nucl. Instrum. Methods Phys. Res. Sect. A Accel. Spectrom. Detect. Assoc. Equip.* **2007**, *583*, 516–525. [CrossRef]

85. Eland, J.H.D.; Vieuxmaire, O.; Kinugawa, T.; Lablanquie, P.; Hall, R.I.; Penent, F. Complete Two-Electron Spectra in Double Photoionization: The Rare Gases Ar, Kr, and Xe. *Phys. Rev. Lett.* **2003**, *90*, 53003. [CrossRef] [PubMed]

86. Mazza, T.; Ilchen, M.; Rafipoor, A.J.; Callegari, C.; Finetti, P.; Plekan, O.; Prince, K.C.; Richter, R.; Demidovich, A.; Grazioli, C.; et al. Angular distribution and circular dichroism in the two-colour XUV + NIR above-threshold ionization of helium. *J. Mod. Opt.* **2016**, *63*, 367–382. [CrossRef]

87. Gauthier, D.; Allaria, E.; Coreno, M.; Cudin, I.; Dacasa, H.; Danailov, M.B.; Demidovich, A.; Di Mitri, S.; Diviacco, B.; Ferrari, E.; et al. Chirped pulse amplification in an extreme-ultraviolet free-electron laser. *Nat. Commun.* **2016**, *7*, 13688. [CrossRef] [PubMed]

88. Meyer, M.; Radcliffe, P.; Tschentscher, T.; Costello, J.T.; Cavalieri, A.L.; Grguras, I.; Maier, A.R.; Kienberger, R.; Bozek, J.; Bostedt, C.; et al. Angle-Resolved Electron Spectroscopy of Laser-Assisted Auger Decay Induced by a Few-Femtosecond X-ray Pulse. *Phys. Rev. Lett.* **2012**, *108*, 63007. [CrossRef] [PubMed]

89. Mazza, T.; Ilchen, M.; Rafipoor, A.J.; Callegari, C.; Finetti, P.; Plekan, O.; Prince, K.C.; Richter, R.; Danailov, M.B.; Demidovich, A.; et al. Determining the polarization state of an extreme ultraviolet free-electron laser beam using atomic circular dichroism. *Nat. Commun.* **2014**, *5*, 3648. [CrossRef] [PubMed]

90. Hartmann, G.; Lindahl, A.O.; Knie, A.; Hartmann, N.; Lutman, A.A.; MacArthur, J.P.; Shevchuk, I.; Buck, J.; Galler, A.; Glownia, J.M.; et al. Circular dichroism measurements at an x-ray free-electron laser with polarization control. *Rev. Sci. Instrum.* **2016**, *87*, 83113. [CrossRef] [PubMed]

91. Düsterer, S.; Rading, L.; Johnsson, P.; Rouzée, A.; Hundertmark, A.; Vrakking, M.J.J.; Radcliffe, P.; Meyer, M.; Kazansky, A.K.; Kabachnik, N.M. Interference in the angular distribution of photoelectrons in superimposed XUV and optical laser fields. *J. Phys. B At. Mol. Opt. Phys.* **2013**, *46*, 164026. [CrossRef]

92. Krausz, F.; Ivanov, M.Y. Attosecond physics. *Rev. Mod. Phys.* **2009**, *81*, 163–234. [CrossRef]

93. Yakovlev, V.S.; Gagnon, J.; Karpowicz, N.; Krausz, F. Attosecond Streaking Enables the Measurement of Quantum Phase. *Phys. Rev. Lett.* **2010**, *105*, 3–6. [CrossRef] [PubMed]

94. Gagnon, J.; Goulielmakis, E.; Yakovlev, V.S. The accurate FROG characterization of attosecond pulses from streaking measurements. *Appl. Phys. B* **2008**, *92*, 25–32. [CrossRef]

95. Grguraš, I. Time Resolved Photoelectron Spectroscopy for Femtosecond Characterization of X-ray Free-Electron Laser Pulses. Ph.D. Thesis, Universität Hamburg, Hamburg, Germany, 2015. Available online: https://www.physnet.uni-amburg.de/services/biblio/dissertation/dissfbPhysik/___Kurzfassungen/Ivanka___Grguras (accessed on 1 September 2017).

96. Helml, W. Development & Characterization of Sources for High-Energy, High-Intensity Coherent Radiation. Ph.D. Thesis, Technische Universität München, München, Germany, 2012. Available online: http://nbn-resolving.de/urn/resolver.pl?urn:nbn:de:bvb:91-diss-20120723--1110298--1-6 (accessed on 1 September 2017).

97. Bozek, J.D. AMO instrumentation for the LCLS X-ray FEL. *Eur. Phys. J. Spec. Top.* **2009**, *169*, 129–132. [CrossRef]

98. Kruit, P.; Read, F.H. Magnetic field paralleliser for 2π electron-spectrometer and electron-image magnifier. *J. Phys. E* **1983**, *16*, 313–324. [CrossRef]

99. Gagnon, J.; Yakovlev, V.S. The robustness of attosecond streaking measurements. *Opt. Express* **2009**, *17*, 17678–17693. [CrossRef] [PubMed]

100. Krinsky, S.; Gluckstern, R. Analysis of statistical correlations and intensity spiking in the self-amplified spontaneous-emission free-electron laser. *Phys. Rev. Spec. Top. Accel. Beams* **2003**, *6*, 50701. [CrossRef]

101. Frühling, U.; Wieland, M.; Gensch, M.; Gebert, T.; Schütte, B.; Krikunova, M.; Kalms, R.; Budzyn, F.; Grimm, O.; Rossbach, J.; et al. Single-shot terahertz-field-driven X-ray streak camera. *Nat. Photonics* **2009**, *3*, 523–528. [CrossRef]

102. Yeh, K.-L.; Hoffmann, M.C.; Hebling, J.; Nelson, K.A. Generation of 10 µJ ultrashort terahertz pulses by optical rectification. *Appl. Phys. Lett.* **2007**, *90*, 171121. [CrossRef]

103. Grguraš, I.; Maier, A.R.; Behrens, C.; Mazza, T.; Kelly, T.J.; Radcliffe, P.; Düsterer, S.; Kazansky, A.K.; Kabachnik, N.M.; Tschentscher, T.; et al. Ultrafast X-ray pulse characterization at free-electron lasers. *Nat. Photonics* **2012**, *6*, 852–857. [CrossRef]

104. Gagnon, J.; Yakovlev, V.S. The direct evaluation of attosecond chirp from a streaking measurement. *Appl. Phys. B Lasers Opt.* **2011**, *103*, 303–309. [CrossRef]

105. Gouy, L.G. Sur Une Propriete Nouvelle Des Ondes Lumineuses. *C. R. Acad. Sci.* **1890**, *110*, 1251–1253.

106. Wellhöfer, M.; Hoeft, J.T.; Martins, M.; Wurth, W.; Braune, M.; Viefhaus, J.; Tiedtke, K.; Richter, M. Photoelectron spectroscopy as a non-invasive method to monitor SASE-FEL spectra. *J. Instrum.* **2008**, *3*, P02003. [CrossRef]

107. Bionta, M.R.; Lemke, H.T.; Cryan, J.P.; Glownia, J.M.; Bostedt, C.; Cammarata, M.; Castagna, J.-C.; Ding, Y.; Fritz, D.M.; Fry, A.R.; et al. Spectral encoding of X-ray/optical relative delay. *Opt. Express* **2011**, *19*, 21855–21865. [CrossRef] [PubMed]

108. Henke, B.L.; Gullikson, E.M.; Davis, J.C. X-ray Interactions: Photoabsorption, Scattering, Transmission, and Reflection at E = 50–30,000 eV, Z = 1–92. *At. Data Nucl. Data Tables* **1993**, *54*, 181–342. [CrossRef]

109. Juranić, P.N.; Stepanov, A.; Ischebeck, R.; Schlott, V.; Pradervand, C.; Patthey, L.; Radović, M.; Gorgisyan, I.; Rivkin, L.; Hauri, C.P.; et al. High-precision X-ray FEL pulse arrival time measurements at SACLA by a THz streak camera with Xe clusters. *Opt. Express* **2014**, *22*, 30004. [CrossRef] [PubMed]

110. Gorgisyan, I.; Ischebeck, R.; Erny, C.; Dax, A.; Patthey, L.; Pradervand, C.; Sala, L.; Milne, C.; Lemke, H.T.; Hauri, C.P.; et al. THz streak camera method for synchronous arrival time measurement of two-color hard X-ray FEL pulses. *Opt. Express* **2017**, *25*, 2080. [CrossRef]

111. Irimia, D.; Dobrikov, D.; Kortekaas, R.; Voet, H.; van den Ende, D.A.; Groen, W.A.; Janssen, M.H.M. A short pulse (7 µs FWHM) and high repetition rate (dc-5 kHz) cantilever piezovalve for pulsed atomic and molecular beams. *Rev. Sci. Instrum.* **2009**, *80*, 113303. [CrossRef] [PubMed]

112. Hagena, O.F.; Obert, W. Cluster Formation in Expanding Supersonic Jets: Effect of Pressure, Temperature, Nozzle Size, and Test Gas. *J. Chem. Phys.* **1972**, *56*, 1793–1802. [CrossRef]

113. Juranić, P.N.; Stepanov, A.; Peier, P.; Hauri, C.P.; Ischebeck, R.; Schlott, V.; Radović, M.; Erny, C.; Ardana-Lamas, F.; Monoszlai, B.; et al. A scheme for a shot-to-shot, femtosecond-resolved pulse length and arrival time measurement of free electron laser X-ray pulses that overcomes the time jitter problem between the FEL and the laser. *J. Instrum.* **2014**, *9*, P03006. [CrossRef]

114. Thompson, N.R.; McNeil, B.W.J. Mode Locking in a Free-Electron Laser Amplifier. *Phys. Rev. Lett.* **2008**, *100*, 203901. [CrossRef] [PubMed]

115. Kur, E.; Dunning, D.J.; McNeil, B.W.J.; Wurtele, J.; Zholents, A.A. A wide bandwidth free-electron laser with mode locking using current modulation. *New J. Phys.* **2011**, *13*, 63012. [CrossRef]

116. Xiang, D.; Ding, Y.; Raubenheimer, T.; Wu, J. Mode-locked multichromatic x rays in a seeded free-electron laser for single-shot X-ray spectroscopy. *Phys. Rev. Spec. Top. Accel. Beams* **2012**, *15*, 50707. [CrossRef]

117. Ding, Y.; Decker, F.-J.; Emma, P.; Feng, C.; Field, C.; Frisch, J.; Huang, Z.; Krzywinski, J.; Loos, H.; Welch, J.; et al. Femtosecond X-ray Pulse Characterization in Free-Electron Lasers Using a Cross-Correlation Technique. *Phys. Rev. Lett.* **2012**, *109*, 254802. [CrossRef] [PubMed]

118. Riedel, R.; Al-Shemmary, A.; Gensch, M.; Golz, T.; Harmand, M.; Medvedev, N.; Prandolini, M.J.; Sokolowski-Tinten, K.; Toleikis, S.; Wegner, U.; et al. Single-shot pulse duration monitor for extreme ultraviolet and X-ray free-electron lasers. *Nat. Commun.* **2013**, *4*, 1731. [CrossRef] [PubMed]

119. Maxwell, T.J.; Behrens, C.; Ding, Y.; Fisher, A.S.; Frisch, J.; Huang, Z.; Loos, H. Coherent-Radiation Spectroscopy of Few-Femtosecond Electron Bunches Using a Middle-Infrared Prism Spectrometer. *Phys. Rev. Lett.* **2013**, *111*, 184801. [CrossRef] [PubMed]

120. Hartmann, N.; Helml, W.; Galler, A.; Bionta, M.R.; Grünert, J.; Molodtsov, S.L.; Ferguson, K.R.; Schorb, S.; Swiggers, M.L.; Carron, S.; et al. Sub-femtosecond precision measurement of relative X-ray arrival time for free-electron lasers. *Nat. Photonics* **2014**, *8*, 706–709. [CrossRef]
121. Löhl, F.; Arsov, V.; Felber, M.; Hacker, K.; Jalmuzna, W.; Lorbeer, B.; Ludwig, F.; Matthiesen, K.-H.; Schlarb, H.; Schmidt, B.; et al. Electron Bunch Timing with Femtosecond Precision in a Superconducting Free-Electron Laser. *Phys. Rev. Lett.* **2010**, *104*, 144801. [CrossRef] [PubMed]
122. Allaria, E.; Diviacco, B.; Callegari, C.; Finetti, P.; Mahieu, B.; Viefhaus, J.; Zangrando, M.; De Ninno, G.; Lambert, G.; Ferrari, E.; et al. Control of the Polarization of a Vacuum-Ultraviolet, High-Gain, Free-Electron Laser. *Phys. Rev. X* **2014**, *4*, 41040. [CrossRef]
123. Kazansky, A.K.; Bozhevolnov, A.V.; Sazhina, I.P.; Kabachnik, N.M. Interference effects in angular streaking with a rotating terahertz field. *Phys. Rev. A* **2016**, *93*, 13407. [CrossRef]
124. Kazansky, A.K.; Sazhina, I.P.; Nosik, V.L.; Kabachnik, N.M. Angular streaking and sideband formation in rotating terahertz and far-infrared fields. *J. Phys. B At. Mol. Opt. Phys.* **2017**, *50*, 105601. [CrossRef]
125. Constant, E.; Taranukhin, V.; Stolow, A.; Corkum, P. Methods for the measurement of the duration of high-harmonic pulses. *Phys. Rev. A* **1997**, *56*, 3870–3878. [CrossRef]
126. Eckle, P.; Smolarski, M.; Schlup, P.; Biegert, J.; Staudte, A.; Schöffler, M.; Muller, H.G.; Dörner, R.; Keller, U. Attosecond angular streaking. *Nat. Phys.* **2008**, *4*, 565–570. [CrossRef]
127. Ueda, K.; Ishikawa, K.L. Attosecond science: Attoclocks play devil's advocate. *Nat. Phys.* **2011**, *7*, 371–372. [CrossRef]

applied
sciences

MDPI

Communication

Measurement of the X-ray Spectrum of a Free Electron Laser with a Wide-Range High-Resolution Single-Shot Spectrometer

Yuichi Inubushi [1,2,*], Ichiro Inoue [2], Jangwoo Kim [3,4], Akihiko Nishihara [3], Satoshi Matsuyama [3], Hirokatsu Yumoto [1,2], Takahisa Koyama [1,2], Kensuke Tono [1,2], Haruhiko Ohashi [1,2], Kazuto Yamauchi [3] and Makina Yabashi [1,2]

1   Japan Synchrotron Radiation Research Institute, 1-1-1 Kouto, Sayo-cho, Sayo-gun, Hyogo 679-5198, Japan; yumoto@spring8.or.jp (H.Y.); koyama@spring8.or.jp (T.K.); tono@spring8.or.jp (K.T.); hohashi@spring8.or.jp (H.O.); yabashi@spring8.or.jp (M.Y.)
2   RIKEN SPring-8 Center, 1-1-1 Kouto, Sayo-cho, Sayo-gun, Hyogo 679-5148, Japan; inoue@spring8.or.jp
3   Department of Precision Science and Technology, Graduate School of Engineering, Osaka University, 2-1 Yamada-oka, Suita, Osaka 565-0871, Japan; jkpal@postech.ac.kr (J.K.); nishihara@up.prec.eng.osaka-u.ac.jp (A.N.); matsuyama@prec.eng.osaka-u.ac.jp (S.M.); yamauchi@prec.eng.osaka-u.ac.jp (K.Y.)
4   Pohang Accelerator Laboratory (PAL), POSTECH, 127-80 Jigokro, Nam-gu, Pohang, Gyeongbuk 37673, Korea
*   Correspondence: inubushi@spring8.or.jp; Tel.: +81-791-58-0802

Academic Editor: Kiyoshi Ueda
Received: 31 March 2017; Accepted: 30 May 2017; Published: 6 June 2017

Abstract: We developed a single-shot X-ray spectrometer for wide-range high-resolution measurements of Self-Amplified Spontaneous Emission (SASE) X-ray Free Electron Laser (XFEL) pulses. The spectrometer consists of a multi-layer elliptical mirror for producing a large divergence of 22 mrad around 9070 eV and a silicon (553) analyzer crystal. We achieved a wide energy range of 55 eV with a fine spectral resolution of 80 meV, which enabled the observation of a whole SASE-XFEL spectrum with fully-resolved spike structures. We found that a SASE-XFEL pulse has around 60 longitudinal modes with a pulse duration of 7.7 ± 1.1 fs.

Keywords: XFEL; spectroscopy; ultra-shot pulse

1. Introduction

Recently, X-ray Free Electron Lasers (XFELs), such as the Linac Coherent Light Source (LCLS) [1] and the SPring-8 Angstrom Compact Free Electron Laser (SACLA) [2], have successfully generated brilliant, femtosecond X-ray pulses, achieving an ultrahigh resolving power in both space and time that opens up new frontiers in various scientific fields [3–5]. Currently, the Self-Amplified Spontaneous Emission (SASE) scheme [6,7] is widely utilized for generating XFEL pulses. SASE-XFEL pulses have many narrow spike structures originating from the stochastic nature of the electron beam used as light source. The spike structures reflect important characteristics of the XFEL beam: the width and the number of spikes in the whole spectrum closely relate with the XFEL pulse duration and the longitudinal mode number, respectively. To measure spike structures, single-shot measurements with a dispersive X-ray spectrometer are required, because spike structures change shot by shot [8–11]. However, it was difficult to fully resolve the spike structure with a typical width of ~100 meV over the entire pulse spectrum covering a few tens of eV. In this paper, we report on the development of a single-shot spectrometer covering the whole XFEL spectrum with fully-resolved spike structures.

So far, two methods have been developed for designing single-shot spectrometers of well-collimated XFEL beams. One uses a thin, bent analyzer crystal [9,11]. The total spectral range can be simply tuned by controlling the bending radius, while possible lattice distortions induced by the bending can degrade the energy resolution. The other one combines a flat analyzer crystal and an elliptical total-reflection mirror to increase the angular divergence of X-rays. In this case, one can achieve high resolution of 14 meV at around 10 keV photon energy by using a higher-order diffraction for the analyzer, although a spectral range is limited to 4 eV due to the small angular divergence of 2.7 mrad [10]. In this study, we utilized a multi-layer mirror to achieve a large divergence of 22 mrad to measure a wide spectral range of 55 eV, while keeping the resolution sufficient to resolve the spike structures of a few hundred meV.

2. Design of Spectrometer

The basic design of the dispersive X-ray spectrometer in this study consists of an elliptical mirror, a flat analyzer crystal, and a position sensitive detector [8]. First, the spectral resolution dE/E is described by

$$\frac{dE}{E} = \frac{\Psi}{\tan \theta_B}. \tag{1}$$

where E and θ_B are the photon energy and the Bragg angle, respectively. Ψ is determined by parameters of the X-ray source and optics Ψ_1 and the angular resolution of the detector Ψ_2, as given by

$$\Psi = \sqrt{\Psi_1^2 + \Psi_2^2}, \tag{2}$$

$$\Psi_1 = \frac{\sqrt{\sigma^2 + L^2\omega^2}}{L}, \tag{3}$$

and

$$\Psi_2 = \frac{p}{L}. \tag{4}$$

Here, σ, ω, L, and p are the X-ray source size, the Darwin width of the Bragg reflection, the distance from the X-ray source to the detector, and the pixel size of the detector, respectively.

Second, the spectral range ΔE is described by

$$\frac{\Delta E}{E} = \frac{\Delta \theta}{\tan \theta_B}. \tag{5}$$

Here, $\Delta\theta$ is the angular divergence of the X-ray beam on the analyzer crystal. Equation (5) shows that a larger divergence is required for expanding the covered spectral range. Although an elliptical mirror with total reflection was utilized in our previous experiment [10], the mirror cannot provide a sufficient divergence angle for measuring the entire spectrum in one shot due to its small reflection angle of a few mrad.

To overcome the limitation, we used platinum/carbon multi-layer mirrors [12], which were installed as the last optics of the two-stage focusing system at SACLA [13], to produce a large divergence $\Delta\theta$ of 22 mrad with a central photon energy of 9070 eV and a central incident angle of 14 mrad. We note that the reflectivity of the single mirror with 30 layers is ~70% at a photon energy range between 8900 and 9300 eV. To obtain a high spectral resolution in the dispersive spectrometer, a large Bragg angle is required from Equation (1). For the distance L of 1.5 m and a pixel size p of 50 μm for a Multi-Port Charge Coupled Device (MPCCD) detector [14] in the present geometry, we derive both the spectral range of 55 eV and the resolution of 80 meV at around 9070 eV, simultaneously, with a Bragg angle of 75.2° using a silicon (553) analyzer crystal.

3. Experiment and Discussion

Our experiment was conducted at BL3 of SACLA [15]. A schematic of the experimental setup is shown in Figure 1. XFEL pulses with a central photon energy of 9070 eV were focused with the two-stage focusing system using platinum/carbon multi-layer mirrors. The silicon (553) analyzer crystal was set at 1.2 m from the focal point, and an MPCCD detector was placed on a 0.3-m-long detector arm. Thus, the total length from the focal point to the detector L was 1.5 m, as used in the estimation in Section 2. Figure 2a shows a typical single-shot spectrum of an XFEL pulse, demonstrating that the spectrometer is capable of covering the complete pulse spectrum. In Figure 2b, we show the enlarged spectrum around the peak maximum of Figure 2a. The typical width of observed spikes is 390 ± 30 meV in full width at half maximum (FWHM), which is sufficiently larger with respect to the resolution of the spectrometer of 80 meV. Based on this result, we estimated the XFEL pulse duration with a method described in [10]. Here, an energy chirp of the electron beam of 6×10^{-5} fs^{-1} was used as the simulation parameter. By comparing the measurement with a simulation, the pulse duration was estimated to be 7.7 ± 1.1 fs in FWHM. From the evaluation of Figure 2a, the longitudinal mode number was around 60.

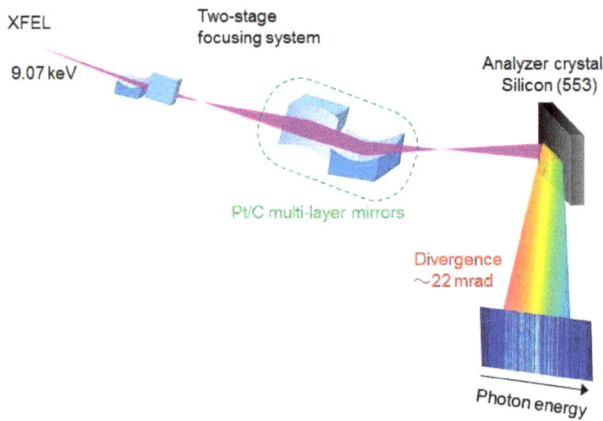

Figure 1. Experimental setup. The X-ray divergence was increased with the two-stage focusing optics, in which multi-layer elliptical mirrors were used as a second K-B optics. The spectral profile analyzed by a silicon (553) crystal is recorded with a MPCCD detector.

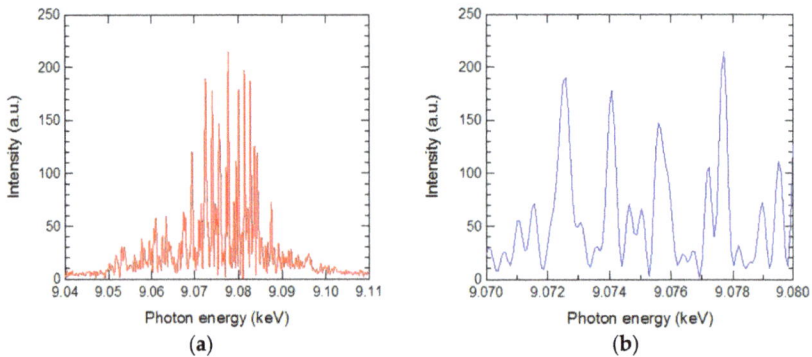

Figure 2. (a) Spectrum of the XFEL beam measured with the spectrometer; (b) enlarged spectrum around the central region of Figure 2a.

Appl. Sci. **2017**, *7*, 584

4. Summary

We demonstrated the measurement of the single-shot entire XFEL spectrum with fully resolved spike structures. In addition to the duration of the XFEL pulse, we obtained the longitudinal mode number. The results could provide new information to a theoretical study of XFEL physics. Moreover, the method is applicable to studies of X-ray nonlinear optics [5,16–18] because the mirrors in our spectrometer are a part of an XFEL two-stage focusing system. In particular, the quality of spectrum measurements for an absorption spectroscopy [18] and stimulated emissions [5] will be improved with a higher resolution over a wider energy range.

Acknowledgments: We would like to acknowledge the supporting members of SACLA facility. We also thank Takashi Tanaka and Ryota Kinjo for advice regarding spectral analysis. This research was partially supported by JSPS KAKENHI Grant Numbers 15H05434 and 23226004. The experiment was performed at BL3 of SACLA with the approval of the Japan Synchrotron Radiation Research Institute (JASRI) (proposal Nos. 2015A8009, 2015B8005 and 2015B8013).

Author Contributions: Y.I. conceived and designed the experiment. J.K., A.N., S.M., H.Y., T.K., H.O. and K.Y. developed the X-ray optics. Y.I., I.I. and K.T. performed the experiment. Y.I. and M.Y. wrote the paper.

Conflicts of Interest: The authors declare no conflict of interest.

References

1. Emma, P.; Akre, R.; Arthur, J.; Bionta, R.; Bostedt, C.; Bozek, J.; Brachmann, A.; Bucksbaum, P.; Coffee, R.; Decker, F.-J.; et al. First lasing and operation of an ångstrom-wavelength free-electron laser. *Nat. Photonics* **2010**, *4*, 641–647. [CrossRef]

2. Ishikawa, T.; Aoyagi, H.; Asaka, T.; Asano, Y.; Azumi, N.; Bizen, T.; Ego, H.; Fukami, K.; Fukui, T.; Furukawa, Y.; et al. A compact X-ray free-electron laser emitting in the sub-ångstrom region. *Nat. Photonics* **2012**, *6*, 540–544. [CrossRef]

3. Chapman, H.N.; Fromme, P.; Barty, A.; White, T.A.; Kirian, R.A.; Aquila, A.; Hunter, M.S.; Schulz, J.; DePonte, D.P.; Weierstall, U.; et al. Femtosecond X-ray protein nanocrystallography. *Nature* **2011**, *470*, 73–77. [CrossRef] [PubMed]

4. Suga, M.; Akita, F.; Hirata, K.; Ueno, G.; Murakami, H.; Nakajima, Y; Shimizu, T.; Yamashita, K.; Yamamoto, M.; Ago, H.; et al. Native structure of photosystem II at 1.95 Å resolution viewed by femtosecond X-ray pulses. *Nature* **2015**, *517*, 99–103. [CrossRef] [PubMed]

5. Yoneda, H.; Inubushi, Y.; Nagamine, K.; Michine, Y.; Ohashi, H.; Yumoto, H.; Yamauchi, K.; Mimura, H.; Kitamura, H.; Katayama, T.; et al. Atomic inner-shell laser at 1.5-ångström wavelength pumped by an X-ray free-electron laser. *Nature* **2015**, *524*, 446–449. [CrossRef] [PubMed]

6. Bonifacio, R.; Pellegrini, C.; Narducci, L.M. Collective instabilities and high-gain regime in a free electron laser. *Opt. Commun.* **1984**, *50*, 373–378. [CrossRef]

7. Saldin, E.L.; Schneidmiller, E.A.; Shvyd'ko, Y.V.; Yurkov, M.V. X-ray FEL with a meV bandwidth. *Nucl. Instrum. Methods Phys.* **2001**, *475*, 357–362. [CrossRef]

8. Yabashi, M.; Hastings, J.B.; Zolotorev, M.S.; Mimura, H.; Yumoto, H.; Matsuyama, S.; Yamauchi, K.; Ishikawa, T. Single-Shot Spectrometry for X-ray Free-Electron Lasers. *Phys. Rev. Lett.* **2006**, *97*, 084802. [CrossRef] [PubMed]

9. Zhu, D.; Cammarata, M.; Feldkamp, J.M.; Fritz, D.M.; Hastings, J.B.; Lee, S.; Lemke, H.Y.; Robert, A.; Turner, J.L.; Feng, Y. A single-shot transmissive spectrometer for hard X-ray free electron lasers. *Appl. Phys. Lett.* **2012**, *101*, 034103. [CrossRef]

10. Inubushi, Y.; Tono, K.; Togashi, T.; Sato, T.; Hatsui, T.; Kameshima, T.; Togawa, K.; Hara, T.; Tanaka, T.; Tanaka, H.; et al. Determination of the Pulse Duration of an X-ray Free Electron Laser Using Highly Resolved Single-Shot Spectra. *Phys. Rev. Lett.* **2012**, *109*, 144801. [CrossRef] [PubMed]

11. Rich, D.; Zhu, D.; Turner, J.; Zhang, D.; Hill, B.; Feng, Y. The LCLS variable-energy hard X-ray single-shot spectrometer. *J. Synchrotron Rad.* **2015**, *23*, 3–9. [CrossRef] [PubMed]

12. Kim, J.; Nagahira, A.; Koyama, T.; Matsuyama, S.; Sano, Y.; Yabashi, M.; Ohashi, H.; Ishikawa, T.; Yamauchi, K. Damage threshold of platinum/carbon multilayers under hard X-ray free-electron laser irradiation. *Opt. Express* **2015**, *23*, 29032. [CrossRef] [PubMed]

13. Mimura, H.; Yumoto, H.; Matsuyama, S.; Koyama, T.; Tono, K.; Inubushi, Y.; Togashi, T.; Sato, T.; Kim, J.; Fukui, R.; et al. Generation of 10^{20} W cm^{-2} hard X-ray laser pulses with two-stage reflective focusing system. *Nat. Commun.* **2014**, *5*, 3539. [CrossRef] [PubMed]

14. Kameshima, T.; Ono, S.; Kudo, T.; Ozaki, K.; Kirihara, Y.; Kobayashi, K.; Inubushi, Y.; Yabashi, M.; Horigome, T.; Holland, A.; et al. Development of an X-ray pixel detector with multi-port charge-coupled device for X-ray free-electron laser experiments. *Rev. Sci. Instrum.* **2014**, *85*, 033110. [CrossRef] [PubMed]

15. Tono, K.; Togashi, T.; Inubushi, Y.; Sato, T.; Katayama, T.; Ogawa, K.; Ohashi, H.; Kimura, H.; Takahashi, S.; Takeshita, K.; et al. Beamline, experimental stations and photon beam diagnostics for the hard X-ray free electron laser of SACLA. *New J. Phys.* **2013**, *15*, 083035. [CrossRef]

16. Tamasaku, K.; Nagasono, M.; Iwayama, H.; Shigemasa, E.; Inubushi, Y.; Tanaka, T.; Tono, K.; Togashi, T.; Sato, T.; Katayama, T.; et al. Double Core-hole Creation by Sequential Attosecond Photo-ionization. *Phys. Rev. Lett.* **2013**, *111*, 043001. [CrossRef] [PubMed]

17. Tamasaku, K.; Shigemasa, E.; Inubushi, Y.; Katayama, T.; Sawada, K.; Yumoto, H.; Ohashi, H.; Mimura, H.; Yabashi, M.; Yamauchi, K.; et al. X-ray two-photon absorption competing against single and sequential multiphoton processes. *Nat. Photonics* **2014**, *8*, 313. [CrossRef]

18. Yoneda, H.; Inubushi, Y.; Yabashi, M.; Katayama, T.; Ishikawa, T.; Ohashi, H.; Yumoto, H.; Yamauchi, K.; Mimura, H.; Kitamura, H. Saturable absorption of intense hard X-rays in iron. *Nat. Commun.* **2014**, *5*, 5080. [CrossRef] [PubMed]

![applied sciences logo]

applied
sciences

MDPI

Article

A Dispersive Inelastic X-ray Scattering Spectrometer for Use at X-ray Free Electron Lasers

Jakub Szlachetko [1,2,*], Maarten Nachtegaal [1], Daniel Grolimund [1], Gregor Knopp [1], Sergey Peredkov [3], Joanna Czapla–Masztafiak [4] and Christopher J. Milne [1,*] ![ORCID icon]

[1]   Paul Scherrer Institut, 5232 Villigen, Switzerland; maarten.nachtegaal@psi.ch (M.N.); daniel.grolimund@psi.ch (D.G.); gregor.knopp@psi.ch (G.K.)
[2]   Institute of Physics, Jan Kochanowski University, 25-001 Kielce, Poland
[3]   Max-Planck-Institute for Chemical Energy Conversion, 45470 Mülheim an der Ruhr, Germany; sergey.peredkov@cec.mpg.de
[4]   The Henryk Niewodniczanski Institute of Nuclear Physics, Polish Academy of Sciences, 31342 Kraków, Poland; joanna.czapla@ifj.edu.pl
*   Correspondence: jakub.szlachetko@ujk.edu.pl (J.S.); chris.milne@psi.ch (C.J.M.); Tel.: +48-41-349-6440 (J.S.); +41-56-310-5477 (C.J.M.)

Academic Editor: Kiyoshi Ueda
Received: 14 July 2017; Accepted: 26 August 2017; Published: 1 September 2017

Abstract: We report on the application of a short working distance von Hamos geometry spectrometer to measure the inelastic X-ray scattering (IXS) signals from solids and liquids. In contrast to typical IXS instruments where the spectrometer geometry is fixed and the incoming beam energy is scanned, the von Hamos geometry allows measurements to be made using a fixed optical arrangement with no moving parts. Thanks to the shot-to-shot capability of the spectrometer setup, we anticipate its application for the IXS technique at X-ray free electron lasers (XFELs). We discuss the capability of the spectrometer setup for IXS studies in terms of efficiency and required total incident photon flux for a given signal-to-noise ratio. The ultimate energy resolution of the spectrometer, which is a key parameter for IXS studies, was measured to the level of 150 meV at short crystal radius thanks to the application of segmented crystals for X-ray diffraction. The short working distance is a key parameter for spectrometer efficiency that is necessary to measure weak IXS signals.

Keywords: dispersive X-ray spectrometer; von Hamos geometry; inelastic X-ray scattering; X-ray free electron laser; SwissFEL; segmented crystal

---

1. Introduction

With the continuous increase in X-ray flux available from accelerator and lab-based sources, many demanding techniques are now being explored and applied in a variety of research fields. One class of techniques that has recently received significant attention has been the application of dispersive X-ray spectrometer geometries to measure inelastic X-ray scattering signals from a wide-range of samples. Dispersive X-ray spectrometry consists of using a crystal to simultaneously energy-resolve a broad spectral range of the X-rays scattered or emitted from the sample after exposure to an X-ray source. This experimental description covers everything from non-resonant and resonant X-ray emission spectroscopy (XES [1] and RXES [2], respectively) to X-ray Raman scattering (XRS [3,4]). In general, these dispersive spectrometers cover a finite range of X-ray bandwidth in a fixed geometry. This allows them to measure a range of the scattered X-ray spectrum in a single measurement, avoiding the necessity of scanning any part of the spectrometer or the incident beam energy. Here, we will describe the characterization and application of such a spectrometer based on the von Hamos geometry using segmented cylindrically bent crystals with 25 cm radius of curvature. This spectrometer is an evolution of our previous design [5], which has been in operation at the SuperXAS beamline at the Swiss Light Source (Paul Scherrer Institute, Switzerland) since 2012, as well as being used in temporary installations [6–9] at the Advanced Photon Source [10] (Argonne National Labs, Lemont, IL, USA), the SACLA XFEL [11] (Spring-8, Hyogo Prefecture, Japan), and the LCLS XFEL [12] (SLAC National Accelerator Laboratory, Menlo Park, CA, USA). Similar spectrometers will also be available at Experimental Station Alvra [13] at the SwissFEL XFEL [14] when it begins user operation in 2018.

The inelastic X-ray scattering technique (IXS), also called X-ray Raman scattering (XRS), is based on a photon-in photon-out scattering process of hard X-rays from low Z elements [4,15–18]. Through this scattering process, the core electron is excited to an unoccupied electronic state just above the Fermi level, and the energy loss shifts the scattered photon energy to lower values. The energy conservation for the IXS process is expressed by $E_1 = E_{initial} + E_{electron} + E_2$, where E_1 and E_2 are the energies of incoming and scattered X-rays, respectively. The sum of $E_{initial}$ and $E_{electron}$ represent the total energy loss of the scattered X-ray, where $E_{initial}$ stands for the binding energy of an electron and $E_{electron}$ represents its energy above the Fermi level. By changing the energy of either E_1 or E_2, and monitoring E_1 and E_2 energies, the unoccupied electronic states probed by the scattering electron may thus be determined.

Compared to soft X-ray absorption spectroscopy (XAS) measurements (<1 keV), which are typically employed for electronic state and structure determination in low Z elements, the IXS technique has both advantages and limitations. The primary drawback of applying the IXS technique is the low cross section for the scattering process. As compared to resonant X-ray scattering, the IXS yields are lower by 10^{-4}–10^{-6}. On the other hand, the IXS technique provides capabilities that are complementary when compared to soft-XAS techniques. The IXS technique uses hard X-rays for incoming and emitted X-ray energies; therefore, the penetration depth is much larger, making the IXS technique bulk-sensitive. Thanks to this advantage, the IXS technique may thus be applied to probe the electronic structure of low Z elements in liquids, concentrated heterogeneous materials, or matter under in situ conditions. To date, the IXS technique has been applied in many scientific areas where soft X-ray techniques could either not be applied [19,20] or could not probe the desired state of matter [21,22].

The IXS technique is generally implemented at synchrotron sources at dedicated beamlines, equipped with instruments ensuring good energy resolution and efficient X-ray detection [3,23–25]. In order to obtain good quality spectra with relatively short acquisition times, the IXS signal is often recorded by scanning the incident X-ray energy rather than scanning the emission spectrometer. Such an approach, called inverse-IXS, allows the efficiency of the IXS experiments to be improved by building X-ray spectrometers equipped with multiple analyzer crystals. Such spectrometers, arranged in the Johann geometry, are operated at a fixed Bragg angle close to 90 degrees. A large solid angle can be achieved by using even a few tens of analyzer crystals. To date, this experimental approach is the best possible approach for use at 3rd-generation storage rings, which are characterized by very stable beam parameters and allow for fast, reproducible scanning of the incident X-ray beam energy using fixed-exit double-crystal monochromators synchronized to insertion device gaps. This approach cannot be applied at XFELs, where the pulse and photon energy stability are poor and vary significantly from pulse to pulse [26,27]. This means the application of the IXS technique at an XFEL should avoid, if possible, the necessity of scanning the incident photon energy. An alternative approach to scanning the incident X-ray photon energy is to use a dispersive X-ray spectrometer at a fixed incident X-ray beam energy. This allows the full and direct IXS spectrum to be obtained in a single acquisition, and each acquisition can then be simply summed up, removing the necessity of both signal normalization and changing the incident beam photon energy. This approach is clearly advantageous for sources where the stability of the incident X-ray beam parameters varies from pulse to pulse.

Due to the properties of X-ray radiation delivered by XFEL sources, the implementation of X-ray spectroscopy techniques commonly applied and established at synchrotron sources is not straightforward due to the inherent instabilities of the XFEL beam. In the present work, we report on the application of a dispersive von Hamos spectrometer to measure the IXS signals from solid and liquid samples. Thanks to the dispersive detection, the entire IXS signal may be measured on a shot-to-shot basis. Moreover, the shape of the IXS signal does not depend on the incident beam energy, and only requires shifting the IXS spectra according to the energy conservation rule to be centered on the spectrometer acquisition bandwidth. Therefore, combining a dispersive-type spectrometer with the self-seeded operation of an XFEL source [26,28] will allow one to record IXS signals for every X-ray pulse, improving the efficiency of the experiment.

2. Materials and Methods

The experiments were performed at the SuperXAS and microXAS beamlines of the Swiss Light Source (Paul Scherrer Institute, Villigen, Switzerland). We employed the von Hamos spectrometer described in detail in [5], thus only the general characteristics and operational details specific to its application to inelastic X-ray scattering studies will be presented here.

The schematic drawing of the von Hamos geometry employing segmented crystals is presented in Figure 1. The von Hamos setup consists of three main components: the X-ray source located at the sample position, the analyzer crystal and a position-sensitive detector. The crystal and sample/detector axes are separated at a distance equal to the radius of curvature (R) of the analyzer crystal, and the detector axis is positioned along the interaction point of the incoming X-rays on the sample (see Figure 1). The X-rays scattered from the sample will undergo a diffraction process from the crystal if the Bragg law criteria is met. In the von Hamos geometry, the Bragg angle is defined by the X-ray direction with respect to the crystal position (along crystal axis CRY_{POS}) and is described by the following formula:

$$\theta_B = tan^{-1}\left(\frac{R}{CRY_{POS}}\right). \tag{1}$$

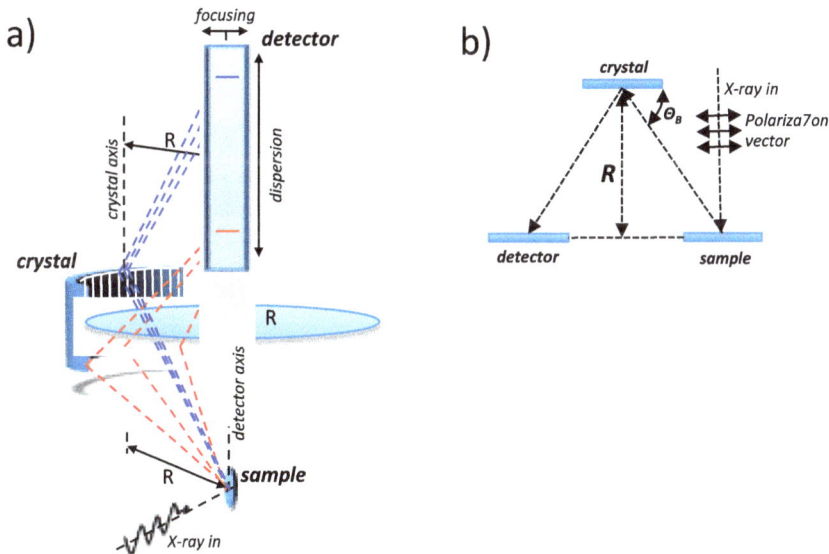

Figure 1. (a) Schematic of the von Hamos spectrometer layout employing segmented crystals for X-ray diffraction in a vertical scattering geometry; **(b)** Schematic view of the spectrometer geometry along the dispersive axis as applied to inelastic X-ray scattering (IXS) studies.

The formula implies that the Bragg angle range, and hence the energy range, covered by the spectrometer is limited only by the length, along the dispersive axis, of the crystal or detector. In the typical von Hamos geometry, the crystal is continuously bent in the focusing plane in order to direct the diffracted X-rays onto a single spot on the detector. In the present setup, we used a segmented crystal design that allows one to maintain the energy resolution at the level of the Darwin width of the reflection provided by a perfect, flat crystal. Indeed, crystal segmentation does not introduce strain in the crystal when the crystal segments are attached to the curved crystal support. The strain induced by the crystal curvature is the major source of the poor energy resolution obtained using bent crystals. The application of a segmented crystal leads to similar quasi-focusing properties of the von Hamos setup as when bent crystals are used. As schematically shown in Figure 1 by the blue dashed line, after diffraction on one crystal segment, the X-rays will focus onto the detector plane. This focus will have a size equal to two times the size of the crystal segment. Since all the segments are placed on a common radius, the total spot size on the detector will be the same, independently of the number of segments used (Figure 1, red dashed line).

In the present experiment, we used an Si(111) crystal glued to a cylindrically shaped support. The crystal consists of 100 segments, each 1×50 mm (focusing $\times$ dispersion) size, and a radius of curvature of the support of 250 mm. The diffracted X-rays were measured by means of either a two-dimensional Pilatus 100 K detector [29,30] consisting of 195×490 pixels with 172×172 μm^2 size, or a Mythen strip detector consisting of a linear array of 1280 pixels of dimension 50 $\mu m \times 8$ mm [31,32]. Both detectors are photon-counting, so all reported signals are directly in counted photons. To measure the IXS signals, the spectrometer was operated at a Bragg angle (marked as Θ_B in Figure 1b) of around 80 degrees and Si(444) diffraction, and arranged in the backscattering geometry. Using such a setup, the spectrometer could record X-rays of energy around 8030 eV and a dispersive energy bandwidth of 80 eV. The incident X-ray beam was delivered by an Si(311) monochromator and focused down to a size of 100×100 μm^2 onto the sample position by means of a Pt-coated mirror. The higher energy X-rays were rejected by primary Si mirror operated at an angle of 3 mrad. The experimental resolution, measured from the full-width at half maximum (FWHM) of elastically scattered photons from a piece of solid Pb, was found to be 300 meV (see Figure 2), indicating that the main broadening contribution is from the Si(311) monochromator bandwidth (250 meV), while the influence from the spectrometer is substantially smaller. The experimental data were fitted with the convolution of two Gauss functions representing the contributions of experimental broadening from the incidence beam and spectrometer, respectively. In the fit, the FWHM of the incidence beam was fixed to 250 meV and the width of the spectrometer contribution was left as a free parameter. From this procedure, the FWHM of the spectrometer contribution was found to be 164 meV, which includes all other contributions to the experimental energy resolution [5]. The primary reason for this improved spectrometer resolution, in comparison to previous results, is due to the decreased segment size from 5 to 1 mm, which reduces the geometric contribution to the energy resolution [5].

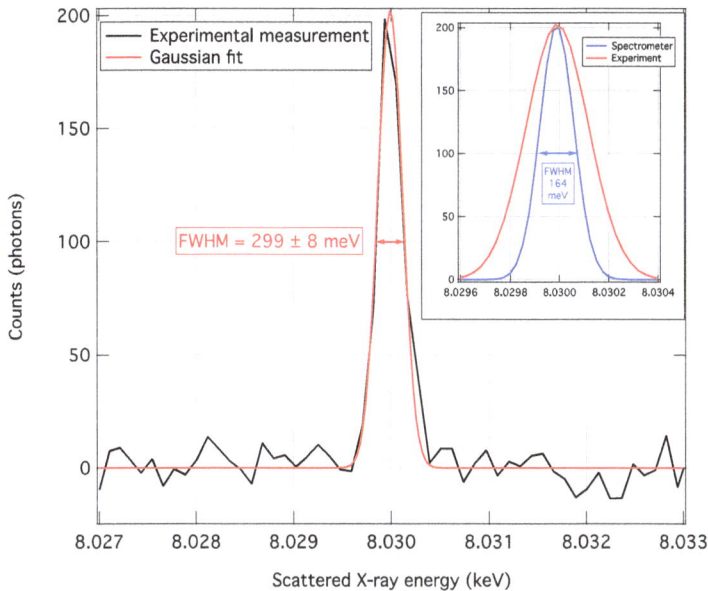

Figure 2. Elastically scattered X-rays measured using the von Hamos spectrometer with the Si(444) diffraction signal at a Bragg angle of around 80 degrees and an X-ray energy of 8030 eV. The incident beam was monochromatized using an Si(311) channel-cut monochromator, resulting in a 250 meV incident X-ray bandwidth. Inset: A Gaussian fit to the experimental measurement (red) overlaid with the de-convolved spectrometer energy resolution (blue).

3. Results

The IXS measurement of the C K-edge recorded from a chemical vapor deposition (CVD) diamond sample with the von Hamos spectrometer is plotted in Figure 3. The incident beam energy was set to 8350 eV. The spectrum is shifted in energy according to the energy conservation rule, in order to be compared with experimental data recorded by means of Electron Loss Near Edge Spectroscopy (ELNES) [33,34]. As shown, excellent agreement is obtained between the spectra. A total incident photon flux of around 10^{15}–10^{16} photons was needed to record a good quality IXS signal. Following the ELNES and XAS experimental and theoretical interpretation, the first peak at around 292 eV corresponds to 1s→σ* excitation, while the features from 295–310 eV relate to σ-type unoccupied states, and the peak at 328 eV is the first EXAFS feature.

The IXS technique has the capability to probe not only K-edges of low Z elements but also the absorption spectra of higher electronic levels. As an example, we performed the IXS measurements on Ti to detect the 3p and 3s absorption spectra by means of IXS. As sample, we used a 5-μm-thick Ti foil. The spectrometer settings were the same as in the case of the IXS C K-edge measurements with the incident beam energy tuned to energy of 8070 eV. The resulting spectrum is plotted in Figure 4. Two distinct features are observed at an energy transfer range between 20 and 80 eV. The first feature located at 30–45 eV corresponds to the 3p absorption spectrum. A sharp peak at 38 eV is detected that corresponds to 3p→d excitation. The second feature at energies above 55 eV relates to 3s absorption. Resonance excitation is observed at 60 eV, that corresponds to dipole 3s→p transition. Note that these transitions probe the same final electronic states as Ti K- and L-edge spectroscopies, and are difficult to address without using electron spectroscopic techniques.

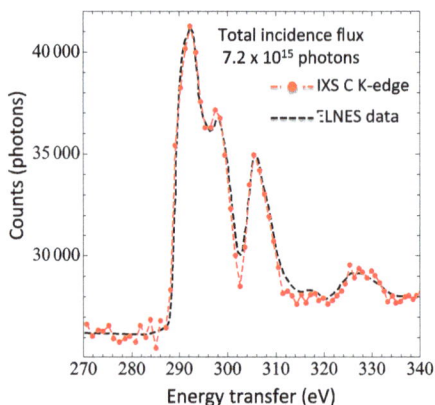

Figure 3. IXS signal of the carbon K-edge of a chemical vapor deposition (CVD) diamond sample. Electron Loss Near Edge Spectroscopy (ELNES) data adapted from Reference [33] with permission from the Royal Society of Chemistry.

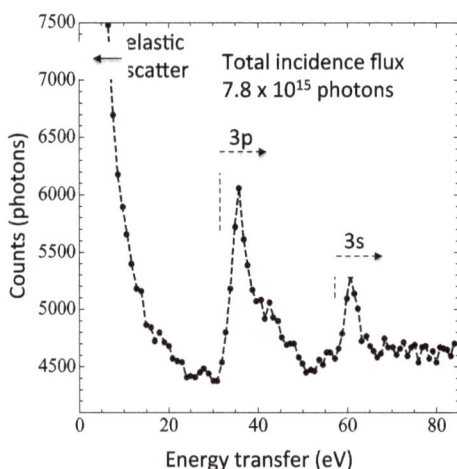

Figure 4. IXS signal of Ti foil showing the excitation signatures of 3p→d and 3s→p.

One of the significant advantages of the IXS technique is the ability to probe bulk samples, such as liquid water. The oxygen IXS signals from water are shown in Figure 5. The O K-edge measurement (Figure 5 right) shows the characteristic XAS and EXAFS of liquid water [35–37], with no contribution from the gas-phase species [38], as expected for a bulk measurement. As for the Ti sample, the dispersive spectrometer can also easily access higher-lying excitations, in this case excitations from the $1b_2$ and $2a_1$ molecular orbitals of the water molecule [39]. In general, these types of excitations are measured using photoemission, which requires vacuum techniques to be applied [40]. Here, we measure the signals from the bulk liquid under ambient atmospheric conditions.

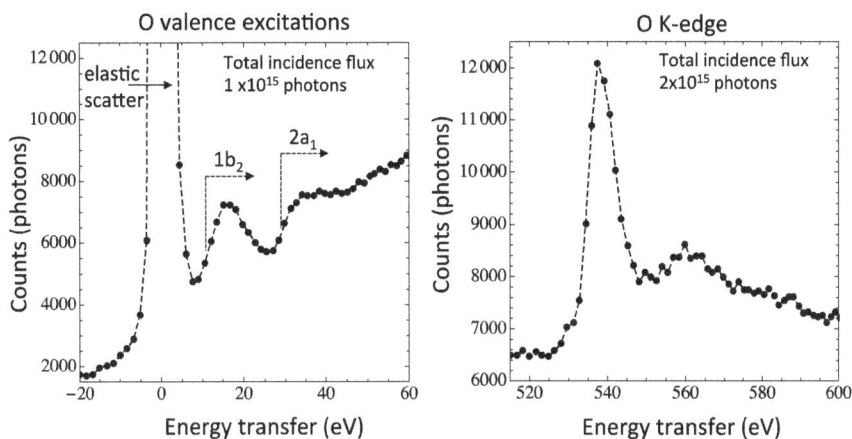

Figure 5. IXS signal of liquid water. Left: high-lying excitations from the molecular orbitals of the water molecule sitting on top of the Compton scattering signal. Right: The O K-edge IXS signal, showing the characteristics of bulk water.

4. Discussion

The presented spectrometer design has been shown to be ideal for a range of different types of X-ray experiments including both off- and on-resonant X-ray techniques [5,8,41–44]. Here, we have demonstrated its application to inelastic X-ray scattering to probe low-energy electronic excitations in condensed matter. Due to its unique combination of large solid angle and high-energy resolution, it can measure IXS signals within realistic timescales of several hours at a bending magnet beamline (X-ray flux 10^{11} photons/second/0.015% bandwidth) at a third-generation storage ring X-ray source. Increasing the number of crystals used is a straightforward way of increasing the X-ray signals.

In terms of the application of a short working distance von Hamos spectrometer to IXS measurements at XFEL sources, the required incident photon flux for the measured IXS signals as shown in Figures 3–5 are in the range of 10^{15}–10^{16} photons. Assuming an XFEL pulse intensity of 10^{11} photons and 100 Hz operation would translate to about 1000 s of total acquisition time. Therefore, a large margin is left for improved spectra quality by increasing the acquisition time to several hours. For example at a total of 2×10^{15} incident photons for the liquid water sample, we note the statistical error for the maximum white line intensity to be around 1.4%, which can be further diminished to 0.14% for 24 h acquisition at an XFEL. Such a level of uncertainty is sufficient for many pump-probe experiments where signal differences on the level of a few percent are detected by means of X-ray absorption and X-ray emission spectroscopies [45]. We would like to emphasize that the present studies include only one analyzer crystal; thus, further signal enhancement may be achieved by application of multi-crystal arrangements, as commonly applied for X-ray emission spectroscopy setups [46]. We anticipate that IXS can be used to probe ultrafast structural and electronic dynamics in samples such as graphite [47] and liquid water [48–50], with a possible extension to probing more dilute species at higher repetition rate XFEL sources in the future [51].

Acknowledgments: The authors would like to acknowledge the contributions of Jörg Schneider and Konrad Vogelsang to the manufacture and assembly of the segmented crystals. Furthermore, we would like to acknowledge the contributions of Beat Meyer, Urs Vogelsang, and Lorenz Baeni for their technical support during the measurements.

Author Contributions: Jakub Szlachetko and Christopher J. Milne conceived and designed the experiments; Jakub Szlachetko, Christopher J. Milne, Joanna Czapla-Masztafiak, Sergey Peredkov, Gregor Knopp, Maarten Nachtegaal, and Daniel Grolimund performed the experiments; Jakub Szlachetko and Christopher J. Milne. analyzed the data; Jakub Szlachetko and Christopher J. Milne wrote the paper with contributions from all authors.

Conflicts of Interest: The authors declare no conflict of interest.

References

1. Bergmann, U.; Glatzel, P. X-ray emission spectroscopy. *Photosynth. Res.* **2009**, *102*, 255–266. [CrossRef] [PubMed]
2. Glatzel, P.; Bergmann, U. High resolution 1s core hole X-ray spectroscopy in 3d transition metal complexes—Electronic and structural information. *Coord. Chem. Rev.* **2005**, *249*, 65–95. [CrossRef]
3. Huotari, S.; Sahle, C.J.; Henriquet, C.; Al-Zein, A.; Martel, K.; Simonelli, L.; Verbeni, R.; Gonzalez, H.; Lagier, M.C.; Ponchut, C.; et al. A large-solid-angle X-ray Raman scattering spectrometer at ID20 of the European Synchrotron Radiation Facility. *J. Synchrotron Rad.* **2017**, *24*, 521–530. [CrossRef] [PubMed]
4. Tohji, K.; Udagawa, Y. X-ray Raman scattering as a substitute for soft-X-ray extended X-ray-absorption fine structure. *Phys. Rev. B* **1989**, *39*, 7590–7594. [CrossRef]
5. Szlachetko, J.; Nachtegaal, M.; de Boni, E.; Willimann, M.; Safonova, O.; Sa, J.; Smolentsev, G.; Szlachetko, M.; van Bokhoven, J.A.; Dousse, J.C.; et al. A von Hamos X-ray spectrometer based on a segmented-type diffraction crystal for single-shot X-ray emission spectroscopy and time-resolved resonant inelastic X-ray scattering studies. *Rev. Sci. Instrum.* **2012**, *83*, 103105. [CrossRef] [PubMed]
6. Milne, C.J.; Szlachetko, J.; Penfold, T.; Santomauro, F.; Britz, A.; Gawelda, W.; Doumy, G.; March, A.M.; Southworth, S.H.; Rittmann, J.; et al. Time-resolved X-ray absorption and emission spectroscopy on ZnO nanoparticles in solution. In *Optical Society of America: Okinawa*; OSA Publishing: Washington, DC, USA, 2014.
7. Penfold, T.J.; Szlachetko, J.; Gawelda, W.; Santomauro, F.G.; Britz, A.; van Driel, T.B.; Sala, L.; Ebner, S.; Southworth, S.H.; Doumy, G.; et al. Femtosecond X-ray Absorption and Emission Spectroscopy on ZnO Nanoparticles in Solution. In Proceedings of the Optical Society of America, Santa Fe, NM, USA, 17–22 July 2016.
8. Szlachetko, J.; Milne, C.J.; Hoszowska, J.; Dousse, J.C.; Błachucki, W.; Sa, J.; Kayser, Y.; Messerschmidt, M.; Abela, R.; Boutet, S.; et al. Communication: The electronic structure of matter probed with a single femtosecond hard X-ray pulse. *Struct. Dyn.* **2014**, *1*, 021101. [CrossRef] [PubMed]
9. Szlachetko, J.; Hoszowska, J.; Dousse, J.-C.; Nachtegaal, M.; Błachucki, W.; Kayser, Y.; Sa, J.; Messerschmidt, M.; Boutet, S.; Williams, G.J.; et al. Establishing nonlinearity thresholds with ultraintense X-ray pulses. *Sci. Rep.* **2016**, *6*, 33292. [CrossRef] [PubMed]
10. March, A.M.; Stickrath, A.; Doumy, G.; Kanter, E.P.; Krässig, B.; Southworth, S.H.; Attenkofer, K.; Kurtz, C.A.; Chen, L.X.; Young, L. Development of high-repetition-rate laser pump/X-ray probe methodologies for synchrotron facilities. *Rev. Sci. Instrum.* **2011**, *82*, 073110. [CrossRef] [PubMed]
11. Ishikawa, T.; Aoyagi, H.; Asaka, T.; Asano, Y.; Azumi, N.; Bizen, T.; Ego, H.; Fukami, K.; Fukui, T.; Furukawa, Y.; et al. A compact X-ray free-electron laser emitting in the sub-ångström region. *Nat. Photonics* **2012**, *6*, 540–544. [CrossRef]
12. Emma, P.J.; Akre, R.; Arthur, J.; Bionta, R.; Bostedt, C.; Bozek, J.; Brachmann, A.; Bucksbaum, P.; Coffee, R.; Decker, F.J.; et al. First lasing and operation of an ångstrom-wavelength free-electron laser. *Nat. Photonics* **2010**, *4*, 641–647. [CrossRef]
13. Milne, C.J.; Beaud, P.; Deng, Y.; Erny, C.; Follath, R.; Flechsig, U.; Hauri, C.P.; Ingold, G.; Juranić, P.; Knopp, G.; et al. Opportunities for Chemistry at the SwissFEL X-ray Free Electron Laser. *CHIMIA* **2017**, *71*, 299–307. [CrossRef] [PubMed]
14. Milne, C.J.; Schietinger, T.; Aiba, M.; Alarcon, A.; Alex, J.; Anghel, A.; Arsov, V.; Beard, C.; Beaud, P.; Bettoni, S.; et al. SwissFEL: The Swiss X-ray Free Electron Laser. *Appl. Sci.* **2017**, *7*, 720. [CrossRef]
15. Mizuno, Y.; Ohmura, Y. Theory of X-Ray Raman Scattering. *J. Phys. Soc. Jpn.* **1967**, *22*, 445–449. [CrossRef]
16. Suzuki, T. X-Ray Raman Scattering Experiment. I. *J. Phys. Soc. Jpn.* **1967**, *22*, 1139–1150. [CrossRef]
17. Tohji, K.; Udagawa, Y. Observation of X-ray Raman scattering. *Phys. B Condens. Matter* **1989**, *158*, 550–552. [CrossRef]
18. Bergmann, U.; Glatzel, P.; Cramer, S.P. Bulk-sensitive XAS characterization of light elements: From X-ray Raman scattering to X-ray Raman spectroscopy. *Microchem. J.* **2002**, *71*, 221–230. [CrossRef]
19. Miedema, P.S.; Ngene, P.; van der Eerden, A.M.J.; Sokaras, D.; Weng, T.-C.; Nordlund, D.; Au, Y.S.; de Groot, F.M.F. Experimental section. *Phys. Chem. Chem. Phys.* **2014**, *16*, 22651–22658. [CrossRef] [PubMed]

20. Miedema, P.S.; Ngene, P.; van der Eerden, A.M.J.; Weng, T.-C.; Nordlund, D.; Sokaras, D.; Alonso-Mori, R.; Juhin, A.; de Jongh, P.E.; de Groot, F.M.F. In situ X-ray Raman spectroscopy of LiBH$_4$. *Phys. Chem. Chem. Phys.* **2012**, *14*, 5581–5587. [CrossRef] [PubMed]

21. Bergmann, U.; Di Cicco, A.; Wernet, P.; Principi, E.; Glatzel, P.; Nilsson, A. Nearest-neighbor oxygen distances in liquid water and ice observed by X-ray Raman based extended X-ray absorption fine structure. *J. Chem. Phys.* **2007**, *127*. [CrossRef] [PubMed]

22. Bergmann, U.; Wernet, P.; Glatzel, P.; Cavalleri, M.; Pettersson, L.G.M.; Nilsson, A.; Cramer, S.P. X-ray raman spectroscopy at the oxygen K edge of water and ice: Implications on local structure models. *Phys. Rev. B* **2002**, *66*, 921071–921074. [CrossRef]

23. Verbeni, R.; Kocsis, M.; Huotari, S.; Krisch, M.; Monaco, G.; Sette, F.; Vankó, G. Advances in crystal analyzers for inelastic X-ray scattering. *J. Phys. Chem. Solids* **2005**, *66*, 2299–2305. [CrossRef]

24. Verbeni, R.; Pylkkaenen, T.; Huotari, S.; Simonelli, L.; Vankó, G.; Martel, K.; Henriquet, C.; Monaco, G. Multiple-element spectrometer for non-resonant inelastic X-ray spectroscopy of electronic excitations. *J. Synchrotron Rad.* **2009**, *16*, 469–476. [CrossRef] [PubMed]

25. Sokaras, D.; Nordlund, D.; Weng, T.C.; Mori, R.A.; Velikov, P.; Wenger, D.; Garachtchenko, A.; George, M.; Borzenets, V.; Johnson, B.; et al. A high resolution and large solid angle X-ray Raman spectroscopy end-station at the Stanford Synchrotron Radiation Lightsource. *Rev. Sci. Instrum.* **2012**, *83*, 043112. [CrossRef] [PubMed]

26. Amann, J.; Berg, W.; Blank, V.; Decker, F.J.; Ding, Y.; Emma, P.J.; Feng, Y.; Frisch, J.; Fritz, D.; Hastings, J.; et al. Demonstration of self-seeding in a hard-X-ray free-electron laser. *Nat. Photonics* **2012**, *6*, 693–698. [CrossRef]

27. Lemke, H.T.; Bressler, C.; Chen, L.X.; Fritz, D.M.; Gaffney, K.J.; Galler, A.; Gawelda, W.; Haldrup, K.; Hartsock, R.W.; Ihee, H.; et al. Femtosecond X-ray Absorption Spectroscopy at a Hard X-ray Free Electron Laser: Application to Spin Crossover Dynamics. *J. Phys. Chem. A* **2013**, *117*, 735–740. [CrossRef] [PubMed]

28. Ratner, D.; Abela, R.; Amann, J.; Behrens, C.; Bohler, D.; Bouchard, G.; Bostedt, C.; Boyes, M.; Chow, K.; Cocco, D.; et al. Experimental demonstration of a soft X-ray self-seeded free-electron laser. *Phys. Rev. Lett.* **2015**, *114*. [CrossRef] [PubMed]

29. Broennimann, C.; Eikenberry, E.F.; Henrich, B.; Horisberger, R.; Huelsen, G.; Pohl, E.; Schmitt, B.; Schulze-Briese, C.; Suzuki, M.; Tomizaki, T.; et al. The PILATUS 1M detector. *J. Synchrotron Rad.* **2006**, *13*, 120–130. [CrossRef] [PubMed]

30. Henrich, B.; Bergamaschi, A.; Broennimann, C.; Dinapoli, R.; Eikenberry, E.F.; Johnson, I.; Kobas, M.; Kraft, P.; Mozzanica, A.; Schmitt, B. PILATUS: A single photon counting pixel detector for X-ray applications. *Nucl. Inst. Methods Phys. Res. Sect. A* **2009**, *607*, 247–249. [CrossRef]

31. Bergamaschi, A.; Cervellino, A.; Dinapoli, R.; Gozzo, F.; Henrich, B.; Johnson, I.; Kraft, P.; Mozzanica, A.; Schmitt, B.; Shi, X. The MYTHEN detector for X-ray powder diffraction experiments at the Swiss Light Source. *J. Synchrotron Rad.* **2010**, *17*, 653–668. [CrossRef] [PubMed]

32. Schmitt, B.; Brönnimann, C.; Eikenberry, E.F.; Gozzo, F.; Hörmann, C.; Horisberger, R.; Patterson, B. Mython detector system. *Nucl. Instrum. Method A* **2003**, *501*, 267–272. [CrossRef]

33. Hamon, A.-L.; Verbeeck, J.; Schryvers, D.; Benedikt, J.; vd Sanden, R.M. ELNES study of carbon K-edge spectra of plasma deposited carbon films. *J. Mater. Chem.* **2004**, *14*, 2030. [CrossRef]

34. Merchant, A.R.; McCulloch, D.G.; Brydson, R. A comparison of experimental and calculated electron-energy loss near-edge structure of carbon, and the nitrides of boron, carbon and silicon using multiple scattering theory. *Diam. Relat. Mater.* **1998**, *7*, 1303–1307. [CrossRef]

35. Näslund, L.Å.; Luning, J.; Ufuktepe, Y.; Ogasawara, H.; Wernet, P.; Bergmann, U.; Pettersson, L.G.M.; Nilsson, A. X-ray Absorption Spectroscopy Measurements of Liquid Water. *J. Phys. Chem. B* **2005**, *109*, 13835–13839. [CrossRef] [PubMed]

36. Yang, B.; Kirz, J. Extended X-ray-absorption fine structure of liquid water. *Phys. Rev. B* **1987**, *36*, 1361–1364. [CrossRef]

37. Fukui, H.; Huotari, S.; Andrault, D.; Kawamoto, T. Oxygen K-edge fine structures of water by X-ray Raman scattering spectroscopy under pressure conditions. *J. Chem. Phys.* **2007**, *127*, 134502–134504. [CrossRef] [PubMed]

38. Myneni, S.; Luo, Y.; Näslund, L.Å.; Cavalleri, M.; Ojamae, L.; Ogasawara, H.; Pelmenschikov, A.; Wernet, P.; Väterlein, P.; Heske, C.; et al. Spectroscopic probing of local hydrogen-bonding structures in liquid water. *J. Phys.-Condens Matter* **2002**, *14*, L213–L219. [CrossRef]

39. Winter, B.; Weber, R.; Widdra, W.; Dittmar, M.; Faubel, M.; Hertel, I.V. Full Valence Band Photoemission from Liquid Water Using EUV Synchrotron Radiation. *J. Phys. Chem. A* **2004**, *108*, 2625–2632. [CrossRef]
40. Winter, B.; Faubel, M. Photoemission from liquid aqueous solutions. *Chem. Rev.* **2006**, *106*, 1176–1211. [CrossRef] [PubMed]
41. Szlachetko, J.; Nachtegaal, M.; Sa, J.; Dousse, J.-C.; Hoszowska, J.; Kleymenov, E.; Janousch, M.; Safonova, O.V.; König, C.; van Bokhoven, J.A. High energy resolution off-resonant spectroscopy at sub-second time resolution: (Pt(acac)2) decomposition. *Chem. Commun.* **2012**, *48*, 10898. [CrossRef] [PubMed]
42. Szlachetko, J.; Sa, J.; Safonova, O.V.; Smolentsev, G.; Szlachetko, M.; van Bokhoven, J.A.; Nachtegaal, M. In situ hard X-ray quick rixs to probe dynamic changes in the electronic structure of functional materials. *J. Electron. Spectrosc.* **2013**, *188*, 161–165. [CrossRef]
43. Szlachetko, J.; Ferri, D.; Marchionni, V.; Kambolis, A.; Safonova, O.V.; Milne, C.J.; Kröcher, O.; Nachtegaal, M.; Sa, J. Subsecond and in Situ Chemical Speciation of Pt/Al$_2$O$_3$ during Oxidation–Reduction Cycles Monitored by High-Energy Resolution Off-Resonant X-ray Spectroscopy. *J. Am. Chem. Soc.* **2013**, *135*, 19071–19074. [CrossRef] [PubMed]
44. Błachucki, W.; Szlachetko, J.; Hoszowska, J.; Dousse, J.C.; Kayser, Y.; Nachtegaal, M.; Sa, J. High Energy Resolution Off-Resonant Spectroscopy for X-Ray Absorption Spectra Free of Self-Absorption Effects. *Phys. Rev. Lett.* **2014**, *112*, 173003. [CrossRef] [PubMed]
45. Milne, C.J.; Penfold, T.J.; Chergui, M. Recent experimental and theoretical developments in time-resolved X-ray spectroscopies. *Coord. Chem. Rev.* **2014**, *277*, 44–68. [CrossRef]
46. Alonso-Mori, R.; Kern, J.; Sokaras, D.; Weng, T.-C.; Nordlund, D.; Tran, R.; Montanez, P.; Delor, J.; Yachandra, V.K.; Yano, J.; et al. A multi-crystal wavelength dispersive X-ray spectrometer. *Rev. Sci. Instrum.* **2012**, *83*, 073114. [CrossRef] [PubMed]
47. Van der Veen, R.M.; Penfold, T.J.; Zewail, A.H. Ultrafast core-loss spectroscopy in four-dimensional electron microscopy. *Struct. Dyn.* **2015**, *2*, 24302–24313. [CrossRef] [PubMed]
48. Wen, H.; Huse, N.; Schoenlein, R.W.; Lindenberg, A.M. Ultrafast conversions between hydrogen bonded structures in liquid water observed by femtosecond X-ray spectroscopy. *J. Chem. Phys.* **2009**, *131*, 234505. [CrossRef] [PubMed]
49. Huse, N.; Wen, H.; Nordlund, D.; Szilagyi, E.; Daranciang, D.; Miller, T.A.; Nilsson, A.; Schoenlein, R.W.; Lindenberg, A.M. Probing the hydrogen-bond network of water via time-resolved soft X-ray spectroscopy. *Phys. Chem. Chem. Phys.* **2009**, *11*, 3951–3957. [CrossRef] [PubMed]
50. Wernet, P.; Gavrila, G.; Godehusen, K.; Weniger, C.; Nibbering, E.T.J.; Elsaesser, T.; Eberhardt, W. Ultrafast temperature jump in liquid water studied by a novel infrared pump-X-ray probe technique. *Appl. Phys. A* **2008**, *92*, 511–516. [CrossRef]
51. Tschentscher, T.; Bressler, C.; Grünert, J.; Madsen, A.; Mancuso, A.; Meyer, M.; Scherz, A.; Sinn, H.; Zastrau, U. Photon Beam Transport and Scientific Instruments at the European XFEL. *Appl. Sci.* **2017**, *7*, 592. [CrossRef]

applied
sciences

MDPI

Article

Split-And-Delay Unit for FEL Interferometry in the XUV Spectral Range

Sergey Usenko [1,2], Andreas Przystawik [1], Leslie Lamberto Lazzarino [3],
Markus Alexander Jakob [2,3], Florian Jacobs [3], Christoph Becker [3], Christian Haunhorst [4],
Detlef Kip [4] and Tim Laarmann [1,2,*]

[1] Photon Science Division, Deutsches Elektronen-Synchrotron DESY, 22607 Hamburg, Germany;
     sergey.usenko@desy.de (S.U.); andreas.przystawik@desy.de (A.P.)
[2] The Hamburg Centre for Ultrafast Imaging CUI, 22761 Hamburg, Germany; markus.jakob@desy.de
[3] Department of Physics, University of Hamburg, 22761 Hamburg, Germany;
     leslie.lamberto.lazzarino@desy.de (L.L.L.); florian.jacobs@desy.de (F.J.);
     cbecker@physik.uni-hamburg.de (C.B.)
[4] Faculty of Electrical Engineering, Helmut Schmidt University, 22043 Hamburg, Germany;
     ceh@hsu-hh.de (C.H.); kip@hsu-hh.de (D.K.)
* Correspondence: tim.laarmann@desy.de; Tel.: +49-40-8998-4940

Academic Editor: Kiyoshi Ueda
Received: 29 March 2017; Accepted: 22 May 2017; Published: 25 May 2017

Abstract: In this work we present a reflective split-and-delay unit (SDU) developed for interferometric
time-resolved experiments utilizing an (extreme ultraviolet) XUV pump–XUV probe scheme with
focused free-electron laser beams. The developed SDU overcomes limitations for phase-resolved
measurements inherent to conventional two-element split mirrors by a special design using two
reflective lamellar gratings. The gratings produce a high-contrast interference signal controlled by
the grating displacement in every diffraction order. The orders are separated in the focal plane of
the focusing optics, which enables one to avoid phase averaging by spatially selective detection of a
single interference state of the two light fields. Interferometry requires a precise relative phase control
of the light fields, which presents a challenge at short wavelengths. In our setup the phase delay is
determined by an in-vacuum white light interferometer (WLI) that monitors the surface profile of the
SDU in real time and thus measures the delay for each laser shot. The precision of the WLI is 1 nm as
determined by optical laser interferometry. In the presented experimental geometry it corresponds to
a time delay accuracy of 3 as, which enables phase-resolved XUV pump–XUV probe experiments at
free-electron laser (FEL) repetition rates up to 60 Hz.

Keywords: white light interferometry; split-and-delay unit; XUV optics characterization; pump-probe

1. Introduction

The recent fast development of table-top high harmonic generation (HHG) [1,2] and free-electron
laser (FEL) light sources [3–5] made short and intense light pulses available in the extreme ultraviolet
(XUV) and soft X-ray wavelength ranges. This opens new opportunities for ultrafast science, in
particular new time-resolved studies. Currently, FELs have two major advantages over HHG sources:
the wavelength tunability and the high intensity. Multidimensional spectroscopy [6], coherent
control [7] and coherent nonlinear optics [8] require multiple intense pulses in the XUV or soft
X-ray spectral range with a precisely defined relative phase.

The simplest interferometric experiment requires two phase-locked light fields. A sequence
of two phase-locked pulses can either be directly generated or produced from a single pulse by
physical manipulation of the beam using a beam splitter and a delay stage. Direct generation and

attosecond control of phase-locked pulse sequences were recently demonstrated at a seeded FEL [9,10]. Nevertheless, mechanical split-and-delay units (SDUs) offer a number of advantages over the direct methods. The first advantage of SDUs lies in their versatility as they are not tied to a particular light source, making the experimental setup more flexible and mobile. The second advantage is a wide continuous range of the pulse replica delays SDUs can provide. Today, the direct methods are limited to either very short (<1 fs) [9] or relatively long (>300 fs) pulse delays [10]. This leaves uncovered a large 0.3–300 fs gap interesting for investigation of ultrafast dynamics of light–matter interaction. By contrast, the delay window of an SDU is limited solely by the travel magnitude of its driving mechanism and can continuously cover a larger range. For a long time the working horse for XUV pump–XUV probe experiments was a double-mirror split-and-delay unit [11–13] which impedes phase-resolved measurements due to the nature of its wavefront splitting (see [14–16] and Section 2). However, recent successful adaptations of ideas originally invented for far-infrared interferometry [17–20] to UV [21] and XUV [22] frequencies open new opportunities for phase-resolved (interferometric) experiments at short wavelength FELs with a typical spectral bandwidth of 1%.

One of the major challenges presented by interferometry at short wavelengths is the need for a precise control of the light phase. Any interferometric experiment requires the delay between the light fields to be controlled within a fraction of their optical cycle. For XUV frequencies, these times lie in the attosecond domain. When applied to an SDU, this requires the geometry of the device to be manipulated with nanometer precision, which imposes very high demands on the precision of mechanics and electronics. Yet sometimes the complexity of the task can be reduced. Precise control over some variable involves two steps: (1) measuring this variable and (2) steering the variable to the desired value, i.e., controlling it. Usually the second step—the actual control—requires much more effort than the measurement. However, in many parameter dependence studies the genuine control can be omitted if one can measure the independent variable and then appropriately sort the dependent observables. All interferometric studies investigate dependence of some observable (photon flux, ion or electron yield) on the phase delay between the light fields. With respect to an SDU, it is sufficient to measure the delay in real time and later sort the data accordingly, provided that some sort of "coarse" delay control is available for the desired time scale. In this way, the question how precisely one can control the delay is transformed into how precisely one can measure it. Typically, high-precision piezo actuators employed in SDUs possess measurement and control systems based on strain-gauge and capacitive sensors. However, often the control electronics fail to measure and maintain the actuator position in environments where vibrations are present and one has to rely on different methods. Interferometric profilometry is an attractive choice as it provides a non-contact, quick and precise measurement of surface profiles with nanometer resolution. White light interferometry is one of the most powerful techniques in this family because it overcomes the phase ambiguity from which monochromatic sources suffer [23–28] by using broadband light sources with short coherence lengths.

In this paper we present a reflective split-and-delay unit that allows for interferometric pump–probe experiments with short XUV laser pulses provided, for instance, by FEL facilities. The SDU is paired with a diagnostics system based on a white light interferometer that measures the generated pump–probe delay with attosecond precision. The performance of the WLI system is characterized using monochromatic interferometry in the optical spectral range.

2. Split-And-Delay Unit for XUV Interferometry

To produce a double pulse sequence from a single pulse, one requires two key components: a beam splitter to divide the initial pulse in two and a translation stage to introduce a phase shift to one of them. Common tools used to split beams in the visible wavelength range are semi-transparent mirrors (amplitude beam splitters). These devices divide the incident wavefront in two replicas—one transmitted and one reflected—each carrying half the energy (for a 50:50 split) and preserving the shape of the original beam. One of the replicas is delayed in time and then both of them are recombined in another (or the same) beam splitter. The advantage of amplitude beamsplitters is that they inherently allow for collinear beam mixing. Collinear propagation puts the pulse replicas in a single interference state as their phase difference is solely defined by the introduced delay. Phase-resolved signal detection thus becomes easy, provided that the delay stage is capable of sub-wavelength movement control. However, amplitude beam splitters are not available below 140 nm [29] and for shorter wavelengths, one has to rely on reflective optics.

A common design of a reflective split-and-delay unit for XUV wavelengths is a double split mirror shown in Figure 1a. Its reflective surface is divided into two halves, one of which can be displaced along the normal and thus delay part of the incident wavefront. The two partial beams are later overlapped by tilting the parts of the split mirror or by focusing optics as shown in the figure. Since the two half-beams are spatially separated, they are superimposed at a small skew angle with their wavefronts tilted in opposite directions. This means that the relative phase between the beams varies continuously along the intersection plane. The simulated light distribution created by the split mirror in the focus is shown in Figure 1a as a function of phase delay between the two partial beams. As the delay changes, the interference pattern continuously "scrolls" inside the intensity envelope defined by the transverse focus size. One can see from the fringe spacing that the phase difference between the beams changes across the focal spot on the scale of 2π. In experiments demanding high photon fluxes, the XUV beams are focused into spots that do not exceed several microns in size. Therefore, to select a region with a fixed phase relation between the two light fields required for interferometry, one would need a detector with a submicron spatial resolution. This is a very challenging task. Instead, the signal is generally integrated over the whole focal volume and the phase information is lost due to phase averaging. This is a major obstacle for phase-resolved measurements with split mirror SDUs.

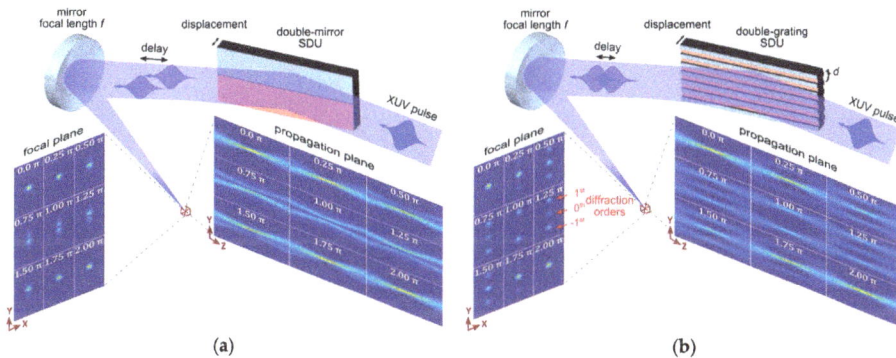

Figure 1. Comparison of a conventional double-mirror SDU (**a**) and a lamellar-grating SDU (**b**). The insets with blue background show simulated light intensity distributions in the focal (XY) and the propagation (YZ) planes for nine different phase delays ranging from 0 to 2π. (**a**) The double-mirror SDU generates two spatially separated beams which are superimposed by the focusing optics at a small angle with their wavefronts tilted in opposite directions. As a result, their relative phase varies continuously along the Y-axis. Hence, a detector with infinitely high resolution would be required to discriminate single interference states. (**b**) Each grating of the SDU diffracts the incident beam. Partial beams from the two gratings propagate collinearly in every diffraction order, which is justified by the complete destructive interference in the zeroth order for $\Delta\phi = \pi$. Spatial selection of a single order for detection allows for high-contrast interferometric measurement.

Limitations of the double split mirror can be overcome using diffractive optics. The idea to use a diffraction grating as a beam splitter is not new [30,31]. For example, a Michelson interferometer based on a transmission grating was successfully employed to characterize femtosecond UV pulses and was proposed for applications in XUV wavelength range [32,33]. However, the setup utilizing a transmission grating is difficult to align, requires additional mirrors and has a low efficiency of the order of several percent [32]. An alternative approach using reflective optics was suggested by Strong and Vanasse in their lamellar-grating interferometer developed for Fourier spectrometry in the far-infrared [17]. The authors overcame the problem of the two-element split mirror—generation of two spatially separated beams—by using two interleaved multi-mirror arrays (lamellar gratings) shown in Figure 1b. Each grating represents a sequence of narrow rectangular facets separated by gaps wide enough to house the facets of another grating. When interleaved, the gratings form a sequence of alternating lamellae with neighboring elements belonging to different gratings. Optical performance of the lamellar-grating beam splitter was thoroughly investigated [17,21,34,35]. In short, each grating diffracts the incident beam in a number of diffraction orders. The partial beams from both gratings propagate collinearly in every order which ensures that their interference depends only on the longitudinal offset between the gratings. This property makes the lamellar-grating SDU a reflective analog of a Michelson interferometer. As an example, for a π phase delay, diffraction from each lamellar grating results in constructive interference in odd orders and destructive interference in even orders. Upon focusing, the angular distribution of the diffracted light translates into a sequence of spots in the focal plane separated by:

$$\Delta r = \frac{\lambda f}{d}, \tag{1}$$

where d is the grating period, f is the focal length of the focusing optics and λ is the mean wavelength of the incident light. The intensity of the zeroth order recorded as a function of grating displacement yields the interferogram of the light source. Provided that Δr is larger than the focus size and can be resolved by the detection system, a single order can be selected for the signal accumulation. Compared to a transmission grating setup, the lamellar-grating SDU has fewer optical elements and offers higher efficiency.

Experimental Design of the SDU

The SDU designed for XUV experiments utilizes two lamellar gratings of different but complementary design. The first grating is a $60 \times 35 \times 1$ mm^3 Si wafer with the central 10×20 mm^2 area processed as a slotted grid with a circular diamond saw. The structure of 250 μm period comprises 150 μm wide slits separated by 100 μm wide reflective facets. The second grating comprises 100 μm wide, rectangular ridges projecting (protruding) from the substrate for 1.25 mm. The ridges are spaced 150 μm apart and fit into the slits of the first grating. When interleaved, the gratings form an alternating pattern of 100 μm facets separated by 25 μm gaps giving a fill factor of 0.8 for the assembled device.

The interleaved gratings are installed on a specially designed mount equipped with motors necessary for grating alignment and translation as shown in Figure 2. The slotted grating is rigidly fixed to the mount while the ridged grating has three degrees of freedom driven by piezo actuators. One of them is used to translate the grating and thus delay the reflected beam. The two others control the grating rotation in two planes and keep both gratings parallel. The delay piezo stage has a travel range of 250 μm in the closed loop control mode, which translates in the delay ranging from −50 to +574 fs at the incidence angle of 22° with respect to the surface. We note in passing that the resulting peak fluence of unfocused FEL beams on the optics surface in this geometry is well below the melting and ablation thresholds of bulk Si.

The described SDU is designed for time-resolved ionization experiments of low-density gas targets by XUV photons with detection of ionization products, i.e., electrons and ions. It follows from Equation (1) that diffraction orders from an XUV beam generated by a grating with a 250 μm period in the focal plane will be separated only by some tens of microns. For example, $\Delta r = 46$ μm for $\lambda = 38$ nm and $f = 300$ mm [22]. Therefore the detection system must have high spatial resolution in order to distinguish signals from individual orders. The ionization volume can be readily imaged with a resolution of few microns by an elongated velocity map imaging (VMI) spectrometer operated in the spatial imaging mode, which proved to be sufficient for 38 nm wavelength [22]. Higher resolution required for shorter wavelengths can be achieved by extending the focal length of the mirror and using a spectrometer purposely designed for magnified spatial imaging of charged particles, i.e., an ion microscope [36].

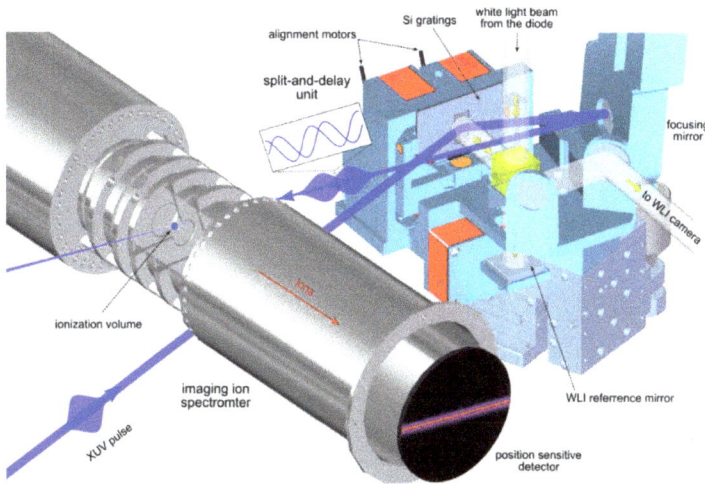

Figure 2. Layout of the experimental setup inside the vacuum chamber. As an illustration, the image on the position-sensitive detector (PSD) displays the spatial distribution of Xe^+ ions produced by 38 nm photons [22]. The image shows a characteristic triple foci structure generated by the SDU. The image is an accumulation of ~5000 shots including many different pulse-replica delays.

3. White Light Interferometer for Time Delay Diagnostics

In our experimental setup we determine the time delay introduced by the SDU using a white light interferometer that monitors the topography of the split-and-delay unit. The typical task of a WLI system is to characterize the height profile $h(x, y)$ of a sample surface placed in one of the interferometer arms by finding zero optical path difference (OPD) for each point of the surface. Usually an experiment is performed by mechanically scanning the length z of one arm and recording the light intensity $I(x, y)$ for different z positions by a camera. The recorded intensity $I(x, y; z)$ is a collection of z-dependent interferograms for all camera pixels. The optical path difference s between the two arms can be expressed as a sum of the surface heightmap $h(x, y)$ relative to some reference plane and a translation z of this plane along the interferometer arm: $s(x, y) = 2(z + h(x, y))$. By scanning z, the positions where $s = 0$ (interferogram maximum) are found for every pixel yielding thus the surface profile $h(x, y)$.

A mechanical z-scan requires the setup to be static on the timescale of at least few seconds. This condition is often difficult to fulfill. Equipment present in a typical laboratory environment (e.g., vacuum pumps or ventilation system) operates at frequencies from several tens to several hundred Hz and generates vibrations which may be transmitted to the experimental setup. For interferometric experiments in the XUV, even small vibration-induced displacements on a scale of tens of nm become crucial. With respect to an SDU, the important aspect to take into account is not the vibration of the system as the whole but the relative jitter of the two components of the device. These oscillations transform into the jitter of the time delay between the generated pulse replicas and thus have a direct impact on the outcome of an interferometric measurement. For example, a displacement of just 50 nm between the SDU reflectors translates into a delay which may comprise several optical cycles for XUV wavelengths. This makes interferometric, i.e., sub-cycle resolved, experiments impossible if the signal is accumulated over time without considering the exact topography of the SDU for each laser shot. Under such a scenario, a mechanical z-scan of the WLI is not applicable and one has to rely on a single-frame technique. Therefore, we used the interferometer in the so-called "electronically scanned" configuration [37,38] with non-collinear beams. If one of the reflective surfaces is tilted to some small angle β, the two beams intersect at the skew angle 2β. As a result, the OPD between the beams varies continuously along the intersection plane and a single camera image actually presents an OPD scan. In this way, the interferogram $I(z)$ is mapped onto the axis perpendicular to the beam propagation: $I(z) \rightarrow I(x)$ as shown in Figure 3. A shift of the interference pattern Δx along the horizontal axis of the image will correspond to some longitudinal displacement Δz of the reflective surface along the interferometer arm. The calibration factor dz/dx for a fixed tilt angle can be calculated from the spacing of interference fringes knowing the average wavelength of the white light. Using this factor, Δz can be derived for every single frame.

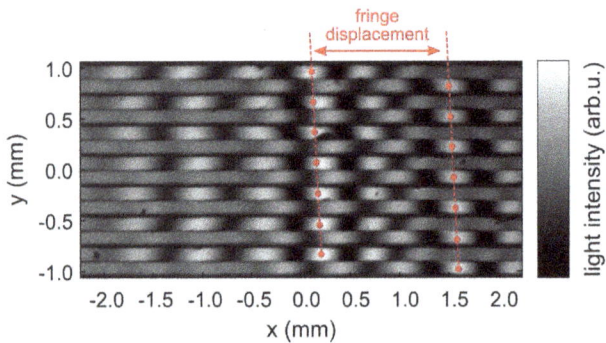

Figure 3. Surface fragment of the SDU seen through the WLI camera. The image clearly shows horizontal facets of the gratings with white light interference fringes on their surface. Neighboring facets belong to different gratings. One of the gratings is displaced along the surface normal (z-direction) as is indicated by the shift of the interference pattern.

Experimental Design of the WLI

The diagnostics setup for a precise determination of the delay introduced by the SDU is based on a Michelson white light interferometer. All of its components except the white light source and the imaging system are placed in a vacuum chamber in close proximity to the SDU (see Figure 2). The interferometer employs a white light diode with a mean wavelength $\lambda_D = 593$ nm as the illumination source. The diode is mounted on top of the vacuum chamber and illuminates the interferometer through a viewport. The beam from the diode is split equally into two arms by a 20 mm broad-band beam splitting cube. The SDU is placed at the end of the horizontal arm. The vertical arm terminates with a super-polished reference Si mirror mounted on a translational piezo stage. After reflection, the two beams recombine in the same beam splitter and pass through a vacuum window to a camera placed outside the chamber. The camera can image a 10×10 mm^2 area of the SDU surface with a lateral resolution of 8 µm at a full frame rate up to 60 Hz.

4. Results and Discussion

4.1. Analysis of the SDU Displacement Jitter

Using the WLI, we found that the movable grating is affected by environmental vibrations transmitted to the experimental chamber. The vibrations cause the grating to oscillate around the set position with amplitudes up to 100 nm. The frequency spectrum of this displacement jitter was analyzed as follows. A movie of the jittering interference pattern was recorded by the WLI camera at a high frame rate of 1380 Hz. In order to increase the frame rate from the maximum 60 Hz available for the full frame data acquisition, the frame was sized down to a small area of interest comprising only two 100 µm wide lamellae (one from each grating). The positions of interferogram maxima x_0 corresponding to $s = 0$ were found for each grating in every movie frame. The relative displacements Δz between the gratings were derived from these data as described in Section 2. The Fourier transform of $\Delta z(t)$ directly gives the vibrational spectrum shown in Figure 4. As seen from the figure, the major frequency components lay below 100 Hz. The spectrum demonstrates that typical jitter periods (>10 ms) are much longer than the typical exposure time of the camera (~1 ms). In an experiment with femtosecond laser pulses, this allows the SDU displacement to be determined on a single-shot basis at a repetition rate limited solely by the camera characteristics (up to 60 Hz for the full frame acquisition in our case).

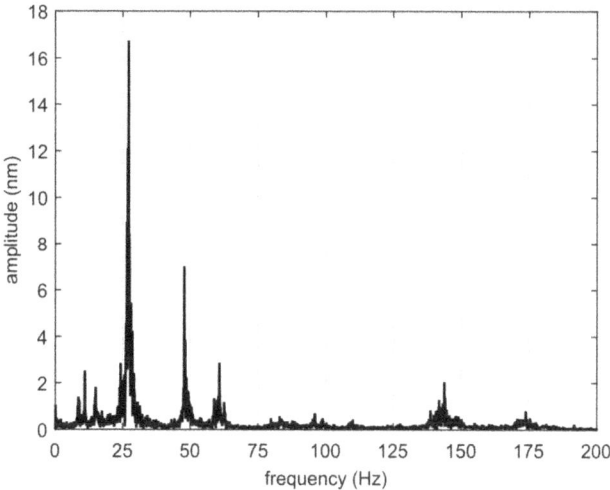

Figure 4. Vibrational spectrum of the SDU displacement. The spectrum was obtained by recording a movie with the WLI camera at a frame rate of 1380 Hz and then making the Fourier transform of the central fringe position.

4.2. WLI Single-Shot Precision Determination

To characterize the performance of the WLI system, we used the split-and-delay unit as a second independent interferometer. The SDU conveniently combines a beam splitter and two interferometer arms in a single device and requires only a light source and a detector to complete the setup. Hence, the second interferometric experiment can be performed simultaneously with the WLI measurement and used to gauge the precision of the latter. In the experiment, we used a small continuous wave (CW) diode laser to illuminate the SDU and imaged the reflected intensity in the focal plane with the second camera in synchronization with the WLI measurement. The interferogram of the laser recorded as a function of SDU displacement was then used to characterize the performance of the WLI system.

The laser used in the experiment has a narrow spectrum with the central wavelength $\lambda_L = 638.4$ nm determined by a grating spectrometer. The laser beam was collimated and shaped with a 1.5×7 mm² rectangular aperture in order to restrict the illuminated area of the SDU to the structured region. After reflection from the SDU, the laser beam was focused by a toroidal mirror with an effective focal length $f = 317$ mm. The irradiance distribution in the focal plane was imaged by a camera placed behind the focus. The spatial distribution of the laser intensity for different OPDs between the gratings is shown in Figure 5. According to Equation (1), the separation between diffraction orders in the focal plane for our experimental geometry is $\Delta r = 801$ µm and could be easily resolved by the camera.

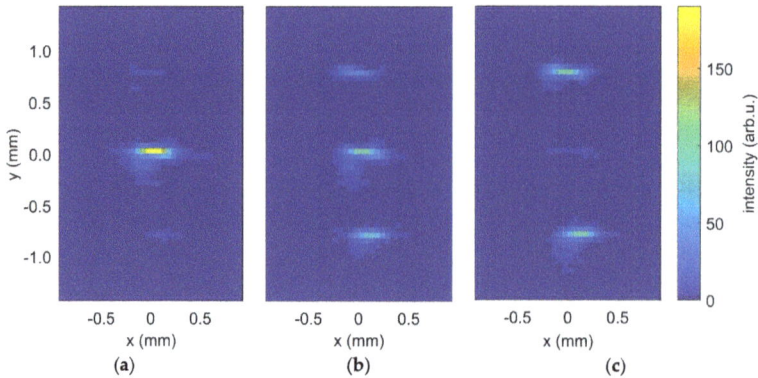

Figure 5. Irradiance distribution of the laser diode in the focal plane after reflection from the SDU for three different phase delays: (**a**) 0, (**b**) $\pi/2$, (**c**) π. The foci have line shapes because the laser had a 1.5×7 mm^2 rectangular aperture to match the beam size with the structured area of the SDU.

We scanned the OPD by translating the movable grating. The WLI and the laser cameras took images synchronously with equal exposure time of 1 ms at a rate of 8 Hz. The OPD s between the partial laser beams is connected to the displacement Δz between the gratings by the relation:

$$s = 2\Delta z \sin \alpha \tag{2}$$

where is the angle of incidence on the SDU counted from its surface. We performed a 2130 nm long z-scan which corresponds to an OPD window of 1600 nm considering $\alpha = 22°$ in our experiment. The actual displacement between the gratings for each exposure cycle was determined by the WLI according to the procedure described in Section 3. The laser intensity in the zeroth diffraction order as a function of derived OPD s is shown in Figure 6. The obtained interferogram is sinusoidal without any noticeable envelope as is expected for a highly monochromatic light source. The oscillation period of 640 nm agrees well with $\lambda_L = 638.4$ nm measured previously by the grating spectrometer.

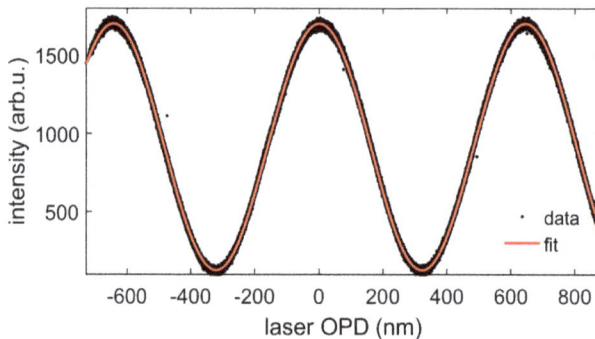

Figure 6. Intensity of the diode laser in the zeroth diffraction order as a function of optical path difference between the gratings of the SDU derived from the WLI images. The red curve is a sine fit to the data.

The deviation of the experimental data from the fit in the interferogram plot has two main contributions: (1) the noise of the camera imaging the focal plane and (2) the error in the OPD derived by the WLI. The latter error defines the precision of the WLI system and we will quantify it using its dependence on the interferogram shape in the following. The contribution of the WLI error to the standard deviation (SD) of the data is proportional to dI/ds, where I is the interferogram intensity. Consequently, the contribution is the smallest for data points located on crests of the sinusoid and the largest for points of the highest gradient. Therefore, the SD plotted as a function of s oscillates at twice the frequency of the interferogram and is shifted by $\pi/2$. To compare the contribution of the WLI error with the camera noise we bin the interferogram data in OPD bins of different size (see Figure 7). The figure clearly demonstrates that the ratio of the oscillation amplitude to the noise level decreases as the bin size gets smaller. At the bin size $\delta s = 1$ nm, the oscillation amplitude becomes comparable to the noise level and we consider this value to be the effective precision of our WLI system. In our experimental geometry, $s = 1$ nm corresponds to the SDU displacement $\Delta z = 0.75$ nm and translates into 3 attoseconds of time delay. This value is sufficient to perform interferometric experiments with laser pulses in the XUV wavelength range as was already demonstrated [22].

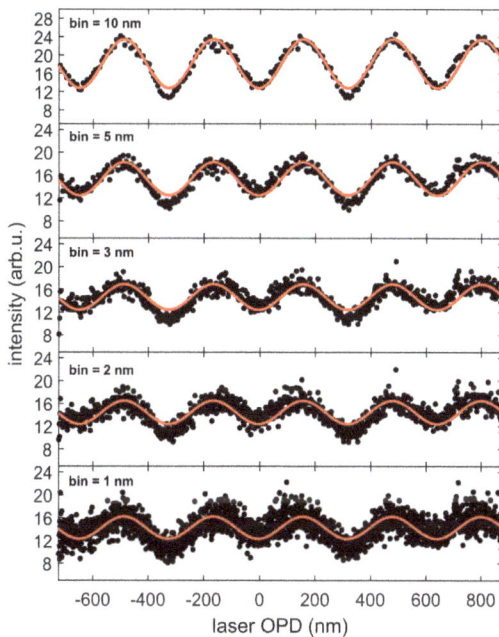

Figure 7. Standard deviation of the laser intensity data from the fit plotted as a function of optical path difference s. The black scatter represents the data, and the red curves are fits of a form $\sin 2\frac{2\pi}{\lambda_L}s$, where λ_L is the laser wavelength. Each data point is the center of a bin, the size of which is indicated in the left top corner of each plot.

5. Conclusions

In summary, we have described a split-and-delay unit suitable for interferometric XUV pump–XUV probe experiments at coherent FELs. The SDU consists of two interleaved lamellar gratings. The intensity of the zeroth diffraction order monitored as a function of delay is analogous to that in a Michelson interferometer. The signal from individual orders can be selected using spatially-resolved imaging of the ionization volume with a VMI spectrometer or an ion microscope. The time delay produced by the SDU is precisely determined by an in-vacuum white light interferometer. The WLI in "electronically scanned" configuration records the topography of the SDU on a single-shot basis. These data are used to determine the pump–probe delay generated by the SDU for each laser shot. The WLI was characterized with an interferometric measurement using a CW laser to have a precision of 1 nm.

Appl. Sci. **2017**, *7*, 544

This allows for interferometric experiments in the XUV spectral range with a delay resolution down to 3 attoseconds at repetition rates up to 60 Hz.

Towards short-wavelength FEL applications, the key aspect defining the usability of the described SDU for phase-resolved studies is whether the individual diffraction orders can be resolved under the particular experimental conditions. For a given light wavelength, this depends on the grating surface quality and period, focusing geometry and the charged particle detector, all of which can be improved. High surface quality of the optics becomes an especially important aspect when dealing with short wavelengths. The major challenge here is to structure the gratings with small period maintaining the flatness of the original substrate. Distortions of the substrate induced by the sawing process lead to phase errors in the reflected wavefront, which reduce the interference contrast in the ionization region. This effect becomes more pronounced the shorter the wavelength is. To some extent, the phase errors produced by the surface inhomogeneity can be scaled down by reducing the angle of incidence of the XUV beam on the SDU, which is required anyway for shorter wavelengths to maintain high reflectivity. Another aspect to take into account is the order separation equal to $\lambda f / d$ in the focal area. For a fixed wavelength, the free parameters are the mirror focal length and the grating period. Manufacturing lamellar gratings with shorter periods and high surface quality is challenging; therefore, extending the focal length is a feasible option. The imaging detector is yet another key element that defines the resolution of the setup. In the experiment with 38 nm wavelength and a focal length of 300 mm, the resolution of the VMI spectrometer operated in the spatial-imaging mode with a moderate magnification of 20 was sufficient to distinguish individual diffraction orders. A spectrometer purposely designed for magnified spatial imaging of charged particles, i.e., an ion microscope, can reach magnification factors up to 100 [36] and significantly improve the resolution of the setup. In conclusion, we are confident that the lamellar-grating concept can be extended to the few nm-wavelength range for advanced applications with soft X-ray FEL pulses.

Acknowledgments: This work was supported by the Deutsche Forschungsgemeinschaft through the excellence cluster "The Hamburg Centre for Ultrafast Imaging (CUI)—Structure, Dynamics and Control of Matter at the Atomic Scale" (DFG-EXC1074), the collaborative research center "Light-induced Dynamics and Control of Correlated Quantum Systems" (SFB925), the GrK 1355 and by the Federal Ministry of Education and Research of Germany under contract No. 05K16GU4.

Author Contributions: Sergey Usenko, Andreas Przystawik, Leslie Lamberto Lazzarino, Markus Alexander Jakob, Florian Jacobs, Christoph Becker, Christian Haunhorst, Detlef Kip and Tim Laarmann contributed to the design of the experimental setup. Andreas Przystawik performed the experiment with the diode laser. Florian Jacobs performed the vibration measurement. Sergey Usenko analyzed the data. Sergey Usenko wrote the paper with contributions from Andreas Przystawik and Tim Laarmann and input from all the other authors.

Conflicts of Interest: The authors declare no competing financial interests.

References

1. Krausz, F.; Ivanov, M. Attosecond physics. *Rev. Mod. Phys.* **2009**, *81*, 163–234. [CrossRef]
2. Kohler, M.C.; Pfeifer, T.; Hatsagortsyan, K.Z.; Keitel, C.H. Frontiers of Atomic High-Harmonic Generation. In *Advances in Atomic, Molecular, and Optical Physics*; Elsevier: Amsterdam, The Netherlands, 2012; Volume 61, pp. 159–208.
3. Ackermann, W.; Asova, G.; Ayvazyan, V.; Azima, A.; Baboi, N.; Bähr, J.; Balandin, V.; Beutner, B.; Brandt, A.; Bolzmann, A.; et al. Operation of a free-electron laser from the extreme ultraviolet to the water window. *Nat. Photonics* **2007**, *1*, 336–342. [CrossRef]
4. Emma, P.; Akre, R.; Arthur, J.; Bionta, R.; Bostedt, C.; Bozek, J.; Brachmann, A.; Bucksbaum, P.; Coffee, R.; Decker, F.-J.; et al. First lasing and operation of an ångstrom-wavelength free-electron laser. *Nat. Photonics* **2010**, *4*, 641–647. [CrossRef]
5. Allaria, E.; Appio, R.; Badano, L.; Barletta, W.A.; Bassanese, S.; Biedron, S.G.; Borga, A.; Busetto, E.; Castronovo, D.; Cinquegrana, P.; et al. Highly coherent and stable pulses from the FERMI seeded free-electron laser in the extreme ultraviolet. *Nat. Photonics* **2012**, *6*, 699–704. [CrossRef]
6. Cundiff, S.T.; Mukamel, S. Optical multidimensional coherent spectroscopy. *Phys. Today* **2013**, *66*, 44–49. [CrossRef]

7. Adams, B.W.; Buth, C.; Cavaletto, S.M.; Evers, J.; Harman, Z.; Keitel, C.H.; Pálffy, A.; Picón, A.; Röhlsberger, R.; Rostovtsev, Y.; et al. X-ray quantum optics. *J. Mod. Opt.* **2013**, *60*, 2–21. [CrossRef]
8. Bencivenga, F.; Cucini, R.; Capotondi, F.; Battistoni, A.; Mincigrucci, R.; Giangrisostomi, E.; Gessini, A.; Manfredda, M.; Nikolov, I.P.; Pedersoli, E.; et al. Four-wave mixing experiments with extreme ultraviolet transient gratings. *Nature* **2015**, *520*, 205–208. [CrossRef] [PubMed]
9. Prince, K.C.; Allaria, E.; Callegari, C.; Cucini, R.; Ninno, G.D.; Mitri, S.D.; Diviacco, B.; Ferrari, E.; Finetti, P.; Gauthier, D.; et al. Coherent control with a short-wavelength free-electron laser. *Nat. Photonics* **2006**, *10*, 985–990. [CrossRef]
10. Gauthier, D.; Ribič, P.R.; Ninno, G.D.; Allaria, E.; Cinquegrana, P.; Danailov, M.B.; Demidovich, A.; Ferrari, E.; Giannessi, L. Generation of phase-locked pulses from a seeded free-electron laser. *Phys. Rev. Lett.* **2016**, *116*, 024801. [CrossRef] [PubMed]
11. Sorgenfrei, F.; Schlotter, W.F.; Beeck, T.; Nagasono, M.; Gieschen, S.; Meyer, H.; Foehlisch, A.; Beye, M.; Wurth, W. The extreme ultraviolet split and femtosecond delay unit at the plane grating monochromator beamline PG2 at FLASH. *Rev. Sci. Instrum.* **2010**, *81*, 043107. [CrossRef] [PubMed]
12. Wöstmann, M.; Mitzner, R.; Noll, T.; Roling, S.; Siemer, B.; Siewert, F.; Eppenhoff, S.; Wahlert, F.; Zacharias, H. The XUV split-and-delay unit at beamline BL2 at FLASH. *J. Phys. B* **2013**, *46*, 164005. [CrossRef]
13. Campi, F.; Coudert-Alteirac, H.; Miranda, M.; Rading, L.; Manschwetus, B.; Rudawski, P.; L'Huillier, A.; Johnsson, P. Design and test of a broadband split-and-delay unit for attosecond XUV-XUV pump-probe experiments. *Rev. Sci. Instrum.* **2016**, *87*, 023106. [CrossRef]
14. Mashiko, H.; Suda, A.; Midorikawa, K. All-reflective interferometric autocorrelator for the measurement of ultra-short optical pulses. *Appl. Phys. B* **2003**, *76*, 525–530. [CrossRef]
15. Faucher, O.; Tzallas, P.; Benis, E.P.; Kruse, J.; Conde, A.P.; Kalpouzos, C.; Charalambidis, D. Four-dimensional investigation of the 2nd order volume autocorrelation technique. *Appl. Phys. B* **2009**, *97*, 505–510. [CrossRef]
16. Tzallas, P.; Charalambidis, D.; Papadogiannis, N.A.; Witte, K.; Tsakiris, G.D. Direct observation of attosecond light bunching. *Nature* **2003**, *426*, 267. [CrossRef] [PubMed]
17. Strong, J.; Vanasse, G.A. Lamellar grating far-infrared interferometer. *J. Opt. Soc. Am.* **1960**, *50*, 113. [CrossRef]
18. Hall, R.T.; Vrabec, D.; Dowling, J.M. A high-resolution, far infrared double-beam lamellar grating interferometer. *Appl. Opt.* **1966**, *5*, 1147. [CrossRef] [PubMed]
19. Milward, R.C. A small lamellar grating interferometer for the very far-infrared. *Infrared Phys.* **1969**, *9*, 59–74. [CrossRef]
20. Henry, R.L.; Tanner, D.B. A lamellar grating interferometer for the far-infrared. *Infrared Phys.* **1979**, *19*, 163–174. [CrossRef]
21. Gebert, T.; Rompotis, D.; Wieland, M.; Karimi, F.; Azima, A.; Drescher, M. Michelson-type all-reflective interferometric autocorrelation in the VUV regime. *New J. Phys.* **2014**, *16*, 239–310. [CrossRef]
22. Usenko, S.; Przystawik, A.; Jakob, M.; Lazzarino, L.L.; Brenner, G.; Toleikis, S.; Haunhorst, C.; Kip, D.; Laarmann, T. Attosecond interferometry with self-amplified spontaneous emission of a free-electron laser. *Nat. Commun.* **2017**, *8*, 15626.
23. Flournoy, P.A.; McClure, R.W.; Wyntjes, G. White-light interferometric thickness gauge. *Appl. Opt.* **1972**, *11*, 1907. [CrossRef] [PubMed]
24. Lee, B.S.; Strand, T.C. Profilometry with a coherence scanning microscope. *Appl. Opt.* **1990**, *29*, 3784–3788. [CrossRef] [PubMed]
25. Chim, S.S.C.; Kino, G.S. Correlation microscope. *Opt. Lett.* **1990**, *15*, 579. [CrossRef] [PubMed]
26. Kino, G.S.; Chim, S.S.C. Mirau correlation microscope. *Appl. Opt.* **1990**, *29*, 3775–3783. [CrossRef] [PubMed]
27. Danielson, B.L.; Boisrobert, C.Y. Absolute optical ranging using low coherence interferometry. *Appl. Opt.* **1991**, *30*, 2975. [CrossRef] [PubMed]
28. Dresel, T.; Häusler, G.; Venzke, H. Three-dimensional sensing of rough surfaces by coherence radar. *Appl. Opt.* **1992**, *31*, 919–925. [CrossRef] [PubMed]
29. Thorne, A. Fourier transform spectrometry in the vacuum ultraviolet: Applications and progress. *Phys. Scr.* **1996**, *T65*, 31–35. [CrossRef]
30. Ronchi, V. Forty years of history of a grating interferometer. *Appl. Opt.* **1964**, *3*, 437. [CrossRef]
31. Munnerlyn, C.R. A Simple Laser Interferometer. *Appl. Opt.* **1969**, *8*, 827. [CrossRef] [PubMed]

32. Goulielmakis, E.; Nersisyan, G.; Papadogiannis, N.A.; Charalambidis, D.; Tsakiris, G.D.; Witte, K. A dispersionless Michelson interferometer for the characterization of attosecond pulses. *Appl. Phys. B* **2002**, *74*, 197–206. [CrossRef]
33. Papadogiannis, N.A.; Nersisyan, G.; Goulielmakis, E.; Rakitzis, T.P.; Hertz, E.; Charalambidis, D.; Tsakiris, G.D.; Witte, K. Temporal characterization of short-pulse third-harmonic generation in an atomic gas by a transmission-grating Michelson interferometer. *Opt. Lett.* **2002**, *27*, 1561. [CrossRef] [PubMed]
34. Möller, K.D. Wavefront dividing interferometers. *Infrared Phys.* **1991**, *32*, 321–331. [CrossRef]
35. Yin, H.; Wang, M.; Ström, M.; Nordgren, J. Study of a wave-front-dividing interferometer for Fourier transform spectroscopy. *Nucl. Instrum. Methods Phys. Res. Sect. A* **2000**, *451*, 529–539. [CrossRef]
36. Schultze, M.; Bergüs, B.; Schroeder, H.; Krausz, F.; Kompa, K.L. Spatially resolved measurement of ionization yields in the focus of an intense laser pulse. *New J. Phys.* **2011**, *13*, 033001. [CrossRef]
37. Bosselmann, T.; Ulrich, R. High-accuracy position-sensing with fiber-coupled white-light interferometers. In *Proceedings SPIE 0514*; Kersten, R.T., Kist, R., Eds.; SPIE: Bellingham, WA, USA, 1984; pp. 361–364.
38. Chen, S.; Meggitt, B.T.; Rogers, A.J. Electronically scanned optical-fiber Young's white-light interferometer. *Opt. Lett.* **1991**, *16*, 761. [CrossRef] [PubMed]

applied
sciences

MDPI

Article

Terawatt-Isolated Attosecond X-ray Pulse Using a Tapered X-ray Free Electron Laser

Sandeep Kumar [1,2], **Alexandra S. Landsman** [2,3] **and Dong Eon Kim** [1,2,*]

[1] Department of Physics, Center for Attosecond Science and Technology,
Pohang University of Science and Technology, Pohang 790-784, Korea; skiitd@gmail.com
[2] Max Planck Center for Attosecond Science, Pohang 790-784, Korea; landsmanster@gmail.com
[3] Max Planck Institute for the Physics of Complex Systems, Noethnitzer Street 38, 01187 Dresden, Germany
[*] Correspondence: kimd@postech.ac.kr; Tel.: +82-054-279-2089

Academic Editor: Kiyoshi Ueda
Received: 30 March 2017; Accepted: 8 June 2017; Published: 13 June 2017

Abstract: High power attosecond (as) X-ray pulses are in great demand for ultrafast dynamics and high resolution microscopy. We numerically demonstrate the generation of a ~230 attosecond, 1.5 terawatt (TW) pulse at a photon energy of 1 keV, and a 115 attosecond, 1.2 TW pulse at a photon energy of 12.4 keV, using the realistic electron beam parameters such as those of Korean X-ray free electron laser (XFEL) in a tapered undulator configuration. To compensate the energy loss of the electron beam and maximize its radiation power, a tapering is introduced in the downstream section of the undulator. It is found that the tapering helps in not only amplifying a target radiation pulse but also suppressing the growth of satellite radiation pulses. Tapering allows one to achieve a terawatt-attosecond pulse only with a 60 m long undulator. Such an attosecond X-ray pulse is inherently synchronized to a driving optical laser pulse; hence, it is well suited for the pump-probe experiments for studying the electron dynamics in atoms, molecules, and solids on the attosecond time-scale. For the realization of these experiments, a high level of synchronization up to attosecond precision between optical laser and X-ray pulse is demanded, which can be possible by using an interferometric feedback loop.

Keywords: ultraviolet (UV); extreme ultraviolet (EUV); X-ray lasers; attosecond pulses; free-electron laser; undulator radiation

1. Introduction

The advent of X-ray free electron laser (XFEL) [1–5] sources has set a new frontier in X-ray science due to remarkable advances in its characteristics. Synchrotron sources produce insufficient brightness and picosecond-long X-ray pulses, which yield only blurred images of atoms and molecules in motion. XFELs are coherent, ultra-brilliant tunable laser pulses. The current XFEL pulse is characterized by its peak power of 10–50 GW and its pulse duration of a few to about 100 femtoseconds (fs).

Attosecond science is a new exciting frontier born with the subfemtosecond extreme ultraviolet light pulses via high harmonic generation (HHG) based on femtosecond lasers [6,7]. This field will enrich even more with the development of isolated attosecond (as) XFEL pulses. XFEL presently exceeds HHG sources [8,9], with a greater power and a shorter wavelength. It is these properties that still suggest XFEL pulse as a future light source because even a shorter pulse duration and higher power is possible [10]. By the development of such high-intensity attosecond X-ray sources, the realm of ultrafast processes that can be explored will be greatly extended. The imaging of a single molecule [11], the real time tracking of electron distribution around atoms and molecules [12,13], and the dynamical investigation of X-ray nonlinear processes [14] are a few examples among the investigations that will set milestones. Therefore, the generation of attosecond X-ray pulses has drawn attention. Such high-intensity X-ray pulses cause radiation damage by depositing energy directly into the sample. However, the estimation of radiation damage as a function of photon energy, pulse length, integrated pulse intensity, and sample size shows that experiments using very high X-ray dose rates and ultrashort exposures provide useful structural information before radiation damage destroys the sample [15,16]. Novel techniques have been proposed to further shorten the pulse length, mostly employing one or more external lasers [17–26] and generating short electron bunches [27,28].

Recently, Tanaka [29] proposed a scheme to produce high peak powers of up to the terawatt range. His design uses a combination of slotted foil [27], enhanced self-amplified spontaneous emission (SASE) [20], and optical and electron beam delay between undulator sections. In another work, Prat and Reiche [30] suggested a simpler scheme for a TW-attosecond free electron laser (FEL) in which a multiple slot foil is used to effectively divide an electron bunch into many different parts by preserving the emittance in some parts. The same authors in another work [31] show a TW-attosecond pulse by introducing a transverse tilt to an electron beam while properly delaying and correcting the trajectory of the electron beam between certain undulator modules. Kumar et al. [32] devised a new idea where only one single electron spike is used repeatedly. The radiation amplification is based on the superradiant behavior of short pulses [33,34], where the power level can significantly exceed the saturation power of an XFEL while shortening its pulse length.

In this paper, the multiple electron spike scheme [29] is adopted for realistic simulations. For this purpose, the beam parameters of the XFEL in Pohang Accelerator Laboratory (PAL-XFEL) are taken. We studied both cases of hard and soft TW-attosecond XFEL. The role of tapered undulator [35] on the output radiation pulse characteristics is also discussed for both cases. This is the first numerical demonstration for a TW, isolated, attosecond, soft X-ray pulse.

2. Scheme

The basic scheme is shown in Figure 1. A slotted foil [27] is inserted in the last bunch compressor section of the PAL-XFEL linac. The foil spoils the electron beam emittance in the head and tail parts of the bunch and suppresses lasing there. Note that the slot is set relatively wide in our scheme because its function is not to shorten the pulse length as in the original proposal but to set a defined temporal window of lasing and define the lasing domain in the electron bunch [29]. The next is the ESASE [20] section that consists of an optical laser, a modulator, and a magnetic chicane for the density modulation of an unspoiled part of the electron beam at the end of a linac section (Figure 1a). In this ESASE section, the electron beam is energy-modulated in a strong magnetic field modulator via the interaction with a co-propagating laser. The shape of the optical cycles of the pulse is replicated in the energy distribution of the electron beam. Later, a magnetic chicane converts the energy modulation to a density modulation so that the flat current profile changes into a profile of current spikes. The electron beam with these current spikes is sent to an undulator (Figure 1b) for radiation generation. The SASE undulator is made by three undulator modules (UMs). The length of the undulator is kept short to avoid saturation and to keep the electron-beam energy spread at a minimum.

Figure 1c represents a chicane–mirror system composed of four dipole magnets and a set of mirrors [36]. The magnetic chicane is used for delaying an electron beam and diluting the microbunching developed in an undulator. A set of reflecting mirrors is used to delay an X-ray pulse with respect to a particular electron spike in the electron beam. The X-ray pulses are delayed in such a way that the leading radiation spike (target pulse) acts as a seed to trailing current spikes (more description below along with Figure 2) [29]. Figure 1d shows the second, even shorter undulator (1 UM only) for a radiation amplification. This undulator is used to amplify the target pulse. For a good amplification, the radiation seed always needs a fresh electron spike. Figure 1e shows a small magnetic chicane followed by a small undulator section (1 UM only). A small chicane is used to create the delay of the electron beam by a laser wavelength λ_L to align the radiation spike to the next current spike. The same unit shown in Figure 1e is used several times in the downstream undulator for repeated electron beam delays and repeated amplifications of a target pulse to obtain an isolated TW-attosecond radiation pulse at the end.

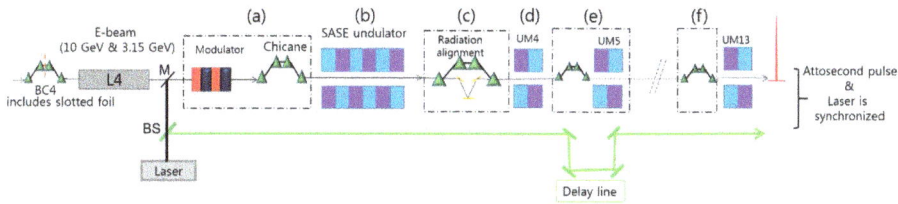

Figure 1. (Color online) Schematic layout for the terawatt (TW)-attosecond X-ray free electron laser (XFEL) [29], (**a**) ESASE section; (**b**) SASE undulator; (**c**) Chicane-mirror system; (**d**) single undulator module (UM4); (**e**) a small magnet chicane for e-beam delay followed by single undulator module (UM5); and (**f**) a small magnet chicane for e-beam delay.

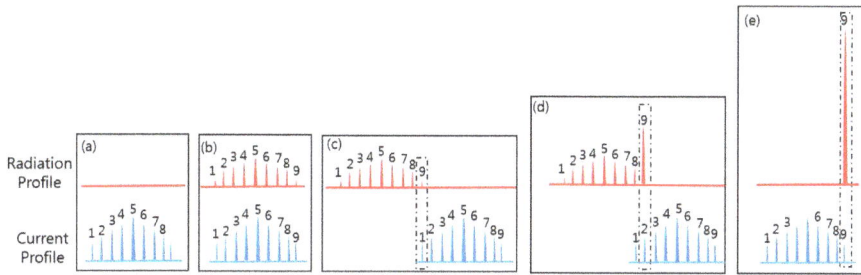

Figure 2. Working principle of the scheme: the alignment process between the current profile of electron bunch (consisting of nine blue spikes produced by ESASE in Figure 1a) and the radiation profile (consisting of nine red spikes) at different locations of Figure 1.

Figure 2 shows the working principle of the scheme. First, Figure 2a shows that a few current spikes with different magnitudes are generated by an optical laser's interaction with electron beam inside the modulator, followed by a density modulation in a magnetic chicane (see Figure 1a). The separation between the current spikes is in the order of the modulation laser wavelength λ_L. Figure 2b shows the current and radiation profile after the first stage of amplification (Figure 1b), and the same number of radiation spikes (a total of 9) are generated. Now, in Figure 2c, the 9th radiation spike is selected as a target pulse and aligned with the tail current peak (1st current peak) via a chicane–mirror system (Figure 1c). Figure 2d shows the current and radiation profile after 2 UM amplification. The electron beam is delayed by λ_L to align the 2nd current spike with the target radiation pulse for next amplification (Figure 1e). This procedure is repeated and the target pulse is aligned sequentially with all the current spikes to yield a solitary TW-attosecond radiation pulse, as shown in Figure 2e, at the undulator exit (Figure 1f).

3. Results and Discussion

We illustrate two cases of TW-attosecond XFEL: one in a hard X-ray regime (12.4 keV or 0.1 nm) and the other in the soft X-ray region (~1 keV or 1.25 nm). To consider a more realistic situation, we use the parameters of Korean XFEL [5].

3.1. Terawatt-Attosecond Hard X-ray Pulse Generation

First, an electron beam with an energy of 10 GeV, a 200 pC bunch charge, an average current of 2.0 kA, 66 fs in length with a normalized emittance of 0.5 mm-mrad, and an un-correlated energy spread of 0.01% is generated using ELEGANT code [37] by taking into account the space-charge effects and the microbunching instabilities [38]. The popular ELEGANT code has been used to track the particle dynamics in six dimensional phase space coordinates.

For the energy modulation of an electron beam (Figure 1a), an 800 nm, 30 fs, and 50 GW laser is used inside a two period modulator with a wiggler period of 50 cm; the density modulation is generated by a magnetic chicane with a time-of-flight parameter R_{56} of ~0.16 mm. The generated current distribution after these two processes is shown in Figure 3a: several current spikes with different magnitudes at an average background current of 2 kA. Figure 3b shows the corresponding plot in the longitudinal phase space after the energy and current modulation of a 10 GeV electron beam. Note that the current spikes are larger by a factor of approximately 8 compared to the background current. Note also that the window for the lasing section of the current profile can be controlled by the slotted foil system [27]. The red line in Figure 3a is the unspoiled section of the electron bunch, which is fed to UMs to generate an X-ray pulse at a photon energy of 12.4 keV. The nine current spikes in total are considered. These spikes radiate strongly in an undulator. Usually, the advantage of operating an FEL with a high peak current is the increase in X-ray output power. However, high current spikes also suffer from the energy spread, which is taken into account in this calculation. In Figure 3b, one can see that the increase in peak currents is accompanied by corresponding increases in the peak–peak energy modulation depth.

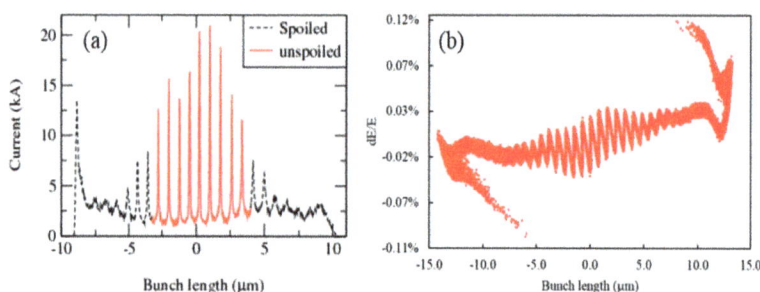

Figure 3. (a) The current modulation of a 10 GeV electron beam for hard X-ray amplification; the black-dashed line shows the spoiled section of the current profile and the red-dashed line shows the unspoiled section. (b) The longitudinal phase space plot showing energy modulation of the electrons along the electron bunch after the interaction with the optical laser (Figure 1a).

Now, the electron beam is sent to UMs to generate an X-ray pulse at a 12.4 keV photon energy. For undulator radiation, simulations are performed using three-dimensional time-dependent FEL code GENESIS [39]. Note that, due to large energy modulation, high peak current spikes suffer from the energy chirp [40] due to short-range and long-range space-charge effects. Short range space-charge effects acting against microbunching have been included in the GENESIS code [39] but the long-range space-charge effect (or longitudinal debunching effect) is overlooked in the simulation code [39]. Here, it is worth acknowledging that according to [41], longitudinal debunching causes a reduction in the peak current of the current spikes during radiation amplification and the reduction is of the order of only a few percents, not significant for high-electron beam energies (i.e., >3 GeV).

Figure 4a shows the temporal profile of the radiation pulse after the 3rd UM. The first three UMs play a role in generating initial SASE radiation. Each UM is 5 m long with an undulator period of 2.6 cm (Figure 1b). The drift space between two UMs is long enough to accommodate the magnetic chicane used for the electron beam delay. After the first amplification stage, the optical delay is adjusted before the next UM to align the target pulse (i.e., the 9th pulse in the radiation profile) with the tail current spike (i.e., the 1st current spike in the current profile) (Figure 2c). At each of the following amplification stages, the target pulse is aligned with a fresh current spike (i.e., the 2nd, 3rd, 4th, and so on) using the electron-beam delay unit. Figure 4b shows the snapshot of the radiation pulse after 8 UMs. After 8 UMs, we get one dominant radiation spike with a few weak neighboring radiation spikes. The snapshot of the radiation after 13 UMs in a uniform undulator is shown in Figure 4c. After the 13th UM, we obtain a 0.55 TW and 130 attosecond radiation pulse at a photon energy of 12.4 keV in a 75 m long uniform undulator. This power corresponds to 0.1 mJ energy (or 5×10^{10} photons) per pulse. As the amplification progresses, the resonant condition is gradually broken. The undulator strength parameter K may need to be tapered to maintain the resonance condition as the electron beam loses its energy [35]. Initially, up to 47.4 m long, the energy loss was not severe. Thus, we considered tapering in the last part of the undulator. For each undulator module, K value was optimized to maximize the power. Figure 4d shows the result of tapering optimization: the radiation profile after 13 UMs in a tapered undulator. The radiation power was doubled, up to 1.2 TW, by applying the tapering from UM9 to UM13, within a 75 m long undulator without any other modification.

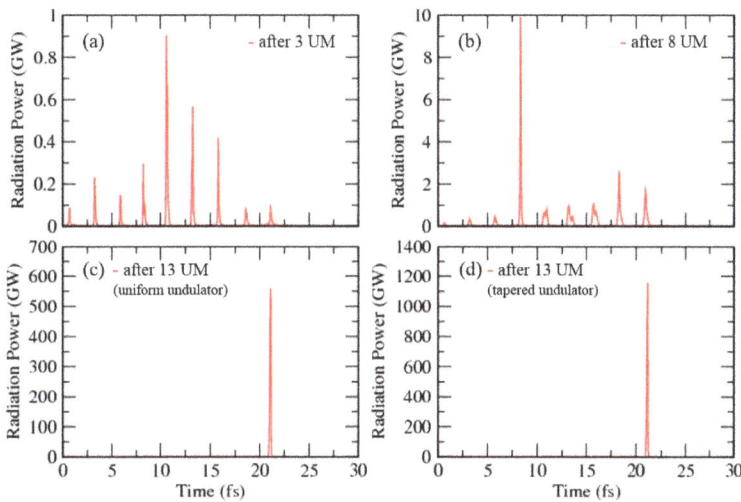

Figure 4. The radiation amplification along the undulator for hard X-ray case. Snapshot of the radiation pulse (**a**) after 3 UMs; (**b**) after 8 UMs; (**c**) after 13 UMs in a uniform undulator; and (**d**) after 13 UMs when a tapering is considered in UM9-UM13. A single isolated 130 attosecond FWHM, 1.2 TW radiation pulse is obtained in a tapered undulator.

High peak currents suffer from the energy spread and the major energy loss occurs at the current spikes to which the target radiation spike is seeded. By choosing a tapering to compensate the degradation of the seeded current spike, we preserve the resonance condition in that region and partially suppress the amplification at the neighboring current spikes due to the off-resonance condition. The ideal tapering profile is a smoothly varying function of undulator length. However, a stepwise undulator tapering is considered here along UM9-UM13 modules, maximizing the output power. The stepwise tapering is shown in Figure 5a. First, UM1-UM8 undulator modules have the same K parameter K_0. The modules UM9-UM13 belongs to the exponential regime of radiation amplification. Therefore, K is adjusted to decrease sharply along the undulator. The undulator parameter K_n of the nth undulator module can be described as

$$K_n = \begin{cases} K_0, \ n \leq k \\ K_0 - a(n-k) - b(n-k)^2, \ n > k \end{cases} \tag{1}$$

where K_0 is the initial undulator parameter, k is the segment number at which the tapering starts, and a and b are the coefficients obtained by multidimensional scans that maximize the radiation power. The optimal values for hard X-ray case are plotted in Figure 5a with $K_0 = 1.9728$ and $k = 8$.

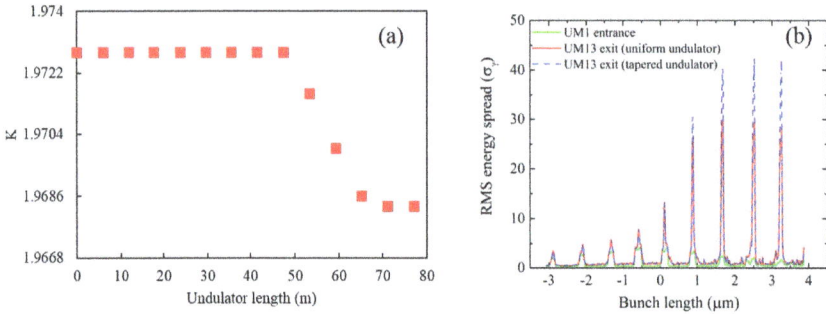

Figure 5. (**a**) Undulator parameter K for a stepwise tapering, and (**b**) the electron beam energy spread at the undulator entrance and exit for hard X-ray case.

Figure 5b shows the growth of the electron-beam energy spread. The green curve shows the initial energy spread due to the current modulation by the ESASE section prior to the undulator entrance. The red curve shows the growth of the energy spread after 13 UMs in the uniform undulator case and the blue-dotted curve in the tapered undulator case. One can see that some of the current spikes are still reusable due to the small energy spread. Therefore, the radiation power can be further increased if more units are added.

For the robustness of the scheme, the shot noise effect on the X-ray pulse peak power and the pulse duration have been tested. The simulation was performed for 10 different seed values. Figure 6a shows the average power for uniform and tapered configuration. For the tapered undulator system, the average power after 13 UM stage amplification is 0.9 ± 0.2 TW (red curve), while the average pulse duration is 100 ± 15 attosecond FWHM. For the uniform undulator system, the average power is restricted to a maximum of 0.42 ± 0.1 TW.

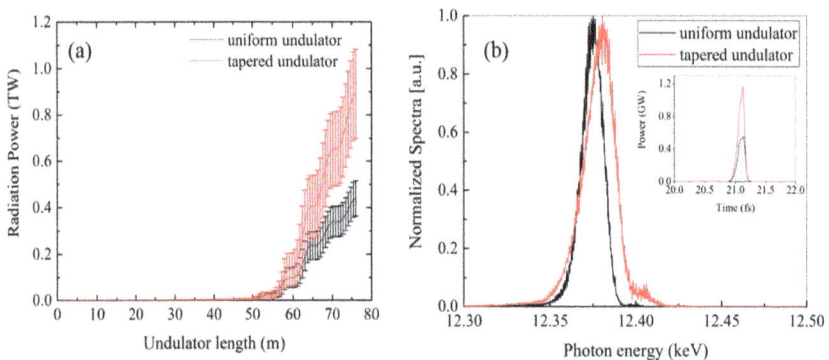

Figure 6. (**a**) The average radiation power and the power fluctuations for 10 different random seeds for a uniform and a tapered undulator. (**b**) Power spectrum of the radiation output for hard X-ray case (inset shows the temporal profiles of the radiation powers).

Figure 6b shows the clean radiation spectrum for both the uniform (black line) and the tapered undulator (red line). For the uniform undulator, the FWHM pulse width is 130 as (inset) and the spectrum bandwidth is 15 eV wide. From the frequency–time bandwidth relation, $\Delta\nu\Delta t = 0.441$, the bandwidth turns out to be 14 eV wide. In case of tapered undulator, the pulse width is 115 as, and the spectrum width is 20 eV wide. Using $\Delta\nu\Delta t = 0.441$, we obtain a bandwidth 15.8 eV wide. These estimates indicate that there is a chirp induced in the amplification. The chirp is larger in the tapered case than in the uniform case. The radiation spectrum of the tapered undulator is slightly broader compared to that of the uniform undulator. This spectral bandwidth increases due to the sideband growth in the tapered undulator case.

3.2. Terawatt-Attosecond Soft X-ray Pulse Generation Tables and Schemes

The generation of attosecond soft X-ray XFEL pulse has also been investigated. In the soft X-ray beamline of PAL-XFEL, the electron beam has an energy of 3.15 GeV with a bunch charge of 200 pC and an average current of 2.0 kilo-ampere (kA). The electron-bunch is 66 fs long with a normalized emittance of 0.3 mm-mrad and an uncorrelated energy spread of 0.01%.

A series of simulations similar to those for the hard X-ray XFEL has been carried out. Figure 7a shows the current distribution after the modulator-chicane section. The black-dotted line represents the spoiled part by a slotted foil, and the central part (red line) shows the unspoiled part of the electron beam, which is chosen for radiation amplification. To generate this comb-like current profile, an optical laser at 1200 nm with a 65 fs pulse duration and 26 GW power is used for energy modulation in electron beam using a two-period modulator of 50 cm wiggler period. For the current modulation, a magnetic chicane of R_{56} ~0.11 mm is used. In total, 13 UMs are used for amplification. Each UM is 4.9 m long with a 3.4 cm undulator period. Figure 7b shows the temporal profile of the output radiation power after the first 3 UMs. The first three UMs without the delay of the electron beam are optimized, at the expense of radiation power, for minimum energy spread and a short pulse length. Figure 7c shows a solitary radiation pulse of 0.75 TW and 258 attosecond pulse duration in a 60 m long uniform undulator that includes 13 UMs and the drift sections. The radiation power can be increased up to 1.3 TW in a tapered undulator and to a 230 attosecond pulse duration, as shown in Figure 7d. This power corresponds to 0.3 mJ of energy (or 1×10^{12} photons) per pulse.

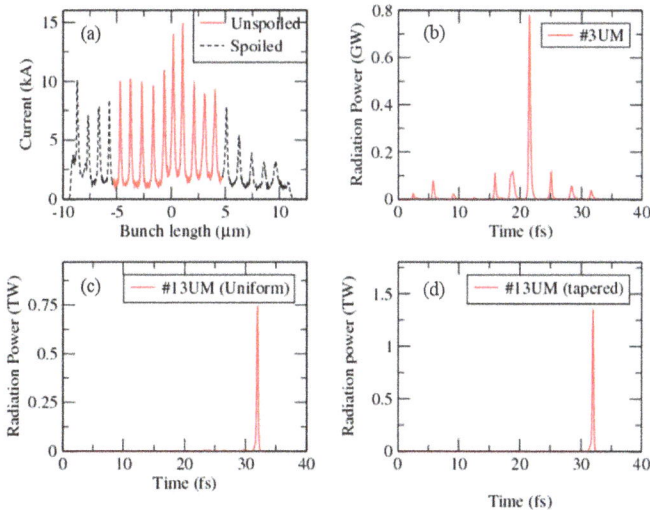

Figure 7. (**a**) Current modulation of an electron beam of 3.15 GeV energy. The dashed-black line shows the spoiled section, and the red line is the unspoiled section of the current profile. The snapshot of the radiation pulse amplification (**b**) after 3 UMs; (**c**) after 13 UMs in a uniform undulator; and (**d**) after 13 UMs in a tapered undulator.

A stepwise tapering (Equation (1)) is applied along UM7-UM13 modules of the soft X-ray undulator to maximize the radiation power. Figure 8a shows the tapering of K parameter applied along the undulator in the soft X-ray case. Optimal values for the soft X-ray case are as follows: $K_0 = 1.8940$, k = 6, and a and b are optimized by multi scans to maximize the output radiation power. Figure 8b shows the electron-beam energy spread at various places: the black line at the entrance of the undulator, the blue-dotted line at the undulator exit in the uniform undulator, and the red line at the undulator exit in the tapered undulator.

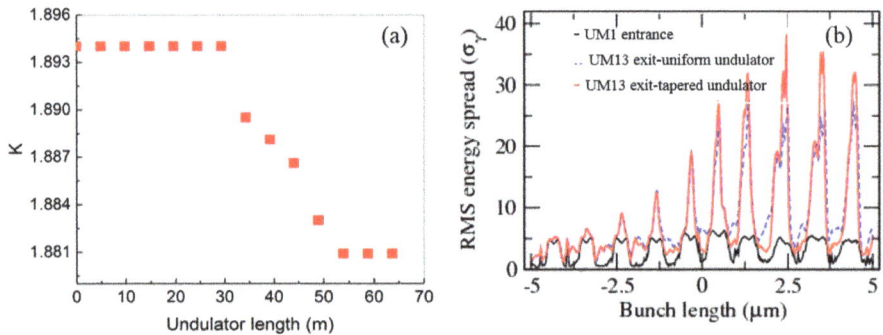

Figure 8. (**a**) Stepwise tapering in the undulator parameter K for the attosecond-TW XFEL in soft X-ray region, and (**b**) variation of the electron beam energy spread along the undulator.

The effect of the electron beam's shot noise on the peak power of soft X-ray pulse and the pulse length is tested. The simulation was performed for 10 different random seeds, whose results are shown in Figure 9a. The black curve with errors shows the radiation power for 10 different random seeds in a uniform undulator. The average power (black curve) after 13 UM stage amplification is 0.8 ± 0.1 TW, and the average pulse duration is 260 ± 20 attosecond FWHM. Simulations in a tapered undulator show that the average power (red curve and errors) after 13 UM stage amplification is 1.4 ± 0.2 TW, while the average pulse duration is 230 ± 20 attosecond FWHM. From Figure 9a, it is clear that the radiation growth is not saturated even after 13 UMs. Therefore, the radiation power can be further increased if more UMs are added. Figure 9b shows the radiation spectra for a uniform (black curve) and a tapered (red curve) undulator. Note that the radiation spectrum in case of a tapered undulator (red curve) is broader and shows one additional mode due to the tapering of the K parameter.

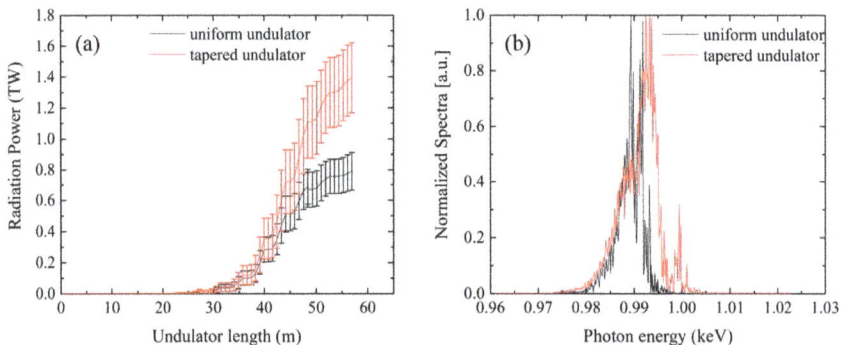

Figure 9. (**a**) The average radiation power for 10 different random seeds along the uniform undulator (black line) and along the tapered undulator (red line) in the soft X-ray case, and (**b**) the power spectra for the uniform undulator (black line) and for the tapered undulator (red line).

The realization of X-ray delay is important in this scheme. A soft X-ray mirror typically suffers from transmission losses while hard X-ray Bragg reflectors are only narrow-band, which cannot be applied to this scheme due to the broad-band nature of the radiation spikes [42]. Hence, the design of an X-ray delay unit using grazing incidence geometry has been discussed in detail in [32]. According to [32], a chicane–mirror system using mirrors with a length of 5~6 cm and a deflection angle of 0.1° in grazing incidence geometry would be suitable for obtaining the required optical delay. A similar X-ray delay unit would also be applicable in soft X-ray cases. Moreover, the mirror stages with a 10 nm resolution are commercially available. Therefore, the delay can be controlled with a high degree of precision.

For meaningful pump-probe experiments at the XFEL end stations, the synchronization of the X-ray pulse and optical laser has to be sustained over the full legnth of the experimental configuration including the delay chicanes and other transport optics. Therefore, it is required that all electron beam and optical components are placed onto the same support system. The mechanical vibration leads to the instability in optical path lengths. However, using a feedback loop [43], such an instability can be suppressed and the time delay between pump and probe pulses can be controlled within 20 as RMS. Optical transport of ultrashort pulses generated by optical laser from its optical table to the experimental stations, which are ~100 m away, requires ultra-high vacuum transport system. S. Schulz et al. [44] and P. Cinquegrana et al. [45] have demonstrated that a carefully designed laser beam transport, including a free space propagation of 150 m and a number of beam folding mirrors allows one to keep the timing fluctuations to less than 4 femtoseconds RMS.

4. Conclusions

We have carried out simulations for the hard X-ray beamline (10 GeV e-beam) and for the soft X-ray beamline (3.15 GeV e-beam) of Korean XFEL to assess the performance of the multi-electron spike scheme. The simulation results indicate that an isolated attosecond radiation pulse of 0.6 TW power and a 130 attosecond duration can be achieved at a photon energy of 12.4 keV in a ~75 m long uniform undulator, which can be further scaled up to 1 TW and 115 attosecond duration in a 75 m long tapered undulator. Similarly, in the soft X-ray case, a radiation pulse of 0.75 TW power and 258 attosecond duration can be achieved at a photon energy of ~1 keV (1.25 nm) in a ~60 m long uniform undulator, which can be further enhanced up to 1.35 TW and a duration of 230 attosecond via tapering to a few downstream undulator sections within a 60 m long undulator. We noticed that tapering helps in suppressing the amplification of satellite pulses. Only a target pulse (seed pulse) is amplified due to on-resonance amplification. Moreover, the undulator tapering allows one to achieve a two-fold increase in the peak power within the same undulator length. Additional work on the careful optimization of electron beams, lasing domains, and ESASE parameters may further improve the performance of this scheme in SASE FELs. Such high power and short X-ray pulses will be useful in such research fields as bioimaging, mostly for single-molecule imaging [11], as well as ultrafast science.

Acknowledgments: This research has been supported in part by the Global Research Laboratory Program [Grant No. 2009-00439], by the Max Planck POSTECH/KOREA Research Initiative Program [2016K1A4A4A01922028], through the National Research Foundation of Korea (NRF) funded by Ministry of Science, ICT Future Planning, and by the Nuclear Research Foundation of Korea (NRF) grant funded by the Korea government (MEST) (No. 2012027506). Alexandra S. Landsman acknowledges the support of the Max Planck Center for Attosecond Science (MPC-AS).

Author Contributions: S.K. and D.E.K. conceived and designed the simulation method; S.K. performed the simulations and analyzed the data; S.K., A.S.L. and D.E.K. wrote the paper.

Conflicts of Interest: The authors declare no conflict of interest.

References

1. Emma, P.; Akre, R.; Arthur, J.; Bionta, R. First lasing and operation of an angstrom wavelength free-electron laser. *Nat. Photonics* **2010**, *4*, 641–647. [CrossRef]

2. Altarelli, M.; Brinkmann, R.; Chergui, M.; Decking, W. *DESY Report*; The European X-ray Free Electron Laser: Schenefeld, Germany, 2006.
3. Ishikawa, T.; Aoyagi, H.; Asaka, T.; Asano, Y.A. Compact X-ray free-electron laser emitting in the sub-angstrom region. *Nat. Photonics* **2012**, *6*, 540–544. [CrossRef]
4. Amann, J.; Berg, W.; Blank, V.; Decker, F.J. Demonstration of self-seeding in a Hard-X-ray free-electron laser. *Nat. Photonics* **2012**, *6*, 693–698. [CrossRef]
5. Ko, I.S.; Han, J.H. Current status of PAL-XFEL. In Proceedings of the 2014, 27th Linear Accelerator Conference, Geneva, Switzerland, 31 August–5 September 2014.
6. Paul, P.M.; Toma, E.S.; Breger, P.; Mullot, G. Observation of a train of attosecond pulses from high harmonic generation. *Science* **2001**, *292*, 1689–1692. [CrossRef] [PubMed]
7. Hentschel, M.; Kienberger, R.; Spielmann, C.H.; Reider, G.A. Attosecond Metrology. *Nature* **2001**, *414*, 509–513. [CrossRef] [PubMed]
8. Krausz, F.; Ivanov, M. Attosecond Physics. *Rev. Mod. Phys.* **2009**, *81*, 163–234. [CrossRef]
9. Corkum, P.B.; Krausz, F. Attosecond Science. *Nat. Phys.* **2007**, *3*, 381. [CrossRef]
10. Dunning, D.J.; McNeil, B.W.J.; Thompson, N.R. Towards Zeptosecond-scale pulses from X-ray free-electron lasers. *Phys. Procedia* **2014**, *52*, 62–67. [CrossRef]
11. Fratalocchi, A.; Ruocco, G. Single-molecules imaging with X-ray free-electron lasers: Dream or reality. *Phys. Rev. Lett.* **2011**, *106*, 105504. [CrossRef] [PubMed]
12. Goulielmakis, E.; Loh, Z.H.; Wirth, A.; Santra, R. Real-time observation of valence electron motion. *Nature* **2010**, *466*, 739–743. [CrossRef] [PubMed]
13. Geiseler, H.; Rottke, H.; Zhavoronkov, N.; Sandner, W. Real-Time observation of interference between atomic one-electron and two-electron Excittaions. *Phys. Rev. Lett.* **2012**, *108*, 123601. [CrossRef] [PubMed]
14. Fuchs, M.; Trigo, M.; Chen, J.; Ghimire, S. Anomalous nonlinear X-ray Compton scattering. *Nat. Phys.* **2015**, *11*, 964–970. [CrossRef]
15. Neutze, R.; Wouts, R.; Spoel, D.V.D.; Weckert, E. Potential for biomolecular imaging with femtosecond X-ray pulses. *Nature* **2000**, *406*, 752–757. [CrossRef] [PubMed]
16. Inoue, I.; Inubushi, Y.; Sato, T.; Tono, K. Observation of femtosecond X-ray Interactions with matter using an X-ray-X-ray pump-probe scheme. *Proc. Natl. Acad Sci. USA* **2016**, *113*, 1492–1497. [CrossRef] [PubMed]
17. Saldin, E.; Schneidmiller, A.; Yurkov, M.V. A new technique to generate 100 GW-level attosecond X-ray pulses from the X-ray SASE FELs. *Opt. Commun.* **2004**, *239*, 161–172. [CrossRef]
18. Zholents, A.A.; Fawley, W.M. Proposal for intense attosecond radiation from an X-ray free-electron laser. *Phys. Rev. Lett.* **2004**, *92*, 224801. [CrossRef] [PubMed]
19. Saldin, E.L.; Schneidmiller, E.A.; Yurkov, M.V. Self-amplified spontaneous emission FEL with energy-chirped electron beam and its application for generation of attosecond X-ray pulses. *Phys. Rev. ST Accel. Beams* **2006**, *9*, 050702. [CrossRef]
20. Zholents, A.A. Method of an enhanced self-amplified spontaneous emission for X-ray free electron lasers. *Phys. Rev. ST Accel. Beams* **2005**, *8*, 040701. [CrossRef]
21. Zholents, A.A.; Zolotorev, M.S. Attosecond X-ray pulses produced by ultra-short transverse slicing via laser electron beam interaction. *New J. Phys.* **2008**, *10*, 025005. [CrossRef]
22. Xiang, D.; Huang, Z.; Stupakov, G. Generation of intense attosecond X-ray pulses using ultraviolet laser induced microbunching in electron beams. *Phys. Rev. ST Accel. Beams* **2009**, *12*, 060701. [CrossRef]
23. Ding, Y.; Huang, Z.; Ratner, D.; Bucksbaum, P.; Merdji, H. Generation of attosecond X-ray pulses with a multicycle two-color enhanced self-amplified spontaneous emission scheme. *Phys. Rev. ST Accel. Beams* **2009**, *12*, 060703. [CrossRef]
24. Kumar, S.; Kang, H.S.; Kim, D.E. Generation of isolated attosecond hard X-ray pulse in enhanced self-amplified spontaneous emission scheme. *Opt. Express* **2011**, *19*, 7537. [CrossRef] [PubMed]
25. Kumar, S.; Kang, H.S.; Kim, D.E. Tailoring the amplification of attosecond pulse through detuned X-ray FEL undulator. *Opt. Express* **2015**, *23*, 2808. [CrossRef] [PubMed]
26. Chung, S.Y.; Yoon, M.; Kim, D.E. Generation of attosecond X-ray and gamma-ray via Compton backscattering. *Opt. Express* **2009**, *17*, 7853–7861. [CrossRef] [PubMed]
27. Emma, P.; Bane, K.; Cornacchia, M.; Huang, Z. Femtosecond and subfemtosecond X-ray pulses from a self-amplified spontaneous emission-based free electron laser. *Phys. Rev. Lett.* **2004**, *92*, 074801. [CrossRef] [PubMed]

Appl. Sci. **2017**, *7*, 614

28. Reiche, S.; Musumeci, P.; Pellegrini, C.; Rosenzweig, J.B. Development of ultra-short pulse, single coherent spike for SASE X-ray FELs. *Nucl. Instrum. Methods Phys. Res. Sect. A* **2008**, *593*, 45. [CrossRef]
29. Tanaka, T. Proposal for a Pulse-compression scheme in X-ray free electron lasers to generate a multiterawatt, attosecond X-ray pulse. *Phys. Rev. Lett.* **2013**, *110*, 084801. [CrossRef] [PubMed]
30. Prat, E.; Reiche, S. Simple method to generate terawatt-attosecond X-ray free electron laser pulses. *Phys. Rev. Lett.* **2015**, *114*, 244801. [CrossRef] [PubMed]
31. Prat, E.; Lohl, F.; Reiche, S. Efficient generation of short and high power X-ray free electron laser pulses based on superradiance with a transversely tilted beam. *Phys. Rev. ST Accel. Beams* **2015**, *18*, 100701. [CrossRef]
32. Kumar, S.; Parc, Y.W.; Landsman, A.S.; Kim, D.E. Temporally-coherent terawatt attosecond XFEL synchronized with a few cycle laser. *Sci. Rep.* **2016**, *6*, 37700. [CrossRef] [PubMed]
33. Bonifacio, R.; Souza, L.D.S.; Pierini, P.; Piovella, N. The superradiant regime of an FEL: Analytical and numerical results. *Nucl. Instrum. Methods Phys. Res. Sect. A* **1990**, *296*, 358–367. [CrossRef]
34. Bonifacio, R.; Piovella, N.; McNeil, B.W.J. Superradiant evolution of radiation pulses in a free electron laser. *Phys. Rev. A* **1991**, *44*, R3441. [CrossRef] [PubMed]
35. Kroll, N.M.; Morton, P.L.; Rosenbluth, M.N. Free-electron lasers with variable parameter wigglers. *IEEE J. Quant. Electron.* **1981**, *17*, 1436–1468. [CrossRef]
36. Geloni, G.; Kocharyan, V.; Saldin, E. A simple method for controlling the line width of SASE X-ray FELs. In *DESY Report*; No. 10-004; Cornell University Library: Ithaca, NY, USA, 2010.
37. Borland, M. *Elegant: A Flexible SDDS-Compliant Code for Accelerator Simulation*; Report No. LS-287; Cornell University Library: Ithaca, NY, USA, 2000; pp. 1–11.
38. Lee, J.H.; Han, J.H.; Lee, S.; Hong, J.; Kim, C.H.; Min, C.K.; Ko, I.S. PAL-XFEL laser heater commissioning. *Nucl. Instrum. Methods Phys. Res. Sect. A* **2017**, *843*, 39. [CrossRef]
39. Reiche, S. GENESIS 1.3: A fully 3D time-dependent FEL simulation code. *Nucl. Instrum. Methods Phys. Res. Sect. A* **1999**, *429*, 243–248. [CrossRef]
40. Geloni, G.; Saldin, E.; Schneidmiller, E.; Yurkov, M. Longitudinal impedence and wake from XFEL undulators. Impact on current-enhanced SASE schemes. *Nucl. Instrum. Methods Phys. Res. Sect. A* **2007**, *583*, 228. [CrossRef]
41. Gruner, F.J.; Schroeder, C.B.; Maier, A.R.; Becker, S.; Mikhailova, J.M. Space-charge effects in ultrahigh electron bunches generated by laser-plasma accelerators. *Phys. Rev. ST Accel. Beams* **2009**, *12*, 020701. [CrossRef]
42. Feldhaus, J.; Saldin, E.L.; Schneider, J.R.; Schneidmiller, E.A.; Yurkov, M.V. Possible application of X-ray optical elements for reducing the spectral bandwidth of an X-ray SASE FEL. *Opt. Commun.* **1997**, *140*, 341. [CrossRef]
43. Chini, M.; Mashiko, H.; Wang, H.; Chen, S. Delay control in attosecond pump-probe experiments. *Opt. Express* **2009**, *17*, 21459–21464. [CrossRef] [PubMed]
44. Schulz, S.; Grguras, I.; Behrens, C.; Bromberger, H. Femtosecond all optical synchronization of an X-ray free electron laser. *Nat. Commun.* **2015**, *6*, 5938. [CrossRef] [PubMed]
45. Cinquegrana, P.; Cleva, S.; Demidovich, A.; Gaio, G. Optical beam transport to a remote location for low jitter pump-probe experiments with a free electron laser. *Phys. Rev. ST Accel. Beams* **2014**, *17*, 040702. [CrossRef]

applied
sciences

MDPI

Review

Molecular Dynamics of XFEL-Induced Photo-Dissociation, Revealed by Ion-Ion Coincidence Measurements

Edwin Kukk [1,*], Koji Motomura [2], Hironobu Fukuzawa [2,4], Kiyonobu Nagaya [3,4] and Kiyoshi Ueda [2,4]

[1]   Department of Physics and Astronomy, University of Turku, FI-20014 Turku, Finland
[2]   Institute of Multidisciplinary Research for Advanced Materials, Tohoku University, Sendai 980-8577, Japan; motomura@tagen.tohoku.ac.jp (K.M.); fukuzawa@tagen.tohoku.ac.jp (H.F.); ueda@tagen.tohoku.ac.jp (K.U.)
[3]   Department of Physics, Graduate School of Science, Kyoto University, Kyoto 606-8502, Japan; nagaya@scphys.kyoto-u.ac.jp
[4]   RIKEN Spring-8 Center, Sayo, Hyogo 679-5148, Japan
*   Correspondence: edwin.kukk@utu.fi

Academic Editor: Malte C. Kaluza
Received: 24 March 2017; Accepted: 12 May 2017; Published: 19 May 2017

Abstract: X-ray free electron lasers (XFELs) providing ultrashort intense pulses of X-rays have proven to be excellent tools to investigate the dynamics of radiation-induced dissociation and charge redistribution in molecules and nanoparticles. Coincidence techniques, in particular multi-ion time-of-flight (TOF) coincident experiments, can provide detailed information on the photoabsorption, charge generation, and Coulomb explosion events. Here we review several such recent experiments performed at the SPring-8 Angstrom Compact free electron LAser (SACLA) facility in Japan, with iodomethane, diiodomethane, and 5-iodouracil as targets. We demonstrate how to utilize the momentum-resolving capabilities of the ion TOF spectrometers to resolve and filter the coincidence data and extract various information essential in understanding the time evolution of the processes induced by the XFEL pulses.

Keywords: free electron laser; Coulomb explosion; radiation damage; molecular dynamics; X-ray absorption; ultrafast dissociation; coincidence; photoion-photoion coincidence (PIPICO); ion mass spectroscopy; time-of-flight

---

1. Introduction

Nuclear motion in molecules and clusters takes place in a timescale starting from a few femtoseconds. Photoinduced nuclear dynamics—isomerization, bond breakage and dissociation, Coulomb explosion of molecules and clusters, chemical reactions—have long been of particular interest from both the fundamental and applied points of view [1–3]. In order to accurately observe such motion and trace its development, a sufficiently short initiating pulse of radiation is needed as not to obscure the real dynamics. Lasers have provided such an excitation source. The advent of free electron lasers (FELs) operating in the UV- and X-ray regime, however, opened up completely new avenues, providing access to tunable atomic inner-shell multiphoton excitation and ionization in femtosecond timescales [4–11]. In the present paper, we focus on the unimolecular dissociation reactions induced by intense femtosecond-range X-ray pulses of FEL radiation. Specifically, we present examples on how to extract detailed information about the nuclear, and also the related electron dynamics by using coincident ion spectroscopy and momentum filtering analysis methods. The full results of these experiments and their interpretation are published or submitted for publication elsewhere, in [12] for CH_3I, in [13] for CH_2I_2, and in [14,15] for 5-iodouracil.

In order to reconstruct the dissociation process, as much information as possible should be gathered from each dissociation event; ideally this means detecting all atomic and molecular fragments (of all charges), as well as emitted electrons, and measuring their momenta. In practice, it would be a formidable task for all but the simplest systems and processes. Various experimental set-ups, as well as analysis methods, have been developed towards this goal. Most usable at FEL sources so far have been electron and ion time-of-flight (TOF) spectrometers and their various developments, such as COLTRIMS for ion-electron coincidences [16,17], and magnetic bottle spectrometers for electrons [18–23].

The results presented in this article are based on experiments performed at SACLA XFEL in Japan, using momentum-resolving ion TOF spectrometers with multi-ion detection capabilities per FEL pulse [24]. The multi-ion detection offers the opportunity for performing *coincidence* (specifically ion-ion or multi-ion coincidence) analysis. In coincidence methods, each event is analyzed separately, tracing back the detected ions to a single source—an ionized and subsequently dissociating molecule, cluster, or nanoparticle. By its nature, it is one of the most accurate approaches in experimentally reconstructing the quantum mechanical event [25–29].

Coincidence experiments at FEL sources also face major challenges. Firstly, the repetition rates of FEL sources have been relatively low, even as low as 10 Hz for SACLA, which means that collecting sufficient coincidence data is a lengthy process during the extremely valuable beamtime (once referred to as a "heroic experiment" by a referee). This limitation is gradually becoming less severe with, for example, the Linac Coherent Light Source (LCLS) operating at 120 Hz and the European XFEL shortly starting operations at the planned 27,000 pulses per second. The second limitation is more fundamental, namely that for efficient coincidence analysis it must be ensured that all detected fragment ions indeed originate from the same event. Otherwise, the combinations of ions from different sources would be meaningless and create erroneous data, often referred to as "false coincidences". It can be a particularly severe problem at FEL sources, since the pulses have very high photon density (often desirable for a specific process to occur), creating many ionization events per pulse while, for a coincidence experiment, the ideal would be a single ionized particle per pulse. As a simplified example, in an (N^+, N^+) coincidence analysis of the photodissociation of N_2, only 20% are proper ("true") coincidences if three molecules are ionized by the FEL pulse; and only 2% if 25 molecules are ionized. In the latter regime (which is not uncommon in a FEL experiment), coincidence analysis would be overwhelmed by the false coincidences, without additional means of distinguishing them from the true ones.

One solution to this predicament is to perform *covariance* analysis, in which statistically significant relationships between various detected particles are looked for, not on an event-by event basis, but by using statistical analysis methods [30–34]. Covariance analysis has been shown to extract useful information even from datasets with high ionization rates per pulse. The second solution is to utilize additional information in sorting the raw data into coincidence events. Modern ion TOF spectrometers equipped with accurate position-sensitive detectors are capable of recording the data for complete ion momentum reconstruction. On the same example of N_2 above, one can confidently assume that the true coincident ions should have the momentum vectors antiparallel and of equal length, as to satisfy the momentum conservation. That condition can be used as a very efficient filter, dramatically increasing the true/false coincidence ratio and making a coincidence analysis possible at seemingly too high ionization rates. Naturally, the above example is trivial, but similar physically justified momentum filtering (hereafter referred to as MF) conditions can often also be found in significantly larger systems.

In the present paper, we present three systems of increasing complexity (Figure 1)—iodomethane (CH_3I) possessing a three-fold symmetry axis, diiodomethane (CH_2I_2) with two symmetry planes, and a planar molecule 5-iodouracil ($C_4H_3IN_2O_2$). We show how multi-ion coincidence analysis can be applied in combination with the MF as a powerful tool for extracting information on femtosecond-scale molecular dynamics.

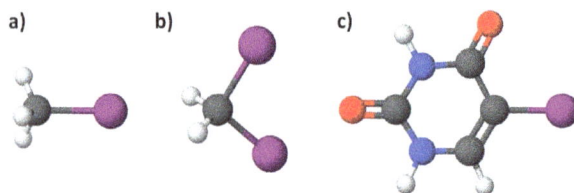

Figure 1. The molecules studied: (**a**) iodomethane, (**b**) diiodomethane, and (**c**) 5-iodouracil. Pink—iodine, blue—nitrogen, red—oxygen, gray—carbon, and light gray—hydrogen.

2. Materials and Methods

The experiments were carried out at beam line 3 (BL3) of the SACLA X-ray free electron laser (XFEL) facility [35] and are described in detail in [12–15]. The XFEL beam is focused by the Kirkpatrick-Baez (KB) mirror system to a focal size of about 1 μm (FWHM) diameter. The photon energy was set to 5.5 keV and the photon bandwidth was about 40 eV (FWHM). The repetition rate of the XFEL pulses was 10 (for CH_3I and 5-iodouracil) or 30 Hz (for CH_2I_2). The pulse duration was estimated to be about 10 fs. The peak fluence was estimated to be 26 μJ/μm² in the CH_3I and 5-iodouracil experiment, and 26 μJ/μm² in the CH_2I_2 experiment.

The gas phase samples were introduced to the focal point of the XFEL pulses as a pulsed supersonic gas jet seeded by helium gas. The ions were detected by a multi-coincidence recoil ion momentum spectrometer to measure three-dimensional momenta of each fragment ion (see Figure 2).

The molecular beam was crossed with the focused XFEL beam at the reaction point and the emitted ions were projected by electric fields onto a microchannel plate (MCP) detector in front of a delay-line anode (Roentdek HEX80, by RoentDek Handels GmbH, Kelkheim, Germany). We used velocity-map-imaging (VMI) electric field conditions. Signals from the delay-line anode and MCP were recorded by a digitizer and analyzed by a software discriminator. The arrival time and the arrival position of each ion were determined and allowed us to extract the three-dimensional momentum of each ion.

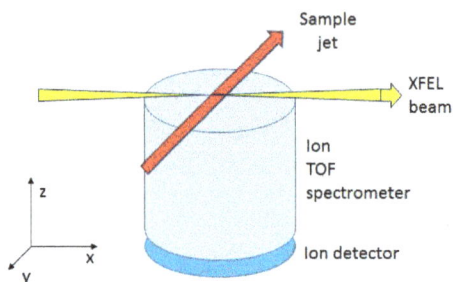

Figure 2. Sketch of the experimental setup at SPring-8 Angstrom Compact free electron LAser (SACLA).

An essential element in interpreting the observations in all three cases was a comparison with a rather simple Coulomb explosion model, predicting the fragments' momentum vector correlations, magnitudes and geometry. Such quick, empirical models proved an invaluable aid in data analysis and we, thus, summarize its main properties. The classical point charge model assumes the source of the total charge creation to be the heavy (in this case iodine) atom, due to its much higher photoionization cross-section at around 5 keV. The charge build-up is described statistically as a smooth function:

$$Q_{tot}(t) = Q_{tot,final}\left(1 - \exp\left(-\frac{t}{\tau}\right)\right),$$ (1)

where τ is the empirical, adjustable charge build-up constant. In comparison with the experiments, we found τ close to 10 fs in all cases. The positive charge is multiplied by additional absorption and by Auger cascades, transferring charge from iodine (Q_I) to the rest of the molecule (Q_X):

$$\frac{dQ_X(t)}{dt} = R \times Q_I(t), \tag{2}$$

where R is the charge transfer rate—another empirical constant in the model. The fragment momenta and their correlations is a sensitive function of the charge dynamics and the above-given model parameters.

3. Results and Discussion

3.1. Iodomethane

CH$_3$I molecules were irradiated with ultrashort ($\sim$ 10 fs) XFEL pulses of 5.5 keV energy. The experimental results, as well as the Coulomb explosion modeling, was published in [12]; here we summarize them as far as necessary for the coincidence analysis' background.

The molecules absorb X-rays predominantly in the iodine atomic site, where electrons from inner shells down to I 2p are emitted. Multiphoton absorption occurs during the pulse and the charge is also further increased by Auger cascades, as demonstrated in the case of xenon—an atom with a very similar electronic structure [36,37]. The Auger cascades are also the most likely mechanism of transferring charge to the methyl group of the molecule. The highly-charged molecule undergoes Coulomb explosion, producing the I^{n+}, C^{m+}, and $H^{0,1+}$ fragments. Ion-ion coincidence analysis of these fragments was detected by the momentum-resolving ion TOF spectrometer, which then provided the data to reconstruct the dynamics of the dissociation events.

The ion count per XFEL pulse was quite high, with the average of 7.25 heavy (excluding H$^+$) ions/pulse. Assuming an ion collection and detection efficiency of roughly 50% means that in average of about 7 molecules/pulse were ionized; quite unfavorable conditions for coincidence analysis. Under such conditions, the most informative approach was, mainly, to perform a two-particle ion-ion coincidence (PIPICO) analysis with the help of MF.

A possible pitfall in the MF-enhanced PIPICO lies in carrying over the filtering conditions to the result and erroneously interpreting the apparent correlations as coincidence information, while they were just created by the filtering process. To avoid this, we used several strategies. First, the MF can be performed using only two momentum vector components on the detector plane perpendicular to the TOF spectrometer's axis (See Figure 2). That leaves the third axial component undisturbed and, if a clear PIPICO pattern emerges in the first ion's TOF versus the second ion's TOF plot, it is a confirmation that true coincidence events were found. Second, the measured events can be "scrambled" by redistributing the individual ions between the events. Scrambling ensures that there are no true coincident ion pairs left in the data and, thus, all remaining apparent correlations, if any, must be assigned to the MF action and discarded.

We first present an example of the analysis of a coincident pair (I^{5+}, C^{2+}) that has the best counting statistics. At first, ion pairs are formed from all events with no filtering nor bias—if, for example, one iodine and two carbon ions are detected in a pulse, a pair is formed by randomly choosing one of the carbons. A momentum sum plot is then generated and is shown in Figure 3a, where the direction of the iodine's momentum is always pointing along the x-axis. It is seen from Figure 3a, that there is a diffuse background of uncorrelated momenta and a bright spot that corresponds to well-correlated momenta. However, the bright spot is not centered at zero; its shift indicates that the iodine's momentum is larger than the carbon's. This is easily understood from the molecular geometry: both the H$^+$ and C^{2+} ions balance the momentum of I^{5+}, but the H$^+$ momenta are not included in the sum. This "dark" momentum is manifested as the shift of the correlated spot. However, it is also seen that the spot is well defined in the y-direction, meaning that the three hydrogens carry a concerted momentum that is directed closely along the C-I bond. In short, the momentum sharing between I and C ions can be, based on Figure 3a, approximated by a two-body dissociation if the C momentum is corrected by a dark momentum coefficient k:

$$p_{I,x} = -(p_{C,x} + 3p_{H,x}) = -\frac{1}{k}p_{C,x}.$$

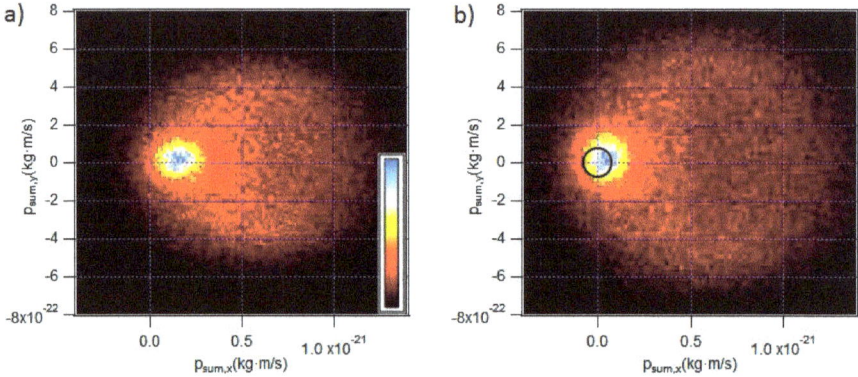

Figure 3. Momentum sum plots on the detector plane of the two ions in the (I^{5+}, C^{2+}) coincident pairs of iodomethane. The pairs were rotated to point the momentum of I^{5+} in the x-direction. (**a**) Applying the correction coefficient $k = 1$, (**b**) $k = 0.78$. The black circle illustrates the strength of the MF, applied later.

As the next step, the correct value of this coefficient is determined by applying a two-dimensional MF and restricting the momentum sum of acceptable pairs on the detector plane to a small circle around zero. The value of the coefficient k is scanned and the filtered coincident pair count obtained for each value, with the result shown in Figure 4. It shows a clear maximum at $k = 0.78 \pm 0.02$. When we now regenerate the momentum sum plot with this coefficient, but without applying the MF (Figure 3b), it has the characteristics of a two-body process: a circular diffuse false coincidence background and the true coincidence spot close to zero. Misleadingly, the spot still appears not exactly centered at zero. This is an artifact of the rotation to point $\mathbf{p_I}$ always in the x-direction, and warrants a brief discussion in Appendix A.

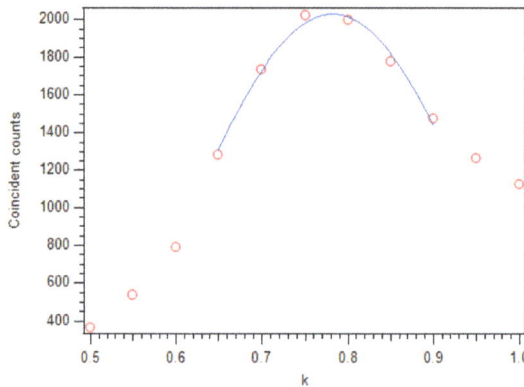

Figure 4. Coincident pair (I^{5+}, C^{2+}) counts as a function of the dark momentum coefficient k.

To verify that we have obtained true coincidences, a PIPICO plot is shown in Figure 5a, obtained with the correction coefficient $k = 0.78$, but not applying MF. Here one can see both the true coincidences as a tilted rectangular pattern and the false coincidence background as the fainter rectangle. The tilted pattern is the result of correlation in the axial momentum component, not used for MF. This pattern can be improved by applying the MF to the detector plane, as shown in Figure 5b, where a much sharper PIPICO pattern with almost no background is seen. The strength of the MF is indicated in Figure 3 by the black circle, although the events were chosen from within this circle in the *unrotated* image, not in the rotated one (see Appendix A) (there was a technical problem in this experiment that resulted in an increased ion source size along the TOF axis. The strong filtering applied resulted also in ions accepted only from part of the source and created a narrower tilted pattern).

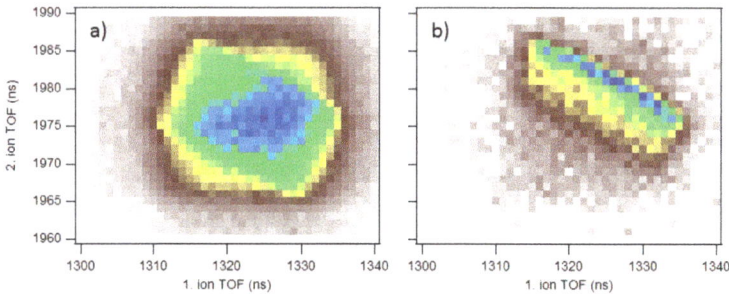

Figure 5. Photoion-photoion coincidence (PIPICO) maps of the (I^{5+}, C^{2+}) coincidences, (**a**) without momentum filtering (MF), and (**b**) with two-dimensional MF applied.

We also demonstrate an application of the MF to a difficult case where no PIPICO pattern is initially discernible at all—the (I^{11+},C^{3+}) coincidences. Here, the I^{11+} ions (with m/q of 11.5) are completely buried under the strong and broad C^+ peak in the ion TOF spectrum. The initial creation of the unfiltered and unbiased ion pairs produces a momentum sum plot in Figure 6a that is dominated by false coincidences (mostly between C^+ and C^{3+}), but a weak coincident spot of I^{11+},C^{3+} is also seen as a slight enhancement of the intensity just right of the (0,0) position in the plot.

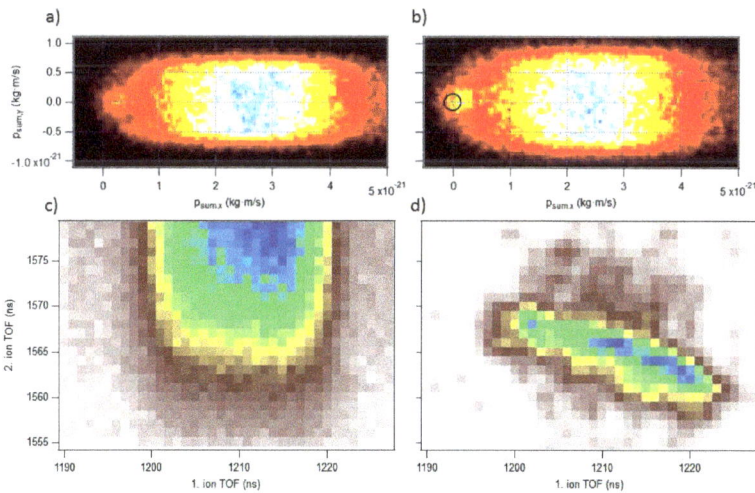

Figure 6. Momentum sum plots (**a**) and (**b**) and PIPICO plots (**c**) and (**d**) of (I^{11+}, C^{3+}) coincidences.

The PIPICO plot in Figure 6c does not show any typical tilted pattern. Repeating the procedure described before, we first obtain the correlation coefficient, $k = 0.825$, and recreate the momentum sum plot (Figure 6b). For this plot, the lowest-strength MF is applied—the momentum sum is not restricted to a certain maximum length, but a set of ion pairs was formed that gave the best momentum correlation, instead of forming pairs randomly. Finally, strong MF is applied on the detector plane (eliminating 97.8% of the ion pairs), as indicated with the black circle in Figure 6b. A PIPICO pattern now emerges clearly, as seen in Figure 6d.

Since the coincidences are extractable from the (I^{11+}, C^{3+}) pairs, let us attempt to also derive the kinetic energy distributions of the ions from this most difficult case in our dataset. The results are shown in Figure 7, where first the curves were extracted for the (I^{11+}, C^{3+}) set using MF to remove false coincidences. However, even the filtered data still contains a number of false coincidences from the diffuse background (see Figure 6b, for example). In order to remove that, the filtering was also applied to a *scrambled* dataset, where iodine ions were always paired with carbon ions from a different event (XFEL pulse) and which contained only false coincidences. The final kinetic energy distributions were obtained by subtracting the scrambled distributions. In the case of C^{3+}, for example, it is seen from Figure 7 that the false coincidences mostly contribute to the lower kinetic energies: this is since most of the C^{3+} ions come from other events with iodine charges lower than 11+ and, thus, with lower energy release.

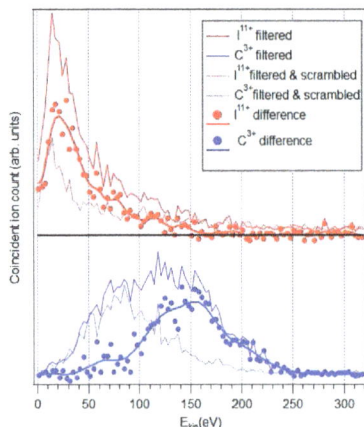

Figure 7. Kinetic energy distributions of ions in the (I^{11+}, C^{3+}) coincidences. Top panel: I^{11+}, bottom panel: C^{3+}.

Briefly, the complete analysis similar to the procedures described here allowed us to draw important conclusions on the interplay of the electron and nuclear dynamics during short XFEL pulses. We found that the C-I bond has only time to elongate only 10% or even less during the 10 fs pulse, whereas the C-H distances are nearly tripled [12]. These results are rather different from the ones predicted by instantaneous charge creation and redistribution.

3.2. Diiodomethane

In diiodomethane that contains three heavier atoms, one does not expect to find two-ion coincidences that can be well-approximated by a two-body dissociation process. However, the geometry suggests that, at least in the cases where the charge distribution between the two iodine atoms is not very unbalanced, a similar approach than for CH_3I could be useful. We, therefore, assume that the C ion is well momentum correlated with the *combined* momenta of the two I ions, but that some momentum is removed by the hydrogens along the direction of the carbon momentum. This is supported by point-charge Coulomb explosion simulations.

A test of this approach is presented on the example of a triple-coincidence set, (I^{3+}, I^{3+}, C^{2+}). In a dark momentum coefficient scan, such as in CH_3I, the triple coincident count reaches maximum at around $k = 0.75$, although in this case the maximum is not as sharply defined as in CH_3I. The triple momentum sum plot, after applying the correction coefficient to the carbon momentum, is shown in Figure 8. An alternative would be to apply the dark momentum correction to the combined momentum of the two iodine atoms, reducing it instead of increasing the carbon momentum. Both approaches gave very similar results. In Figure 8 one can see a rather well defined true coincidence region (again there is a slight displacement of it from the origin, for the same reason as discussed above).

For triple coincidences, a more informative representation than PIPICO plots is, for example, a Newton plot, as shown in Figure 9. The plot was obtained by applying MF with the strength as indicated by the circle in Figure 9, but in order to further reduce the false coincidence contributions, the filter was applied in all three dimensions (restricting the full vector sum to within a sphere around zero). The MF, itself, does not define any structures in the Newton plot, as it only restricts the three-vector sum values not, for example, the angle between the iodine ions. In generating Figure 9, all momenta of the first I^{3+} ion in the set were rescaled to 1×10^{-21} kg·m/s and rotated onto the x-axis. The red spot is the distribution of the second iodine ion's momenta and the blue spot shows the momenta of carbon ions. The black dots are the calculated momentum vectors from a point-charge Coulomb explosion simulation. For the simulation, the initial positions of the charges were fixed according to the ground state equilibrium geometry of diiodomethane, full charges were assigned instantaneously to the atoms and then their classical trajectories were followed. The comparison with the experimental momentum islands in Figure 9 shows a very good agreement in the angular distribution, but a clearly shorter vector length for the carbon ions, indicating that in the Coulomb explosion they obtain a lesser share of the kinetic energy release than predicted by the simplest model. Detailed investigations of the Coulomb explosion's dynamics in CH_2I_2 can be found in [13].

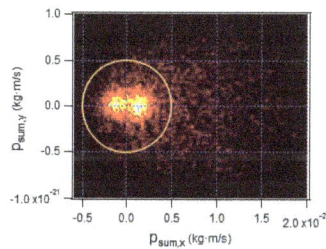

Figure 8. Momentum sum plot of (I^{3+}, I^{3+}, C^{2+}) triple coincidences. The circle illustrates the strength of the three-dimensional MF applied afterwards. The sum of the iodine momenta is directed along the x-axis.

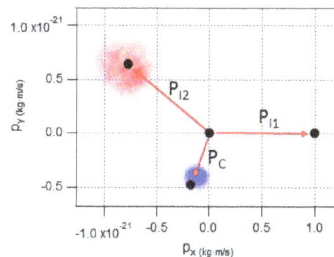

Figure 9. Newton plot of the (I^{3+}, I^{3+}, C^{2+}) triple coincidences, with the first I^{3+} ion's momentum in the x-direction and normalized to 1×10^{-21} kg·m/s. The red-coloured area marks the second I^{3+} ion's momenta and the blue-colored area the C^{2+} ion's momenta. Black dots are from point-charge Coulomb explosion simulation.

Finally, Figure 10 displays an ion-ion coincidence plot between the C^{2+} and I^{3+} ions in the (I^{3+}, I^{3+}, C^{2+}) triple coincidence set, after applying the same MF as for the Newton plot. The emission angle of about 107° between the two ions (see Figure 9) results in an elliptical PIPICO island, which is also quite well represented by a simple modeling of isotropic dissociation with the angular correlation of 107° between the two ions, and using a simple linear dependency of the ion TOF from the axial velocity component. The shape of the PIPICO island is also affected by the "dark" momentum of hydrogens and by the nonlinearity of the dependency of ion TOF from the axial velocity.

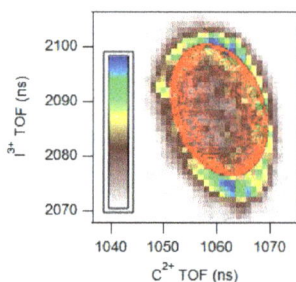

Figure 10. (C^{2+}, I^{3+}) PIPICO plot from the (I^{3+}, I^{3+}, C^{2+}) coincidence set in CH_2I_2. Color map: experiment, red dots: model.

3.3. 5-Iodouracil

The last molecule in the series studied at SACLA is 5-iodouracil, a cyclic planar molecule. Extracting high-quality coincidence datasets by MF is, on one hand, more difficult in this case because no clear axis of momentum distribution exists and it is shared by a number of ions. On the other hand, the ionization rate/pulse was relatively low in this experiment [14,15]. Here, again, preliminary Coulomb explosion simulations are valuable in deciding the strategy of applying MF. Modeling suggested, first, that the ion trajectories are significantly affected by their initial vibrational motion, which should, therefore, be added to the model. Second, the ion momenta are not expected to remain planar (due to vibrations) but strong deviations can occur. Planarity is, therefore, not the best criterion for the MF. Third, the carbon ions, being surrounded by heavier elements, have very variable trajectories that bear little resemblance to the original geometry and, therefore, are not the best candidates to be included in MF. On the other hand, oxygen ions, being on the outer positions, have trajectories that reflect well the geometry of the ring and also have sufficiently high momenta for efficient filtering. However, the lack of axial symmetry means that the filtering should be done based on the *angle* between the momenta, rather than the length of their sum. We will demonstrate this type of MF on the cases of triple coincidence sets containing the two oxygen ions. Figure 11 shows the angular correlations between the momenta of the ions in the (O^+, O^{2+}, H^+) triple coincidence set and confirms the predictions of the simulations: the O^+, O^{2+} angle has a strong correlation maximum at around 115°, while the planarity of the three vectors is not sharply defined (although the dissociation is preferentially planar).

Next, we extract triple coincidence sets with angular correlation MF for the two oxygen ions. Figure 12 shows a Newton plot for the (O^+, O^+, N^+) dataset. There are two possible origins for the oxygen defining the x-axis; these molecular orientations are depicted next to the figure. Then there are four initial positions for the N^+ ions, two of them degenerate. Indeed, the N^+ island in the figure corresponding to the degenerate position shows approximately twice the intensity of the other two. In extracting this plot, the background was slightly improved by subtracting a Newton plot generated from a scrambled coincidence set (not showing any defined islands). The conical sharp limits of the O^{2+} island is the result of applying the angular MF.

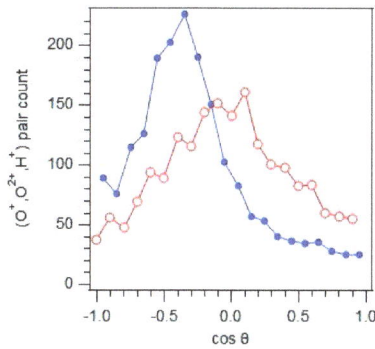

Figure 11. Distribution of angles between the momentum vectors in the (O^+,O^{2+},H^+) coincidence dataset in 5-iodouracil. Blue curve with filled circles: $\cos(p(O^+), p(O^{2+}))$; red curve with open circles: cosine of $p(H^+)$ from the (O^+,O^{2+}) momentum plane.

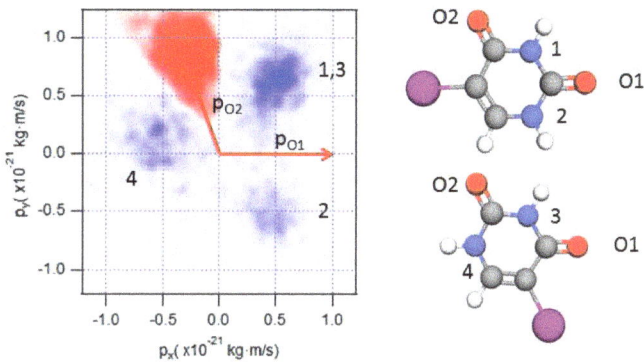

Figure 12. Newton plot of the (O^+,O^+,N^+) coincidences in 5-iodouracil, with the first O^+ ion's momentum in the x-direction and normalized to 1×10^{-21} kg·m/s, the red-colored area marking the second O^+ ion's momenta and the blue-colored area marking the N^+ ion's momenta. Possible molecular orientations and labeling of the nitrogen atoms and islands are shown to the right.

The second example is the (O^+, O^+, H^+) dataset. Due to the relatively low total charge in this experiment and the larger size of the molecule, the kinetic energies of the H^+ ions were sufficiently low for efficient detection together with the heavier ions. After applying angular MF for the two oxygens and subtracting the scrambled coincidence background, again a clear island structure emerges for the third (H^+) ion in the dataset, as shown in Figure 13. Indeed, all different initial positions of the hydrogens are discernible, indicating that the Coulomb explosion involving these ions has a common center of force in a good approximation. In contrast, we also generated Newton plots for (O^+,O^+,C^+) datasets, which showed no visible island structures for the carbon ions, in agreement with the simulations.

Interestingly, hydrogens emitted to the directions 1,4 appear to have larger momenta than the rest. Unfortunately, the number of events are not high enough to perform useful quantitative analysis of the energy distributions of ions originating from specific positions in a molecule in a particular triple coincidence dataset. Such analysis would await for FEL sources with high repetition rates. Additionally, there are indications that the probability of hydrogens from different sites acquiring charge varies somewhat, but with insufficient statistics for firm conclusions.

Appl. Sci. **2017**, *7*, 531

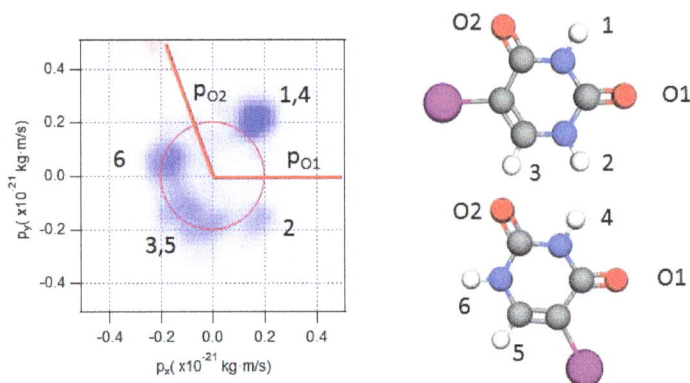

Figure 13. Newton plot of the (O^+, O^+, H^+) coincidences in 5-iodouracil. The image is slightly smoothed for better visibility. Blue-colored islands correspond to the H^+ momenta. The red circle is a guide for the eye.

4. Conclusions

We presented three cases of successful extraction and analysis of coincidence information in molecular dissociation studies with ultrashort XFEL pulses from the SACLA facility in Japan. The molecular samples possessed quite different symmetry properties, and also an increasing number of atoms from five in iodomethane to 12 in 5-iodouracil. We showed how efficient extraction of true coincidence data is achieved by applying ion momentum filtering. The exact implementation of these techniques must be flexible and adapted to the particular geometries of the samples, but is seen to work efficiently even in the case of complete atomization of a 12-atom molecule.

Coincidence techniques, proven to be powerful methods for extracting detailed information on molecular dynamics from experiments conducted at more conventional light sources, are equally valuable in FEL-based experiments. Their full potential is likely to be realized with high repetition rate sources, where the required long collection time becomes less of a restriction.

Acknowledgments: First and foremost, we are most grateful for all of the co-authors of [11,12,14,15] for their contributions in making these demanding experiments a success. The experiments were performed at SACLA with the approval of JASRI and the program review committee. They were supported by the X-ray Free Electron Laser Utilization Research Project and the X-ray Free Electron Laser Priority Strategy Program of the Ministry of Education, Culture, Sports, Science, and Technology of Japan (MEXT), by the Japan Society for the Promotion of Science (JSPS), by the Proposal Program of SACLA Experimental Instruments of RIKEN, by the IMRAM project and by the Cooperative Research Program of "Network Joint Research Center for Materials and Devices: Dynamic Alliance for Open Innovation Bridging Human, Environment and Materials". E.K. acknowledges financial support by the Academy of Finland.

Author Contributions: E.K. participated in the experiments, performed the analysis reported here and wrote the manuscript, K.M. prepared and participated in the experiments and analyzed data, H.F. planned, prepared and participated in the experiments and contributed to the analysis, K.N. prepared and participated in the experiments and analyzed the 5-iodouracil data, K.U. supervised the experimental program, participated in experiments and in the preparation of the manuscript.

Conflicts of Interest: The authors declare no conflict of interest.

Appendix A. Two-Dimensional Ion Momentum Sum Plots

Let us first assume a perfect measurement of a two-body dissociation, with $\mathbf{p}_C = -\mathbf{p}_I$. Then, before, as well as after, the rotation, $\mathbf{p}_{sum}$ is exactly zero. However, the imperfect measurement adds an uncertainty $\Delta\mathbf{p}_I$ and $\Delta\mathbf{p}_C$ to both vectors. We can make a *gedankenexperiment* where we first choose the x-axis according to the accurate vector $\mathbf{p}_I$ and then add the uncertainty $\Delta\mathbf{p}_I$. (see Figure A1). For the sake of simplicity we do not add the uncertainty $\Delta\mathbf{p}_C$. This uncertainly is directly transferred to $\mathbf{p}_{sum}$:

$\mathbf{p}_{sum} = \Delta\mathbf{p}_I$. and moves in a circle of radius $\Delta\mathbf{p}_I$ around the origin. In order to bring the vector $\mathbf{p}_I$ back on the x-axis, a rotation by an angle α relative to the x-axis needs to be performed, and with a reasonable assumption that $\Delta p_I < p_I$, $|\alpha| < 90°$. The vector $\mathbf{p}_C$ and the sum vector $\mathbf{p}_{sum}$ are naturally rotated by α as well and, thus, $\mathbf{p}_{sum}$ *always shifts in the direction of the x-axis* in rotation as a consequence of random uncertainty in the vector defining the x-axis. Thus, although the rotated sum momentum plots give a good visualization of the correlations and quality in the coincident dataset, they should not be used to find the correction coefficient by shifting the center of the true coincidence spot to zero—therefore, the scans, such as in Figure 4, for example, always operated on the *unrotated* momenta.

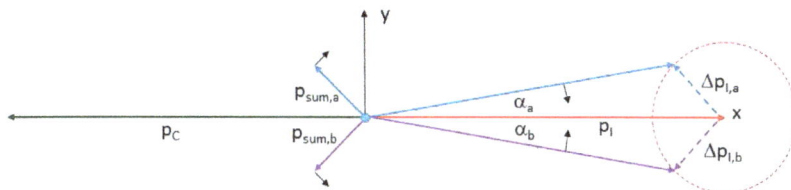

Figure A1. Sketch of the effect of the measurement error of $\mathbf{p}_I$ on the $\mathbf{P}_{sum}$ plot, with two choices a, b of the measurement error vector.

References

1. Prince, K.C.; Bolognesi, P.; Feyer, V.; Plekan, O.; Avaldi, L. Study of complex molecules of biological interest with synchrotron radiation. *J. Electron Spectrosc. Relat. Phenom.* **2015**. [CrossRef]
2. Piancastelli, M.N.; Simon, M.; Ueda, K. Present trends and future perspectives for atomic and molecular physics at the new X-ray light sources. *J. Electron Spectrosc. Relat. Phenom.* **2010**, *181*, 98–110. [CrossRef]
3. Continetti, R.E. COINCIDENCE SPECTROSCOPY. *Annu. Rev. Phys. Chem.* **2001**, *52*, 165–192. [CrossRef] [PubMed]
4. Ullrich, J.; Rudenko, A.; Moshammer, R. Free-electron lasers: New avenues in molecular physics and photochemistry. *Annu. Rev. Phys. Chem.* **2012**, *63*, 635–660. [CrossRef] [PubMed]
5. Liekhus-Schmaltz, C.E.; Tenney, I.; Osipov, T.; Sanchez-Gonzalez, A.; Berrah, N.; Boll, R.; Bomme, C.; Bostedt, C.; Bozek, J.D.; Carron, S.; et al. Ultrafast isomerization initiated by X-ray core ionization. *Nat. Commun.* **2015**, *6*, 8199. [CrossRef] [PubMed]
6. Milne, C.J.; Penfold, T.J.; Chergui, M. Recent experimental and theoretical developments in time-resolved X-ray spectroscopies. *Coord. Chem. Rev.* **2014**, *277–278*, 44–68. [CrossRef]
7. Berrah, N. Molecular dynamics induced by short and intense X-ray pulses from the LCLS. *Phys. Scr.* **2016**, *T169*, 14001. [CrossRef]
8. Bucksbaum, P.H.; Coffee, R.; Berrah, N. The first atomic and molecular experiments at the linac coherent light source X-ray free electron laser. *Adv. At. Mol. Opt. Phys.* **2011**, *60*, 239–289.
9. McFarland, B.K.; Farrell, J.P.; Miyabe, S.; Tarantelli, F.; Aguilar, A.; Berrah, N.; Bostedt, C.; Bozek, J.D.; Bucksbaum, P.H.; Castagna, J.C.; et al. Ultrafast X-ray auger probing of photoexcited molecular dynamics. *Nat. Commun.* **2014**, *5*. [CrossRef] [PubMed]
10. Minitti, M.P.; Budarz, J.M.; Kirrander, A.; Robinson, J.S.; Ratner, D.; Lane, T.J.; Zhu, D.; Glownia, J.M.; Kozina, M.; Lemke, H.T.; et al. Imaging molecular motion: Femtosecond X-ray scattering of an electrocyclic chemical reaction. *Phys. Rev. Lett.* **2015**, *114*. [CrossRef] [PubMed]
11. Tenboer, J.; Basu, S.; Zatsepin, N.; Pande, K.; Milathianaki, D.; Frank, M.; Hunter, M.; Boutet, S.; Williams, G.J.; Koglin, J.E.; et al. Time-resolved serial crystallography captures high-resolution intermediates of photoactive yellow protein. *Science* **2014**, *346*, 1242–1246. [CrossRef] [PubMed]
12. Motomura, K.; Kukk, E.; Fukuzawa, H.; Wada, S.-I.; Nagaya, K.; Ohmura, S.; Mondal, S.; Tachibana, T.; Ito, Y.; Koga, R.; et al. Charge and nuclear dynamics induced by deep inner-shell multiphoton ionization of CH3I molecules by intense X-ray free-electron laser pulses. *J. Phys. Chem. Lett.* **2015**, *6*, 2944–2949. [CrossRef] [PubMed]

13. Takanashi, T.; Nakamura, K.; Kukk, E.; Motomura, K.; Fukuzawa, H.; Nagaya, K.; Wada, S.I.; Kumagai, Y.; Iablonskyi, D.; Ito, Y.; et al. Ultrafast coulomb explosion of the diiodomethane molecule induced by an X-ray free-electron laser pulse. *Phys. Chem. Chem. Phys.* **2017**, submitted. [CrossRef]

14. Nagaya, K.; Motomura, K.; Kukk, E.; Fukuzawa, H.; Wada, S.; Tachibana, T.; Ito, Y.; Mondal, S.; Sakai, T.; Matsunami, K.; et al. Ultrafast dynamics of a nucleobase analogue illuminated by a short intense X-ray free electron laser pulse. *Phys. Rev. X* **2016**, *6*. [CrossRef]

15. Nagaya, K.; Motomura, K.; Kukk, E.; Takahashi, Y.; Yamazaki, K.; Ohmura, S.; Fukuzawa, H.; Wada, S.; Mondal, S.; Tachibana, T.; et al. Femtosecond charge and molecular dynamics of I-containing organic molecules induced by intense X-ray free-electron laser pulses. *Faraday Discuss.* **2016**, *194*, 537–562. [CrossRef] [PubMed]

16. Dörner, R.; Mergel, V.; Jagutzki, O.; Spielberger, L.; Ullrich, J.; Moshammer, R.; Schmidt-Böcking, H. Cold Target Recoil Ion Momentum Spectroscopy: A "momentum microscope" to view atomic collision dynamics. *Phys. Rep.* **2000**, *330*, 95–192. [CrossRef]

17. Ullrich, J.; Moshammer, R.; Dorn, A.; Dörner, R.; Schmidt, L.P.H.; Schmidt-Böcking, H. Recoil-ion and electron momentum spectroscopy: Reaction-microscopes. *Rep. Prog. Phys.* **2003**, *66*, 1463–1545. [CrossRef]

18. Tsuboi, T.; Xu, E.Y.; Bae, Y.K.; Gillen, K.T. Magnetic bottle electron spectrometer using permanent magnets. *Rev. Sci. Instrum.* **1988**, *59*, 1357–1362. [CrossRef]

19. Rijs, A.M.; Backus, E.H.G.; de Lange, C.A.; Westwood, N.P.C.; Janssen, M.H.M. "Magnetic bottle" spectrometer as a versatile tool for laser photoelectron spectroscopy. *J. Electron Spectrosc. Relat. Phenom.* **2000**, *112*, 151–162. [CrossRef]

20. Mucke, M.; Förstel, M.; Lischke, T.; Arion, T.; Bradshaw, A.M.; Hergenhahn, U. Performance of a short "magnetic bottle" electron spectrometer. *Rev. Sci. Instrum.* **2012**, *83*, 63106. [CrossRef] [PubMed]

21. Penent, F.; Palaudoux, J.; Lablanquie, P.; Andric, L.; Feifel, R.; Eland, J.H.D. Multielectron spectroscopy: The xenon 4D hole double auger decay. *Phys. Rev. Lett.* **2005**, *95*. [CrossRef] [PubMed]

22. Murphy, B.F.; Osipov, T.; Jurek, Z.; Fang, L.; Son, S.-K.; Mucke, M.; Eland, J.H.D.; Zhaunerchyk, V.; Feifel, R.; Avaldi, L.; et al. Femtosecond X-ray-induced explosion of C60 at extreme intensity. *Nat. Commun.* **2014**, *5*. [CrossRef] [PubMed]

23. Frasinski, L.J.; Zhaunerchyk, V.; Mucke, M.; Squibb, R.J.; Siano, M.; Eland, J.H.D.; Linusson, P.; Vd Meulen, P.; Salén, P.; Thomas, R.D.; et al. Dynamics of hollow atom formation in intense X-ray pulses probed by partial covariance mapping. *Phys. Rev. Lett.* **2013**, *111*. [CrossRef] [PubMed]

24. Motomura, K.; Foucar, L.; Czasch, A.; Saito, N.; Jagutzki, O.; Schmidt-Böcking, H.; Dörner, R.; Liu, X.-J.; Fukuzawa, H.; Prümper, G.; et al. Multi-coincidence ion detection system for EUV–FEL fragmentation experiments at SPring-8. *Nucl. Instrum. Methods Phys. Res. Sect. A* **2009**, *606*, 770–773. [CrossRef]

25. Arion, T.; Hergenhahn, U. Coincidence spectroscopy: Past, present and perspectives. *J. Electron Spectrosc. Relat. Phenom.* **2015**, *200*, 222–231. [CrossRef]

26. Eland, J.H.D.; Wort, F.S.; Lablanquie, P.; Nenner, I. Mass spectrometric and coincidence studies of double photoionization of small molecules. *Z. Phys. D Atoms Mol. Clust.* **1986**, *4*, 31–42. [CrossRef]

27. Eland, J.H.D.; Linusson, P.; Mucke, M.; Feifel, R. Homonuclear site-specific photochemistry by an ion–electron multi-coincidence spectroscopy technique. *Chem. Phys. Lett.* **2012**, *548*, 90–94. [CrossRef]

28. Rolles, D.; Pešić, Z.D.; Perri, M.; Bilodeau, R.C.; Ackerman, G.D.; Rude, B.S.; Kilcoyne, A.L.D.; Bozek, J.D.; Berrah, N. A velocity map imaging spectrometer for electron–ion and ion–ion coincidence experiments with synchrotron radiation. *Nucl. Instrum. Methods Phys. Res. Sect. B* **2007**, *261*, 170–174. [CrossRef]

29. Miron, C.; Morin, P. High-resolution inner-shell coincidence spectroscopy. *Nucl. Instrum. Methods Phys. Res. Sect. A* **2009**, *601*, 66–77. [CrossRef]

30. Card, D.A.; Folmer, D.E.; Sato, S.; Buzza, S.A.; Castleman, A.W. Covariance mapping of ammonia clusters: Evidence of the connectiveness of clusters with coulombic explosion. *J. Phys. Chem. A* **1997**, *101*, 3417–3423. [CrossRef]

31. Frasinski, L.J.; Codling, K.; Hatherly, P.A. Covariance mapping: A correlation method applied to multiphoton multiple ionization. *Science* **1989**, *246*, 1029–1031. [CrossRef] [PubMed]

32. Kornilov, O; Eckstein, M.; Rosenblatt, M.; Schulz, C.P.; Motomura, K.; Rouzée, A.; Klei, J.; Foucar, L.; Siano, M.; Lübcke, A.; et al. Coulomb explosion of diatomic molecules in intense XUV fields mapped by partial covariance. *J. Phys. B* **2013**, *46*, 164028. [CrossRef]

33. Zhaunerchyk, V.; Frasinski, L.J.; Eland, J.H.D.; Feifel, R. Theory and simulations of covariance mapping in multiple dimensions for data analysis in high-event-rate experiments. *Phys. Rev. A* **2014**, *89*. [CrossRef]
34. Mucke, M.; Zhaunerchyk, V.; Frasinski, L.J.; Squibb, R.J.; Siano, M.; Eland, J.H.D.; Linusson, P.; Salén, P.; Vd Meulen, P.; Thomas, R.D.; et al. Covariance mapping of two-photon double core hole states in C2H2 and C2H6 produced by an X-ray free electron laser. *New J. Phys.* **2015**, *17*, 73002. [CrossRef]
35. Yabashi, M.; Tanaka, H.; Ishikawa, T. Overview of the SACLA Facility. *J. Synchrotron Radiat.* **2015**, *22*, 477–484. [CrossRef] [PubMed]
36. Fukuzawa, H.; Motomura, K.; Son, S.-K.; Mondal, S.; Tachibana, T.; Ito, Y.; Kimura, M.; Nagaya, K.; Sakai, T.; Matsunami, K.; et al. Sequential multiphoton multiple ionization of Ar and Xe by X-ray free electron laser pulses at SACLA. *J. Phys.* **2014**, *488*, 32034. [CrossRef]
37. Motomura, K.; Fukuzawa, H.; Son, S.-K.; Mondal, S.; Tachibana, T.; Ito, Y.; Kimura, M.; Nagaya, K.; Sakai, T.; Matsunami, K.; et al. Sequential multiphoton multiple ionization of atomic argon and xenon irradiated by X-ray free-electron laser pulses from SACLA. *J. Phys. B* **2013**, *46*, 164024. [CrossRef]

Article

Observing Femtosecond Fragmentation Using Ultrafast X-ray-Induced Auger Spectra

Thomas J. A. Wolf [1] ![ORCID], Fabian Holzmeier [2,3], Isabella Wagner [3], Nora Berrah [4],
Christoph Bostedt [5,6,7], John Bozek [2,5], Phil Bucksbaum [1,8], Ryan Coffee [5], James Cryan [1],
Joe Farrell [1], Raimund Feifel [9], Todd J. Martinez [1,10] ![ORCID], Brian McFarland [1], Melanie Mucke [11],
Saikat Nandi [2,12], Francesco Tarantelli [13] ![ORCID], Ingo Fischer [3] ![ORCID] and Markus Gühr [1,14,*]

[1]   Stanford PULSE Institute, SLAC National Accelerator Laboratory, 2575 Sand Hill Road, Menlo Park,
      CA 94025, USA; thomas.wolf@stanford.edu (T.J.A.W.); phbuck@stanford.edu (P.B.);
      jcryan@slac.stanford.edu (J.C.); joepfar@gmail.com (J.F.); todd.martinez@stanford.edu (T.J.M.);
      bigbmac@gmail.com (B.M.)
[2]   Synchrotron SOLEIL, L'Orme des Merisiers, Saint-Aubin, BP 48, 91192 Gif-sur-Yvette Cedex, France;
      fabian.holzmeier@synchrotron-soleil.fr (F.H.); john.bozek@synchrotron-soleil.fr (J.B.);
      saikat.nandi@fysik.lth.se (S.N.)
[3]   Institut für Physikalische & Theoretische Chemie, Universität Würzburg, D-97074 Würzburg, Germany;
      isabella.wagner@stud-mail.uni-wuerzburg.de (I.W.); ingo.fischer@uni-wuerzburg.de (I.F.)
[4]   Department of Physics, University of Connecticut, 2152 Hillside Road, Storrs, CT 06269, USA;
      nora.berrah@wmich.edu
[5]   Linac Coherent Light Source, SLAC National Accelerator Laboratory, 2575 Sand Hill Road, Menlo Park,
      CA 94720, USA; cbostedt@anl.gov (C.B.); coffee@slac.stanford.edu (R.C.)
[6]   Argonne National Laboratory, 9700 Cass Ave, Lemont, IL 60439, USA
[7]   Department of Physics and Astronomy, 2145 Sheridan Road, Northwestern University, Evanston,
      IL 60208, USA
[8]   Departments of Physics and Applied Physics, Stanford University, 382 Via Pueblo Mall, Stanford,
      CA 94305, USA
[9]   Department of Physics, University of Gothenburg, SE-412 96 Gothenburg, Sweden;
      raimund.feifel@physics.gu.se
[10] Department of Chemistry, Stanford University, 333 Campus Drive, Stanford, CA 94305, USA
[11] Department of Physics and Astronomy, Uppsala University, Box 516, SE-751 20 Uppsala, Sweden;
      melanie.mucke@physics.uu.se
[12] Department of Physics, Lund University, P. O. Box 118, SE-221 00 Lund, Sweden
[13] Department of Chemistry, Biology and Biotechnology, University of Perugia, 06123 Perugia, Italy;
      francesco.tarantelli@unipg.it
[14] Institut für Physik und Astronomie, Universität Potsdam, 14476 Potsdam, Germany
*   Correspondence: mguehr@uni-potsdam.de; Tel.: + 49-331-977-5571

Academic Editor: Kiyoshi Ueda
Received: 11 June 2017; Accepted: 28 June 2017; Published: 1 July 2017

Abstract: Molecules often fragment after photoionization in the gas phase. Usually, this process can only be investigated spectroscopically as long as there exists electron correlation between the photofragments. Important parameters, like their kinetic energy after separation, cannot be investigated. We are reporting on a femtosecond time-resolved Auger electron spectroscopy study concerning the photofragmentation dynamics of thymine. We observe the appearance of clearly distinguishable signatures from thymine's neutral photofragment isocyanic acid. Furthermore, we observe a time-dependent shift of its spectrum, which we can attribute to the influence of the charged fragment on the Auger electron. This allows us to map our time-dependent dataset onto the fragmentation coordinate. The time dependence of the shift supports efficient transformation of the excess energy gained from photoionization into kinetic energy of the fragments. Our method is broadly applicable to the investigation of photofragmentation processes.

Appl. Sci. **2017**, *7*, 681

Keywords: ultrafast dynamics; Auger electron spectroscopy; photofragmentation; photochemistry

1. Introduction

The speed of a photoexcited chemical reaction is determined by gradients of potential energy surfaces and the mass of reaction products or precursors. Generally, the decisive steps—the making or breaking of chemical bonds—occur on a femtosecond timescale. Numerous experimental methods have been implemented to approach this topic. The development of femtosecond laser pulses brought an extreme wealth of ultrafast techniques to life, among them visible and ultraviolet pump-probe spectroscopy [1] and multidimensional spectroscopy [2]. For isolated molecules, photoelectron spectroscopy [3] and electron, as well as X-ray scattering [4–6], have developed as high fidelity tools that can be directly compared to highest level ab initio calculations on isolated molecular dynamics.

In this article, we focus on investigating a specific type of reaction, photoion fragmentation as it occurs in the presence of ionizing radiation in the upper atmosphere or in space [7]. Organic molecules can be photoionized by absorbing either one or multiple photons, which combined can overcome the ionization potential (typically on the order of 9 eV [8]). When sufficient energy is deposited in the molecule, it is placed on a cationic potential energy surface with repulsive character in at least one internal degree of freedom. Thus, the nuclei are sped up towards rapid bond dissociation and fragmentation by a steep gradient in the Franck–Condon region. The positive charge localizes on one of the fragments. Charged and neutral species separate with considerable velocity.

Due to strongly different ionization potentials in charged and neutral species, such a process can only be investigated by probing the parent photocation and the fragments with photon energies considerably beyond what is readily available from table-top laser systems in the vacuum ultraviolet and soft X-ray regime.

Short wavelength pulses in the femtosecond domain with unprecedented pulse energy became available with the advent of X-ray free electron lasers [9–12]. For isolated molecules, this allowed for X-ray scattering experiments in extremely diluted targets [13–15] as well as the development of spectroscopic methods in the soft X-ray domain [16–22]. Due to the deeply-bound core electrons, X-ray spectroscopy is element-, as well as site-, selective [23].

In the present study, we demonstrate that unique insight into ionic photofragmentation can be gained by probing the dynamics using Auger electron spectroscopy. Here, the photoion and its fragments are core-ionized by a femtosecond soft X-ray pulse. They undergo Auger decay by filling the core hole with a valence electron and simultaneously ejecting another valence electron for energy conservation reasons, which is schematically shown in Figure 1. We measured the kinetic energy spectrum of the ejected Auger electron as a function of the time delay between the start of the photofragmentation reaction by the optical UV pulse and the X-ray ionization. By choosing the non-resonant soft X-ray photon energy to be high enough, Auger spectra of charged and neutral fragments can be observed simultaneously and the method is insensitive to small fluctuations of the photon energy at free electron lasers starting from noise. Furthermore, as shown in [16], Auger spectroscopy methods provide the unique opportunity to follow fragmentation dynamics beyond the typical bond distances of 1–2 Å, since molecular Auger decay channels, which involve valence electrons situated at different atoms, still function at considerably larger atomic distances. We show in the following that the Auger spectral features reveal signatures of fragmentation dynamics at even longer distances. This allows for determination of e.g., their kinetic energies and can, therefore, give important mechanistic information about the redistribution of energy during the fragmentation process.

We demonstrate the concept on a specific example, the multi-photon induced fragmentation process of the nucleobase thymine (see Figure 1). Ultraviolet light of 4.65 eV photon energy induces a resonant three photon transition to an excited ionic state, from where the molecule is predicted to fragment into neutral isocyanic acid (HNCO) and a charged $C_4H_5NO^+$ fragment.

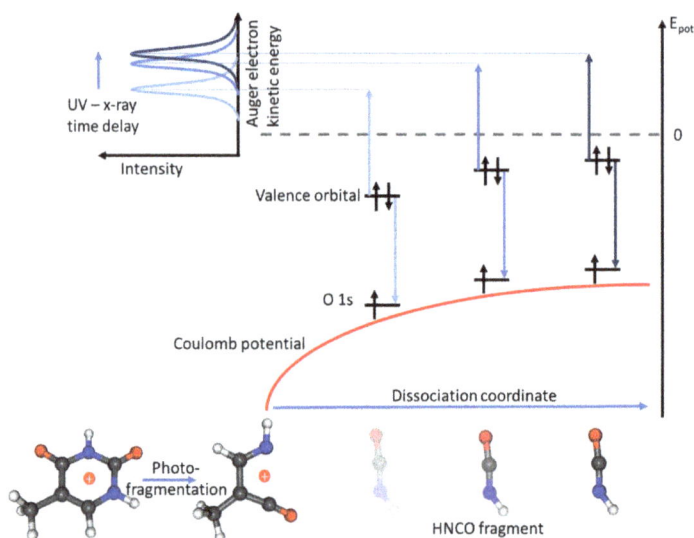

Figure 1. Probing molecular photofragmentation by soft X-ray-induced Auger spectra. We exemplify the method on the thymine molecular cation, which breaks apart into neutral isocyanic acid (HNCO) and singly-charged $C_4H_5NO^+$ after three-photon UV ionization (bottom). The distance of the HNCO fragment is interrogated by a time-delayed soft X-ray pulse creating core ionized molecules. The subsequent Auger decay leads to the emission of an additional electron, which does not only have to overcome the ionization potential intrinsic to the molecule, but also the external Coulomb potential (red) of the nearby cationic fragment. The Auger emission from the small neutral HNCO in the vicinity of the positively-charged large fragment leads to a red-shifted Auger kinetic energy with respect to the Auger electron spectrum of the isolated molecule. As the dissociation coordinate increases, the redshift asymptotically disappears.

Charged fragments have been observed in classical mass spectroscopy under high energy electron or photon excitation. The onset of the $C_4H_5NO^+$ fragment channel has been determined by threshold spectroscopy [24] to be above 10 eV. Thus, three photons of 4.65 eV photon energy need to be absorbed to observe the described fragmentation. The fragmentation of cationic thymine into HNCO and $C_4H_5NO^+$ has been theoretically investigated [25]. Although the reaction can be formally understood as a retro-Diels–Alder reaction, which should be concerted, a stepwise mechanism cleaving the two involved C–N bonds, as sketched in Figure 1, was found.

Large molecular systems possess a relatively unstructured Auger spectrum after creation of 1s core holes in carbon, nitrogen, or oxygen. Many different decay paths are energetically allowed and broad bands originating from the 2s2s, 2s2p, and 2p2p valence electron orbitals are involved in the Auger decay [17,26,27]. In terms of potential energy surfaces, one can rephrase this assertion by stating that the density of final states of the Auger decay process is high and the Auger lines overlap to form large bands.

After fragmentation and sufficient separation, however, electrons from the cationic fragment cannot take part in Auger decays on the neutral fragments, because such decay processes have distance dependences of r^{-6} [28]. Small molecules show relatively narrow lines with a width on the few 100 meV scale [29] because the density of final states is much lower as the number of electrons in the molecule is smaller. Different fragments can therefore be easily distinguished in the Auger spectra by characteristic sharp fingerprints. It is energetically more favorable to delocalize a positive charge on a large fragment than on a small one. Therefore, as a rule of thumb, neutral fragments will be observable as sharper lines on top of a background from the cationic system with higher density of final states and, thus, less structural.

The electrons, which are emitted during Auger decay of the neutral fragment, still experience the Coulomb potential of the cationic fragment with considerably longer-range characteristic (r^{-1}) than the above mentioned joint neutral-charged fragment Auger channels (r^{-6}). The effect is observable by a shift of the Auger spectrum to lower kinetic energies.

Probing the fragmentation process just after the optical excitation still interrogates the large molecular ion. The Auger spectrum will be showing a broad unstructured band. After some time, however, fragmentation has taken place and one can treat the molecular fragments as small molecules with a reduced dicationic state density. The resulting spectrum will be structured in much sharper lines. These lines will first appear shifted with respect to their steady-state position (see Figure 1). The increasing distance of neutral and cationic fragments can then be followed in the reduction of this shift.

The article is structured in the following way: In the next section, we will introduce the experimental setups used at the free electron laser (FEL) and the synchrotron. The FEL setup was used to perform the time resolved dissociation experiment on the nucleobase thymine. Since the theoretically proposed fragmentation paths suggest splitting off neutral isocyanic acid (HNCO), we separately measured its Auger spectrum at a synchrotron to assign the neutral fragment.

We show the results and discuss spectral shape, as well as the spectral position of the fragment Auger lines. From the latter, we calibrate the distances of the fragments on a sub-picosecond timescale. We validate our calibration with a simple mechanical model based on full transformation of excess energy into the kinetic energy of the fragments.

2. Experimental Methods and Theory

The time-resolved fragmentation experiments on thymine were performed at the AMO (Atomic, Molecular and Optical physics) instrument of the LCLS (Linac Coherent Light Source) free electron laser. The instrument and some of its science are described in detail in [30,31]. Details of the setup we used for this particular experiment can be found in [17,32]. A sketch of the most important parts is given in Figure 2a. In short, thymine powder was evaporated in an oven at 140 °C with an outlet to the interaction region in the form of a heated capillary, thus forming an effusive jet in the collection volume of a magnetic bottle electron spectrometer.

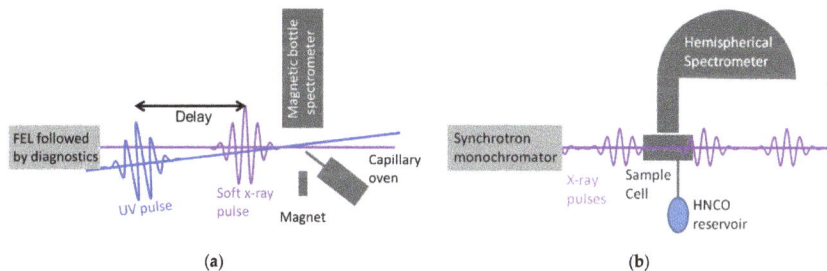

Figure 2. Experimental setups. (**a**) Sketch of the setup at the free electron laser. Molecules are inserted into the interaction region by evaporation in a capillary oven. A first UV pulse ionizes and fragments the molecules. The fragments are then probed by a delayed soft X-ray pulse via core electron ionization. The Auger electrons are collected and analyzed in a magnetic bottle spectrometer. (**b**) Sketch of the setup at the synchrotron beamline. Isocyanic acid is produced close to the beamline and then transported to the spectrometer under cryogenic conditions. The sample is inserted in a differentially pumped cell. Soft X-ray radiation from the monochromator beamline core ionizes the molecules and Auger electrons of HNCO are analyzed via a hemispherical electron analyzer.

A first UV pulse (hv = 4.65 eV, 70 fs duration) triggered the fragmentation process via resonance-enhanced three-photon ionization. A time-delayed free electron laser (FEL) soft X-ray pulse (hv = 570 eV, 70 fs duration) was used to create an oxygen core hole in the photoionized molecules. Auger electrons from O 1s core holes were detected with the magnetic bottle spectrometer as a function of time delay between UV excitation and X-ray probe pulses.

The time resolution of the experiment is about 330 fs, which is dominated by the jitter between the optical excitation and the FEL probe pulses. This is much larger than the time resolution observed in our study of the non-adiabatic dynamics in the neutral excited states of thymine [17], since we had not implemented the single shot optical X-ray cross-correlator [33] to compensate for the jitter between optical and FEL pulses in the data analysis for the fragmentation study reported here.

The steady state experiments on isolated HNCO were performed at the high-resolution photoelectron spectroscopy instrument of the PLEIADES beamline at the synchrotron SOLEIL [34]. A schematic overview is given in Figure 2b. Synchrotron light at a photon energy of hv = 570 eV and a bandwidth of 750 meV ionized the HNCO molecules contained in a differentially-pumped sample cell with entrance and exit windows for the synchrotron beam. The emitted O 1s Auger electrons in the direction of the light polarization were dispersed by a hemispherical analyzer. The monochromator, as well as the hemispherical analyzer were calibrated by measuring the photoabsorption and Auger spectra of CO_2.

The isocyanic acid target was prepared according to the method described in [35]. After synthesis, the molecular sample was stored at liquid nitrogen temperatures to avoid polymerization. The vapor pressure at around −40 °C was sufficient to obtain the necessary molecular density inside the gas cell attached to the hemispherical analyzer.

The O1s core decay spectrum of HNCO has been calculated using the second-order Algebraic Diagrammatic Construction (ADC(2)), which is a well-established Green's function-based direct method to compute double ionization and Auger spectra [36,37] in combination with a cc-pVDZ basis [38]. The details of the simulation are described in [17].

3. Results and Discussion

Figure 3a shows Auger spectra of fragmenting thymine ions as a function of kinetic energy. The different spectra are measured at various time delays between the UV ionization and X-ray probe pulse as given in the figure. At negative delays, the ultra-short X-ray pulses hit molecules before the ultraviolet photoionization, thus, those spectra reflect the core hole decay of thymine in its electronic ground state. The spectrum is only weakly structured, centered around 504 eV photon energy, and possesses a width of around 10 eV. The Auger spectra of thymine have been published by us in another context, however, we realized during the work on this publication, that the energy calibration in [17] must be shifted up by 3 eV. The ADC(2) theory in [17] was too high in energy by about 1.5 eV; now it is too low compared to the experimental thymine spectrum by about the same amount.

The overlap of UV and X-ray pulses in the pump-probe experiment manifests itself by a strong bleach of the main Auger line due to the removal of the population from the electronic ground state of neutral thymine [17]. This feature has also been used to set the zero-delay time for this experiment. As shown in Figure 3a, the Auger spectrum broadens considerably and develops a pronounced wing towards lower kinetic energies after the UV pulse excitation. The underlying reason for this wing are additional states newly populated by absorption of the UV pump pulse. In part, this is the desired, dissociative cationic state. Since the photoionization efficiency is enhanced by resonant excited states in the neutral molecule, there are certainly additional contributions from population in those states. From our previous study [17], we know that photoexcitation to the lowest valence state of the neutral results in a shift of the Auger spectrum to lower energies by about 5 eV within 500 fs. We can expect the Auger electron spectrum of the cationic state to shift more strongly in the same direction, simply based on the argument that Auger electron emission must work against one more charge in the molecular ion compared to the case of Auger decay from a core-ionized neutral molecule. The combination of energy shifts to the spectrum from excited and ionized states can explain unambiguously the appearance of a broad signature at lower electron kinetic energies.

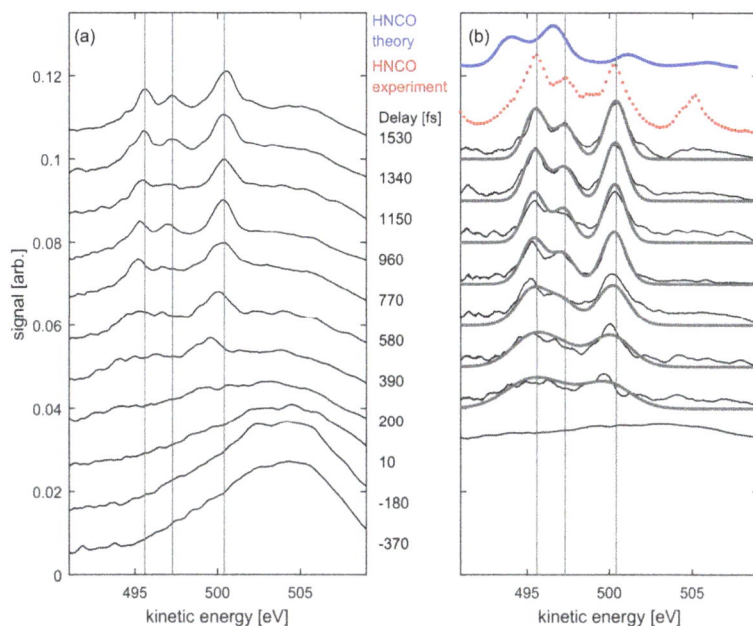

Figure 3. Results of the transient Auger studies on fragmenting thymine ions. (**a**) The Auger spectra of the UV-fragmented molecules are shown as a function of delay between the UV and soft X-ray pulse. Narrow lines appear on a broadening baseline with increasing delay. The three vertical grey lines indicate the position of peak maxima for long delays, showing that the narrow lines shift towards higher energies with increasing delay. The time-resolved data is synthesized from single FEL shots. The relative error of the spectral intensity on is on the order of 5–10%, which was determined in another paper on data from the same lincac coherent light source experiment using bootstrap statistical analysis [17]. (**b**) The same transient spectra as in (**a**), corrected by a background (see text). For large delays, the transient spectrum fits the reference HNCO spectrum taken at the synchrotron (red dots). The relative intensity error in the significant range is on the order of a few percent. The calculation indicated in blue reproduces all the essential features of the HNCO reference spectrum, although with slightly different peak distances. The thick grey lines model the transient spectra by shifting and broadening three Gaussian lines according to a one-dimensional fragmentation process as described in the text.

Two new features, having a width smaller than the broad underlying band, appear in the spectrum after about 400 fs. At a UV-X-ray delay of 600 fs, three features stand out in the lower kinetic energy wing. They continue to sharpen and shift towards higher kinetic energies as the delay progresses. At 1200 fs after UV excitation, the features have reached a steady state.

We now compare the time-resolved spectra with steady-state Auger spectra of HNCO. The red dotted line in Figure 3b is the steady-state spectrum of HNCO as taken at the PLEIADES beamline of SOLEIL. The spectrum is characterized by 4 dominant peaks in the region between 494 and 508 eV. To further understand the character of these signatures, we also calculated the Auger decay features using the ADC(2) method. The method has been described in the context of thymine in [17]. The result of the calculation is shown as the solid blue spectrum on top of Figure 3b. The simulation reproduces the essential features of the steady-state experimental spectrum. The four lower energy maxima are reproduced by the ADC(2) calculation. The upper energy maximum is only poorly modulated in the simulation. In addition, we observe energy shifts in the four lower maxima. Shifts on the order of 1 eV are typical for the ADC(2) method and the deviations will be the topic of future investigations.

In order to discern the details of the sharp features in the time-resolved spectra we model a background from a scaled spectrum at a 200 fs delay and subtract it from the spectra at all delays. The scaling factor is fitted such that the low and high energy wings of the spectra after this delay essentially have zero background. We plot the background-corrected spectra in Figure 3b. The Auger spectra of UV-photoionized thymine for delays 1150, 1340, and 1530 fs show a very good agreement with the steady-state Auger spectrum of HNCO. The three lower peaks at positions 495.6, 497.3 and 500.4 eV (in the remainder of the paper labeled as Peak 1–3 in ascending Auger electron kinetic energy) are perfectly reproduced both in position and width. The highest energy peak at 505.2 eV in the steady-state spectrum is not well reproduced in the time-dependent spectra. This can be attributed to additional contributions from excited state dynamics in the neutral thymine and the ground state bleach, which are known to both appear in this regime [17]. Inspecting the transient trend in these peaks, one realizes that the peak width as well as position follows a similar trend for all three features. The peaks sharpen after the UV excitation to reach their stationary shape at about 1 ps. In addition, all peaks experience a shift towards higher kinetic energies. Around 400 fs, peak 3 appears at 499.7 eV, which is 0.7 eV shifted from the steady state. Around 770 fs, this shift diminishes to 0.3 eV. At the same delay, peak 1 is also red-shifted by about 0.3 eV from its steady state value of 495.6 eV. Peak 2 needs a longer time to fully form and also has a larger shift than the high and low energy features. At 770 fs, it is shifted by 0.6 eV lower from the steady state value of 497.3 eV.

In the following, we provide an explanation of the three main observations in the experimental data, the appearance of sharp new features, their spectroscopic shift, and their sharpening. From our calculations of the Auger spectra of thymine, we know that both oxygen atoms of thymine give rise to very dense spectral features, resulting from the high density of final states (DOFS) in this rather large molecule [17]. The high DOFS is reflected in a broad and unstructured Auger spectrum even for only one of the two oxygen atoms, demonstrated by calculations in the supporting material in [17]. From photoion mass spectrometry, it was suggested that photoionized thymine fragments into neutral HNCO and a positively-charged residual with mass 83 [24]. As the neutral HNCO fragment splits off the positively-charged residual, the electronic overlap between the two constituents gets smaller. This in turn reduces the density of dicationic states, in particular for the small HNCO fragment with only 22 electrons compared to the 43 electrons of the $C_4H_5NO^+$ fragment. The appearance of narrow lines within 600 fs, therefore, directly reflects the formation of neutral HNCO fragments.

The transient shift in the peak positions stems from the interaction of the outgoing Auger electron of HNCO with the positive charge of the $C_4H_5NO^+$ fragment. If the Auger electron is emitted in the vicinity of a positive fragment, it has to overcome the Coulomb potential of that charge. It is, thus, detected with lower kinetic energy compared to a HNCO Auger electron being emitted far away from the charged fragment. With the assumption of the Coulomb interaction resulting in a shift in atomic units being directly 1/distance to be responsible for the peak shifts, we can directly determine distances between charged and neutral fragments from the energy shifts. The distances are listed in Table 1. The evaluated distances for different peaks in part differ considerably, which can be mainly attributed to the limited accuracy to which the peak positions could be extracted from the spectra.

Table 1. Fragment distances calculated based on the transient peak shifts in the Auger electron spectra and a Coulomb interaction. The confidence intervals are calculated assuming a constant error in relative peak position of 0.1 eV. For comparison, the expected distances under the model assumption of one-dimensional fragmentation that the entire absorbed excess energy is transformed into kinetic energy are included.

Delay (fs)	Peak 1, Distance (Å)	Peak 2, Distance (Å)	Peak 3, Distance (Å)	Model, Distance (Å)
380	-	-	21 ± 3	17
580	-	-	36 ± 9	27
770	48 ± 16	24 ± 4	48 ± 16	35
960	72 ± 36	72 ± 36	144 ± 144	44

We confirm the validity of our interpretation by comparing the values in Table 1 with time-dependent distances calculated under the model assumption that the entire excess energy from the UV photoionization is transferred into kinetic energy. It is clear that three UV photons of 4.65 eV are needed to overcome the appearance energy of 10.7 eV for HNCO fragments in thymine [24]. This leaves approximately 3 eV of excess kinetic energy. The masses of HNCO and the $C_4H_5NO^+$ fragment are $m_1 = 43$ a.u. and $m_2 = 83$ a.u., respectively, and the velocity of their separation approximately 4600 m/s. The distance of the two fragments x as a function of time t is given in Equation (1):

$$x(t) = v_2 t + v_1 t = t \left(\sqrt{\frac{2E_{kin}}{m_2 \left(\frac{m_2}{m_1} + 1 \right)}} + \sqrt{\frac{2E_{kin}}{m_1 \left(\frac{m_1}{m_2} + 1 \right)}} \right) \tag{1}$$

Time-dependent distances from this model agree reasonably well with those from evaluating the peak shifts, which strongly supports our assignment to the effect of Coulomb interactions with the cationic fragment.

We are using the expression for time-dependent distances to model the evolution of the HNCO Auger spectra. We are describing the HNCO peaks as Gaussian functions with 1.5 eV full width at half maximum as extracted from the steady-state HNCO spectra. We simulate the time-dependent shifts by assuming a pure Coulombic $1/x$ potential (using atomic units). A trivial explanation for the peak sharpening would be that it is due to the limited time resolution of 330 fs. Since the gradient of the Coulomb potential has an inverse dependence on the distance to the charged fragment, sampling over the 330 fs time evolution would broaden the peaks more strongly at earlier times than at later times. Convolving our modeled time-dependent spectra, however, does not completely account for the peak sharpening. To achieve good agreement with experimental spectra, we had to include an additional time-dependent broadening to the Gaussian functions, which decreases from 4.7 to 1.5 eV during the first picosecond. The Gaussian peaks are normalized according to their area and scaled by the same universal factor for all time delays. The resulting simulated spectra are given as the grey thick lines in Figure 3b. The shift as well as the position fits well within the noise on the experimental Auger spectra.

The additional broadening most likely stems from the finite width of the nuclear wavepacket. From [24], we know that the dissociating nuclear wavepacket has to overcome two potential barriers which will broaden it in the dissociation coordinate. The broad wavepacket in this degree of freedom will have a similar effect as the time resolution on the Auger electron peak width. At small distances/early times, there might, however, be an additional contribution from long-range interactions during Auger decay, which can have a distance-dependent effect on the DOFS, even at separations of several angstroms substantially beyond the van der Waals radii [16]. Future experiments with improved time resolution can easily investigate this early phase of photofragmentation. This should allow us to follow the reduction in DOFS during separation of the fragments in real-time and provide important information about the nature of the fragmentation process.

Appl. Sci. **2017**, *7*, 681

4. Conclusions

We conclude by pointing out the unique long-range sensitivity of Auger electron spectroscopy to the presence of charged particles, which allows one to map time-dependent signatures of fragments onto a dissociation coordinate. The important observable, the spectral shift with respect to field-free Auger decay, can be easily interpreted based on a Coulomb potential without recourse to more complicated theory. In principle, any probe process generating and detecting charged particles is sensitive to the Coulomb potential of a nearby charged fragment. However, the sensitivity is strongly reduced in the case of ion detection due to their orders of magnitude larger masses. Valence electron spectroscopy should, in principle, show the same effect. However, due to strong vibrational substructure in valence photoelectron spectra, the effect would be difficult to observe, Auger electron spectroscopy on the other hand has the dynamic range to observe both the signatures of the molecular ground state and of the fragmentation process. Moreover, due to the strong difference in DOFS, it provides distinct signatures of fragment generation. Future experiments are planned to exploit this powerful X-ray induced Auger electron technique.

Acknowledgments: This work was supported by the AMOS program within the Chemical Sciences, Geosciences, and Biosciences Division of the Office of Basic Energy Sciences, Office of Science, U.S. Department of Energy. Parts of this research were carried out at the Linac Coherent Light Source (LCLS) at the SLAC National Accelerator Laboratory. LCLS is an Office of Science User Facility operated for the U.S. Department of Energy Office of Science by Stanford University. Parts of this research were carried out at the PLEIADES beamline, Soleil Synchrotron, France. MG acknowledges funding via the Office of Science Early Career Research Program through the Office of Basic Energy Sciences, U.S. Department of Energy and NB under grant no. DE-SC0012376. MG is funded by a Lichtenberg Professorship from the Volkswagen foundation. IF acknowledges DFG, Project FI 575/7-3 for funding. TJAW thanks the German National Academy of Sciences Leopoldina for a fellowship (Grant No. LPDS2013-14). R.F. acknowledges funding from the Swedish Research Council (VR), the Göran Gustafsson Foundation and the Knut and Alice Wallenberg Foundation, Sweden. We thank S. Miyabe, A. Aguilar, J. C. Castagna, L. Fang, J. M. Glownia, B. Murphy, A. Natan, T. Osipov, V. S. Petrovic, S. Schorb, T. Schultz, L. S. Spector, M. Swiggers, I. Tenney, S. Wang, and J. L. White for contributions to the thymine experimental data.

Author Contributions: Markus Gühr, Nora Berrah, Christoph Bostedt, John Bozek, Phil Bucksbaum, Ryan Coffee, James Cryan, Joseph Farrell, Raimund Feifel, Todd J. Martinez, Brian McFarland, and Melanie Mucke prepared and conducted the time-resolved experiment at LCLS and analyzed the results. Thomas J. A. Wolf, Fabian Holzmeier, Isabella Wagner, Saikat Nandi, Iingo Fischer, and Markus Gühr synthesized HNCO and performed the experiment at Synchrotron Soleil. Francesco Tarantelli performed the simulations. Markus Gühr analyzed the experimental data. Thomas J. A. Wolf, Markus Gühr, Fabian Holzmeier, and Ingo Fischer analyzed the results and wrote the paper.

Conflicts of Interest: The authors declare no conflict of interest.

References

1. Zewail, A.H. Femtochemistry: Atomic-Scale Dynamics of the Chemical Bond Using Ultrafast Lasers (Nobel Lecture). *Angew. Chem. Int. Ed.* **2000**, *39*, 2586–2631. [CrossRef]
2. Mukamel, S. *Principles of nonlinear optical spectroscopy*; Oxford university press: Oxford, UK, 1995.
3. Blanchet, V.; Zgierski, M.Z.; Seideman, T.; Stolow, A. Discerning vibronic molecular dynamics using time-resolved photoelectron spectroscopy. *Nature* **1999**, *401*, 52–54. [CrossRef]
4. Srinivasan, R.; Lobastov, V.A.; Ruan, C.-Y.; Zewail, A.H. Ultrafast Electron Diffraction (UED). *Helv. Chim. Acta* **2003**, *86*, 1761–1799. [CrossRef]
5. Yang, J.; Guehr, M.; Vecchione, T.; Robinson, M.S.; Li, R.; Hartmann, N.; Shen, X.; Coffee, R.; Corbett, J.; Fry, A.; et al. Diffractive imaging of a rotational wavepacket in nitrogen molecules with femtosecond megaelectronvolt electron pulses. *Nat. Commun.* **2016**, *7*, 11232. [CrossRef] [PubMed]
6. Yang, J.; Guehr, M.; Shen, X.; Li, R.; Vecchione, T.; Coffee, R.; Corbett, J.; Fry, A.; Hartmann, N.; Hast, C.; et al. Diffractive Imaging of Coherent Nuclear Motion in Isolated Molecules. *Phys. Rev. Lett.* **2016**, *117*, 153002. [CrossRef] [PubMed]
7. Dishoeck, E.F. Astrochemistry of dust, ice and gas: introduction and overview. *Faraday Discuss.* **2014**, *168*, 9–47. [CrossRef] [PubMed]
8. Koch, M.; Wolf, T.J.A.; Gühr, M. Understanding the modulation mechanism in resonance-enhanced multiphoton probing of molecular dynamics. *Phys. Rev. A* **2015**, *91*, 031403. [CrossRef]

9.   Ackermann, W.; Asova, G.; Ayvazyan, V.; Azima, A.; Baboi, N.; Bähr, J.; Balandin, V.; Beutner, B.; Brandt, A.; Bolzmann, A.; et al. Operation of a free-electron laser from the extreme ultraviolet to the water window. *Nat. Photonics* **2007**, *1*, 336–342. [CrossRef]

10. Emma, P.; Akre, R.; Arthur, J.; Bionta, R.; Bostedt, C.; Bozek, J.; Brachmann, A.; Bucksbaum, P.; Coffee, R.; Decker, F.-J.; et al. First lasing and operation of an angstrom-wavelength free-electron laser. *Nat. Photonics* **2010**, *4*, 641–647. [CrossRef]

11. Ishikawa, T.; Aoyagi, H.; Asaka, T.; Asano, Y.; Azumi, N.; Bizen, T.; Ego, H.; Fukami, K.; Fukui, T.; Furukawa, Y.; et al. A compact X-ray free-electron laser emitting in the sub-ångström region. *Nat. Photonics* **2012**, *6*, 540–544. [CrossRef]

12. Allaria, E.; Appio, R.; Badano, L.; Barletta, W.A.; Bassanese, S.; Biedron, S.G.; Borga, A.; Busetto, E.; Castronovo, D.; Cinquegrana, P.; et al. Highly coherent and stable pulses from the FERMI seeded free-electron laser in the extreme ultraviolet. *Nat. Photonics* **2012**, *6*, 699–704. [CrossRef]

13. Minitti, M.P.; Budarz, J.M.; Kirrander, A.; Robinson, J.S.; Ratner, D.; Lane, T.J.; Zhu, D.; Glownia, J.M.; Kozina, M.; Lemke, H.T.; et al. Imaging Molecular Motion: Femtosecond X-Ray Scattering of an Electrocyclic Chemical Reaction. *Phys. Rev. Lett.* **2015**, *114*, 255501. [CrossRef] [PubMed]

14. Küpper, J.; Stern, S.; Holmegaard, L.; Filsinger, F.; Rouzée, A.; Rudenko, A.; Johnsson, P.; Martin, A.V.; Adolph, M.; Aquila, A.; et al. X-Ray Diffraction from Isolated and Strongly Aligned Gas-Phase Molecules with a Free-Electron Laser. *Phys. Rev. Lett.* **2014**, *112*, 083002. [CrossRef]

15. Glownia, J.M.; Natan, A.; Cryan, J.P.; Hartsock, R.; Kozina, M.; Minitti, M.P.; Nelson, S.; Robinson, J.; Sato, T.; van Driel, T.; et al. Self-Referenced Coherent Diffraction X-Ray Movie of Ångstrom and Femtosecond-Scale Atomic Motion. *Phys. Rev. Lett.* **2016**, *117*, 153002. [CrossRef] [PubMed]

16. Erk, B.; Boll, R.; Trippel, S.; Anielski, D.; Foucar, L.; Rudek, B.; Epp, S.W.; Coffee, R.; Carron, S.; Schorb, S.; et al. Imaging charge transfer in iodomethane upon x-ray photoabsorption. *Science* **2014**, *345*, 288–291. [CrossRef] [PubMed]

17. McFarland, B.K.; Farrell, J.P.; Miyabe, S.; Tarantelli, F.; Aguilar, A.; Berrah, N.; Bostedt, C.; Bozek, J.D.; Bucksbaum, P.H.; Castagna, J.C.; et al. Ultrafast X-ray Auger probing of photoexcited molecular dynamics. *Nat. Commun.* **2014**, *5*, 4235. [CrossRef] [PubMed]

18. Liekhus-Schmaltz, C.E.; Tenney, I.; Osipov, T.; Sanchez-Gonzalez, A.; Berrah, N.; Boll, R.; Bomme, C.; Bostedt, C.; Bozek, J.D.; Carron, S.; et al. Ultrafast isomerization initiated by X-ray core ionization. *Nat. Commun.* **2015**, *6*, 8199. [CrossRef] [PubMed]

19. Petrovic, V.S.; Siano, M.; White, J.L.; Berrah, N.; Bostedt, C.; Bozek, J.D.; Broege, D.; Chalfin, M.; Coffee, R.N.; Cryan, J.; et al. Transient X-ray Fragmentation: Probing a Prototypical Photoinduced Ring Opening. *Phys. Rev. Lett.* **2012**, *108*, 253006. [CrossRef] [PubMed]

20. Berrah, N.; Fang, L.; Murphy, B.; Osipov, T.; Ueda, K.; Kukk, E.; Feifel, R.; van der Meulen, P.; Salen, P.; Schmidt, H.T.; et al. Double-core-hole spectroscopy for chemical analysis with an intense X-ray femtosecond laser. *Proc. Natl. Acad. Sci. U. S. A.* **2011**, *108*, 16912–16915. [CrossRef] [PubMed]

21. Wolf, T.J.A.; Myhre, R.H.; Cryan, J.P.; Coriani, S.; Squibb, R.J.; Battistoni, A.; Berrah, N.; Bostedt, C.; Bucksbaum, P.; Coslovich, G.; et al. Probing ultrafast ππ*/nπ* internal conversion in organic chromophores via K-edge resonant absorption. *Nat. Commun.* **2017**, *8*, 29. [CrossRef] [PubMed]

22. Ullrich, J.; Rudenko, A.; Moshammer, R. Free-Electron Lasers: New Avenues in Molecular Physics and Photochemistry. *Annu. Rev. Phys. Chem.* **2012**, *63*, 635–660. [CrossRef] [PubMed]

23. Siegbahn, K. *ESCA Applied to Free Molecules*; North-Holland Pub. Co: Amsterdam, The Netherlands, 1969.

24. Jochims, H.-W.; Schwell, M.; Baumgärtel, H.; Leach, S. Photoion mass spectrometry of adenine, thymine and uracil in the 6–22 eV photon energy range. *Chem. Phys.* **2005**, *314*, 263–282. [CrossRef]

25. Improta, R.; Scalmani, G.; Barone, V. Radical cations of DNA bases: some insights on structure and fragmentation patterns by density functional methods. *Int. J. Mass Spectrom.* **2000**, *201*, 321–336. [CrossRef]

26. Storchi, L.; Tarantelli, F.; Veronesi, S.; Bolognesi, P.; Fainelli, E.; Avaldi, L. The Auger spectroscopy of pyrimidine and halogen-substituted pyrimidines. *J. Chem. Phys.* **2008**, *129*, 154309. [CrossRef] [PubMed]

27. Rennie, E.E.; Hergenhahn, U.; Kugeler, O. A core-level photoionization study of furan. *J. Chem. Phys.* **2002**, *117*, 6524–6532. [CrossRef]

28. Sisourat, N.; Kryzhevoi, N.V.; Kolorenč, P.; Scheit, S.; Jahnke, T.; Cederbaum, L.S. Ultralong-range energy transfer by interatomic Coulombic decay in an extreme quantum system. *Nat. Phys.* **2010**, *6*, 508–511. [CrossRef]

29. Moddeman, W.E.; Carlson, T.A.; Krause, M.O.; Pullen, B.P.; Bull, W.E.; Schweitzer, G.K. Determination of the K-LL Auger Spectra of N2, O2, CO, NO, H2O, and CO2. *J. Chem. Phys.* **1971**, *55*, 2317–2336. [CrossRef]
30. Ferguson, K.R.; Bucher, M.; Bozek, J.D.; Carron, S.; Castagna, J.-C.; Coffee, R.; Curiel, G.I.; Holmes, M.; Krzywinski, J.; Messerschmidt, M.; et al. The Atomic, Molecular and Optical Science instrument at the Linac Coherent Light Source. *J. Synchrotron Radiat.* **2015**, *22*, 492–497. [CrossRef] [PubMed]
31. Bozek, J.D. AMO instrumentation for the LCLS X-ray FEL. *Eur. Phys. J.-Spec. Top.* **2009**, *169*, 129–132. [CrossRef]
32. McFarland, B.K.; Berrah, N.; Bostedt, C.; Bozek, J.; Bucksbaum, P.H.; Castagna, J.C.; Coffee, R.N.; Cryan, J.P.; Fang, L.; Farrell, J.P.; et al. Experimental strategies for optical pump: Soft X-ray probe experiments at the LCLS. *J. Phys. Conf. Ser.* **2014**, *488*, 12015. [CrossRef]
33. Schorb, S.; Gorkhover, T.; Cryan, J.P.; Glownia, J.M.; Bionta, M.R.; Coffee, R.N.; Erk, B.; Boll, R.; Schmidt, C.; Rolles, D.; et al. X-ray–optical cross-correlator for gas-phase experiments at the Linac Coherent Light Source free-electron laser. *Appl. Phys. Lett.* **2012**, *100*, 121107. [CrossRef]
34. Lindblad, A.; Söderström, J.; Nicolas, C.; Robert, E.; Miron, C. A multi-purpose source chamber at the PLEIADES beamline at SOLEIL for spectroscopic studies of isolated species: Cold molecules, clusters, and nanoparticles. *Rev. Sci. Instrum.* **2013**, *84*, 113105. [CrossRef] [PubMed]
35. Fischer, G.; Geith, J.; Klapötke, T.M.; Krumm, B. Synthesis, Properties and Dimerization Study of Isocyanic Acid. *Z. Für Naturforschung B* **2002**, *57*, 19–24. [CrossRef]
36. Schirmer, J.; Barth, A. Higher-order approximations for the particle-particle propagator. *Z. Phys. At. Nucl.* **1984**, *317*, 267–279. [CrossRef]
37. Tarantelli, F. The calculation of molecular double ionization spectra by Green's functions. *Chem. Phys.* **2006**, *329*, 11–21. [CrossRef]
38. Kendall, R.A.; Dunning, T.H.; Harrison, R.J. Electron affinities of the first-row atoms revisited. Systematic basis sets and wave functions. *J. Chem. Phys.* **1992**, *96*, 6796–6806. [CrossRef]

![applied sciences logo] *applied sciences*

MDPI

Article

X-ray Pump–Probe Investigation of Charge and Dissociation Dynamics in Methyl Iodine Molecule

Li Fang [1,*], Hui Xiong [2], Edwin Kukk [3] and Nora Berrah [2,*]

1   Center for High Energy Density Science, University of Texas, Austin, TX 78712, USA
2   Department of Physics, University of Connecticut, Storrs, CT 06269, USA; alexionghui@hotmail.com
3   Department of Physics and Astronomy, University of Turku, FI-20014 Turku, Finland; ekukk@utu.fi
*   Correspondence: lifang@austin.utexas.edu (L.F.); nora.berrah@uconn.edu (N.B.);
    Tel.: +1-512-471-3675 (L.F.); +1-860-486-0439 (N.B.)

Academic Editor: Kiyoshi Ueda
Received: 26 April 2017; Accepted: 15 May 2017; Published: 19 May 2017

Abstract: Molecular dynamics is of fundamental interest in natural science research. The capability of investigating molecular dynamics is one of the various motivations for ultrafast optics. We present our investigation of photoionization and nuclear dynamics in methyl iodine (CH_3I) molecule with an X-ray pump X-ray probe scheme. The pump–probe experiment was realized with a two-mirror X-ray split and delay apparatus. Time-of-flight mass spectra at various pump–probe delay times were recorded to obtain the time profile for the creation of high charge states via sequential ionization and for molecular dissociation. We observed high charge states of atomic iodine up to 29+, and visualized the evolution of creating these high atomic ion charge states, including their population suppression and enhancement as the arrival time of the second X-ray pulse was varied. We also show the evolution of the kinetics of the high charge states upon the timing of their creation during the ionization-dissociation coupled dynamics. We demonstrate the implementation of X-ray pump–probe methodology for investigating X-ray induced molecular dynamics with femtosecond temporal resolution. The results indicate the footprints of ionization that lead to high charge states, probing the long-range potential curves of the high charge states.

Keywords: free electron laser; X-ray pump–probe; molecular dynamics; charge dynamics

1. Introduction

Molecular dynamics, including nuclear motion and electronic motion, are one of the fundamental aspects in scientific investigations, the information of which provides microscopic insights into chemical reactions and biological processes. Photo-induced dynamics in molecules can be essential to natural processes as well as in applications: electron transportation fuels the photo-chemical energy transformation in photosynthesis [1], photoabsorption, and the subsequent molecular fragmentation leads to application in bioimaging [2] and radiation therapy [3]. Knowledge concerning the dynamics of the charge and the nuclei in a molecule, underlying phenomena—such as photoionization, charge migration, and molecular dissociation—underpins the potential for photo-control of these processes and therefore, the possibility for photo-steering chemical reactions at an atomic level, as well as for optimization of photo-production.

Due to the ultrafast nature of molecular dynamics, e.g., tens to a few hundred femtoseconds for nuclear motion and sub-femtosecond to a few femtoseconds for electronic motion, ultrafast optical tools are needed for the investigation of these dynamics. Tabletop optical lasers in near infrared, VUV, and XUV regimes have been widely applied in work regarding the evolution of nuclear wavepackets and electronic wavepackets, typically with a pump–probe scheme [4,5]. However, pump–probe experiments in the X-ray regime with tabletop lasers are still a challenge, due to the low photon output in the X-ray photon range, e.g., from high harmonic generation [6]. With the recent developments in free electron lasers (FELs), short wavelength pump–probe experiments have become practical, due to the high flux of FELs and have been realized with various X-ray split and delay (XRSD) designs [7,8]. For accelerator-based X-ray sources, pulse splitting and delay can also be implemented in the electron accelerator stage, before creating the X-ray pulses [9].

X-ray photons can access specific atomic sites in a molecule by selective inner-shell ionization. This atomic-site specificity allows for creating localized charge and hence, tracing charge transfer even with a single pulse or with a pulse duration that is longer than the actual time scale of the dynamical process [10,11]. In a previous work [10], charger transfer from iodine to the methyl group in methyl iodine molecule (CH_3I) was investigated using near-infrared (NIR) and FEL pulses. Therein, the INR pulse produced singly charged molecular ions and initiated the dissociation and the FEL pulse created positive charge localized at the iodine which transferred to the methyl group at a later time. In this work, the vanished atomic iodine ion population at short pump–probe delay time indicates the charge transfer from the iodine to the methyl group. Single FEL pulse was also used to investigate molecular dynamics in CH_3I and methylselenol (CH_3SeH) [11,12], where ion coincidence mapping was used for correlation of carbon and iodine charge states indicating the charge redistribution within the molecule. In single pulse experiments, the extraction of the timing information delay on theoretical models, typically involving Coulomb explosion [11,13].

In the present work, we used a two-mirror XRSD apparatus to generate two X-ray pulses for the pump–probe experiment where we investigated the molecular dynamics in CH_3I molecule. Using a single-color X-ray pump and X-ray probe scheme, we were able to probe with an independent clock X-ray induced molecular dynamics, e.g., the sequential ionization and relaxation of highly excited molecular ionic states via fragmentation. Using time-of-flight mass spectroscopy, we obtained the time profile of high charge states of atomic iodine and the evolution map of the kinetic energy (KE) distribution for selected charge states. The results show suppression of high charge states at small pump–probe delay time, a dramatic increase till a delay time comparable to the pulse duration, and an enhancement at relatively longer delay times, all associated with modifications in charge state branching ratios. The KE distribution shows a monotonic width narrowing down as the pump–probe delay time increases, as a consequence of increasing abundance of slow ions.

2. Materials and Methods

The experiment was carried out at the atomic, molecular and optical physics (AMO) end station at the LCLS free electron laser facility (SLAC National Laboratory, Menlo Park, CA, USA). The photon energy used was 1619 eV which dominantly ionizes the M1 subshell of iodine. The pulse duration for both pump and probe was 50 fs and the pulse energy measured upstream the focusing mirrors and the XRSD was 1.3 mJ. Considering an X-ray transportation loss of 85%, the actual total pulse energy at the interaction region is estimated to be ~0.2 mJ, corresponding to an intensity of 4×10^{16} W/cm^2 with a focal size of ~10 μm^2. Gas phase CH_3I was delivered using a pulsed valve jet. Ion Time of flight (TOF) signal was detected and recorded at different pump–probe delay times. The details of the experimental setup are described in [8].

The XRSD apparatus (see [8] for details) consists of two plane mirrors, is designed for soft X-ray photon energy range (250 eV–1800 eV) and is located downstream of the X-ray focusing mirrors of the AMO hutch. The positions (angles and height) of the two mirrors are adjusted by actuators to achieve the desired pump–probe delay time with a minimum time-resolution of 100 as and the maximum delay time of about 400 fs. The high position reproducibility of the actuators using position sensors make possible reliable pre-experiment calibration of the actuator settings for any particular pump–probe delay-time with optimal spatial overlap.

To extract the KE distribution of the ions, we first obtained with SIMION (v5, SIS Inc., Ringoes, NJ, USA) simulation the peak shapes D (E_j, t_i) of the TOF response for a sequence of discrete energy release, E_j ($j = 1, 2, \ldots, n$). These peak profiles are regarded as a set of base functions. The experimental TOF was fitted by the following function:

$$TOF(t_i) = \sum_{j=1}^{n} I_j B\left(E_j, t_i\right) \Delta E_j$$

where ΔE_j is the energy interval and I_j are the fitting coefficients.

We also used a point-charge Coulomb explosion model to calculate the time evolution for the various ionic charges. The model is described in detail in [11], where it was applied to a single-pulse scenario. The model is designed to represent, in average, the time evolution into any given set of final atomic fragment charges, including charge build-up, redistribution, and the concurrent Coulomb explosion. The sequence starts by single-photon absorption by the M1 subshell of iodine, with subsequent positive charge created by Auger cascades and possibly by additional photoabsorption. Due to large disparity in photoionization cross-sections in iodine and in the methyl group at this photon energy, only iodine is assumed to absorb the photons and charge build-up in the methyl group that occurs via electron dynamics, such as Auger cascades involving molecular orbitals. The statistical nature of the molecule is reflected in the smooth charge build-up function (instead of stepwise increase at random intervals for each single event)

$$Q_{tot}(t) = Q_{tot,\,final}\left(1 - \exp\left(-\frac{t}{\tau}\right)\right),$$

where τ is the empirical, adjustable charge build-up constant.

3. Results

3.1. Timeline for Generation of Highly Charged Atomic Iodine Ions

With a photon energy of 1619 eV, the dominant ionization at the first ionization step by a single FEL pulse is the photoionization of the M_1 shell electrons of the iodine atom which have a binding energy of ~1072 eV. The subsequent cascade Auger decay and sequential decay will lead to high charge states of iodine. The second laser pulse, arriving at a particular delay time, will further ionize the molecular ions, likely dissociating after being generated by the first pulse. At short delay time, the ionization by the second pulse kicks in while the ionization by the first pulse is still ongoing. At longer delay, the second pulse will ionize established charge states during molecular dissociation.

We observed high charge states up to 29+ for iodine and 3+ for carbon, as shown in Figure 1a (C^{4+} ion peaks marked in Figure 1a cannot be confirmed in the current experiment, because the estimated KE is ~230 eV, an increase from that for 3+ is much higher than expected or measured in previous work [11]). In the case of 'no overlap', we moved one of the two split mirrors to avoid spatial overlap of the two X-ray pulses to experimentally mimic the situation of single pulse with the same total pulse energy as two separate pulses. To ensure 'no overlap', we also set a long time delay of 150 fs. This single pulse condition will be presented in figures as the case of zero delay. The true time zero, the zero reference for all pump–probe delays, was set by mechanically making the two split mirrors co-plane [8]. With a single pulse ('no overlap'), the overall production of highly charged iodine ions is minimum even though the interaction region volume doubled, in comparison with the high charge state yield by two pulses. In contrast, the production of high charge states of carbon is at maximum in the single pulse scenario, which is similar to the production at long delay time in the two-pulse scenario. The spectral peaks for I^{21+} to I^{23+} overlap with those for C^{2+}. Their contribution, however, broadened the C^{2+} peaks and even leads to a split structure on top of the fast C^{2+} peak at the delay time of 150 fs (see Figure 1a). The I^{29+} peak also overlaps with a carbon peak, but is obviously seen on top of the slow C^{3+} peak. It is possible that charge states of iodine higher than 29+ were produced but buried under the C^{3+} peak, as barely seen at the valley between the two C^{3+} peaks (see Figure 1a).

Similarly, the I^{21+} appears between the two C^{2+} peaks. The abundance of high charge states decreases for higher charges. The seemingly higher yield of I^{16+} compared to the adjacent peaks is due to the contribution of O^{2+} from the background gas in the vacuum chamber. The production of low charge states (I^+ to I^{6+}) is favored at long delay time, reaching a maximum with a single pulse as shown in Figure 1b. At the low charge states, the ion yields at short (20 fs) and intermediate (~40 fs) delay times are similar.

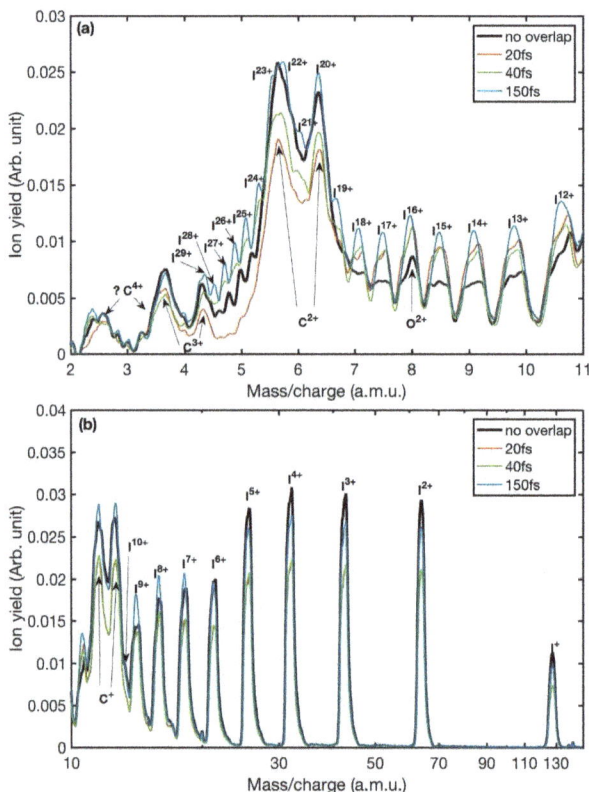

Figure 1. Ion Time of flight (TOF) spectra of CH_3I at a photon energy of 1619 eV and a pulse duration of 50 fs. Different curves (color coded) correspond to selected pump–probe delay times (see legend). The pulse energy is estimated to be ~0.2 mJ, based on 1.3 mJ upstream value and a transportation loss of 85%. 'no overlap' is considered to represent the case at zero delay time (see text). (**a**) Spectral range: <11 amu; (**b**) Spectral range: >10 amu. the x-axis is plotted in log scale.

Figure 2 shows the TOF spectra as a function of pump–probe delay with a finer step size (see figure y-axis tick labels). The time zero spectrum corresponds to that with single pulse mentioned above. As seen in Figure 2, there is an overall drop in the ionization yield at short pump–probe delay time (20 fs). Particularly, the very high charge states (>24+) are almost vanished. Most CH_3I molecular ions went through dissociation, since only a small amount of CH_3I^+ remains (see Figure 1b). The first significant peak is I^+, which is much smaller compared to the dominant I^{2+} peak, because Auger decay preferentially generates evenly charged ions and the charges are mainly localized. As the pump–probe delay time increases, the ion yield for high charge states increases within the pulse duration. The very high charge state (>24+) peaks reached their maxima at ~100 fs time delay, meanwhile the abundance of mid and low high charge states increases (see Figure 2). In Figure 2a, one clearly sees the rising peaks of I^{19+}, I^{21+}, and I^{24+} above the shoulders of the dominant C^{2+} peaks. Also, becoming clear are the peaks of I^{22+} and I^{23+} which appear as a split branch on top of the fast C^{2+} peak.

Charge states of carbon ions up to 3+ were observed. The highest charge state reached does not depend on the delay time. As seen in Figure 1, C^{3+} was produced with a single pulse or at short delay time. However, the abundance of these carbon ions shows a delay-time dependence, increasing for longer delay time. The contributions of iodine ions under the carbon peaks are not large enough to explain the changes in the carbon peaks which are dominant. With a single pulse, the carbon ion signal at all charge states reaches around the maxima, similar to the values at long delay time. This could be an indication of charge transfer from the iodine to the methyl group that was thwarted by the ionization with the probe pulse [10].

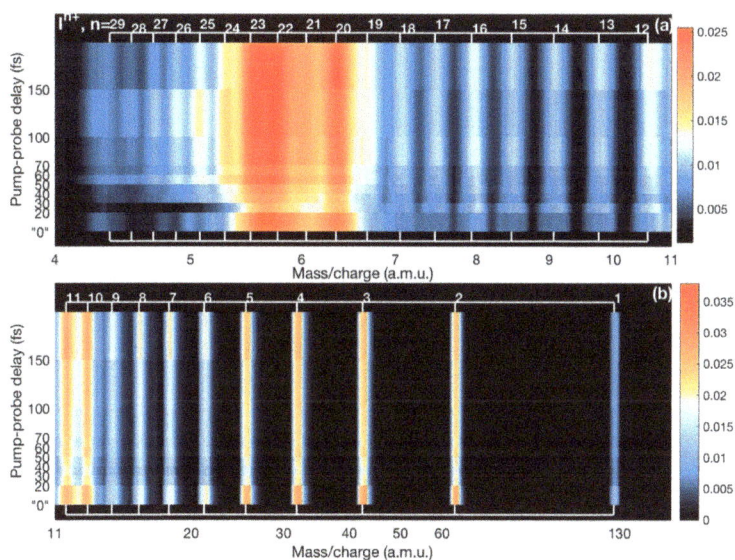

Figure 2. Ion TOF spectra as a function of pump–probe delay time at a photon energy of 1619 eV and a pulse duration of 50 fs. The color represents the ion signal per laser shot (arb. unit) (see color bar). The "0" delay corresponds to the case of no overlap between the two X-ray beams (see text). The positons for various charge states of iodine atomic ions are marked and labeled at the top of the figure. (**a**) Spectral range: mass/charge < 11 amu; (**b**) Spectral range: mass/charge > 10 amu. The *x*-axes are plotted in log scale.

To illustrate the overall picture of the evolution of the iodine charge states, we integrated the signal under the spectral peaks of various charge states at different pump–probe times, as shown in Figure 3. The integral signal with a single pulse was subtracted from the integrated signal at all other delay times, i.e., using the single pulse spectrum as a reference. Clearly seen in Figure 3 is: (1) high charge states were enhanced and low charge states were partially depleted by the second FEL pulse, shown as the trend from negative to positive ion yield difference for increasing charge states; (2) at short delay time, the depletion of low charge states were stronger compared to that at long delay, shown as the trend of decreasing absolute ion yield difference for increasing delay time within the block for each different charge state; (3) long delay time favors the production of very high charge states, consistent with the theme that these charge states were suppressed at 20 fs delay (negative values for I^{25+} to I^{28+}); (4) 50 fs delay, which is equal to the pulse duration, is an odd point at which the low and very high charges are dramatically enhanced, but not the intermediate charge states, compared to the case at adjacent delay times.

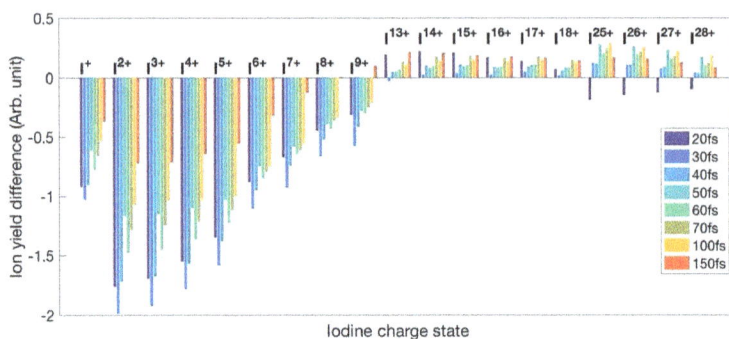

Figure 3. Ion yield as a function of pump–probe delay time for various iodine charge states. Data were extracted from TOF spectra in Figure 2. The ion yield with a single X-ray pulse was used as a reference spectrum and subtracted from the iodine charge states at all other delay times.

As roughly seen in Figure 2a, there seems to be a phase shift between the spectrum with single pulse and those at delays larger than 50 fs for charge states I^{25+} to I^{28+}. The spectra of the same charge state region are shown in Figure 4, where a decay base line of the spectrum at 20 fs delay was subtracted from all spectra and the spectrum at 20 fs delay was amplified for clarity of comparison. The 'phase shift' actually reflects the fact that at single pulse or small delay times, ions with large KE were produced; as the delay time increases, low KE ions were produced, filling the valley between the forward and backward peaks of the high KE ion and even forming a new peak for large delay time. At intermediate delay times, around 30 fs, the transient case is shown as multiple peaks from both above origins are seen.

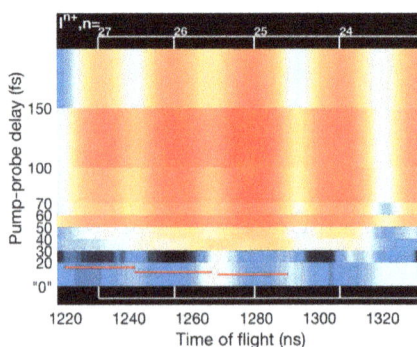

Figure 4. Ion TOF spectra as a function of pump–probe delay time. The data were from Figure 2a with the *x*-axis plotted in TOF. A background trace obtained for the spectrum at 20 fs was subtracted from all spectra. The spectrum at 20 fs delay was magnified by ×6 for purpose of clarity for presentation. The '0' delay corresponds to the case of no overlap between the two X-ray beams (see text). At the top of the figure, the positions for the charge states of atomic iodine were marked and labeled. The red horizontal lines indicate the TOF spread of the forward and backward peaks of I^{25+}, I^{26+}, and I^{27+}, identified in the spectrum at 20 fs or single pulse.

3.2. Evolution of KE Distribution

As seen in Figure 1a, with a single pulse, the KE for high charge states (I^{12+}–I^{18+}) has a broad distribution, shown as a flat top shape in the TOF spectrum. As the delay time increases, the abundance of ions with small KE increased, adding signal on top of the flat top, eventually leading to a sharp dome shape, as seen at long delay times. As to extremely high charge states of iodine (I^{20+}–I^{29+}), a sharp distribution starts at short delay time. For low charge states (shown in Figure 1b), the distribution is relatively sharp even at single pulse or short delay time, indicating small average KEs. Similarly seen in Figure 2a, the broad KE distribution with a single pulse is represented by an evenly colored rectangular region at around zero delay time. As more low KE ions were detected at longer delay times, the shape of distribution becomes sharper, forming slim columns (see Figure 2).

We extract the ion KE from the TOF spectra with the method described in Section 2. Figure 5 shows the KE of I^{14+} as a function of pump–probe delay time. Each KE spectrum was normalized to its maximum. The map was interpreted without significant distortion of the distribution to smooth out the uneven delay-time steps. The ion species of I^{14+} was selected as an example to show the evolution of the KE distribution and other high charge state species show a similar trend (see Figure 1). The difference between the distribution obtained from the forward ions and backward ions is mainly due to the uneven base line caused by spectral overlap with the adjacent peaks. Figure 6 shows the weighted KE averaged over the forward and backward ions. At small delay time, the KE of I^{14+} shows a broad distribution with an average KE of 10 eV and a width of ~15 eV. At about 60 fs delay time, the KE distribution approaches an asymptotic value of ~8 eV with a width of ~8 eV.

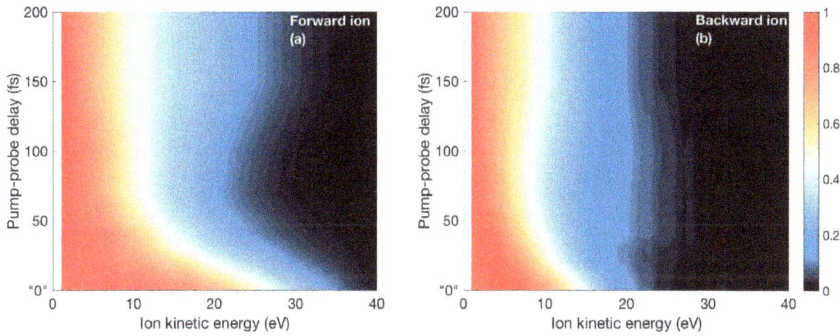

Figure 5. Ion KE distribution of I^{14+} as a function of pump–probe delay time at a photon energy of 1619 eV and a pulse duration of 50 fs. The color represents the ion signal strength which is normalized to the signal maximum at each delay time. The '0' delay corresponds to the case of no overlap between the two X-ray beams (see text). (**a**) Ions with the direction of the initial velocity towards the detector ('forward ion'); (**b**) Ions with the direction of the initial velocity against the detector ('backward ion').

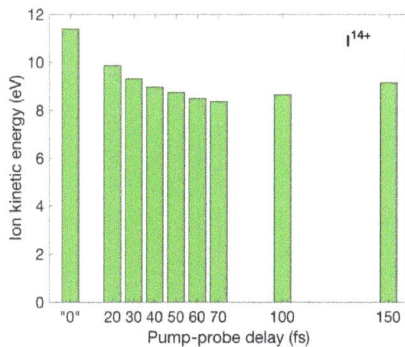

Figure 6. Weighted KE averaged over the forward and backward I^{14+} ions as a function of pump–probe delay time, extracted from data shown in Figure 5. The '0' delay corresponds to the case of no overlap between the two X-ray beams (see text).

4. Discussion

As mentioned above, two pulses (with spatial overlap) produced more high charge state iodine ions than a single pulse even with the same total pulse energy. The reason is that the ions generated by the first pulse and further ionized by the second pulse are associated with higher ionization potentials which are closer to the photon energy and therefore have larger ionization cross sections compared to the neutral particles. Since the life time of the iodine M1 shell is only ~435 as [14], much shorter than the X-ray pulse duration, the suppression of every high charge states at short delay time is likely due to the depletion of valence shell electrons which hinders the Auger decay, rather than X-ray induced photoabsorption frustration caused by inner-shell electron depletion which hinders sequential photoionization [15,16]. Indeed, with a photon energy of 1916 eV, photoionization of M1 shell electron alone can only lead to up to I^{26+}, since the binding energy of a 3s electron of I^{26+} (with all valence vacancies) is ~1680 eV [17]. At long delay times, the population in iodine high charge states increased, because the long delay times allow for the completion of the electron transfer from carbon sites to the methyl group, restocking the valence electrons at iodine needed for the Auger decay.

We obtained the average charge of the atomic iodine ions for different pump–probe delay times, as shown in Figure 7a. With a single FEL pulse, the average charge of iodine is ~7 amu; with the second pulse, a charge of ~9 amu was established around 60 fs. The experimental results show a charge saturation at ~30 fs. We simulated the charge evolution in a pump–probe experiment, as shown in Figure 7b. The charge by a single X-ray pulse was set to be ~7 amu and a charge built up time of ~30 fs was used. Since the photoionization cross section for carbon is more than one order of magnitude less than that of iodine at 1619 eV, photoionization of carbon was excluded in the simulation. Hence, the carbon charge mainly resulted from the charge transfer between the iodine and the methyl group, which can charge the carbon site up to 4+ [10,11]. As seen in Figure 7b, the charge on carbon is not affected by the probe pulse arriving 50 fs later after the pump pulse, because the charge transfer was hindered due to the large C–I bond separation at the latter time of the molecular dissociation.

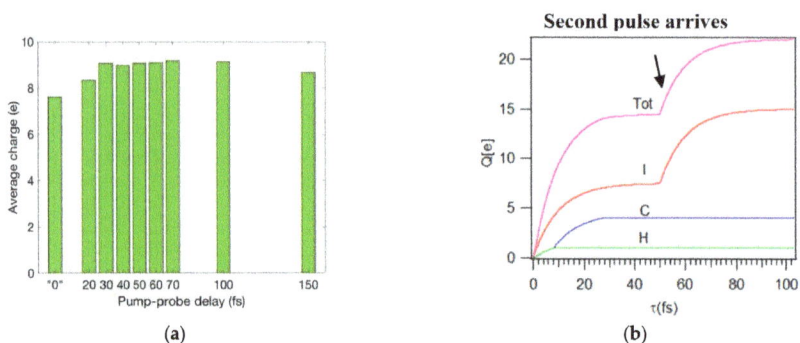

Figure 7. (a) Experimental average charge of iodine as a function of pump–probe delay time. The '0' delay corresponds to the case of no overlap between the two X-ray beams (see text); (b) Simulation of charges as a function of time for different ionic species. Pulse duration of 50 fs and delay time of 50 fs were used. The *x*-axis is time with zero position at the onset of the first pulse.

From the TOF spectral peaks with single pulse, we estimated the KEs for I^{25+}, I^{26}, and I^{27+} (corresponding to horizontal lines in Figure 4, see figure caption), as well as the KEs for the carbon ions and their associated iodine ions on average. The results are presented in Table 1. Also shown in Table 1 for comparison are the KEs of carbon ions estimated from the results reported in [11]. Since the C-I bond breaks within the first 10 fs of the X-ray pulse [11], we expect the KE for C^+ which was dominantly produced at an early time during the FEL pulse to be similar between the experiments in current work and in [11]. This is indeed the case as shown in Table 1. However, for higher carbon charge states

(2+, 3+), the KEs in the current work are much lower than those in [11]. A plausible explanation is: the high charge states observed in [11] were produced within 10 fs of a single pulse, where one can consider a scenario of internal clock pump–probe experiment [13,18]; at the very short effective delay time, the dissociation at short range of the internuclear separation was probed, leading to large KEs due to stronger Coulomb repulsion; whereas in the current work where an external independent clock was applied, allowing for relatively larger delay times, the potential curve at long range was probed, leading to lower dissociation energy at the end. As an example, for low charge states of iodine, we estimated the KE for I^{4+} to be 4.5 eV which is nearly constant for different delay times, indicating that the C^{3+} ions are likely to be associated with I^{4+}. With a similar reasoning, the very high charge states of iodine are likely associated with high charge states of carbon, but due to their small abundance, the contribution of the associated carbon ions is also small.

Table 1. Kinetic energy (KE) of carbon and iodine ions. The values for the current work were obtained from the spectrum with single pulse. The charge states in the parenthesis are the species associated those for which the KEs are presented.

Species	KE (eV) [1]	Species	KE (eV) [1]	Species	KE (eV) [1]	Species	KE (eV) [2]
C^+ ($-I^{n+}$)	5.0 ± 0.9	I^{n+} ($-C^+$)	0.5 ± 0.3	I^{25+}	49 ± 9	C^+ ($-I^+$)	~6
C^{2+} ($-I^{n+}$)	19.1 ± 2.3	I^{n+} ($-C^{2+}$)	1.8 ± 0.2	I^{26+}	61 ± 10	C^{2+} ($-I^{5+}$)	~47
C^{3+} ($-I^{n+}$)	42.5 ± 7.9	I^{n+} ($-C^{3+}$)	4.0 ± 0.4	I^{27+}	55 ± 10	C^{3+} ($-I^{7+}$)	~100

[1] Current work. The KEs of I^{n+} are extracted from that of C ion, assuming a two-body break-up. [2] Reference [11]. Single pulse experiment with FEL pulses of 10 fs duration and 5.5 keV photon energy.

As mentioned above, the KE approaches a constant value at ~60 fs delay and the ion yield at 50 fs delay is an odd point in the overall trend. It remains a question whether this tip point around 50 fs for the kinetics is a pulse duration factor or a species-dependent factor. Seeking the answer will be for future investigations with different FEL pulse durations. The pulse duration was not directly measured but derived based on electron bunch duration measurements [15]. The actual pulse duration has been reported to be about half of the cited value [19]. Considering an actual pulse duration of 25 fs for a nominal value of 50 fs, the interference between the pump and probe pulses is not expected to be an issue in the current work.

5. Conclusions

We presented an investigation of the ionization and dissociation dynamics in CH_3I molecule with an X-ray pump and X-ray probe scheme. To illustrate the charge dynamics and nuclear kinetics, we showed the charge state and the KE distribution for selected charge states as a function of pump–probe delay time. We observed the creation of ions with low KE at a relatively long delay time, probing the long-range potential curves of molecular high charge states. We also observed the suppression of high charge states at short delay time and an increase of the high charge states at long delay time, indicating the time scale of the C–I charge transfer. X-ray pump–probe experimental approach combined with coincidence measurements for resolving fragmentation channels could hold the promise for unambiguous time-resolved investigation of molecular dynamics, including decoupled dissociation and charge transfer dynamics.

Acknowledgments: We acknowledge the XRSD commissioning team: Brendan F. Murphy, Timur Y. Osipov, Ryan Coffee, Ken R. Ferguson, Jean-Charles Castagna, Vladmire S. Petrovic, Sebastian Carron Montero, and John D. Bozek. This work was funded by the Department of Energy, Office of Science, Basic Energy Sciences (BES), Division of Chemical Sciences, Geosciences, and Biosciences under a SISGR program1 No. DE-SC0002004, under grant No. DE-SC0012376 and for the Linac Coherent Light Source (LCLS), SLAC National Accelerator Laboratory, under Contract No. DE-AC02-76SF00515. E.K. acknowledges the support by the Academy of Finland. L.F. acknowledges the support by Defense Advanced Research Project Agency Contract 12-63-PULSE-FP014, and by National Nuclear Security Administration Cooperative Agreement DE-NA0002008.

Appl. Sci. **2017**, *7*, 529

Author Contributions: N.B. conceived the experiment; L.F., H.X., E.K. and N.B. performed the experiments; H.X. and L.F. analyzed the data; E.K. did the simulation; L.F. wrote the paper.

Conflicts of Interest: The authors declare no conflict of interest.

References

1. Lewis, K.L.M.; Ogilvie, J.P. Probing photosynthetic energy and charge transfer with two-dimensional electronic spectroscopy. *J. Phys. Chem. Lett.* **2012**, *3*, 503–510. [CrossRef] [PubMed]
2. Neustetter, M.; Aysina, J.; da Silva, F.F.; Denifl, S. The Effect of Solvation on Electron Attachment to Pure and Hydrated Pyrimidine Clusters. *Angew. Chem. Int. Ed. Engl.* **2015**, *54*, 9124–9126. [CrossRef] [PubMed]
3. Khan, F.M. Part I 5. Interactions of Ionizing radiation. In *The Physics of Radiation Therapy*, 3rd ed.; Lippincott Williams&Wilkins: Philadelphia, PA, USA, 2003.
4. Magrakvelidze, M.; Herrwerth, O.; Jiang, Y.H.; Rudenko, A.; Kurka, M.; Foucar, L.; Kühnel, K.U.; Kübel, M.; Johnson, N.G.; Schröter, C.D.; et al. Tracing nuclear-wave-packet dynamics in singly and doubly charged states of N_2 and O_2 with XUV-pump–XUV-probe experiments. *Phys. Rev. A* **2012**, *86*, 013415. [CrossRef]
5. Leone, S.R.; McCurdy, C.W.; Burgdörfer, J.; Cederbaum, L.S.; Chang, Z.; Dudovich, N.; Feist, J.; Greene, C.H.; Ivanov, M.; Kienberger, R.; et al. What will it take to observe processes in 'real time'? *Nat. Photonics* **2014**, *8*, 162–166. [CrossRef]
6. Hädrich, S.; Rothhardt, J.; Krebs, M.; Demmler, S.; Klenke, A.; Tünnermann, A.; Limpert, J. Single-pass high harmonic generation at high repetition rate and photon flux. *J. Phys. B At. Mol. Opt. Phys.* **2016**, *49*, 1–26. [CrossRef]
7. Woestmann, M.; Mitzner, R.; Noll, T.; Roling, S.; Siemer, B.; Siewert, F.; Eppenhoff, S.; Wahlert, F.; Zacharias, H. The XUV split-and-delay unit at beamline BL2 at FLASH. *J. Phys. B At. Mol. Opt. Phys.* **2013**, *46*, 164005. [CrossRef]
8. Berrah, N.; Fang, L.; Murphy, B.F.; Kukk, E.; Osipov, T.Y.; Coffee, R.; Ferguson, K.R.; Xiong, H.; Castagna, J.C.; Petrovic, V.S.; et al. Two mirror X-ray pulse split and delay instrument for femtosecond time resolved investigations at the LCLS free electron laser facility. *Opt. Express* **2016**, *24*, 11768–11781. [CrossRef] [PubMed]
9. Ding, Y.; Decker, F.J.; Emma, P.; Feng, C.; Field, C.; Frisch, J.; Huang, Z.; Krzywinski, J.; Loos, H.; Welch, J.; et al. Femtosecond X-ray pulse characterization in free-electron lasers using a cross-correlation technique. *Phys. Rev. Lett.* **2012**, *109*, 254802. [CrossRef] [PubMed]
10. Erk, B.; Boll, R.; Trippel, S.; Anielski, D.; Foucar, L.; Rudek, B.; Epp, S.W.; Coffee, R.; Carron, S.; Schorb, S.; et al. Imaging charge transfer in iodomethane upon X-ray photoabsorption. *Science* **2014**, *345*, 288–291. [CrossRef] [PubMed]
11. Motomura, K.; Kukk, E.; Fukuzawa, H.; Wada, S.; Nagaya, K.; Ohmura, S.; Mondal, S.; Tachibana, T.; Ito, Y.; Koga, R.; et al. Charge and Nuclear Dynamics Induced by Deep Inner-Shell Multiphoton Ionization of CH_3I Molecules by Intense X-ray Free-Electron Laser Pulses. *J. Phys. Chem. Lett.* **2015**, *6*, 2944–2949. [CrossRef] [PubMed]
12. Erk, B.; Rolles, D.; Foucar, L.; Rudek, B.; Epp, S.W.; Cryle, M.; Bostedt, C.; Schorb, S.; Bozek, J.; Rouzee, A.; et al. Ultrafast Charge Rearrangement and Nuclear Dynamics upon Inner-Shell Multiple Ionization of Small Polyatomic Molecules. *Phys. Rev. Lett.* **2013**, *110*, 053003. [CrossRef] [PubMed]
13. Fang, L.; Osipov, T.; Murphy, B.; Tarantelli, F.; Kukk, E.; Cryan, J.P.; Glownia, M.; Bucksbaum, P.H.; Coffee, R.N.; Chen, M.; et al. Multiphoton ionization as a clock to reveal molecular dynamics with intense short X-FEL pulses. *Phys. Rev. Lett.* **2012**, *109*, 263001. [CrossRef] [PubMed]
14. Dampbell, J.L.; Papp, T. Widths of the atomic K-N7 levels. *Atom. Data Nucl. Data* **2001**, *77*, 1–56. [CrossRef]
15. Young, L.; Kanter, E.P.; Krässig, B.; Li, Y.; March, A.M.; Pratt, S.T.; Santra, R.; Southworth, S.H.; Rohringer, N.; DiMauro, L.F.; et al. Femtosecond Electronic Response of Atoms to Ultra-Intense X-rays. *Nature* **2010**, *466*, 56–61. [CrossRef] [PubMed]
16. Hoener, M.; Fang, L.; Kornilov, O.; Gessner, O.; Pratt, S.T.; Gühr, M.; Kanter, E.P.; Blaga, C.; Bostedt, C.; Bozek, J.D.; et al. Ultraintense X-ray Induced Ionization, Dissociation, and Frustrated Absorption in Molecular Nitrogen. *Phys. Rev. Lett.* **2010**, *104*, 253002. [CrossRef] [PubMed]
17. Estimated Using Los Alamos Atomic Physics Codes. Available online: http://aphysics2.lanl.gov/cgi-bin/ION/runlanl08f.pl (accessed on 26 April 2017).

18. Fang, L.; Osipov, T.; Murphy, B.F.; Rudenko, A.; Rolles, D.; Petrovic, V.; Bostedt, C.; Bozek, J.D.; Bucksbaum, P.H.; Berrah, N. Probing ultrafast electronic and molecular dynamics with free electron lasers. *J. Phys. B At. Mol. Opt. Phys.* **2014**, *47*, 124006. [CrossRef]

19. Murphy, B.F.; Osipov, T.; Jurek, Z.; Fang, L.; Son, S.K.; Mucke, M.; Eland, J.H.D.; Zhaunerchyk, V.; Feifel, R.; Avaldi, L.; et al. Bucky ball explosion by intense femtosecond X-ray pulses: A model system for complex molecules. *Nat. Commun.* **2014**, *5*, 4281. [CrossRef] [PubMed]

applied
sciences

MDPI

Article

Application of Matched-Filter Concepts to Unbiased Selection of Data in Pump-Probe Experiments with Free Electron Lasers

Carlo Callegari [1,2,*,†], Tsukasa Takanashi [3,†], Hironobu Fukuzawa [3], Koji Motomura [3,‡],
Denys Iablonskyi [3], Yoshiaki Kumagai [3,§], Subhendu Mondal [3], Tetsuya Tachibana [3],
Kiyonobu Nagaya [4], Toshiyuki Nishiyama [4], Kenji Matsunami [4], Per Johnsson [5], Paolo Piseri [6],
Giuseppe Sansone [7,8], Antoine Dubrouil [7,||], Maurizio Reduzzi [7], Paolo Carpeggiani [7],
Caterina Vozzi [7], Michele Devetta [7], Davide Faccialà [7], Francesca Calegari [7,9],
Mattea Carmen Castrovilli [7], Marcello Coreno [1,2], Michele Alagia [10], Bernd Schütte [11],
Nora Berrah [12], Oksana Plekan [1], Paola Finetti [1], Eugenio Ferrari [1,¶], Kevin Charles Prince [1,10]
and Kiyoshi Ueda [3]

1   Elettra-Sincrotrone Trieste S.C.p.A., Strada Statale 14–km 163.5 in Area Science Park,
    34149 Basovizza, Trieste, Italy; marcello.coreno@elettra.eu (M.C.); oksana.plekan@elettra.eu (O.P.);
    paola.finetti12@gmail.com (P.F.); eugenio.ferrari@psi.ch (E.F.); kevin.prince@elettra.eu (K.C.P.)
2   CNR-ISM, Area Science Park, 34149 Basovizza, Trieste, Italy
3   Institute of Multidisciplinary Research for Advanced Materials, Tohoku University, Sendai 980-8577, Japan;
    tsukasat@mail.tagen.tohoku.ac.jp (Ts.T.); fukuzawa@tagen.tohoku.ac.jp (H.F.);
    motomura@spring8.or.jp (Ko.M.); denys@tagen.tohoku.ac.jp (D.I.); kumagay@anl.gov (Y.K.);
    justsm@gmail.com (S.M.); tach.ymn@gmail.com (Te.T.); ueda@tagen.tohoku.ac.jp (K.U.)
4   Department of Physics, Graduate School of Science, Kyoto University, Kyoto 606-8502, Japan;
    nagaya@scphys.kyoto-u.ac.jp (K.N.); t-nishiyama@scphys.kyoto-u.ac.jp (T.N.);
    matsunami.yaolab.kyoto@gmail.com (Ke.M.)
5   Department of Physics, Lund University, P.O. Box 118, 22100 Lund, Sweden; per.johnsson@fysik.lth.se
6   CIMAINA and Dipartimento di Fisica, Università degli Studi di Milano, Via Celoria 16, 20133 Milano, Italy;
    paolo.piseri@unimi.it
7   CNR-IFN, Piazza Leonardo da Vinci 32, 20133 Milano, Italy; giuseppe.sansone@physik.uni-freiburg.de (G.S.);
    dubrouil@femtoeasy.eu (A.D.); mauriziobattista.reduzzi@polimi.it (M.R.);
    paolo.carpeggiani@tuwien.ac.at (P.C.); caterina.vozzi@ifn.cnr.it (C.V.); michele.devetta@mail.polimi.it (M.D.);
    davide.facciala@polimi.it (D.F.); francesca.calegari@desy.de (F.C.); matteac@gmail.com (M.C.C.)
8   Physikalisches Institut, Albert-Ludwigs-Universität, 79104 Freiburg, Germany
9   Center for Free-Electron Laser Science, DESY, Notkestrasse 85, 22607 Hamburg, Germany
10 CNR-IOM, Area Science Park, 34149 Basovizza, Trieste, Italy; alagiam@elettra.eu
11 Max-Born-Institut, Max-Born-Strasse 2 A, 12489 Berlin, Germany; schuette@mbi-berlin.de
12 Department of Physics, University of Connecticut, 2152 Hillside Road, Storrs, CT 06269, USA;
    nora.berrah@uconn.edu
*   Correspondence: carlo.callegari@elettra.eu; Tel.: +39-040-375-8844
†   These authors contributed equally to this work.
‡   Current address: RIKEN SPring-8 Center, Kouto, Sayo, Hyogo 679-5148, Japan.
§   Current address: Argonne National Laboratory, 9700 S. Cass Avenue, Argonne, IL 60439, USA.
|| Current address: Femto Easy, Parc scientifique Laseris 1, 33114 Le Barp, France.
¶   Current address: Particle Accelerator Physics Laboratory, École Polytechnique Fédérale de Lausanne EPFL,
    CH-1015 Lausanne, Switzerland.

Academic Editor: Malte Kaluza
Received: 27 April 2017; Accepted: 9 June 2017; Published: 16 June 2017

Abstract: Pump-probe experiments are commonly used at Free Electron Lasers (FEL) to elucidate the femtosecond dynamics of atoms, molecules, clusters, liquids and solids. Maximizing the signal-to-noise ratio of the measurements is often a primary need of the experiment, and the aggregation of repeated, rapid, scans of the pump-probe delay is preferable to a single long-lasting scan.

Appl. Sci. **2017,** *7,* 621

The limited availability of beamtime makes it impractical to repeat measurements indiscriminately, and the large, rapid flow of single-shot data that need to be processed and aggregated into a dataset, makes it difficult to assess the quality of a measurement in real time. In post-analysis it is then necessary to devise unbiased criteria to select or reject datasets, and to assign the weight with which they enter the analysis. One such case was the measurement of the lifetime of Intermolecular Coulombic Decay in the weakly-bound neon dimer. We report on the method we used to accomplish this goal for the pump-probe delay scans that constitute the core of the measurement; namely we report on the use of simple auto- and cross-correlation techniques based on the general concept of "matched filter". We are able to unambiguously assess the signal-to-noise ratio (*SNR*) of each scan, which then becomes the weight with which a scan enters the average of multiple scans. We also observe a clear gap in the values of *SNR*, and we discard all the scans below a *SNR* of 0.45. We are able to generate an average delay scan profile, suitable for further analysis: in our previous work we used it for comparison with theory. Here we argue that the method is sufficiently simple and devoid of human action to be applicable not only in post-analysis, but also for the real-time assessment of the quality of a dataset.

Keywords: correlation; matched filter; pump-probe; free electron laser; data processing; statistical weight

---

1. Introduction

Recently we described the results of a pump-probe experiment, in which the lifetimes of doubly excited states of neon dimers were measured [1]. The dimers were excited by absorption of two EUV photons from the Free Electron Laser (FEL) FERMI-1 [2] and probed via ionization by a UV laser pulse.

The excited dimers decayed by Interatomic Coulombic Decay to stable dimer cations Ne_2^+, which were detected by a time-of-flight (TOF) mass spectrometer. Ionization by the UV pulse led to a repulsive state of the dimer, which dissociated, so that the yield of dimer ions was reduced, and the yield of Ne^+ increased. The dimer sample used in that work was very dilute with a large atomic background, as it was produced in a supersonic expansion of neon gas, with a yield of about 1%. The data were therefore noisy, and a substantial effort was needed to analyze and filter them. In this paper, we describe the methods used for that analysis.

The data consisted of a number of temporal scans, each lasting approximately 30 min, in which the ion-TOF signal was measured as a function of the delay between pump and probe pulses, and these scans had poor signal-to-noise ratios. Simply averaging them did not give good results, probably because some of them suffered from poor FEL conditions or poor alignment of the pump and probe pulses. One should reject those scans that do not contribute significant signal, while avoiding human bias. A possible approach is to exploit the difference between signal and noise with respect to auto- and cross-correlation. This is an instance of the more general concept of "matched filter", i.e., the linear filter maximizing the signal-to-noise ratio (SNR) of a measured noisy sample [3]. Under the assumption of white noise, the matched filter is the complex-conjugate time-reversal of the signal sought, i.e., application of the matched filter returns the autocorrelation of the signal [4]. The concept is widely applied in signal processing, where one is primarily interested in extracting a burst of periodic signal from a noisy sample. As far as peak detection is concerned, a popular field of application is chromatography, but the general results obtained there apply equally well to our case; Ref. [5] explicitly discusses the use of a matched filter to determine amplitude and time shift of the peak being sought. We consider the use of auto- and cross-correlation for three purposes:

1. validate the scans to be included or excluded from averaging
2. determine the weight with which each scan should enter the average
3. determine, if desired, by how much to temporally shift a scan prior to averaging

2. Results

For a set of delays $\{d_i\}$, $i = 0..n$, let us consider two delay scans $\mathbf{R} = \{R_0, \ldots, R_n\}$ and $\mathbf{S} = \{S_0, \ldots, S_n\}$ each consisting of a sequence of Ne_2^+ ion-TOF signals, reduced as explained in Section 4, and padded with zeros outside of the delay range scanned. We will consider $\mathbf{R}$ our reference scan, which for convenience is assumed to be noise-free.

The definition of cross-correlation without normalization, $\Gamma_{\mathbf{RS},q}$ is:

$$\Gamma_{\mathbf{RS},q} = \sum_i R_i S_{i+q} \equiv (\mathbf{R} \star \mathbf{S})_q \tag{1}$$

Note that Γ is linear in the two sequences (scaling either one by a factor α scales Γ by the same factor), and that $\Gamma_{\mathbf{SS},q}$ is the autocorrelation of $\mathbf{S}$, which has a maximum for $q = 0$:

$$(\mathbf{S} \star \mathbf{S})_0 = \sum_i (S_i)^2 \tag{2}$$

We now assume that any scan $\mathbf{N} = \{N_0, \ldots, N_n\}$ consisting of pure noise has zero (negligible) cross-correlation with any other scan (including itself except, obviously, at zero-shift), thus:

$$(\mathbf{N} \star \mathbf{N})_q = \delta_{q0} \sum_i (N_i)^2 \tag{3}$$

$$(\mathbf{R} \star \mathbf{N})_q = 0 \tag{4}$$

with δ_{q0} the Kronecker delta; we use Equation (3) to define a variance $\sigma^2 = (\mathbf{N} \star \mathbf{N})_0 = \sum_i (N_i)^2$. Equations (3) and (4) strictly hold for white noise and infinite n; in a real-life situation we can expect that noise correlation just decays much faster than signal correlation: thus all scans whose autocorrelation is sharply peaked are probably pure noise, and will also have poor cross-correlation with the reference scan.

Any noisy scan $\mathbf{S}$ can be written in terms of the reference scan $\mathbf{R}$ and a pure-noise sequence $\mathbf{N}$ as: $\mathbf{S} = \alpha \mathbf{R} + \mathbf{N}$, with α a real number. In a worse-case scenario, S_i may be shifted (for simplicity by an integer index r), that is: $S_i = \alpha R_{i+r} + N_i$. Then:

$$(\mathbf{R} \star \mathbf{S})_q = \alpha (\mathbf{R} \star \mathbf{R})_{q-r} + (\mathbf{R} \star \mathbf{N})_q = \alpha (\mathbf{R} \star \mathbf{R})_{q-r} \tag{5}$$

note that $\mathbf{R} \star \mathbf{S}$ is proportional to $\mathbf{R} \star \mathbf{R}$ but shifted by r as expected; this is the argument invoked in Ref. [5] to associate the position of the maximum of the cross-correlation function to the shift of the peak being sought. In our work [1] it was not necessary to include a shift; while Equation (5) could be used to estimate the signal amplitude α, it is preferable to use the autocorrelation instead:

$$(\mathbf{S} \star \mathbf{S})_q = \alpha^2 (\mathbf{R} \star \mathbf{R})_q + (\mathbf{N} \star \mathbf{N})_q = \alpha^2 (\mathbf{R} \star \mathbf{R})_q + \sigma^2 \delta_{q0} \tag{6}$$

(note that the shift by r cancels, here). Equation (6) tells us in particular that all autocorrelation sequences should be the same except for a scale factor α^2 and a sharp noise peak at $q = 0$. Visual inspection shows that this is true for a number of scans which we consider good; other scans exhibit a narrower, or structured, sometimes negative, autocorrelation, probably indicating correlated noise, i.e., a drift during the measurement; the rest only exhibits the noise peak, indicating no signal at all (Figure 1). From now on we will consider $(\mathbf{R} \star \mathbf{R})$ to be noiseless and, as empirically found, satisfactorily approximated by a gaussian; we derive the width of the gaussian from our best scan (Figure 1).

Our choice of a gaussian is purely empirical, and our method does not critically depend on it: to the extent that all good scans have the same shape of the autocorrelation function, Equation (6), the best fit parameters will be the same for all good scans, except for a scale factor for the height. What is important is that the fitting function provides a reasonable approximation of α in Equation (6), i.e., of the peak value of the signal component of the autocorrelation.

It is nevertheless instructive to discuss some limiting cases for the shape of the autocorrelation function: we begin by noting that the expected shape of a pump-probe signal such as that of our experiment is an exponential decay (we ignore the possible complication brought on by the presence of more than one decay constant) convoluted with the instrumental resolution (in our case a gaussian, coming from the finite duration of the FEL and UV pulses). The respective autocorrelation functions are an exponential and a gaussian. For the case of non-negligible instrumental resolution the autocorrelation will resemble a gaussian near $q = 0$, but will have broader wings decaying as $\exp(-|q|)$ rather than $\exp(-q^2)$. An early truncation of the scan will clip the wings of the autocorrelation function. In our case an early truncation does somewhat contribute to determining the shape of the autocorrelation function, but does so equally for all scans (the pump-probe delay was scanned in reverse, and point index $i = 0$ in Figure 1 consistently corresponds to the maximum value of the delay).

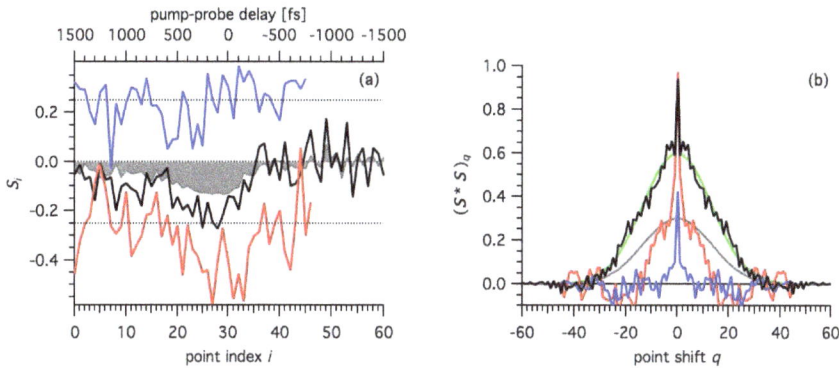

Figure 1. (a) Three representative delay scans and (b) their autocorrelation curves. We use the autocorrelation curves ($q = 0$ excluded) as a criterion to classify scans as good (black), drifting (red), or pure noise (blue). The autocorrelation curve of a good scan is well approximated by a gaussian (green). In panel (a), traces have been offset for clarity, and the shaded curve is the weighed average Equation (7); in panel (b), the green curve is a gaussian fit of the black autocorrelation trace; the gray curve is the same, scaled to the height of the red autocorrelation trace (white-noise peak excluded). The calculated signal-to-noise ratio (SNR) for the three scans are 1.30, 0.52, and 0.20.

Let us finally consider two possible sources of artefacts, namely a constant, or a linearly drifting, baseline (remembering that the scans are padded with zeros outside of the delay range scanned). This would primarily contribute a slowly decaying component to the autocorrelation signal (of width comparable to that of the scan itself). Let us however note that because the scans are acquired by rapid double-background subtraction (Equation (9) and Figure 4), we expect a complete baseline cancellation. Finally, a drifting signal would alter the shape of the peak and consequently of its autocorrelation function: assuming an even drift that causes a loss of the optimal experimental conditions (temporal or spatial overlap; quality of the focus; resonance wavelength) one can speculate that its main effect would be a reduction of the measured width.

3. Discussion

Given a set of scans $\left\{ \mathbf{S}^{(k)} \right\}$, their weighted average is

$$\langle \mathbf{S} \rangle = \frac{\sum_k w_k \mathbf{S}^{(k)}}{\sum_k w_k} \tag{7}$$

$\langle \mathbf{S} \rangle$ is the quantity reported in our work ([1] note for the sake of exactness that in Figure 2a therein, the unity baseline was not subtracted); we want to determine which scans to include in the average, and their weights w_k.

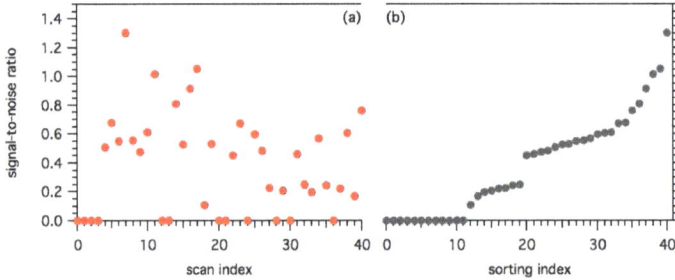

Figure 2. (**a**) Signal-to-noise ratio plotted versus scan index. (**b**) Same data as (**a**) sorted by increasing value, to highlight the gap between $SNR > 0.25$ and $SNR < 0.45$.

From Equation (6) and Figure 1b we can estimate the signal-to-noise ratio as:

$$SNR_k = \sqrt{\frac{(\mathbf{S}^{(k)} \star \mathbf{S}^{(k)})_0}{(\mathbf{N}^{(k)} \star \mathbf{N}^{(k)})_0} - 1} = \sqrt{\frac{\alpha_k^2 (\mathbf{R} \star \mathbf{R})_0}{(\mathbf{N}^{(k)} \star \mathbf{N}^{(k)})_0}} = \frac{\alpha_k}{\sigma_k} \tag{8}$$

A simple analysis of its trend over the course of the experiment reveals some regularities that we exploit to qualitatively classify scans, and to define a quantitative criterion that we adopt to accept or reject them. When the scans are ordered in the sequence they were acquired (Figure 2a) no obvious trend is visible for the SNR, but one does note a large number of scans with $SNR = 0$: they are either those for which a gaussian fit of the autocorrelation was not successful, or those which have been excluded *a priori* (e.g., because the scan was aborted). The ordering of the scans by increasing SNR (Figure 2b) reveals a gap between scans with $SNR < 0.25$ and $SNR > 0.45$. We cannot find an obvious reason for this gap; because this was one of the first resonant two-photon experiments performed at FERMI we can speculate a threshold behavior of some of the less-controlled parameters of the FEL (peak intensity; second harmonic content). In any case, we decided to use the condition $SNR > 0.45$ as a discriminant to include a scan in the average.

Let us now come to the weight with which each accepted scan enters the average. We show in Appendix A that the weight which maximizes the SNR of $\langle \mathbf{S} \rangle$ is $w_k = \alpha_k / \sigma_k^2 = SNR_k / \sigma_k$; in Ref. [1] we used $w_k = SNR_k$, which gives a slightly worse result, although the difference is not significant. We presume that the latter fact depends on the limited number of samples, the predominance of few of them (see Figure 2a), and the low dispersion of the σ_k. In both cases we observe an improvement of the signal-to-noise ratio of $\langle \mathbf{S} \rangle$ by a factor ≈ 2.25 relative to the best single scan.

Unfortunately at the time of the experiment we were not anticipating the need for this test, so we are not able to perform further checks on which parameters and events mostly affected our experiment. Likewise our primary goal was to identify a simple unbiased method of data selection, and we do not attempt a systematic analysis of its merits and limitations, for which we refer the reader to the vast existing literature ([6,7] and references therein); we simply note that the method is most easily applicable to the case of white noise and single- or well-separated peaks, although generalizations to other noise [6] or multiple peaks [8] have been discussed. Let us finally note that because the autocorrelation is the Fourier Transform of the power spectrum (Wiener-Khinchin Theorem), the same determination of the signal-to-noise ratio could have been accomplished by Fourier Transform of each scan (in the frequency domain, the white-noise component appears as a constant).

Despite its simplicity and limitations, we believe that our simple test can be further characterized and profitably applied in future experiments. Let us note that the original and most immediate application of the method addresses a situation common to the beginning of many experiments, namely the need of making a weak signal visible when the expected signal shape is not precisely known in advance.

4. Materials and Methods

4.1. Experimental

The experiment was performed at the Low Density Matter (LDM) beamline [9] at FERMI FEL-1 [2]. The EUV pulses had an average energy of 16 μJ, duration estimated between 60 and 80 fs FWHM, circular polarization, focal size 30 μm FWHM, and were tuned to the wavelength resonant with the two-photon doubly excited target state, 75.65 nm. The UV pulses had an energy of 35 μJ, estimated duration 200 fs, focal size 80 μm FWHM, fixed wavelength (261 nm). The Ne dimers were produced by adiabatic expansion of Ne gas at a pressure of 0.8 MPa and temperature of 190 K through a 100 μm nozzle. The target ions were detected by the time-of-flight mass spectrometer of the LDM endstation [10].

4.2. Data Acquisition

The Free Electron Laser FERMI currently operates at a repetition rate of 50 Hz (10 Hz at the time of the experiment). This is determined by the operation rate of the LINAC that accelerates the electron bunches used to generate the FEL pulses. Single-shot ion TOF traces (Figure 3) are acquired, tagged with a progressive integer ("bunchnumber") and stored along with a wide selection of experimental parameters and machine parameters, for post-processing. The relevant single-shot quantities for us are the FEL energy per pulse ($I0$), the UV energy per pulse (I_{UV}), and the integrated area a of the TOF peak of interest (specifically the $^{20}Ne_2^+$ peak).

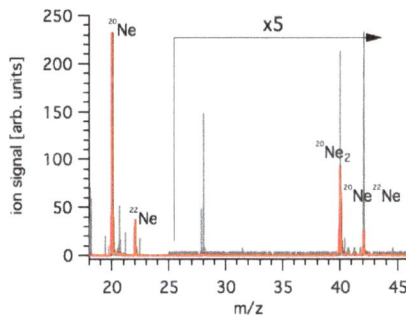

Figure 3. Single-shot ion-TOF trace (gray) and average over 12,200 shots (red). One in three shots is a blank shot (no gas sample), which is subtracted from the average to eliminate spurious signals, such as that at $m/z = 28$ in the single shot spectrum, due to residual nitrogen gas. Note that for $m/z > 25$ both spectra are magnified by a factor 5.

A preset number of single-shot values a_j at nominally identical conditions are acquired and aggregated into a datapoint A_i; one in three shots is a blank shot (no gas sample), which is subtracted from the average to eliminate spurious signals (Equation (9)). Although in reality the bunchnumber is never reset, i.e., each bunchnumber is unique since the inception of FERMI, for the purpose of this work we will think of it as spanning a range $j = \{1..m\}$ for each datapoint. A set of datapoints at a sequence of delay values constitutes a delay scan. Delay scans are averaged into a global delay scan, where each scan enters with the weight determined in Equation (8).

4.3. Data Reduction

From the single-shot values we build a datapoint as follows:

$$A_i = \frac{\sum\limits_{j=\text{gas}} a_j}{\sum\limits_{j=\text{gas}} \left(I0_j\right)^p} - \frac{\sum\limits_{j=\text{blank}} a_j}{\sum\limits_{j=\text{blank}} \left(I0_j\right)^p} \tag{9}$$

with the new index i labelling sets of nominally identical measurements at a series of pump-probe delays d_i. $\tilde{A}_i$ is the analogue of A_i with UV off. We consider equally spaced delays, so $d_i = \Delta t \times i$ $(i = 0..n)$. Let us note that single-shot data can be excluded from the sum by applying in post-analysis suitable filter conditions (e.g., a threshold on $I0$, or on the quality of the FEL spectrum); for this reason the number of single-shots aggregated into a datapoint is not strictly constant. For each i we define the quantity $S_i = A_i / \tilde{A}_i - 1$: this is the base datapoint of a delay scan (Figure 4). The integer p in Equation (9) is the normalization order; for the data presented in Ref. [1] we chose $p = 0$, i.e., no normalization. Let us also note that the intensity of the UV pulse was kept constant and is very stable, so it does not appear in the data processing. Let us finally note that at negative delays, and large positive delays we expect the UV to have no effect, i.e., $S_i = 0$.

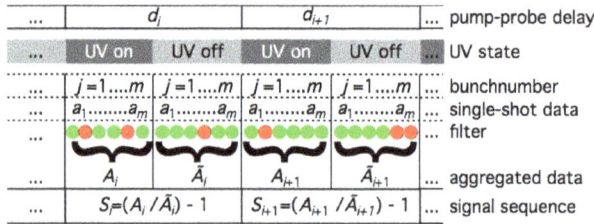

Figure 4. Schematic representation of the generation of a delay scan $\{S_i\}$, as a function of pump-probe delay $\{d_i\}$ from single-shot data a_j. The green (red) dots symbolize shots that pass (fail) the filter conditions.

5. Conclusions

Although rather simple, the method proposed has many advantages and applications. It provides a quick and reliable method to evaluate the quality of a measurement in real time, even when the shape of the expected signal is unknown. In fact, in the approximation of white noise, and of an unstructured pump-probe signal, it provides a first coarse estimate of the signal shape, and thus of the delay range that needs to be scanned. In post-analysis it provides an unbiased criterion to include each scan in an averaged measurement.

Acknowledgments: This work was supported by the X-ray Free Electron Laser Priority Strategy Program of the Ministry of Education, Culture, Sports, Science, and Technology of Japan (MEXT); by JSPS and CNR under the Japan - Italy Research Cooperative Program; by the Grant-in-Aid for the Global COE Program 'the Next Generation of Physics, Spun from Universality and Emergence' from the MEXT; by the Grants-in-Aid (No. 20310055 and No. 21244062) from the Japan Society for the Promotion of Science (JSPS); by the JSPS KAKENHI Grant Number JP 16J02270; by the Swedish Research Council and the Swedish Foundation for Strategic Research; by the ERC Starting Research Grant UDYNI No. 307964; by the European Union Horizon 2020 research and innovation programme under the Marie Skłodowska-Curie grant agreement No. 641789 "MEDEA" (Molecular Electron Dynamics investigated by IntensE Fields and Attosecond Pulses); by the DOE-SC-BES under Award No. DE-SC0012376; and by the Italian Ministry of Education, Universities and Research (MIUR) (PRIN 2012 - NOXSS)

Author Contributions: K.U. conceived and designed the experiment; A.D., B.S., C.C., C.V., D.F., D.I., E.F., F.C., G.S., H.F., K.C.P., Ke.M., Ko.M., K.N., K.U., M.A., M.C., M.C.C, M.D., M.R., N.B., O.P., P.C., P.F., P.J., P.P., S.M., T.N., Ts.T., Te.T., Y.K. performed the experiments; C.C., H.F., Ko.M., Ts.T. analyzed the data; A.D., C.C., P.J., P.P. contributed analysis tools; C.C. and K.C.P. drafted the paper, which was completed in consultation with all authors.

Conflicts of Interest: The authors declare no conflict of interest. The founding sponsors had no role in the design of the study; in the collection, analyses, or interpretation of data; in the writing of the manuscript, and in the decision to publish the results.

Abbreviations

The following abbreviations are used in this manuscript:

EUV	Extreme Ultra Violet
FEL	Free Electron Laser
FERMI	Free Electron laser Radiation for Multidisciplinary Investigations
	(this acronym identifies the Free Electron Laser facility in Trieste, Italy)
FWHM	Full Width at Half Maximum
I0	FEL energy per pulse
LINAC	Linear Particle Accelerator
SNR	Signal-to-noise Ratio
TOF	Time-of-flight
UV	Ultra Violet

Appendix A

We demonstrate that within a proportionality factor the optimum value of the weights appearing in Equation (7) is $w_j = \frac{\alpha_j}{\sigma_j^2}$; all three quantities α_j, σ_j, w_j are assumed positive; note that we use j to indicate a specific index, and tacitly replace it with k whenever it becomes a dummy index in a sum. We wish to maximize

$$\frac{\left(\sum_k w_k \alpha_k\right)^2}{\sum_k w_k^2 \sigma_k^2} - \lambda \left(1 - \sum_k w_k\right) \tag{A1}$$

where λ is a Lagrange multiplier. The condition of zero partial derivatives with respect to each of the w_j and of λ yields:

$$2\alpha_j \left(\sum_k w_k \alpha_k\right)\left(\sum_k w_k^2 \sigma_k^2\right) - 2w_j\sigma_j^2\left(\sum_k w_k\alpha_k\right)^2 + \lambda\left(\sum_k w_k^2\sigma_k^2\right)^2 = 0 \tag{A2}$$

$$\sum_k w_k = 1 \tag{A3}$$

Multiplying (A2) by w_j and summing over j, one has:

$$2\left(\sum_k w_k\alpha_k\right)^2\left(\sum_k w_k^2\sigma_k^2\right) - 2\left(\sum_k w_k^2\sigma_k^2\right)\left(\sum_k w_k\alpha_k\right)^2 + \lambda\left(\sum_k w_k\right)\left(\sum_k w_k^2\sigma_k^2\right)^2 = 0 \tag{A4}$$

hence $\lambda = 0$; and Equation (A2), divided by $2\sigma_j^2 \neq 0$, simplifies to

$$\frac{\alpha_j}{\sigma_j^2}\left(\sum_k w_k^2\sigma_k^2\right) - w_j\left(\sum_k w_k\alpha_k\right) = 0 \tag{A5}$$

Summing over j, and using Equation (A3) yields

$$\sum_k w_k\alpha_k = \left(\sum_k \frac{\alpha_k}{\sigma_k^2}\right)\left(\sum_k w_k^2\sigma_k^2\right) \tag{A6}$$

which replaced in Equation (A5) finally gives

$$w_j = \frac{\alpha_j/\sigma_j^2}{\sum_k \alpha_k/\sigma_k^2} \tag{A7}$$

Note that in the special case when α_k is the same for all k, one has the familiar result that the minimum variance of an average is the one with $w_j = \sigma_j^{-2} / \sum_k \sigma_k^{-2}$ [11].

References

1. Takanashi, T.; Golubev, N.V.; Callegari, C.; Fukuzawa, H.; Motomura, K.; Iablonskyi, D.; Kumagai, Y.; Mondal, S.; Tachibana, T.; Nagaya, K.; et al. Time-Resolved Measurement of Interatomic Coulombic Decay Induced by Two-Photon Double Excitation of Ne$_2$. *Phys. Rev. Lett.* **2017**, *118*, 33202.
2. Allaria, E.; Appio, R.; Badano, L.; Barletta, W.A.; Bassanese, S.; Biedron, S.G.; Borga, A.; Busetto, E.; Castronovo, D.; Cinquegrana, P.; et al. Highly coherent and stable pulses from the FERMI seeded free-electron laser in the extreme ultraviolet. *Nat. Photonics* **2012**, *6*, 699–704.
3. Green, P. Preface to the matched filter issue. *IRE Trans. Inf. Theory* **1960**, *6*, 310.
4. North, D. An Analysis of the factors which determine signal/noise discrimination in pulsed-carrier systems. *Proc. IEEE* **1963**, *51*, 1016–1027.
5. Van den Heuvel, E.; van Malssen, K.; Smit, H. Optimal estimation of intensity of noisy peaks by matched filtering with application to chromatography. *Anal. Chim. Acta* **1990**, *235*, 343–353.
6. Smit, H. Specification and estimation of noisy analytical signals: Part I. Characterization, time invariant filtering and signal approximation. *Chemom. Intell. Lab. Syst.* **1990**, *8*, 15–27.
7. Smit, H. Specification and estimation of noisy analytical signals: Part II. Curve fitting, optimum filtering and uncertainty determination. *Chemom. Intell. Lab. Syst.* **1990**, *8*, 29–41.
8. Van den Bogaert, B.; Boelens, H.F.M.; Smit, H.C. Quantification of overlapping chromatographic peaks using a matched filter. *Chemom. Intell. Lab. Syst.* **1994**, *25*, 297–311.
9. Svetina, C.; Grazioli, C.; Mahne, N.; Raimondi, L.; Fava, C.; Zangrando, M.; Gerusina, S.; Alagia, M.; Avaldi, L.; Cautero, G.; et al. The Low Density Matter (LDM) beamline at FERMI: optical layout and first commissioning. *J. Synchrotron Radiat.* **2015**, *22*, 538–543.
10. Lyamayev, V.; Ovcharenko, Y.; Katzy, R.; Devetta, M.; Bruder, L.; LaForge, A.; Mudrich, M.; Person, U.; Stienkemeier, F.; Krikunova, M.; et al. A modular end-station for atomic, molecular, and cluster science at the low density matter beamline of FERMI@Elettra. *J. Phys. B* **2013**, *46*, 164007.
11. Bevington, P.R.; Robinson, D.K. *Data Reduction and Error Analysis for the Physical Sciences*, 2nd ed.; McGraw-Hill: New York, NY, USA, 1992.

applied
sciences

MDPI

Review

Measurement of the Resonant Magneto-Optical Kerr Effect Using a Free Electron Laser

Shingo Yamamoto and Iwao Matsuda *

Institute for Solid State Physics, The University of Tokyo, Kashiwa, Chiba 277-8581, Japan;
shingo.yamamoto@issp.u-tokyo.ac.jp
* Correspondence: imatuda@issp.u-tokyo.ac.jp; Tel.: +81-(0)4-7136-3402

Academic Editor: Kiyoshi Ueda
Received: 1 June 2017; Accepted: 21 June 2017; Published: 27 June 2017

Abstract: We present a new experimental magneto-optical system that uses soft X-rays and describe its extension to time-resolved measurements using a free electron laser (FEL). In measurements of the magneto-optical Kerr effect (MOKE), we tune the photon energy to the material absorption edge and thus induce the resonance effect required for the resonant MOKE (RMOKE). The method has the characteristics of element specificity, large Kerr rotation angle values when compared with the conventional MOKE using visible light, feasibility for M-edge, as well as L-edge measurements for $3d$ transition metals, the use of the linearly-polarized light and the capability for tracing magnetization dynamics in the subpicosecond timescale by the use of the FEL. The time-resolved (TR)-RMOKE with polarization analysis using FEL is compared with various experimental techniques for tracing magnetization dynamics. The method described here is promising for use in femtomagnetism research and for the development of ultrafast spintronics.

Keywords: magneto-optical Kerr effect (MOKE); free electron laser; ultrafast spin dynamics

1. Introduction

Femtomagnetism, which refers to magnetization dynamics on a femtosecond timescale, has been attracting research attention for more than two decades because of its fundamental physics and its potential for use in the development of novel spintronic devices [1]. The ultrafast dynamics of femtomagnetism can be accessed using ultrashort laser pulses to perturb magnetic materials via thermal and nonthermal effects. This produces a system of strongly-nonequilibrium states [2]. Immediately after a sudden disturbance, magnetic systems show demagnetization due to a transition from ferro-/ferri-magnetism to paramagnetism that is caused by impulsive heating. Additionally, in certain magnetic compounds, such as ferromagnetic alloys and ferrimagnets composed of rare-earth and transition metals, the magnetization is reversed on a femtosecond timescale [3]. Laser-induced phase transitions from antiferromagnetic to ferromagnetic phases [4] and optical control of the spin precession have also been reported [5]. These ultrafast demagnetization and magnetization reversal phenomena, and particularly the mechanism of spin-flips, have been interpreted from various phenomenological viewpoints [1,6,7]. Interactions between the electron, spins, lattices and photon fields have been incorporated in some models. In addition, superdiffusive spin transport [8] and interactions with collective excitations such as phonons and magnons [9] have also been considered in other models.

While the magnetization dynamics, especially on the femtosecond timescale, remain controversial from the microscopic perspective, a number of experimental techniques for probing of the transient characteristics have emerged to meet the demand for probing methods that are suitable for a wide variety of magnetic materials. One of these techniques is the magneto-optical Kerr effect (MOKE) method, which has been applied to investigations of static bulk/surface magnetism, spin transport and magnetization dynamics using linearly-polarized light [10]. These experiments have mainly

Appl. Sci. **2017**, *7*, 662

been conducted in the visible range. In previous studies, MOKE in the soft X-ray range, which involves tuning of the photon energy to the absorption edge of a target magnetic element, has been investigated, and we call this method resonant MOKE (RMOKE) (see Equation (1)) in this review [11,12]. However, when it comes to the static magnetization measurements, the usefulness of RMOKE has not been focused on until now by the availability of other convenient soft X-ray magneto-optical measurement methods, such as X-ray magnetic circular dichroism (XMCD) spectroscopy [13,14]. Following the development of femtosecond light sources such as synchrotron radiation (SR) sources using a laser slicing technique, the free electron laser (FEL) and the high harmonic generation (HHG) laser, the RMOKE technique used in combination with these state-of-the-art light sources is becoming increasingly important, particularly for temporal domain measurements in the subpicosecond timescale. We first demonstrated time-resolved RMOKE (TR-RMOKE) measurements using the FEL in 2015 [15]. In this review, we clarify the importance of the RMOKE technique when using an extreme ultraviolet (EUV) FEL from the methodological view point and demonstrate time-resolved RMOKE measurements using a seeded FEL at FERMI@Elettra (FERMI-I is generally called X-ray ultraviolet (XUV) FEL).

In the next section, the MOKE is introduced in terms of two of its aspects: rotation and ellipticity. The experimental geometry of the MOKE is explained with respect to the magnetization components that can be detected when using each configuration. In Section 3, we give an overview of the magnetic probing techniques, including the MOKE, that have been used to investigate static magnetism and laser-induced magnetization dynamics in previous research. In this section, the characteristics of the TR-RMOKE technique are clarified. In Section 4, methods used to perform MOKE measurements in the visible and soft X-ray ranges are explained. In Section 5, the results of RMOKE measurements based on rotating analyzer ellipsometry are shown for the ferrimagnetic alloy GdFeCo. In Section 6, the TR-RMOKE measurement method is demonstrated using a soft X-ray FEL. In the final section, we summarize this review.

2. MOKE Phenomena

The MOKE can be characterized based on its rotation and ellipticity properties. Linearly-polarized light can be decomposed into left and right circularly-polarized components. Each of these components has the same phase velocity and amplitude. When linearly-polarized light interacts with magnetized materials, the polarization state of the light will change in two ways. The first way involves the rotation of the polarization axis, and the second is a change from linearly-polarized to elliptically-polarized light that is characterized by the ellipticity η_k. Figure 1 shows an example of the (polar) MOKE measurement. The phase (amplitude) variations between the right and left circularly-polarized light components is responsible for the rotation, denoted by θ_K (ellipticity, η_K).

The measurement geometry is shown in Figure 2. There are two main schemes that are used for MOKE measurements. One scheme is based on polarization analysis, because the polarization of the reflected light rotates, and its ellipticity changes (see Figure 2a,b). The other scheme involves intensity measurements and is based on the fact that the polarization states of the reflected light do not change from the corresponding states of the incident light (see Figure 2c) This is called the transverse MOKE, or T-MOKE. Schemes in the first group are classified based on the magnetization direction **M**. When **M** is perpendicular to the sample surface, the method is called polar MOKE (P-MOKE; see Figure 2a). If **M** is parallel to the sample surface and the reflection plane defined by the incident and reflected light beams, it is called longitudinal MOKE (L-MOKE; see Figure 2b). Historically, the schemes shown in Figure 2a,b are recognized separately, but the MOKE signals that involve polarization analysis in the cases of Figure 2a,b both originate from $\mathbf{k} \cdot \mathbf{M}$, and there is thus no intrinsic difference between the two geometries. These measurement geometries correspond to the measurement under applying a magnetic field to a specific direction with respect to the incident plane as shown in Figure 2. In the case of the arbitrary direction of the external magnetic field, magneto-optical effects observed in the reflected light are contributed from any of the three MO geometries. The phenomenological and analytical expressions in such cases can be found in the preceding studies [16,17]. θ_k and η_k are defined with Fresnel coefficients, which depend on the frequency of incident light (ω) as follows [16].

$$
\begin{aligned}
\theta_{ks} &= -\mathrm{Re}\left(\frac{\tilde{r}_{ps}(\omega)}{\tilde{r}_{ss}(\omega)}\right) \\
\theta_{kp} &= \mathrm{Re}\left(\frac{\tilde{r}_{sp}(\omega)}{\tilde{r}_{pp}(\omega)}\right) \\
\eta_{ks} &= \mathrm{Im}\left(\frac{\tilde{r}_{ps}(\omega)}{\tilde{r}_{ss}(\omega)}\right)\mathrm{Re}\left(\frac{\tilde{r}_{ps}(\omega)}{\tilde{r}_{ss}(\omega)}\right) \\
\eta_{kp} &= \frac{\mathrm{Im}\left(\frac{\tilde{r}_{sp}(\omega)}{\tilde{r}_{pp}(\omega)}\right)}{\mathrm{Re}\left(\frac{\tilde{r}_{sp}(\omega)}{\tilde{r}_{pp}(\omega)}\right)} \\
\omega &\approx \omega_{\mathrm{res}} \ (\text{RMOKE definition})
\end{aligned}
\tag{1}
$$

ω_{res} corresponds to an absorption edge of the target magnetic element. θ_{ks} (η_{ks}) and θ_{kp} (η_{ks}) are the Kerr rotation angle (ellipticity) for *s*- and *p*-polarized incident light, respectively. *i* and *j* in the $\tilde{r}_{ij}$ are the electric-field component of the reflected and of the incident light, respectively. The tilde shows that the quantity is complex. The specific derivatives depends on the MOKE configurations, which are found in the earlier study [16].

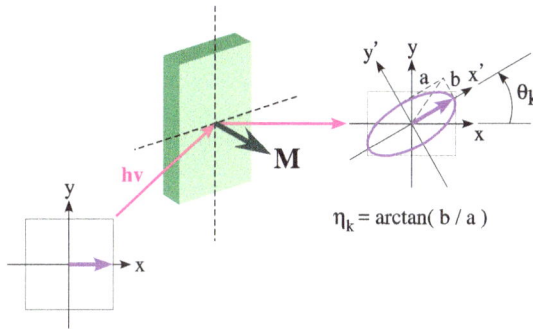

Figure 1. Illustration of the (polar) magneto-optical Kerr effect (MOKE). Linearly-polarized light with photon energy of *hν* becomes elliptically polarized after reflection from a magnetized material with magnetization (**M**), and the main polarization plane is tilted by a small angle θ_k with respect to that of the incident light. The ellipticity of the reflected light is quantified using $\eta_k = \arctan\frac{b}{a}$.

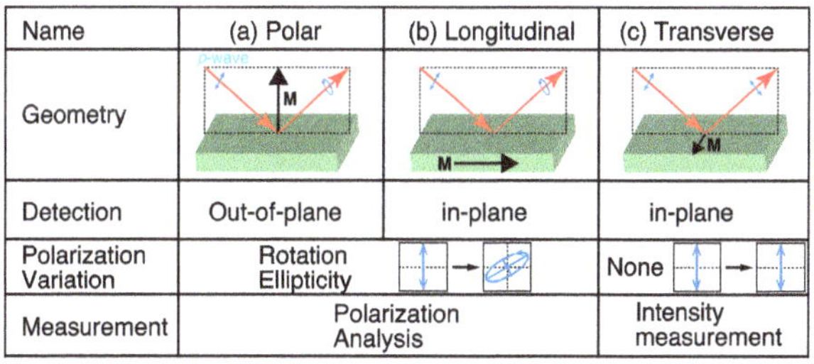

Name	(a) Polar	(b) Longitudinal	(c) Transverse
Geometry			
Detection	Out-of-plane	in-plane	in-plane
Polarization Variation	Rotation Ellipticity		None
Measurement	Polarization Analysis		Intensity measurement

Figure 2. Schematic diagram of MOKE measurement geometry for *p*-polarized incident light. The dashed line in the geometry row expresses the incident plane. In the polarization variation row, the changes in the polarization states that are projected in the plane that lies perpendicular to the direction of the travel of the light are shown for both incident (left) and reflected light (right).

3. Techniques for Magnetization Dynamics Capture

Figure 3 presents the various experimental methods that have been used to trace magnetization dynamics in previous studies, including MOKE. Each method is shown with respect to the energy required and the timescale of the operation, together with the relevant optical transitions and magnetization dynamics phenomena. The optical transitions are classified into two regimes: one from the perspective of the valence bands for energies of a few eV and the other from the perspective of the core levels for energies higher than those of the EUV range. Three light sources with sub-picosecond pulses, comprising the FEL, the HHG laser and the laser slicing source, are shown based on the energy ranges in which each source can emit.

In the microwave and millimeter-wave range, magnetic resonance, such as ferromagnetic resonance (FMR) [18] and antiferromagnetic resonance (AFMR) [19], has been used to investigate the precession frequency and the magnetic collective excitation or magnon. Inelastic light scattering, for example Brillouin (BLS) and Raman light scattering methods, is also employed for exploring the nature of collective spin excitation modes. These occur on a timescale of hundreds of picoseconds. Terahertz time-domain spectroscopy (THz-TDS) can directly visualize the electric field waveforms of ultrashort pulses [20]. Additionally, in time-resolved measurements, THz pulses can excite the system without being absorbed by the target materials, which means that heating effects can be avoided during the analysis of the magnetization dynamics [21]. In the energy range from the infrared to the ultraviolet, the magneto-optical effects in both reflection (Kerr) and transmission (Faraday) regimes have been used. In some magnetic compounds composed of rare-earths and transition-metals, element-specific measurements were reported by probing different wavelength for each magnetic element [22]. Furthermore, a nonlinear magneto-optical effect, magnetization-induced second harmonic generation (MSHG), has also been used to detect magnetic systems without inversion symmetry, such as systems involving surface and interface magnetism [23,24]. While these methods can be used easily with laboratory-based lasers, most of them lack elemental selectivity because they are involved with the optical transitions that occur between delocalized states.

Above the EUV range, magnetic probe measurements are generally performed by tuning the photon energy to a specific absorption edge, which then enables element-specific measurements. In static measurements, the Kerr and Faraday effects in the EUV and soft X-ray ranges were particularly investigated from the late 1990s until the mid-2000s. Magneto-optical effects that are sensitive to $<M^2>$ are also used, including the Voigt effect [25] and X-ray magnetic linear dichroism (XMLD) [26]. These methods can probe both the ferromagnetic and antiferromagnetic orders. However, these magneto-optical effects, which are proportional to $<M^2>$ are much smaller than those proportional to $<M>$ [25,27]. Additionally, in XMLD measurements, it is difficult to separate the magnetic and nonmagnetic contributions [28]. XMCD is the most commonly-used technique for probing of ferro-/ferri-magnets, particularly in the soft X-ray range, and uses circularly-polarized light. This method can extract the spin and the orbital magnetic moment separately using magneto-optical sum-rules. While the XMCD signals in the EUV range are quite small when compared with the signal in the soft X-ray range, M-edge XMCD for $3d$ transition metals has also been demonstrated using a circularly-polarized HHG laser. The resonant inelastic X-ray scattering (RIXS) method is relatively new when compared with the other magnetic probing techniques and is complementary to the inelastic neutron scattering method [29–32]. This method could become more easily available as a result of the increasing brilliance of third generation SR sources and the advent of the FEL. Furthermore, through its enhancement of the energy resolution, RIXS has an advantage in that it can be used to detect collective magnetic excitations. Unlike similar techniques, RIXS can also detect momentum-resolved information. Small-angle X-ray scattering (SAXS) is similar to resonant X-ray scattering and can be used to determine magnetic structures on the nanometer scale [33,34]. Recently, this method has been used to investigate the topological spin textures of the skyrmion. Fourier transform holography (FTH) in the EUV and soft X-ray ranges has been measured using coherent light sources such as HHG lasers and FELs [35,36]. Conventional magnetic imaging has been performed by magnetic transmission X-ray microscopy using zone plates. The FTH scheme of holography measurements can be achieved without a lens, such as zone plate. In SAXS, the average correlation length can be extracted, while the FTH reveals the element-selective real-space magnetic distribution. The resonant soft X-ray diffraction (RSXD) technique has been used to investigate the charge, spin, orbital order and structural information of specific elements, particularly in strongly-correlated systems, such as transition metal oxides. This technique can determine the magnetic structures of antiferromagnets and helimagnets. Resonant X-ray diffraction is also measured in the hard X-ray range for detection of charges and orbital orders; however, the magnetic scattering cross-section is extremely small in the hard X-ray range when compared with that in the soft X-ray range [30,37]. Additionally, because the $3d$ states can be accessed directly in the soft X-ray range, resonant magnetic scattering is mainly used in the soft X-ray range. The following must be kept in mind for RSXD measurements: (1) the target ordering structure is limited, with a typically long periodic length of >10 Å, because the wavelength of the soft X-ray is longer than that in the hard X-ray range; (2) the attenuation length is smaller ($\sim$100–200 nm) than that in the hard X-ray and neutron scattering ranges.

When transient magnetization measurements have been performed above the EUV range, HHG lasers, FELs and laser slicing light sources have been combined with the experimental methods described above. Transverse RMOKE (T-RMOKE) has recently been combined with an HHG laser to reveal in-plane magnetization dynamics on a femtosecond timescale [8,38,39]. When a grating is used, the time-resolved T-RMOKE spectrum can be measured using an HHG laser that covers the energy range around the M-edge of the $3d$ transition metals. Only recently, a time-resolved T-RMOKE technique has been combined with FEL at FERMI@Elettra [40]. Time-resolved XMCD (TR-XMCD) measurements have mainly been performed in the soft X-ray range, particularly for the L-edges of $3d$ transition metals when using a laser slicing light source [3,41]. Recently, TR-XMCD measurements of the M-edge of a $3d$ transition metal have been demonstrated using an HHG laser [42] and an FEL [43]. These time-resolved reflectivity measurements using circular (XMCD) and linear polarization (T-RMOKE) have had an importance in tracing in-plane magnetization dynamics in the previous studies. A time-resolved RIXS (TR-RIXS) method with a femtosecond timescale has also been demonstrated using an FEL [9]. This enables tracing of the magnetic correlation with nanometer-scale momentum resolution, which provides information with regard to magnetic melting on various length scales. A time-resolved SAXS (TR-SAXS) method has been used to reveal the magnetic spatial response on a nanometer scale during demagnetization or magnetization reversal processes using FELs or HHG lasers [44–46]. Time-resolved FTH (TR-FTH) has been performed using both SR sources [47,48] and an FEL [49]. Because of the higher brilliance and shorter pulse durations of FELs when compared with those of SR sources, holographic images can be obtained in much shorter times when the FEL is used. Time-resolved RSXD (TR-RSXD) for tracing of magnetic orders with q $\neq$ 0 (where q is the wave number) has been implemented using a laser slicing source [50] and FELs [51,52].

In Figure 3, we focused only on photon-in and photon-out schemes. In these regimes, the measurement system is not influenced by the existence of an external field. This scheme is crucial for measurement of magnetic systems under operando conditions, such as application of an electric field and a magnetic field for both insulators and metals. Direct spin detection schemes such as time-resolved spin-polarized photoemission and scanning tunneling microscopy are also considered to be important experimental options, although these are not photon-out techniques.

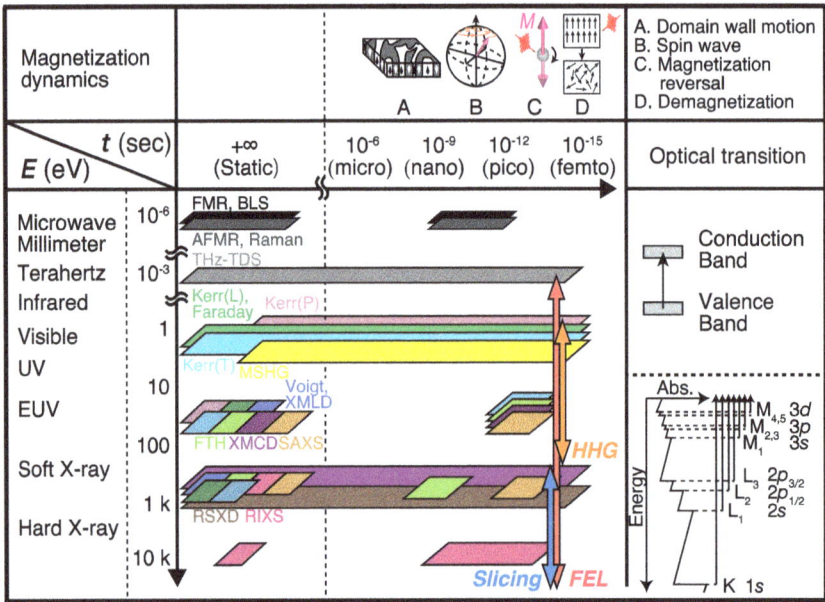

Figure 3. Experimental methods that have been used to trace magnetization dynamics over a time scale ranging from $+\infty$ (corresponding to static measurements) to 10^{-15} s (femtosecond scale) are presented with respect to the energies and temporal ranges with which each technique has been used. The methods are limited to photon-in and photon-out schemes. With regard to the Kerr methods, longitudinal (L), polar (P) and transverse (T) represent the longitudinal, polar and transverse geometries, respectively, which are explained in the text. Above the temporal axis, the magnetization dynamics that occur for each timescale are shown. On the right side, the related optical transitions are depicted schematically. The core-state positions are those for the $3d$ transition metals. Use of femtosecond X-ray pulses, free electron laser (FEL), high harmonic generation (HHG) laser and laser slicing sources, are indicated according to their pulse durations and energy ranges.

Characteristics of the TR-RMOKE Technique

Below the UV energy range, magnetization probing techniques detect the average information of target materials, whereas element selectivity is added in techniques that are involved with the core levels of target materials. We extract the features of the (TR-)RMOKE technique with polarization analysis, such as the features for polar and longitudinal geometries, through comparison with the other element-selective methods that are described in Figure 3. Comparisons are made with the following notable techniques: TR-RMOKE in a transverse geometry, TR-XMCD, TR-RIXS, TR-SAXS, TR-FTH and TR-RSXD. The comparisons with TR-RMOKE in the transverse geometry and TR-XMCD are described in Features (1)–(3), which are presented in the next paragraph. TR-RIXS has only recently been used in combination with TR-RSXD. This technique focuses on tracing of the magnetic low-energy excitation mode. Therefore, if one is interested in the possibility that a demagnetization or magnetization reversal is affected by the magnetic correlated modes, this technique provides a deeper insight than the other methods. However, the laser-induced macroscopic magnetization dynamics, which are revealed by elastic scattering (including reflection), should be measured using other techniques, including TR-XMCD, TR-RMOKE for ferro-/ferrimagnetic orders and TR-RSXD for antiferromagnetic orders, along with TR-RIXS. TR-SAXS and TR-FTH are powerful tools for determination of the magnetic structure on the nanometer scale. However, these techniques are mainly used in transmission geometries. During sample preparation, when trying to arrange a transmission-type experiment, the target magnetic system must be thin enough for a sizable transmission to be obtained. TR-RSXD is mainly used for materials with long-range magnetic orders with q $\neq$ 0, such as antiferromagnets. In the cases where ferro-/ferri-magnets are targeted, TR-XMCD and TR-RMOKE techniques are preferred.

The TR-RMOKE technique with polarization analysis using an FEL offers the following characteristics: (1) M-edge measurement feasibility, (2) measurements of both rotation and ellipticity, (3) use of linearly-polarized pulses, (4) giant Kerr rotation when compared with that of conventional visible MOKE and (5) element selectivity. We explain (1)–(3) in detail here. Characteristic (4) is discussed in Section 5.

(1) M-edge measurement feasibility:

In TR-RMOKE measurements performed in the soft X-ray range, the M-edge region is preferred to the L-edge region when the target materials contain $3d$ transition metals and when one is interested in the out-of-plane magnetization dynamics. In the L-edge range (100 $\sim$ a few keV), the experimental setup in the reflection geometry is limited because the light in this energy range is strongly absorbed by the materials. To detect the out-of-plane component of the magnetization, it is important to ensure that the angle of incidence is as close to the sample surface normal as possible. However, the reflectivity drops off dramatically with decreasing angle of incidence θ (with respect to the sample normal) and $R \sim 10^{-10} - 10^{-11}$ at $\theta = 45°$ [53]. To use the reflection setup in the L-edge region, grazing incidence is required to enhance the reflectivity, and this reduces the magnetic contrast in the out-of-plane magnetization components. TR-XMCD is usually used in the L-edge range and is detected using a reflection geometry with grazing incidence [54]. There is also the possibility that out-of-plane magnetization dynamics may be detected using TR-XMCD in the L-edge range via transmission [3] and total fluorescence yield (TFY) [55] modes. However, for the transmission measurements, the sample thickness must be thin enough to allow high transmission, and the detection system for the TFY measurements must be carefully constructed to prevent the pump laser from entering the detectors and collecting a low photon flux. In contrast, around the M absorption edge (50$\sim$70 eV), the light is less absorbed by the materials than it is in the L-edge range. Therefore, sizable reflectivity can be obtained at all angles of incidence. In the polar MOKE geometry, which is used to detect out-of-plane magnetization, the effect is maximized at normal incidence. However, even under this condition, there is a detectable reflectance that ensures that a better signal-to-noise ratio is obtained in the M-edge range than in the L-edge range. These facts mean that the freedom of the experimental setup is greater in the M-edge range than in the L-edge range for the detection of out-of-plane magnetization dynamics on a sub-picosecond timescale. When these advantages are used, it is expected that it will be possible to measure the depth-dependent RMOKE signal by varying the angle of incidence. In addition to the high degree of measurement freedom, the magnitude of the RMOKE for the M-edge is of the same order as that for the L-edge that was reported in earlier studies [56,57], whereas the magnitude of the M-edge XMCD is much smaller than that of the L-edge XMCD. In addition, the M-edge XMCD lacks the advantage that exists in the L-edge range, in which the spin and the orbital magnetic moment can be extracted using the sum-rule. Because the spin-orbit splitting in the M-edge range is smaller than that in the L-edge range, the assumption made for formalization of the sum-rule is not fully met in the M-edge range. This causes a discrepancy between the spin/orbital magnetic moments that were extracted from the M-edge and L-edge XMCD measurements [58].

(2)   Measurement of both rotation and ellipticity:

In principle, it is possible to measure both rotation and ellipticity in MOKE measurements, whereas XMCD measures only the ellipticity. Through simple calculations, the non-diagonal component of the permittivity tensor can be determined based on the rotation and the ellipticity [10]. In earlier studies, the determination of the permittivity over the EUV to soft X-ray energy range was conducted indirectly with a certain number of errors. However, by performing RMOKE measurements in this energy range, it is possible to determine the permittivity component that carries magnetic information directly.

There is another advantage to be obtained from measuring both the rotation and the ellipticity. In the femtomagnetism field, the magnetic response in transient magneto-optical signals is not trivial, and this has led to controversial discussions in earlier studies. There are both optical and magnetic contributions to the magneto-optical response. A recent theoretical study suggested that analysis of time-dependent magneto-optical signals in terms of both their rotation and ellipticity is important for the extraction of the intrinsic magnetization dynamics on a sub-picosecond timescale [59]. In TR-RMOKE techniques for use in a transverse geometry, the transient reflected intensity is measured. However, two physical quantities, i.e., the rotation and the ellipticity, can be extracted in TR-RMOKE with polarization analysis, as in the cases of polar and longitudinal geometries.

(3)   Use of linearly-polarized pulses:

In TR-RMOKE measurements, linearly-polarized light in the EUV ~ soft X-ray range is used. The polarization state is analyzed through ellipsometry in the polar RMOKE (P-RMOKE) and the longitudinal RMOKE (L-RMOKE). In contrast, TR-XMCD measurements use circularly-polarized light. Most of the TR-XMCD measurements on the femtosecond timescale are conducted using a laser slicing source with circular polarization. However, these measurements suffer from extremely low photon flux when compared with those using SR and FEL sources [60]. Among the FEL facilities that are currently in operation, circularly-polarized light can be obtained in the EUV to soft X-ray range at the FERMI@Elettra and Linac Coherent Light Source (LCLS) [61–63] facilities. However, to extract the magnetization dynamics, accurate preliminary determination of the degree of circular polarization, which is dependent on the energy, is required. In this sense, measurements using linearly-polarized light are straightforward for use in analysis of the magnetic response from the transient magneto-optical signals. In HHG lasers, circularly-polarized EUV light has recently been produced using various techniques, including use of a phase-shifter and production of two circularly- or linearly-polarized light beams at slightly different wavelengths [64,65]. However, HHG lasers have smaller output intensities when compared with FELs, and thus, an FEL source is preferred when attempting to detect the dynamics of weak magnetic signals.

The combination of the RMOKE with an FEL provides another possibility for nonlinear RMOKE signal detection, while the nonlinear regime of XMCD is not expected to be useful. In the visible range, MSHG is often used, as shown in Figure 3. Through a polarization analysis of the SHG signal, a large-scale rotation was observed in the visible range [66]. Because second-order nonlinear signals only appear from systems without inversion symmetry, the nonlinear RMOKE technique can provide element-selective and surface/interface-selective measurements, which is important for investigation of spintronic materials, such as magnetoresistive, spin-valve and magnetic topological insulator systems.

4. MOKE Measurement Scheme

4.1. Visible MOKE

In the visible region, the MOKE is typically measured using the polarization plane modulation technique shown in Figure 4. A Faraday cell set between a polarizer and an analyzer compensates for the rotation of the polarization plane due to the Kerr effect in the sample. In the magneto-optic measurement field, the polarizer that is positioned after the sample is called the "analyzer". To enhance the sensitivity to the Kerr rotation angle, an AC current for modulation is added to the DC current, and the output signal from a detector is input to a lock-in amplifier; its output signal is then given as feedback to the current for the Faraday cell. Anisotropic materials such as LiF and MgF_2 can be used for the polarizer because they either show birefringence or dichroism [67,68]. Figure 5 shows an experimental example of the Ta (2 nm)/Gd_{21}($Fe_{90}Co_{10}$)$_{79}$ (20 nm)/Ta (10 nm)/SiO_2 sample. The material is ferrimagnetic; its magnetization saturates at ± 0.1 kOe (0.01 T), and the magneto-optical Kerr rotation angle reaches $+0.24°$. Data from the sample are used as examples throughout this article, and the sample is described in detail in the next section.

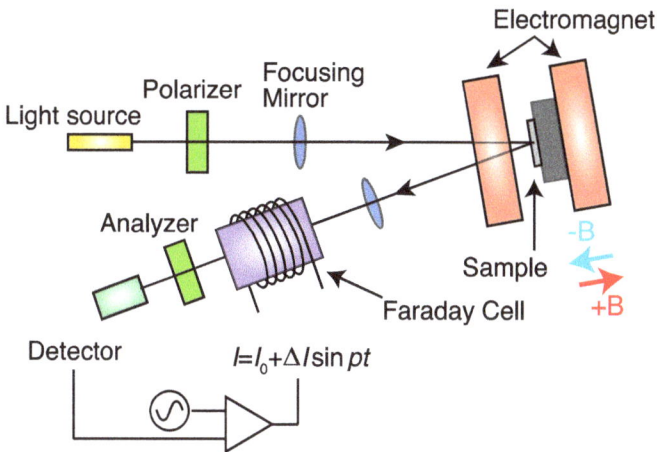

Figure 4. Measurement system for polar MOKE in the visible range using a polarization plane modulation technique. A Faraday cell is used to compensate for the Kerr angle that originated from the sample.

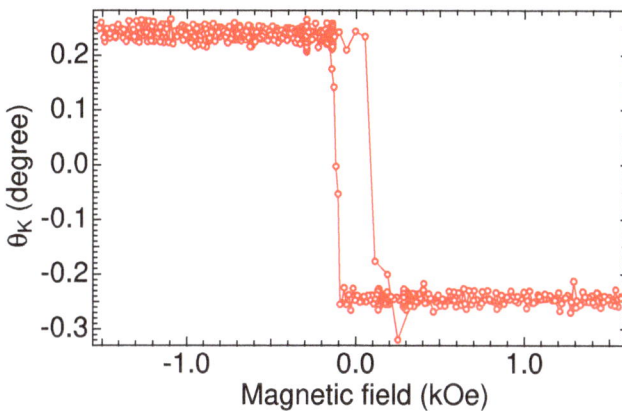

Figure 5. Polar Kerr hysteresis loop measured at a wavelength of 700 nm. The sample is a thin-film structure composed of Ta (2 nm)/$Gd_{21}(Fe_{90}Co_{10})_{79}$ (20 nm)/Ta (10 nm)/SiO_2. The measurements were taken at room temperature.

4.2. RMOKE

In the soft X-ray region, the method described above cannot be applied because of a lack of appropriate optical components, such as Faraday cells and transmission-type polarizers. This is because of the extremely low transmission or the high extinction coefficient k of the optical constant in this photon energy range. Additionally, the refractive index n of the optical constant of a soft X-ray is nearly one, and is thus close to that of air, which means that the intensity of the reflected beam is also very low. Under these severe experimental conditions, researchers have tended to choose reflection-type optical components because there are very few candidates for the transparent materials for soft X-ray light.

To improve the reflectivity of these components, the constructive interference effects from multilayer structures have been useful in guaranteeing sizable reflectivity in the soft X-ray range. Multilayer mirrors can reflect light within a specific energy range that is roughly defined by Bragg's law ($\lambda = 2d\sin\theta$, where d is the thickness of a single period of the layer and θ is the angle of incidence with respect to the surface normal). The constituents of the multilayer structures are materials with low k values and large differences in n. In the soft X-ray region, the first material is selected to have an absorption edge at a higher energy than a specific target energy, and a counterpart material with low k and n in the energy region is then chosen. If light is incident on the multilayer mirror at the Brewster angle, which is defined by the angle where the p-component of the reflected light is suppressed when compared with the s-component, the multilayer mirror can then be used as a polarizer. In the L-edge region, the value of n is almost one, which means that the Brewster angle is approximately 45°. On the other hand, in the M-edge region, n deviates from one, so that the Brewster angle is not around 45°. In this case, the polarizance, which is defined as the ratio between the s- and p-reflected intensities r_s/r_p, under 45° incident angle is of the order of 10–100 using Mo/Si multilayer mirrors. In the L-edge range, the polarizance is around 1000–100,000 with the same reflection geometry, although the multilayer combination is different (e.g., W (tungsten)/B_4C (boron carbide), W/C (carbon)). Figure 6 shows a collection of polarizers for photons at wavelengths ranging from visible light to hard X-rays [69]. Even with the order of the polarizance in the EUV range, it is able to determine a principal axis of reflected polarization for extracting Kerr angles.

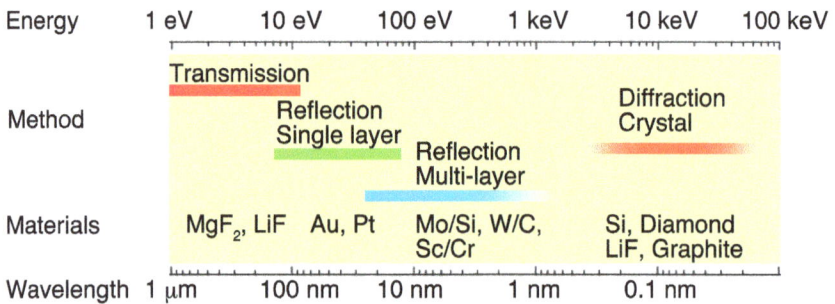

Figure 6. Methods for polarization analysis for wavelengths ranging from visible light to hard X-rays. Typical materials that are used as polarizers are shown for each energy region.

When the multilayer mirror described above is used, the Kerr rotation angle θ_k and the Kerr ellipticity η_k can be determined experimentally using the rotating analyzer ellipsometry (RAE) technique, which is shown in Figure 7. In RMOKE measurements using this setup, the intensity of the light that is reflected by the analyzer is monitored at the detector as a function of χ, as shown in Figure 7. The incident light impinges on the sample, and the reflected light is then transferred to the ellipsometry unit, which is used to determine the Kerr rotation angles. After passing through two pinholes that are positioned to ensure accurate alignment of the reflected light, the light reaches an analyzer with an angle of incidence that is roughly equal to the Brewster angle. The reflected light from the analyzer is then detected by a detector represented by the microchannel plate (MCP). The MCP is used in current detection mode. During extraction of the Kerr rotation angle, the section that is indicated by the rectangle of broken yellow lines in Figure 7 is rotated with respect to the axis of the light reflected from the sample using a rotary flange. The reflected light intensity from the magnetic sample is dependent on the ellipticity angle η_K and the azimuthal angle of the major axis of the polarization ellipse θ_K of the beam. It should be noted that the RMOKE measurement system does not require the lock-in amplification that was adopted in the visible MOKE setup shown in Figure 4.

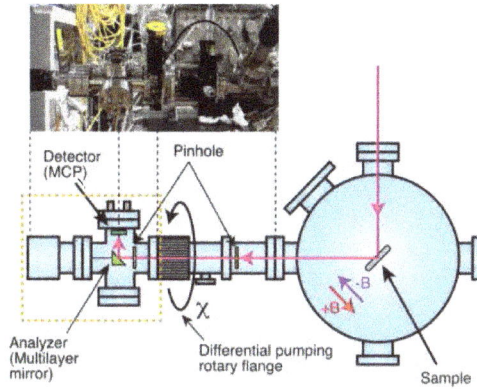

Figure 7. Measurement system for polar resonant MOKE that uses a rotating analyzer ellipsometry (RAE) technique. The section enclosed by the dotted rectangular line is rotated together. The RAE unit is shown in the inset photograph.

We describe the resulting intensity that is obtained by rotating the ellipsometry unit indicated by the yellow dashed rectangle in Figure 7 using a Mueller formalism [70,71], in which the polarization states are expressed using the Stokes vector. The Stokes vector consists of four elements, S_0, S_1, S_2 and S_3, which express the sum of the vertical and horizontal components of the polarization (= total intensity), the difference between the vertical and horizontal components, the difference between the $\pi/4$ and $-\pi/4$ components and the difference between the right-handed and left-handed components, respectively. It is given by:

$$S = \begin{bmatrix} S_0 \\ S_1 \\ S_2 \\ S_3 \end{bmatrix} \tag{2}$$

The degree of polarization, V, can be expressed as follows:

$$V = \frac{\sqrt{S_1^2 + S_2^2 + S_3^2}}{S_0} \tag{3}$$

The absolute intensity is less important than the polarization state in most real cases, so it is convenient to define the normalized Stokes parameter that is divided by S_0, which means that $S_0 = 1$. The resulting normalized Stokes parameters can then be expressed as follows.

$$S_0 = 1 \tag{4}$$
$$S_1 = V\cos2\eta_K\cos2\theta_K \tag{5}$$
$$S_2 = V\cos2\eta_K\sin2\theta_K \tag{6}$$
$$S_3 = V\sin2\eta_K \tag{7}$$

In the Mueller scheme, the optical components are expressed by the Mueller matrix. Using A_R, which is the analyzer, and $R(\chi)$, which is the coordinate rotation, the Stokes vector $S'(\chi)$ of the light that is reflected by the analyzer and reaches the detector is a function of χ. It is expressed using the Mueller matrices as:

$$S'(\chi) = R(\chi) \cdot A_R \cdot R(-\chi) \cdot S \tag{8}$$

where S' is the Stokes vector of the light that reaches the detector after reflection from the analyzer.

The intensity I(χ) of the reflected light that reaches the detector is derived from Equation (8):

$$I(\chi) = S'_0(\chi)\frac{r_p^2}{2}\{S_0(\alpha^2 + 1) + S_1(\alpha^2 - 1)\cos2\chi + S_2(\alpha^2 - 1)\sin2\chi\} \tag{9}$$

where α is the ratio of the reflectance amplitudes, r_s/r_p, for the s- and p-components.

Using the relationship between the Stokes vector and V, η_K, and θ_K (from Equation (7)), the intensity can be written as:

$$I(\chi) = \frac{r_p^2}{2}[2V(\alpha^2 - 1)\cos2\eta_K \cdot \cos^2(\theta_K - \chi) + \alpha^2 + 1 - V(\alpha^2 - 1)\cos2\eta_K] \tag{10}$$

The phase-shift in $I(\chi)$ corresponds to θ_K. The value of V is unity if the light is perfectly polarized. The dependence of the cosine square function on χ is Malus' law [72,73]. If we assume that $V = 1$, the intensity $I(\chi)$ can simply be expressed using a cosine function with η_K and θ_K as:

$$I(\chi) = C_1(\eta_K)\cos2(\chi - \theta_K) + C_2(\eta_K) \tag{11}$$

where the values of C_1 and C_2 determine η_K. Figure 8 shows an example of the experimental results using SR. The vertical axis shows the signals detected at MCP as shown in Figure 7. The Kerr rotation angle θ_k can be determined using this RAE technique. For example, the phase difference that appeared in curves that were measured under the application of +B and −B fields corresponds to double the Kerr rotation angle, i.e., $2\theta_k$. It should be noted that in the RAE technique, V affects the value of the ellipticity. If the light is not completely polarized and unpolarized components are included, there is then a reduction in the amplitude that leads to a decrease in $C_1(\eta_K)$ in Equation (11). Unpolarized or fully-circularly-polarized light gives no intensity variations in the RAE measurement. The degree of polarization is dependent on the energy and the optical components in the SR beamline. The polarizance of an analyzer is also taken into account for extraction of the ellipticity from the RAE technique. In the remainder of this review paper, we focus solely on θ_K for revealing the first demonstration of TR-RMOKE with FEL. θ_K is independent of V, and we thus take $V = 1$.

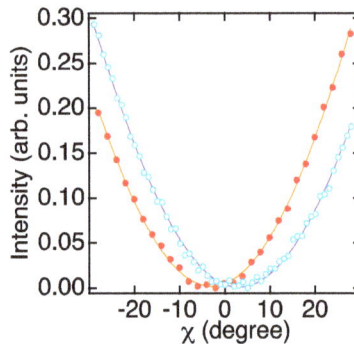

Figure 8. Intensity variation with rotation angle χ. Red (blue) circles show the results for +H (−H). The orange and purple curves show fitted cosine functions. The sample is a thin-film structure of Ta (2 nm)/$Gd_{21}(Fe_{90}Co_{10})_{79}$ (20 nm)/Ta (10 nm)/SiO_2. Measurements were taken at room temperature. The photon energy was tuned to 53 eV, which corresponds to the M edge of iron.

5. Static RMOKE Measurement

In the visible region, the MOKE measurements essentially probe the average magnetizations of target samples. In contrast, in the shorter wavelength range around the soft X-ray region, light interacts with the electronic states in the core levels of matter, such as the $2p$ or $3p$ absorption edges for $3d$ transition metals. These core level states are relatively localized to each of the element atoms when compared with the states near the Fermi level for magnetic metals, and there is a resonant effect in the MOKE data. Therefore, the RMOKE measurements detect the magnetization of materials with element selectivity. In this section, we review the research on the various types of RMOKE, including transverse RMOKE (T-RMOKE), longitudinal RMOKE (L-RMOKE) and polar RMOKE (P-RMOKE).

In the soft X-ray region, T-RMOKE has been used earlier and adopted more widely than the other RMOKE geometries. This seems to be because T-RMOKE requires only intensity measurements, as shown in Figure 2, and it is more technically feasible than the polarization analysis. The T-RMOKE measurements were initiated by measurement of a Ni sample at the Ni K-edge that led to observation of 0.2% as a peak-to-peak value of the asymmetry [74]. The research itself was motivated by the study of an interference effect between X-ray magnetic Bragg scattering and the electric scattering in the X-ray diffraction measurements [75]. Tuning of the photon energy at the K-edge means that the 1 s $\rightarrow$ $4p$ dipole transition occurs and the magnetic signal, then mainly originates from p-d hybridization [76], which was similar to the visible MOKE case [77]. Subsequently, T-RMOKE measurements were carried out on a Fe sample at the L-edge [78]. In this case, the dipole transition is $2p \rightarrow 3d$ and thus directly probed the itinerant $3d$ bands. The asymmetry ratio increases by up to $\sim$20% as a peak-to-peak value, and the data obtained allow us to extract the total width of an excited state and the magnitude of the exchange splitting when combined with theoretical resonant scattering calculations [79]. Triggered by this research, T-RMOKE experiments were reported in [80–83] for the Co M-edge, in [82–84], for the Co L-edge, in [82,85–87], for the Fe M-edge, in [82,88–91], for the Fe L-edge, in [82,87,92], for the Ni M-edge, in [82,87], for the Ni L-edge, in [93], for the Pt L-edge, and in [94], for the Mn L-edge. Additionally, the L-edge T-MOKE experiment was applied to the investigation of the magneto-crystalline anisotropy energy of an ultra-thin transition metal (Co) film [84]. It should be noted that the T-RMOKE signal basically originates from pure-charge and pure-magnetic signals; however, when the angle of incidence is tuned to the Brewster angle, components of the charge scattering are suppressed, and a pure-magnetic signal can be obtained within this geometry [86,89,95].

Moving onto the L-RMOKE research, polarization analyses of L-RMOKE were reported in [11,96]. The Kerr rotation angle was measured for a Fe/Cr multilayer structure at the individual photon energies of the Fe and Cr absorption edges. The Kerr rotation angles were two orders of magnitude higher than the corresponding values in the visible region. The L-RMOKE measurements were found to enable depth analysis by varying the photon energy, including analysis of the pre-edge region. In [91,97–99], depth profile measurements using Fe, Co and Ni L-edge L-RMOKE were demonstrated by selecting appropriate angles of incidence and energies. The $L_{2,3}$ spectra of the rotation angles and the ellipticities were measured for the Fe, Co and Ni metal samples [12,100]. The L-RMOKE Kerr rotation angle was related to the T-RMOKE signals, while the L-MOKE ellipticities corresponded to the XMCD spectra of the reflected beam. The researchers also discovered notable interference effects from the light reflected from the surface and the interfaces in the L-RMOKE spectra. Recently, M-edge L-RMOKE measurements of Co and Ni were reported in [83,101].

A P-RMOKE experiment has recently been the subject of intense attention from researchers because it has the only measurement geometry that can detect perpendicular magnetization (Figure 2), which is significant for the development of modern storage devices. The P-RMOKE was predicted in 1975 when Ni metal showed a θ_k value at the $M_{2,3}$-edge that was 10-times larger than that in the visible region [102]. Then, P-RMOKE measurements were carried out at a synchrotron radiation bending-magnet beamline and confirmed that θ_k was 50-times larger at the Ni $M_{2,3}$ edges [56]. Basically, there is no apparent physical difference in the RMOKE signals between the longitudinal and polar MOKE geometries. Either technique can be used, depending on whether the magnetization direction is parallel or perpendicular.

We now provide an example of RMOKE for a thin-film structure of Ta (2 nm)/Gd$_{21}$(Fe$_{90}$Co$_{10}$)$_{79}$ (20 nm)/Ta (10 nm) that was fabricated on thermally-oxidized silicon wafers using a radio-frequency

Appl. Sci. **2017**, *7*, 662

(RF) magnetron sputtering process. The $Gd_{21}(Fe_{90}Co_{10})_{79}$ alloy is ferrimagnetic. The magnetic moment of the transition metal (Fe) sublattice at room temperature is higher than that of the rare earth (Gd) metal, and the direction of the magnetic moment of the Fe atom is parallel to that of the external magnetic field. The Ta (2 nm) capping layer prevents oxidation of the $Gd_{21}(Fe_{90}Co_{10})_{79}$, and the Ta (10 nm) underlayer helps with adhesion to the Si substrate. Figure 9a shows a set of soft X-ray absorption spectra for the $Gd_{21}(Fe_{90}Co_{10})_{79}$ sample that were measured under saturated magnetization conditions (Figure 5), where the external field of ± 0.47 T was applied perpendicularly to the sample surface. The measurements were performed at bending-magnet beamline BL5B at the Ultraviolet Synchrotron Orbital Radiation facility (UVSOR, Okazaki, Japan). The degree of linear polarization is at least 0.98 [101]. The spectra were obtained using the total electron yield mode. The vertical axis in Figure 9a was normalized by incident intensity measured by a gold mesh. The peaks at 33 eV, 42 eV, 53 eV and 66 eV were assigned to the Ta $5p_{3/2}$, $5p_{1/2}$, Fe and Co $3p$ absorptions, respectively.

Figure 9b shows the experimental RMOKE spectra obtained around the Fe M-absorption edge. The RMOKE measurements were conducted in a polar geometry (P-RMOKE), and the Kerr rotation angle values were obtained by the RAE method, which was shown in Figure 7. At the Fe $3p$ absorption, θ_K was approximately $3°$, while at the Co $3p$ absorption, θ_K was approximately $5°$, and these values were approximately 10-times larger than those obtained from the visible MOKE, as shown in Figure 5. From a set of the absorption and RMOKE spectra, the photon energy of 53 eV is found to have sufficiently large values for both the absorption peak and θ_K, and the Fe atom is thus the most suitable candidate for tracing of the magnetization dynamics in $Gd_{21}(Fe_{90}Co_{10})_{79}$ during ultrafast switching for the time-resolved measurements.

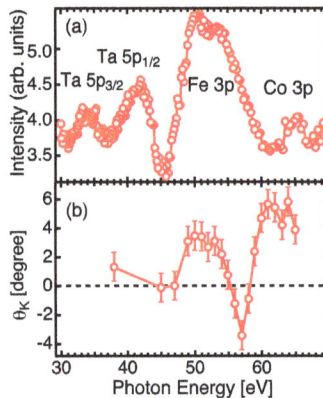

Figure 9. (a) Absorption spectrum and (b) θ_K variation with photon energy measured using a polar geometry. The sample is a thin-film structure of Ta (2 nm)/$Gd_{21}(Fe_{90}Co_{10})_{79}$ (20 nm)/Ta (10 nm)/SiO_2. The measurements were taken at room temperature.

6. Demonstration of TR-RMOKE with a Soft X-ray FEL

Figure 10 shows a schematic of the TR-RMOKE experiment that was carried out on the Diffraction and Projection Imaging (DiProI) beamline [103] at the seeded FERMI FEL at the Elettra laboratory in Italy. We used the FEL-I, which is normally operated at 60–20 nm (where hv = 12.4–62 eV) with an electron beam energy of 1.2 GeV [104,105]. The optical pumping was performed using the infrared (IR) lasers that are used for seeding of the FEL, and the pump pulses were thus intrinsically synchronized with the FEL probe pulses with practically jitter-free time resolution [106]. It should be noted that the seeded FEL at FERMI has excellent longitudinal coherence and spectral purity when compared with the commonly-used self-amplified spontaneous emission (SASE) scheme, and it also offers a multiple polarized pulse capability [61].

Measurements were performed using 23.6-nm (52.5 eV) FEL pulses with widths of 80–100 fs and a 780-nm IR laser. The pump-probe method was used with a repetition rate of 10 Hz. The FEL and IR laser beam spot sizes on the sample were 420 μm and 530 μm in diameter, respectively. The time resolution was limited by the pulse width of the pump laser of 150 fs. The fluences of the pump beam and the probe pulses were tuned to 14 mJ/cm^2 and 3 mJ/cm^2, respectively. The temporal overlap of these pulses was determined with 250 fs resolution by monitoring the reflectivity changes of Si_3N_4 in the FEL pump/IR laser probe experiment [106,107]. The spatial overlap was checked using a YAG crystal.

The measurement configuration was set such that a linear-horizontally-polarized IR beam and the linear-vertically polarized FEL beam were coaxially incident on the sample. The degree of linear polarization of the FEL, which was nominally called vertical polarization, was ~0.97 [108]. The FEL was used to irradiate the sample in the *s*-polarization configuration with an angle of incidence of 45° with respect to the surface normal. The ellipsometry was conducted by rotation of the RAE unit, in which the reflected FEL beam traveled through a rotary flange, which was then reflected at a Mo/Si multilayer mirror (10 periods of 19.1-nm layers) and was finally detected at the MCP. An Al filter was used to attenuate the reflected IR laser beam. The fluctuating pulse-to-pulse intensity of the FERMI pulses was monitored in a shot-by-shot manner using an intensity monitor composed of a gas cell.

Figure 11 shows the results of time-resolved measurements of the $Gd_{21}(Fe_{90}Co_{10})_{79}$ sample (using the Ta (2 nm)/$Gd_{21}(Fe_{90}Co_{10})_{79}$ (20 nm)/Ta (10 nm)/SiO_2 structure) that has been known to enable ultrafast spin switching. Each panel shows the intensity variation when normalized with respect to the incident intensity, which was monitored in a shot-by-shot manner with χ at each delay time. The vertical axis denotes the normalized intensity, i.e., the intensity that was detected at the MCP divided by the incident intensity. The intensity of the light that was reflected by the analyzer is monitored at the detector as a function of χ, as shown in the inset of Figure 10. The two curves (blue and red) correspond to measurements performed under applied magnetic fields in the up and down directions along the sample surface normal, respectively. The solid lines indicate cosine fitting (Equation (11)) to the experimental results. The zero angle was set at the middle angle between the minima of the curves that were obtained under application of +H and −H. The position of the middle angle did not change with respect to the delay time to within 0.5 degrees, which was the resolution used to determine θ_K in this experiment. We see that the apparent shifts in the entire curves increase the reliability and the accuracy of the change, even with the possible existence of fluctuations between the individual data points. The Kerr rotation angle at each delay time can be extracted from the phase difference between the two curves, 3.2°, 2.5°, −1.1° and −0.7° for the delay times of −100, 100, 200 and 600 fs, respectively. The Kerr rotation angle of 3.1°, which was measured from the static RMOKE (see Figure 9), was reproduced in the time-resolved measurements by tuning the FEL energy to 53 eV at −100 fs, at which the Kerr rotation angle is 3.2°. Because a polar geometry is used in the TR-RMOKE measurements, the Kerr rotation angle indicates the out-of-plane magnetic moment of Fe in the $Gd_{21}(Fe_{90}Co_{10})_{79}$ structure. Figure 11 shows a schematic depiction of the magnetization dynamics with respect to the external field that result from the time-dependent Kerr rotation angle. The lengths of the arrows in the figure correspond to the Kerr rotation angle magnitude for each delay time. At 200 fs after the high-intensity laser irradiation, the changes in the sign of the Kerr rotation angle indicate reversal of the Fe magnetization. Because −1.1° at 200-fs delay time was larger than the resolution for determining Kerr angle in these measurements, the results showed magnetization reversal rather than the demagnetization. When the FERMI-FEL repetition rate of 10 Hz is considered, the Fe magnetic moment is recovered within at least 100 ms after pumping. This reversal mechanism is classified as a thermal process, unlike the non-thermal effect, such as the process involving the inverse Faraday effect that was observed in previous studies using circularly-polarized beams [109]. Because a linearly-polarized IR pump laser was used, there is no coupling in terms of the exchange of angular momentum between the photons and the spins in the material, although there is a possibility of coherent interaction between spin and photon field [110], which occurs on a much faster timescale than this experimental resolution. Therefore, the angular momentum transfer path is closed between the Gd and Fe sub-lattices [111]. In Figure 12, the time-dependent results that were obtained from this TR-RMOKE experiment with the FEL are shown. The magnetization reversal timescale of the Fe sublattice is the same as that in a previous study that used TR-XMCD with laser slicing [3]. Although the microscopic mechanism of the magnetization dynamics should be explored with various experimental conditions and materials, which is beyond the scope of this review paper, here we demonstrated that the TR-RMOKE scheme is useful for tracing of light-induced magnetization dynamics on a subpicosecond timescale. It should be noted that in time-resolved MOKE measurements, it has been argued that the MOKE signal is modified by a nonequilibrium state that is generated during the femtosecond pulse, meaning that the MOKE signal does not reflect the sample magnetization [112,113]. While this effect is negligible for the time delays shown in Figures 11 and 12, this should be considered when the delay becomes close to the dephasing time of the coherent correlation between the photons and the spins.

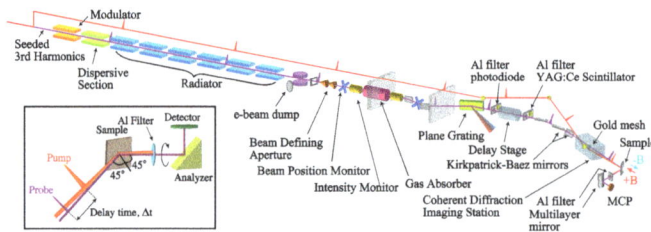

Figure 10. Schematic diagram of seeded FEL at Free Electron laser Radiation for Multidisciplinary Investigations (FERMI)@Elettra. Femtosecond pulses are used as seeds for the FEL and as a pump source for the TR-RMOKE measurements. The measurement chamber was directly connected on the downstream side of the chamber for the Diffraction and Projection Imaging (DiProI) station. (Inset) Details of the TR-RMOKE measurements. An Al filter was inserted to prevent the pump pulses from entering the unit composed of the analyzer (Mo/Si multilayer) and the detector (MCP).

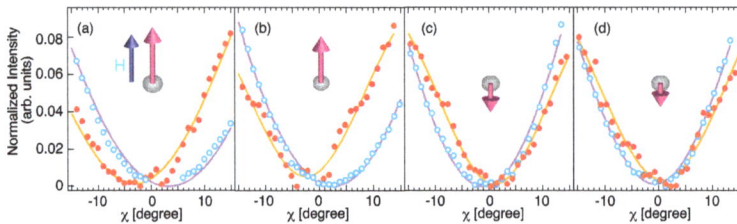

Figure 11. Intensity variation with rotation angle χ at delay times of (**a**) -100 fs, (**b**) 100 fs, (**c**) 200 fs and (**d**) 600 fs under application of the external magnetic fields +H (red circles) and $-$H (blue circles). These characteristics were obtained using a photon energy of 53 eV. The sample is a thin-film structure composed of Ta (2 nm)/$Gd_{21}(Fe_{90}Co_{10})_{79}$ (20 nm)/Ta (10 nm)/SiO_2. Solid lines show cosine fitting using Equation (11) for the data points that were obtained under application of +H (orange) and $-$H (purple). Inside the graphs, schematic diagrams of the time evolution of the Fe magnetic moment that were deduced from the time-dependent Kerr rotation angle are shown. At the delay time of -100 fs, **H** pointed in the same direction as the magnetic moment of the Fe sublattice.

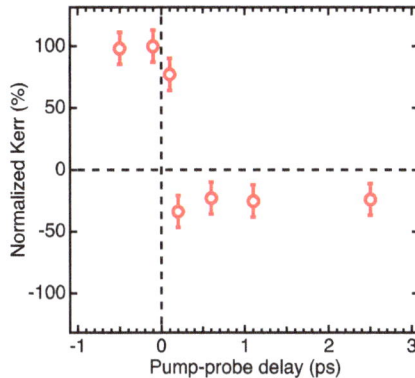

Figure 12. Time evolution of normalized Kerr rotation angle obtained from TR-RMOKE measurements. The sample used was a thin film structure composed of Ta (2 nm)/Gd$_{21}$(Fe$_{90}$Co$_{10}$)$_{79}$ (20 nm)/Ta (10 nm)/SiO$_2$. The vertical axis represents the Kerr rotation angles divided by the average of the Kerr angles for negative delay times of -500 fs and -100 fs.

7. Conclusions and Outlook

In summary, we reviewed experimental measurements of the polar resonant magneto-optical Kerr effect (P-RMOKE) using soft X-rays and the extension of the method to time-resolved measurements using a free electron laser (FEL). We showed that the method has various advantages, including: (1) element specificity, (2) large Kerr rotation angle, (3) probing of the out-of-plane magnetization with various experimental setup at M-edge range for 3d transition metals, (4) use of the linearly-polarized light and (5) femtosecond-scale time resolution. Because the experiments were performed using the photon-in and photon-out configuration, we expect that the measurements could also be taken under magnetic and electric fields. This type of operando experimental method will be a promising tool for research into femtomagnetism and for the development of ultrafast spintronics.

Acknowledgments: This work was partly supported by the MEXT program "X-ray Free Electron Laser Priority Strategy Program", by the Hyogo Science and Technology Association and by the Japan Society for the Promotion of Science with a grant-in-aid for Scientific Research (C) (Grant No. 26400328). Sh. Yamamoto was supported by a Grant-in-Aid for JSPS Fellows and by the Program for Leading Graduate Schools (Materials Education program for the future leaders in Research, Industry, and Technology, MERIT). The supporting experiments were performed using the facilities of the Synchrotron Radiation Research Organization of the University of Tokyo, Nos. 2014A7401, 2014B7401, 2014B7473, 2015A7401 and 2015B7401. The authors are grateful to S. Iwata and T. Kato of Nagoya University for synthesizing the GdFeCo sample and measuring the visible MOKE. E. Shigemasa and M. Hasumoto supported the static RMOKE measurement experiments at Ultraviolet Synchrotron Orbital Radiation (UVSOR). M. Kiskinova, F. Capotondi, E. Pedersoli and M. Manfredda supported the TR-RMOKE experiments at FERMI@Elettra. H. Wadati, M. Fujisawa and T. Someya conducted the TR-RMOKE experiments together.

Conflicts of Interest: The authors declare no conflict of interest.

Abbreviations

The following abbreviations are used in this manuscript:

AFMR	Antiferromagnetic resonance
BLS	Brillouin light scattering
EUV	Extreme ultraviolet
FEL	Free electron laser
FMR	Ferromagnetic resonance
FTH	Fourier transform holography

HHG	High harmonic generation
L-RMOKE	Longitudinal resonant magneto-optical Kerr effect
MSHG	Magnetization-induced second harmonic generation
P-RMOKE	Polar resonant magneto-optical Kerr effect
RAE	Rotating analyzer ellipsometry
RMOKE	Resonant magneto-optical Kerr effect
RIXS	Resonant inelastic X-ray scattering
RSXD	Resonant soft X-ray diffraction
SAXS	Small angle X-ray scattering
SR	Synchrotron radiation
TFY	Total fluorescence yield
THz-TDS	Terahertz time domain spectroscopy
TR	Time-resolved
XMCD	X-ray magnetic circular dichroism
T-RMOKE	Transverse resonant magneto-optical Kerr effect
UV	Ultraviolet
XMCD	X-ray magnetic circular dichroism
XMLD	X-ray magnetic linear dichroism

References

1. Kirilyuk, A.; Kimel, A.V.; Rasing, T. Ultrafast optical manipulation of magnetic order. *Rev. Mod. Phys.* **2010**, *82*, 2731–2784.
2. Beaurepaire, E.; Merle, J.C.; Daunois, A.; Bigot, J.Y. Ultrafast spin dynamics in ferromagnetic nickel. *Phys. Rev. Lett.* **1996**, *76*, 4250–4253.
3. Radu, I.; Vahaplar, K.; Stamm, C.; Kachel, T.; Pontius, N.; Dürr, H.; Ostler, T.; Barker, J.; Evans, R.; Chantrell, R.; et al. Transient ferromagnetic-like state mediating ultrafast reversal of antiferromagnetically coupled spins. *Nature* **2011**, *472*, 205–208.
4. Thiele, J.U.; Buess, M.; Back, C.H. Spin dynamics of the antiferromagnetic-to-ferromagnetic phase transition in FeRh on a sub-picosecond time scale. *Appl. Phys. Lett.* **2004**, *85*, 2857–2859.
5. Yamaguchi, K.; Nakajima, M.; Suemoto, T. Coherent control of spin precession motion with impulsive magnetic fields of half-cycle terahertz radiation. *Phys. Rev. Lett.* **2010**, *105*, 237201.
6. Schellekens, A.; Koopmans, B. Comparing ultrafast demagnetization rates between competing models for finite temperature magnetism. *Phys. Rev. Lett.* **2013**, *110*, 217204.
7. Bigot, J.Y.; Vomir, M. Ultrafast magnetization dynamics of nanostructures. *Annalen der Physik* **2013**, *525*, 2–30.
8. Rudolf, D.; Chan, L.O.; Battiato, M.; Adam, R.; Shaw, J.M.; Turgut, E.; Maldonado, P.; Mathias, S.; Grychtol, P.; Nembach, H.T.; et al. Ultrafast magnetization enhancement in metallic multilayers driven by superdiffusive spin current. *Nat. Commun.* **2012**, *3*, 1037.
9. Dean, M.; Cao, Y.; Liu, X.; Wall, S.; Zhu, D.; Mankowsky, R.; Thampy, V.; Chen, X.; Vale, J.; Casa, D.; et al. Ultrafast energy-and momentum-resolved dynamics of magnetic correlations in the photo-doped Mott insulator Sr2IrO4. *Nat. Mater.* **2016**, *15*, 601–605.
10. Oppeneer, P. Magneto-optical Kerr spectra. *Handb. Magn. Mater.* **2001**, *13*, 229–422.
11. Kortright, J.; Rice, M. Soft X-ray magneto-optic Kerr rotation and element-specific hysteresis measurement. *Rev. Sci. Instrum.* **1996**, *67*, 3353.
12. Mertins, H.C.; Valencia, S.; Abramsohn, D.; Gaupp, A.; Gudat, W.; Oppeneer, P.M. X-ray Kerr rotation and ellipticity spectra at the 2 p edges of Fe, Co, and Ni. *Phys. Rev. B* **2004**, *69*, doi:10.1103/PhysRevB.69.064407.
13. Chen, C.; Idzerda, Y.; Lin, H.J.; Smith, N.; Meigs, G.; Chaban, E.; Ho, G.; Pellegrin, E.; Sette, F. Experimental confirmation of the X-ray magnetic circular dichroism sum rules for iron and cobalt. *Phys. Rev. Lett.* **1995**, *75*, 152–155.
14. Stöhr, J.; Padmore, H.; Anders, S.; Stammler, T.; Scheinfein, M. Principles of X-ray magnetic dichroism spectromicroscopy. *Surf. Rev. Lett.* **1998**, *5*, 1297–1308.

15. Yamamoto, Sh.; Taguchi, M.; Someya, T.; Kubota, Y.; Ito, S.; Wadati, H.; Fujisawa, M.; Capotondi, F.; Pedersoli, E.; Manfredda, M.; et al. Ultrafast spin-switching of a ferrimagnetic alloy at room temperature traced by resonant magneto-optical Kerr effect using a seeded free electron laser. *Revi. Sci. Instrum.* **2015**, *86*, doi:10.1063/1.4927828.

16. Yang, Z.; Scheinfein, M. Combined three-axis surface magneto-optical Kerr effects in the study of surface and ultra-thin-film magnetism. *J. Appl. Phys.* **1993**, *74*, 6810–6823.

17. Strelniker, Y.M.; Bergman, D.J. Magneto-optical response of a periodic metallic nano-structure. In Proceedings of the SPIE 9574, International Society for Optics and Photonics, San Diego, CA, USA, 9 August 2015; p. 954705.

18. Nozaki, T.; Shiota, Y.; Miwa, S.; Murakami, S.; Bonell, F.; Ishibashi, S.; Kubota, H.; Yakushiji, K.; Saruya, T.; Fukushima, A.; et al. Electric-field-induced ferromagnetic resonance excitation in an ultra-thin ferromagnetic metal layer. *Nat. Phys.* **2012**, *8*, 491–496.

19. Milano, J.; Steren, L.; Grimsditch, M. Effect of dipolar interaction on the antiferromagnetic resonance spectra of NiO. *Phys. Rev. Lett.* **2004**, *93*, doi:10.1103/PhysRevLett.93.077601.

20. Hangyo, M.; Tani, M.; Nagashima, T. Terahertz time-domain spectroscopy of solids: A review. *Int. J. Infrared Millim. Waves* **2005**, *26*, 1661–1690.

21. Parchenko, S.; Satoh, T.; Yoshimine, I.; Stobiecki, F.; Maziewski, A.; Stupakiewicz, A. Non-thermal optical excitation of terahertz-spin precession in a magneto-optical insulator. *Appl. Phys. Lett.* **2016**, *108*, doi:10.1063/1.4940241.

22. Khorsand, A.; Savoini, M.; Kirilyuk, A.; Kimel, A.; Tsukamoto, A.; Itoh, A.; Rasing, T. Element-specific probing of ultrafast spin dynamics in multisublattice magnets with visible light. *Phys. Rev. Lett.* **2013**, *110*, 107205.

23. Ogawa, Y.; Kaneko, Y.; He, J.; Yu, X.; Arima, T.; Tokura, Y. Magnetization-induced second harmonic generation in a polar ferromagnet. *Phys. Rev. Lett.* **2004**, *92*, 047401.

24. Regensburger, H.; Vollmer, R.; Kirschner, J. Time-resolved magnetization-induced second-harmonic generation from the Ni (110) surface. *Phys. Rev. B* **2000**, *61*, 14716–14722.

25. Mertins, H.C.; Oppeneer, P.; Kuneš, J.; Gaupp, A.; Abramsohn, D.; Schäfers, F. Observation of the X-ray magneto-optical Voigt effect. *Phys. Rev. Lett.* **2001**, *87*, 047401.

26. Höchst, H.; Rioux, D.; Zhao, D.; Huber, D.L. Magnetic linear dichroism effects in reflection spectroscopy: A case study at the Fe M 2, 3 edge. *J. Appl. Phys.* **1997**, *81*, 7584–7588.

27. Schwickert, M.; Guo, G.; Tomaz, M.; O'Brien, W.; Harp, G. X-ray magnetic linear dichroism in absorption at the L edge of metallic Co, Fe, Cr, and V. *Phys. Rev. B* **1998**, *58*, R4289.

28. Finazzi, M.; Duo, L.; Ciccacci, F. Magnetic properties of interfaces and multilayers based on thin antiferromagnetic oxide films. *Surf. Sci. Rep.* **2009**, *64*, 139–167.

29. Ament, L.J.; Van Veenendaal, M.; Devereaux, T.P.; Hill, J.P.; Van Den Brink, J. Resonant inelastic X-ray scattering studies of elementary excitations. *Rev. Mod. Phys.* **2011**, *83*, 705–767.

30. Fabbris, G.; Meyers, D.; Xu, L.; Katukuri, V.; Hozoi, L.; Liu, X.; Chen, Z.Y.; Okamoto, J.; Schmitt, T.; Uldry, A.; et al. Doping Dependence of Collective Spin and Orbital Excitations in the Spin-1 Quantum Antiferromagnet $La_{2-x}Sr_xNiO_4$ Observed by X Rays. *Phys. Rev. Lett.* **2017**, *118*, 156402.

31. Dean, M.; Springell, R.; Monney, C.; Zhou, K.; Pereiro, J.; Božović, I.; Dalla Piazza, B.; Rønnow, H.; Morenzoni, E.; Van Den Brink, J.; et al. Spin excitations in a single La_2CuO_4 layer. *Nat. Mater.* **2012**, *11*, 850–854.

32. Minola, M.; Di Castro, D.; Braicovich, L.; Brookes, N.; Innocenti, D.; Sala, M.M.; Tebano, A.; Balestrino, G.; Ghiringhelli, G. Magnetic and ligand field properties of copper at the interfaces of $(CaCuO_2)n/(SrTiO_3)n$ superlattices. *Phys. Rev. B* **2012**, *85*, 235138.

33. Müller, L.; Gutt, C.; Pfau, B.; Schaffert, S.; Geilhufe, J.; Büttner, F.; Mohanty, J.; Flewett, S.; Treusch, R.; Düsterer, S.; et al. Breakdown of the X-ray resonant magnetic scattering signal during intense pulses of extreme ultraviolet free-electron-laser radiation. *Phys. Rev. Lett.* **2013**, *110*, 234801.

34. Yamasaki, Y.; Sudayama, T.; Okamoto, J.; Nakao, H.; Kubota, M.; Murakami, Y. Diffractometer for small angle resonant soft X-ray scattering under magnetic field. *J. Phys. Conf. Ser.* **2013**, *425*, doi:10.1088/1742-6596/425/13/132012.

35. Günther, C.M.; Guehrs, E.; Schneider, M.; Pfau, B.; von Korff Schmising, C.; Geilhufe, J.; Schaffert, S.; Eisebitt, S. Experimental evaluations of signal-to-noise in spectro-holography via modified uniformly redundant arrays in the soft X-ray and extreme ultraviolet spectral regime. *J. Opt.* **2017**, *19*, doi:10.1088/2040-8986/aa6380.

36. Eisebitt, S.; Lüning, J.; Schlotter, W.; Lörgen, M.; Hellwig, O.; Eberhardt, W.; Stöhr, J. Lensless imaging of magnetic nanostructures by X-ray spectro-holography. *Nature* **2004**, *432*, 885–888.

37. Tanaka, A.; Chang, C.; Buchholz, M.; Trabant, C.; Schierle, E.; Schlappa, J.; Schmitz, D.; Ott, H.; Metcalf, P.; Tjeng, L.; et al. Analysis of charge and orbital order in Fe_3O_4 by Fe $L_{2,3}$ resonant X-ray diffraction. *Phys. Rev. B* **2013**, *88*, 195110.

38. Chan, L.O.; Siemens, M.; Murnane, M.M.; Kapteyn, H.C.; Mathias, S.; Aeschlimann, M.; Grychtol, P.; Adam, R.; Schneider, C.M.; Shaw, J.M.; et al. Ultrafast demagnetization dynamics at the M edges of magnetic elements observed using a tabletop high-harmonic soft X-ray source. *Phys. Rev. Lett.* **2009**, *103*, 257402.

39. Turgut, E.; Shaw, J.M.; Grychtol, P.; Nembach, H.T.; Rudolf, D.; Adam, R.; Aeschlimann, M.; Schneider, C.M.; Silva, T.J.; Murnane, M.M.; et al. Controlling the competition between optically induced ultrafast spin-flip scattering and spin transport in magnetic multilayers. *Phys. Rev. Lett.* **2013**, *110*, 197201.

40. Ferrari, E.; Spezzani, C.; Fortuna, F.; Delaunay, R.; Vidal, F.; Nikolov, I.; Cinquegrana, P.; Diviacco, B.; Gauthier, D.; Penco, G.; et al. Element Selective Probe of the Ultra-Fast Magnetic Response to an Element Selective Excitation in Fe-Ni Compounds Using a Two-Color FEL Source. *Photonics* **2017**, *4*, 6.

41. Stamm, C.; Kachel, T.; Pontius, N.; Mitzner, R.; Quast, T.; Holldack, K.; Khan, S.; Lupulescu, C.; Aziz, E.; Wietstruk, M.; et al. Femtosecond modification of electron localization and transfer of angular momentum in nickel. *Nat. Mater.* **2007**, *6*, 740–743.

42. Willems, F.; Smeenk, C.; Zhavoronkov, N.; Kornilov, O.; Radu, I.; Schmidbauer, M.; Hanke, M.; von Korff Schmising, C.; Vrakking, M.; Eisebitt, S. Probing ultrafast spin dynamics with high-harmonic magnetic circular dichroism spectroscopy. *Phys. Rev. B* **2015**, *92*, 220405.

43. Higley, D.J.; Hirsch, K.; Dakovski, G.L.; Jal, E.; Yuan, E.; Liu, T.; Lutman, A.A.; MacArthur, J.P.; Arenholz, E.; Chen, Z.; et al. Femtosecond X-ray magnetic circular dichroism absorption spectroscopy at an X-ray free electron laser. *Rev. Sci. Instrum.* **2016**, *87*, doi:10.1063/1.4944410.

44. Pfau, B.; Schaffert, S.; Müller, L.; Gutt, C.; Al-Shemmary, A.; Büttner, F.; Delaunay, R.; Düsterer, S.; Flewett, S.; Frömter, R.; et al. Ultrafast optical demagnetization manipulates nanoscale spin structure in domain walls. *Nat. Commun.* **2012**, *3*, 1100.

45. Vodungbo, B.; Gautier, J.; Lambert, G.; Sardinha, A.B.; Lozano, M.; Sebban, S.; Ducousso, M.; Boutu, W.; Li, K.; Tudu, B.; et al. Laser-induced ultrafast demagnetization in the presence of a nanoscale magnetic domain network. *Nat. Commun.* **2012**, *3*, 999.

46. Graves, C.; Reid, A.; Wang, T.; Wu, B.; De Jong, S.; Vahaplar, K.; Radu, I.; Bernstein, D.; Messerschmidt, M.; Müller, L.; et al. Nanoscale spin reversal by non-local angular momentum transfer following ultrafast laser excitation in ferrimagnetic GdFeCo. *Nat. Mater.* **2013**, *12*, 293–298.

47. Büttner, F.; Moutafis, C.; Schneider, M.; Krüger, B.; Günther, C.; Geilhufe, J.; Schmising, C.v.K.; Mohanty, J.; Pfau, B.; Schaffert, S.; et al. Dynamics and inertia of skyrmionic spin structures. *Nat. Phys.* **2015**, *11*, 225–228.

48. Bukin, N.; McKeever, C.; Burgos-Parra, E.; Keatley, P.; Hicken, R.; Ogrin, F.; Beutier, G.; Dupraz, M.; Popescu, H.; Jaouen, N.; et al. Time-resolved imaging of magnetic vortex dynamics using holography with extended reference autocorrelation by linear differential operator. *Sci. Rep.* **2016**, *6*, doi:10.1038/srep36307.

49. Von Korff Schmising, C.; Pfau, B.; Schneider, M.; Günther, C.; Giovannella, M.; Perron, J.; Vodungbo, B.; Müller, L.; Capotondi, F.; Pedersoli, E.; et al. Imaging ultrafast demagnetization dynamics after a spatially localized optical excitation. *Phys. Rev. Lett.* **2014**, *112*, 217203.

50. Holldack, K.; Pontius, N.; Schierle, E.; Kachel, T.; Soltwisch, V.; Mitzner, R.; Quast, T.; Springholz, G.; Weschke, E. Ultrafast dynamics of antiferromagnetic order studied by femtosecond resonant soft X-ray diffraction. *Appl. Phys. Lett.* **2010**, *97*, doi:10.1063/1.3474612.

51. Pontius, N.; Kachel, T.; Schüßler-Langeheine, C.; Schlotter, W.; Beye, M.; Sorgenfrei, F.; Chang, C.; Foehlisch, A.; Wurth, W.; Metcalf, P.; et al. Time-resolved resonant soft X-ray diffraction with free-electron lasers: Femtosecond dynamics across the Verwey transition in magnetite. *Appl. Phys. Lett.* **2011**, *98*, 182504.

52. Först, M.; Caviglia, A.; Scherwitzl, R.; Mankowsky, R.; Zubko, P.; Khanna, V.; Bromberger, H.; Wilkins, S.; Chuang, Y.D.; Lee, W.; et al. Spatially resolved ultrafast magnetic dynamics initiated at a complex oxide heterointerface. *Nat. Mater.* **2015**, *14*, 883–888.

53. Henke, B.L.; Gullikson, E.; Davis, J.C. X-ray Interactions: Photoabsorption, Scattering, Transmission, and Reflection at E = 50–30,000 eV, Z = 1–92. *At. Data Nucl. Data Tables* **1993**, *54*, 181–342.

54. Tsuyama, T.; Chakraverty, S.; Macke, S.; Pontius, N.; Schüßler-Langeheine, C.; Hwang, H.; Tokura, Y.; Wadati, H. Photoinduced Demagnetization and Insulator-to-Metal Transition in Ferromagnetic Insulating BaFeO₃ Thin Films. *Phys. Rev. Lett.* **2016**, *116*, 256402.

55. Takubo, K.; Yamamoto, K.; Hirata, Y.; Yokoyama, Y.; Kubota, Y.; Yamamoto, Sh.; Yamamoto, S.; Matsuda, I.; Shin, S.; Seki, T.; et al. Capturing ultrafast magnetic dynamics by time-resolved soft x-ray magnetic circular dichroism. *Appl. Phys. Lett.* **2017**, *110*, 162401.

56. Yamamoto, Sh.; Taguchi, M.; Fujisawa, M.; Hobara, R.; Yamamoto, S.; Yaji, K.; Nakamura, T.; Fujikawa, K.; Yukawa, R.; Togashi, T.; et al. Observation of a giant Kerr rotation in a ferromagnetic transition metal by M-edge resonant magneto-optic Kerr effect. *Phys. Rev. B* **2014**, *89*, 064423.

57. Valencia, S.; Gaupp, A.; Gudat, W.; Mertins, H.C.; Oppeneer, P.; Abramsohn, D.; Schneider, C. Faraday rotation spectra at shallow core levels: 3p edges of Fe, Co, and Ni. *New J. Phys.* **2006**, *8*, 254.

58. Miyahara, T.; Park, S.Y.; Hanyu, T.; Hatano, T.; Moto, S.; Kagoshima, Y. Comparison between 3 p and 2 p magnetic circular dichroism in transition metals and alloys: Is the sum rule applicable to itinerant magnetic systems? *Rev. Sci. Instrum.* **1995**, *66*, 1558–1560.

59. Razdolski, I.; Alekhin, A.; Martens, U.; Bürstel, D.; Diesing, D.; Münzenberg, M.; Bovensiepen, U.; Melnikov, A. Analysis of the Time-Resolved Magneto-Optical Kerr Effect for Ultrafast Magnetization Dynamics in Ferromagnetic Thin Films. *J. Phys. Condens. Matter* **2017**, *29*, 174002.

60. Holldack, K.; Bahrdt, J.; Balzer, A.; Bovensiepen, U.; Brzhezinskaya, M.; Erko, A.; Eschenlohr, A.; Follath, R.; Firsov, A.; Frentrup, W.; et al. FemtoSpeX: A versatile optical pump–soft X-ray probe facility with 100 fs X-ray pulses of variable polarization. *J. Synchrotron. Radiat.* **2014**, *21*, 1090–1104.

61. Allaria, E.; Diviacco, B.; Callegari, C.; Finetti, P.; Mahieu, B.; Viefhaus, J.; Zangrando, M.; De Ninno, G.; Lambert, G.; Ferrari, E.; et al. Control of the polarization of a vacuum-ultraviolet, high-gain, free-electron laser. *Phys. Rev. X* **2014**, *4*, 041040.

62. Roussel, E.; Allaria, E.; Callegari, C.; Coreno, M.; Cucini, R.; Mitri, S.D.; Diviacco, B.; Ferrari, E.; Finetti, P.; Gauthier, D.; et al. Polarization Characterization of Soft X-ray Radiation at FERMI FEL-2. *Photonics* **2017**, *4*, 29.

63. Lutman, A.A.; MacArthur, J.P.; Ilchen, M.; Lindahl, A.O.; Buck, J.; Coffee, R.N.; Dakovski, G.L.; Dammann, L.; Ding, Y.; Dürr, H.A.; et al. Polarization control in an X-ray free-electron laser. *Nat. Photonics* **2016**, *10*, 468–472.

64. Vodungbo, B.; Sardinha, A.B.; Gautier, J.; Lambert, G.; Valentin, C.; Lozano, M.; Iaquaniello, G.; Delmotte, F.; Sebban, S.; Lüning, J.; et al. Polarization control of high order harmonics in the EUV photon energy range. *Opt. Express* **2011**, *19*, 4346–4356.

65. Lambert, G.; Vodungbo, B.; Gautier, J.; Mahieu, B.; Malka, V.; Sebban, S.; Zeitoun, P.; Luning, J.; Perron, J.; Andreev, A.; et al. Towards enabling femtosecond helicity-dependent spectroscopy with high-harmonic sources. *Nat. Commun.* **2015**, *6*, doi:10.1038/ncomms7167.

66. Rasing, T.; Koerkamp, M.G.; Koopmans, B.; Berg, H.V. Giant nonlinear magneto-optical Kerr effects from Fe interfaces. *J. Appl. Phys.* **1996**, *79*, 6181–6185.

67. Williams, P.; Rose, A.; Wang, C. Rotating-polarizer polarimeter for accurate retardance measurement. *Appl. Opt.* **1997**, *36*, 6466–6472.

68. Aspnes, D.; Studna, A. High precision scanning ellipsometer. *Appl. Opt.* **1975**, *14*, 220–228.

69. Suzuki, M.; Hirono, T. Control the light polarization. *Jpn. Soc. Synchrotron Radiat. Res.* **2006**, *19*, 444–453.

70. Fujiwara, H. *Spectroscopic Ellipsometry: Principles and Applications*; John Wiley & Sons: Chichester, UK, 2007.

71. Schäfers, F.; Mertins, H.C.; Gaupp, A.; Gudat, W.; Mertin, M.; Packe, I.; Schmolla, F.; Di Fonzo, S.; Soullié, G.; Jark, W.; et al. Soft-X-ray polarimeter with multilayer optics: Complete analysis of the polarization state of light. *Appl. Opt.* **1999**, *38*, 4074–4088.

72. Kimura, H.; Hirono, T.; Tamenori, Y.; Saitoh, Y.; Salashchenko, N.; Ishikawa, T. Transmission type Sc/Cr multilayer as a quarter-wave plate for near 400 eV. *J. Electron. Spectrosc. Relat. Phenom.* **2005**, *144*, 1079–1081.

73. Hirono, T.; Kimura, H.; Muro, T.; Saitoh, Y.; Ishikawa, T. Full polarization measurement of SR emitted from twin helical undulators with use of Sc/Cr multilayers at near 400 eV. *J. Electron. Spectrosc. Relat. Phenom.* **2005**, *144*, 1097–1099.

74. Namikawa, K.; Ando, M.; Nakajima, T.; Kawata, H. X-ray resonance magnetic scattering. *J. Phys. Soc. Jpn.* **1985**, *54*, 4099–4102.

75. De Bergevin, F.; Brunel, M. Diffraction of X-rays by magnetic materials. I. General formulae and measurements on ferro-and ferrimagnetic compounds. *Acta Crystallogr. Sect. A* **1981**, *37*, 314–324.

76. Igarashi, J.I.; Hirai, K. Magnetic circular dichroism at the K edge of nickel and iron. *Phys. Rev. B* **1994**, *50*, 17820.

77. Krinchik, G.; Gushchin, V. Investigation of Interband Transitions in Ferromagnetic Metals and Alloys by the Magneto-optical Method. *Sov. Phys. JETP* **1969**, *29*, 984.

78. Kao, C.; Hastings, J.; Johnson, E.; Siddons, D.; Smith, G.; Prinz, G. Magnetic-resonance exchange scattering at the iron L II and L III edges. *Phys. Rev. Lett.* **1990**, *65*, 373.

79. Hannon, J.; Trammell, G.; Blume, M.; Gibbs, D. X-ray resonance exchange scattering. *Phys. Rev. Lett.* **1988**, *61*, 1245–1248.

80. Hillebrecht, F.; Kinoshita, T.; Spanke, D.; Dresselhaus, J.; Roth, C.; Rose, H.; Kisker, E. New magnetic linear dichroism in total photoelectron yield for magnetic domain imaging. *Phys. Rev. Lett.* **1995**, *75*, 2224–2227.

81. Kinoshita, T.; Rose, H.B.; Roth, C.; Spanke, D.; Hillebrecht, F.U.; Kisker, E. A new type of magnetic linear dichroism at Fe and Co $M_{2,3}$ edges. *J. Electron. Spectrosc. Relat. Phenom.* **1996**, *78*, 237–240.

82. Grychtol, P.; Adam, R.; Kaiser, A.; Cramm, S.; Bürgler, D.; Schneider, C. Layer-selective studies of an anti-ferromagnetically coupled multilayer by resonant magnetic reflectivity in the extreme ultraviolet range. *J. Electron. Spectrosc. Relat. Phenom.* **2011**, *184*, 287–290.

83. Tesch, M.; Gilbert, M.; Mertins, H.C.; Bürgler, D.; Berges, U.; Schneider, C. X-ray magneto-optical polarization spectroscopy: An analysis from the visible region to the X-ray regime. *Appl. Opt.* **2013**, *52*, 4294–4310.

84. Kleibert, A.; Senz, V.; Bansmann, J.; Oppeneer, P. Thickness dependence and magnetocrystalline anisotropy of the X-ray transverse magneto-optical Kerr effect at the Co 2 p edges of ultra-thin Co films on W (110). *Phys. Rev. B* **2005**, *72*, 144404.

85. Pretorius, M.; Friedrich, J.; Ranck, A.; Schroeder, M.; Voss, J.; Wedemeier, V.; Spanke, D.; Knabben, D.; Rozhko, I.; Ohldag, H.; et al. Transverse magneto-optical Kerr effect of Fe at the Fe 3p threshold. *Phys. Rev. B* **1997**, *55*, 14133.

86. Hecker, M.; Oppeneer, P.M.; Valencia, S.; Mertins, H.C.; Schneider, C.M. Soft X-ray magnetic reflection spectroscopy at the 3p absorption edges of thin Fe films. *J. Electron. Spectrosc. Relat. Phenom.* **2005**, *144*, 881–884.

87. Zaharko, O.; Oppeneer, P.; Grimmer, H.; Horisberger, M.; Mertins, H.C.; Abramsohn, D.; Schäfers, F.; Bill, A.; Braun, H.B. Exchange coupling in Fe/NiO/Co film studied by soft X-ray resonant magnetic reflectivity. *Phys. Rev. B* **2002**, *66*, 134406.

88. Knabben, D.; Weber, N.; Raab, B.; Koop, T.; Hillebrecht, F.; Kisker, E.; Guo, G. Transverse magneto-optical Kerr effect of Fe at the 2p excitation threshold. *J. Magn. Magn. Mater.* **1998**, *190*, 349–356.

89. Mertins, H.C.; Abramsohn, D.; Gaupp, A.; Schäfers, F.; Gudat, W.; Zaharko, O.; Grimmer, H.; Oppeneer, P. Resonant magnetic reflection coefficients at the Fe 2 p edge obtained with linearly and circularly-polarized soft x rays. *Phys. Rev. B* **2002**, *66*, 184404.

90. Senz, V.; Kleibert, A.; Bansmann, J. Transverse magneto-optical Kerr effect in the soft X-ray regime of ultra-thin iron films and islands on W (110). *Surf. Rev. Lett.* **2002**, *9*, 913–919.

91. Kortright, J.; Kim, S.K.; Fullerton, E.; Jiang, J.; Bader, S. X-ray magneto-optic Kerr effect studies of spring magnet heterostructures. *Nucl. Instrum. Methods Phys. Res. A* **2001**, *467*, 1396–1403.

92. Sacchi, M.; Panaccione, G.; Vogel, J.; Mirone, A.; van der Laan, G. Magnetic dichroism in reflectivity and photoemission using linearly polarized light: 3 p core level of Ni (110). *Phys. Rev. B* **1998**, *58*, 3750–3754.

93. De Bergevin, F.; Brunel, M.; Gale, R.; Vettier, C.; Elkaim, E.; Bessière, M.; Lefebvre, S. X-ray resonant scattering in the ferromagnet CoPt. *Phys. Rev. B* **1992**, *46*, 10772.

94. Valencia, S.; Gaupp, A.; Gudat, W.; Abad, L.; Balcells, L.; Martinez, B. Surface degradation of magnetic properties in manganite thin films proved with magneto-optical techniques in reflection geometry. *Appl. Phys. Lett.* **2007**, *90*, 252509.

95. Kortright, J.B. Resonant soft X-ray and extreme ultraviolet magnetic scattering in nanostructured magnetic materials: Fundamentals and directions. *J. Electron. Spectrosc. Relat. Phenom.* **2013**, *189*, 178–186.

96. Kortright, J.; Rice, M.; Kim, S.K.; Walton, C.; Warwick, T. Optics for element-resolved soft X-ray magneto-optical studies. *J. Magn. Magn. Mater.* **1999**, *191*, 79–89.

97. Hellwig, O.; Kortright, J.; Takano, K.; Fullerton, E.E. Switching behavior of Fe-Pt/Ni-Fe exchange-spring films studied by resonant soft-X-ray magneto-optical Kerr effect. *Phys. Rev. B* **2000**, *62*, 11694.

98. Lee, K.S.; Kim, S.K.; Kortright, J. Atomic-scale depth selectivity of soft X-ray resonant Kerr effect. *Appl. Phys. Lett.* **2003**, *83*, 3764–3766.

99. Kim, S.K.; Lee, K.S.; Kortright, J.; Shin, S.C. Soft X-ray resonant Kerr rotation measurement and simulation of element-resolved and interface-sensitive magnetization reversals in a NiFe/FeMn/Co trilayer structure. *Appl. Phys. Lett.* **2005**, *86*, 102502.

100. Valencia, S.; Mertins, H.C.; Abramsohn, D.; Gaupp, A.; Gudat, W.; Oppeneer, P.M. Interference effects in the X-ray Kerr rotation spectrum at the Fe 2p edge. *Phys. B Condensed Matter* **2004**, *345*, 189–192.
101. Saito, K.; Igeta, M.; Ejima, T.; Hatano, T.; Watanabe, M. Faraday and Magnetic Kerr Rotation Measurements on Co and Ni Films Around M2, 3 Edges. *Surf. Rev. Lett.* **2002**, *9*, 943–947.
102. Erskine, J.; Stern, E. Calculation of the m 23 magneto-optical absorption spectrum of ferromagnetic nickel. *Phys. Rev. B* **1975**, *12*, 5016.
103. Capotondi, F.; Pedersoli, E.; Mahne, N.; Menk, R.; Passos, G.; Raimondi, L.; Svetina, C.; Sandrin, G.; Zangrando, M.; Kiskinova, M.; et al. Invited Article: Coherent imaging using seeded free-electron laser pulses with variable polarization: First results and research opportunities. *Rev. Sci. Instrum.* **2013**, *84*, 051301.
104. Allaria, E.; Appio, R.; Badano, L.; Barletta, W.; Bassanese, S.; Biedron, S.; Borga, A.; Busetto, E.; Castronovo, D.; Cinquegrana, P.; et al. Highly coherent and stable pulses from the FERMI seeded free-electron laser in the extreme ultraviolet. *Nat. Photonics* **2012**, *6*, 699–704.
105. Allaria, E.; Battistoni, A.; Bencivenga, F.; Borghes, R.; Callegari, C.; Capotondi, F.; Castronovo, D.; Cinquegrana, P.; Cocco, D.; Coreno, M.; et al. Tunability experiments at the FERMI@ Elettra free-electron laser. *New J. Phys.* **2012**, *14*, 113009.
106. Danailov, M.B.; Bencivenga, F.; Capotondi, F.; Casolari, F.; Cinquegrana, P.; Demidovich, A.; Giangrisostomi, E.; Kiskinova, M.P.; Kurdi, G.; Manfredda, M.; et al. Towards jitter-free pump-probe measurements at seeded free electron laser facilities. *Opt. Express* **2014**, *22*, 12869–12879.
107. Casolari, F.; Bencivenga, F.; Capotondi, F.; Giangrisostomi, E.; Manfredda, M.; Mincigrucci, R.; Pedersoli, E.; Principi, E.; Masciovecchio, C.; Kiskinova, M. Role of multilayer-like interference effects on the transient optical response of Si3N4 films pumped with free-electron laser pulses. *Appl. Phys. Lett.* **2014**, *104*, 191104.
108. Finetti, P.; Allaria, E.; Diviacco, B.; Callegari, C.; Mahieu, B.; Viefhaus, J.; Zangrando, M.; De Ninno, G.; Lambert, G.; Ferrari, E.; et al. Polarization measurement of free electron laser pulses in the VUV generated by the variable polarization source FERMI. In Proceedings of the SPIE 9210, International Society for Optics and Photonics, San Diego, CA, USA, 17 August 2014; p. 92100K.
109. Stanciu, C.; Hansteen, F.; Kimel, A.; Kirilyuk, A.; Tsukamoto, A.; Itoh, A.; Rasing, T. All-optical magnetic recording with circularly-polarized light. *Phys. Rev. Lett.* **2007**, *99*, 047601.
110. Bigot, J.Y.; Vomir, M.; Beaurepaire, E. Coherent ultrafast magnetism induced by femtosecond laser pulses. *Nat. Phys.* **2009**, *5*, 515–520.
111. Mentink, J.; Hellsvik, J.; Afanasiev, D.; Ivanov, B.; Kirilyuk, A.; Kimel, A.; Eriksson, O.; Katsnelson, M.; Rasing, T. Ultrafast spin dynamics in multisublattice magnets. *Phys. Rev. Lett.* **2012**, *108*, 057202.
112. Zhang, G.; Hübner, W.; Lefkidis, G.; Bai, Y.; George, T.F. Paradigm of the time-resolved magneto-optical Kerr effect for femtosecond magnetism. *Nat. Phys.* **2009**, *5*, 499–502.
113. Carva, K.; Battiato, M.; Oppeneer, P.M. Is the controversy over femtosecond magneto-optics really solved? *Nat. Phys.* **2011**, *7*, 665.

![applied sciences logo] *applied sciences*

MDPI

Review

Current Status of Single Particle Imaging with X-ray Lasers

Zhibin Sun [1,2,3], **Jiadong Fan** [3], **Haoyuan Li** [2,4] **and Huaidong Jiang** [3,*]

1 State Key Laboratory of Crystal Materials, Shandong University, Jinan 250100, China; zbsun@slac.stanford.edu
2 Linac Coherent Light Source, SLAC National Accelerator Laboratory, Menlo Park, CA 94025, USA; haoyuan@slac.stanford.edu
3 School of Physical Science and Technology, ShanghaiTech University, Shanghai 201210, China; fanjd@shanghaitech.edu.cn
4 Department of Physics, Stanford University, Stanford, CA 94305, USA
* Correspondence: jianghd@shanghaitech.edu.cn

Received: 20 December 2017; Accepted: 10 January 2018; Published: 22 January 2018

Abstract: The advent of ultrafast X-ray free-electron lasers (XFELs) opens the tantalizing possibility of the atomic-resolution imaging of reproducible objects such as viruses, nanoparticles, single molecules, clusters, and perhaps biological cells, achieving a resolution for single particle imaging better than a few tens of nanometers. Improving upon this is a significant challenge which has been the focus of a global single particle imaging (SPI) initiative launched in December 2014 at the Linac Coherent Light Source (LCLS), SLAC National Accelerator Laboratory, USA. A roadmap was outlined, and significant multi-disciplinary effort has since been devoted to work on the technical challenges of SPI such as radiation damage, beam characterization, beamline instrumentation and optics, sample preparation and delivery and algorithm development at multiple institutions involved in the SPI initiative. Currently, the SPI initiative has achieved 3D imaging of rice dwarf virus (RDV) and coliphage PR772 viruses at ~10 nm resolution by using soft X-ray FEL pulses at the Atomic Molecular and Optical (AMO) instrument of LCLS. Meanwhile, diffraction patterns with signal above noise up to the corner of the detector with a resolution of ~6 Ångström (Å) were also recorded with hard X-rays at the Coherent X-ray Imaging (CXI) instrument, also at LCLS. Achieving atomic resolution is truly a grand challenge and there is still a long way to go in light of recent developments in electron microscopy. However, the potential for studying dynamics at physiological conditions and capturing ultrafast biological, chemical and physical processes represents a tremendous potential application, attracting continued interest in pursuing further method development. In this paper, we give a brief introduction of SPI developments and look ahead to further method development.

Keywords: X-ray free-electron lasers; XFEL; coherent diffraction imaging; single particle imaging; resolution

---

1. Coherent Diffraction Imaging Using Synchrotron Light Source

High-resolution structure determination of macromolecular and biological particles is an important tool for the life science and biological community [1–4]. Apart from X-ray crystallography and spectroscopy, microscopy is the most widely used technique for structure determination in physical, material and biological sciences. Multiple imaging methods, such as optical microscopy (including general optical microscopy, laser confocal microscopy and super-resolution microscopy) [5–9], electron microscopy [10–13] and X-ray microscopy [14–16] have been employed by researchers. Nevertheless, the resolution of an optical microscope is always limited by the wavelength of the radiation used (~200 nm for visible light). For super-resolution microscopy, the sample has to be labeled, which will limit the applications of photoactivated localization microscopy (PALM) [17], stochastic optical reconstruction microscopy (STROM) [18] and stimulated emission depletion (STED) [19]. For scanning electron microscopy (SEM), it can only provide surface information [20]. Regarding transmission electron microscopy (TEM) [21], the penetration ability of electrons is a critical issue. The sample thickness has to be less than 50 nm [22], otherwise contrast is reduced due to multiple scattering of the electrons. To obtain structural information below the surface for thick samples, the sample has to be sliced, introducing artifacts in the final images. Compared to electrons, X-rays have a larger penetration depth, making them an ideal probe for most applications in structure determination. However, the scattering cross-section of X-rays is smaller than that of electrons [23]. The scattering efficiency is worse than that of electrons. Actually, neither X-rays nor electrons can achieve atomic resolution due to the limitation of radiation damage [24]. Only when the X-ray or electron pulse is shorter than the atomic motion timescale, such as in the order of 10 fs, could the valid snapshot be captured [25]. Under the same conditions, X-rays are better for thick samples without slicing and electrons are better for weak scattering samples with cryogenic cooling. Classical X-ray imaging methods, such as transmission X-ray microscopy (TXM), scanning transmission X-ray microscopy (STXM) and X-ray fluorescence imaging are based on X-ray optics [26] which will limit the highest obtainable resolution. At the moment, the achievable resolution of focusing optics has broken the 10 nm barrier [27,28]. For X-ray crystallography, the samples have to be crystallized typically to a certain minimum size to obtain high resolution [29]. In addition to crystal growth and phasing, radiation damage is an extra challenge, especially when radiation sensitive samples such membrane or metal containing proteins.

In 1952, D. Sayre underlined in a half-page paper that by measuring the diffraction intensity between the Bragg peaks, the phase problem of non-crystalline samples might be resolved [30]. Since then, oversampling theory [31,32] and different types of phase retrieval algorithms [33–41] have developed rapidly. In 1999, J. Miao and colleagues performed the first successful coherent diffraction imaging experiment [42]. This newly developed coherent diffraction imaging method is a form of lensless microscopy rather than relying on X-ray optics. The theoretical resolution is only limited by the incident light wavelength and the maximum diffraction angle recorded by the detector. The past 18 years have witnessed a great success of coherent diffraction imaging, especially under the use of high flux X-ray light source facilities, such as advanced 3rd generation synchrotron facility Super Photon ring-8 GeV (Spring-8, Hyogo, Japan), European Synchrotron Radiation Facility (ESRF, Grenoble, France) and Advanced Photon Source (APS, Lemont, IL, USA). With developing X-ray instrumentation and phase retrieval algorithms, coherent X-ray diffraction microscopy has been extended to nanoparticles [43–48], mineral materials [49,50], biological cells, bacteria and viruses [51–63]. Meanwhile, different CDI methods also developed quickly, such as Bragg-CDI [43,44,46,48], Fresnel-CDI [64], Keyhole-CDI [65], Reflective-CDI [66] and Ptychography [67,68]. These methods greatly increased the diversity and breadth of applications. Some technologies were incorporated from Cyro-EM, for example, Cryo-CDI [55,56] and wet-CDI [60]. These borrowed technologies make high-resolution imaging without obvious radiation damage possible.

In order to summarize the areas of attention for coherent X-ray diffraction microscopy, ~200 published papers were analyzed. Fifty of them [24,31,32,38,42–87] were selected to generate the word cloud of Figure 1 which shows that reconstruction, diffraction, sample, resolution, structure and coherent are the most used terms during the last 18 years of synchrotron-based CDI papers. Besides, exposure, photons and radiation are also frequently used words. These suggest that published efforts have focused intensely on phase retrieval. It seems the reconstructed real space image was the main aspect of a publication a decade ago. Furthermore, the type of samples and the achieved resolution are also some of the most important parts of the selected papers. For biological specimen, radiation damage has also been a concern. From M. Howells's summarizing information [24], it can be found that one should be able to image a frozen-hydrated biological sample with CDI at a 10 nm resolution. To our knowledge, the currently published results showed that the best 3D image of biological samples' resolution is ~30 nm [61], which is a little far away from the 10 nm resolution under the maximum radiation dose tolerance using a third-generation synchrotron light source. This lags behind the atomic resolution achieved by electron microscopy. For the similar samples, such as human chromosome, the resolution of 3D structures with CDI and Cryo-EM is ~120 nm [57] and ~50 nm [88], respectively.

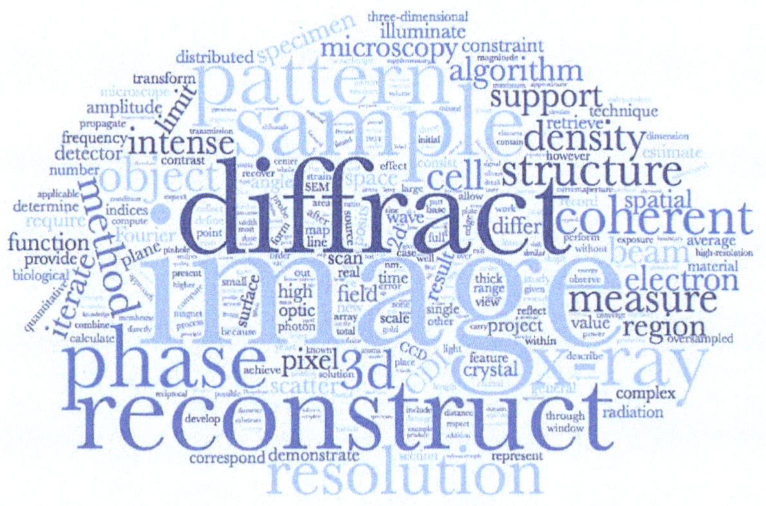

Figure 1. Word cloud of synchrotron-based coherent diffraction imaging published research papers. About 50 widely-read published papers were downloaded. These papers are published during the last 18 years since the first coherent diffraction imaging experiment was conducted successfully in 1999. The main text of the papers was analyzed by open python codes. The frequency of each meaningful word was calculated and then demonstrated by different size and colors. The bigger the size means the larger the frequency. The image shows that reconstruction, sample, phase and resolution are the most used terms for the past 18 years of coherent diffraction imaging research. At the same time, structure, intensity and radiation have also been key topics for biological imaging since 1999. (Original datas were collected from References [24,31,32,38,42–87]. Figure was generated under the help of Github open codes [89].).

The advent of a fourth-generation source, diffraction-limit storage ring (DLSR), opens a promising pathway for CDI [90–92]. The low transverse emittance and beam divergence improve the coherent fraction of the X-ray beam, especially for hard X-rays. The direct result is that the coherent photon of DLSR will be higher than that of current synchrotron light source. The obtainable resolution for CDI scales directly with the spatial coherence (as long as damage can be avoided). At the moment, many facilities around the world intend to update or build DLSR facilities [93], such as ESRF Upgrade Phase II [94], Spring-8 [95], MAX-IV [96], Sirius [97], etc. Looking forward, the investment X-ray facilities are putting into DLSR operations will yield new science possibilities for CDI.

2. Coherent Diffraction Imaging with X-ray Lasers

During the rapid development of CDI, the X-ray light source technologies have also made great progress. Thanks to the advent of ultra-intense and ultrafast X-ray free-electron lasers [98,99], the femtosecond pulses can outrun main radiation damage processes [100] using the so-called "diffraction-before-destruction" strategy [101]. Under proper conditions, XFELs can limit radiation damage and provide clear images of molecules that can otherwise be damaged using a continuous illumination. The state-of-the-art serial femtosecond crystallography (SFX) [102] and single particle imaging (SPI) methods [103] have proven to be promising ideas for native structure determination, with the additional benefit of time-resolution for making movies of reactions. Due to noise, detector limitation and demand of large amount dataset for atomic resolution, big data and automatic data analysis are needed to achieve the challenging goal.

Owing to the ultra-intense and ultrafast X-ray pulses generated by FLASH (Hamburg, Germany) [104,105], LCLS (Menlo Park, CA, USA) [106,107], SACLA (Hyogo, Japan) [108,109] and FERMI (Trieste, Italy) [110,111], many exciting CDI results have been obtained. It is worthwhile mentioning that the European-XFEL [112] and PAL-XFEL [113] have recently been commissioned, and that Swiss-FEL [114] is under construction and soon to be commissioned. The Shanghai Soft XFEL [115,116] is under construction, and a hard XFEL facility has also been proposed [117]. We believe that the newly-built facilities will continue to be a great impetus to ultra-small and ultrafast research.

Coherent diffractive imaging experiments with X-ray free-electron lasers (XFEL-CDI) have studied a wide range of samples including man-made objects, nano-materials, clusters, and aerosol particles in the gas phase [101,118–132]. Many of these were used to test the validity of XFEL-CDI and included measurements of the morphology of complex structures, functional imaging method development, and dynamic changes under a stimulus (pump-probe). On the biological side, viruses, bacteria, organelles and associated biological components have been imaged at various resolution levels [133–142] and have helped further develop high throughput imaging methods.

Figure 2 shows typical results from non-biological particles using XFEL-CDI. The first experiment demonstrating the "diffraction-before-destruction" principle using the FLASH soft X-ray free-electron laser by H. Chapman and colleagues [101] is shown on Figure 2a. An intense pulse with 10^{12} photons at 32 nm wavelength was transported to a nanostructured non-periodic object. A speckled pattern was obtained before the beam completely destroyed the nanostructure. The diffraction pattern and the reconstructed real-space image showed no measurable effect from radiation damage. The final imaging resolution was estimated to be 62 nm, where the phase retrieval transfer function (PRTF) drops to a value of $1/e$. After this successful demonstration of the principle, pump-probe methodologies were applied to XFEL-CDI. Figure 2b,c [120,128] show time-resolved imaging of transforming materials under a visible light stimulus at FLASH and LCLS, respectively. Dynamic changes in man-made structures (Figure 2b) and nanoparticles (Figure 2c) were observed. Such measurements represent examples of the truly unique power of XFELs to study dynamic systems.

Other methods can also be combined with XFEL-CDI to yield further information. Figure 2d [126] shows the functional imaging of multiple nanoparticles. Time-of-flight mass spectrometry was used simultaneously with imaging, which allowed the morphology and chemical content to be captured at the same time by a single shot. Other methods such as fluorescence and emission spectroscopies can also be applied to reveal specific chemical information and correlate it to the structure revealed by imaging.

Three-dimensional imaging is typically not only desired but also necessary to fully understand a system being studied. However, focused XFEL pulses are too intense and destroy the sample typically in one pulse, limiting the information available to two-dimensions in one pulse. Without attenuation, the ultra-intense pulses will destroy almost all particles interacting with the X-ray laser [143]. Three-dimensional images typically cannot be obtained by rotating samples using an X-ray FEL which makes computed tomography (CT) and transmission X-ray microscopy (TXM) regularly performed at synchrotron light sources not particularly suitable for XFELs. Three-dimensional information is obtained using an XFEL by using re-producible particles delivered to the X-ray laser pulses one at a time. In this manner, thousands to hundreds of thousands of diffraction patterns with random orientations are recorded by large pixel array detectors. Using methods such as diffusion maps or threshold methods, valid hits and single-particle hits can be selected and their relative orientation can be determined. After phase retrieval, high-resolution real-space images can be obtained.

Worthy of noted, J. Miao and colleagues developed a single-shot 3D imaging method for high-symmetry objects and demonstrated the idea on gold nanoparticles with X-ray lasers [131]. Figure 2e shows a 3D image of a nanoparticle from SACLA. By using the symmetry of gold nanoparticles, 48 diffraction patterns were generated. The 3D image was obtained after the finding of common arcs and phase retrieval. The final resolution was determined to be ~5.5 nm.

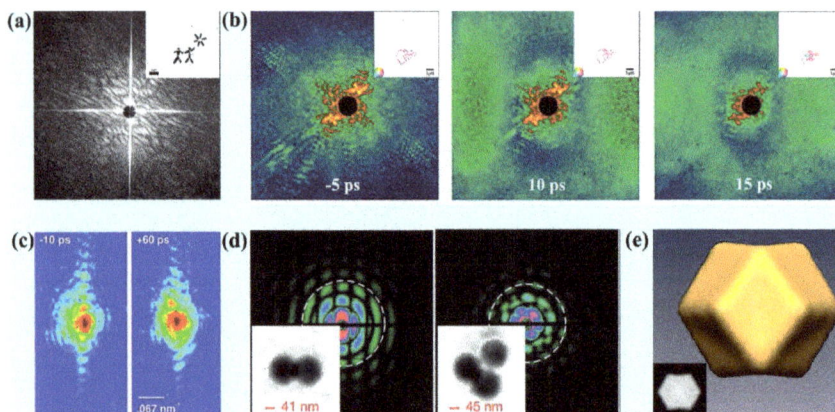

Figure 2. Coherent diffraction imaging of materials with or without pump-probe using X-ray free-electron lasers. (**a**) The first demonstration of coherent diffraction imaging with X-ray lasers at FLASH in 2006. The concept of "diffraction-before-destruction" was demonstrated at the limited resolution available; (**b**) Time-resolved diffractive imaging of a man-made sample. Laser-pump X-ray-probe was applied at different time delays. In this case, the sample is seen to get destroyed under the high pump laser intensity and the features are disappearing after 15 ps; (**c**) Bragg coherent diffraction imaging of gold nanoparticles with hard X-ray lasers was employed at the Linac Coherent Light Source (LCLS). For different time delays, the changes in the diffraction patterns can be used to image phonon dynamics; (**d**) Morphology and chemical components of aerosol particles were captured by coherent diffraction imaging combined with time-of-flight mass spectrometry at LCLS; (**e**) Three-dimensional imaging of nanoparticle from SACLA. By using the symmetry of gold nanoparticles, 48 diffraction patterns were generated. The 3D image was obtained after the finding of common arcs and phase retrieval. The resolution as determined by Fourier shell correlation (FSC) is ~5.5 nm. (Figure 2a was reproduced with permission from [101]. Copyright Springer Nature, 2006. Figure 2b was reproduced with permission from [120]. Copyright Springer Nature, 2008. Figure 2c was reproduced with permission from [128]. Copyright The American Association for the Advancement of Science, 2013. Figure 2d was reproduced with permission from [126]. Copyright Springer Nature, 2012. Figure 2e was reproduced with permission from [131]. Copyright Springer Nature, 2014.).

Compared to non-bioparticles, the coherent scattering cross-section of bioparticles is typically much smaller as they are composed of low atomic number elements primarily carbon, oxygen, nitrogen and hydrogen. In other words, the scattering signal of bioparticles is less than other particles when the particle size and photon parameters are the same. For example, the coherent scattering cross-section of gold (19.32 g/cm^3) and DNA ($C_{10}H_{11}N_4O_6P$, 1.70 g/cm^3) are 6.5 cm^2/g and 0.4 cm^2/g under 7 keV. However, the scattering ability of gold is over 100 times stronger than that of DNA. The weak scattering and low signal-to-noise-ratio (SNR) make high-resolution imaging of biological samples more challenging. Figure 3a [135] shows the first coherent diffraction imaging of biological samples at LCLS. High-quality diffraction data were obtained with a single X-ray laser pulse from a mimivirus. The reconstructed exit wavefront (image) shows a full-period resolution of 32 nm. Single cell, virus or bacteria imaging is of potential great importance to disease detection and drug design. It enables the collections of separate cell images from individual cells and high-throughput imaging over populations of cells can sample unique occurrences. In this big data era, images from populations of genetically identical cells often reveal heterogeneous phenotypes, which can be helpful to group behavior research. Figure 3b [137] shows the high-throughput imaging of heterogeneous cell organelles with 120 Hz X-ray laser pulses from LCLS. Morphology and inner electron density were reconstructed by phase retrieval automatically. This result could pave the way for further biological functional imaging with XFEL. Figure 3c [136] demonstrates the combination of synchrotron XFEL together for one sample where one facility could not solve all challenges. This combined experiment illustrates the differences between facilities. Figure 3d [141] shows the successful demonstration of living cell imaging of cyanobacteria. Living cyanobacteria were injected into the X-ray beam using an aerosol injector. During the short delivery time from solution to the vacuum of the experimental chamber, the cyanobacteria remain in a living state and they were then imaged by coherent diffraction. This demonstrates the capability for the real structure determination of biological samples under near native state. Figure 3e shows the first 3D imaging of biological samples using an XFEL [139]. Hundreds of diffraction patterns with different SNR and random orientations were recorded. After hit finding, pattern classifications, orientation determination and phasing, a 3D real-space volume with a full-period resolution of 120 nm was generated. This demonstration paves the way for 3D single-particle imaging, with the ultimate goal of achieving atomic resolution requiring further effort.

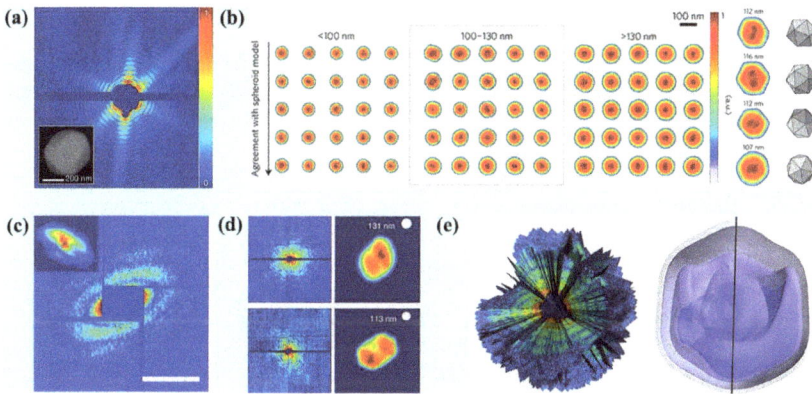

Figure 3. Coherent diffractive imaging of biological samples with X-ray free-electron lasers. (**a**) First coherent diffraction imaging of biological samples with hard X-ray lasers at LCLS. The sample, mimivirus, is one of the largest known viruses. The total diameter of the particle, including fibrils, is about 750 nm. The full-period resolution of this first demonstration is ~32 nm; (**b**) High-throughput imaging of heterogeneous cell organelles. Around 70,000 low-noise diffraction patterns were collected within 12 min at LCLS at a frequency of 120 Hz. Real space images with different organelles' size and shape were phased automatically; (**c**) Structure determination of RNAi microsponge combining single-shot X-ray free-electron laser (XFEL) diffraction with synchrotron-based coherent diffraction imaging; (**d**) Living cell imaging of cyanobacteria with XFEL. Living cyanobacteria were injected to the X-ray laser using an aerosol particle injector; (**e**) The first demonstration of three-dimensional imaging of biological samples. Mimivirus was delivered to the X-ray laser beam, hundreds of random orientation diffraction patterns were collected. After hit finding, single-shot classification and orientation determination, a three-dimensional diffraction volume was obtained. The spatial resolution was estimated to ~125 nm. (Figure 3a was reproduced with permission from [135]. Copyright Springer Nature, 2011. Figure 3b was reproduced with permission from [137]. Copyright Springer Nature, 2014. Figure 3c was reproduced with permission from [136]. Copyright Springer Nature, 2014. Figure 3d was reproduced with permission from [141]. Copyright Springer Nature, 2015. Figure 3e was reproduced with permission from [139]. Copyright American Physical Society, 2015).

3. Single-Particle Imaging with X-ray Lasers

As mentioned above, coherent diffraction imaging with X-ray lasers has undergone a fast development and many high-profile results have been achieved. However, the published resolution has been limited, and there is a huge gap between the current resolution and atomic-scale resolution. In order to deeply investigate the limitations and move forward, start-to-end considerations have to be taken. The light source, such as the photon energy, pulse duration and pulse energy need to be optimized. X-ray instruments, such as optics, focusing mirrors, front-end slits and apertures before the sample, also need to be optimized for the measurement. Sample delivery systems, such as aerosol injectors, liquid-jet injectors and fixed targets, need development, optimization and suitable design for different experiments. Data acquisition system, such as high quantum efficiency (QE) and high dynamic range detectors, online data monitoring and analysis software, phase retrieval algorithms, will greatly affect the efficiency and validity of final images.

To speed up the progress and take full advantage of the scientific community around the world, LCLS launched an international cooperation team in December 2014, the Single-Particle Imaging initiative, which includes more than 100 scientists from 21 institutions over eight countries. A step-by-step roadmap was published where a host of challenges was outlined to guide the required developments. A series of experiments has been dedicated to systematically address the challenges of SPI.

3.1. The Road Map

To achieve atomic-scale resolution, such as 3 Å, the SPI initiative analyzed all the factors that could limit the resolution from the start to the end of the experiment. To initiate this work, workshops and brainstorming sessions were arranged to clarify the steps to undertake. Following the published roadmap [103], the following key challenges were identified:

(1) Radiation damage;
(2) Start-to-end simulation pipeline;
(3) Samples issues;
(4) Sample delivery system;
(5) Characterization of parasitic scattering and noise;
(6) Beam diagnostics and characterization;
(7) Data analysis and phase retrieval algorithms development.

3.2. Setup and the Experiment Procedure

Many aspects are required to be specified prior to conducting an SPI experiment. First, the proposed resolution, sample type and size should be defined. Special diagnostics required for the experiment, such as the ion/e^- TOF, high speed camera, fluorescence detector and pumping laser system and the space they require should also be taken into consideration. Secondly, the ideal sample-to-detector-distance, photon energy, pulse duration, and pulse energy should be determined. Then, careful alignment of all optics, slits, KB mirrors and large-panel pixel array detectors must carefully be accomplished at the start of the beamtime. Simultaneously, beam characterization, such as the beam profile, the focused beam spot size, beam intensity and position fluctuations, wavefront or arrival time should be diagnosed for further reference. The high harmonic content of the beam should also ideally be tested. Finally, selection of a proper sample delivery method will be of great importance to the background and hit rate. The simple flow chart of the experiment and data process is shown in Figure 4. After alignment of the instrumentation, samples are delivered into the X-ray laser beam path. The forward scattering exit wavefront will propagate downstream and then to be recorded by the detector and this pattern must be oversampled properly. Data processing involves center determination, background subtraction, analog-to-digital units (ADU) to photons conversion, hit finding, pattern classification, and orientation determination. Finally, phasing can produce a real-space 3D image. By using PRTF, FSC or other criteria, the spatial resolution could be determined.

3.3. Current Status

Eight experiments have been performed since the start of the SPI initiative. Different samples, such as FIB Au nanostructures, rice dwarf virus (RDV), coliphage PR772 viruses and bacteriophage MS2 virus, were used for the various experiments. The reasons for the sample selection will be discussed in the sample selection section. The main purpose of each experiment was different. For example, amo86615 aimed primarily to collect a lot of diffraction patterns of RDV and PR772. However, the main target for experiment amo11416 was to test sample delivery and collect diffraction data from MS2. To reduce the background from sample delivery, different sample methods were tested.

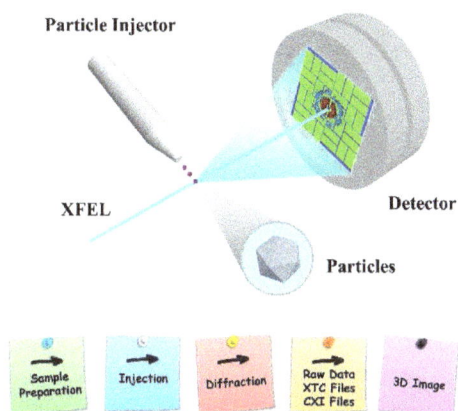

Figure 4. Schematic layout of a single-particle imaging measurement using femtosecond XFEL pulses. Nanoparticles and viruses are injected to the XFEL focus by a particle injector or a fixed target system using thin membranes such as Si_3N_4 and graphene. High dynamic range, high quantum efficiency and large panel detectors are employed to record patterns. In general, five main steps are required for a single-particle imaging experiment. After sample selection and preparation, they are delivered to the XFEL focus and the patterns are collected by the detector. This produces large data sets requiring advanced automatic processing. Hit finding, background subtraction and detector calibrations have to be applied to reduce the data. By using threshold analysis or diffusion maps, single-shot patterns are selected. After orientation determination and phase retrieval, a real-space three-dimensional image can in principle be achieved.

3.4. Light Source and Instruments

There are seven X-ray instruments at the LCLS. The Coherent X-ray Imaging (CXI) instrument and the Atomic, Molecular & Optical Science (AMO) instrument are the instruments suitable for SPI. The CXI instrument is located in the Far Experimental Hall of LCLS. The hutch is located 440 m far away from the source. The instrument makes use of hard X-ray pulses to perform coherent X-ray imaging and serial femtosecond crystallography experiments. The primary operating photon energy range is 5~11 keV, with some capability for higher harmonics up to 25 keV. Three sample chambers exist with foci of ~5 µm, 1 µm, and 0.1 µm being possible. The AMO instrument is located in the Near Experimental Hall of LCLS and the photon energy range is 280~2000 eV. The focusing optics for the AMO instrument are bendable Kirkpatrick-Baez (KB) mirrors and the focused beam size is estimated to be 1.5~5 µm. In both the CXI and AMO cases, a high-power laser synchronized with the X-ray laser is available. Figure 5 shows a simple layout of AMO and CXI instruments. More detailed information can be found from reference [144–146] and from the LCLS facility website [147,148].

As mentioned in the setup and experiment procedure section, photon parameters should be determined before the experiment. The selection of the photon energy is very important to single particle imaging. As shown by momentum transfer $q = 4\pi \sin\theta/\lambda$ (2θ is the scattering angle), to obtain higher resolution, higher photon energy is needed. On the contrary, for higher photon energy, the wavelength is shorter and the coherent scattering cross-section is weaker. In order to estimate the scattering intensity per shot, a sphere model with uniform electron density can be used [149],

$$I(q) = I_0 r_e^2 \Delta\Omega |\Delta\rho|^2 (4\pi \frac{\sin(Rq) - Rq\cos(Rq)}{q^3})^2 \tag{1}$$

where I_0 is the incident intensity, r_e is the classical electron radius, $\Delta\rho$ is the difference of the complex electron density of the sphere to the medium and R is the sphere radius. If we take the complex refractive index into the above equation, we can get the maximum ring intensity [150],

$$I_{max}(q) = I_0 r_e^2 \Delta\Omega |\Delta\rho|^2 16\pi^2 R^2 q^{-4} \tag{2}$$

The intensity for high q will decrease rapidly. For the same sample size and scattering angle, shorter wavelength will contribute to higher resolution but lower intensity. Therefore, the optimized photon energy should be chosen depends on proposed resolution and sample types.

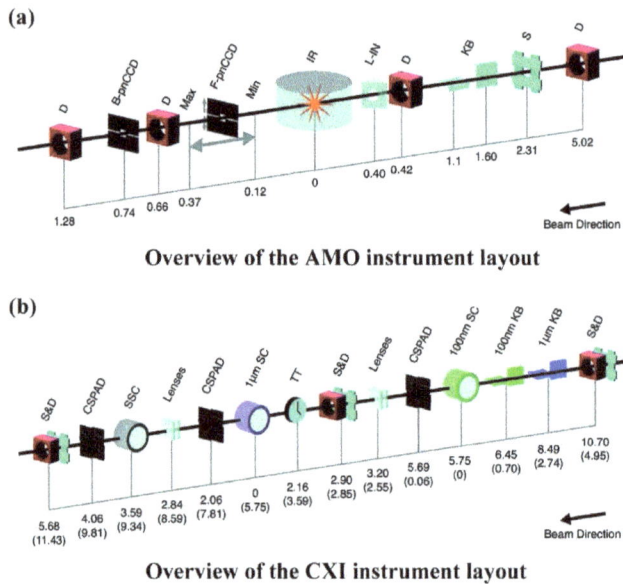

(a) Overview of the AMO instrument layout

(b) Overview of the CXI instrument layout

Figure 5. Overview of Atomic, Molecular & Optical Science (AMO) and Coherent X-ray Imaging (CXI) instrument layout. Distances are indicated in meters from the interaction region (IR) for AMO, 1 μm sample chamber (SC) and in parentheses from the 100 nm SC for CXI. The X-ray beam enters the hutch and passes through the diagnostics (D) and slits (S) and is then focused by KB mirrors. (**a**) An optical laser in-coupling (L-IN) is located 0.4m upstream of the IR. Two pairs of pnCCD detectors are located in different positions downstream of the IR; (**b**) For the CXI instrument, each chamber is colored to match its designed KB mirrors. A timing tool (TT) is designed to obtain the fine time between X-ray laser and pump laser. (Figures were reproduced from Reference [145,146] with permission from International Union of Crystallography/IUCr).

Another principle for the photon energy selection is the optimized X-ray energy range for specified experiment instruments. For different X-ray instruments at LCLS or different beamlines at SACLA, the optimized photon energy varies. The commonly used photon energy for SPI is 1700 eV that just below the Si K-edge at AMO, 7 keV at CXI that below the Fe edge of 7.1 keV, LCLS. Also, the photon count ability of detectors has also been considered during the photon energy selection. Other photon energies, such as 1250 eV, 1800 eV, 5 keV and 10 keV have also been used and tested, with varying degrees of success. A balance has to be struck between the resolution and scattering efficiency, and the ability for the various detectors to count photons with a reasonable signal to noise ratio.

High harmonics are an inevitable aspect of XFEL sources. Radiation at the fundamental wavelength dominates but appreciable levels of high harmonics also present at the 1% level. These harmonics may extend the capability of instruments but also may contribute background to experiments. The Soft X-ray Offset Mirror System (SOMS) and Hard X-ray Offset Mirror System (HOMS) located in the Front End Enclosure of LCLS have harmonic cutoffs at ~2.5 and ~25 keV respectively. The SOMS operate in the FEL energy range from 0.20 keV to 2.0 keV and the reflectivity is above 90% [151]. The system reflectivity above 2.48 keV is below 20%, which contributes to suppressing the third harmonic FEL peak [152]. The fundamental operation energy range for HOMS is 2.0~12.0 keV, and be able to deliver 2~25 keV photons up to the third harmonic. From former measurement, it is known that the second harmonic content in normal operation at soft X-ray region (near 1keV) is 0.04~0.1%, and the third harmonic is as high as 2.0~2.5% [153]. The third harmonic for hard X-ray wavelength (6~8 keV) is 0.2~2%. The reflectivity for the second and third harmonics are different for the downstream KB mirrors located at the experiment station, which provide some level of harmonic rejection. The third harmonic can produce undesired noise, for example via fluorescence by crossing above an absorption edge such as iron, a material making up most of the vacuum chamber. It is also less efficiently block by apertures and slits and can again increase the background noise.

Knowing and optimizing the focused beam size and the wavefront is another key aspect of SPI experiments [154]. The position and angles of KB mirrors should ideally be aligned for every SPI experiments. Imprint methods have been employed for the AMO and CXI instruments to find the best focusing position [155,156]. This is a time-consuming procedure for the current experiments. Two to three hours are usually required. According to the nature of SASE mode, the beam position and intensity fluctuates shot-to-shot. The SPI initiative has also taken the phenomenon into consideration. A backscattering position and intensity monitor was used for diagnostic for the CXI instruments [157].

The parasitic scattering of particle injectors is the primary source of scatter, the team is still working on this problem. Apart from particle injector scattering, the left stray scattering comes from the upstream components of the beamline. For the AMO instrument, the parasitic scattering comes from the roughness of KB mirrors. For the CXI instrument, the stray scattering comes from the diamond window that is located downstream of the KB mirrors to protect the mirrors from particles. The detailed information will come out soon by the SPI team. For SPI experiments, the alignment of apertures is also of great importance. Ge, Si, Si_3N_4 and B_4C apertures are typically used depending on the instrument and photon energy. Background levels from apertures of different materials have been investigated with varying performance depending on the material and thickness. For example, cleaved and etched Ge was found to be a good choice for hard X-rays around 7 keV.

The typical LCLS pulse energy for SPI is 2~5 mJ, equals to 7.80×10^{12}~1.95×10^{13} photons/pulse@1.6 keV and 1.78~4.46×10^{12} photons/pulse@7.0 keV. The calculation of photon number per pulse was based on the 100% efficiency assumption of optical system. Actually, the efficiency is closer to 50%. More photons typically mean higher SNR with stability being desirable. A single diffraction pattern is just one slice from the 3D diffraction volume. The number of photons interacting with the sample is different due to the intensity fluctuations shot-to-shot. The side effects of intensity fluctuations to the SPI 3D diffraction volume also need further study. Figure 6 shows the pulse intensity fluctuation with shots (time), the common fluctuation of intensity is ~10%. Also, the focused beam position fluctuates. The relative positions between X-ray laser beam and the sample are also different shot to shot. This leads to different incident photons onto samples and means that for SPI, is not so easy to normalize every frame.

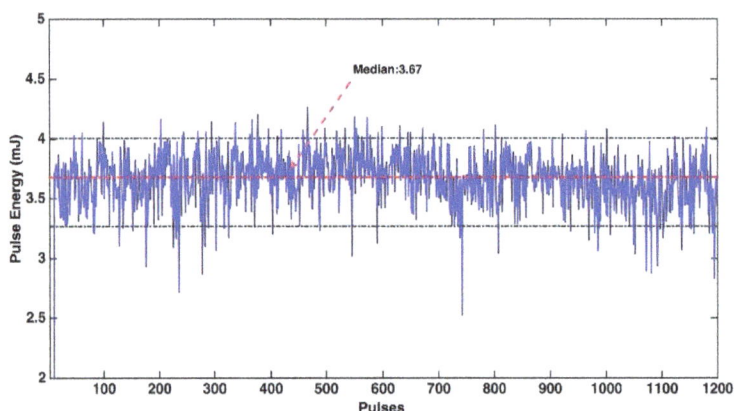

Figure 6. Pulse energy fluctuation with time. The intensity and position fluctuations come from the origin of SASE. The common fluctuations are typically 10% but sometimes as high as 30%.

The pulse duration is another main factor for XFEL experiments. Short pulse duration is always required to outrun radiation damage. However, since shorter pulses usually come with a reduced number of photons, a balance is needed between pulse duration and pulse energy, a balance between damage and signal. For biological samples and molecules, Auger decay is predominant in carbon, nitrogen and oxygen [158]. The photoelectric effects will lead to a removal of two electrons from these elements. The energy and lifetime of these two electrons are different [159–161]. To figure out the exact Auger emission time scale and non-elastic scattering for specified samples is complex. However, the classic simulation and crystalline experiments have demonstrated that the radiation damage could be reduced or eliminated by using shorter pulses, such as <10 fs [100]. For SPI, to guarantee enough photons per pulse, the commonly used pulse duration is 30~70 fs. At the current limited resolution, these pulse durations do not lead to noticeable damage. For higher spatial resolution, radiation damage will be a limit for the development of SPI [25,162,163].

3.5. Sample Selection

For the development phase of SPI, the selection of samples is key. To make things easier, choosing a high Z material is tempting, but this is not very representative of samples that are the ultimate goal of the method with the scattering strength of high Z nanoparticles being stronger than bioparticles and biomolecules. Nevertheless, as initial steps towards developing aspects of the SPI methods, such artificial test samples can prove useful. Nanofabricated samples present the challenge of producing particles with regular shapes and sizes, a key requirement for putting together a consistent 3D diffraction volume from multiple copies of an object. This is also a challenge for bioparticles but less so.

The sample size should be smaller than the focused beam size. If the X-ray pulse only illuminates part of a single particle, one does not get a diffraction pattern representative of the whole object and therefore cannot merge this data reliably. As the sample quantities needed are large, the cost, mass production capacity should also be considered. High concentration should also be easy to achieve to improve the hit rate. Last but not least, the particles should be able to be aerosolized stably for injection to the XFEL beam. Under these principles, gold octahedra with size range from 2 μm to 100 nm were produced by a polyol process [164]. For bioparticles, RDV, PR772 and MS2 were selected. RDV (Rice

Dwarf Virus, Figure 7a) is an icosahedral RNA virus of about 70 nm in diameter [165]. A 3D structure of the capsid was determined by X-ray crystallography with a resolution of 3.5 Å. The associated PDB ID is 1UF2 [166]. PR772 (Figure 7b) is a lipid-containing bacteriophage that infects *E. coli* [167]. The diameter of this icosahedral virus is about 69 nm. Unfortunately, the inner arrangement and structures of lipid layer and DNA are also not clearly known. MS2 (Figure 7c) is a bacteriophage that infects F + *E. coli*. The diameter of this icosahedral virus is about 26 nm [168]. The internal arrangement of the complex is unknown.

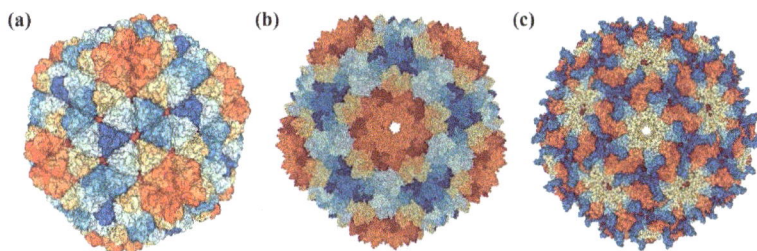

Figure 7. Projections of bioparticles used for SPI experiments. (**a**) Biological assembly of double-shelled rice dwarf virus (RDV). RDV is a member of the genus Phytoreovirus in the family Reoviridae. The atomic structure was determined by X-ray crystallography at 3.5 Å resolution [166]; (**b**) Quasi-atomic structure resolution model of bacteriophage PRD1 wild type virion. PR772 belongs to the PRD1 family. The dsDNA bacteriophage has a membrane inside its icosahedral capsid. The atomic structure was determined by combined cryo-EM and X-ray crystallography at 25 Å resolution [169]; (**c**) Crystal structures of MS2 coat protein mutants in complex with wild-type RNA operator fragments. MS2 is a group I RNA bacteriophage that infects *Escherichia coli*. The atomic structure was resolved by X-ray crystallography at a resolution of 2.8 Å [170]. The projections were generated by NSL Viewer, a WebGL based 3D viewer [171,172].

3.6. Sample Delivery

Sample delivery is a key component of XFEL experiments. There are two mainly used methods: (a) fix the sample by solid support, fixed target method; (b) Inject the particle streams into X-ray laser beam path, injection method. For the first method, Si_3N_4 membrane or amorphous carbon film are regularly used. The method has been used for many 2D crystalline samples and synchrotron-based CDI experiments. However, the diffraction background is strong and can overwhelm the signal from the sample if Si_3N_4 is used. For particle injectors, there are many injector devices commonly used by scientists. The gas dynamic virtual nozzle (GDVN) injector [173], delivers samples inside an unbroken liquid stream with tunable diameter by changing nozzle diameter and gas flow rate. The main general drawback of the GDVN is that they don't form droplets that are small enough. Also, non-volatile components in the buffer solution trend to stick to the particle of interest, which will make valid single particle shot more challenging. Another drawback of the GDVN is the high rate of sample consumption. This is not a problem when the sample is easy to achieve or synthesis. But this will be a big problem when the sample is valuable and hard to get. For SPI specifically, the liquid carrier medium scatters X-rays much more than the sample itself and makes it unsuitable for measuring single particle diffraction patterns of biological sampels. To solve this problem, the GDVN is always used combine with aerodynamic lens stack or electrospray during the SPI experiments.

Another mainly used injector device is the aerosol injector. The particles can be focused by gradient pressure in the aerodynamic lens stack. The sample delivery method of choice for SPI is typically to transport the purified samples into the buffer solution and then aerosolized with helium gas using a GDVN and then introduced to the X-ray laser beam path via an aerodynamic lens. The purified samples could also be aerosolized by a gas nebulizer, such as electrospray [174], and then injected by the aerodynamic lens stack [175]. These unfortunately typically lead to low hit rates. For CXI nanofocusing, the hit rate is just ~1% or lower. The number of diffraction patterns needed for high resolution is large, requiring a lot of beamtime, beyond what is reasonably possible. This is an issue that will need to be resolved for SPI to become a broadly useful method. Another challenge for SPI is the scattering background from residual gas of aerosol injector. Different gases, such as He and CO_2, have been tested and there are indications that the background level is now limited by the aerosol carrier gas.

3.7. Detector and Data Analysis

Detector systems are critical important to all scientific experiments and SPI is not an exception. The 2017 Chemistry Nobel Prize was awarded to Cryo-EM, a powerful technology for single-particle protein crystal structures determination [176]. The current resolution is around 3 Å, comparable with the X-ray crystallography [177]. About 20 years ago, the first 3D reconstruction at the sub-nanometer resolution of icosahedra virus was around 10 Å. Since then, thousands of structures have been determined by the Cryo-EM at a resolution from 30 Å to 3 Å [3]. A breakthrough was suddenly made around 2014. Great progress was made with the direct electron detectors and single particle analysis methods [178]. From that moment, Cryo-EM has attracted much attention from biologists and chemists. Compared with the Cryo-EM, there is a long way to go for SPI. The current 3D resolution of SPI is about 10 nm. The CXI nanofocusing data showed a post-sample aperture [179] limited resolution close to 5.9 Å [180]. Figure 8 shows the calculated photons number vs resolution (Q). The integrated single hit signal is higher than the background. A real-space image could not be obtained from this data due to the lack of enough hits. While the main limitation currently is hit rate, improved detectors with lower noise and easier identification of single scattered photons from the noise will be required to achieve higher resolution imaging. The SPI detector of the future will need to combine the best of dark current, detector panel size, gain, readout noise, dynamic range and quantum efficiency, with an increased readout speed. For LCLS, the maximum pulse repetition is 120 Hz. At the increased repetition rates of the European-XFEL and LCLS-II, new detectors are needed. Millions or more diffraction patterns are needed to obtain a high-resolution image. At the current hit rates of about 1%, higher repetition rate X-ray FELs will have a profound impact. The main characteristics of detectors used or planned for SPI related experiments at LCLS, LCLS-II, European-XFEL, SACLA and SwissFEL are listed in Table 1 for reference.

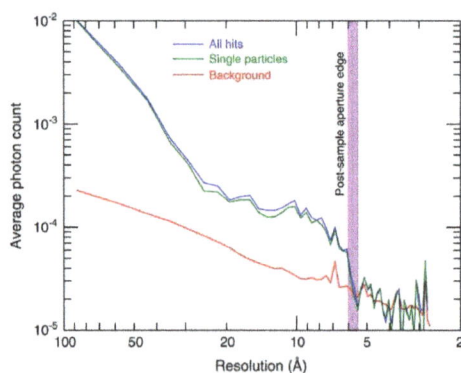

Figure 8. Radial average of signals using hard X-rays. Elevated photon counts from the sample are visible up to an angle commensurate with 5.9 Å resolution, this being the resolution limit set by the angular acceptance of a post-sample aperture. (reproduced with permission from [180]. Copyright Springer Nature, 2016).

Table 1. Main characteristics of detectors used for LCLS, LCLS-II, European-XFEL and SACLA.

	AMO, LCLS	CXI, LCLS	TXI, LCLS-II	SPB/SFX, European-XFEL	BL 2, SACLA	SwissFEL
Detector	pnCCD	CSPAD [b]	ePix×10k	AGIPD [d]	MPCCD	JUNGFRAU [f]
Pixel Size (μm^2)	75 × 75	110 × 110	100 × 100	200 × 200	50 × 50	75 × 75
Single Photon Sensitivity	Yes	Yes	Yes	Yes	Yes [e]	Yes
Quantum Efficiency	>80%@0.3–12 keV	~97%@8 keV	~85%@5 keV	>80%@0.3–25 keV	~85%@5.5keV	Up to 85%@1 keV
Dynamic Range	10^3@2 keV	3.5×10^2@8 keV	~10^4@8 keV	>10^4@12.5 keV	1.2×10^3@6 keV	>10^4@12 keV
Noise (e^-)	20/2 [a]	300	120	265	300	100
Frame Rate (kHz)	0.12	0.12	0.48/2~4/10~20 [c]	4500	0.06	0.1~2.4

[a] For pnCCD, the readout noise is 20 e^- for low gain, 2 e^- for high gain [181]. [b] The dynamic range and ASIC noise for CSPAD under high gain mode are 350 photons@8 keV and 300 e^- r.m.s. For low gain mode, they're 2700 photons@8 keV and 1000 e^- r.m.s [182]. [c] The current version is about 480 Hz. For digital domain multiplexing, the frame rate will be 2~4 kHz. For fast ADCs, the frame rate will be 10~20 kHz [183,184]. [d] See reference [185,186] for detailed information. [e] Keeping the single photon detection capability for X-ray photon energy higher than 6 keV [187]. [f] See reference [188] for detailed information.

Data analysis is obviously key to the final results. As XFEL beamtime is rare and precious around the world, to optimize the experimental parameters before the formal experiment and data collection period is of great importance to guarantee the success and save beamtime. Simulations based on physical theories, actual light source and instrument specifications will be a great help. The SPB/SFX instrument of the European XFEL created a start-to-end simulation (S2Esim) software (European-XFEL, Hamburg, Germany) [189] to help researchers figure out the optimized parameters for SPI experiments based on European XFEL accelerator and undulator data bank. Recently, a powerful program was designed for SPI to demonstrate single particle imaging under ultra-shot X-ray pulses [190]. LCLS is planning to leverage these developments from the European XFEL. Other open sources, such as condor [191], provided by Uppsala University can simulate speckle patterns, reconstructed images and evaluated resolution that based on photon energy, focusing beam spot size, sample type and detector, etc. These will help researchers design and optimize their experiments.

The quantity of raw data of XFEL experiments makes it fall under the big data category. Raw data files contain electron and photon beams parameters on a shoit by shot basis, instrument settings and motor positions, diagnostics and DAQ information. How to extract useful information from the raw data efficiently is challenging. To provide a user-friendly solution, LCLS, European-XFEL and SACLA created their own DAQ systems [192]. At the LCLS, data analysis department developed psana module that a python-script interfaces could be used online (real-time) offline and parallelized over many machines [193]. This data analysis strategy makes SPI data analysis easier by using building blocks.

Online data monitoring is also an important part of SPI experiments. To give an instant feedback from the SPI data collection, a Python-based software named *Hummgbird* (Uppsala University, Uppsala, Sweden) [194] was developed. This gives the needed fast feedback for the particle injector and instruments alignment or optimization.

Offline analysis of such big data sets is challenging. The LCLS data analysis department developed psocake [195], a GUI program that makes SPI data analysis simple. The input of experiment name, run number and detector ID will make diffraction data visible. Different masks, different algorithms for hit finding and multiple types of pixel readout (gain corrected ADU, common mode corrected ADU, pedestal corrected ADU, raw ADU and photon counts) make diffraction patterns preliminary analysis (e.g., ADUs to photons, background subtraction, hit finding) powerful and effective.

Apart from the shot-to-shot intensity variation, non-linear detector response, stray scattering from optics and injectors, sample heterogeneity and conformations mentioned above, multiple particle hits are also a challenge factor for offline data analysis. Significant progress has been made in classification algorithms by using diffusion maps [196]. Experimental data contains only intensities with no phase information. Diffusion map algorithms provide mathematical links to a cloud of points via the Laplace-Beltrami operator. Each point on a diffusion map represents a diffraction pattern as shown on Figure 9 [197]. Single hit, multiple-particle hit, water diffraction, background and defects from the detector are captured in such mapping. Diffusion map approaches will fail in the limit of strong background and low SNR for small particles (e.g., MS2). For that situation, the classification can be conducted by hand. This is time-consuming and the precision will be non-reproducible for different operators. To solve this problem, automatic and supervised machine learning methods are being developed. The convolutional neural network is based on VGG16 net. For training, simulated data and experiments data are both used. Further study is still needed but this represents a promising future approach.

After classification, orientation determination and phase retrieval are needed. The commonly used methods are generative topographic mapping (GTM) [198], expectation maximization compression (EMC) [199,200], correlation maximization [201], manifold [202,203] and angular correlation methods [204]. The current method of choice for orientation determination is EMC. The EMC algorithm is implemented in *Dragonfly* [205], which is a powerful python-script program. Figure 10 shows a 3D diffraction volume after orientation determination and assembly [206]. The associated 3D real-space image is reconstructed by error-reduction (ER) method. Many other phase retrieval algorithms could be used, such as hybrid-input-output (HIO), Guided-hybrid-input-output (GHIO), relaxed averaged alternating reflections (RAAR) and difference map (DM).

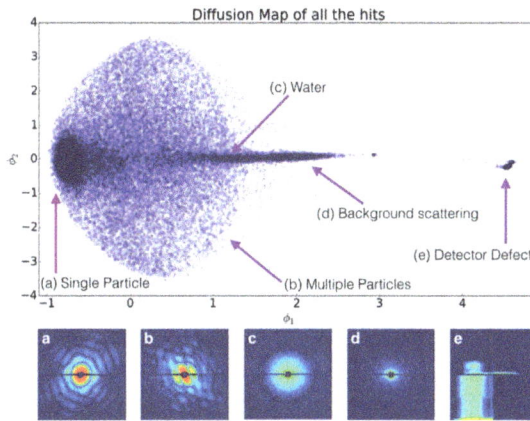

Figure 9. Diffusion map analysis of all hits. Large data sets were reduced to single-particle hit patterns. (**a**) Single-particle hit; (**b**) multiple-particle hit; (**c**) water diffraction that may contain contaminant residue from samples and buffer solutions; (**d**) background diffraction from detector; (**e**) detector defect pattern. (reproduced with permission from [197]. Copyright Springer Nature, 2017).

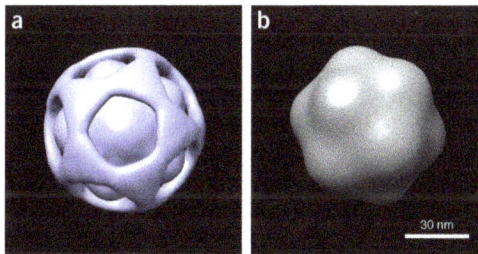

Figure 10. Three-dimensional diffraction pattern and reconstructed structure obtained without conformational analysis. (**a**) Diffraction pattern extracted from 37,550 2D single-particle diffraction snapshots of the PR772 virus, obtained at LCLS; (**b**) Corresponding 3D reconstructed image with a resolution of 9 nm, corresponding to scattering to the edge of the detector. (reproduced with permission from [206]. Copyright Springer Nature, 2017).

Recently, great progress has been made in real-space image reconstruction phase retrieval algorithms, multi-tiered iterative phasing (M-TIP) [207]. M-TIP is an extension of standard iterative phasing algorithms, which recover the 3D internal intensity directly from fluctuation X-ray scattering data [208]. The angular cross-correlations method makes a valuable statistical tool for structural analysis. The approach in the case of scattering also offers a valuable opportunity for multi-particle analysis. Figure 11 shows the recovered 3D structures of RDV and PR772 by MTIP. A non-uniform distribution of internal structures was obtained.

4. Summary and Future Prospects

Since the start of the SPI initiative, eight experiments have been completed to try to overcome the technical challenges in achieving atomic resolution in single particles imaging with XFELs at the

Linac Coherent Light Source. Two data papers have been published to describing experimental details and progress to date. Progress was also made by the SPI initiative members developing conformation change detection and new phase retrieval algorithm. The goals of past SPI experiments mainly focused on light source and instruments optimization, sample preparations and delivery development, background reduction and data analysis methods. The team achieved measurement signals at ~6 Å resolution, to the corner of the CSPAD detector used with the 100 nm-focusing system of the CXI instrument. At the moment, the data is insufficient to reconstruct the 3D real-space image. On the other hand, hundreds of thousands of diffraction patterns were recorded by a far-field pnCCD at the AMO instrument using soft X-rays. Different team members have successfully reconstructed 3D real-space images using different methods at a full-period resolution of ~10 nm. Some of the SPI beamtime have been dedicated to the improvement of sample injection. Different injection methods, such as GDVN and electrospray for aerosolization, have been tested to reduce the background. This is still a challenge for further experiments. For the data analysis, classification has made rapid progress via the use of diffusion maps and machine learning. Powerful programs, such as *psana, Condor, Hummingbird, Ondata, Dragonfly* have made great contributions to SPI.

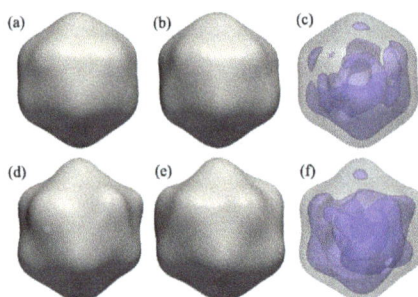

Figure 11. Three-dimensional reconstructed images of RDV (top row) and PR772 virus (bottom row). Two different views of the reconstructed RDV (**a,b**) and PR772 (**d,e**) particles, as well as density plots showing nonuniformities in the internal distribution of material inside the viruses (**c,f**). (reproduced with permission from [207]. Copyright American Physical Society, 2017).

For higher resolution in the near future, such as 5 Å or better, more challenges lay ahead. The reproducibility of samples and homogeneity of internal structures are hard to guarantee. The hit rate and sample contaminations will have a great influence on the validity of patterns and numbers of valid frames. Higher resolution means more frames, this also requires more semi- and automatic data analysis methods. Not only are enough frames needed for 3D reconstructions but they also need enough photons in each frame for interpretation above noise. To totally outrun the radiation damage, the pulse duration should be reduced from 40 fs to below 10 fs without too much loss in flux. High QE, high dynamic range and large array pixel area detector will also be a bottleneck for higher resolution approach using higher repetition rate machines. Fortunately, the European XFEL and LCLS-II are both developing such detectors.

A new tendency in scientific research is to share the data, sources and codes to the public [209–211]. With data sharing, the validity and repeatability of the analysis and interpretation can be tested by different institutes and personnel [212–216]. "Open science, data sharing, software sharing is the future of science", Carly Strasser says [209]. For SPI, this concept has also been accepted and will be extended to a wider range [217–220]. At the moment, most published data has been uploaded to the Coherent X-ray Imaging Data Bank (CXIDB) [221,222]. The community can download the raw data and try to reproduce the results. Commonly used tools can also be obtained easily from Github [223] or facility websites [224]. The move to make scientific findings transparent will speed up the challenging project and improve the scientific outcome.

Moving forward and overcoming the technical challenges can be painful, but this necessary method development effort is improving the prospects of single particle imaging with XFEL. High resolution X-ray based single particle imaging will make dynamic changes, light-induced phenomena, phase transition and femtosecond chemistry & catalysis valuable possibility. This will provide a new horizon to the scientific community to explore the ultra-small and ultrafast world. The potential rewards are worth the effort and will only benefit from the advent of high repetition rate of state-of-the-art X-ray FELs such as the European-XFEL and LCLS-II.

Acknowledgments: We would like to thank the funding support from the National Natural Science Foundation of China (grant No. 31430031) and the Major State Basic Research Development Program of China (grant No. 2014CB910401). The Linac Coherent Light Source (LCLS) at the SLAC National Accelerator Laboratory is an Office of Science User Facility operated for the U.S. Department of Energy Office of Science by Stanford University. This work was: supported by the U.S. Department of Energy, Office of Science, Basic Energy Sciences under Contract No. DEAC02-76SF00515. We also would like to sincerely thank the stimulated discussions and help from Sébastien Boutet and Andrew L. Aquila at the Linac Coherent Light Source. Z.S. acknowledges funding from China Scholarship Council.

Author Contributions: Z.S. wrote the paper. All authors reviewed and critiqued the manuscript.

Conflicts of Interest: The authors declare no conflict of interest.

References

1. Opella, S.J. Structure Determination of Membrane Proteins by Nuclear Magnetic Resonance Spectroscopy. *Annu. Rev. Anal. Chem.* **2013**, *6*, 305–328. [CrossRef] [PubMed]
2. Garman, E.F. Developments in X-ray Crystallographic Structure Determination of Biological Macromolecules. *Science* **2014**, *343*, 1102–1108. [CrossRef] [PubMed]
3. Elmlund, D.; Elmlund, H. Cryogenic Electron Microscopy and Single-Particle Analysis. *Annu. Rev. Biochem.* **2015**, *84*, 499–517. [CrossRef] [PubMed]
4. Marsh, J.A.; Teichmann, S.A. Structure, Dynamics, Assembly, and Evolution of Protein Complexes. *Annu. Rev. Biochem.* **2015**, *84*, 551–575. [CrossRef] [PubMed]
5. Rajadhyaksha, M.; Grossman, M.; Esterowitz, D.; Webb, R.H.; Rox Anderson, R. In Vivo Confocal Scanning Laser Microscopy of Human Skin: Melanin Provides Strong Contrast. *J. Investig. Dermatol.* **1995**, *104*, 946–952. [CrossRef] [PubMed]
6. Suzuki, T.; Fujikura, K.; Higashiyama, T.; Takata, K. DNA Staining for Fluorescence and Laser Confocal Microscopy. *J. Histochem. Cytochem.* **1997**, *45*, 49–53. [CrossRef] [PubMed]
7. Ntziachristos, V. Going deeper than microscopy: The optical imaging frontier in biology. *Nat. Methods* **2010**, *7*, 603–614. [CrossRef] [PubMed]
8. Willig, K.I.; Rizzoli, S.O.; Westphal, V.; Jahn, R.; Hell, S.W. STED microscopy reveals that synaptotagmin remains clustered after synaptic vesicle exocytosis. *Nature* **2006**, *440*, 935–939. [CrossRef] [PubMed]
9. Pawley, J.B. Fundamental Limits in Confocal Microscopy. In *Handbook of Biological Confocal Microscopy*; Springer: Boston, MA, USA, 2006; pp. 20–42, ISBN 978-03-8-745524-2.
10. Binnig, G.; Rohrer, H. Scanning tunneling microscopy. *Surf. Sci.* **1983**, *126*, 236–244. [CrossRef]
11. Binnig, G.; Quate, C.F.; Gerber, C. Atomic Force Microscope. *Phys. Rev. Lett.* **1986**, *56*, 930–933. [CrossRef] [PubMed]
12. Williams, D.B.; Carter, C.B. *The Transmission Electron Microscope*; Springer: Boston, MA, USA, 1996; ISBN 978-14-7-572519-3.
13. Reichelt, R. *Scanning Electron Microscopy*; Springer: New York, NY, USA, 2007; pp. 133–272. ISBN 978-03-8-749762-4.
14. Larson, B.C.; Yang, W.; Ice, G.E.; Budai, J.D.; Tischler, J.Z. Three-dimensional X-ray structural microscopy with submicrometre resolution. *Nature* **2002**, *415*, 887–890. [CrossRef] [PubMed]
15. Chao, W.; Harteneck, B.D.; Liddle, J.A.; Anderson, E.H.; Attwood, D.T. Soft X-ray microscopy at a spatial resolution better than 15 nm. *Nature* **2005**, *435*, 1210–1213. [CrossRef] [PubMed]
16. Sakdinawat, A.; Attwood, D. Nanoscale X-ray imaging. *Nat Photon.* **2010**, *4*, 840–848. [CrossRef]

17. Betzig, E.; Patterson, G.H.; Sougrat, R.; Lindwasser, O.W.; Olenych, S.; Bonifacino, J.S.; Davidson, M.W.; Lippincott-Schwartz, J.; Hess, H.F. Imaging Intracellular Fluorescent Proteins at Nanometer Resolution. *Science* **2006**, *313*, 1642–1645. [CrossRef] [PubMed]

18. Rust, M.J.; Bates, M.; Zhuang, X. Sub-diffraction-limit imaging by stochastic optical reconstruction microscopy (STORM). *Nat. Methods* **2006**, *3*, 793–796. [CrossRef] [PubMed]

19. Klar, T.A.; Jakobs, S.; Dyba, M.; Egner, A.; Hell, S.W. Fluorescence microscopy with diffraction resolution barrier broken by stimulated emission. *Proc. Natl. Acad. Sci. USA* **2000**, *97*, 8206–8210. [CrossRef] [PubMed]

20. Hofer, W.A.; Foster, A.S.; Shluger, A.L. Theories of scanning probe microscopes at the atomic scale. *Rev. Mod. Phys.* **2003**, *75*, 1287–1331. [CrossRef]

21. Kohl, L.R.H. *Transmission Electron Microscopy*, 5th ed.; Springer: New York, NY, USA, 2008; ISBN 978-03-8-740093-8.

22. Spence, J.C.H. *High-Resolution Electron Microscopy*, 4th ed.; Oxford University Press: New York, NY, USA, 2017; ISBN 978-01-9-879583-4.

23. Breedlove, J.R., Jr.; Trammell, G.T. Molecular Microscopy: Fundamental Limitations. *Science* **1970**, *170*, 1310–1313. [CrossRef] [PubMed]

24. Howells, M.R.; Beetz, T.; Chapman, H.N.; Cui, C.; Holton, J.M.; Jacobsen, C.J.; Kirz, J.; Lima, E.; Marchesini, S.; Miao, H.; et al. An assessment of the resolution limitation due to radiation-damage in X-ray diffraction microscopy. *J. Electron. Spectrosc. Relat. Phenom.* **2009**, *170*, 4–12. [CrossRef] [PubMed]

25. Barty, A.; Caleman, C.; Aquila, A.; Timneanu, N.; Lomb, L.; White, T.A.; Andreasson, J.; Arnlund, D.; Bajt, S.; Barends, T.R.M.; et al. Self-terminating diffraction gates femtosecond X-ray nanocrystallography measurements. *Nat. Photon.* **2012**, *6*, 35–40. [CrossRef] [PubMed]

26. Attwood, D. *Soft X-rays and Extreme Ultraviolet Radiation: Principles and Applications*, 1st ed.; Cambridge University Press: New York, NY, USA, 2007; ISBN 978-0521029971.

27. Mimura, H.; Handa, S.; Kimura, T.; Yumoto, H.; Yamakawa, D.; Yokoyama, H.; Matsuyama, S.; Inagaki, K.; Yamamura, K.; Sano, Y.; et al. Breaking the 10nm barrier in hard-X-ray focusing. *Nat. Phys.* **2010**, *6*, 122–125. [CrossRef]

28. Döring, F.; Robisch, A.L.; Eberl, C.; Osterhoff, M.; Ruhlandt, A.; Liese, T.; Schlenkrich, F.; Hoffmann, S.; Bartels, M.; Salditt, T.; et al. Sub-5 nm hard X-ray point focusing by a combined Kirkpatrick-Baez mirror and multilayer zone plate. *Opt. Express* **2013**, *21*, 19311–19323. [CrossRef] [PubMed]

29. Drenth, J. *Principles of Protein X-ray Crystallography*, 3rd ed.; Springer: New York, NY, USA, 2007; ISBN 978-0-387-33334-2.

30. Sayre, D. Some implications of a theorem due to Shannon. *Acta. Crystallogr.* **1952**, *5*, 843. [CrossRef]

31. Miao, J.; Sayre, D.; Chapman, H.N. Phase retrieval from the magnitude of the Fourier transforms of nonperiodic objects. *J. Opt. Soc. Am. A* **1998**, *15*, 1662–1669. [CrossRef]

32. Miao, J.; Ishikawa, T.; Anderson, E.H.; Hodgson, K.O. Phase retrieval of diffraction patterns from noncrystalline samples using the oversampling method. *Phys. Rev. B* **2003**, *67*, 174104. [CrossRef]

33. Fienup, J.R. Reconstruction of an object from modulus of Its Fourier-transform. *Opt. Lett.* **1978**, *3*, 27–29. [CrossRef] [PubMed]

34. Fienup, J.R. Phase retrieval algorithms: A comparison. *Appl. Opt.* **1982**, *21*, 2758–2769. [CrossRef] [PubMed]

35. Fienup, J.R. Lensless coherent imaging by phase retrieval with an illumination pattern constraint. *Opt. Express* **2006**, *14*, 498–508. [CrossRef] [PubMed]

36. Bauschke, H.H.; Combettes, P.L.; Luke, D.R. Hybrid projection–reflection method for phase retrieval. *J. Opt. Soc. Am. A* **2003**, *20*, 1025–1034. [CrossRef]

37. Elser, V. Solution of the crystallographic phase problem by iterated projections. *Acta Crystallogr. Sect. A Found. Crystallogr.* **2003**, *59*, 201–209. [CrossRef]

38. Marchesini, S. X-ray image reconstruction from a diffraction pattern alone. *Phys. Rev. B* **2003**, *68*, 140101. [CrossRef]

39. Luke, D.R. Relaxed averaged alternating reflections for diffraction imaging. *Inverse Probl.* **2005**, *21*, 37–50. [CrossRef]

40. Chen, C.-C.; Miao, J.; Wang, C.W.; Lee, T.K. Application of optimization technique to noncrystalline x-ray diffraction microscopy: Guided hybrid input-output method. *Phys. Rev. B* **2007**, *76*, 064113. [CrossRef]

41. Marchesini, S. Invited Article: A unified evaluation of iterative projection algorithms for phase retrieval. *Rev. Sci. Instrum.* **2007**, *78*, 011301. [CrossRef] [PubMed]

42. Miao, J.; Charalambous, P.; Kirz, J.; Sayre, D. Extending the methodology of X-ray crystallography to allow imaging of micrometre-sized non-crystalline specimens. *Nature* **1999**, *400*, 342–344. [CrossRef]
43. Robinson, I.K.; Vartanyants, I.A.; Williams, G.J.; Pfeifer, M.A.; Pitney, J.A. Reconstruction of the Shapes of Gold Nanocrystals Using Coherent X-Ray Diffraction. *Phys. Rev. Lett.* **2001**, *87*, 195505. [CrossRef] [PubMed]
44. Williams, G.J.; Pfeifer, M.A.; Vartanyants, I.A.; Robinson, I.K. Three-Dimensional Imaging of Microstructure in Au Nanocrystals. *Phys. Rev. Lett.* **2003**, *90*, 175501. [CrossRef] [PubMed]
45. Miao, J.; Chen, C.-C.; Song, C.; Nishino, Y.; Kohmura, Y.; Ishikawa, T.; Ramunno-Johnson, D.; Lee, T.-K.; Risbud, S.H. Three-dimensional GaN-Ga2O3 core shell structure revealed by x-ray diffraction microscopy. *Phys. Rev. Lett.* **2006**, *97*, 215503. [CrossRef] [PubMed]
46. Pfeifer, M.A.; Williams, G.J.; Vartanyants, I.A.; Harder, R.; Robinson, I.K. Three-dimensional mapping of a deformation field inside a nanocrystal. *Nature* **2006**, *442*, 63–66. [CrossRef] [PubMed]
47. Song, C.; Bergstrom, R.; Ramunno-Johnson, D.; Jiang, H.; Paterson, D.; de Jonge, M.D.; McNulty, I.; Lee, J.; Wang, K.L.; Miao, J. Nanoscale Imaging of Buried Structures with Elemental Specificity Using Resonant X-ray Diffraction Microscopy. *Phys. Rev. Lett.* **2008**, *100*, 025504. [CrossRef] [PubMed]
48. Robinson, I.; Harder, R. Coherent X-ray diffraction imaging of strain at the nanoscale. *Nat. Mater.* **2009**, *8*, 291–298. [CrossRef] [PubMed]
49. Jiang, H.; Ramunno-Johnson, D.; Song, C.; Amirbekian, B.; Kohmura, Y.; Nishino, Y.; Takahashi, Y.; Ishikawa, T.; Miao, J. Nanoscale imaging of mineral crystals inside biological composite materials using X-ray diffraction microscopy. *Phys. Rev. Lett.* **2008**, *100*, 038103. [CrossRef] [PubMed]
50. Jiang, H.; Xu, R.; Chen, C.-C.; Yang, W.; Fan, J.; Tao, X.; Song, C.; Kohmura, Y.; Xiao, T.; Wang, Y.; et al. Three-dimensional coherent X-ray diffraction imaging of molten iron in mantle olivine at nanoscale resolution. *Phys. Rev. Lett.* **2013**, *110*, 205501. [CrossRef] [PubMed]
51. Miao, J.; Hodgson, K.O.; Ishikawa, T.; Larabell, C.A.; LeGros, M.A.; Nishino, Y. Imaging whole Escherichia coli bacteria by using single-particle x-ray diffraction. *Proc. Natl. Acad. Sci. USA* **2003**, *100*, 110–112. [CrossRef] [PubMed]
52. Shapiro, D.; Thibault, P.; Beetz, T.; Elser, V.; Howells, M.; Jacobsen, C.; Kirz, J.; Lima, E.; Miao, H.; Neiman, A.M.; et al. Biological imaging by soft x-ray diffraction microscopy. *Proc. Natl. Acad. Sci. USA* **2005**, *102*, 15343–15346. [CrossRef] [PubMed]
53. Song, C.; Jiang, H.; Mancuso, A.; Amirbekian, B.; Peng, L.; Sun, R.; Shah, S.S.; Zhou, Z.H.; Ishikawa, T.; Miao, J. Quantitative imaging of single, unstained viruses with coherent x-rays. *Phys. Rev. Lett.* **2008**, *101*, 158101. [CrossRef] [PubMed]
54. Williams, G.J.; Hanssen, E.; Peele, A.G.; Pfeifer, M.A.; Clark, J.; Abbey, B.; Cadenazzi, G.; de Jonge, M.D.; Vogt, S.; Tilley, L.; et al. High-resolution X-ray imaging of Plasmodium falciparum-infected red blood cells. *Cytom. Part A* **2008**, *73A*, 949–957. [CrossRef] [PubMed]
55. Huang, X.; Nelson, J.; Kirz, J.; Lima, E.; Marchesini, S.; Miao, H.; Neiman, A.M.; Shapiro, D.; Steinbrener, J.; Stewart, A.; et al. Soft X-Ray Diffraction Microscopy of a Frozen Hydrated Yeast Cell. *Phys. Rev. Lett.* **2009**, *103*, 198101. [CrossRef] [PubMed]
56. Lima, E.; Wiegart, L.; Pernot, P.; Howells, M.; Timmins, J.; Zontone, F.; Madsen, A. Cryogenic X-Ray Diffraction Microscopy for Biological Samples. *Phys. Rev. Lett.* **2009**, *103*, 198102. [CrossRef] [PubMed]
57. Nishino, Y.; Takahashi, Y.; Imamoto, N.; Ishikawa, T.; Maeshima, K. Three-dimensional visualization of a human chromosome using coherent X-ray diffraction. *Phys. Rev. Lett.* **2009**, *102*, 018101. [CrossRef] [PubMed]
58. Nelson, J.; Huang, X.; Steinbrener, J.; Shapiro, D.; Kirz, J.; Marchesini, S.; Neiman, A.M.; Turner, J.J.; Jacobsen, C. High-resolution x-ray diffraction microscopy of specifically labeled yeast cells. *Proc. Natl. Acad. Sci. USA* **2010**, *107*, 7235–7239. [CrossRef] [PubMed]
59. Lima, E.; Diaz, A.; Guizar-Sicairos, M.; Gorelick, S.; Pernot, P.; Schleier, T.; Menzel, A. Cryo-scanning X-ray diffraction microscopy of frozen-hydrated yeast. *J. Microsc.* **2013**, *249*, 1–7. [CrossRef]
60. Nam, D.; Park, J.; Gallagher-Jones, M.; Kim, S.; Kim, S.; Kohmura, Y.; Naitow, H.; Kunishima, N.; Yoshida, T.; Ishikawa, T.; et al. Imaging fully hydrated whole cells by coherent X-ray diffraction microscopy. *Phys. Rev. Lett.* **2013**, *110*, 098103. [CrossRef] [PubMed]
61. Song, C.; Takagi, M.; Park, J.; Xu, R.; Gallagher-Jones, M.; Imamoto, N.; Ishikawa, T. Analytic 3D Imaging of Mammalian Nucleus at Nanoscale Using Coherent X-rays and Optical Fluorescence Microscopy. *Biophys. J.* **2014**, *107*, 1074–1081. [CrossRef] [PubMed]

62. Fan, J.; Sun, Z.; Zhang, J.; Huang, Q.; Yao, S.; Zong, Y.; Kohmura, Y.; Ishikawa, T.; Liu, H.; Jiang, H. Quantitative Imaging of Single Unstained Magnetotactic Bacteria by Coherent X-ray Diffraction Microscopy. *Anal. Chem.* **2015**, *87*, 5849–5853. [CrossRef] [PubMed]

63. Rodriguez, J.A.; Xu, R.; Chen, C.-C.; Huang, Z.; Jiang, H.; Chen, A.L.; Raines, K.S.; Pryor, A., Jr.; Nam, D.; Wiegart, L.; et al. Three-dimensional coherent X-ray diffractive imaging of whole frozen-hydrated cells. *IUCrJ* **2015**, *2*, 575–583. [CrossRef] [PubMed]

64. Williams, G.J.; Quiney, H.M.; Dhal, B.B.; Tran, C.Q.; Nugent, K.A.; Peele, A.G.; Paterson, D.; Jonge, M.D.D. Fresnel coherent diffractive imaging. *Phys. Rev. Lett.* **2006**, *97*, 025506. [CrossRef] [PubMed]

65. Abbey, B.; Nugent, K.A.; Williams, G.J.; Clark, J.N.; Peele, A.G.; Pfeifer, M.A.; de Jonge, M.; McNulty, I. Keyhole coherent diffractive imaging. *Nat. Phys.* **2008**, *4*, 394–398. [CrossRef]

66. Marathe, S.; Kim, S.S.; Kim, S.N.; Kim, C.; Kang, H.C.; Nickles, P.V.; Noh, D.Y. Coherent diffraction surface imaging in reflection geometry. *Opt. Express* **2010**, *18*, 7253–7262. [CrossRef] [PubMed]

67. Rodenburg, J.M.; Hurst, A.C.; Cullis, A.G.; Dobson, B.R.; Pfeiffer, F.; Bunk, O.; David, C.; Jefimovs, K.; Johnson, I. Hard-X-ray lensless imaging of extended objects. *Phys. Rev. Lett.* **2007**, *98*, 034801. [CrossRef] [PubMed]

68. Dierolf, M.; Menzel, A.; Thibault, P.; Schneider, P.; Kewish, C.M.; Wepf, R.; Bunk, O.; Pfeiffer, F. Ptychographic X-ray computed tomography at the nanoscale. *Nature* **2010**, *467*, 436–439. [CrossRef] [PubMed]

69. Miao, J.; Hodgson, K.O.; Sayre, D. An approach to three-dimensional structures of biomolecules by using single-molecule diffraction images. *Proc. Natl. Acad. Sci. USA* **2001**, *98*, 6641–6645. [CrossRef] [PubMed]

70. Miao, J.; Ishikawa, T.; Johnson, B.; Anderson, E.H.; Lai, B.; Hodgson, K.O. High Resolution 3D X-ray Diffraction Microscopy. *Phys. Rev. Lett.* **2002**, *89*, 088303. [CrossRef] [PubMed]

71. Nugent, K.A.; Peele, A.G.; Chapman, H.N.; Mancuso, A.P. Unique phase recovery for nonperiodic objects. *Phys. Rev. Lett.* **2003**, *91*, 203902. [CrossRef] [PubMed]

72. Chapman, H.N.; Barty, A.; Marchesini, S.; Noy, A.; Hau-Riege, S.P.; Cui, C.; Howells, M.R.; Rosen, R.; He, H.; Spence, J.C.H.; et al. High-resolution ab initio three-dimensional x-ray diffraction microscopy. *J. Opt. Soc. Am. A* **2006**, *23*, 1179–1200. [CrossRef]

73. Quiney, H.M.; Peele, A.G.; Cai, Z.; Paterson, D.; Nugent, K.A. Diffractive imaging of highly focused X-ray fields. *Nat. Phys.* **2006**, *2*, 101–104. [CrossRef]

74. Thibault, P.; Dierolf, M.; Menzel, A.; Bunk, O.; David, C.; Pfeiffer, F. High-Resolution Scanning X-ray Diffraction Microscopy. *Science* **2008**, *321*, 379–382. [CrossRef] [PubMed]

75. Chapman, H.N.; Nugent, K.A. Coherent lensless X-ray imaging. *Nat. Photon.* **2010**, *4*, 833–839. [CrossRef]

76. Giewekemeyera, K.; Thibaultb, P.; Kalbfleischa, S.; Beerlinka, A.; Kewishc, C.M.; Dierolfb, M.; Pfeifferb, F.; Salditta, T. Quantitative biological imaging by ptychographic X-ray diffraction microscopy. *Proc. Natl. Acad. Sci. USA* **2010**, *107*, 529–534. [CrossRef] [PubMed]

77. Jiang, H.; Song, C.; Chen, C.-C.; Xu, R.; Raines, K.S.; Fahimian, B.P.; Lu, C.-H.; Lee, T.-K.; Nakashima, A.; Urano, J.; et al. Quantitative 3D imaging of whole, unstained cells by using X-ray diffraction microscopy. *Proc. Natl. Acad. Sci. USA* **2010**, *107*, 11234–11239. [CrossRef] [PubMed]

78. Raines, K.S.; Salha, S.; Sandberg, R.L.; Jiang, H.; Rodriguez, J.A.; Fahimian, B.P.; Kapteyn, H.C.; Du, J.; Miao, J. Three-dimensional structure determination from a single view. *Nature* **2010**, *463*, 214–217. [CrossRef] [PubMed]

79. Abbey, B.; Whitehead, L.W.; Quiney, H.M.; Vine, D.J.; Cadenazzi, G.A.; Henderson, C.A.; Nugent, K.A.; Balaur, E.; Putkunz, C.T.; Peele, A.G.; et al. Lensless imaging using broadband X-ray sources. *Nat. Photon.* **2011**, *5*, 420–424. [CrossRef]

80. Roy, S.; Parks, D.; Seu, K.A.; Su, R.; Turner, J.J.; Chao, W.; Anderson, E.H.; Cabrini, S.; Kevan, S.D. Lensless X-ray imaging in reflection geometry. *Nat. Photon.* **2011**, *5*, 243–245. [CrossRef]

81. Tripathi, A.; Mohanty, J.; Dietze, S.H.; Shpyrko, O.G.; Shipton, E.; Fullerton, E.E.; Kim, S.S.; McNulty, I. Dichroic coherent diffractive imaging. *Proc. Natl. Acad. Sci. USA* **2011**, *108*, 13393–13398. [CrossRef] [PubMed]

82. Clark, J.N.; Huang, X.; Harder, R.; Robinson, I.K. High-resolution three-dimensional partially coherent diffraction imaging. *Nat. Commun.* **2012**, *3*, 993. [CrossRef] [PubMed]

83. Miao, J.; Sandberg, R.L.; Song, C. Coherent X-ray Diffraction Imaging. *IEEE J. Sel. Top. Quantum Electron.* **2012**, *18*, 399–410. [CrossRef]

84. Sun, T.; Jiang, Z.; Strzalka, J.; Ocola, L.; Wang, J. Three-dimensional coherent X-ray surface scattering imaging near total external reflection. *Nat. Photon.* **2012**, *6*, 586–590. [CrossRef]

85. Szameit, A.; Shechtman, Y.; Osherovich, E.; Bullkich, E.; Sidorenko, P.; Dana, H.; Steiner, S.; Kley, E.B.; Gazit, S.; Cohen-Hyams, T.; et al. Sparsity-based single-shot subwavelength coherent diffractive imaging. *Nat. Mater.* **2012**, *11*, 455–459. [CrossRef] [PubMed]

86. Seaberg, M.D.; Zhang, B.; Gardner, D.F.; Shanblatt, E.R.; Murnane, M.M.; Kapteyn, H.C.; Adams, D.E. Tabletop nanometer extreme ultraviolet imaging in an extended reflection mode using coherent Fresnel ptychography. *Optica* **2014**, *1*, 39–44. [CrossRef]

87. Miao, J.; Ishikawa, T.; Robinson, I.K.; Murnane, M.M. Beyond crystallography: Diffractive imaging using coherent X-ray light sources. *Science* **2015**, *348*, 530–535. [CrossRef] [PubMed]

88. Chen, B.; Yusuf, M.; Hashimoto, T.; Estandarte, A.K.; Thompson, G.; Robinson, I. Three-dimensional positioning and structure of chromosomes in a human prophase nucleus. *Sci. Adv.* **2017**, *3*. [CrossRef] [PubMed]

89. HTML5 Word Cloud. Available online: https://github.com/timdream/wordcloud (accessed on 15 October 2017).

90. Eriksson, M.; van der Veen, J.F.; Quitmann, C. Diffraction-limited storage rings—A window to the science of tomorrow. *J. Synchrotron Radiat.* **2014**, *21*, 837–842. [CrossRef] [PubMed]

91. Hitchcock, A.P.; Toney, M.F. Spectromicroscopy and coherent diffraction imaging: Focus on energy materials applications. *J. Synchrotron Radiat.* **2014**, *21*, 1019–1030. [CrossRef] [PubMed]

92. Thibault, P.; Guizar-Sicairos, M.; Menzel, A. Coherent imaging at the diffraction limit. *J. Synchrotron Radiat.* **2014**, *21*, 1011–1018. [CrossRef] [PubMed]

93. Hettel, R. DLSR design and plans: An international overview. *J. Synchrotron Radiat.* **2014**, *21*, 843–855. [CrossRef] [PubMed]

94. Susini, J.; Barrett, R.; Chavanne, J.; Fajardo, P.; Gotz, A.; Revol, J.-L.; Zhang, L. New challenges in beamline instrumentation for the ESRF Upgrade Programme Phase II. *J. Synchrotron Radiat.* **2014**, *21*, 986–995. [CrossRef] [PubMed]

95. Tanaka, H.; Ishikawa, T. SPring-8 upgrade project. In Proceedings of the IPAC2016, Busan, Korea, 8–13 May 2016.

96. Tavares, P.F.; Leemann, S.C.; Sjostrom, M.; Andersson, A. The MAX IV storage ring project. *J. Synchrotron Radiat.* **2014**, *21*, 862–877. [CrossRef] [PubMed]

97. Liu, L.; Milas, N.; Mukai, A.H.C.; Resende, X.R.; de Sa, F.H. The Sirius project. *J. Synchrotron Radiat.* **2014**, *21*, 904–911. [CrossRef] [PubMed]

98. Madey, J.M.J. Stimulated Emission of Bremsstrahlung in a Periodic Magnetic Field. *J. Appl. Phys.* **1971**, *42*, 1906–1913. [CrossRef]

99. McNeil, B.W.J.; Thompson, N.R. X-ray free-electron lasers. *Nat. Photon.* **2010**, *4*, 814–821. [CrossRef]

100. Neutze, R.; Wouts, R.; Spoel, D.; Weckert, E.; Hajdu, J. Potential for biomolecular imaging with femtosecond X-ray pulses. *Nature* **2000**, *406*, 752–757. [CrossRef] [PubMed]

101. Chapman, H.N.; Barty, A.; Bogan, M.J.; Boutet, S.; Frank, M.; Hau-Riege, S.P.; Marchesini, S.; Woods, B.W.; Bajt, S.; Benner, W.H.; et al. Femtosecond diffractive imaging with a soft-X-ray free-electron laser. *Nat. Phys.* **2006**, *2*, 839–843. [CrossRef]

102. Chapman, H.N. Femtosecond X-ray protein nanocrystallography. *Nature* **2011**, *470*, 73–77. [CrossRef] [PubMed]

103. Aquila, A. The linac coherent light source single particle imaging road map. *Struct. Dyn.* **2015**, *2*, 041701. [CrossRef] [PubMed]

104. Tiedtke, K.; Azima, A.; Bargen, N.V.; Bittner, L.; Bonfigt, S.; Düsterer, S.; Faatz, B.; Frühling, U.; Gensch, M.; Gerth, C.; et al. The soft X-ray free-electron laser FLASH at DESY: Beamlines, diagnostics and end-stations. *New J. Phys.* **2009**, *11*, 023029. [CrossRef]

105. Free-Electron Laser FLASH. Available online: https://flash.desy.de (accessed on 15 October 2017).

106. Emma, P.; Akre, R.; Arthur, J.; Bionta, R.; Bostedt, C.; Bozek, J.; Brachmann, A.; Bucksbaum, P.; Coffee, R.; Decker, F.J.; et al. First lasing and operation of an angstrom-wavelength free-electron laser. *Nat. Photon.* **2010**, *4*, 641–647. [CrossRef]

107. Linac Coherent Light Source. Available online: https://lcls.slac.stanford.edu (accessed on 15 December 2017).

108. Ishikawa, T.; Aoyagi, H.; Asaka, T.; Asano, Y.; Azumi, N.; Bizen, T.; Ego, H.; Fukami, K.; Fukui, T.; Furukawa, Y.; et al. A compact X-ray free-electron laser emitting in the sub-angstrom region. *Nat. Photon.* **2012**, *6*, 540–544. [CrossRef]

109. SACLA XFEL. Available online: http://xfel.riken.jp/eng/ (accessed on 15 October 2017).

110. Allaria, E.; Appio, R.; Badano, L.; Barletta, W.A.; Bassanese, S.; Biedron, S.G.; Borga, A.; Busetto, E.; Castronovo, D.; Cinquegrana, P.; et al. Highly coherent and stable pulses from the FERMI seeded free-electron laser in the extreme ultraviolet. *Nat. Photon.* **2012**, *6*, 699–704. [CrossRef]

111. Elettra and FERMI Lightsources. Available online: https://www.elettra.trieste.it/lightsources/fermi.html (accessed on 15 October 2017).

112. European XFEL. Available online: https://www.xfel.eu (accessed on 15 October 2017).

113. Pohang Accelerator Laboratory. Available online: http://pal.postech.ac.kr/paleng/ (accessed on 15 October 2017).

114. SwissFEL. Available online: https://www.psi.ch/swissfel/ (accessed on 15 October 2017).

115. Wang, D. Soft X-ray Free-electron Laser at SINAP. In Proceedings of the IPAC2016, Busan, Korea, 8–13 May 2016.

116. Zhao, Z.; Wang, D.; Gu, Q.; Yin, L.; Gu, M.; Leng, Y.; Liu, B. Status of the SXFEL Facility. *Appl. Sci.* **2017**, *7*, 607. [CrossRef]

117. Zhu, Z.; Zhao, Z.; Wang, D.; Liu, Z.; Li, R.; Yin, L.; Yang, Z.H. SCLF: An 8-GeV CW SCRF linac-based X-ray FEL facility in Shanghai. In Proceedings of the FEL2017, Santa Fe, NM, USA, 20–25 August 2017.

118. Gaffney, K.J.; Chapman, H.N. Imaging atomic structure and dynamics with ultrafast X-ray scattering. *Science* **2007**, *316*, 1444–1448. [CrossRef] [PubMed]

119. Vartanyants, I.A.; Robinson, I.K.; McNulty, I.; David, C.; Wochner, P.; Tschentscher, T. Coherent X-ray scattering and lensless imaging at the European XFEL Facility. *J. Synchrotron Radiat.* **2007**, *14*, 453–470. [CrossRef] [PubMed]

120. Barty, A.; Boutet, S.; Bogan, M.J.; Hau-Riege, S.; Marchesini, S.; Sokolowski-Tinten, K.; Stojanovic, N.; Tobey, R.A.; Ehrke, H.; Cavalleri, A.; et al. Ultrafast single-shot diffraction imaging of nanoscale dynamics. *Nat. Photon.* **2008**, *2*, 415–419. [CrossRef]

121. Bogan, M.J.; Benner, W.H.; Boutet, S.; Rohner, U.; Frank, M.; Barty, A.; Seibert, M.M.; Maia, F.; Marchesini, S.; Bajt, S.; et al. Single particle X-ray diffractive imaging. *Nano Lett.* **2008**, *8*, 310–316. [CrossRef] [PubMed]

122. Mancuso, A.P.; Schropp, A.; Reime, B.; Stadler, L.-M.; Singer, A.; Gulden, J.; Streit-Nierobisch, S.; Gutt, C.; Bel, G.G.; Feldhaus, J.; et al. Coherent-pulse 2D crystallography using a free-electron laser X-ray source. *Phys. Rev. Lett.* **2009**, *102*, 035502. [CrossRef] [PubMed]

123. Bogan, M.J.; Boutet, S.; Chapman, H.N.; Marchesini, S.; Barty, A.; Benner, W.H.; Rohner, U.; Frank, M.; Hau-Riege, S.P.; Bajt, S.; et al. Aerosol Imaging with a Soft X-Ray Free Electron Laser. *Aerosol Sci. Technol.* **2010**, *44*, i–vi. [CrossRef]

124. Hau-Riege, S.P.; Boutet, S.; Barty, A.; Bajt, S.; Bogan, M.J.; Frank, M.; Andreasson, J.; Iwan, B.; Seibert, M.M.; Hajdu, J.; et al. Sacrificial Tamper Slows Down Sample Explosion in FLASH Diffraction Experiments. *Phys. Rev. Lett.* **2010**, *104*, 064801. [CrossRef] [PubMed]

125. Kassemeyer, S.; Steinbrener, J.; Lomb, L.; Hartmann, E.; Aquila, A.; Barty, A.; Martin, A.V.; Hampton, C.Y.; Bajt, S.; Barthelmess, M.; et al. Femtosecond free-electron laser x-ray diffraction data sets for algorithm development. *Opt. Express* **2012**, *20*, 4149–4158. [CrossRef] [PubMed]

126. Loh, N.D.; Hampton, C.Y.; Martin, A.V.; Starodub, D.; Sierra, R.G.; Barty, A.; Aquila, A.; Schulz, J.; Lomb, L.; Steinbrener, J.; et al. Fractal morphology, imaging and mass spectrometry of single aerosol particles in flight. *Nature* **2012**, *486*, 513–517. [CrossRef] [PubMed]

127. Starodub, D.; Aquila, A.; Bajt, S.; Barthelmess, M.; Barty, A.; Bostedt, C.; Bozek, J.D.; Coppola, N.; Doak, R.B.; Epp, S.W.; et al. Single-particle structure determination by correlations of snapshot X-ray diffraction patterns. *Nat. Commun.* **2012**, *3*, 1276. [CrossRef] [PubMed]

128. Clark, J.N.; Beitra, L.; Xiong, G.; Higginbotham, A.; Fritz, D.M.; Lemke, H.T.; Zhu, D.; Chollet, M.; Williams, G.J.; Messerschmidt, M.; et al. Ultrafast Three-Dimensional Imaging of Lattice Dynamics in Individual Gold Nanocrystals. *Science* **2013**, *341*, 56–59. [CrossRef] [PubMed]

129. Takahashi, Y.; Suzuki, A.; Zettsu, N.; Oroguchi, T.; Takayama, Y.; Sekiguchi, Y.; Kobayashi, A.; Yamamoto, M.; Nakasako, M. Coherent Diffraction Imaging Analysis of Shape-Controlled Nanoparticles with Focused Hard X-ray Free-Electron Laser Pulses. *Nano Lett.* **2013**, *13*, 6028–6032. [CrossRef] [PubMed]

130. Andreasson, J.; Martin, A.V.; Liang, M.; Timneanu, N.; Aquila, A.; Wang, F.; Iwan, B.; Svenda, M.; Ekeberg, T.; Hantke, M.; et al. Automated identification and classification of single particle serial femtosecond X-ray diffraction data. *Opt. Express* **2014**, *22*, 2497–2510. [CrossRef] [PubMed]

131. Xu, R.; Jiang, H.; Song, C.; Rodriguez, J.A.; Huang, Z.; Chen, C.-C.; Nam, D.; Park, J.; Gallagher-Jones, M.; Kim, S.; et al. Single-shot three-dimensional structure determination of nanocrystals with femtosecond X-ray free-electron laser pulses. *Nat. Commun.* **2014**, *5*, 4061. [CrossRef] [PubMed]

132. Clark, J.N.; Beitra, L.; Xiong, G.; Fritz, D.M.; Lemke, H.T.; Zhu, D.; Chollet, M.; Williams, G.J.; Messerschmidt, M.M.; Abbey, B.; et al. Imaging transient melting of a nanocrystal using an X-ray laser. *Proc. Natl. Acad. Sci. USA* **2015**, *112*, 7444–7448. [CrossRef] [PubMed]

133. Mancuso, A.P.; Gorniak, T.; Staier, F.; Yefanov, O.M.; Barth, R.; Christophis, C.; Reime, B.; Gulden, J.; Singer, A.; Pettit, M.E.; et al. Coherent imaging of biological samples with femtosecond pulses at the free-electron laser FLASH. *New J. Phys.* **2010**, *12*, 035003. [CrossRef]

134. Seibert, M.M.; Boutet, S.; Svenda, M.; Ekeberg, T.; Maia, F.R.N.C.; Bogan, M.J.; Tîmneanu, N.; Barty, A.; Hau-Riege, S.; Caleman, C.; et al. Femtosecond diffractive imaging of biological cells. *J. Phys. B At. Mol. Opt. Phys.* **2010**, *43*, 194015. [CrossRef]

135. Seibert, M.M.; Ekeberg, T.; Maia, F.R.N.C.; Svenda, M.; Andreasson, J.; Jonsson, O.; Odic, D.; Iwan, B.; Rocker, A.; Westphal, D.; et al. Single mimivirus particles intercepted and imaged with an X-ray laser. *Nature* **2011**, *470*, 78–81. [CrossRef] [PubMed]

136. Gallagher-Jones, M.; Bessho, Y.; Kim, S.; Park, J.; Kim, S.; Nam, D.; Kim, C.; Kim, Y.; Noh, D.Y.; Miyashita, O.; et al. Macromolecular structures probed by combining single-shot free-electron laser diffraction with synchrotron coherent X-ray imaging. *Nat. Commun.* **2014**, *5*, 3798. [CrossRef] [PubMed]

137. Hantke, M.F.; Hasse, D.; Maia, F.R.N.C.; Ekeberg, T.; John, K.; Svenda, M.; Loh, N.D.; Martin, A.V.; Timneanu, N.; Larsson, D.S.D.; et al. High-throughput imaging of heterogeneous cell organelles with an X-ray laser. *Nat. Photon.* **2014**, *8*, 943–949. [CrossRef]

138. Kimura, T.; Joti, Y.; Shibuya, A.; Song, C.; Kim, S.; Tono, K.; Yabashi, M.; Tamakoshi, M.; Moriya, T.; Oshima, T.; et al. Imaging live cell in micro-liquid enclosure by X-ray laser diffraction. *Nat. Commun.* **2014**, *5*, 3052. [CrossRef] [PubMed]

139. Ekeberg, T.; Svenda, M.; Abergel, C.; Maia, F.R.N.C.; Seltzer, V.; Claverie, J.-M.; Hantke, M.; Jönsson, O.; Nettelblad, C.; Van Der Schot, G. Three-dimensional reconstruction of the giant mimivirus particle with an x-ray free-electron laser. *Phys. Rev. Lett.* **2015**, *114*, 098102. [CrossRef] [PubMed]

140. Takayama, Y.; Inui, Y.; Sekiguchi, Y.; Kobayashi, A.; Oroguchi, T.; Yamamoto, M.; Matsunaga, S.; Nakasako, M. Coherent X-Ray Diffraction Imaging of Chloroplasts from Cyanidioschyzon merolae by Using X-Ray Free Electron Laser. *Plant Cell Physiol.* **2015**, *56*, 1272–1286. [CrossRef] [PubMed]

141. Schot, G.V.D.; Svenda, M.; Maia, F.R.N.C.; Hantke, M.; DePonte, D.P.; Seibert, M.M.; Aquila, A.; Schulz, J.; Kirian, R.; Liang, M.; et al. Imaging single cells in a beam of live cyanobacteria with an X-ray laser. *Nat. Commun.* **2015**, *6*, 5704. [CrossRef] [PubMed]

142. Fan, J.; Sun, Z.; Wang, Y.; Park, J.; Kim, S.; Gallagher-Jones, M.; Kim, Y.; Song, C.; Yao, S.; Zhang, J.; et al. Single-pulse enhanced coherent diffraction imaging of bacteria with an X-ray free-electron laser. *Sci. Rep.* **2016**, *6*, 34008. [CrossRef] [PubMed]

143. Bostedt, C.; Thomas, H.; Hoener, M.; Eremina, E.; Fennel, T.; Meiwes-Broer, K.H.; Wabnitz, H.; Kuhlmann, M.; Plönjes, E.; Tiedtke, K.; et al. Multistep Ionization of Argon Clusters in Intense Femtosecond Extreme Ultraviolet Pulses. *Phys. Rev. Lett.* **2008**, *100*, 133401. [CrossRef] [PubMed]

144. Sébastien, B.; Garth, J.W. The Coherent X-ray Imaging (CXI) instrument at the Linac Coherent Light Source (LCLS). *New J. Phys.* **2010**, *12*, 035024. [CrossRef]

145. Ferguson, K.R.; Bucher, M.; Bozek, J.D.; Carron, S.; Castagna, J.-C.; Coffee, R.; Curiel, G.I.; Holmes, M.; Krzywinski, J.; Messerschmidt, M.; et al. The Atomic, Molecular and Optical Science instrument at the Linac Coherent Light Source. *J. Synchrotron Radiat.* **2015**, *22*, 492–497. [CrossRef] [PubMed]

146. Liang, M.; Williams, G.J.; Messerschmidt, M.; Seibert, M.M.; Montanez, P.A.; Hayes, M.; Milathianaki, D.; Aquila, A.; Hunter, M.S.; Koglin, J.E.; et al. The Coherent X-ray Imaging instrument at the Linac Coherent Light Source. *J. Synchrotron Radiat.* **2015**, *22*, 514–519. [CrossRef] [PubMed]

147. Coherent X-ray Imaging (CXI). Available online: https://lcls.slac.stanford.edu/instruments/cxi (accessed on 15 December 2017).

148. Atomic, Molecular & Optical Science (AMO). Available online: https://lcls.slac.stanford.edu/instruments/amo (accessed on 15 December 2017).

149. Starodub, D.; Rez, P.; Hembree, G.; Howells, M.; Shapiro, D.; Chapman, H.N.; Fromme, P.; Schmidt, K.; Weierstall, U.; Doak, R.B.; et al. Dose, exposure time and resolution in serial X-ray crystallography. *J. Synchrotron Radiat.* **2008**, *15*, 62–73. [CrossRef] [PubMed]

150. Spence, J.C.H. X-ray Lasers in Biology: Structure and Dynamics. *Adv. Imag. Electron Phys.* **2017**, *200*, 103–152. [CrossRef]

151. Pivovaroffa, M.J.; Biontaa, R.M.; Mccarvillea, T.J.; Souflia, R.; Stefanb, P.M. Soft X-ray Mirrors for the Linac Coherent Light Source. *Proc. SPIE* **2007**, *6705*, 670500. [CrossRef]

152. Soufli, R.; Pivovaroff, M.J.; Baker, S.L.; Robinson, J.C.; Gullikson, E.M.; McCarville, T.J.; Stefan, P.M.; Aquila, A.L.; Ayers, J.; McKernan, M.A.; et al. Development, characterization and experimental performance of X-ray optics for the LCLS free-electron laser. *Proc. SPIE* **2008**, *7077*, 707716. [CrossRef]

153. Ratner, D.; Brachmann, A.; Decker, F.J.; Ding, Y.; Dowell, D.; Emma, P.; Fisher, A.; Frisch, J.; Gilevich, S.; Huang, Z.; et al. Second and third harmonic measurements at the linac coherent light source. *Phys. Rev. ST Accel. Beams* **2011**, *14*, 060701. [CrossRef]

154. Barty, A.; Soufli, R.; McCarville, T.; Baker, S.L.; Pivovaroff, M.J.; Stefan, P.; Bionta, R. Predicting the coherent X-ray wavefront focal properties at the Linac Coherent Light Source (LCLS) X-ray free electron laser. *Opt. Express* **2009**, *17*, 15508–15519. [CrossRef] [PubMed]

155. Hau-Riege, S.P.; Pardini, T. The effect of electron transport on the characterization of X-ray free-electron laser pulses via ablation. *Appl. Phys. Lett.* **2017**, *111*, 144102. [CrossRef]

156. David, C.; Gorelick, S.; Rutishauser, S.; Krzywinski, J.; Vila-Comamala, J.; Guzenko, V.A.; Bunk, O.; Färm, E.; Ritala, M.; Cammarata, M.; et al. Nanofocusing of hard X-ray free electron laser pulses using diamond based Fresnel zone plates. *Sci. Rep.* **2011**, *1*. [CrossRef] [PubMed]

157. Feng, Y.; Feldkamp, J.M.; Fritz, D.M.; Cammarata, M.; Robert, A.; Caronna, C.; Lemke, H.T.; Zhu, D.; Lee, S.; Boutet, S.; et al. A single-shot intensity-position monitor for hard X-ray FEL sources. *Proc. SPIE* **2011**, *8140*, 81400Q. [CrossRef]

158. Hau-Riege, S.P.; London, R.A.; Szoke, A. Dynamics of biological molecules irradiated by short x-ray pulses. *Phys. Rev. B* **2004**, *69*, 051906. [CrossRef] [PubMed]

159. Ziaja, B.; van der Spoel, D.; Szöke, A.; Hajdu, J. Auger-electron cascades in diamond and amorphous carbon. *Phys. Rev. B* **2001**, *64*, 214104. [CrossRef]

160. Ziaja, B.; de Castro, A.R.B.; Weckert, E.; Möller, T. Modelling dynamics of samples exposed to free-electron-laser radiation with Boltzmann equations. *Eur. Phys. J. D* **2006**, *40*, 465–480. [CrossRef]

161. Caleman, C.; Ortiz, C.; Marklund, E.; Bultmark, F.; Gabrysch, M.; Parak, F.G.; Hajdu, J.; Klintenberg, M.; Tîmneanu, N. Radiation damage in biological material: Electronic properties and electron impact ionization in urea. *EPL* **2009**, *88*, 29901. [CrossRef]

162. Young, L.; Kanter, E.P.; Krässig, B.; Li, Y.; March, A.M.; Pratt, S.T.; Santra, R.; Southworth, S.H.; Rohringer, N.; DiMauro, L.F.; et al. Femtosecond electronic response of atoms to ultra-intense X-rays. *Nature* **2010**, *466*, 56–61. [CrossRef] [PubMed]

163. Garman, E.F.; Weik, M. Radiation damage to biological macromolecules: Some answers and more questions. *J. Synchrotron Radiat.* **2013**, *20*, 1–6. [CrossRef] [PubMed]

164. Yuan, L.; Yang, M.; Qu, F.; Shen, G.; Yu, R. Seed-mediated growth of platinum nanoparticles on carbon nanotubes for the fabrication of electrochemical biosensors. *Electrochim. Acta* **2008**, *53*, 3559–3565. [CrossRef]

165. Zhong, B.; Kikuchi, A.; Moriyasu, Y.; Higashi, T.; Hagiwara, K.; Omura, T. A minor outer capsid protein, P9, of Rice dwarf virus. *Arch. Virol.* **2003**, *148*, 2275–2280. [CrossRef] [PubMed]

166. Nakagawa, A.; Miyazaki, N.; Taka, J.; Naitow, H.; Ogawa, A.; Fujimoto, Z.; Mizuno, H.; Higashi, T.; Watanabe, Y.; Omura, T.; et al. The Atomic Structure of Rice dwarf Virus Reveals the Self-Assembly Mechanism of Component Proteins. *Structure* **2003**, *11*, 1227–1238. [CrossRef] [PubMed]

167. Coetzee, J.N.; Lecatsas, G.; Coetzee, W.F.; Hedges, R.W. Properties of R plasmid R772 and the corresponding pilus-specific phage PR772. *J. Gen. Microbiol.* **1979**, *110*, 263–273. [CrossRef] [PubMed]

168. Toropova, K.; Basnak, G.; Twarock, R.; Stockley, P.G.; Ranson, N.A. The Three-dimensional Structure of Genomic RNA in Bacteriophage MS2: Implications for Assembly. *J. Mol. Biol.* **2008**, *375*, 824–836. [CrossRef] [PubMed]

169. Martín, C.S.; Burnett, R.M.; de Haas, F.; Heinkel, R.; Rutten, T.; Fuller, S.D.; Butcher, S.J.; Bamford, D.H. Combined EM/X-Ray Imaging Yields a Quasi-Atomic Model of the Adenovirus-Related Bacteriophage PRD1 and Shows Key Capsid and Membrane Interactions. *Structure* **2001**, *9*, 917–930. [CrossRef]

170. Worm, S.H.E.V.D.; Stonehouse, N.J.; Valegård, K.; Murray, J.B.; Walton, C.; Fridborg, K.; Stockley, P.G.; Liljas, L. Crystal structures of MS2 coat protein mutants in complex with wild-type RNA operator fragments. *Nucleic Acids Res.* **1998**, *26*, 1345–1351. [CrossRef] [PubMed]

171. Rose, A.S.; Bradley, A.R.; Valasatava, Y.; Duarte, J.M.; Prlić, A.; Rose, P.W. Web-based molecular graphics for large complexes. In Proceedings of the 21st International Conference on Web3D Technology, Anaheim, CA, USA, 22–24 July 2016; pp. 185–186.

172. Rose, A.S.; Hildebrand, P.W. NGL Viewer: A web application for molecular visualization. *Nucleic Acids Res.* **2015**, *43*, W576–W579. [CrossRef] [PubMed]

173. DePonte, D.P. Gas dynamic virtual nozzle for generation of microscopic droplet streams. *J. Phys. D Appl Phys.* **2008**, *41*, 195505. [CrossRef]

174. Sierra, R.G. Nanoflow electrospinning serial femtosecond crystallography. *Acta Cryst. D* **2012**, *68*, 1584–1587. [CrossRef] [PubMed]

175. Kirian, R.A.; Awel, S.; Eckerskorn, N.; Fleckenstein, H.; Wiedorn, M.; Adriano, L.; Bajt, S.; Barthelmess, M.; Bean, R.; Beyerlein, K.R.; et al. Simple convergent-nozzle aerosol injector for single-particle diffractive imaging with X-ray free-electron lasers. *Struct. Dyn.* **2015**, *2*, 041717. [CrossRef] [PubMed]

176. The Nobel Prize in Chemistry 2017. Available online: https://www.nobelprize.org/nobel_prizes/chemistry/laureates/2017/ (accessed on 15 October 2017).

177. Fernandez-Leiro, R.; Scheres, S.H.W. Unravelling biological macromolecules with cryo-electron microscopy. *Nature* **2016**, *537*, 339–346. [CrossRef] [PubMed]

178. Cheng, Y. Single-particle cryo-EM at crystallographic resolution. *Cell* **2015**, *161*, 450–457. [CrossRef] [PubMed]

179. Wiedorn, M.O.; Awel, S.; Morgan, A.J.; Barthelmess, M.; Bean, R.; Beyerlein, K.R.; Chavas, L.M.G.; Eckerskorn, N.; Fleckenstein, H.; Heymann, M.; et al. Post-sample aperture for low background diffraction experiments at X-ray free-electron lasers. *J. Synchrotron Radiat.* **2017**, *24*, 1296–1298. [CrossRef] [PubMed]

180. Munke, A.; Andreasson, J.; Aquila, A.; Awel, S.; Ayyer, K.; Barty, A.; Bean, R.J.; Berntsen, P.; Bielecki, J.; Boutet, S.; et al. Coherent diffraction of single Rice Dwarf virus particles using hard X-rays at the Linac Coherent Light Source. *Sci. Data* **2016**, *3*, 160064. [CrossRef] [PubMed]

181. Strüder, L.; Epp, S.; Rolles, D.; Hartmann, R.; Holl, P.; Lutz, G.; Soltau, H.; Eckart, R.; Reich, C.; Heinzinger, K.; et al. Large-format, high-speed, X-ray pnCCDs combined with electron and ion imaging spectrometers in a multipurpose chamber for experiments at 4th generation light sources. *Nucl. Instrum. Methods Phys. Res. A* **2010**, *614*, 483–496. [CrossRef]

182. Blaj, G.; Caragiulo, P.; Carini, G.; Carron, S.; Dragone, A.; Freytag, D.; Haller, G.; Hart, P.; Hasi, J.; Herbst, R.; et al. X-ray detectors at the Linac Coherent Light Source. *J. Synchrotron Radiat.* **2015**, *22*, 577–583. [CrossRef] [PubMed]

183. Blaj, G.; Caragiulo, P.; Carini, G.; Dragone, A.; Haller, G.; Hart, P.; Hasi, J.; Herbst, R.; Kenney, C.; Markovic, B.; et al. Future of ePix detectors for high repetition rate FELs. *AIP Conf. Proc.* **2016**, *1741*, 040012. [CrossRef]

184. Blaj, G.; Caragiulo, P.; Carini, G.; Carron, S.; Dragone, A.; Freytag, D.; Haller, G.; Hart, P.; Herbst, R.; Herrmann, S.; et al. Detector Development for the Linac Coherent Light Source. *Synchrotron Radiat. News* **2014**, *27*, 14–19. [CrossRef]

185. Greiffenberg, D. The AGIPD detector for the European XFEL. *J. Instrum.* **2012**, *7*, C01103. [CrossRef]

186. Allahgholi, A.; Becker, J.; Bianco, L.; Delfs, A.; Dinapoli, R.; Goettlicher, P.; Graafsma, H.; Greiffenberg, D.; Hirsemann, H.; Jack, S.; et al. AGIPD, a high dynamic range fast detector for the European XFEL. *J. Instrum.* **2015**, *10*, C01023. [CrossRef]

187. Kameshima, T.; Ono, S.; Kudo, T.; Ozaki, K.; Kirihara, Y.; Kobayashi, K.; Inubushi, Y.; Yabashi, M.; Horigome, T.; Holland, A.; et al. Development of an X-ray pixel detector with multi-port charge-coupled device for X-ray free-electron laser experiments. *Rev. Sci. Instrum.* **2014**, *85*, 033110. [CrossRef] [PubMed]

188. Jungmann-Smith, J.H.; Bergamaschi, A.; Bruckner, M.; Cartier, S.; Dinapoli, R.; Greiffenberg, D.; Huthwelker, T.; Maliakal, D.; Mayilyan, D.; Medjoubi, K.; et al. Towards hybrid pixel detectors for energy-dispersive or soft X-ray photon science. *J. Synchrotron Radiat.* **2016**, *23*, 385–394. [CrossRef] [PubMed]

189. Yoon, C.H.; Yurkov, M.V.; Schneidmiller, E.A.; Samoylova, L.; Buzmakov, A.; Jurek, Z.; Ziaja, B.; Santra, R.; Loh, N.D.; Tschentscher, T.; et al. A comprehensive simulation framework for imaging single particles and biomolecules at the European X-ray Free-Electron Laser. *Sci. Rep.* **2016**, *6*, 24791. [CrossRef] [PubMed]

190. Fortmann-Grote, C.; Buzmakov, A.; Jurek, Z.; Loh, N.-T.D.; Samoylova, L.; Santra, R.; Schneidmiller, E.A.; Tschentscher, T.; Yakubov, S.; Yoon, C.H.; et al. Start-to-end simulation of single-particle imaging using ultra-short pulses at the European X-ray Free-Electron Laser. *IUCrJ* **2017**, *4*, 560–568. [CrossRef] [PubMed]

191. Hantke, M.F.; Ekeberg, T.; Maia, F.R.N.C. Condor: A simulation tool for flash X-ray imaging. *J. Appl. Cryst.* **2016**, *49*, 1356–1362. [CrossRef] [PubMed]

192. Thayer, J.; Damiani, D.; Ford, C.; Gaponenko, I.; Kroeger, W.; O'Grady, C.; Pines, J.; Tookey, T.; Weaver, M.; Perazzo, A. Data systems for the Linac Coherent Light Source. *J. Appl. Cryst.* **2016**, *49*, 1363–1369. [CrossRef]

193. Damiani, D. Linac Coherent Light Source data analysis using psana. *J. Appl. Cryst.* **2016**, *49*, 672–679. [CrossRef]

194. Daurer, B.J.; Hantke, M.F.; Nettelblad, C.; Maia, F.R.N.C. Hummingbird: Monitoring and analyzing flash X-ray imaging experiments in real time. *J. Appl. Cryst.* **2016**, *49*, 1042–1047. [CrossRef] [PubMed]

195. Psocake SPI Tutorial. Available online: https://confluence.slac.stanford.edu/display/PSDM/Psocake+SPI+tutorial (accessed on 15 October 2017).

196. Coifman, R.R.; Lafon, S. Diffusion maps. *Appl. Comput. Harmon. Anal.* **2006**, *21*, 5–30. [CrossRef]

197. Reddy, H.K.N.; Yoon, C.H.; Aquila, A.; Awel, S.; Ayyer, K.; Barty, A.; Berntsen, P.; Bielecki, J.; Bobkov, S.; Bucher, M.; et al. Coherent soft X-ray diffraction imaging of coliphage PR772 at the Linac coherent light source. *Sci. Data* **2017**. [CrossRef] [PubMed]

198. Fung, R.; Shneerson, V.; Saldin, D.K.; Ourmazd, A. Structure from fleeting illumination of faint spinning objects in flight. *Nat. Phys.* **2008**, *5*, 64. [CrossRef]

199. Loh, N.D.; Bogan, M.J.; Elser, V.; Barty, A.; Boutet, S.; Bajt, S.; Hajdu, J.; Ekeberg, T.; Maia, F.R.N.C.; Schulz, J.; et al. Cryptotomography: Reconstructing 3D Fourier Intensities from Randomly Oriented Single-Shot Diffraction Patterns. *Phys. Rev. Lett.* **2010**, *104*, 225501. [CrossRef] [PubMed]

200. Moths, B.; Ourmazd, A. Bayesian algorithms for recovering structure from single-particle diffraction snapshots of unknown orientation: A comparison. *Acta Cryst. A* **2011**, *67*, 481–486. [CrossRef] [PubMed]

201. Tegze, M.; Bortel, G. Atomic structure of a single large biomolecule from diffraction patterns of random orientations. *J. Struct. Biol.* **2012**, *179*, 41–45. [CrossRef] [PubMed]

202. Giannakis, D.; Schwander, P.; Ourmazd, A. The symmetries of image formation by scattering. I. Theoretical framework. *Opt. Express* **2012**, *20*, 12799–12826. [CrossRef] [PubMed]

203. Schwander, P.; Giannakis, D.; Yoon, C.H.; Ourmazd, A. The symmetries of image formation by scattering. II. Applications. *Opt. Express* **2012**, *20*, 12827–12849. [CrossRef] [PubMed]

204. Saldin, D.K.; Poon, H.C.; Schwander, P.; Uddin, M.; Schmidt, M. Reconstructing an icosahedral virus from single-particle diffraction experiments. *Opt. Express* **2011**, *19*, 17318–17335. [CrossRef] [PubMed]

205. Ayyer, K.; Lan, T.-Y.; Elser, V.; Loh, N.D. Dragonfly: An implementation of the expand-maximize-compress algorithm for single-particle imaging. *J. Appl. Cryst.* **2016**, *49*, 1320–1335. [CrossRef] [PubMed]

206. Hosseinizadeh, A.; Mashayekhi, G.; Copperman, J.; Schwander, P.; Dashti, A.; Sepehr, R.; Fung, R.; Schmidt, M.; Yoon, C.H.; Hogue, B.G.; et al. Conformational landscape of a virus by single-particle X-ray scattering. *Nat. Methods* **2017**, *14*, 877–881. [CrossRef] [PubMed]

207. Kurta, R.P.; Donatelli, J.J.; Yoon, C.H.; Berntsen, P.; Bielecki, J.; Daurer, B.J.; DeMirci, H.; Fromme, P.; Hantke, M.F.; Maia, F.R.N.C.; et al. Correlations in Scattered X-Ray Laser Pulses Reveal Nanoscale Structural Features of Viruses. *Phys. Rev. Lett.* **2017**, *119*, 158102. [CrossRef] [PubMed]

208. Donatelli, J.J.; Zwart, P.H.; Sethian, J.A. Iterative phasing for fluctuation X-ray scattering. *Proc. Natl. Acad. Sci. USA* **2015**, *112*, 10286–10291. [CrossRef] [PubMed]

209. Gewin, V. Data sharing: An open mind on open data. *Nature* **2016**, *529*, 117–119. [CrossRef] [PubMed]

210. Ince, D.C.; Hatton, L.; Graham-Cumming, J. The case for open computer programs. *Nature* **2012**, *482*, 485–488. [CrossRef] [PubMed]

211. Anonymous. Announcement: Transparency upgrade for Nature journals. *Nature* **2017**, *543*, 288. [CrossRef]

212. Baker, M. Independent labs to verify high-profile papers. *Nature* **2012**. [CrossRef]

213. Anonymous. Announcement: Reducing our irreproducibility. *Nature* **2013**, *496*, 398. [CrossRef]

214. McNutt, M. Reproducibility. *Science* **2014**, *343*, 229. [CrossRef] [PubMed]

215. Allison, D.B.; Brown, A.W.; George, B.J.; Kaiser, K.A. Reproducibility: A tragedy of errors. *Nature* **2016**, *530*, 27–29. [CrossRef] [PubMed]

216. Loeb, A. Good data are not enough. *Nature* **2016**, *539*, 23–25. [CrossRef] [PubMed]

217. Daurer, B.J.; Okamoto, K.; Bielecki, J.; Maia, F.R.N.C.; Muhlig, K.; Seibert, M.M.; Hantke, M.F.; Nettelblad, C.; Benner, W.H.; Svenda, M.; et al. Experimental strategies for imaging bioparticles with femtosecond hard X-ray pulses. *IUCrJ* **2017**, *4*, 251–262. [CrossRef] [PubMed]

218. Ekeberg, T.; Svenda, M.; Seibert, M.M.; Abergel, C.; Maia, F.R.N.C.; Seltzer, V.; DePonte, D.P.; Aquila, A.; Andreasson, J.; Iwan, B.; et al. Single-shot diffraction data from the Mimivirus particle using an X-ray free-electron laser. *Sci. Data* **2016**, *3*, 160060. [CrossRef] [PubMed]

219. Hantke, M.F.; Hasse, D.; Ekeberg, T.; John, K.; Svenda, M.; Loh, D.; Martin, A.V.; Timneanu, N.; Larsson, D.S.D.; van der Schot, G.; et al. A data set from flash X-ray imaging of carboxysomes. *Sci. Data* **2016**, *3*, 160061. [CrossRef] [PubMed]

220. Van der Schot, G.; Svenda, M.; Maia, F.R.N.C.; Hantke, M.F.; DePonte, D.P.; Seibert, M.M.; Aquila, A.; Schulz, J.; Kirian, R.A.; Liang, M.; et al. Open data set of live cyanobacterial cells imaged using an X-ray laser. *Sci. Data* **2016**, *3*, 160058. [CrossRef] [PubMed]

221. Maia, F.R.N.C. The Coherent X-ray Imaging Data Bank. *Nat. Methods* **2012**, *9*, 854–855. [CrossRef] [PubMed]

222. Coherent X-Ray Imaging Data Bank (CXIDB). Available online: http://www.cxidb.org (accessed on 15 October 2017).

223. Dragonfly. Available online: https://github.com/duaneloh/Dragonfly/wiki (accessed on 15 October 2017).

224. Condor (Online Tool for CXI Pattern Simulation). Available online: http://lmb.icm.uu.se/condor/simulation/ (accessed on 15 October 2017).

applied sciences

MDPI

Article

Probing Dynamics in Colloidal Crystals with Pump-Probe Experiments at LCLS: Methodology and Analysis

Nastasia Mukharamova [1], Sergey Lazarev [1,2], Janne-Mieke Meijer [3,9], Matthieu Chollet [4], Andrej Singer [5], Ruslan P. Kurta [1,10], Dmitry Dzhigaev [1], Oleg Yu. Gorobtsov [1], Garth Williams [4,11], Diling Zhu [4], Yiping Feng [4], Marcin Sikorski [4,10], Sanghoon Song [4], Anatoly G. Shabalin [1,12], Tatiana Gurieva [1], Elena A. Sulyanova [6], Oleksandr M. Yefanov [7] and Ivan A. Vartanyants [1,8,*]

[1] Deutsches Elektronen-Synchrotron DESY, Notkestrasse 85, D-22607 Hamburg, Germany; nastasia.mukharamova@desy.de (N.M.); sergey.lazarev@desy.de (S.L.); ruslan.kurta@xfel.eu (R.P.K.); dmitry.dzhigaev@desy.de (D.D.); oleg.gorobtsov@desy.de (O.Y.G.); a.shabalin.r@gmail.com (A.G.S.); tatyana.guryeva@desy.de (T.G.)
[2] National Research Tomsk Polytechnic University (TPU), Pr. Lenina 30, 634050 Tomsk, Russia
[3] Van 't Hoff Laboratory for Physical and Colloid Chemistry, Debye Institute for Nanomaterials Science, Utrecht University, Padualaan 8, 3508 TB Utrecht, The Netherlands; janne-mieke.meijer@uni-konstanz.de
[4] SLAC National Accelerator Laboratory, 2575 Sand Hill Rd, Menlo Park, CA 94025, USA; mchollet@slac.Stanford.edu (M.C.); gwilliams@bnl.gov (G.W.); dlzhu@slac.stanford.edu (D.Z.); yfeng@slac.stanford.edu (Y.F.); marcin.sikorski@xfel.eu (M.S.); sanghoon@slac.stanford.edu (S.S.)
[5] UC San Diego, 9500 Gilman Dr., La Jolla, CA 92093, USA; singer.andrej@gmail.com
[6] Shubnikov Institute of Crystallography RAS, Leninskii Pr. 59, 119333 Moscow, Russia; sulyanova@gmail.com
[7] Center for Free-Electron Laser Science, DESY, Notkestraße 85, D-22607 Hamburg, Germany; oleksandr.yefanov@cfel.de
[8] National Research Nuclear University MEPhI, Kashirskoye Ch. 31, 115409 Moscow, Russia
[9] Present address: Department of Physics, University of Konstanz, D-78457 Konstanz, Germany
[10] Present address: European XFEL GmbH, Holzkoppel 4, D-22869 Schenefeld, Germany
[11] Present address: NSLS-II, Brookhaven National Laboratory, Upton, NY 11973-5000, USA
[12] Present address: UC San Diego, 9500 Gilman Dr., La Jolla, CA 92093, USA
* Correspondence: ivan.vartaniants@desy.de; Tel.: +49-040-8998-2653

Academic Editor: Kiyoshi Ueda
Received: 4 April 2017; Accepted: 8 May 2017; Published: 19 May 2017

Abstract: We present results of the studies of dynamics in colloidal crystals performed by pump-probe experiments using an X-ray free-electron laser (XFEL). Colloidal crystals were pumped with an infrared laser at a wavelength of 800 nm with varying power and probed by XFEL pulses at an energy of 8 keV with a time delay up to 1000 ps. The positions of the Bragg peaks, and their radial and azimuthal widths were analyzed as a function of the time delay. The spectral analysis of the data did not reveal significant enhancement of frequencies expected in this experiment. This allowed us to conclude that the amplitude of vibrational modes excited in colloidal crystals was less than the systematic error caused by the noise level.

Keywords: free-electron laser; pump-probe; colloidal crystal; dynamics

1. Introduction

Acoustic motion in nanoscale objects induced by light is of importance for both fundamental science and applications. This interest is due to various potential applications in acousto-optical devices which can be used for ultrafast manipulation and control of electromagnetic waves by hypersonic (GHz) acoustic waves [1]. Colloidal crystals, formed by the self-assembly method, have shown phononic band gaps in the GHz frequency range [2–7]. Due to the recent progress in the fabrication of high-quality colloids, this provides an opportunity to produce inexpensive phononic crystals.

Dynamics in submicrometer colloidal crystals was investigated using different techniques such as Raman scattering [8], Brillouin light scattering [2], and optical pump-probe spectroscopy [4–7]. However, these methods have limited spatial resolution and are not suitable for the observation of the detailed structure of the sample. Furthermore, for some materials obtaining a sufficient refractive index is a challenging task, if not impossible.

At the same time, newly-developed X-ray free-electron lasers (XFELs) produce X-rays with unprecedented brightness, pulse duration and coherence, which are ideal for performing time-resolved experiments on various materials with a time resolution that outperforms that of third-generation synchrotron sources [9–11]. Moreover, the femtosecond X-ray pulses from free-electron laser sources allow an investigation of transient states of time-dependent processes [12–17]. In this respect, XFELs are especially well suited for investigating ultrafast structural dynamics of colloidal crystals.

The vibrations of the colloidal crystals were recently measured at free-electron laser FLASH in Hamburg [18]. The polystyrene colloidal crystals were pumped by an infrared (IR) laser of 800 nm and probed by an X-ray pulse with a wavelength of 8 nm. This pump-probe experiment indicated the possibility of observation of the dynamics in the colloidal crystals by diffraction with a FEL. To give a reliable interpretation of the observed results, it was necessary to investigate the dynamics with a larger time interval and better statistics.

Here we present the results of a time-resolved pump-probe diffraction experiment in which the ultrafast dynamics in colloidal crystals induced by an IR laser was investigated. A theoretical analysis of the vibrations of the colloidal crystals was performed, and the vibrational frequency was calculated. This analysis was applied to the experimental results.

2. Experiment

The pump-probe experiments were performed at the Linac Coherent Light Source (LCLS) [9] in Stanford, USA at the X-ray Pump Probe (XPP) beamline [19]. LCLS was operated in the Self Amplified Spontaneous Emission (SASE) mode [20]. A sketch of the experiment is shown in Figure 1. A double-crystal diamond (111) monochromator with the thicknesses of the monochromator crystals of 100 µm and 300 µm, enabled splitting of the primary X-ray beam into a pink (transmitted) and a monochromatic (reflected) branches and simultaneous operation of two beamlines (see Figure 1) [21]. We employed the monochromatic regime of LCLS with the energy of a single XFEL pulse of 8 keV (1.5498 Å), energy bandwidth of 4.3×10^{-5} [22], and a pulse duration of about 50 fs as provided by electron bunch measurements at a repetition rate of 120 Hz.

The X-ray beam was focused using the compound refractive lenses (CRLs) on the sample down to 50 µm full width at half maximum (FWHM). The photon flux of the beam at the sample position was about 10^9 ph/pulse. This allowed us to perform measurements in a non-destructive regime. In the experiment, the colloidal crystal was oriented vertically, with its surface perpendicular to the incoming XFEL pulse in the transmission diffraction geometry.

Colloidal crystal films were prepared using the vertical deposition method [23] from spherical colloids consisting of two different materials, either polystyrene (PS) or silica. The crystalline films consisted of 30–40 particle layers, depending on the position on the film along the growth direction (the number of crystal layers increases at later growth times). After crystal growth, some of the films were covered with a 20–50 nm thick aluminum layer to increase interaction with the IR laser. In total, three types of samples were investigated: PS colloidal crystals, PS colloidal crystals covered by Al, and silica crystals, with different sphere diameters of 420 ± 9 nm, 376 ± 8 nm and 238 ± 7 nm, respectively. Due to the growing method, the dried colloidal crystal films used in our experiment contain cracks between single-crystal regions of about 50 µm in size. Microscopic images of these samples are shown in Figure 2.

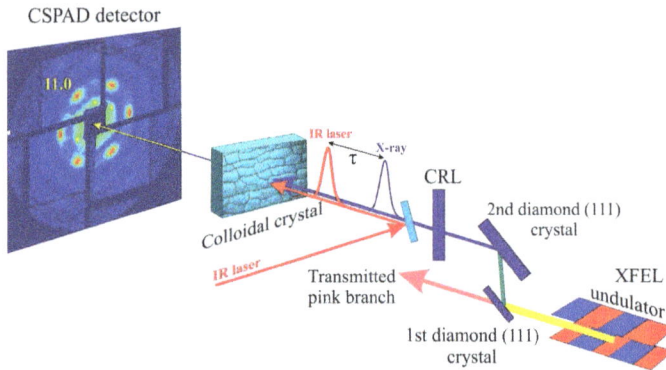

Figure 1. The scheme of the pump-probe experiment. X-ray free-electron laser (XFEL) pulses are generated by the undulator and reflected by the first diamond crystal. The second diamond crystal reflects the beam in the direction to the sample position. The beam is focused by the compound refractive lenses (CRLs) to the size of 50 µm at the sample position. The Cornell-SLAC Pixel Array Detector (CSPAD) is positioned 10 m downstream from the colloidal sample.

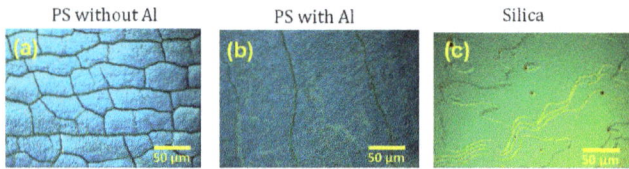

Figure 2. Microscopic images of colloidal crystals made from (**a**) polystyrene (PS); (**b**) PS covered with an Al layer; and (**c**) silica samples measured at Linac Coherent Light Source (LCLS).

Series of X-ray diffraction images were recorded using the Cornell-SLAC Pixel Array Detector (CSPAD) megapixel X-ray detector [24], positioned 10 m downstream from the sample, that is comprised of 32 silicon sensors with the pixel size of 110×110 µm^2 covering an area approximately 17×17 cm^2. This experimental arrangement provided resolution of 0.5 µm^{-1} per pixel in reciprocal space. The 2.3 megapixels of the detector were read out at 120 Hz, encoded at 14 bits per pixel. In order to prevent damage of the detector, a 5×5 mm^2 Al beamstop was placed in front of it. To reduce air scattering, a 10 m long flight tube filled with He was mounted between the sample and the detector.

A Ti:sapphire IR laser was used to excite the colloidal crystal film. The pump pulses were generated with a wavelength $\lambda = 800$ nm (1.5497 eV) and duration about 50 fs (FWHM). The size of the laser footprint on the sample was 100 µm (FWHM) and, thus, twice larger in comparison to the X-ray beam. The corresponding intensity of the laser pulses was on the order of 10^{13} W/cm^2" The IR laser pulses were propagating along the XFEL pulses and were synchronized to the pulses from the XFEL with less than 0.5 ps jitter.

The IR laser energy was controlled by rotation of the optical axis of a waveplate (see Figure 1), and was calibrated by power sensor at the position of the sample. The measured calibration curve is presented in Figure 3. Zero degrees of waveplate angle corresponds to the minimum and 10 degrees correspond to the maximum calibrated energy and intensity of the IR laser.

A series of pump-probe experiments were performed with the variation of the IR laser energy from 195 µJ to 691 µJ (3.2×10^{13}–1.1×10^{14} W/cm^2) for PS and silica samples and from 195 µJ to

326 μJ (3.2×10^{13}–5.4×10^{13} W/cm^2) for PS films covered by Al. Measurements with different IR laser energies were performed at the new position of the sample to avoid sample damage. The upper limit of IR laser energies was set due to the damage threshold of the colloidal crystals and was lower for Al-covered samples due to enhanced absorption of IR light by the Al film. The results for the highest non-destructive IR laser energies, which are marked with arrows on the calibration curve, are presented in this work. At energies higher than 1 mJ, ultrafast melting of the colloidal crystals was observed and was investigated in a separate work [25].

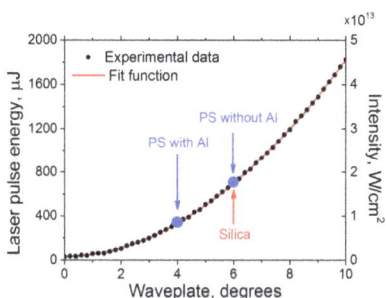

Figure 3. The infrared (IR) laser energy calibration curve. The maximum of the IR laser energy used in the experiments for three samples is marked with arrows. For the PS colloidal crystal covered with Al layer, the maximum IR laser energy was 326 μJ, and for the pure PS and silica samples, 691 μJ.

The pump-probe experiment was performed with a time delay variation τ from −50 ps to +1000 ps, with a 10 ps time increment. This time delay region and the time interval were considered in order to resolve GHz frequency dynamics of colloidal crystals. In order to obtain sufficient statistics of the measured data, for each time delay a large number of diffraction patterns (600 for polystyrene and 120 for silica) with and without the IR laser were measured. Typical single-shot diffraction patterns for three different samples are shown in Figure 4. The six-fold symmetry of the diffraction pattern is due to the hexagonal close-packed structure of the colloidal crystals. Two orders of Bragg peaks can be seen in single diffraction patterns shown in Figure 4. A family of 11.0 Bragg reflections, indicated in Figure 4 by squares, was used in further analysis. Indices of the structure are adopted from the hexagonal structure and the point symbolizes the fourth index equal to the negative sum of the first two indices (11.0 equals to 11$\bar{2}$0).

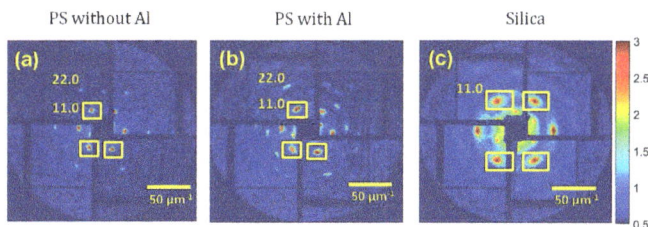

Figure 4. Single-shot diffraction patterns (in logarithmic scale) from colloidal crystals made from (**a**) PS; (**b**) PS covered with Al layer, and (**c**) silica samples measured at LCLS. Squares indicate Bragg peaks considered in the further analysis.

3. Data Analysis

The following strategy was implemented for the data analysis. First, for the patterns with sufficiently high intensity (more than 5000 counts per single diffraction pattern) of the XFEL pulse,

Appl. Sci. **2017**, 7, 519

we selected Bragg peaks located far away from the detector gaps (see Figure 4). Patterns with a lower signal were not treated. Using this procedure, typically three to four Bragg peaks were selected for the analysis (see Figure 4). Next, the Bragg peak positions were extracted using the center of mass of their intensity distributions as well as the relative distances between the opposite Bragg peaks. The change of the relative positions of the Bragg peaks as a function of the time delay allowed us to probe the lattice dynamics along specific crystal directions. The following parameters as a function of time delay were analyzed in order to reveal the dynamics of the colloidal crystals, induced by the IR laser: Bragg peak position $q(\tau)$, as well as FWHMs in the radial $w_q(\tau)$ and azimuthal $w_\phi(\tau)$ directions in reciprocal space. The temporal variation of the momentum transfer vector $q(\tau)$ is related to the dynamics in the inter-particle spacing, while $w_q(\tau)$ corresponds to the dynamics of the average particle size, and $w_\phi(\tau)$ defines angular misorientation of coherent scattering domains [26].

Due to the nature of the SASE process, each XFEL pulse has a unique fine spatial structure that was mapped by the intensity distribution at each Bragg position. A multi-spiked structure of the Bragg peaks varying from pulse to pulse was observed in our experiment (see Figure 5). From these diffraction patterns spatial and temporal coherence properties of hard XFEL could be determined by spatial correlation analysis [27]. Due to the different shape of each FEL pulse (see Figure 5), it was not possible to perform deconvolution of each scattered pulse shape. For the same reason (non-Gaussian structure of Bragg peaks) fitting of the peaks with the two-dimensional Gaussian function was not reliable (see Figure 6a). In order to improve Bragg peak characterization, projections of their intensities on azimuthal and radial directions were performed. These data were fitted with the one-dimensional Gaussian function (see Figure 6b), and peak broadening in the radial (w_q) and azimuthal (w_ϕ) directions was determined.

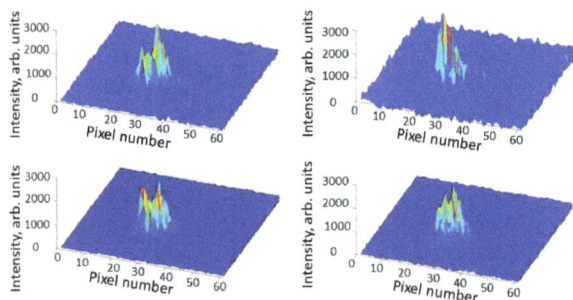

Figure 5. Four single-shot diffraction patterns measured at the same Bragg peak position. The spike-shaped peaks demonstrate an influence of the individual structure of each XFEL pulse on the Bragg peak.

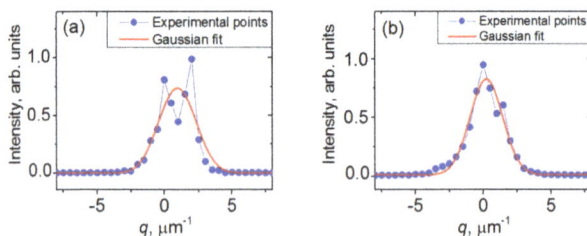

Figure 6. (**a**) The cross-section of a two-dimensional Bragg peak along azimuthal direction and (**b**) the projection of the same Bragg peak on the same direction . It is seen that the cross-section dataset (**a**) deviates from the Gaussian fit (red line), whereas the projection points are fitted with the Gaussian function with higher accuracy.

In order to compare dynamics of the collected data as a function of the time delay τ for different samples and IR laser energies, the following dimensionless parameters were used:

$$\frac{\Delta q(\tau)}{q} = \frac{\langle q_{on}(\tau)\rangle - \langle q_{off}\rangle}{\langle q_{off}\rangle} \tag{1}$$

$$\frac{\Delta w(\tau)}{w} = \frac{\langle w_{on}(\tau)\rangle - \langle w_{off}\rangle}{\langle w_{off}\rangle} \tag{2}$$

where $q(\tau)$ is the distance between the opposite Bragg peaks, and $w(\tau)$ is the radial or azimuthal FWHM of the Bragg peak. Subscripts "on" and "off" define measurements with and without the IR laser, respectively. Brackets $< \ldots >$ denote ensemble averaging of the chosen Bragg peak parameter for each time delay τ over all XFEL pulses. Time dependencies of the momentum transfer vector $q(\tau)$, as well as radial $w_q(\tau)$ and azimuthal $w_\phi(\tau)$ broadening of the Bragg peaks for three measured samples, are shown in Figure 7. Some points in this figure are excluded due to the unexpected drops of intensity of the XFEL (such drops of intensity were observed when operation of some clystrons in the accelerator complex was failing). The statistical error of these parameters was determined using the standard approach for the error determination in the multi-parameter equation [28]. As clearly seen in Figure 7, an error of the $\Delta q(\tau)/q$ values is one order of magnitude lower than for the Bragg peaks broadening parameters. This could be explained by the influence of the XFEL pulse shape on the shape of the Bragg peaks with a relatively stable position of each pulse. Due to this, the Gaussian fit of the Bragg peaks shape was not accurate, while the influence of the XFEL pulse shape on the variation of the center of mass was insignificant. The error values for the silica sample are higher than for the PS due to lower statistics (for the PS crystals 600 shots and for the silica crystals 120 shots were collected with the IR laser).

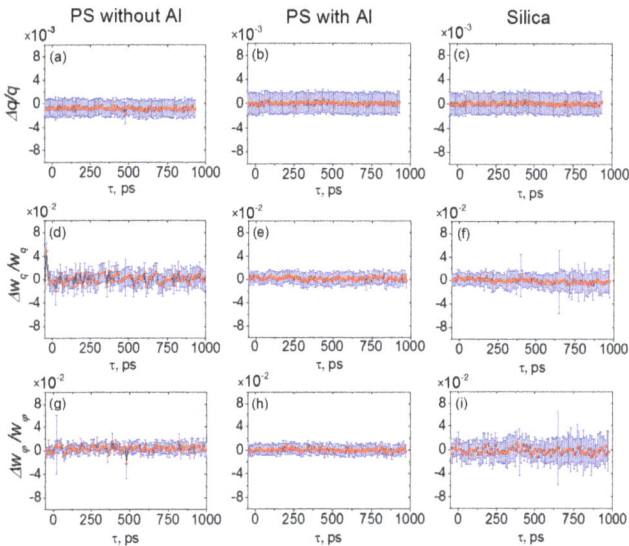

Figure 7. Time dependence of the relative change of the distance between the two opposite Bragg peaks $\Delta q(\tau)/q$ (**a–c**) and Bragg peaks widths in radial $\Delta w_q(\tau)/w_q$ (**d–f**) and azimuthal $\Delta w_\phi(\tau)/w_\phi$ (**g–i**) directions for three selected samples. Analysis of the data was performed for the IR pump laser intensities shown in Figure 3 and 11.0 Bragg peaks indicated by a square in Figure 4.

Figure 7 also clearly shows that variations of all parameters are around zero and in all cases, the signal is lower than the statistical error. To reveal characteristic frequencies excited by the IR laser, we performed the Fourier analysis of the time dependencies of the Bragg peak parameters described above. The corresponding Fourier spectra are shown in Figure 8. Due to the more precise measurements of the momentum transfer vector $q(\tau)$, the average value of the Fourier components of $q(\tau)$ is one order of magnitude lower than for the Bragg peaks broadening $w(\tau)$. There was no significant enhancement of any particular Fourier component, for all Fourier spectra. We determined an average value of the Fourier components and the standard deviation σ from the distribution of these Fourier components. For all Fourier spectra shown in Figure 8, there is no Fourier component higher than 3σ above the average Fourier component value, which indicates that the periodic signal could not be reliably detected in these conditions.

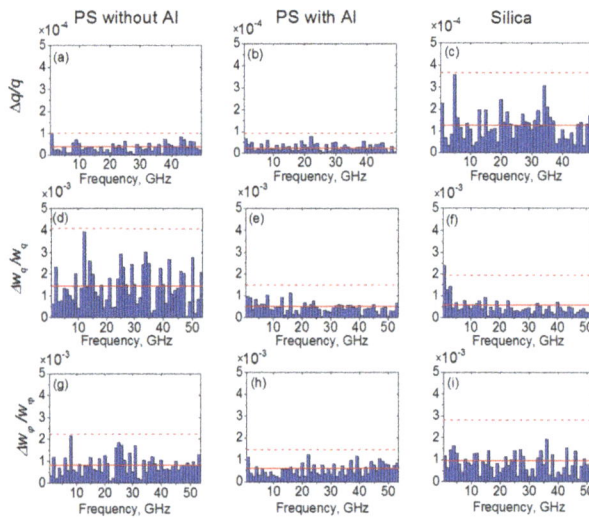

Figure 8. Fourier spectra from the distance between the two opposite Bragg peaks $\Delta q(\tau)/q$ (**a–c**) and Bragg peaks widths in radial $\Delta w_q(\tau)/w_q$ (**d–f**) and azimuthal $\Delta w_\phi(\tau)/w_\phi$ (**g–i**) directions for three selected samples. Fourier spectra calculated from the results of diffraction pattern analysis shown in Figure 7. Average value of all Fourier components is indicated by the red line. The dashed red line corresponds to the 3σ level above the average value, where σ is the standard deviation and is determined from the distribution of Fourier components.

4. Model Simulations

To have better understanding of the obtained results, we performed simulations of the colloidal crystal dynamics induced by an external IR pulse excitation. In our model colloidal crystals consist of isotropic homogeneous spheres. The characteristic frequencies of vibration of a colloidal particle were determined using the Lamb theory describing vibrations of an isotropic elastic sphere [29]. Theoretically, two families of Lamb modes, namely spheroidal and torsional ones, can be derived from the equations of motion. The spheroidal modes cause the change of the sphere volume, and the torsional modes leave the sphere volume unperturbed. During the IR pulse propagation through a sphere, the changes of the sphere volume are expected and, therefore, we used spheroidal modes in our simulations. We determined the frequencies of the first spheroidal mode for all three measured samples using reference [30]. Sphere diameter, longitudinal c_L and transverse c_T sound velocities of PS and silica were obtained from reference [31]. These parameters, as well as the frequencies of the first spheroidal mode for the colloidal crystals, are summarized in Table 1.

Table 1. Parameters of PS and silica nanoparticles used in our model. The corresponding frequencies of the first spheroidal mode of colloidal particles were obtained from reference [30]. The noise level was derived from the experimental data. Relative and absolute deviation in the interparticle distance ($\Delta d/d_0$ and Δd), which could be detected in our experiment with the probability of 60%, were calculated from the experimental data and simulations.

	PS without Al	PS with Al	Silica
Sphere diameter, nm	420	376	238
Longitudinal sound velocity c_L, m/s	2350	2350	5950
Transverse sound velocity c_T, m/s	1210	1210	3760
Frequency of the breathing mode, GHz	4.97	5.54	19.06
Noise level $N \times 10^{-3}$	1.4	1.8	1.8
Relative deviation in interparticle distance $\Delta d/d_0 \times 10^{-3}$	0.7	0.9	0.9
Deviation in interparticle distance Δd, nm	0.29	0.35	0.22

The values of $\Delta q(\tau)/q$ were obtained from the experimental data with the smallest error in comparison to other Bragg peak parameters for all investigated samples (see Figure 7a–c). Therefore, the parameter $\Delta q(\tau)/q$ was considered for our simulations. Assuming an ideal close-packed crystal with the interparticle distances equal to the particles diameter, the unperturbed momentum transfer modulus of the 11.0 Bragg reflection can be described as $q_0 = 2\pi/d_{11.0} = 4\pi/d_0$, where d_0 is the unperturbed particle diameter. We simulated vibrations of the colloidal spheres with the diameter d_0 as a periodic sinusoidal signal:

$$d(\tau) = d_0(1 - S \cdot \sin(\omega\tau)) \tag{3}$$

where S is the relative amplitude and ω is the characteristic frequency of the vibration. The dynamics of the experimentally-observed 11.0 diffraction peak can be described as:

$$q(\tau) = \frac{4\pi}{d(\tau)} \approx \frac{4\pi}{d_0}(1 + S \cdot \sin(\omega\tau)) \tag{4}$$

where we used the Taylor series expansion of the momentum transfer vector $q(\tau)$ due to the small values of the amplitude $S \ll 1$.

As a result, measured changes in the Bragg peak positions and $\Delta q(\tau)/q$ can be simulated as a periodic sinusoidal signal. We should also take into account systematic errors of the measurement as an additional noise. Finally, simulations of the parameter $\Delta q(\tau)/q$ in our model were described using the following expression:

$$\frac{\Delta q(\tau)}{q_0} = \frac{q(\tau) - q_0}{q_0} = S \cdot \sin(\omega\tau) + N \cdot \eta_{noise}(\tau) \tag{5}$$

where N is the amplitude of a noise and $\eta_{noise}(\tau)$ is a random noise function with normal distribution and standard deviation value equal to one. The signal-to-noise ratio in this case is assumed to be equal to the ratio of the amplitudes S/N.

Next, we defined criterion when the periodic signal is detected by the algorithm in our simulations. Due to the random behavior of the noise function $\eta_{noise}(\tau)$, each realization of the signal $\Delta q(\tau)/q_0$ (see Equation (5)) for a constant S/N is different. As an example, we demonstrate this for the experimental data of the polystyrene sample without aluminum. This sample was considered as one with the lowest noise level in our experiment. The time interval and increment of the simulated signal $\Delta q(\tau)/q_0$ were chosen the same as in our pump-probe experiment. One of the typical realizations of the $\Delta q(\tau)/q_0$ with the signal level $S = 0.7 \times 10^{-3}$ and noise level $N = 1.4 \times 10^{-3}$ is shown in Figure 9a. From this figure it is difficult to conclude on the presence or absence of the signal.

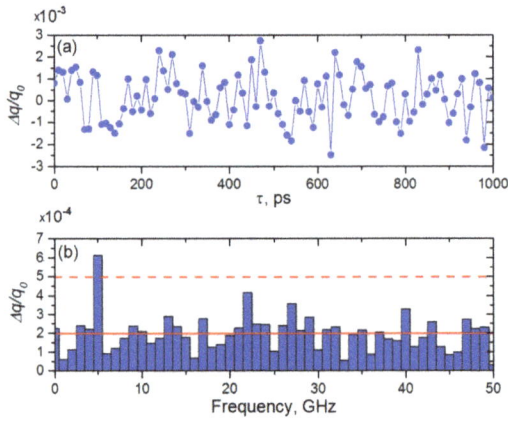

Figure 9. (a) Simulation of the signal $\Delta q(\tau)/q_0$ and **(b)** Fourier transform of this signal. The average value of all Fourier components is indicated by red line. The dashed red line corresponds to level of 3σ above the average value, where σ is a standard deviation and is determined from the distribution of Fourier components.

Similar to the analysis of the experimental data, the Fourier spectra were calculated for the simulated signal with the noise (see Figure 9b). Next, the behavior of the Fourier component amplitude (A_ω) corresponding to the first spheroidal mode frequency $\omega = 4.97$ GHz was studied. The criterion of detecting the signal was considered as:

$$A_\omega > \overline{A} + 3\sigma \tag{6}$$

where $\overline{A}$ is the mean and σ is the standard deviation of all Fourier components in the spectrum. The Fourier spectrum of the simulated data (see Figure 9a), as well as its 3σ level are shown in Figure 9b. It can be seen that the contribution of the vibrational mode in the Fourier spectrum is higher than the criterion stipulated by Equation (6). Therefore, the algorithm detected the presence of the 4.97 GHz frequency signal.

We calculated the probability of successful signal detection as a function of signal-to-noise ratio S/N in the range of (S/N) from 0 to 1.5 with 0.01 increment. During the simulations the amplitude of the signal S was changing, while the noise level N was kept constant. The probability of detecting signal $P(S/N)$ was defined as:

$$P(S/N) = \lim_{n \to \infty} \left(\frac{n_{detected}(S/N)}{n} \right) \tag{7}$$

where $n_{detected}(S/N)$ is a number of simulations with successful detection of signal (according to the criterion stipulated by Equation (6)) and $n = 10{,}000$ is the total number of simulations for each S/N step. Results of these simulations for the experimental data of the polystyrene sample without aluminum are shown in Figure 10. As one can see from Figure 10 that, according to our criterion stipulated by Equation (6), the signal is 100% detectable if it is on the level of the noise ($S/N = 1$).

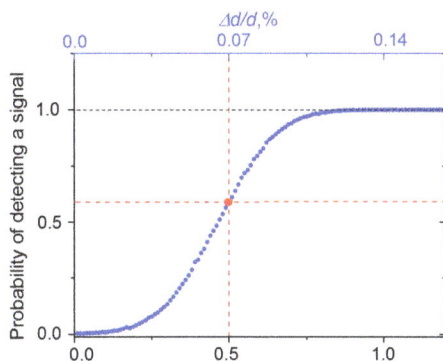

Figure 10. The probability of successful signal detection as a function of signal-to-noise ratio (S/N). On the top horizontal axis, the change of the corresponding interparticle distance for the polystyrene sample without Al is shown. The red lines correspond to the value of $S/N = 0.5$ and $P(S/N) = 0.6$.

5. Results and Discussion

Simulations, described in the previous section, were performed for each sample (polystyrene sample without aluminum, polystyrene sample with aluminum, and silica samples) to derive the probability function $P(S/N)$ of detecting the periodic signal (see Figure 10) in each case. It is clearly visible from Figure 10 that even the value of the signal-to-noise ratio $S/N = 0.5$ gives 60% probability of detecting the signal. This condition was considered as a criterion for the minimum measurable signal. Using well-known relations between the deviations of the momentum transfer vectors and interparticle distances $\Delta q(\tau)/q_0 = -\Delta d(\tau)/d_0$, as well as Equation (5), we can determine the relative (and absolute) values of the changes of the interparticle distances $\Delta d(\tau)/d_0$ ($\Delta d(\tau)$) for the given values of the experimental noise N. As a result of this analysis, we can obtain the minimum values of these parameters which could be detected in our experimental conditions with the probability of 60%. For all measured samples results of this analysis are summarized in Table 1.

As it follows from Table 1, the minimum relative changes of the interparticle distance $\Delta d/d_0$ that can be detected with the probability of 60% for all samples were about 10^{-3}. Such values correspond to relatively large amplitudes of the deviations in interparticle distances of about 0.2–0.4 nm. Clearly, in our experimental conditions excited vibrations were lower than the noise level of the experiment that prevented us from their experimental determination.

For the successful measurement of vibrations of colloidal crystals at the FEL sources, some further improvements of the experiment may be suggested. First, the accuracy of the measurement of the incoming X-ray beam intensity from pulse to pulse should be improved, as this would allow measurement of the Bragg peak intensity with higher precision. In our experiment such an accurate measurement of the Bragg peak intensity variation was not possible due to the complicated structure of each XFEL pulse and a relatively low accuracy (about 3%) in the measurement of the incoming X-ray beam intensity from pulse to pulse. Second, colloidal crystals may be covered by a thicker aluminum layer which could lead to a more efficient energy transfer from an IR laser to the colloidal crystal [32]. This would lead to a higher amplitude of vibrations of a PS colloidal crystal, and enable their detection in IR pump and X-ray probe experiments.

6. Conclusions

We performed experiments at the LCLS to investigate dynamics in the colloidal crystals. The diffraction patterns from the colloidal crystals, which were pumped by IR laser pulses and probed by XFEL radiation, were collected. The measurements were performed in a non-destructive regime at the different positions of the samples. The pump-probe experiments were performed in the wide range of the IR laser energies, from 195 µJ up to the upper limit of the non-destructive threshold of 691 µJ.

Through the analysis of the Bragg peaks, extracted from the diffraction patterns, changes in the colloidal crystals caused by the IR laser were investigated. The dynamics at different timescales was studied by the Fourier analysis of parameters associated with the momentum transfer and Bragg peak broadening in the radial and angular directions. Simulations of the colloidal crystal dynamics were performed and the probability function $P(S/N)$ was derived. Relative and absolute deviations in the interparticle distance ($\Delta d/d_0$ and Δd), which could be detected in such an experiment with the probability of 60% (corresponding to the signal to noise level of 0.5), were calculated for all measured samples. From the experiment and simulations, we were able to conclude that, for all measured samples, the amplitudes of vibrational modes excited in colloidal crystals were less than the systematic error caused by the noise level.

In future experiments with XFEL sources different scattering geometries may be foreseen. For example, performing scattering experiments on colloidal crystals in grazing incidence small-angle X-ray scattering (GISAXS) [33,34] may increase the sensitivity of the scattered radiation to changes induced by the IR pump laser.

Acknowledgments: We acknowledge fruitful discussions and support of the project by Edgar Weckert as well as help and support by Andrei V. Petukhov. We acknowledge the help of Ralf Röhlsberger in preparation of Al-covered colloidal crystals. We acknowledge careful reading of the manuscript by Tim Laarman. This work was supported by the Virtual Institute VH-VI-403 of the Helmholtz Association. The experimental work was carried out at the Linac Coherent Light Source (LCLS), a National User Facility operated by Stanford University on behalf of the U.S. Department of Energy, Office of Basic Energy Sciences. Use of the LCLS, SLAC National Accelerator Laboratory, is supported by the U.S. Department of Energy, Office of Science, and Office of Basic Energy Sciences under contract number DE-AC02-76SF00515. A.S. was supported by U.S. Department of Energy, Office of Science, Office of Basic Energy Sciences, under Contracts No. DE- SC0001805. For G.W. portions of this research were carried out at Brookhaven National Laboratory, operated under Contract No. DE-SC0012704 from the U.S. Department of Energy (DOE) Office of Science.

Author Contributions: M.C., A.S., R.P.K., O.M.Y., D.D., O.Y.G., G.W., D.Z., Y.F., M.S., S.S., and I.A.V. performed the experiments; J.-M.M. and T.G. prepared the colloidal crystal samples; I.A.V. supervised N.M. and the project; N.M., A.G.S., E.A.S., and S.L. developed the code; N.M. analyzed the data; and N.M. and S.L. wrote the paper.

Conflicts of Interest: The authors declare no conflict of interest.

References

1. Gorishnyy, T.; Ullal, C.K.; Maldovan, M.; Fytas, G.; Thomas, E. Hypersonic phononic crystals. *Phys. Rev. Lett.* **2005**, *94*, 115501. [CrossRef] [PubMed]
2. Cheng, W.; Wang, J.; Jonas, U.; Fytas, G.; Stefanou, N. Observation and tuning of hypersonic bandgaps in colloidal crystals. *Nat. Mater.* **2006**, *5*, 830. [CrossRef] [PubMed]
3. Still, T.; Cheng, W.; Retsch, M.; Sainidou, R.; Wang, J.; Jonas, U.; Stefanou, N.; Fytas, G. Simultaneous occurrence of structure-directed and particle-resonance-induced phononic gaps in colloidal films. *Phys. Rev. Lett.* **2008**, *100*, 194301. [CrossRef] [PubMed]
4. Thomas, E.L.; Gorishnyy, T.; Maldovan, M. Phononics: Colloidal crystals go hypersonic. *Nat. Mater.* **2006**, *5*, 773–774. [CrossRef] [PubMed]
5. Akimov, A.; Tanaka, Y.; Pevtsov, A.; Kaplan, S.; Golubev, V.; Tamura, S.; Yakovlev, D.; Bayer, M. Hypersonic modulation of light in three-dimensional photonic and phononic band-gap materials. *Phys. Rev. Lett.* **2008**, *101*, 033902. [CrossRef] [PubMed]
6. Salasyuk, A.S.; Scherbakov, A.V.; Yakovlev, D.R.; Akimov, A.V.; Kaplyanskii, A.A.; Kaplan, S.F.; Grudinkin, S.A.; Nashchekin, A.V.; Pevtsov, A.B.; Golubev, V.G.; et al. Filtering of elastic waves by opal-based hypersonic crystal. *Nano Lett.* **2010**, *10*, 1319–1323. [CrossRef] [PubMed]
7. Mazurenko, D.; Shan, X.; Stiefelhagen, J.; Graf, C.; Van Blaaderen, A.; Dijkhuis, J. Coherent vibrations of submicron spherical gold shells in a photonic crystal. *Phys. Rev. B* **2007**, *75*, 161102. [CrossRef]
8. Duval, E.; Boukenter, A.; Champagnon, B. Vibration eigenmodes and size of microcrystallites in glass: Observation by very-low-frequency Raman scattering. *Phys. Rev. Lett.* **1986**, *56*, 2052. [CrossRef] [PubMed]

9. Emma, P.; Akre, R.; Arthur, J.; Bionta, R.; Bostedt, C.; Bozek, J.; Brachmann, A.; Bucksbaum, P.; Coffee, R.; Decker, F.J.; et al. First lasing and operation of an ångstrom-wavelength free-electron laser. *Nat. Photonics* **2010**, *4*, 641–647. [CrossRef]

10. Altarelli, M.; Brinkmann, R.; Chergui, M.; Decking, W.; Dobson, B.; Dusterer, S.; Grubel, G.; Graeff, W.; Graafsma, H.; Hajdu, J.; et al. *XFEL The European X-ray Free-Electron Laser*; Technical Design Report, DESY Report No. 2006–097; DESY XFEL Project Group, European XFEL Project Team, Deutsches Elektronen-Synchrotron, Member of the Helmholtz Association: Hamburg, Germany, 2006; Available online: http://xfel.desy.de/technical_information/tdr/tdr/ (accessed on 17 May 2017).

11. Ishikawa, T.; Aoyagi, H.; Asaka, T.; Asano, Y.; Azumi, N.; Bizen, T.; Ego, H.; Fukami, K.; Fukui, T.; Furukawa, Y.; et al. A compact X-ray free-electron laser emitting in the sub-ångstrom region. *Nat. Photonics* **2012**, *6*, 540–544. [CrossRef]

12. McFarland, B.; Farrell, J.; Miyabe, S.; Tarantelli, F.; Aguilar, A.; Berrah, N.; Bostedt, C.; Bozek, J.; Bucksbaum, P.; Castagna, J.; et al. Ultrafast X-ray Auger probing of photoexcited molecular dynamics. *Nat. Commun.* **2014**, *5*. [CrossRef] [PubMed]

13. Trigo, M.; Fuchs, M.; Chen, J.; Jiang, M.; Cammarata, M.; Fahy, S.; Fritz, D.; Gaffney, K.; Ghimire, S.; Higginbotham, A.; et al. Fourier-transform inelastic X-ray scattering from time-and momentum-dependent phonon-phonon correlations. *Nat. Phys.* **2013**, *9*, 790–794. [CrossRef]

14. Kupitz, C.; Basu, S.; Grotjohann, I.; Fromme, R.; Zatsepin, N.A.; Rendek, K.N.; Hunter, M.S.; Shoeman, R.L.; White, T.A.; Wang, D.; et al. Serial time-resolved crystallography of photosystem II using a femtosecond X-ray laser. *Nature* **2014**, *513*, 261–265. [CrossRef] [PubMed]

15. Glownia, J.; Natan, A.; Cryan, J.; Hartsock, R.; Kozina, M.; Minitti, M.; Nelson, S.; Robinson, J.; Sato, T.; van Driel, T.; et al. Self-referenced coherent diffraction X-ray movie of Ångstrom-and femtosecond-scale atomic motion. *Phys. Rev. Lett.* **2016**, *117*, 153003. [CrossRef] [PubMed]

16. Abbey, B.; Dilanian, R.A.; Darmanin, C.; Ryan, R.A.; Putkunz, C.T.; Martin, A.V.; Wood, D.; Streltsov, V.; Jones, M.W.; Gaffney, N.; et al. X-ray laser-induced electron dynamics observed by femtosecond diffraction from nanocrystals of Buckminsterfullerene. *Sci. Adv.* **2016**, *2*, e1601186. [CrossRef] [PubMed]

17. Clark, J.; Beitra, L.; Xiong, G.; Higginbotham, A.; Fritz, D.; Lemke, H.; Zhu, D.; Chollet, M.; Williams, G.; Messerschmidt, M.; et al. Ultrafast three-dimensional imaging of lattice dynamics in individual gold nanocrystals. *Science* **2013**, *341*, 56–59. [CrossRef] [PubMed]

18. Dronyak, R.; Gulden, J.; Yefanov, O.; Singer, A.; Gorniak, T.; Senkbeil, T.; Meijer, J.M.; Al-Shemmary, A.; Hallmann, J.; Mai, D.; et al. Dynamics of colloidal crystals studied by pump-probe experiments at FLASH. *Phys. Rev. B* **2012**, *86*, 064303. [CrossRef]

19. Chollet, M.; Alonso-Mori, R.; Cammarata, M.; Damiani, D.; Defever, J.; Delor, J.T.; Feng, Y.; Glownia, J.M.; Langton, J.B.; Nelson, S.; et al. The X-ray pump–probe instrument at the Linac Coherent Light Source. *J. Synchrotron Radiat.* **2015**, *22*, 503–507. [CrossRef] [PubMed]

20. Saldin, E.L.; Schneidmiller, E.A.; Yurkov, M.V. *The Physics of Free Electron Lasers*; Springer: Berlin, Germany, 2000.

21. Stoupin, S.; Terentyev, S.; Blank, V.; Shvyd'ko, Y.V.; Goetze, K.; Assoufid, L.; Polyakov, S.; Kuznetsov, M.; Kornilov, N.; Katsoudas, J.; et al. All-diamond optical assemblies for a beam-multiplexing X-ray monochromator at the Linac Coherent Light Source. *J. Appl. Crystallogr.* **2014**, *47*, 1329–1336. [CrossRef] [PubMed]

22. Zhu, D.; Feng, Y.; Stoupin, S.; Terentyev, S.A.; Lemke, H.T.; Fritz, D.M.; Chollet, M.; Glownia, J.; Alonso-Mori, R.; Sikorski, M.; et al. Performance of a beam-multiplexing diamond crystal monochromator at the Linac Coherent Light Source. *Rev. Sci. Instrum.* **2014**, *85*, 063106. [CrossRef] [PubMed]

23. Meijer, J.-M. *Colloidal Crystals of Spheres and Cubes in Real and Reciprocal Space*; Springer: Berlin, Germany, 2015; pp. 23–29.

24. Hart, P.; Boutet, S.; Carini, G.; Dubrovin, M.; Duda, B.; Fritz, D.; Haller, G.; Herbst, R.; Herrmann, S.; Kenney, C.; et al. The CSPAD megapixel X-ray camera at LCLS. In *Proc. SPIE 8504, X-ray Free-Electron Lasers: Beam Diagnostics, Beamline Instrumentation, and Applications, 85040C*; Moeller, S.P., Yabashi, M., Hau-Riege, S.P., Eds.; SPIE: Bellingham, WA, USA, 2012.

25. Lazarev, S.; Mukharamova, N.; Gorobtsov, O.Y.; Chollet, M.; Kurta, R.P.; Meijer, J.-M.; Williams, G.; Zhu, D.; Feng, Y.; Sikorski, M.; et al. Ultrafast melting of the colloidal crystals. 2017, in preparation.

26. Sulyanova, E.A.; Shabalin, A.; Zozulya, A.V.; Meijer, J.M.; Dzhigaev, D.; Gorobtsov, O.; Kurta, R.P.; Lazarev, S.; Lorenz, U.; Singer, A.; et al. Structural evolution of colloidal crystal films in the process of melting revealed by Bragg peak analysis. *Langmuir* **2015**, *31*, 5274–5283. [CrossRef] [PubMed]
27. Gorobtsov, O.Y.; Mukharamova, N.; Lazarev, S.; Chollet, M.; Kurta, R.P.; Meijer, J.-M.; Williams, G.; Zhu, D.; Feng, Y.; Sikorski, M.; et al. Intensity interferometry on crystal diffraction patterns at hard X-ray free electron laser. 2017, in preparation.
28. Hazewinkel, M. *Encyclopedia of Mathematics: An Updated and Annotated Translation of the Soviet Mathematical Encyclopedia*; Springer: Berlin, Germany, 2013; Volume 2.
29. Lamb, H. On the vibrations of an elastic sphere. *Proc. Lond. Math. Soc.* **1881**, *1*, 189–212. [CrossRef]
30. Saviot, L. Vibrational Eigenmodes of an Isotropic Sphere. Available online: http://lucien.saviot.free.fr/lamb/index.en.html (accessed on 30 March 2017).
31. Mark, J.E. *Polymer Data Handbook*; Oxford University Press: Oxford, UK, 2009; p. 834.
32. Tas, G; Maris, H.J. Electron diffusion in metals studied by picosecond ultrasonics. *Phys. Rev. B* **1994**, *49*, 15046. [CrossRef]
33. Huber, P.; Bunk, O.; Pietsch, U.; Textor, M.; Geue, T. Grazing Incidence Small Angle X-ray Scattering on Colloidal Crystals. *J. Phys. Chem. B* **2010**, *114*, 12473–12479. [CrossRef] [PubMed]
34. Zozulya, A.; Zaluzhnyy, I; Lazarev, S.; Meijer, J.-M.; Mukharamova, N.; Shabalin, A.; Kurta, R.P.; Sprung, M.; Petukhov, A.V.; Vartanyants, I.A. Temperature-driven rearrangement of colloidal crystal domains studied in situ by GISAXS. 2017, in preparation.

applied
sciences

MDPI

Review

Probing Physics in Vacuum Using an X-ray Free-Electron Laser, a High-Power Laser, and a High-Field Magnet

Toshiaki Inada [1],*, Takayuki Yamazaki [1],*, Tomohiro Yamaji [2], Yudai Seino [2], Xing Fan [2], Shusei Kamioka [2], Toshio Namba [1] and Shoji Asai [2]

[1] International Center for Elementary Particle Physics, The University of Tokyo, 7-3-1 Hongo, Bunkyo-ku, Tokyo 113-0033, Japan; naniwa@icepp.s.u-tokyo.ac.jp
[2] Department of Physics, Graduate School of Science, The University of Tokyo, 7-3-1 Hongo, Bunkyo-ku, Tokyo 113-0033, Japan; yamaji@icepp.s.u-tokyo.ac.jp (T.Y.); yseino@icepp.s.u-tokyo.ac.jp (Y.S.); xfan@icepp.s.u-tokyo.ac.jp (X.F.); kamioka@icepp.s.u-tokyo.ac.jp (S.K.); Shoji.Asai@cern.ch (S.A.)
* Correspondence: tinada@icepp.s.u-tokyo.ac.jp (T.I.); yamazaki@icepp.s.u-tokyo.ac.jp (T.Y.)

Academic Editor: Kiyoshi Ueda
Received: 31 March 2017; Accepted: 28 June 2017; Published: 29 June 2017

Abstract: A nonlinear interaction between photons is observed in a process that involves charge sources. To observe this process in a vacuum, there are a growing number of theoretical and experimental studies. This process may contain exotic contribution from new physics beyond the Standard Model of particle physics, and is probed by experiments using a high-power laser or a high-field magnet, and more recently using an X-ray free-electron laser (XFEL). Here, we review the present status of our experiments testing various vacuum processes. We describe four experiments with a focus on those using an XFEL: (i) photon–photon scattering in the X-ray region, (ii) laser-induced birefringence and diffraction of X rays, (iii) vacuum birefringence induced by a high-field magnet, and (iv) a dedicated search for axion-like particles using the magnet and X rays.

Keywords: quantum vacuum; strong-field QED; vacuum magnetic birefringence; axion; XFEL

1. Introduction

Photon–photon scattering is a nonlinear interaction between photons, intermediated by a virtual electron loop at the lowest order of quantum electrodynamics (QED) (Figure 1, left). The contribution of this diagram ubiquitously appears in radiative corrections to charged particles at higher orders (Figure 2a), and its agreement with measurements constitutes a solid basis for the Standard Model of particle physics at low energies [1–3]. In addition, some of these photons have been changed into real-state photons (Figure 2b–d) and observed in experiments [4–6]. These processes occur under a Coulomb field of pre-existing, real-state charges, and therefore can be classified as "real-charge processes". By contrast, there are many attempts to observe this diagram in a vacuum by using a PW-class high-power laser or using a high-field magnet. This kind of "vacuum process" occurs without such charge sources and cannot be removed by any effort, thus representing the *core* of all physical processes.

The vacuum process is currently studied in two streams. The first is based in high field science, where the behavior of dense plasma is studied using high-power lasers. They predict many kinds of vacuum phenomena occurring at the focal spot of an intense laser [7,8]. An efficient probe on these phenomena is provided by the recent progress in X-ray free-electron lasers (XFELs) due to their short wavelength, peak brilliance, and more particle nature compared to laser photons [9]. The second is an experimentally ongoing stream that studies the optical birefringence of a magnetized vacuum

(Figure 1, right). The measurement of vacuum magnetic birefringence (VMB) provides the current best sensitivity to the observation of the vacuum process [10].

The theoretical basis describing these vacuum processes can be formulated in terms of the Euler–Heisenberg Lagrangian, which effectively contains only electromagnetic fields. With assumptions that photon energies are well below the electron mass m_e and fields much smaller than the QED critical values ($E \ll 1.3 \times 10^{18}$ V m^{-1}, $B \ll 4.4 \times 10^9$ T), the Lagrangian density is written in natural units ($\hbar = c = 1$) as [11]

$$\mathcal{L} = \frac{1}{2}\left(E^2 - B^2\right) + \frac{2\alpha^2}{45m_e^4}\left[\left(E^2 - B^2\right)^2 + 7\left(\mathbf{E}\cdot\mathbf{B}\right)^2\right],\tag{1}$$

with the fine-structure constant α and electric and magnetic fields $\mathbf{E}$ and $\mathbf{B}$, respectively. The second term of Equation (1) describes four-field interactions, and theoretical studies on these effects have been summarized in References [7,8]. In addition to these nonlinear QED contexts, searches for vacuum processes concern a wider range of research fields, including astrophysics and particle physics. For instance, a practical effect of VMB has been suggested by the recent observation of a strongly magnetized neutron star [12–14]. Furthermore, the vacuum process may be easily affected by the contribution of new physics beyond the Standard Model (BSM) [15]. The conventional benchmark in a low-energy scale is the axion, and more generally, axion-like particles (ALPs) that have an effective coupling to two photons [16]. If such light particles exist, their mass spectrum would increase the diversity of vacuum interactions ascribed to electrons at low energies.

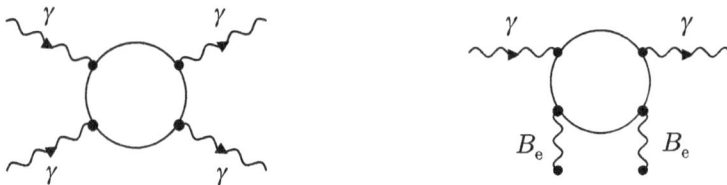

Figure 1. Vacuum processes of photon–photon interaction at the lowest-order quantum electrodynamics (QED). The time goes from left to right. (**Left**) Photon–photon scattering. (**Right**) Interaction of a photon with an external magnetic field that causes vacuum magnetic birefringence. A virtual photon that conveys an electric or magnetic field is shown by a vertical line.

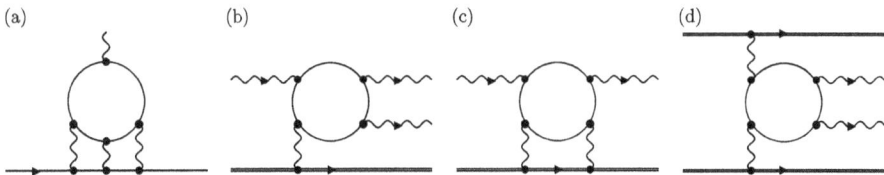

Figure 2. Real-charge processes containing a one-loop photon–photon interaction (see also Reference [9] and the references therein). (**a**) Radiative correction to the electron $g - 2$ at the third order on the fine structure constant, with all relevant photons virtual [1,3]. (**b**) Photon splitting with one virtual photon from a nuclear field [4]. (**c**) Delbrück scattering with two virtual photons [5]. (**d**) Pair production from the crossing of two nuclear fields [6].

To our best knowledge, we are currently the only group that experimentally studies the vacuum physics in both streams. In this article, we review the status and results of our first-phase experiments along with their technical aspects. Section 2 describes experiments probing the vacuum process; two experiments using the XFEL, SACLA (SPring-8 Angstrom Compact Free-Electron Laser) [17,18], and one for VMB using a pulsed magnet. BSM searches by these experiments are discussed in Section 3, particularly in the case of ALPs. A dedicated search using the pulsed magnet and XFEL is also described.

2. Probing the Vacuum Process

So far, no experiments have observed the vacuum process due to its extremely weak signal. Signal photons are characterized by (i) the frequency, (ii) the wave vector, and (iii) the polarization. Any vacuum process is finally detected as the changes in these observables. This simple fact results in a variety of macroscopic phenomena probed in experiments [7,8].

2.1. Elastic Scattering by the Collision of Two X-ray Pulses

This section reviews our experiments at SACLA that provide the first limits on the QED cross-section of photon–photon scattering ($\sigma_{\gamma\gamma}$) in the X-ray region [19,20]. Before the experiments, the scattering was tested by using optical lasers [21]. However, the cross-section is suppressed by the sixth power of the ratio $\hbar\omega/m_ec^2$, where ω is the photon energy and m_e the electron mass [22,23]. Therefore, a significant enhancement can be obtained by using high-energy X-ray photons.

It is not an easy task to collide X-ray pulses if they are delivered as two distinct beams. Thus, we developed a new scheme to divide a single beam into two and then to collide them. This scheme applies a technology established for an X-ray interferometer. A schematic of the beam splitter is shown in Figure 3a. A beam is split into two coherent beams—one is the transmitted beam (T) and the other is the reflected one (R)—by a 0.6 mm-thick blade of Si(440) crystal when the lattice plane matches a diffraction angle of $\theta_B = 36°$ [24]. A similar process occurs at the second blade, creating four beams after the two blades. The TR and RR beams collide obliquely with a crossing angle of $2\theta_B$, and the two-photon system is boosted along θ_B. The change of wave vectors in the center-of-mass system causes energy unbalance of the pairs observed in the laboratory frame. This is an example of a kinematically-induced frequency change occurring in oblique collisions, in contrast to one caused by optical nonlinearities such as four-wave mixing [25]. In this way, the diffraction condition itself ensures the geometrical equivalence of the beam paths split by the symmetric Laue reflection [26].

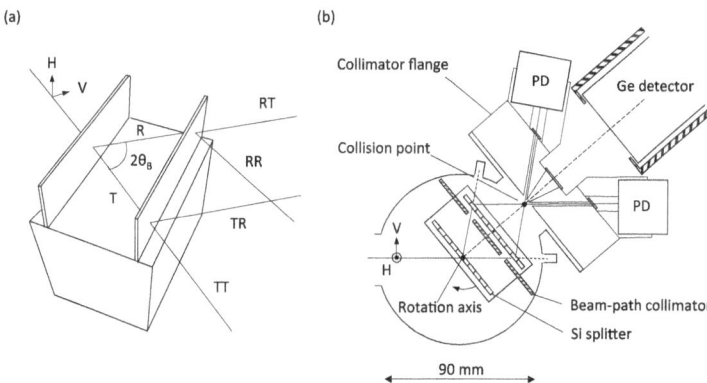

Figure 3. (a) Four beams after the two blades of the silicon beam splitter. (b) Magnified view of the beam-collision chamber. Adapted from Reference [19]. PD: photodiode; R: reflected beam; T: transmitted beam.

A magnified view of the beam-collision chamber is shown in Figure 3b. The splitter is installed in the chamber and fixed to a rotary feedthrough connected to a goniometer. The intensities of the two beams are monitored pulse-by-pulse by silicon PIN photodiodes (PDs). One of the paired signal photons—the forward-scattered one—has higher energy than the original photons

($\omega_0 = 11$ keV), and is easily separated from backgrounds. To select forward-scattered photons, a collimator flange with a conically tapered hole is placed after the collision point. The full angle of the cone is 25°, defining the signal energy region from 18.1 keV to 19.9 keV. The first experiment was performed on 23 and 24 July 2013 at SACLA with a total injection of 6.5×10^5 pulses. A timing cut was applied to select events within ± 1 μs of the arrival of X-ray pulses. Figure 4(left) shows the energy spectrum after the cut. An elastic peak from stray X-rays is observed around ω_0 as well as some events at $2\omega_0$ due to the pileup of two photons. No events are observed in the signal region, and the resulting limit on $\sigma_{\gamma\gamma}$ is shown in Figure 4(right) with other limits.

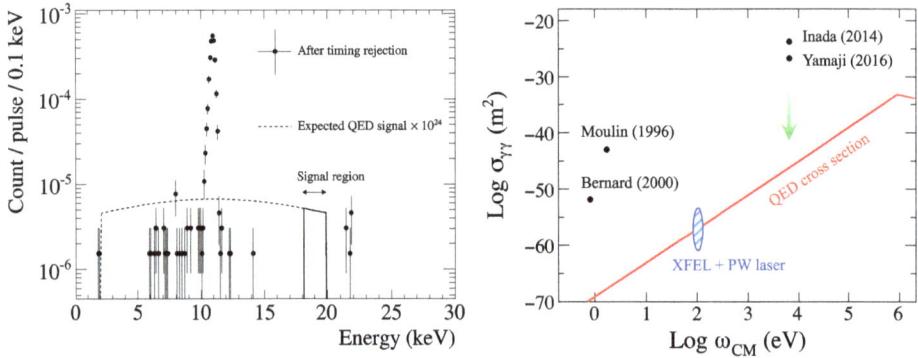

Figure 4. (**Left**) Energ The dotted line shows the expected QED signal shape in the full solid angle (signal strength is $\times 10^{24}$). The solid line corresponds to the signal region from 18.1 keV ($\theta = 12.5°$) to 19.9 keV ($\theta = 0$). Adapted from Reference [19]. (**Right**) Limits on $\sigma_{\gamma\gamma}$ at 95% C.L. obtained by real photon–photon scattering experiments. Moulin (1996) [21] uses two high-power lasers. Bernard (2000) [25] uses three lasers applying four-wave mixing. Inada (2014) [19] is our first search using an X-ray free-electron laser (XFEL), and we upgraded it in Yamaji (2016) [20]. The hatched circle shows a region probed by laser-induced diffraction and birefringence experiments using an XFEL.

After the first experiment, we upgraded it in the second experiment carried out in 2015 [20]. The main improvements are in (i) the efficiency of the beam splitter, (ii) the performance of the XFEL, and (iii) the statistical gain by larger measurement time. The total improvement from the first result is by three orders of magnitude, and gives the most stringent limit in the X-ray region, which still needs to be gained by 20 orders of magnitude to reach the QED cross section. On the basis of these experiences, we are planning to further improve the sensitivity in the next phase. The first approach is to apply a similar splitter in the Bragg case as seen in Reference [27]. This reduces the absorption and beam-branching in the second blade of the Laue case used in the present splitter. Moreover, multiple collisions are possible by bouncing two X-ray pulses many times in the same crystal. The second gain would be obtained by improving the performance of the XFEL. The present mechanism of the X-ray laser is called "self-amplified spontaneous emission (SASE)", where the output beam does not have deterministic monochromaticity, reflecting the stochastic nature of initial electron bunches. An alternative "self-seeding" scheme has been proposed, and is expected to improve the current monochromaticity by about two orders of magnitude, reducing the monochromaticity loss in the splitter [28]. Finally, the direct collision of distinct beams, with each tightly focused at a collision point, can significantly improve the sensitivity. There is a dedicated facility between SACLA and SPring-8 where both beams are available [29]. The temporal synchronization of the two beams is currently underway.

2.2. Laser-Induced Birefringence and Diffraction of X-rays

An XFEL can work as a useful probe on the vacuum process induced at the focal spot of a high-power laser. Since both lasers provide real photons, this is another real process corresponding to a different center-of-mass photon energy $\sqrt{\omega_1\omega_2} = 100$ eV with $\omega_1 = 10$ keV and $\omega_2 = 1$ eV. The high-power laser is often referred to as a pump laser in analogy to a conventional pump-probe experiment where the laser excites a sample and the probe studies the response. The change of the probe's wave vector induced by the scattering is called "vacuum diffraction" [30]. Furthermore, polarization flipping—which occasionally occurs in the scattering—adds tiny ellipticity to the linearly-polarized probe, and is called "vacuum birefringence" [31]. Figure 5(left) shows the principle of the experiment where the pump laser is radiated in a counter-propagating geometry. The probe and pump are focused on the same spot to enhance the diffraction signal that is scattered in a forward direction with a tiny angle. An efficient scheme of detecting photons that have both properties of flipped polarization and a finite diffraction angle was proposed in Reference [32]. In addition, a practical setup using an X-ray focusing lens and crossed polarizers can be found in Reference [9]. While there are many theoretical studies and proposals, no experiments have been put into practice.

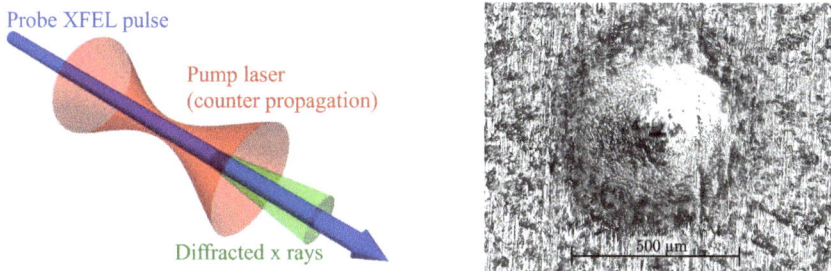

Figure 5. (**Left**) Principle of the experiment probing vacuum diffraction. (**Right**) Microscope image of a zinc film after the temporal and spatial adjustments. The rectangular hole at the center is made by the penetration of X-rays (40(H) × 20(V) μm). The large deformation overlapping the hole is made by a focused 2.5-TW laser (12(H) × 10(V) μm).

Regardless of detecting diffraction or birefringence, the key point of the experiment is the way to synchronize the probe and pump with respect to space and time. As for the timing adjustment, there is a working method using the fast electronic response of a GaAs film that can tune the timing within ~1 ps [33]. However, to our knowledge, there is so far no efficient method to spatially overlap the pump and probe with both focused to ~10 μm, and thus a dedicated method needs to be found and established.

For this purpose, we carried out a test measurement at SACLA in 2016. A probe XFEL and a 2.5-TW pump laser were focused to about 10 μm by composite refractive lenses and an off-axis parabolic mirror, respectively. A zinc film with a thickness of 20 μm was placed at the collision point to check their overlap (the detailed setup and procedures are described in Reference [34]). Their energy deposit on the film created a characteristic pattern of deformation that could be checked later by an optical microscope. Figure 5(right) shows a microscope image of the film after the timing adjustment and some iteration of the spatial overlapping. The XFEL was shot on the film from the back side of the image. The larger hole with a diameter of about 500 μm was made by the focused pump laser, whereas a smaller one at the center was made by the penetration of the probe. The deviation between the two centers was found to be less than ~10 μm, showing the accuracy of this method [34].

A PW-laser facility is under installation at SACLA and is scheduled to be available to users in 2018 [29]. The expected sensitivity to $\sigma_{\gamma\gamma}$ is shown in Figure 4(right). By the time the laser system becomes ready for use, our effort will be directed to the establishment of a synchronization scheme feasible for a higher pump power.

2.3. Birefringence Induced by a High-Field Magnet

Magnet-induced birefringence has long been measured in normal media like gas, and is known as the "Cotton–Mouton effect" [35]. The application of this measurement scheme to a vacuum provides the current best sensitivity to observe the vacuum process. The most recent result of the PVLAS experiment in 2016 provides sensitivity that is smaller by a factor of 20 with respect to the coefficient of the vacuum refractive index [10]. The signal birefringence increases in proportion to the square of the magnetic field strength, and thus a high field of a pulsed magnet is advantageous. In addition, the temporal variation of the pulsed field is essential to distinguish the signal from large static birefringence produced by cavity mirrors. In this scheme, a racetrack pulsed magnet was first used in the BMV experiment [36].

To obtain large statistics along with the field strength, we have been developing a pulsed magnet system suited to a high repetition use. While the typical repetition of a pulsed magnet is about 1 mHz, that of our magnet is higher by about two orders of magnitude [37]. Figure 6 shows a photo of the magnet. The coil part is wound with a copper wire and cooled with liquid nitrogen to reduce the electrical resistivity. Stainless steel metal is used to reinforce the outer strength of the coil that is exposed to an expansion force during the generation of pulsed fields. The peak field reaches 12 T at the magnet center. The magnet has a planar shape to enhance the cooling efficiency. In addition to the magnet, we also developed a discharge unit with capacitors providing pulsed current to the magnet with high repetition [37]. Figure 7(left) shows a cycle of the repetition composed of two successive pulses with alternating field directions. The first pulse produces a field of +10 T at the magnet center with a charged voltage of 4 kV. After a portion of the energy converts into Joule heating in the coil, the second pulse produces −6 T. The residual energy returns to the capacitor that is recharged to 4 kV before the next cycle. Figure 7 (right) shows the distribution of the field strength in a continuous operation. Stable fields are repeatedly produced after the magnets relax to thermal equilibrium. The field length linearly increases the signal birefringence. Up to now, we have tested a simultaneous operation of four magnets, giving a total field length of 0.8 m with a repetition rate of 0.2 Hz and a field strength of 10 T (see also Section 3.2 and Figure 10).

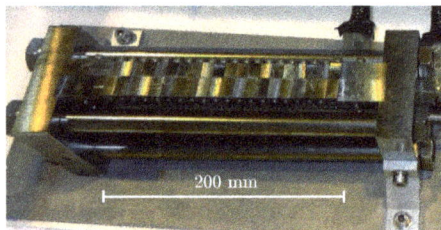

Figure 6. Racetrack pulsed magnet developed for a high-repetition use. The field length is 200 mm.

We have made a setup that combines the magnet with an optical system containing a Fabry–Perot cavity with a finesse of 650,000. The continuous data-taking requires a new scheme that the cavity resonance automatically recovers after the measurements of finesse and static birefringence in each cycle. To verify this scheme, we carried out a test run using a single magnet. The results have been described in Reference [38]. A long-term run of a few months will start after improvements of the optics.

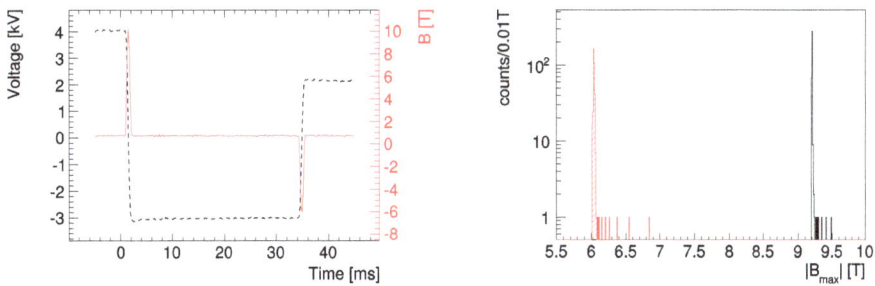

Figure 7. Repetitive operation of the four pulsed magnets. (**Left**) Typical waveform of the capacitor voltage (to left, dashed) and magnetic field (to right, solid). The pulse interval is adjustable and set to 1/30 s here. (**Right**) Peak-field distribution at the magnet center for the first and second pulses, obtained by 1-h operation with 0.2-Hz pulse repetition. Adapted from Reference [37].

3. Searches for Physics Beyond the Standard Model

The subtle signal of the vacuum process would be easily affected by the possible existence of new physics. If a new particle couples to photons, it can be directly searched for by the same vacuum experiment. A good example was provided by the recent discovery of the Higgs boson which decays into two photons ($H \rightarrow \gamma\gamma$), demonstrating that a vacuum is permeated by a scalar field [39,40]. As lighter particles, new pseudoscalar bosons like ALPs provide anomalous contribution to the vacuum process [16]. Moreover, BSM models often predict a dark sector that involves fermions with a tiny fraction of the electric charge, known as millicharged particles (see Reference [15] for a systematic review of BSM in a low energy scale). Among these particles, here we focus on the ALP case.

3.1. Exotic Contribution to Photon–Photon Scattering

As stated before, the VMB measurement provides an efficient strategy for observing the vacuum process, and this is also true for the ALP search for a mass region around 1 meV [10]. Figure 8 shows a diagram where ALPs contribute to VMB. However, it inclusively searches for a wide mass region since the mass information of ALPs is somewhat lost by the virtual off-shell photons conveying magnetic fields. By contrast, the s-channel scattering of real photons provides mass information with a precision of the monochromaticity of the light (Figure 9a). Once the ALP mass is roughly determined, the s-channel search can precisely sweep the relevant mass region by continuously changing the incident photon energy. This is advantageous for real photon–photon processes [19,20], and thus the two methods provide complementary approaches.

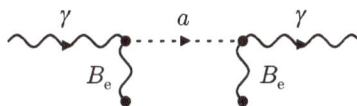

Figure 8. Interaction of a photon with an external magnetic field caused by an axion-like particle, contributing to vacuum magnetic birefringence.

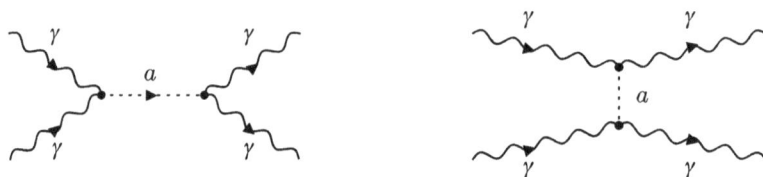

Figure 9. Photon–photon scattering by axion-like particles. (**Left**) *s*-channel. (**Right**) *t*-channel.

Each kind of light source probes a distinct mass region corresponding to its photon energy. As shown in Figure 4(right), we are currently capable of the combinations of XFEL–XFEL, XFEL–laser, and laser–laser, for which the center-of-mass energies are about 10 keV, 100 eV, and 1 eV, respectively. Various combinations of these different light sources widens the experimental opportunities, since the new mass scale is still not clear.

3.2. Dedicated Search for Axion-Like Particles

Compared to the mass region probed by VMB measurements, a heavier region can be studied by using X-rays [41,42]. Especially, the ALP mass around 10–100 meV attracts recent cosmological interests in the inflation, the cosmic microwave background radiation , and dark matter [43]. The photon conversion into a real-state ALP (or vice versa) in a magnetic field is referred to as the "Primakoff effect". Up to now, a large number of laboratory searches for ALPs have been carried out with a "light shining through a wall (LSW)" technique [16]. The first X-ray LSW search was carried out at the European Synchrotron Radiation Facility (ESRF) with superconducting magnets, extending the limit on the ALP-two-photon coupling constant ($g_{a\gamma\gamma}$) up to around 1 eV [44]. While searches for ALP flux from the Sun have also probed this mass region [45], the flux estimation inevitably relies on a solar model and its complex magnetic activity [46], showing the importance of complementary searches using terrestrial and extra-terrestrial X-ray sources [44].

Figure 10 shows a schematic of our LSW setup using four magnets described in Section 2.3. The ALPs generated in the first pair of magnets pass through a beam dump that blocks the unconverted photons. Some of the ALPs then reconvert into detectable photons by an inverse process in the second magnet pair. We carried out the first measurement at SPring-8 BL19LXU [47] in 2015 using a continuous X-ray beam with high intensity. The data obtained with a total of 27,676 pulses produced in a net run time of two days show no events in the signal region (Figure 11, left). This provides a limit on $g_{a\gamma\gamma}$ that is more stringent by a factor of 5.2 compared to the previous X-ray LSW limit for the ALP mass $\lesssim 0.1$ eV (Figure 11, right). The pulsed nature of the magnet provides better sensitivity when used with pulsed X-rays (the repetition is 30 Hz at SACLA; see Figure 7(left) for the corresponding time interval between the magnetic pulses). The expected sensitivity at SACLA with the same magnets and configurations are also shown in the figure.

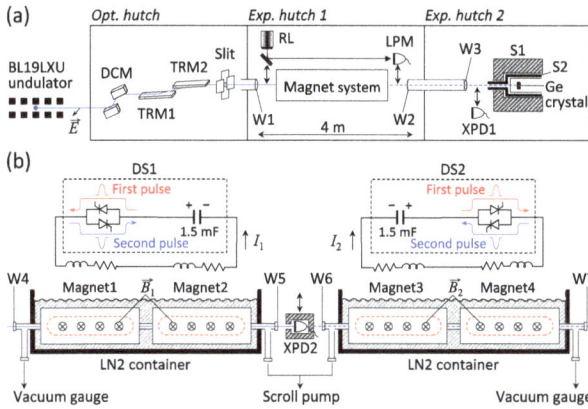

Figure 10. Schematic of the experimental setup. (**a**) Layout of components in the optics hutch and in the two experimental hutches. Retractable components are shown with a vertical arrow. (**b**) Side view of the magnet system. The magnets are placed so as to produce parallel fields with respect to the horizontal polarization of the X-ray beam. Two identical discharge sections (DS1 and DS2) that are contained in the same hutch supply pulsed currents to the magnets by *LC*-discharge circuits. Adapted from Reference [42].

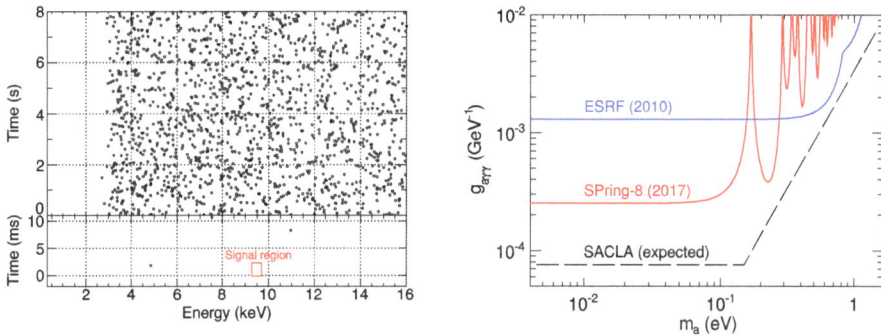

Figure 11. (**Left**) Top: time–energy distribution of X-rays measured with the germanium detector. Each circle represents an event. The horizontal axis is the time from the beginning of the first pulse. The energy threshold was set around 3 keV. Bottom: events around the signal region defined by the beam energy of $\pm 2\sigma$ (σ is the detector resolution) and the time window of 2.1 ms. Adapted from Reference [42]. (**Right**) Upper limit on the axion-like particle (ALP)-two-photon coupling constant $g_{a\gamma\gamma}$ at 95% C.L. as a function of the ALP mass. The X-ray light shining through a wall (LSW) limits obtained at the ESRF (2010) [44] and SPring-8 (2017) [42] are shown along with the expected sensitivity at the SPring-8 Angstrom Compact Free-Electron Laser (SACLA, dashed).

4. Summary

We reviewed the progress of our experiments that probe physics in a vacuum, with a focus on those using an XFEL. Various combinations among an XFEL, a high-power laser, and a high-field magnet provided a test for QED and new physics in a different energy scale. The results obtained by our first-phase experiments are as follows. (i) The photon–photon scattering experiment using an XFEL provides the first limits in the X-ray region [19,20]. The sensitivity needs to be gained by 20 orders of magnitude to reach the QED cross-section and now proceeds to the second phase with significant improvements. (ii) The experiment probing laser-induced birefringence and diffraction of X-rays

Appl. Sci. **2017**, *7*, 671

requires a new synchronization scheme between the pump and probe lasers. A test measurement using a 2.5-TW pump laser succeeded in the temporal and spatial synchronization with the accuracies of ~1 ps and 10 μm, respectively. We are currently preparing for the first measurement using a PW laser in 2018 [34]. (iii) The VMB measurement features a unique field-generation system with high-repetition pulsed magnets. The use of a high-finesse Fabry–Perot cavity verified the automatic resonance recovery scheme required for continuous data-taking and obtained the first results [37,38]. A long-term run for a few months will start after some improvements of the optics. (iv) The dedicated search for ALPs using the magnets and continuous X-rays was carried out at SPring-8 and improved the previous limit on the coupling constant by a factor of 5.2 [42]. The preparation for an experiment at SACLA using pulsed X-rays is underway.

Acknowledgments: This research work at SACLA and SPring-8 BL19LXU is approved and supported by JASRI and RIKEN.

Author Contributions: T.I. wrote the paper. All authors reviewed and critiqued the manuscript.

Conflicts of Interest: The authors declare no conflict of interest.

References

1. Aoyama, T.; Hayakawa, M.; Kinoshita, T.; Nio, M. Tenth-order electron anomalous magnetic moment—Contribution of diagrams without closed lepton loops. *Phys. Rev. D* **2015**, *91*, 033006. [CrossRef]
2. Jegerlehner, F.; Nyffeler, A. The muon g–2. *Phys. Rep.* **2009**, *477*, 1–10. [CrossRef]
3. Hanneke, D.; Fogwell, S.; Gabrielse, G. New Measurement of the Electron Magnetic Moment and the Fine Structure Constant. *Phys. Rev. Lett.* **2008**, *100*, 120801. [CrossRef]
4. Akhmadaliev, S.Z.; Kezerashvili, G.Y.; Klimenko, S.G.; Lee, R.N.; Malyshev, V.M.; Maslennikov, A.L.; Milov, A.M.; Milstein, A.I.; Muchnoi, N.Y.; Naumenkov, A.I.; et al. Experimental Investigation of High-Energy Photon Splitting in Atomic Fields. *Phys. Rev. Lett.* **2002**, *89*, 061802. [CrossRef]
5. Akhmadaliev, S.Z.; Kezerashvili, G.Y.; Klimenko, S.G.; Malyshev, V.M.; Maslennikov, A.L.; Milov, A.M.; Milstein, A.I.; Muchnoi, N.Y.; Naumenkov, A.I.; Panin, V.S.; et al. Delbrück scattering at energies of 140–450 MeV. *Phys. Rev. C* **1998**, *58*, 2844. [CrossRef]
6. ATLAS Collaboration. Light-by-Light Scattering in Ultra-Peripheral Pb+Pb Collisions at sqrt(sNN)=5.02 TeV with the ATLAS Detector at the LHC. *ATLAS-CONF-2016-111.* [CrossRef]
7. Di Piazza, A.; Müller, C.; Hatsagortsyan, K.Z.; Keitel, C.H. Extremely high-intensity laser interactions with fundamental quantum systems. *Rev. Mod. Phys.* **2012**, *84*, 1177. [CrossRef]
8. King, B.; Heinzl, T. Measuring vacuum polarization with high-power lasers. *High Power Laser Sci. Eng.* **2016**, *4*, e5. [CrossRef]
9. Schlenvoigt, H.; Heinzl, T.; Schramm, U.; Cowan, T.; Sauerbrey, R. Detecting vacuum birefringence with X-ray free electron lasers and high-power optical lasers: A feasibility study. *Phys. Scr.* **2016**, *91*, 023010. [CrossRef]
10. Della Valle, F.; Ejlli, A.; Gastaldi, U.; Messineo, G.; Milotti, E.; Pengo, R.; Ruoso, G.; Zavattini, G. The PVLAS experiment: Measuring vacuum magnetic birefringence and dichroism with a birefringent Fabry–Perot cavity. *Eur. Phys. J. C* **2016**, *76*, 24. [CrossRef]
11. Dittrich, W.; Gies, H. *Probing the Quantum Vacuum*; Springer: Berlin, Germany, 2000. [CrossRef]
12. Magnani, R.P.; Testa, V.; González Caniulef, D.; Taverna, R.; Turolla, R.; Zane, S.; Wu, K. Evidence for vacuum birefringence from the first optical-polarimetry measurement of the isolated neutron star RX J1856.5–3754. *MNRAS* **2017**, *465*, 492–500. [CrossRef]
13. Capparelli, L.M.; Maiani, L.; Polosa, A.D. A Note on Polarized Light from Magnetars: QED Effects and Axion-Like Particles. *arXiv* **2017**, arXiv:1705.01540.
14. Turolla, R.; Zane, S.; Taverna, R.; Gonzalez Caniulef, D.; Mignani, R.P.; Testa, V.; Wu, K. A Comment on "A Note on Polarized Light from Magnetars: QED Effects and Axion-Like Particles" by L.M. Capparelli, L. Maiani and A.D. Polosa. *arXiv* **2017**, arXiv:1706.02505.

15. Essig, R.; Jaros, J.A.; Wester, W.; Hansson Adrian, P.; Andreas, S.; Averett, T.; Baker, O.; Batell, B.; Battaglieri, M.; Beacham, J.; et al. Dark Sectors and New, Light, Weakly-Coupled Particles; Report of the Community Summer Study (Snowmass) Intensity Frontier New, Light, Weakly-Coupled Particles Subgroup. *arXiv* 2013, arXiv:1311.0029.

16. Graham, P.W.; Irastorza, I.G.; Lamoreaux, S.K; Lindner, A.; van Bibber, K.A. Experimental Searches for the Axion and Axion-Like Particles. *Annu. Rev. Nucl. Part. Sci.* **2015**, *65*, 485–514. [CrossRef]

17. Ishikawa, T.; Aoyagi, H.; Asaka, T.; Asano, Y.; Asano, N.; Bizen, T.; Ego, H.; Fukami, K.; Fukui, T.; Furukawa, Y.; et al. A compact X-ray free-electron laser emitting in the sub-ångström region. *Nat. Photonics* **2012**, *6*, 540–544. [CrossRef]

18. Tono, K.; Togashi, T.; Inubushi, Y.; Sato, T.; Katayama, T.; Ogawa, K.; Ohashi, H.; Kimura, H.; Takahashi, S.; Takeshita, K.; et al. Beamline, experimental stations and photon beam diagnostics for the hard X-ray free electron laser of SACLA. *New J. Phys.* **2013**, *15*, 083035. [CrossRef]

19. Inada, T.; Yamaji, T.; Adachi, S.; Namba, T.; Asai, S.; Kobayashi, T.; Tamasaku, K.; Tanaka, Y.; Inubushi, Y.; Sawada, K.; et al. Search for photon–photon elastic scattering in the X-ray region. *Phys. Lett. B* **2014**, *732*, 356–359. [CrossRef]

20. Yamaji, T.; Inada, T.; Yamazaki, T.; Namba, T.; Asai, S.; Kobayashi, T.; Tamasaku, K.; Tanaka, Y.; Inubushi, Y.; Sawada, K.; et al. An experiment of X-ray photon–photon elastic scattering with a Laue-case beam collider. *Phys. Lett. B* **2016**, *763*, 454–457. [CrossRef]

21. Moulin, F.; Bernard, D.; Amiranoff, F. Photon-photon elastic scattering in the visible domain. *Z. Phys. C* **1996**, *72*, 607–611. [CrossRef]

22. De Tollis, B. Dispersive approach to photon-photon scattering. *Nuovo Cim.* **1964**, *32*, 757–768. [CrossRef]

23. De Tollis, B. The scattering of photons by photons. *Nuovo Cim.* **1965**, *35*, 1182–1193. [CrossRef]

24. Bonse, U.; Hart, M. Principles and design of Laue-case X-ray interferometers. *Z. Physik* **1965**, *188*, 154–164. [CrossRef]

25. Bernard, D.; Moulin, F.; Amiranoff, F.; Braun, A.; Chambaret, J.P.; Darpentigny, G.; Grillon, G.; Ranc, S.; Perrone, F. Search for stimulated photon-photon scattering in vacuum. *Eur. Phys. J. D* **2000**, *10*, 141–145. [CrossRef]

26. Bonse, U.; Hart, M. An X-ray Interferometer. *Appl. Phys. Lett.* **1965**, *6*, 155–156. [CrossRef]

27. Bonse, U.; Hart, M. An X-ray interferometer with Bragg case beam splitting and beam recombination. *Z. Physik* **1966**, *194*, 1–17. [CrossRef]

28. Amann, J.; Berg, W.; Blank, V.; Decker, F.-J.; Ding, Y.; Emma, P.; Feng, Y.; Frisch, J.; Fritz, D.; Hastings, J.; et al. Demonstration of Self-Seeding in a Hard-X-ray Free-Electron Laser. *Nat. Photonics* **2012**, *6*, 693–698. [CrossRef]

29. SACLA Website. Available online: http://xfel.riken.jp/eng/index.html (accsed on 29 June 2017).

30. Di Piazza, A.; Hatsagortsyan, K.Z.; Keitel, C.H. Light Diffraction by a Strong Standing Electromagnetic Wave. *Phys. Rev. Lett.* **2006**, *97*, 083603. [CrossRef]

31. Heinzl, T.; Liesfeld, B.; Amthor, K.-U.; Schwoerer, H.; Sauerbrey, R.; Wip, A. On the observation of vacuum birefringence. *Opt. Commun.* **2006**, *267*, 318. [CrossRef]

32. Karbstein, F.; Sundqvist, C. Probing vacuum birefringence using X-ray free electron and optical high-intensity lasers. *Phys. Rev. D* **2016**, *94*, 013004. [CrossRef]

33. Katayama, T.; Owada, S.; Togashi, T.; Ogawa, K.; Karvinen, P.; Vartiainen, I.; Eronen, A.; David, C.; Sato, T.; Nakajima, K.; et al. A beam branching method for timing and spectral characterization of hard X-ray free-electron lasers. *Struct. Dyn.* **2016**, *3*, 034301. [CrossRef]

34. Seino, Y. Search for vacuum diffraction using high power laser and X-ray free electron laser SACLA. *Light Driven Nucl.-Part. Phys. Cosmol.* **2017**, in press. [CrossRef]

35. Rizzo, C.; Rizzo, A.; Bishop, D.M. The Cotton-Mouton effect in gases: Experiment and theory. *Int. Rev. Phys. Chem.* **1997**, *16*, 81–111. [CrossRef]

36. Cadène, A; Berceau, P.; Fouché, M.; Battesti, R.; Rizzo, C. Vacuum magnetic linear birefringence using pulsed fields: Status of the BMV experiment. *Eur. Phys. J. D* **2014**, *68*, 16. [CrossRef]

37. Yamazaki, T.; Inada, T.; Namba, T.; Asai, S.; Kobayashi, T.; Matsuo, A.; Kindo, K.; Nojiri, H. Repeating pulsed magnet system for axion-like particle searches and vacuum birefringence experiments. *Nucl. Instrum. Methods A* **2016**, *833*, 122–126. [CrossRef]

38. Fan, X.; Kamioka, S.; Inada, T.; Yamazaki, T.; Namba, T.; Asai, S.; Omachi, J.; Yoshioka, K.; Kuwata-Gonokami, M.; Matsuo, A.; et al. The OVAL experiment: A New Experiment to Measure Vacuum Magnetic Birefringence Using High Repetition Pulsed Magnets. *arXiv* **2017**, arXiv:1705.00495.

39. ATLAS Collaboration. Observation of a new particle in the search for the Standard Model Higgs boson with the ATLAS detector at the LHC. *Phys. Lett. B* **2012**, *716*, 1. [CrossRef]

40. CMS Collaboration. Observation of a new boson at a mass of 125 GeV with the CMS experiment at the LHC. *Phys. Lett. B* **2012**, *716*, 30. [CrossRef]

41. Inada, T.; Namba, T.; Asai, S.; Kobayashi, T.; Tanaka, Y.; Tamasaku, K.; Sawada, K.; Ishikawa, T. Results of a search for paraphotons with intense X-ray beams at SPring-8. *Phys. Lett. B* **2013**, *722*, 301–304. [CrossRef]

42. Inada, T.; Yamazaki, T.; Namba, T.; Asai, S.; Kobayashi, T.; Tamasaku, K.; Tanaka, Y.; Inubushi, Y.; Sawada, K.; Yabashi, M.; et al. Search for Two-Photon Interaction with Axionlike Particles Using High-Repetition Pulsed Magnets and Synchrotron X rays. *Phys. Rev. Lett.* **2017**, *118*, 071803. [CrossRef]

43. Daido, R.; Takahashi, F.; Yin, W. The ALP miracle: Unified inflaton and dark matter. *J. Cosmol. Astropart. Phys.* **2017**, *5*, 44. [CrossRef]

44. Battesti, R.; Fouché, M.; Detlefs, C.; Roth, T.; Berceau, P.; Duc, F.; Frings, P.; Rikken, G.L.J.A.; Rizzo, C. Photon Regeneration Experiment for Axion Search Using X-rays. *Phys. Rev. Lett.* **2010**, *105*, 250405. [CrossRef]

45. Arik, M. Search for Solar Axions by the CERN Axion Solar Telescope with ^{3}He Buffer Gas: Closing the Hot Dark Matter Gap. *Phys. Rev. Lett.* **2014**, *112*, 091302. [CrossRef]

46. Ossendrijver, M. The solar dynamo. *Astron. Astrophys. Rev.* **2003**, *11*, 287. [CrossRef]

47. Yabashi, M.; Mochizuki, T.; Yamazaki, H.; Goto, S.; Ohashi, H.; Takeshita, K.; Ohata, T.; Matsushita, T.; Tamasaku, K.; Tanaka, Y.; et al. Design of a beamline for the SPring-8 long undulator source 1. *Nucl. Instrum. Methods A* **2001**, *467*, 678–681. [CrossRef]

![applied sciences logo] *applied sciences*

MDPI

Article

Two- and Three-Photon Partial Photoionization Cross Sections of Li^+, Ne^{8+} and Ar^{16+} under XUV Radiation

William Hanks, John T. Costello and Lampros A.A. Nikolopoulos *

School of Physical Sciences, Dublin City University, Dublin 9, Ireland; william.hanks2@mail.dcu.ie (W.H.); John.costello@dcu.ie (J.T.C.)
* Correspondence: Lampros.Nikolopoulos@dcu.ie; Tel.: +353-1-7005-300

Academic Editor: Kiyoshi Ueda
Received: 14 February 2017; Accepted: 9 March 2017; Published:17 March 2017

Abstract: In this work, we present the photon energy dependence of the two- and three-photon cross sections of the two-electron Li^+, Ne^{8+} and Ar^{16+} ions, following photoionization from their ground state. The expressions for the cross sections are based on the lowest-order (non-vanishing) perturbation theory for the electric field, while the calculations are made with the use of an ab initio configuration interaction method. The ionization cross section is dominated by pronounced single photon resonances in addition to peaks associated with doubly excited resonances. In the case of two-photon ionization, and in the non-resonant part of the cross section, we find that the ^1D ionization channel overwhelms the ^1S one. We also observe that, as one moves from the lowest atomic number ion, namely Li^+, to the highest atomic number ion, namely Ar^{16+}, the cross sections generally decrease.

Keywords: multiphoton ionization; X-ray radiation; free-electron laser; lowest-order perturbation theory; cross sections

1. Introduction

During the last decade, the advent of new light sources, capable of delivering intense and/or ultrashort, coherent radiation in the soft- and hard-X-ray regime, either directly by free-electron lasers (FEL) or indirectly by high-harmonic generation techniques, has renewed interest in experimental and theoretical photoionization studies of multiply charged ions. The excitation of atoms/molecules involving inner-shell electrons with such light sources offers the possibility of investigating processes at their natural time-scale, such as Auger processes and its variations [1], double core-hole creation and ionization [2], resonant enhanced photoionization, etc. Beyond the viewpoint of accessing the dynamics of these processes, the availability of extremely high flux has made it feasible to experimentally create the required conditions and observe for the first time new processes, for example the two-photon single- and double-(direct) photoionization of helium, the complete stripping of neon [3] and multiphoton inner-shell ionization in noble gases [4] and solid targets [5]. Since a number of recent reviews elaborate on experimental and theoretical studies as well as on their numerous applications in this short wavelength regime [6–9], we will only mention here that typical fluxes of FELs to date range between 5×10^{11} Wcm^{-2} to 5×10^{18} Wcm^{-2}, photon energies range in (0.02–15) KeV, while average full width at half maximum (FWHM) durations vary from 10 fs to 85 fs. It is also worth noting that three new FELs are under development and set to start user operations soon (E-XFEL, Swiss FEL, PAL) with extra pulse parameter specifications compared to the current ones. Notably, the Swiss FEL will provide pulses as short as 2 fs while E-XFEL as long as 100 fs with a repetition rate between 50–100 times higher than any other existing FEL.

From the above discussion, it appears that, nowadays, in these wavelength regimes, excitation/ionization processes involving more than one photon are routinely feasible. Accordingly, their quantitative description requires theoretical approaches capable of coping with the non-linear features of the interaction between X-ray radiation with atomic and molecular systems. Amongst the various theoretical approaches, one is based on a perturbation expansion of the interaction potential with respect to the electric field, known as lowest-order (non-vanishing) perturbation theory (LOPT) [10]. In the case of X-ray quasi-monochromatic radiation, the use of LOPT is well justified for the current available peak intensities. The main requirements are for (a) the ponderomotive potential, $V_p = I_0/4\omega^2$, and (b) the bandwidth, $\Delta\omega$, to be much smaller than the mean central photon frequency, ω, of the radiation. For our case, for example, if we choose a peak intensity at the higher end, $I_0 \simeq 3.51 \times 10^{18}$ W/cm^2, and a mean photon energy at the lowest limit of soft X-ray spectrum, say $\omega \sim 270$ eV, then $V_p/\omega \simeq 0.025 \ll 1$. Since for current FEL sources the bandwidth, $\Delta\omega$, ranges in $(10^{-3}\text{--}10^{-2})\omega$, LOPT is very well suited to provide reliable quantitative information about various quantities of experimental interest, for example, ion and fluorescence yields.

The validity of LOPT to calculate few X-ray photon processes allows the development of an alternative theoretical framework in place of the direct, but much more demanding, integration of a multielectron time-dependent Schrödinger equation. In the present work, we have chosen to apply LOPT to three elements of high experimental interest, namely Li, Ne and Ar. As it is known, these three atomic elements have been the subject of numerous experiments and theoretical studies since the inception of quantum mechanics. Excitation and ionization of the valence shells of the neutral species of these elements have been extensively studied and experimental reports and theoretical calculations about their photoionization cross sections are available in literature. In contrast to this plethora of data, the availability of analogous studies of their ionized species are scarce, especially when few-photon ionization is involved. To provide one example illustrating the need for such data, we mention the case of neon's multiple ionization in the pioneering experiment of Young et al. [3] at LCLS with a FEL pulse at a photon energy circa 1110 eV (pulse peak intensity and FWHM duration were estimated to be 10^{17} W/cm^2 and 100 fs, respectively). In this experiment, the ionization of Ne^{8+}, following the sequential one-photon stripping of the lower charged neon ions, proceeds mainly through a two-photon absorption (the ionization potential of Ne^{8+} is circa 1362 eV). It is needless to say that the same two-photon ionization channel would have been the dominant ionization channel for any photon in the energy range between 681 eV $\leq \omega \leq$ 1362 eV; thus, the need for two-photon ionization cross sections for a range of photon energies. Of course, similar considerations can be carried over to any atomic system and of any degree of charge in the presence of X-ray radiation.

In the simplest case of the one-electron charged ions, multiphoton cross sections can be straightforwardly calculated by scaling the corresponding hydrogen cross sections according to the relation $\sigma_N(Z^2\omega) = \sigma_N^{(H)}/Z^{4N-2}$ [11], where $\sigma_N^{(H)}$ is the N-photon cross section of hydrogen and Z is the atomic number of the element. Throughout the years, a number of reports have appeared in literature with multiphoton cross sections of hydrogen involving quite a high number of photons as, for example, the calculations by Karule [12] and by Potvliege and Shakeshaft ($N = 20$) [13]. Next in line are charged ions with only their K-shell electrons remaining, making them helium-like (two-electron) systems. For helium, systematic studies exist where two, three- and four-photon LOPT ionization cross sections are calculated [14]; it appears, despite the significance of Li, Ne and Ar as experimental and theoretical targets, that no similar detailed study has been extended to these systems. To the best of our knowledge, we are aware of the two- and three-photon total ionization cross sections that have been calculated on Li$^+$ with the use of single-channel quantum-defect theory (SQDT) [15], the work of Novikov and Hopersky in neon ions [16], the more recent work by the group of R. Santra [17] (Ne^{8+}) and the two-photon total cross sections on Ne^{8+} and Ar^{16+}, obtained through a Greens-function method calculation [18].

More specifically, in the present study, we apply an ab initio configuration interaction (CI) approach and make use of LOPT to calculate the two- and three-photon partial photoionization cross sections for Li^+, Ne^{8+} and Ar^{16+}. In addition to this, we have calculated details of their electronic structure. The structure of the text is as follows: in Section 2, we present the theoretical method in sufficient detail for a self-contained formulation of the present study; in Section 3, we present and discuss our results about the calculated energies and the LOPT two- and three-photon ionization partial cross sections. Finally, we conclude with a summary of our findings and a brief discussion of possible further investigations within the present context. In the presentation of the theoretical formulas, we use the atomic-Gaussian unit system ($\hbar = m_e = e = 1/4\pi\varepsilon_0 = 1$). In the figures, the cross sections and the energies are presented in more traditional units, namely, eV for the energies, and cm^4s and cm^6s^2 for the two- and three-photon cross sections, respectively.

2. Theoretical Formulation

The chosen ions have the common feature in that they only have two-electrons, thus allowing a fully ab initio calculation of their atomic structure. The calculation proceeds in two stages. First, we follow a configuration interaction method to calculate the eigenstates and the associated bound and continuum states of these systems, and then we apply LOPT to obtain a generalized multiphoton cross section that can, in turn, be used to calculate two- and three-photon cross sections. These steps are described in greater detail below. Although the computational procedure has been presented in detail in a number of articles [19–23], it is necessary to include some brief presentation, adjusted to the particular case of ionization by linearly polarized X-ray radiation. The latter property of the radiation and the fact that the ground state of the ions, $^1S_0$, is spherically symmetric, combined with the selection rules for electric dipole interactions, restricts the states to those of singlet-symmetry with a total magnetic quantum number value of zero. We take advantage of this from the outset in order to simplify the formulation.

2.1. Atomic Structure Calculation

The non-relativistic two-electron ionic Hamiltonian, H_A, in atomic units, is given by:

$$\hat{H}_A = \hat{h}(\mathbf{r}_1) + \hat{h}(\mathbf{r}_2) + \frac{1}{|\mathbf{r}_1 - \mathbf{r}_2|}, \tag{1}$$

where $\hat{h}(\mathbf{r}_i) = -\nabla_i^2/2 - Z/r_i$ and $\mathbf{r}_i$, $i = 1, 2$, denotes each electron's coordinate. $-\nabla_i^2/2$ represents the electronic kinetic energy operator, $-Z/r_i$ the nucleus-i^{th} electron electrostatic (Coulombic) potential (where it is assumed that the nucleus defines the origin of the working coordinate system) and $1/|\mathbf{r}_1 - \mathbf{r}_2|$ represents the corresponding inter-electronic electrostatic potential. Z is the atomic number, which is 3 for Li^+, 10 for Ne^{8+} and 18 for Ar^{16+}.

First, we numerically solve the Schrödinger equation (SE), $(\hat{h}(\mathbf{r}) - \epsilon)\phi_{\epsilon l m_l}(\mathbf{r}) = 0$, for the mono-electronic systems Li^{2+}, Ne^{9+} and Ar^{17+}. To this end, we adopt a separation of variables approach where the one-electron orbitals are expressed as $\phi_{\epsilon l m_l}(\mathbf{r}) = [P_\epsilon(r)/r]Y_{l m_l}(\theta, \phi)$ where $Y_{l m_l}(\theta, \phi)$, are the spherical harmonic functions. Projection of $\phi_{\epsilon l m_l}(\mathbf{r})$ onto the one-electron SE, followed by angular integration leads to the one-dimensional radial differential equation for the unknown radial orbitals, $P_\epsilon(r)$:

$$\left[-\frac{1}{2}\frac{d^2}{dr^2} + \frac{l(l+1)}{2r^2} - \frac{Z}{r} - \epsilon \right] P_\epsilon(r) = 0. \tag{2}$$

The present method assumes that the radial configuration space of the electron is limited to a finite radius, R, with the boundary conditions chosen as: $P_\epsilon(0) = P_\epsilon(R) = 0$. The immediate consequence of this assumption is to allow for a finite matrix representation of the physical Hamiltonian with a discretized eigenergy spectrum including both the bound ($\epsilon < 0$) and the continuum eigenstates ($\epsilon > 0$). Moreover, we also see that both the eigenenergies and eigenstates are dependent on the particular angular momentum number, l. In the following, we adopt the following discretized notation for the eigenvalues and the eigenfunction: $\varepsilon \to \varepsilon_{nl}$ and $P_e(r) \to P_{nl}$. The positive-energy eigenstates, $\varepsilon_{nl} \geq 0$, exhibit oscillatory asymptotic behaviour, similar to that expected from continuum-like eigenstates. On the other hand, the negative-energy eigenstates, $\varepsilon_{nl} < 0$, have an exponentially decaying asymptotic behaviour associated with the bound spectrum of the finite Hamiltonian. The numerical solution then proceeds by expanding the radial orbitals, $P_{nl}(r) = \sum_i c_i^{(nl)} B_i^{(k_b)}(r)$, on a nonorthogonal set of B-spline polynomials of order k_b and total number n_b defined in the finite interval $[0, R]$ [24]. The particular choice of a B-spline basis, as opposed to a Gaussian or Slater-type basis, is dictated by their superior ability to represent the continuum solutions with great accuracy. This expansion leads to a diagonalization matrix problem, where the solution provides the unknown coefficients, $c_i^{(nl)}$ [25].

Having calculated the radial orbitals, $P_{nl}(r)$, for each partial wave $l = 0, 1, 2, ...$, we proceed with the solution of the SE for the two-electron ions Li^+, Ne^{8+} and Ar^{16+}:

$$\hat{H}_A \Psi_{EL}(\mathbf{r}_1, \mathbf{r}_2) = E_L \Psi_{EL}(\mathbf{r}_1, \mathbf{r}_2). \tag{3}$$

The calculational method proceeds along similar lines to the one-electron case. We shall expand the two-electron eigenstates, Ψ_{EL} (*configuration interaction, CI basis*), on a known two-electron basis set (*configuration basis*), $\Phi_a^{(L)}(\mathbf{r}_1, \mathbf{r}_2)$:

$$\Psi_{EL}(\mathbf{r}_1, \mathbf{r}_2) = \sum_a C_a^{(EL)} \Phi_a^{(L)}(\mathbf{r}_1, \mathbf{r}_2), \tag{4}$$

with the intention to express Equation (3) as an alegbraic equation for the C_a^{EL} coefficients. We choose as configuration states, the eigenstates of the $\hat{H}_A^{(0)}$, $\mathbf{L}^2$, $\hat{L}_z$, $\mathbf{S}^2$, $\hat{S}_z$, and $\hat{\Pi}$ (parity) operators, where $\hat{H}_A^{(0)} = \hat{h}(\mathbf{r}_1) + \hat{h}_2(\mathbf{r}_2)$, $\hat{L} = \hat{l}_1 + \hat{l}_2$ and $\hat{S} = \hat{s}_1 + \hat{s}_2$. $\hat{l}_i$ and $\hat{s}_i$ are the ith electron's angular and spin quantum operators, respectively. $\hat{H}_A^{(0)}$ is the so-called zero-order two-electron Hamiltonian (of a physically fictitious system where the two-electrons are non-interacting), $\hat{L}$ is the total angular momentum operator, $\hat{L}_z$ is its projection onto the quantization axis (chosen to be the z-axis), and $\hat{S}, \hat{S}_z$ are the total spin quantum number and its z-axis projection, respectively. The CI states in Equation (4) are characterized only by their energy, E, and total angular quantum number, L, on the basis that the interaction with the field will not affect the total magnetic quantum number, M_L, as well as the total spin, S, and its z-axis projection, M_S. These values will be equal to those of the initial state. For two electron systems, known to have a ^1S ground state, it is concluded that only states with $M_L = 0$, $S = 0$ and $M_S = 0$ are involved in the photoionization process. Accordingly, the zero-order states are fully determined if the set of L and $a \equiv (n_1 l_1; n_2 l_2)$ parameters is given. The respective zero-order energy is equal to $E_0 = \epsilon_1 + \epsilon_2$. Therefore, the configuration basis set is comprised of singlet ($S = 0$), spatially antisymmetric, angularly coupled, products of one-electron orbitals with $m_l = 0$:

$$\Phi_{n_1 l_1; n_2 l_2}^{(L)}(\mathbf{r}_1, \mathbf{r}_2) = C_{000}^{l_1 l_2 L} \hat{A}_{12} \left[\phi_{n_1 l_1}(\mathbf{r}_1) \phi_{n_2 l_2}(\mathbf{r}_2) \right], \tag{5}$$

where $\hat{A}_{12}$ is the antisymmetrization operator and $C_{000}^{l_1 l_2 L}$ is the Clebsch–Gordan coefficient that ensures $M_L = 0$ [26]. Projection of the above zero-order basis states onto Equation (3) leads again to a matrix diagonalization problem, the solution of which provides the CI energies, E, and CI coefficients, $C_a^{(EL)}$. The physical interpretation of the coefficient is that $|C_a^{(EL)}|^2$ represents the contribution of the (un-correlated) configuration, Φ_a^L (characterized by the set '$a = (n_1 l_1; n_2 l_2)$'), in the formation of the CI state, Ψ_{EL}, with energy E and angular momentum number L.

2.2. Two- and Three-Photon Ionization Cross Section Formulation

According to LOPT, the N-photon partial-wave ionization cross section, following the absorption of N photons of energy ω, from a system in its ground state $|g\rangle$ (of energy E_g) to a final continuum state of energy E and angular momentum L, is given by [10,27]:

$$\sigma_L^{(N)}(\omega) = 2\pi(2\pi\alpha)^N \omega^N |M_{E_L}^{(N)}|^2 \, \delta(E - E_g - N\omega), \tag{6}$$

where α is the fine structure constant. The total N-photon cross section is given by $\sigma_N(\omega) = \sum_L \sigma_L^{(N)}(\omega)$. The N-photon transition amplitude, $M_{E_F L_F}^{(N)}$, in the case of two-photon absorption, $N = 2$, reduces to:

$$M_{E_L}^{(2)} = \sum_{E_P'} \frac{D_{g;E_P'} D_{E_P';E_L}}{E_g + \omega - E_P'}, \qquad L = S, D, \tag{7}$$

while for three-photon absorption, $N = 3$, it is given by:

$$M_{E_P}^{(3)} = \sum_{E_P'} \frac{D_{g;E_P'}}{E_g + \omega - E_P'} \left[\sum_{E_S'} \frac{D_{E_P';E_S'} D_{E_S';E_P}}{E_g + 2\omega - E_S'} + \sum_{E_D'} \frac{D_{E_P';E_D'} D_{E_D';E_P}}{E_g + 2\omega - E_D'} \right], \tag{8}$$

$$M_{E_F}^{(3)} = \sum_{E_P'} \sum_{E_D'} \frac{D_{E_g;E_P'} D_{E_P';E_D'} D_{E_D';E_F}}{(E_g + \omega - E_P')(E_g + 2\omega - E_D')}, \tag{9}$$

where $D_{E_L;E_{L'}'} \equiv \langle \Psi_{EL} | \hat{D} | \Psi_{E'L'} \rangle$ is the two-electron dipole matrix element and $\Psi_{EL} = \Psi_{EL}(\mathbf{r}_1, \mathbf{r}_2)$ is the CI two-electron eigenstate given by Equation (4). The integrals over the angular symmetries include both the bound and continuum states. The length form of the interaction operator is $\hat{D} = -\hat{e}_p \cdot (\mathbf{r}_1 + \mathbf{r}_2)$ and the velocity form is $\hat{D} = \hat{e}_p \cdot (\nabla_1 + \nabla_2)/\omega$, with $\hat{e}_p$ representing the polarization unit vector of the radiation. The detailed expressions of the dipole matrix elements in terms of the calculated CI two-electron wavefunctions can be found in Ref. [23].

At this point, it is appropriate to comment on the adoption of the dipole approximation used in evaluating transition amplitudes. In the present context of the X-ray regime, the use of the dipole approximation is justified by considering the scaling of the dipole transitions of the one-electron (hydrogenic) systems (see Equation (2)). The mean distance of the electron from the ionic core scales with the nuclear charge, Z, as $\langle r \rangle \sim 1/Z$, while the ionization potential scale as $IP(Z) \sim Z^2$. In the present work, we assume transitions with X-ray photons below the first ionization threshold, $\omega < |E(1s) - E(1s^2)|$. Since generally $|E(1s) - E(1s^2)| < IP(Z)$, we thus have $\omega < Z^2$. The dipole approximation is based on the validity of $|\mathbf{k}_\gamma \cdot \mathbf{r}| << 1$, where $\mathbf{k}_\gamma$ is the wavevector of the X-ray photon and $\mathbf{r}$ the electron's position. From the above considerations, we then have $|\mathbf{k}_\gamma \cdot \mathbf{r}| \simeq k_\gamma \langle r \rangle = \omega \langle r \rangle / c \sim Z^2/(Zc) = Z/c << 1$, for all $Z = 3, 10, 18$ that we consider here ($c \simeq 137.036$).

The angular momentum of the final states, following photoabsorption ($L = 0, 2$ for two-photon absorption and $L = 1, 3$ for three-photon absorption), is dictated by the selection rules implicit in the dipole matrix element. For the present case of linearly polarized light, we have $\Delta M_L = 0$ and $\Delta L = \pm 1$ [26]. Since we start from the ion's ground state ($M_L = 0$), we only need to keep the transitions between the $M_L = 0$ states.

3. Results and Discussion

In Table 1, we give information related to the CI basis used for the various symmetries $^1S, ^1P, ^1D$ and ^1F. The configuration states, $\Phi_{n_1 l_1; n_2 l_2}^{(L)}(\mathbf{r}_1, \mathbf{r}_2)$, have been constructed according to Equation (5) by one-electron orbitals with angular momenta given in Table 1 and energies sufficiently high to ensure convergence of the results. The order of B-splines was $k_b = 9$ with the total number of

B-spline polynomials set at $n_b = 110$ for Li^+ and $n_b = 170$ for Ne^{9+} and Ar^{16+}. The box radius varied between $R = 50 - 58$ a.u. for Li^+, $R = 20 - 28$ a.u. for Ne^{8+} and $R = 10 - 15$ a.u. for Ar^{16+}. The knot sequence of the spatial grid for the B-spline basis was linear. The two-electron wavefunctions, $\Phi^{(L)}_{n_1 l_1; n_2 l_2}(\mathbf{r}_1, \mathbf{r}_2)$, have been constructed from the zero-order configurations by one-electron orbitals with angular momenta given in the mentioned table and energies determined by the indices n_1, n_2 in the following ranges: $1 \leq n_1 \leq 6$ and $1 \leq n_2 \leq n_b$. The relationship of the indices n_1, n_2 to the energies of the zero-order wavefunctions depends on the basis size parameters such as the maximum value of the box radius as well as the number of B-spline basis functions used. In summary, the whole basis, for each symmetry, resulted in the inclusion of the following number of functions for each ion: 1650–1940 for Li^+, 1600 for Ne^{8+} and 2040 for Ar^{16+}.

Table 1. (l_1, l_2) electronic configurations included in the configuration interaction (CI) calculations for the ^1L symmetries.

^1S	^1P	^1D	^1F
s^2	sp	sd	sf
p^2	pd	p^2	pd
d^2	df	pf	pg
f^2	fg	d^2	df
g^2		dg	fg
		f^2	

In Table 2, we show the calculated energies for a few lower bound states of the hydrogenic ions Li^{2+}, Ne^{9+} and Ar^{17+}. The degree of the agreement with those reported in the National Institute of Standards and Technology (NIST) atomic spectra database [28] is given in the last row of the table. In all cases, the percentage discrepancies of the ground state energies (between the calculated and those of the NIST database) are of the order 0.1% while the excited states are of similar or even smaller order.

Table 2. Energies of the few lowest states of the one-electron Li^{2+}, Ne^{9+} and Ar^{17+} ions. Energies are given in units of eV. Energies are relative to the single-ionization threshold. $\delta E(1s) \equiv |E(1s) - E_g^{Z+}|/|E_g^{Z+}|$ is the relative discrepancy (between our calculated ground state values with the energies (E_g^{Z+}) listed in the NIST database [28]). The relative discrepancy for excited states is generally less than $\delta E(1s)$.

State	Li^{2+}	Ne^{9+}	Ar^{17+}
$1s$	-122.451	-1360.57	-4408.2
$2s, 2p$	-30.613	-340.14	-1102.1
$3p, 3d$	-13.606	-151.18	-489.8
$4d, 4f$	-7.653	-85.04	-275.5
$5f$	-4.898	-54.42	-176.3
$\delta E(1s)$	2.5×10^{-5}	1.2×10^{-3}	4.1×10^{-3}

In Table 3, we show the energy differences of the few lowest states of Li^+, Ne^{8+} and Ar^{16+} ions with respect to the respective ground state $(1s^2 \, ^1S_0)$ energy of each ion. In all cases, the percentage discrepancies of the ground state energies (between the calculated and those of the NIST) are of the order 0.1% while the excited states are of similar or even of smaller order. This is not surprising as the role of correlation is more important in the lower energy states, where the electrons are on average closer relative to the higher-energy states.

Having examined the reliability of the calculated electronic structure, within the available theoretical and experimental data to compare, we proceed to the main subject of this work: the presentation of ionization cross sections for a range of photon frequencies. In all the following, the horizontal axis in the figures represents the photon energy, given in eV. The cross sections are given in SI units, $cm^{2N}s^{N-1}$, where N is the order of the process (here equivalent with the number of the photons absorbed). The final angular momenta, following two-photon absorption are the $^1S, ^1D$ continua, while in the case of three-photon absorption are the $^1P, ^1F$ continua. The cross sections have been evaluated using both the length and the velocity forms of the dipole operator. They generally have excellent agreement throughout all the spectral regions considered, especially for the non-resonant ones. Relative agreement between the length and the velocity forms is important since it provides strong evidence that the dipole matrix elements, contributing in the multiphoton transition amplitude, have been converged.

Table 3. Energy differences of the few lowest states of Li^+, Ne^{8+} and Ar^{16+} ions with respect to their respective ground state $(1s^2\ ^1S_0)$ energy value. We use boldface for the states' notation to emphasize that they are listed according the dominant configuration in the CI expansion of Equation (4). In the last row, $\delta E(1s^2) \equiv |E(1s^2) - E_g^{Z+}|/|E_g^{Z+}|$ is the relative discrepancy (between our calculated ground state values and the energies (E_g^{Z+}) listed in the NIST database [28]). $E(1s^2)$ is given relative to the double ionization threshold.

State	Li$^+$	Ne^{8+}	Ar^{16+}
1s2s	60.42	914.0	3114.1
1s2p	61.67	920.4	3126.7
1s3s	68.73	1070.1	3666.1
1s3d	69.02	1071.5	3668.7
1s3p	69.09	1072.0	3669.7
1s4s	71.54	1124.5	3858.7
1s4d	71.66	1125.1	3859.9
1s4p	71.67	1125.3	3860.2
1s4f	71.67	1125.1	3859.8
1s5s	72.83	1149.6	3947.7
1s5d	72.89	1148.9	3948.3
1s5p	72.90	1150.0	3948.5
1s5f	72.89	1149.9	3948.3
$E(1s^2)$	-197.518	-2554.53	-8513.84
$\delta E(1s^2)$	2.9×10^{-3}	1.36×10^{-3}	3.87×10^{-3}

In Figures 1–3, we show the photon energy dependence of the calculated two photon partial-ionization cross sections, $\sigma_S^{(2)}(\omega), \sigma_D^{(2)}(\omega)$, of Li^+, Ne^{8+} and Ar^{16+}, respectively, from the ground state, $1s^2(^1S)$, to final states of symmetry $^1S, ^1D$. Summing the latter, we obtain the corresponding total two-photon ionization cross sections, $\sigma_2(\omega)$. For clarity, we have plotted only the length-form results. Generally, for all three ions, the dominant two-photon ionization channel is the ^1D symmetry. The cross sections exhibit strong peak structures, which appear in both the ^1S and ^1D final symmetries, due to one-photon resonance with the intermediate states $1snp\ ^1P, n = 2, 3, ...$ Apart from these intermediate-resonance peaks, there are further peaks due to strong configuration mixing of the type $npn'p, n, n' = 2, 3...$, associated with the $^1S, ^1D$ continua.

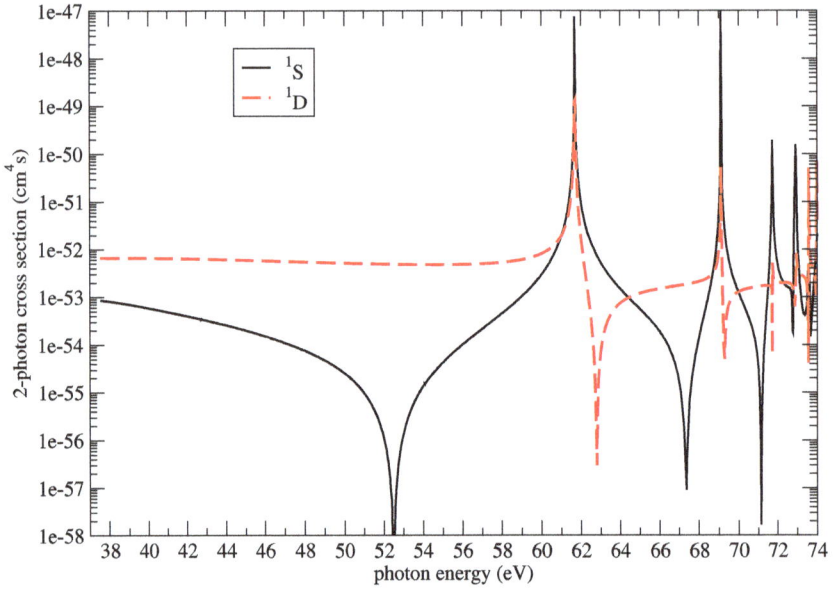

Figure 1. Two-photon partial ionization cross sections of Li^+ from its ground state with linearly polarized light. The characteristic peaks in the cross section are associated with the intermediate bound states of the symmetry ^1P.

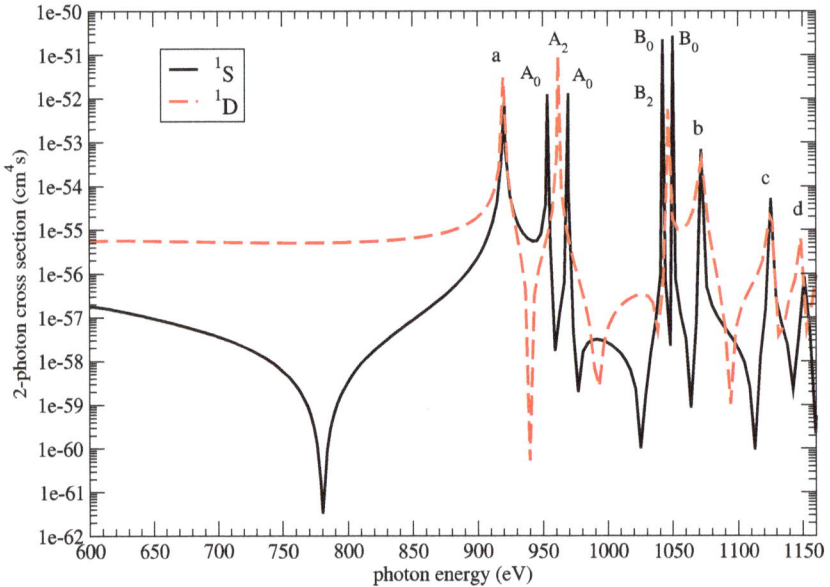

Figure 2. Two-photon partial ionization cross sections of Ne^{8+} from its ground state with linearly polarized light. The small letters (a, b, c) indicate features associated with the intermediate bound states of symmetry ^1P while the capital letters $(A_i, B_i, i = 0, 2)$ are associated with final state correlations.

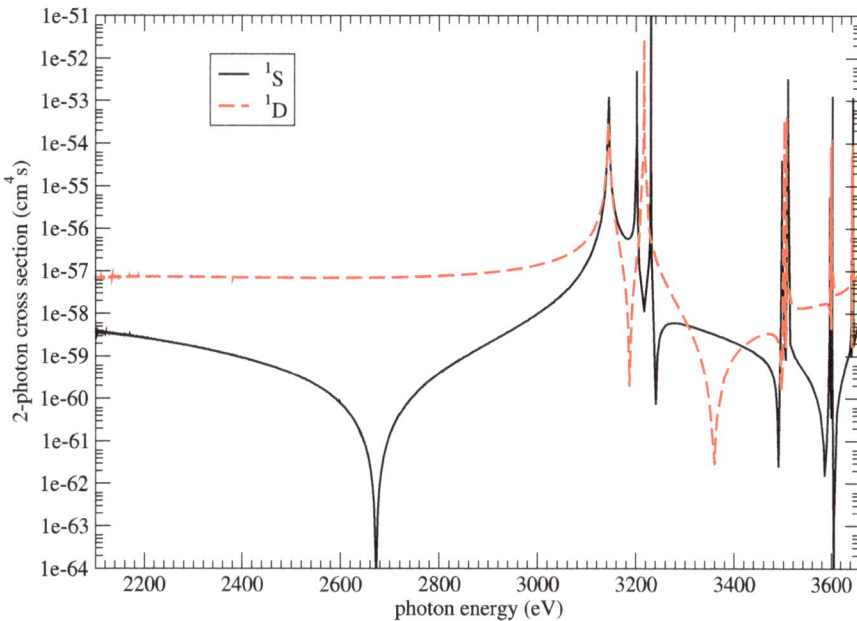

Figure 3. Two-photon partial ionization cross sections of Ar^{16+} from its ground state with linearly polarized light.

The configuration-mixing peaks are absent for two-photon ionization of Li^+ ionization since these would occur at higher photon energy, past about 75 eV, as the lowest post-ionization energy levels are about 150 eV and half that value (two photons) is about 75 eV. When we add the ^1S and ^1D cross sections to obtain the total cross section, we find a relatively good agreement with that of Ref. [15], which models two-photon ionization of Li^+ employing a less elaborate approach, namely, single-channel quantum defect theory; when our shift is accounted for, our peaks occur close to that work (62.2 eV $1s2p(^1P)$, 69.7 eV $1s3p(^1P)$, 72.3 eV $1s4p(^1P)$ and 73.5 eV $1s5p(^1P)$), and our (partial-wave sum) cross section baseline between 50–55 eV is within the same order of magnitude as in Ref. [15] (both between 10^{-53}–10^{-52} cm^4·s); however, our shape is slightly different here, being slightly convex (downward) in this region.

For Ne^{8+} and Ar^{16+}, our values are also in good agreement with the Green-function calculations in Ref. [18]. In addition, for Ne^{8+}, we have also compared our values with the second-order perturbation theory calculation of Novikov and Hopersky [16]; their $1s2p(^1P)$ and $1s3p(^1P)$ one-photon resonance peaks (these are their only peaks) are comparable to ours occurring at around 920 eV and 1070 eV, respectively. Their cross section base-line (non-resonant part) circa 600–800 eV is also close to ours, i.e., within the same order of magnitude (both between 10^{-56}–10^{-55} cm^4·s).

In relation to the intermediate resonance peaks in Table 3, we show the energy differences of the few lowest states of Li^+, Ne^{8+} and Ar^{16+} ions from their respective ground state ($1s^2\,^1S_0$) energy, $E_g = E_{1s^2}$, namely, $\Delta E_P \equiv E_P - E_g$, (i.e., corresponding to peaks (a–d) in Figure 2). Their importance is derived from the fact that these energy differences appear in the denominator of the two-photon cross section expression, Equation (7), i.e., $E_g + \omega - E'_P = \omega - \Delta E_P$. It is then immediately evident that the photon energy detuning from these energy differences generates a series of characteristic features in the cross section. A word of caution is necessary at this point: within the current formulation, the height of these peaks becomes infinite in the exact on-resonance case, $\Delta E_P = \omega$. The first point to note is that we have ignored the inherent spontaneous decay width of the intermediate bound states. This would have served only to remove the unphysical singularities that occur at the resonances positions. However, most importantly, the LOPT cross sections fail to provide the correct ionization yields. In other words, in the resonant case, the LOPT relation for the ionization yield,

$$W_{gf} = \sigma_N F^N, \tag{10}$$

(F is the pulse's flux) becomes invalid. It is well established that, for resonant and near-resonant processes, while perturbation theory is still valid, an alternative formulation is required for the calculation of the expected ionization yields. This formulation, in addition to the ionization of the system directly from the ground state, takes into account the stepwise formation and the subsequent ionization of the intermediate states. Without going into the details, we shall only mention that in such cases the formulation should be developed in terms of a density matrix representation in combination with a proper representation of the spatiotemporal profile of the laser field. At this point, it is worth noting the pulse properties for which the LOPT cross sections presented here are valid. Equation (10) is only just marginally applicable for a pulse with its detuning from a resonance of the same order as its bandwidth, $\Delta E_p \sim \gamma_L$, where γ_L is the bandwidth. If we take as a rough rule that the bandwidth of the pulse is, say, a $1/100$-th of its average photon energy ($\gamma_L \sim \omega/100$), that would mean that the cross section values within the range 61.47 ± 0.61 eV (first peak in the two-photon cross section for Li^+) cannot be safely used in combination with Equation (10). Similar considerations should be assumed for the higher peaks. For completeness, for Ne^{8+} and Ar^{16+}, if based again on the appearance of the first peaks in the two-photon cross section (see the first row for the $1s2p$ state in Table 3), the corresponding intervals are scaled upwards to ± 9.2 eV ($\sim 920/100$) and ± 31 eV ($\sim 3126/100$), respectively. A case that such a discrepancy between the LOPT two-photon ionization cross section [16] and the experimental value [29] is attributed to the bandwidth of the X-ray pulse can be found in Ref. [17]. More specifically, for an X-ray photon energy of 1110 eV (in between (b) and (c) peaks in Figure 2), the reported experimental value was 7×10^{-54} cm^4·s [29] while the theoretical cross section based on a Hartree–Fock–Slater (HFS) model was found to be equal to 4.0×10^{-57} cm^{4}s. Our calculated value for this photon energy is about 4.6×10^{-57} cm^{4}s for the ^1D wave while the ^1S value makes a negligible contribution to the total cross section.

For Ne^{8+} and Ar^{16+}, the 'twin' peaks (see A_0 peaks in Figure 2), exclusive to the ^1S symmetry, are due to the strong coupling between the $2s2s$ and $2p2p$ configurations in the expansion Equation (4). Since in the ^1D symmetry the $2s2s$ configuration is missing, we observe only one peak (A_2 in Figure 2), in between the A_0 ones. To confirm this, we have performed some further tests where, for example, we excluded the Φ_{2s2s}^{1S}, Φ_{2p2p}^{1S} zero-order states (separately each time) from the CI wavefunction, Equation (4). By doing this, we obtain a cross section with only one A_0 peak at the same position where the A_2 peak appears. This suggests that the observed (two) A_0 peaks are the result of strong-mixing of the $2s2s$ and $2p2p$ configurations, mainly due to their proximity in energy.

Similar considerations hold for the peaks B_0, B_2 at higher photon energies. The B_0 twin peaks are due to the mixing of the Φ_{2s3s}, Φ_{2p3p} configurations in the ^1S symmetry, while the B_2 peak is due to the $3p^2$ state exclusively. We mention here that these doubly excited (highly correlated) states are also known as autoionizing states as they are associated with a temporal trap of the two excited electrons in the core's region, eventually leading to the ejection of one of them and the residual (higher-charged ion) to its ground state. In the present (static) context of the CI calculation, these doubly excited states are degenerate with the $1s^2\ ^1S$ or $1sd\ ^1D$ continua, which eventually cause their radiationless (auto)-ionization [30].

In Figures 4–6, we present our calculated three photon partial-ionization cross sections of Li^+, Ne^{8+} and Ar^{16+}, respectively, from the ground state $1s^2(^1S)$. The final angular momentum of the ions, following three-photon absorption, are the $^1P, ^1F$ continua, all being of singlet symmetry. Similarly, as in the two-photon case, the total three-photon ionization cross section is obtained by the addition of the ^1P and ^1F partial-wave cross sections. Again, the final state is dominated by configurations with the residual ion in its ground state and the ejected electron with angular momentum $l = 1$ for the ^1P and $l = 3$ for the ^1F symmetry. The three-photon cross sections exhibit strong peak structures, which appear in both ^1P and ^1F final symmetries, due to two-photon resonances with the intermediate states $1s3d, 1s4d, ...$ (see denominators $(E_g + 2\omega - E_L, L = S, D)$ in Equations (8) and (9)). In the ^1P symmetry, there are additional peaks due to two-photon resonance states of the type $1sns\,^1S, n = 2, 3, ...$ Because the $1s3s$ and $1s3d$ states have slightly different energy positions, the intermediate resonance peaks for the ^1P and ^1F (for example circa 34.4 eV for Li$^+$) do not generally coincide. Note that the ^1P final states are reached by the coherent superposition of two ionization absorption channels: $S \rightarrow P \rightarrow S \rightarrow P$ and $S \rightarrow P \rightarrow D \rightarrow P$. In contrast, the ^1F states are reached only via one ionization channel, namely: $S \rightarrow P \rightarrow D \rightarrow F$.

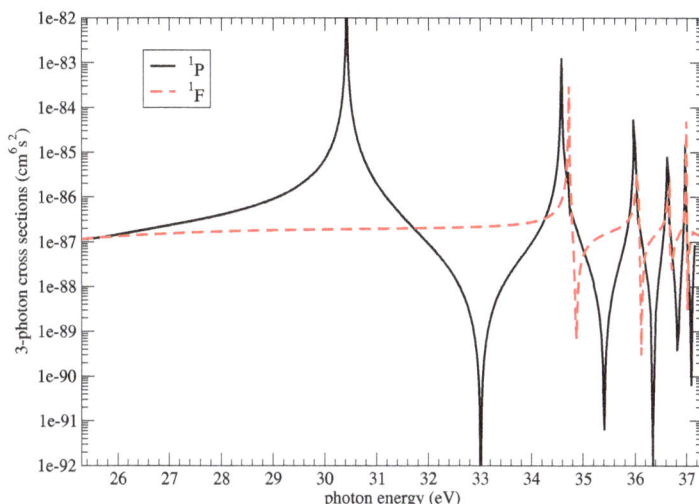

Figure 4. Three-photon ionization partial cross sections of Li$^+$ from its ground state with linearly polarized light.

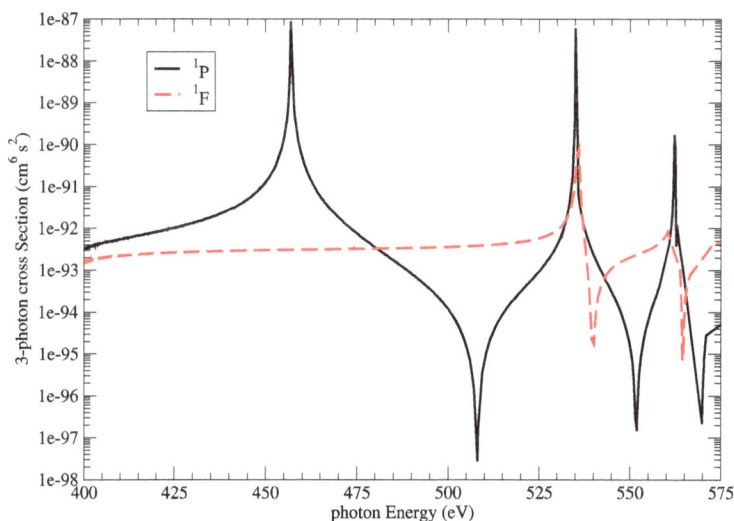

Figure 5. Three-photon ionization partial cross sections of Ne^{8+} from its ground state with linearly polarized light.

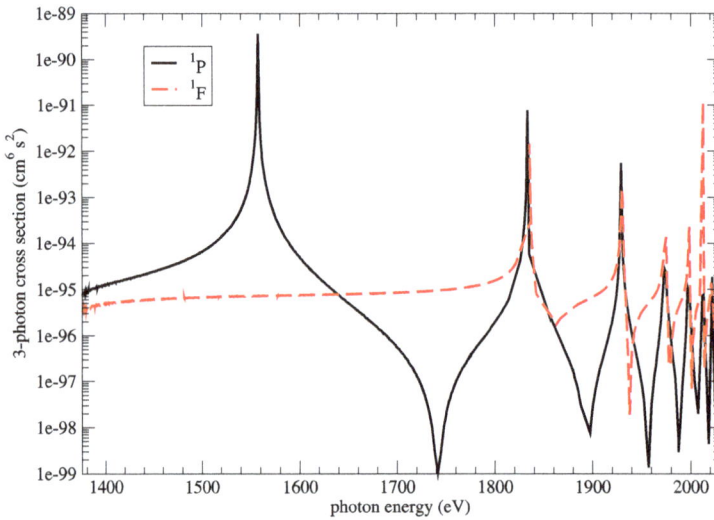

Figure 6. Three-photon ionization partial cross sections of Ar^{16+} from its ground state with linearly polarized light.

At this point, it might be worth comparing the cross sections among the three investigated ions. It is clearly visible from the figures that generally the ionization cross section decreases from Li^+ towards Ar^{16+}. This observation is rather consistent with the (exact) scaling, $1/Z^{4N-2}$, of the N-photon cross section for hydrogenic systems [11]. A second point worth mentioning is that the cross sections ending on the higher symmetry, ^1D for two-photon and ^1F for three-photon ionization, are proportional to the cross sections for circularly polarized light. To be more specific, due to the dipole selection rules, ionization by circularly polarized light will proceed through intermediate states where M_L will change either by +1 or by −1 monotonously. For example, let us assume circularly polarized light which causes a change of the magnetic quantum number by $\Delta M_L = +1$. This means that if we start from the ground state, where $L = 0$ and $M_L = 0$, then the first ionization step will involve only states with $M_L = +1$. Similarly, the next ionization step will involve states that differ by +1 from the previous step, meaning that only states with $M_L = +2$ will be accessed in this step. In this case, these states will necessarily have $L = 2$. Accordingly, if further ionization occurs (three-photon ionization), for the same reason, only states with $M_L = +3$ will be reached, thus ending necessarily with an $L = 3$ total orbital angular momentum. Similar considerations hold had we started by circularly polarized light with opposite helicity, leading to a change of $\Delta M_L = -1$. In short, for two-photon ionization, only the ionization path $S \rightarrow P \rightarrow D$ is allowed. Accordingly, for three-photon ionization, only the $S \rightarrow P \rightarrow D \rightarrow F$ ionization channel will occur. Now the crucial observation is that the transition amplitudes by circularly and linearly polarized light for these ionization paths differ only by the total magnetic quantum number, M_L. For linearly polarized light, it is $\Delta M_L = 0$, while for circularly polarized light, it is $|\Delta M_L| = 1$. Straightforward angular momentum algebra for these transition amplitudes ($T_C(T_L)$ for circular(linear)-polarized light) of two-electron states shows that they differ only by a proportional factor; namely, $|T_C(D)/T_L(D)| = \sqrt{3/2}$ for two-photon ionization and $|T_C(F)/T_L(F)| = \sqrt{5/2}$, for three-photon ionization, where D and F denote the final angular momentum channel for two-photon and three-photon ionization, respectively [31]. Nevertheless, note that circularly polarized light does not guarantee that the total ionization cross section will provide higher rates relative to the ionization by linearly polarized light; the final rate depends on the number of available ionization paths available and the electric field intensity and, in fact, high N-photon ionization by linearly polarized light is more effective.

4. Conclusions

In conclusion, motivated by the wide array of X-ray facilities worldwide that are able to investigate nonlinear interactions with atomic systems on their natural time scales, we have presented calculations for two- and three-photon partial ionization cross sections of Li^+, Ne^{8+} and Ar^{16+} ions. These systems are of high experimental and theoretical interest for the study of interaction of strong and ultrashort X-ray radiation at a fundamental level. We have identified that the ionization cross sections are dominated by a series of intermediate (one or two-photon) resonance peaks in addition to peaks due to doubly excited structures. We have noticed the trend that from the lower Z ion, Li^+, to the higher ones, the ionization cross sections generally decrease.

Acknowledgments: This work was supported by the Science Foundation Ireland under Grant No. 12/IA/1742 and enabled by the EU Education, Audio-visual and Culture Executive Agency (EACEA) Erasmus Mundus Joint Doctorate Programme EXTATIC, Project No. 2013-0033. This work is also associated with the FP7 EU COST Actions MP1203 and CM1204.

Author Contributions: W.H. carried out the calculations and analysed the results. J.T.C. co-supervised W.H. L.A.A.N. supervised W.H. and the project, developed the code and contributed to the analysis of the results. W.H., J.T.C. and L.A.A.N. wrote the paper.

Conflicts of Interest: The authors declare no conflict of interest.

References

1. Glover, T.; Hertlein, M.P.; Southworth, S.H.; Allison, T.K.; van Tilborg, J.; Kanter, E.P.; Krässig, B.; Varma, H.R.; Rude, B.; Santra, R.; et al. Controlling X-rays with light. *Nat. Phys.* **2010**, *6*, 69–74.
2. Tamasaku, K.; Nagasono, M.; Iwayama, H.; Shigemasa, E.; Inubushi, Y.; Tanaka, T.; Tono, K.; Togashi, T.; Sato, T.; Katayama, T.; et al. Double Core-Hole Creation by Sequential Attosecond Photoionization. *Phys. Rev. Lett.* **2013**, *111*, 043001.
3. Young, L.; Kanter, E.P.; Krässig, B.; Li, Y.; March, A.M.; Pratt, S.T.; Santra, R.; Southworth, S.H.; Rohringer, N.; DiMauro, L.F.; et al. Femtosecond electronic response of atoms to ultra-intense X-rays. *Nature* **2010**, *466*, 56–61.
4. Fukuzawa, H.; Son, S.-K.; Motomura, K.; Mondal, S.; Nagaya, K.; Wada, S.; Liu, X.-J.; Feifel, R.; Tachibana, T.; Ito, Y.; et al. Deep Inner-Shell Multiphoton Ionization by Intense X-ray Free-Electron Laser Pulses. *Phys. Rev. Lett.* **2013**, *110*, 173005.
5. Ghimire, S.; Fuchs, M.; Hastings, J.; Herrmann, S.C.; Inubushi, Y.; Pines, J.; Shwartz, S.; Yabashi, M.; Reis, D.A. Nonsequential two-photon absorption from the K shell in solid zirconium. *Phys. Rev. A* **2016**, *94*, 043418.
6. Costello, J.; Kennedy, E.; Nikolopoulos, L. Short wavelength free electron lasers. *J. Mod. Opt.* **2016**, *63*, 285–287.
7. Pellegrini, C.; Marinelli, A.; Reiche, S. The physics of X-ray free-electron lasers. *Rev. Mod. Phys.* **2016**, *88*, 015006.
8. Bostedt, C.; Boutet, S.; Fritz, D.; Huang, Z.; Lee, H.; Lemke, H.; Robert, A.; Schlotter, W.; Turner, J.; Williams, G. Linac Coherent Light Source: The first five years. *Rev. Mod. Phys.* **2016**, *88*, 015007.
9. Falcone, R.; Dunne, M.; Chapman, H.; Yabashi, M.; Ueda, K. Frontiers of free-electron laser science II. *J. Phys. B* **2016**, *49*, 180201.
10. Lambropoulos, P. Topics on multiphoton processes. *Adv. At. Mol. Phys.* **1976**, *12*, 87–164.
11. Madsen, L.; Lambropoulos, P. Scaling of hydrogenic atoms and ions interacting with laser fields: Positronium in a laser field. *Phys. Rev. A* **1999**, *59*, 4574–4579.
12. Karule, E. On the evaluation of transition matrix elements for multiphoton processes in atomic hydrogen. *J. Phys. B* **1971**, *4*, L67.
13. Potvliege, R.M.; Shakeshaft, R. High-order above-threshold ionization of hydrogen in perturbation theory. *Phys. Rev. A* **1989**, *39*, 1545–1548.
14. Saenz, A.; Lambropoulos, P. Theoretical two-, three- and four-photon ionization cross sections of helium in the XUV range. *J. Phys. B* **1999**, *32*, 5629.
15. Emmanouilidou, A.; Hakobyan, V.; Lambropoulos, P. Direct three-photon triple ionization of Li and double ionization of Li^+. *J. Phys. B* **2013**, *46*, 111001.

16. Novikov, S.A.; Hopersky, A.N. Two-photon excitation-ionization of the 1s shell of highly charged positive atomic ions. *J. Phys. B* **2001**, *34*, 4857–4863.
17. Sytcheva, A.; Pabst, S.; Son, S.; Santra, R. Enhanced nonlinear response of Ne8+ to intense ultrafast X-rays. *Phys. Rev. A* **2012**, *85*, 023414.
18. Koval, P. Two-Photon Ionization of Atomic Inner-Shells. Ph.D. Thesis, University of Kassel, Kassel, Germany, 2004.
19. Chang, T.; Kim, Y. Theoretical study of the two-electron interaction in alkaline-earth atoms. *Phys. Rev. A* **1986**, *34*, 2609–2613.
20. Tang, X.; Chang, T.; Lambropoulos, P.; Fournier, S.; DiMauro, L. Multiphoton ionization of magnesium with configuration interaction calculations. *Phys. Rev. A* **1990**, *41*, R5265.
21. Chang, T.; Tang, X. Photoionization of two-electron atoms using a nonvariatonal configuration interaction approach with discretized finite basis. *Phys. Rev. A* **1991**, *44*, 232–238.
22. Chang, T. *Many-Body Theory of Atomic Structure*; Chapter B-Spline Based Configuration-Interaction Approach for Photoionization of Two-Electron and Divalent Atoms; World Scientific: Singapore, 1993; pp. 213–247.
23. Nikolopoulos, L.A.A. A package for the ab initio calculation of one- and two-photon cross sections of two-electron atoms, using a CI B-splines method. *Comput. Phys. Commun.* **2003**, *150*, 140–165.
24. De Boor, C. *A Practical Guide to Splines*; Springer: New York, NY, USA, 1978.
25. Bachau, H.; Cormier, E.; Decleva, P.; Hansen, J.E.; Martin, F. Applications of B-splines in Atomic and Molecular Physics. *Rep. Prog. Phys.* **2001**, *64*, 1815–1942.
26. Sobel'man, I. *Introduction to the Theory of Atomic Spectra*; Pergamon Press Ltd.: Oxford, UK, 1972.
27. Nikolopoulos, L.A.A. Mg in electromagnetic fields: Theoretical partial multiphoton cross sections. *Phys. Rev. A* **2005**, *71*, 033409.
28. Kramida, A.; Yu, R.; Reader, J.; NIST ASD Team. *NIST Atomic Spectra Database (ver. 5.3)*; National Institute of Standards and Technology: Gaithersburg, MD, USA, 2015.
29. Doumy, G.; Roedig, C.; Son, S.-K.; Blaga, C.I.; di Chiara, A.D.; Santra, R.; Berrah, N.; Bostedt, C.; Bozek, J.D.; Bucksbaum, P.H.; et al. Nonlinear atomic response to intense ultrashort X-rays. *Phys. Rev. Lett.* **2011**, *106*, 083002.
30. Fano, U. Effects of Configuration Interaction on Intensities and Phase Shifts. *Phys. Rev.* **1961**, *124*, 1866–1878.
31. Nikolopoulos, L.A.A. Dublin City University, Dublin, Ireland. Unpublished notes.

applied
sciences

MDPI

Article

State-Population Narrowing Effect in Two-Photon Absorption for Intense Hard X-ray Pulses

Krzysztof Tyrała [1], Klaudia Wojtaszek [1], Marek Pajek [1], Yves Kayser [2], Christopher Milne [3], Jacinto Sá [4,5] and Jakub Szlachetko [1,4,*

[1]   Institute of Physics, Jan Kochanowski University, Kielce 25-001, Poland; krzysztof.tyrala1@wp.pl (K.T.); klaudia.wojtaszek@interia.pl (K.W.); m.pajek@ujk.edu.pl (M.P.)
[2]   Physikalisch-Technische Bundesanstalt (PTB), Berlin 10587, Germany; yves.kayser@ptb.de
[3]   Paul Scherrer Institute, Villigen 5232, Switzerland; chris.milne@psi.ch
[4]   Institute of Physical Chemistry, Polish Academy of Sciences, Warsaw 01-224, Poland; jacinto.sa@kemi.uu.se
[5]   Department of Chemistry-Ånsgtröm Laboratory, Uppsala University, Uppsala 752 36, Sweden
*   Correspondence: jakub.szlachetko@ujk.edu.pl; Tel.: +48-41-349-6440

Academic Editor: Kiyoshi Ueda
Received: 29 April 2017; Accepted: 21 June 2017; Published: 24 June 2017

Abstract: We report on studies of state-populations during the two-photon absorption process using intense X-ray pulses. The calculations were performed in a time-dependent manner using a simple three-level model expressed by coupled rate equations. We show that the proposed approach describes well the measured rates of X-rays excited in the one-photon and two-photon absorption processes, and allows detailed investigation of the state population dynamics during the course of the incident X-ray pulse. Finally, we demonstrate that the nonlinear interaction of X-ray pulses with atoms leads to a time-narrowing of state populations. This narrowing-effect is attributed to a quadratic incidence X-ray intensity dependence characteristic for nonlinear interactions of photons with matter.

Keywords: X-ray nonlinear processes; two-photon absorption; rate equations

1. Introduction

In X-ray spectroscopy, the photon absorption and emission processes are usually treated within a weak photon-matter interaction limit, which implies that photon-in and photon-out processes are linearly correlated. However, for strong enough X-ray fields, the nonlinear regime may be accessed and multi-photon processes are possible, leading to a nonlinear dependence of the X-ray photons in/out signals. Such a nonlinear regime, available for many years at optical frequencies in laser spectroscopy [1–4], remained an unexplored area at X-ray wavelengths due to the lack of strong enough X-ray sources. The ability to access nonlinear light-matter interactions at X-ray wavelengths became possible only recently thanks to the development of X-ray Free Electron Lasers (XFELs) [5,6]. In contrast to the optical laser wavelengths, the photon-atom interaction at hard X-ray wavelengths involves bound core-electrons, and leads to the excitation of intermediate electronic states with sub-femtosecond lifetimes [7–10]. Consequently, at femtosecond-durations, the intense pulses made available by XFELs presently allow us to access a thus-far-uninvestigated area of physics, and to probe the physical mechanisms that drive the nonlinear interaction of X-rays with matter [11–13].

With regard to applications, the two-photon absorption (TPA) process using hard X-rays allows access to different excitation states of matter, thanks to the specific selection rules for electronic transitions. In contrast to the one photon absorption (OPA) process, which is determined by dipole-allowed transitions, the TPA process requests a change in the electron orbital quantum number by ± 2 or 0, thus allowing access to quadrupole or forbidden excitations. This opens new avenues for studying unexplored electronic states of matter via two-photon absorption. However, the application of TPA at XFELs for studying electronic excitations is still problematic in practice. Because of low cross-sections, the TPA process is competing with other first- and second-order photon-atom interactions, which have to be considered, in particular, for ultrashort X-ray pulses. Consequently, knowledge of the time-evolution of the population of states, as well as on the subsequent X-ray emission process for ultrashort X-ray pulses, is crucial in order to interpret the experimental results from XFELs.

In the present paper, we combine the results of a TPA/OPA experiment performed in the Linac Coherent Light Source (LCLS) [13], in which the X-rays indicating the TPA and OPA processes were measured, with time-dependent calculations allowing us to interpret the one- and two-photon absorption process for ultrashort hard X-ray pulses. In these calculations, we used a three-level atomic state model to describe the time-evolution of a population of excited electronic states during the course of a femtosecond-duration X-ray pulse. In this way, we could follow the time dependences of population states and subsequent X-ray emission for both linear and nonlinear atomic response. The predictions of the presented model are in good agreement with the measured X-ray rates corresponding to one- and two-photon absorption in copper near the K-absorption edge. We also demonstrate a narrowing effect in population states generated through the nonlinear two-photon absorption mechanism.

2. Materials and Methods

For the calculations of two-photon absorption rates and time dependent population states, a three-level model described by rate equations was assumed. A similar solution was proposed in [13], where a constant intensity for the incident X-rays was assumed and, furthermore, a stationary case was considered. This allowed for an analytical solution of the rates equations. Herein, in contrast, we focus on an exact time-dependent solution of these rate equations for a given initially assumed time-profile of the incident X-ray pulse. Consequently, this approach allows the study of the time evolution of both the state population and the X-ray emission for a given pulse shape, i.e., a full description of the dynamics of the OPA and TPA processes. The schematic representation of atomic states included in the three-level rate equations model is shown in Figure 1.

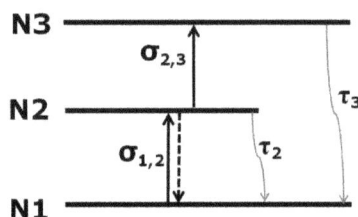

Figure 1. A schematic representation of the three-level model, representing the possible electronic transitions in one- and two-photon absorption processes on atomic core states. N_1 corresponds to the population of the ground state, N_2 to the population of the virtual intermediate state, and N_3 to the ionized state. $\sigma_{1,2}$, and $\sigma_{1,2}$ are cross-sections for the first and second photon absorption. τ_2 and τ_3 corresponds to the lifetimes of the N_2 and N_3 states.

This model is based on the conclusions drawn from the experimental report [13], where the two-photon absorption mechanism leading to sequential ionization through an intermediate virtual atomic state was observed. In this 3-level model (see Figure 1), N_1 corresponds to the population of the ground state, N_2 to the population of the virtual intermediate state, and N_3 to the ionized atom through the sequential ($1 \rightarrow 2 \rightarrow 3$) two-photon absorption process. According to this picture, a subsequent radiative decay of the N_2 state carries information on the direct OPA mechanism and a radiative decay of the N_3 state on the TPA process. We assumed that the excitation of the N_3 state is possible only through the intermediate N_2 state, i.e., only a sequential two-photon absorption process is possible. Direct, simultaneous two-photon absorption is neglected here, because it is expected that the cross-sections for this process are three orders of magnitude lower than the sequential TPA mechanism [9].

Assuming a sequential model for the two-photon absorption mechanism, the following rate equations are considered:

$$\frac{dN_1}{dt} = -I(t) \times (N_1(t) \times \sigma_{1,2} - N_2(t) \times \sigma_{1,2}) + \frac{N_2(t)}{\tau_2} + \frac{N_3(t)}{\tau_3}, \tag{1}$$

$$\frac{dN_2}{dt} = I(t) \times (N_1(t) \times \sigma_{1,2} - N_2(t) \times \sigma_{2,3} - N_2(t) \times \sigma_{1,2}) - \frac{N_2(t)}{\tau_2}, \tag{2}$$

$$\frac{dN_3}{dt} = I(t) \times N_2(t) \times \sigma_{2,3} - \frac{N_3(t)}{\tau_3}, \tag{3}$$

$$N_1(t) + N_2(t) + N_3(t) = N, \tag{4}$$

where $I(t)$ describes the time structure of the incident X-ray pulse and dN_1/dt, dN_2/dt, dN_3/dt are the decay rates of the initial, intermediate, and ionized states of an atom at time t. The cross-sections $\sigma_{1,2}$ and $\sigma_{2,3}$ represent transition probabilities between the N_1 and N_2, and the N_2 and N_3 states, respectively. In Equation (1), the first term describes the transitions between states N_1 and N_2 due to photon absorption and stimulated emission processes, respectively. The N_2/τ_2 and N_3/τ_3 terms in this equation describe the N_1-state decay rates for radiative transitions from the N_2 and N_3 levels with lifetimes τ_2 and τ_3, respectively. Similarly, Equation (2) describes the N_2 state which can be populated by photon absorption from the N_1 level or alternatively depopulated either by absorption of a second photon moving the system to the N_3-state or by stimulated emission as well as by a radiative decay with lifetime τ_2 to the N_1-state. Finally, the N_3 states are populated from the N_2 state through the absorption of a second X-ray photon, and depopulated by a decay onto the N_1 ground state with radiative lifetime τ_3. The number of electrons N in the 3-level system is conserved, and the initial condition $N_1(t = 0) = N$ and $N_2(t = 0) = N_3(t = 0) = 0$ was assumed.

3. Results

3.1. Details on the Computation

In order to compare the predictions of the time-dependent X-ray emission rates obtained from the 3-level model with available experimental data, the calculations were performed for a copper atom and an incident X-ray photon energy close to the Cu-K-shell edge. The details of the experimental setup and the data analysis can be found in [13]. Thanks to a relatively small detuning of only 12 eV with respect to the ionization threshold, both one- and two-photon absorption channels were probed in this experiment by measuring simultaneously the X-rays from a radiative decay of the intermediate state excited through inelastic scattering of the incident photons, as well as the non-resonant $K\alpha_{1,2}$ X-rays resulting from the two-photon absorption process. The X-ray signatures of the TPA and OPA processes were measured with a high-resolution von Hamos diffraction spectrometer. The experiment thus provided information on the flux and incident X-ray energy dependences for the TPA mechanism. In the present calculations, the photon absorption cross sections for the $N_1 \rightarrow N_2$ and $N_2 \rightarrow N_3$ excitations were taken from the reported experimental values [13], which were determined to be $\sigma_{1,2} = 5.5 \times 10^{-22}$ cm^2 and $\sigma_{2,3} = 1.4 \times 10^{-16}$ cm^2. Similarly, the atomic lifetimes for the N_2 and N_3 states were fixed to $\tau_2 = 5.4 \times 10^{-17}$ s and $\tau_3 = 4.42 \times 10^{-17}$ s, respectively.

In the calculations, we assumed a Gaussian-shaped X-ray pulse, which was inserted into Equations (1)–(3) in the following form:

$$I(t) = \frac{I_0}{\sqrt{2\pi}\sigma} \times \exp\left(-\frac{(t)^2}{2\sigma^2}\right) \tag{5}$$

In the above equation, I_0 represent the X-ray pulse intensity, and σ the standard deviation of its time profile. In the computations, we assumed a Gaussian X-ray pulse full width at half maximum value of 30 fs as reported in the experiment. This approximation was made since the experiment took advantage of the self-seeded mode of operation of the LCLS XFEL [6], which transforms the broad, spiky SASE (Self-amplified Spontaneous Emission) spectral structure and reduces it to a quasi-monochromatic spectrum. The resulting X-ray pulse in time will be similarly a single pulse with limited structure [14]. The numerical calculations for the time-dependent population of states were performed in the range of $\Delta t = 100$ fs (i.e., about 3 times the pulse length) and with a time-step (dt) of 10^{-4} fs, i.e., a value being much smaller than the pulse width and the lifetimes of the probed atomic states.

3.2. X-ray Rates for One- and Two-Photon Absorption

Based on the rate equations for the 3-level model described in Section 2, we first performed calculations on the absolute X-ray rates for the OPA and TPA processes versus the incident X-ray flux. With these computations, we could validate the time-dependent approach by comparing the calculated X-ray rates with the previously reported experimental data. The calculated X-ray rates were obtained by integrating the differential decay rates dN_2/τ_2 (OPA) and dN_3/τ_2 (TPA) over the time lapse of the incident X-ray pulse. The results, presented in Figure 2, show a good agreement with the reported experimental results [13].

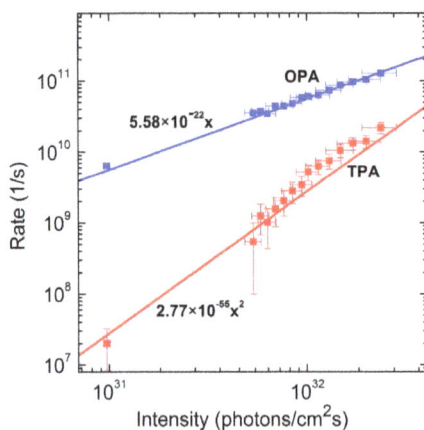

Figure 2. Comparison between calculated (solid line) and measured (squares) X-ray rates at different incidence X-ray flux for one photon absorption (OPA) (blue) and two-photon absorption (TPA) (red) processes. The experimental results were obtained at the Linac Coherent Light Source(LCLS) X-ray Free Electron Laser and are taken from Ref. [13].

In Figure 2, the measured X-ray emission rates corresponding to the OPA and TPA processes are plotted versus the fluence of the incident X-ray photons. The expected linear ($f = ax$) and quadratic dependencies ($f = ax^2$) of the measured rates for OPA and TPA processes, respectively, can be clearly seen. The calculated slope of the linear dependence of the OPA signal provides a one-photon absorption cross-section of 5.6×10^{-22} cm^2. Since the data are plotted on Log-Log scale, the TPA quadratic nature is distinguished from different curve slope as compared to OPA dependence. This value is in agreement with the reported experimental cross-section of 5.5×10^{-22} cm^2. The calculated coefficient of the quadratic function describing the TPA intensity is equal to 2.8×10^{-55} cm^{4}s, hence a qualitative agreement with the fitted experimental value of 4.1×10^{-55} cm^{4}s is given. The obtained results validate, the assumptions made in the theoretical three-level system model, including assumption of a Gaussian shape for the temporal profile of the incident X-ray pulse.

3.3. Time-Dependent Calculations for State Population

The time-dependent approach employing the equations described in Section 2 allows for precise studies of the time evolution of populations in the ground, intermediate and final states during the course of an X-ray pulse. Figure 3 shows the time profiles of the population of the discussed states calculated for an incident X-ray flux of 1.5×10^{32} photons/cm^2·s, which was a typical value used in the experiment discussed in Figure 2. For comparison, the shape of the incident X-ray pulse is also shown in Figure 3.

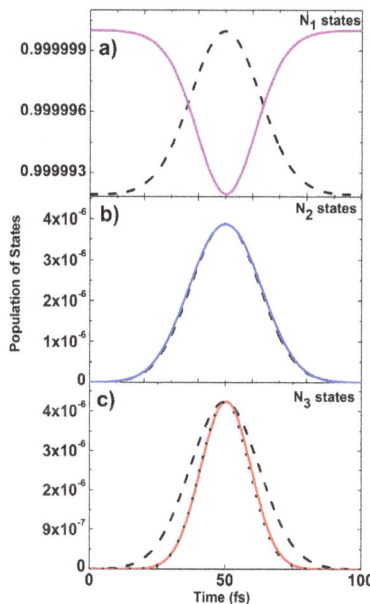

Figure 3. Calculated time-dependent population of the N_1, N_2 and N_3 states. The dashed line represents the shape of the incidence X-ray pulse, and is plotted in (a–c) for comparison. In Figure c, we also plot the Gauss-square function (for details see text).

In Figure 3a, we show the population N_1 of the ground state versus the time, which is anticorrelated with the time evolution of the incident X-ray pulse. Consequently, both curves exhibit the same width and extrema positions. After the incident X-ray pulse ends, the population of the ground state completely recovers its initial value $N_1 = N$ through decays from the N_2 and N_3 states. The population of the intermediate state, plotted in Figure 3b, is fully correlated with the incident X-ray pulse and anticorrelated with the ground state. The biggest difference, shown in Figure 3c, is observed for a continuum state, which is populated by the TPA process. While the peak value in time for the N_3 population state is at the same position as for the peak of the incident X-ray pulse, the width of the N_3 population state evolution is significantly narrower. We find that, while the incident X-ray pulse is 30 fs long, the population of the N_3 state can be characterized with a Gaussian shape of about 21 fs width. Finally, as noticed above for the N_1 state population recovery at the end of X-ray pulse, the populations of both the N_2 and N_3 states vanish to zero through the decays of these states to the ground state.

Using the presented 3-level model approach, we extracted the full widths at half maximum for the populations of the intermediate (N_2) and continuum (N_3) states for X-ray fluxes varying in the range of 10^{30}–10^{33} photons/cm^{2}s. The results are presented in Figure 4, which confirms our previous observations. The calculated population N_3 of the continuum state presents a narrowing effect, which leads to a shorter time distribution of the states populated through the two-photon absorption. The performed calculations demonstrate that at higher X-ray flux, i.e., above 10^{32} photons/cm^{2}s, the widths of the populations of the intermediate and continuum states systematically increase (Figure 4). However, it should be noted that the difference between the widths of these states amounts to about 30% and is constant over the studied incidence X-ray flux range.

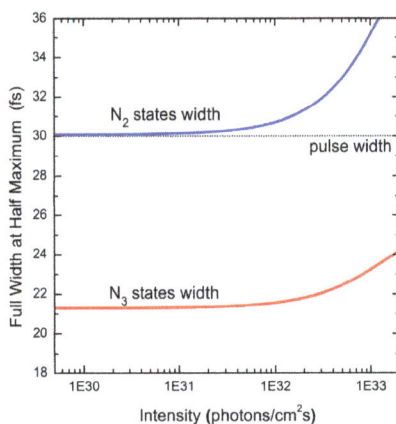

Figure 4. Calculated width (FWHM-full width at half maximum) for the N_2 (blue) and N_3 (red) states population as a function of the incident X-ray flux. The assumed width of the X-ray pulse was 30 fs and is drawn for comparison as a horizontal dotted line.

4. Discussion

The presented calculations regarding the time evolution of the state populations and X-ray emission rates for the OPA and TPA processes can be discussed in different aspects. First, we would like to stress that the presented 3-level model is very simple in terms of number of atomic states, and hence the possible excitations and transitions, that are considered. We are aware that the XFEL beam interaction may lead, in specific cases, to more complicated distributions of atomic states that very often may be difficult or even impossible to treat with such a simple model. Nonetheless, as shown in the present report, the three-level model rate equations can be employed as a general tool for the description of nonlinear phenomena relevant for two-photon absorption via X-ray interaction with core atomic states.

The presented model describes quite well the reported experimental dependences of the X-ray emission rates on the incident X-ray flux for both one- and two-photon absorption processes. The linear and quadratic dependences observed for the OPA and TPA processes were able to be well reproduced by the 3-level model calculations, and the extracted total-cross section values are in agreement with the experiment. The observed underestimation of the TPA cross-section by our calculations will be discussed below, together with the narrowing effect found for the population of the continuum state through the TPA mechanism.

As shown in Figures 3 and 4, the population N_3 of continuum states is about 30% narrower than the population N_2 of the intermediate state and thus also narrower than the applied incident X-ray flux. While the incident X-ray flux width is 30 fs, the estimated width for the population of the continuum state is about 21.4 fs. We found that this effect can be explained by an effective square-dependence of the excitation of these atomic states on the incident photon flux due to the two-photon absorption process. In fact, this dependence manifests, in the occurrence of the nonlinear X-ray interaction with matter. Indeed, the two-photon absorption process scales with the square of the applied X-ray intensity. Therefore, the employed Gaussian-like incident X-ray pulse will effectively act as the squared Gaussian profile. The Gauss-square function, in comparison with the nominal Gauss-shape function, will exhibit a width that is narrower by a factor of $2^{-1/2}$ (i.e., ≈ 1.41). Indeed, this observation is in perfect agreement with the observed narrowing, i.e., the ratio of 30 fs to 21.4 fs is equal to 1.4. This is also confirmed in Figure 3c, where the Gaussian-square function (dotted line) is compared with the calculated population of the continuum state, and both profiles are in perfect agreement. Finally, we would like to note that the discussed narrowing effect should be taken into consideration, in general, for the cross-section determination of nonlinear processes. As reported in [13], the experimental data were normalized while assuming that the incident X-ray pulse has a Gaussian shape. While this normalization is valid for linear processes, it seems that in nonlinear interaction cases the normalization factor resulting from the time distribution of the incident X-ray flux should be, in addition, divided by a factor if $2^{-1/2}$ for the two-photon processes. It is expected that, for this reason, the cross-sections calculated here for TPA mechanism should again be lower, by a factor of about 1.4, than the discussed experimental results.

Finally, we would like to comment on the observed increase of widths of population profiles for excited states, here both N_2 and N_3, at higher incidence X-ray flux. As can be seen in Figure 4, for incident X-ray fluxes above 10^{32} photons/cm^{2}s, both time profiles for these states become wider, however the ratio of their widths remains constant and is equal to about 1.4. The observed increase in the width for the OPA and TPA processes may be considered as a first indication that the saturation regime for nonlinear interaction is approached [15,16]. The atomic state response and state populations no longer follow the incident photon pulse profile, and therefore cannot be evaluated with a simple Gaussian-like approximation. Indeed, the saturation of absorption effects were reported for metallic Fe at X-ray flux intensities of 8.4×10^{34} photons/cm^2·s [16]. At the present point, the studies of such effects need more theoretical investigation, as well as experimental verifications.

Though the present simulations were performed with a simplified Gaussian-like pulse structure, this is still a good approximation for single-shot pulse shapes expected during self-seeded operation of an XFEL [6,14]. In addition, since the experimental data was taken in a cumulative measurement where many pulses were summed together to improve the signal-to-noise, a Gaussian pulse shape is a good representation of the average incident pulse shape. We show that the reported narrowing of atomic states populated through nonlinear interaction relates to a square-dependence of the applied X-ray pulse intensity not to the incident X-ray pulse shape. Further research should address other phenomena that may be related to the single-pulse interaction with SASE-like structure. For example, one may investigate the structure and envelope of the applied X-ray pulse. Moreover, different parameters describing the SASE structure of the incident beam may be evaluated, like spike width and spike separation in the time domain. We anticipate that these effects will start to become significant as they approach the timescales of atomic core-hole lifetimes. For any such single-shot simulations and experiments not only does the experiment require precise measurement of the incident beam properties, but the simulation parameters need to be carefully tuned to match the specific XFEL facility at which the experiment takes place.

Acknowledgments: J. Szlachetko acknowledges the National Science Centre (NCN), Poland for support under grant no. 2015/19/B/ST2/00931. J. Szacknowledges the Polish Ministry of Science and Higher Education for support from the budget for science in 2016–2019 under grant IDEAS Plus II IdPII 2015000164.

Appl. Sci. **2017**, *7*, 653

Author Contributions: J.Sz. conceived and designed the project, M.P. delivered the rate equations and K.T. executed calculations and prepared the figures as well as the initial version of the manuscript. All authors participated in result discussions and manuscript preparation. J.Sz. wrote the manuscript with input from K.T., K.W., Y.K., M.P. C.M. and J.Sá.

Conflicts of Interest: The authors declare no conflict of interest.

References

1. Keller, U. Recent developments in compact ultrafast lasers. *Nature* **2003**, *424*, 831–838. [CrossRef] [PubMed]
2. Schuster, I.; Kubanek, A.; Fuhrmanek, A.; Puppe, T.; Pinkse, P.W.H.; Murr, K.; Rempe, G. Nonlinear spectroscopy of photons bound to one atom. *Nat. Phys.* **2008**, *4*, 382–385. [CrossRef]
3. Srinivasan, K.; Painter, O. Linear and nonlinear optical spectroscopy of a strongly coupled microdisk-quantum dot system. *Nature* **2007**, *450*, 862–865. [CrossRef] [PubMed]
4. Hori, M.; Sótér, A.; Barna, D.; Dax, A.; Hayano, R.S.; Friedreich, S.; Juhász, B.; Pask, T.; Widmann, E.; Horváth, D.; et al. Two-photon laser spectroscopy of antiprotonic helium and the antiproton-to-electron mass ratio. *Nature* **2011**, *475*, 484–488. [CrossRef] [PubMed]
5. Emma, P.; Akre, R.; Arthur, J.; Bionta, R.; Bostedt, C.; Bozek, J.; Brachmann, A.; Bucksbaum, P.; Coffee, R.; Decker, F.-J.; et al. First lasing and operation of an ångstrom-wavelength free-electron laser. *Nat. Photonics* **2010**, *4*, 641–647. [CrossRef]
6. Amann, J.; Berg, W.; Blank, V.; Decker, F.-J.; Ding, Y.; Emma, P.; Feng, Y.; Frisch, J.; Fritz, D.; Hastings, J.; et al. Demonstration of self-seeding in a hard-X-ray free-electron laser. *Nat. Photonics* **2012**, *6*, 693–698. [CrossRef]
7. Young, L.; Kanter, E. P.; Krässig, B.; Li, Y.; March, A.M.; Pratt, S.T.; Santra, R.; Southworth, S.H.; Rohringer, N.; DiMauro, L.F.; et al. Femtosecond electronic response of atoms to ultra-intense X-rays. *Nature* **2010**, *466*, 56–61. [CrossRef] [PubMed]
8. Tamasaku, K.; Nagasono, M.; Iwayama, H.; Shigemasa, E.; Inubushi, Y.; Tanaka, T.; Tono, K.; Togashi, T.; Sato, T.; Katayama, T.; et al. Double Core-Hole Creation by Sequential Attosecond Photoionization. *Phys. Rev. Lett.* **2013**, *111*, 043001. [CrossRef] [PubMed]
9. Tamasaku, K.; Shigemasa, E.; Inubushi, Y.; Katayama, T. X-ray two-photon absorption competing against single and sequential multiphoton processes. *Nature* **2014**, *8*, 313–316. [CrossRef]
10. Rohringer, N.; Ryan, D.; London, R.A.; Purvis, M.; Albert, F.; Dunn, J.; Bozek, J.D.; Bostedt, C.; Graf, A.; Hill, R.; et al. Atomic inner-shell X-ray laser at 1.46 nanometres pumped by an X-ray free-electron laser. *Nature* **2012**, *481*, 488–491. [CrossRef] [PubMed]
11. Beye, M.; Schreck, S.; Sorgenfrei, F.; Trabant, C.; Pontius, N.; Schüßler-Langeheine, C.; Wurth, W.; Föhlisch, A. Stimulated X-ray emission for materials science. *Nature* **2013**, *11*. [CrossRef] [PubMed]
12. Vinko, S.M.; Ciricosta, O.; Cho, B.I.; Engelhorn, K.; Chung, H.-K.; Brown, C.R.D.; Burian, T.; Chalupský, J.; Falcone, R.W.; Graves, C.; et al. Creation and diagnosis of a solid-density plasma with an X-ray free-electron laser. *Nature* **2012**, *482*, 59–62. [CrossRef] [PubMed]
13. Szlachetko, J.; Nachtegaal, M.; De Boni, E.; Willimann, M.; Safonova, O.; Sa, J.; Smolentsev, G.; Szlachetko, M.; Van Bokhoven, J.A.; Dousse, J.C.; et al. Establishing nonlinearity thresholds with ultraintense X-ray pulses. *Sci. Rep.* **2016**, *6*, 33292. [CrossRef] [PubMed]
14. Geloni, G.; Kocharyan, V.; Saldin, E. A novel self-seeding scheme for hard X-ray FELs. *J. Mod. Opt.* **2011**, *58*, 1391–1403. [CrossRef]
15. Nagler, B.; Zastrau, U.; Fäustlin, R.R.; Vinko, S.M.; Whitcher, T.; Nelson, A.J.; Sobierajski, R.; Krzywinski, J.; Chalupsky, J.; Abreu, E.; et al. Turning solid aluminium transparent by intense soft X-ray photoionization. *Nat. Phys.* **2009**, *5*, 693–696. [CrossRef]
16. Yoneda, H.; Inubushi, Y.; Yabashi, M.; Katayama, T.; Ishikawa, T.; Ohashi, H.; Yumoto, H.; Yamauchi, K.; Mimura, H.; Kitamura, H. Saturable absorption of intense hard X-rays in iron. *Nat. Commun.* **2014**, *5*, 5080. [CrossRef] [PubMed]

applied
sciences

MDPI

Article

Modeling Non-Equilibrium Dynamics and Saturable Absorption Induced by Free Electron Laser Radiation

Keisuke Hatada *,† and Andrea Di Cicco

Physics Division, School of Science and Technology, University of Camerino, I-62032 Camerino (MC), Italy; andrea.dicicco@unicam.it

* Correspondence: keisuke.hatada.gm@gmail.com

† Current address: Department Chemie, Ludwig-Maximilians-Universität München, 81377 München, Germany

Received: 31 May 2017; Accepted: 27 July 2017; Published: 9 August 2017

Abstract: Currently available X-ray and extreme ultraviolet free electron laser (FEL) sources provide intense ultrashort photon pulses. Those sources open new exciting perspectives for experimental studies of ultrafast non-equilibrium processes at the nanoscale in condensed matter. Theoretical approaches and computer simulations are being developed to understand the complicated dynamical processes associated with the interaction of FEL pulses with matter. In this work, we present the results of the application of a simplified three-channel model to the non-equilibrium dynamics of ultrathin aluminum films excited by FEL radiation at 33.3, 37 and 92 eV photon energy. The model includes semi-classical rate equations coupled with the equation of propagation of the photon wave packets. X-ray transmission measurements are found to be in agreement with present simulations, which are also able to shed light on temporal dynamics (in the fs range) in nano-sized Al films strongly interacting with the photon pulse. We also expanded our non-linear model, explicitly including the two-photon absorption cross-section and the effect of including electron heating for reproducing transmission measurements.

Keywords: X-ray free electron laser; saturation phenomena; nonlinear optics

---

1. Introduction

Over the past decade, X-ray and extreme ultraviolet free electron laser (FEL) sources have been developed, providing a source of extremely brilliant and ultrafast photon pulses. The present facilities include FELs in the extreme ultra-violet (EUV) and soft X-ray ranges such as FLASH (Hamburg) [1] and FERMI@Elettra (Trieste) [2], and in the hard X-ray range such asLCLS (Stanford) [3], SACLA (Spring-8) [4], and the European XFEL (presently under construction, Hamburg).

Typically, FEL photon pulses show durations in the 10–100 fs range, contain a large number of photons (10^{10}–10^{15}) with a limited wavelength band width, depending on the pulse generation mechanism. Using suitable optics, the pulse spot dimensions can be reduced to 10–100 μm^2 or less. In such extreme conditions, pulse fluences can exceed 100 J/cm^2, leading to the observation of non-linear optical processes in condensed matter such as saturation phenomena, two photon absorption, ultrafast electron, and lattice heating (see [5] and the references therein).

Non-linear effects are quite familiar in optical laser science, while in the EUV and X-ray energy regime, investigations are still in development both experimentally and theoretically. There are several differences among the optical laser and EUV/X-ray radiation interaction with matter, including the penetration depth, the energy deposited, and the lifetime of the excited states. An important feature is exactly the latter, because for X-ray excitations, the lifetime of the excited state is in the femtosecond range (core–hole lifetime). Intense and ultrashort FEL pulses allow us to perform experiments for which the pulse width is of the same order of the core–hole lifetime at fluences for which non-linear effects are not negligible.

Different techniques are used at FEL facilities, including transmission and scattering experiments and pump-and-probe studies using two ultrashort pulses (optical or X-ray). In many cases, simplified approaches are used for modeling the interaction process, basically neglecting the finite time widths and spatial dimensions of the pulses. Although efficient collisional-radiative codes describing dense plasma states are currently available (SCFLY, see [6] and references therein), the complicated dynamical processes associated with the interaction of matter with photon pulses are usually not taken into explicit account. For FEL ultrashort pulses and nanoscale materials, the finite dimensions are expected to play a role, and a detailed modeling of the dynamical pulse–matter interaction appears to be necessary. In particular, reliable theoretical models are needed both for a solid interpretation of the experimental data (transmission, scattering, and so on) and for modeling the evolution of the sample status during the excitation process (transient conditions, local temperature).

In principle, transmission measurements are probably the easiest and cleanest experiments that can be performed using FEL radiation, but they also imply several important difficulties to be overcome both for their practical realization and interpretation of results. The importance of developing proper models can be appreciated by looking first at the results of Nagler et al. [7], in which saturable soft X-ray absorption (92 eV) of an ultrathin aluminum foil was obtained. Those results were followed by other experiments at lower (23.7–37 eV) [8] and higher photon energy (1540–1870 eV) [9], both indicating the importance of accounting for electron heating phenomena at high fluence. Further experiments in the hard X-ray range (7.1 keV) confirmed the existence of important non-linear phenomena with increasing transmission of a factor of 10 and substantial shifts of the absorption edge in solid iron [10].

Those results indicate that the description of photon–matter interaction at high intensities requires specific models to be devised, accounting for the various effects contributing to a modification of the X-ray and EUV absorption cross-section (relaxation of final state, ultrafast electron heating, and so on). There have been several theoretical and computational works on Al EUV and X-ray FEL intense pulse absorption. Examples range from "dynamical" (small cluster) [11,12] to "steady" model [13,14] approaches, including application of the SCFLY code [15].

In previous works, we have developed a simple three-channel model (involving ground, excited and relaxed states) for calculating the transmission of ultrathin metal films for increasing photon fluences using FEL radiation up to the saturation limit [16]. The model was conceived to provide a reliable and physically-intuitive calculation scheme for saturable absorption, also allowing direct inspection of the interaction dynamics including size effects of the pulse width and finite thickness of the absorbing medium at the nanometric scale. The three-channel model was found to provide simulations in excellent agreement with EUV transmission data in an extended range of pulse fluence (0–200 J/cm^2) [8,16]. Simulations allow us direct estimates of time and space profiles for the photon absorption, and a benchmark for calculating observables of interest (transmission) and the deposited energy in nanosized condensed matter for evaluation, for example, of local temperatures. The effects of a local increase of the electron temperature can also be introduced in the calculations as reported in Reference [8]. In this work, we extend previous results for the calculation of EUV transmission of ultrashort pulses through Al films of nanometric thickness. We show the details of the temporal dynamics of the photon–matter interaction at fs resolution, also including possible multi-photon absorption and electron-heating effects.

2. Model for Ultrafast Transmission Measurements

2.1. Three-Channel Model

In our recent work [16], we developed a phenomenological three-state model which is able to reproduce saturation phenomena related to the increased transmission in the high fluence regime. We refer to the original publication [16] for some more details on the computational model. In brief, the model describes the variation of the density of occupation numbers N_1, N_2 and N_3 of three exemplary many-body states (ground $|1\rangle$, excited $|2\rangle$ and an intermediate relaxed state $|3\rangle$) by a set of rate equations with proper constraints:

$$\frac{dN_1(z,t)}{dt} = \frac{g(z,t)I(z,t)}{h\nu} + \frac{N_2(z,t)}{\tau_{21}} + \frac{N_3(z,t)}{\tau_{31}} \tag{1}$$

$$\frac{dN_2(z,t)}{dt} = -\frac{g(z,t)I(z,t)}{h\nu} - \frac{N_2(z,t)}{\tau_{21}} - \frac{N_2(z,t)}{\tau_{23}} \tag{2}$$

$$\frac{dN_3(z,t)}{dt} = \frac{1}{\tau_{23}}N_2(z,t) - \frac{1}{\tau_{31}}N_3(z,t) \tag{3}$$

$$g(z,t) = \sigma(T)\left(N_2(z,t) - N_1(z,t)\right) \tag{4}$$

$$N = N_1(z,t) + N_2(z,t) + N_3(z,t) \tag{5}$$

In this set of equations, the occupation numbers depend thus on the photon field intensity $I(z,t)$ at time t and position z (along the direction of propagation of the pulse), the photon absorption cross-section σ at given photon energy $h\nu$, and on the relaxation times τ between the various states. In this formalism, $g(z,t)$ is an effective time- and space-dependent absorption coefficient—possibly temperature-dependent—that can be considered constant for linear absorption (Lambert–Beer law) at a given temperature T. From the condition of conservation of the total number of the states Equation (5), we just need to solve couples of rate equations.

The set of rate equations are coupled with the transport condition of the incoming pulse, within the classical electrodynamics limit:

$$\frac{dI(z,t)}{dz} + \frac{1}{c}\frac{dI(z,t)}{dt} = g(z,t)I(z,t). \tag{6}$$

Within the model described by the equations above, absorption and stimulated emission by laser radiation involve transitions between ground $|1\rangle$ and excited $|2\rangle$ states. The relaxed state $|3\rangle$ represents the ensemble of all possible relaxed states reached by decay of state $|2\rangle$. It can decay to state $|1\rangle$ by emitting a photon or through other processes.

2.2. Computational Details

We describe here the way of solving the set of coupled equations numerically. For the discretization of Equation (6), we employed the forward differentiation, as the equations can be solved numerically following the dynamics of the pulse, using discretized grids with $\Delta t = 0.4$ as and $\Delta z = 1.2$ nm, which satisfy the *Courant condition*,

$$c \leq \frac{\Delta z}{\Delta t}, \tag{7}$$

so that the *Upward differencing'* method [17] can be applied safely to Equation (6):

$$\frac{I(z_j, t_{n+1}) - I(z_j, t_n)}{c\Delta t} = -\frac{I(z_j, t_n) - I(z_{j-1}, t_n)}{\Delta z} + g(z_j, t_n)I(z_j, t_n). \tag{8}$$

In Equation (8), j runs along the incident direction of the FEL pulse inside the sample and n represents the discrete time grid.

In the rate Equation (5), we just used the *Euler* method; thus, for example, we obtain the following expression for N_2:

$$\frac{N_2(z_j, t_{n+1}) - N_2(z_j, t_n)}{\Delta t} = -\frac{g(z_j, t_n) I(z_j, t_n)}{h\nu} - \left[\frac{1}{\tau_{21}} + \frac{1}{\tau_{23}}\right] N_2(z_j, t_n). \tag{9}$$

We remark that for actual calculations concerning intense ultrafast pulses and nano-sized samples we need extremely small $\Delta t = 0.4$ and $\Delta z = 1.2$ nm, so that efficient algorithms like those mentioned are needed for best accuracy and lower computing times.

For the initial pulse intensity I_0, we assumed a Gaussian time distribution:

$$I_0(z, t) = I_0^{\max} e^{-\frac{(z - c(t + \mu_0))^2}{2(c\sigma_t)^2}}, \tag{10}$$

where μ_0 and σ_t are the mean and standard deviation of the photon pulse time distribution. $I_0^{\max}$ corresponds to the maximum intensity of initial beam. For a given pulse shape, σ_t is estimated looking from the full-width at half maximum FWHM$= 2\sqrt{2\ln 2}\,\sigma_t$. The window of the pulse in time is set at just double of FWHM (namely 30 fs), multiplied by the speed of light. The origin of z is set at the surface of a sample. The three-channel model is implemented in a Fortran 90 code using simple instructions and running on any current operating systems and with any Fortran compiler.

3. Results and Discussion

3.1. Transmission by Ultrashort Pulse of Photons with 92 eV for Al Foil

As mentioned above, saturable absorption using FEL pulses was first observed in a pioneering experiment by Nagler et al. [7], in which single-shot transmission data of a 53 nm Al foil were collected using 92 eV ultra-short (15 fs) photon pulses up to fluences in the 200 J/cm² range. FEL pulses are thus able to excite Al 2p core electrons (binding energies are 72.7 and 73.1 eV), and the lifetime of the excited core–hole state is 40 fs [18].

Several approximations can be used for calculating X-ray transmission using the model reported in the previous section. In particular, fast calculations can be performed considering two opposite limits, namely: (i) a *long pulse* limit in which the pulse width is much longer than the lifetime of the excited state; (ii) a *short pulse* limit in which the lifetime is much longer than the pulse width.

In both limits, the system under the photon field is assumed to be in a steady-state. Hence, the time dependency is neglected; that is to say, the time derivative of the occupation numbers and the intensity (see Equation (5)) become zero. The details of the models are described in Appendixes A and B.

We have performed calculations of Al transmission comparing the results of the long and short pulse limits with those of the three-channel model including the dynamics of the pulse–matter interaction. In Figure 1, we show the comparison between the different models and experimental data [7]. For the dynamical three-channel and *long pulse* models, we included the ground state $|1\rangle$, excited state $|2\rangle$ and relaxed state $|3\rangle$. We assumed that the lifetime of the relaxed state τ_{31} is the same as the one of the excited state τ_{21} (namely, 40 fs), while the relaxation time to the relaxed state τ_{32} is 1 fs, corresponding to the electron mean free path [7]. We also tested a *short pulse* model, which represents just the opposite limit of the *long pulse* model. This model assumes that the pulse width is much shorter than the lifetime of the excited state, so that decay of excited state is neglected. Without decay process, the state cannot access the relaxed state, thus the model reduces to a two channel model.

As shown in Figure 1, all four models show an increase in transmission with the fluence until reaching the saturation at extremely high fluence above 200 J/cm². In the low-fluence limit, the calculations show the convergence to the linear transmission (Lambert–Beer law). The four models result in a completely different transmission at intermediate fluences, and the *long pulse* with and without the relaxed state and *short pulse* limits fail to reproduce the experimental data. These static models imply remarkably increased transmission in the 2–100 J/cm², anticipating the onset of saturable absorption of at least one order of magnitude. On the other hand, the dynamical three-channel model—which explicitly considers the time evolution of the occupation numbers along the pulse direction—is found to be in very good agreement with the experimental data.

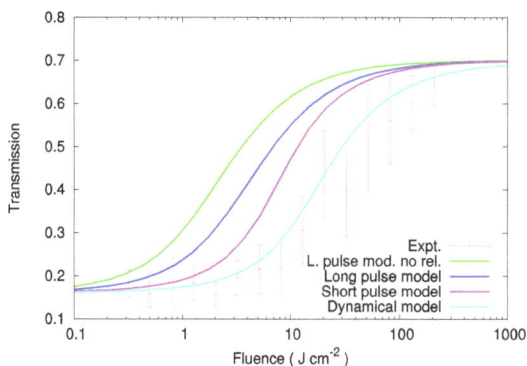

Figure 1. XFEL pulse transmission for Al thin film (53 nm) at 92 eV photon energy. The green and blue solid lines indicate *long pulse* model without (no rel.) and with an intermediate relaxed state; that is, the model of the green line is just a two-channel model. The pink solid line is for the *short pulse* model. The light blue solid line corresponds to the three-channel dynamical model. The dynamical model agrees very well with the experimental (Expt.) data [7].

As will be shown in the next Section, the steady condition of the *long pulse* model is not achieved at the beginning and end of the incident pulse, while a "steady" condition can be obtained only above a threshold intensity of the pulse taking account of its shape.

3.2. Non-Equilibrium Dynamics in Al Thin Film

In the previous section, we have shown that the three-channel model describes the saturation phenomena of X-rays of ultra short pulse well. In this part, we show the detail of the dynamics in the Al thin film during 60 fs. Figure 2 shows the results of time dependence of the density of occupation numbers normalized by the total number N at the end face of Al thin (53 nm) film. The fluences F are 1.5, 15 and 150 J/cm^2 for (a), (b) and (c), respectively. In the calculation, we estimated that the additional absorption by both sides of oxidized Al layers [7] reduces the pulse intensity by 70%. In the figures, the results obtained from the dynamical three-channel model is shown with solid lines, and static *long pulse* model with the dotted line. In Figure 2a, we see that the results of the dynamical model tend asymptotically to reach the static limit. However the time of irradiation is not sufficient to reach those occupation number limits, which is around 10% of the population for the excited state. The dynamical model shows that the system is still highly populated by the ground state. In fact, this can be confirmed looking at Figure 1, where we see that the transmission of *long pulse* model already shows a difference from the linear optical transmission, while the experimental and the dynamical ones are still well below the saturation threshold. For higher fluence at 15 J/cm^2, in Figure 2b, we see that the ratio of the occupation number of the excited state is much increased, such that the linear model (Lambert–Beer law) which works for the ground state fails. This failure is confirmed experimentally by the observation of growth of the transmission in Figure 1. In Figure 2c, we see that by increasing the pulse intensity the occupation of the ground state is strongly decreased, while most electrons are in the excited state. Under these conditions, the increment of the transparency and saturable absorption as the consequence of this are realized for this Al film. In this fluence range, a population inversion corresponding formally to negative temperatures (in equilibrium statistical mechanics terms) is realized. This phenomenon is associated with the existence of the relaxed state $|3\rangle$, which does not participate in the photon absorption and simulated emission. In our calculations, the system returns to the ground state exponentially at the end of the short photon pulse (times larger than 60 fs in Figure 2).

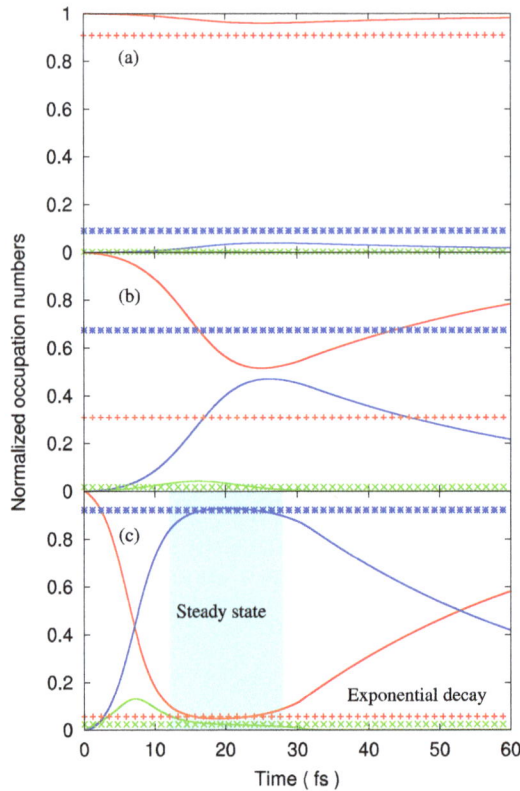

Figure 2. Time dependence of the density of occupation numbers (normalized by the total number N) at the end face of Al thin film. The conditions are the same as in Figure 1. The fluences of incoming Gaussian pulses are 1.5, 15 and 150 J/cm² from the top to the bottom of three figures, respectively. The solid curves and dotted lines are estimated by the three-channel model and *long pulse* model, respectively. The red color indicates the ground state $|1\rangle$, the blue one the excited state $|2\rangle$ and the green one the relaxed state $|3\rangle$. After some time, the curves asymptotically approach the steady state of the *long pulse* model. In the region of the light blue shade, the system has reached a steady state. After about 30 fs (the total pulse width), the excited state decays exponentially. The averaged experimental data are indicated by red points with the errors.

3.3. Simulations for Two Photon Absorption

Another source of non-linear optical phenomena in FEL absorption studies is represented by the presence of two photon absorption (TPA) effects. The presence of TPA can be obtained at extremely large fluences, but the order of two photon absorption cross-section is predicted to be rather small, with typical values in the 10^{-50} cm⁴s range. In order to estimate the size of this effect quantitatively, we apply the following set of equations:

$$\frac{dN_1(z,t)}{dt} = \frac{g(z,t)I(z,t)}{h\nu} + \frac{g_2(z,t)I^2(z,t)}{h\nu} + \frac{N_2(z,t)}{\tau_{21}} + \frac{N_T(z,t)}{\tau_{T1}} \tag{11}$$

$$\frac{dN_2(z,t)}{dt} = -\frac{g(z,t)I(z,t)}{h\nu} - \frac{N_2(z,t)}{\tau_{21}} \tag{12}$$

$$\frac{dN_T(z,t)}{dt} = -\frac{g_2(z,t)I^2(z,t)}{h\nu} - \frac{1}{\tau_{T1}}N_T(z,t) \tag{13}$$

$$\frac{dI(z,t)}{dz} + \frac{1}{c}\frac{dI(z,t)}{dt} = g(z,t)I(z,t) + g_2(z,t)I^2(z,t) \tag{14}$$

$$g(z,t) = \sigma\left(N_2(z,t) - N_1(z,t)\right) \tag{15}$$

$$g_2(z,t) = \frac{\sigma_2}{h\nu}\left(N_T(z,t) - N_1(z,t)\right) \tag{16}$$

$$N = N_1(z,t) + N_2(z,t) + N_T(z,t) \tag{17}$$

where σ_2 is the two photon absorption cross-section (in 10^{-50} cm^{4}s). The new term $g_2(z,t)I^2(z,t)$ in Equation (14) contributes to the increase (or decrease) of the intensity through the two photon absorption and stimulated emission. We introduced a new state $|T\rangle$ which is accessible by TPA. This state is only connected to the ground state $|1\rangle$ by two photon absorption and stimulated emission. The lifetime of state $|T\rangle$ to the ground state $|1\rangle$, τ_{T1} is set to be the same as τ_{21}, which is 40 fs. In order to see the effects of TPA only, we did not introduce the relaxed state. The ionization energy of the ionic state with a hole in 2p orbital is 93 eV [7], which is higher than the photon energy, 92 eV, so we do not need to include the sequential two photon absorption.

We report calculations with four different values of σ_2; namely 0, 10^{-50}, 10^{-49} and 10^{-48} cm^{4}s. $\sigma_2 = 0$ reduces to a single photon absorption two-channel model. For a limit of infinitely long pulse, it corresponds to the *long pulse* model without relaxed state (green line) in Figure 1. $\sigma_2 = 10^{-49}$ and 10^{-48} cm^{4}s represent large cross-sections for TPA and are probably not realistic, but we show the results to check the tendency for TPA effects in X-ray transmission.

In Figure 3, we show the transmission of X-ray pulse for Al thin film with the same condition as in Figure 1. Application of the present TPA models does not seem to reproduce the experimental results. As we see also from the ground state asymptotic model (steady-state TPA, Appendix C, Equation (A19)), TPA plays an important role at high photon intensity, decreasing the total transmission. In the high fluence regime, TPA and saturable absorption are both factors affecting the transmission.

Figure 4a–c show the details of dynamics for the states at the end face of the Al thin film with $\sigma_2 = 10^{-50}$ cm^{4}s for fluences of 1.5, 15 and 150 J/cm^2, respectively. The conditions of the calculations are the same as in Figure 2a–c. For 1 and 10 J/cm^2 in Figure 4a,b, respectively, the results are similar to the previous saturation model (ratio N_1:N_2). For high fluence in Figure 4c, the ratio of the occupation numbers of the states approaches N_1:N_2:N_3 = 1/3:1/3:1/3 in steady conditions (which corresponds to the highest entropy in the sense of the equilibrium statistical mechanics); therefore, the population inversion is not obtained.

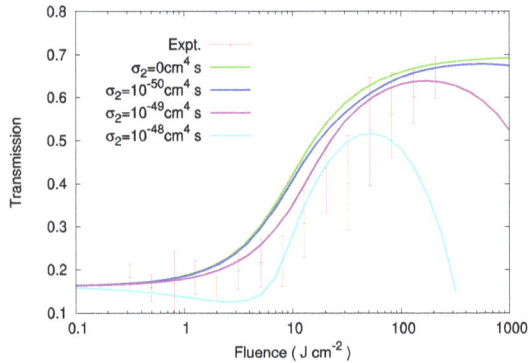

Figure 3. Transmission for Al thin film (53 nm) with two photon absorption (TPA) model at 92 eV photon energy of X-ray pulse, compared with experimental data (Expt.). Four different TPA cross sections—0, 10^{-50}, 10^{-49} and 10^{-48} cm^{4}s—are used to demonstrate a non-linear optical effect.

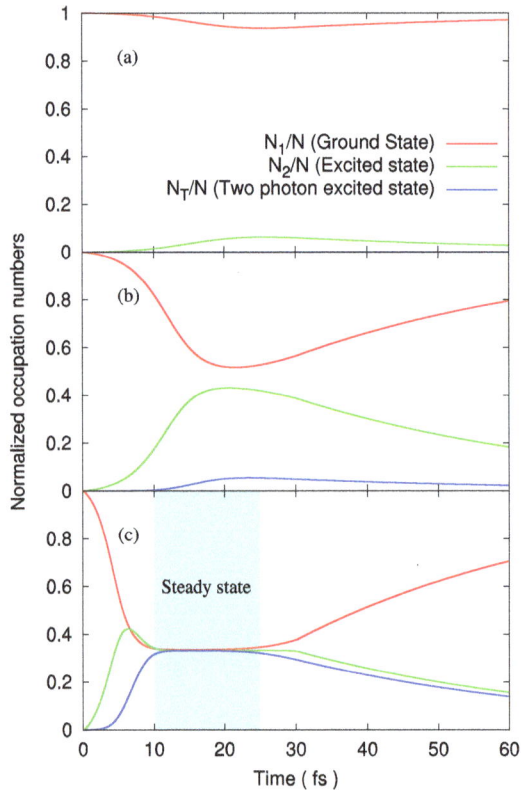

Figure 4. Time dependence of the density of occupation numbers (normalized by the total number N) at the end surface of the Al thin film (53 nm), using the two photon absorption (TPA) contribution at 92 eV photon energy (FEL pulse). The fluences of the incident pulse are 1.5, 15 and 150 J /cm^2 from the top to the bottom, respectively. The TPA cross-section σ_2 is set to 10^{-50} cm^{4}s.

3.4. Transmission by Ultrafast Pulse of Photons with 33.3 and 37 eV for Al Foil

Another EUV transmission experiment [8] was performed at the TIMEX end station [19,20] of the Elastic and Inelastic Scattering (EIS) beamline [21], using the FERMI@Elettra FEL-1 source, a seeded FEL providing clean, tunable [2], and intense subpicosecond (100 fs FWHM) photon pulses in the 19–62 eV photon energy range.

Accurate transmission measurements were carried out using unsupported self-standing 100-nm-thick Al foils. The maximum energy per pulse delivered by the FEL-1 source were 180 µJ and 130 µJ range at 33.3 and 37 eV photon energy, respectively. The actual incoming fluence F at sample position was estimated by measuring the area of the focal spot ($\sigma^2 \approx 100$ µm^2) and accounting for mirror reflectivity and beamline transmission. Data were collected over several decades of fluence F (0.01–20 J/cm^2) through the combined use of filters and a gas attenuator. Appreciable damage of the sample was observed after irradiation with a single FEL pulse with $F > 0.1$ J/cm^2, and single-shot transmittance measurements were carried out in the high fluence regime.

EUV transmission data were measured with unprecedented accuracy, and evidence for a nonmonotonic trend as a function of incoming fluence was found. In particular, an increased transmittance at higher fluence of about 5–10 J/cm^2—associated with saturable absorption effects—was accompanied by an initial decrease at intermediate fluences, as shown in Figure 5.

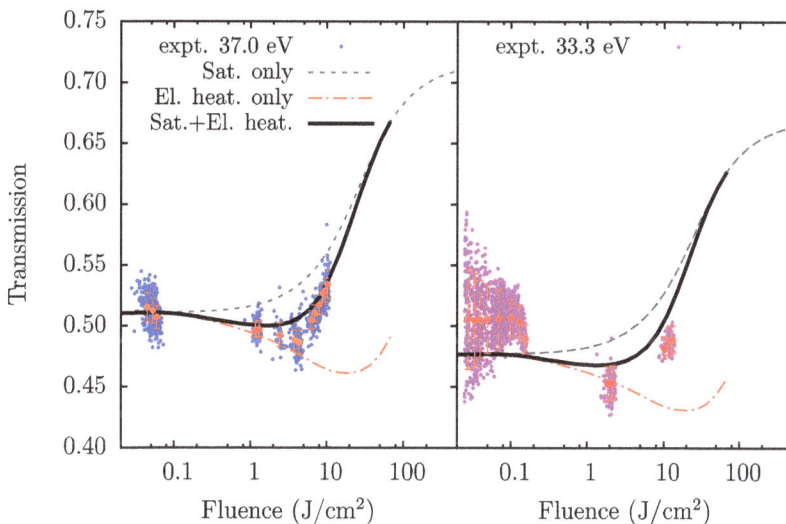

Figure 5. Experimental (expt.) EUV transmission data at 37 eV (**left panel**) and 33.3 eV (**right panel**) compared with different calculations, including: only optical saturation phenomena without accounting for temperature effects in the cross-section (Sat. only); only electron heating but neglecting saturation effects (El. heat. only); both electron heating and saturation effects (Sat.+El. heat.).

As we have seen in previous sections, this decrease can not be reproduced by a pure single-photon optical effect. Two-photon effects can decrease the observed transmittance under suitable conditions. However, we have found that experimental data showing a non-monotonic behaviour cannot be reproduced by realistic TPA models. The cross-section σ_2 for two-photon absorption should be at least on the order of 10^{-48} cm^{4}s, two orders of magnitude larger than expected.

In a recent work done by Vinko et al. [14], detailed calculations of the cross-section for selected photon energy and electron temperatures for aluminium were reported. These calculations show clear changes in photon absorption due to the electron distribution, and in particular, a peculiar increase of absorption in an intermediate range of electron temperatures (1–10 eV) was found. We introduced

Appl. Sci. **2017**, *7*, 814

the results of the calculations through a temperature-dependent absorption coefficient $\sigma(T)$ with effective electron temperature T_e. For each slice of Δz for the simulation, we estimated the local electron temperature from a Maxwell–Boltzmann model as $k_B \Delta T_e = \frac{2}{3n_e} \frac{\Delta F}{\Delta z}$, where n_e is a number of valence electron and ΔF is deposited energy per unit surface. Figure 5 shows experimental (expt.) EUV transmission data at 37 eV (left panel) and 33.3 eV (right panel) compared with different calculations, including: only optical saturation phenomena without accounting for temperature effects in the cross section (Sat. only); only electron heating but neglecting saturation effects (El. heat. only); both electron heating and saturation effects (Sat.+El. heat.). We see that the Sat. only model (no account for electron temperature) does not reproduce the decrease of the transmission between 1 to 5 J/cm². The El. heat. only (temperature-dependent, linear) model shows the decrease in transmittance, but the increase at high fluence is not reproduced. Finally, the Sat.+El. heat. model agrees closely with the experimental results, so for these EUV absorption experiments, electron heating is found to play an important role in understanding the initial decrease in transmittance observed in these accurate experimental data.

4. Conclusions

In this work we have shown the results of the application of a simplified three-channel model to the non-equilibrium dynamics of ultrathin aluminum films excited by FEL radiation at 33.3, 37 and 92 eV photon energy. The theoretical model has been developed and implemented into a Fortran 90 code using semi-classical rate equations coupled with the equation of propagation of the photon wave packets. X-ray transmission measurements for pulses of different photon energy and fluence were found to be in agreement with present simulations. Simulations were shown to be able to shed light on temporal dynamics (in the fs range) for nano-sized Al films strongly interacting with the photon pulse. Simulations of non-equilibrium dynamics were reported for three different maximum intensities of FEL pulses corresponding to typical fluences for a system far from saturation and near the saturation threshold. We remark that the three-channel model including a relaxed state produced a population inversion state, leading to negative temperatures which is a typical consequence of non-equilibrium dynamics induced by laser. We also expanded our non-linear model, including explicitly the two-photon absorption (TPA) cross-section and the effect of including electron heating for reproducing accurate transmission measurements. Within our simplified TPA model, we have obtained the highest entropy state in the limit of high fluence, corresponding to positive temperatures (no inversion). Through the series of numerical calculations using different models reported in this article, we have shown that for the case of short EUV/X-ray pulses, compared to the lifetime of the excited state, we need to employ a dynamical (time-dependent) model. Furthermore, electron heating effects must be taken into account for a deeper level of agreement with experimental data showing saturable absorption at high fluence, as shown by a direct comparison with accurate EUV absorption data.

Acknowledgments: This work has been carried out in the framework of the TIMEX collaboration (time-resolved studies of matter under extreme and metastable conditions: http://gnxas.unicam.it/TIMEX). TIMEX was a project financed by the FERMI@Elettra FEL facility in Trieste in collaboration with the University of Camerino. Keisuke Hatada gratefully acknowledges TIMEX research grants and the authors acknowledge the COST Action MP1306 EUSpec for support.

Author Contributions: Both authors equally contributed to this article.

Conflicts of Interest: The authors declare no conflict of interest.

Appendix A. Long Pulse Limit

The set of rate equations in Equatioins (1)–(5) is simplified by using the condition of conserving the total number of the states; that is to say, substituting N_1 by $N - N_2 - N_3$,

$$\frac{dN_2(z,t)}{dt} = -\frac{g(z,t)I(z,t)}{h\nu} - \frac{N_2(z,t)}{\tau_{21}} - \frac{N_2(z,t)}{\tau_{23}} \tag{A1}$$

$$\frac{dN_3(z,t)}{dt} = \frac{1}{\tau_{23}} N_2(z,t) - \frac{1}{\tau_{31}} N_3(z,t) \tag{A2}$$

$$g(z,t) = \sigma(T) [2N_2(z,t) - N + N_3(z,t)] \tag{A3}$$

For a long pulse limit where the duration of the pulse is much longer than the lifetime of the states, the system reaches a steady state, such that the time derivatives go to zero:

$$\frac{dN_1(z,t)}{dt} = \frac{dN_2(z,t)}{dt} = \frac{dN_3(z,t)}{dt} = 0, \quad \frac{dI(z,t)}{dt} = 0.$$

Thus, we can further simplify the equations

$$\frac{dI(z)}{dz} = \frac{g_0}{1 + \frac{I(z)}{\alpha I_s}} I(z) \tag{A4}$$

$$g_0 = -\sigma(T)N \tag{A5}$$

$$I_s = \frac{h\nu}{2\sigma\tau_{21}} \tag{A6}$$

$$\alpha = 1 + \frac{2\tau_{21} - \tau_{31}}{2\tau_{23} + \tau_{31}} \tag{A7}$$

The time-independent density of occupation numbers are:

$$N_1(z) = N - \left(1 + \frac{\tau_{31}}{\tau_{23}}\right) N_2(z) \tag{A8}$$

$$N_2(z) = \frac{1}{2} \frac{1}{1 + \frac{\tau_{31}}{2\tau_{23}}} \frac{\frac{I(z)}{\alpha I_s}}{1 + \frac{I(z)}{\alpha I_s}} N \tag{A9}$$

$$N_3(z) = \left[1 + \frac{1}{1 + \frac{I(z)}{\alpha I_s}}\right] N. \tag{A10}$$

The solution of Equation (A4) is:

$$I(z) = I_{sat} f_W^{-1} \left(\frac{I_0}{I_{sat}} e^{\frac{I_0}{I_{sat}}} e^{g_0 z}\right) \tag{A11}$$

$$I_{sat} = \alpha I_s \tag{A12}$$

where f_W is Lambert omega function [22], $f_W(W) = We^W$, I_{sat} is the saturation intensity, and I_0 is an initial intensity. When the relaxed state $|3\rangle$ is not considered, namely just two channel model, $\alpha = 1$. In general, the lifetime from the excited state $|2\rangle$ to the relaxed state $|3\rangle$—which is τ_{23}—is much shorter that the others, so the coefficient α becomes greater than 1. This leads to the saturation limit I_{sat} being higher. This situation can be interpreted as due to the quick annihilation of the excited state by the instant relaxation, the number of excited state can be kept small even for high intensity of photons in the pulse. In other words, the relaxed state behaves as a reserver of excited photo electrons. When the intensity I is well above the saturation threshold I_{sat},

$$I = I_0 - \eta, \quad (\eta < 1).$$

The asymptotic behavior of I is,

$$I \sim I_0 + I_{sat} \, g_0 \, z. \tag{A13}$$

We see that the attenuation of intensity is no longer exponential, but linear.

Appendix B. Short Pulse Limit

For a short pulse limit, we employ the Frantz–Nodvik model [23]. In this model, it is supposed that the lifetime of an excited state is much longer than the width of pulse, such that it reduces a limit of lifetime to infinity. In the three-channel model, the decay terms are dropped off and the equations have only absorption and stimulated emission processes. Naturally, it reduces to a two-channel model without an intermediate relaxed state. The detail of the derivation can be found in Reference [23]. The final form of the fluence, $\phi(z) = \int I(z,t)dt$, at a position z in one-dimensional model is

$$\phi(z) = \phi_{sat} \ln \left[1 + e^{g_0 z} \left(e^{\frac{\phi_0}{\phi_{sat}}} - 1 \right) \right] \tag{A14}$$

$$\phi_{sat} = \frac{h\nu}{2\sigma}. \tag{A15}$$

When ϕ_0 is well above ϕ_{sat}, as a *long pulse* model we get the asymptotic linear behavior,

$$\phi(z) = \phi_0 + \phi_{sat} \, g_0 \, z. \tag{A16}$$

Appendix C. Steady State Equation for TPA

The equations and solution with TPA case under steady condition in the ground state are:

$$\frac{dI(z)}{dz} = g_0 \, I(z) + \bar{g}_2 \, I^2(z) \tag{A17}$$

$$I(z) = I_0 \, e^{g_0 z} \frac{1}{1 + \frac{\bar{g}_2}{g_0} I_0 \left(1 - e^{g_0 z} \right)} \tag{A18}$$

$$\bar{g}_2 = -\frac{\sigma_2}{h\nu} N$$

References

1. Ackermann, W.; Asova, G.; Ayvazyan, V.; Azima, A.; Baboi, N.; Bähr, J.; Balandin, V.; Beutner, B.; Brandt, A.; Bolzmann, A.; et al. Operation of a free-electron laser from the extreme ultraviolet to the water window. *Nat. Photonics* **2007**, *1*, 336–342.
2. Allaria, E.; Battistoni, A.; Bencivenga, F.; Borghes, R.; Callegari, C.; Capotondi, F.; Castronovo, D.; Cinquegrana, P.; Cocco, D.; Coreno, M.; et al. Tunability experiments at the FERMI@Elettra free-electron laser. *New J. Phys.* **2012**, *14*, 113009.
3. Emma, P.; Akre, R.; Arthur, J.; Bionta, R.; Bostedt, C.; Bozek, J.; Brachmann, A.; Bucksbaum, P.; Coffee, R.; Decker, F.J.; et al. First lasing and operation of an Ångstrom-wavelength free-electron laser. *Nat. Photonics* **2010**, *4*, 641–647.
4. Ishikawa, T.; Aoyagi, H.; Asaka, T.; Asano, Y.; Azumi, N.; Bizen, T.; Ego, H.; Fukami, K.; Fukui, T.; Furukawa, Y.; et al. A compact X-ray free-electron laser emitting in the sub-ångström region. *Nat. Photonics*, **2012**, *6*, 540–544.
5. Bostedt, C.; Boutet, S.; Fritz, D.M.; Huang, Z.; Lee, H.J.; Lemke, H.T.; Robert, A.; Schlotter, W.F.; Turner, J.J.; Williams, G.J. Linac Coherent Light Source: The first five years. *Rev. Mod. Phys.* **2016**, *88*, 015007.
6. Ciricosta, O.; Chung, H.K.; Lee, R.W.; Wark, J.S. Simulations of neon irradiated by intense X-ray laser radiation. *High Energy Density Phys.* **2011**, *7*, 111–116.

7. Nagler, B.; Zastrau, U.; Faustlin, R.R.; Vinko, S.M.; Whitcher, T.; Nelson, A.J.; Sobierajski, R.; Krzywinski, J.; Chalupsky, J.; Abreu, E.; et al. Turning solid aluminium transparent by intense soft X-ray photoionization. *Nat. Phys.* **2009**, *5*, 693–696.

8. Di Cicco, A.; Hatada, K.; Giangrisostomi, E.; Gunnella, R.; Bencivenga, F.; Principi, E.; Masciovecchio, C.; Filipponi, A. Interplay of electron heating and saturable absorption in ultrafast extreme ultraviolet transmission of condensed matter. *Phys. Rev. B* **2014**, *90*, 220303.

9. Rackstraw, D.S.; Ciricosta, O.; Vinko, S.M.; Barbrel, B.; Burian, T.; Chalupský, J.; Cho, B.I.; Chung, H.K.; Dakovski, G.L.; Engelhorn, K.; et al. Saturable absorption of an X-Ray Free-Electron-Laser heated Solid-Density aluminum plasma. *Phys. Rev. Lett.* **2015**, *114*, 015003.

10. Yoneda, H.; Inubushi, Y.; Yabashi, M.; Katayama, T.; Ishikawa, T.; Ohashi, H.; Yumoto, H.; Yamauchi, K.; Mimura, H.; Kitamura, H. Saturable absorption of intense hard X-rays in iron. *Nat. Commun.* **2014**, *5*, 5080.

11. Kitamura, H. Rate equation for intense core-level photoexcitation and relaxation in metals. *J. Phys. B At. Mol. Opt. Phys.* **2010**, *43*, 115601.

12. Kitamura, H. Rapid energy-level shifts in metals under intense inner-shell photoexcitation. *High Energy Density Phys.* **2012**, *8*, 66–70.

13. Iglesias, C.A. XUV absorption by solid-density aluminum. *High Energy Density Phys.* **2010**, *6*, 311–317.

14. Vinko, S.M.; Gregori, G.; Desjarlais, M.P.; Nagler, B.; Whitcher, T.J.; Lee, R.W.; Audebert, P.; Wark, J.S. Free-free opacity in warm dense aluminum. *High Energy Density Phys.* **2009**, *5*, 124–131.

15. Rackstraw, D.S.; Vinko, S.M.; Ciricosta, O.; Chung, H.K.; Lee, R.W.; Wark, J.S. Simulations of the time and space-resolved x-ray transmission of a free-electron-laser-heated aluminium plasma. *J. Phys. B At. Mol. Opt. Phys.* **2016**, *49*, 035603.

16. Hatada, K.; Di Cicco, A. Modeling saturable absorption for ultra short X-ray pulses. *J. Electron Spectrosc. Relat. Phenom.* **2014**, *196*, 177–180.

17. Courant, R.; Friedrichs, K.; Lewy, H. On the Partial Difference Equations of Mathematical Physics. *IBM J. Res. Dev.* **1967**, *11*, 215–234.

18. Almbladh, C.O.; Morales, A.L.; Grossmann, G. Theory of Auger core-valence-valence processes in simple metals. I. Total yields and core-level lifetime widths. *Phys. Rev. B,* **1989**, *39*, 3489–3502.

19. Di Cicco, A.; Bencivenga, F.; Battistoni, A.; Cocco, D.; Cucini, R.; D'Amico, F.; Fonzo, S.D.; Filipponi, A.; Gessini, A.; Giangrisostomi, E.; et al. Probing matter under extreme conditions at Fermi@Elettra: The TIMEX beamline. In *Damage to VUV, EUV, and X-ray Optics III*; Juha, L., Bajt, S., London, R.A., Eds.; SPIE: Prague, Czech Republic, 2011; Volume 8077, p. 807704.

20. Di Cicco, A.; Masciovecchio, C.; Bencivenga, F.; Principi, E.; Giangrisostomi, E.; Battistoni, A.; Cucini, R.; D'Amico, F.; Di Fonzo, S.; Gessini, A.; et al. Probing matter under extreme conditions at the free-electron-laser facilities: the TIMEX beamline. *Not. Neutroni E Luce Di Sincrotrone* **2013**, *18*, 19–25.

21. Masciovecchio, C.; Battistoni, A.; Giangrisostomi, E.; Bencivenga, F.; Principi, E.; Mincigrucci, R.; Cucini, R.; Gessini, A.; D'Amico, F.; Borghes, R.; et al. EIS: the scattering beamline at FERMI. *J. Synchrotron Radiat.* **2015**, *22*, 553–564.

22. Corless, R.M.; Gonnet, G.H.; Hare, D.E.G.; Jeffrey, D.J.; Knuth, D.E. On the LambertW function. *Adv. Comput. Math.* **1996**, *5*, 329–359.

23. Frantz, L.M.; Nodvik, J.S. Theory of Pulse Propagation in a Laser Amplifier. *J. Appl. Phys.* **1963**, *34*, 2346–2349, doi:10.1063/1.1702744.

*applied
sciences*

Article

Fundamental Limits on Spatial Resolution in Ultrafast X-ray Diffraction

Adam Kirrander [1,*] and Peter M. Weber [2]

[1]   EaStCHEM, School of Chemistry, University of Edinburgh, David Brewster Road, Edinburgh EH9 3FJ, UK
[2]   Department of Chemistry, Brown University, Providence, RI 02912, USA; peter_weber@brown.edu
[*]   Correspondence: Adam.Kirrander@ed.ac.uk; Tel.: +44-(0)131-6504716

Academic Editor: Kiyoshi Ueda
Received: 7 April 2017; Accepted: 17 May 2017; Published: 23 May 2017

Abstract: X-ray Free-Electron Lasers have made it possible to record time-sequences of diffraction images to determine changes in molecular geometry during ultrafast photochemical processes. Using state-of-the-art simulations in three molecules (deuterium, ethylene, and 1,3-cyclohexadiene), we demonstrate that the nature of the nuclear wavepacket initially prepared by the pump laser, and its subsequent dispersion as it propagates along the reaction path, limits the spatial resolution attainable in a structural dynamics experiment. The delocalization of the wavepacket leads to a pronounced damping of the diffraction signal at large values of the momentum transfer vector q, an observation supported by a simple analytical model. This suggests that high-q measurements, beyond 10–15 Å^{-1}, provide scant experimental payback, and that it may be advantageous to prioritize the signal-to-noise ratio and the time-resolution of the experiment as determined by parameters such as the repetition-rate, the photon flux, and the pulse durations. We expect these considerations to influence future experimental designs, including source development and detection schemes.

Keywords: X-ray free-electron lasers; ultrafast dynamics; diffraction; spatial resolution; pump-probe; quantum dynamics; wavepackets; photochemistry

1. Introduction

X-ray and electron scattering have long played an important role in structure determination of matter [1–3]. The recent development of ultrashort pulsed electron and X-ray beams has now expanded the scope of structure determination to the time domain [4,5] (and references therein). Important applications include the determination of molecular structures in excited states and the observation of structural molecular dynamics, i.e., the time-resolved determination of transient molecular structures during chemical reactions [6–13]. The advent of ultrafast pulsed X-ray Free-Electron Lasers (XFELs) in particular has increased the intensity of X-rays while decreasing pulse durations to below 30 fs [14]. Consequently, XFELs are assuming a powerful role in the exploration of gas-phase photochemistry, allowing a direct comparison between experimental results and high-level theory [15]. More generally, ultrafast X-ray scattering will play an important and growing role in our arsenal of ultrafast imaging techniques by providing a structurally-sensitive complement to spectroscopic techniques.

Ultrafast measurements typically involve a pump-probe scheme, where a laser pump pulse prepares an initial wavefunction, whose subsequent time evolution is probed at a sequence of delay-times [15–24]. Depending on the experimental situation, the wavepacket may be of electronic [25,26], nuclear [27], or mixed nuclear-electronic [28] nature. Inevitably, the localization of the initially prepared wavepacket and its subsequent spreading depends not only on the potential energy landscape of the molecular system but also on the exact optical preparation of the initially excited state [29–31]. While in some cases, such as nonradiative transitions in intermediate or statistical limit scenarios, the wave packet is presumed to spread over all receiving modes [32–34]; in other cases, well-defined wave packets persist for at least part of the chemical reaction pathway. In 1,3-cyclohexadiene, for example, the propagation of the wavepacket has been described as 'ballistic', implying that it retains a well-localized form even as it propagates [35]. However, it is inevitable that the nonclassical nature and propagation of the wavepacket leads to a delocalization. It is therefore an important question to consider the extent of this delocalization and how it affects the measurement of the structural dynamics in time-resolved scattering experiments. In the present article, we explore the effective upper limit on the range of the momentum transfer vector (q) that is meaningful to observe in scattering experiments. To do so, we use advanced simulations of ultrafast dynamics in the molecules D_2, ethylene, and 1,3-cyclohexadiene (CHD), and examine the effect of the nuclear wavepacket spreading on the X-ray scattering signals.

2. Methods

2.1. Wavepacket Dynamics

Ultrafast experiments follow the evolution of a time-dependent wavepacket, $|\Psi(\mathbf{r}, \mathbf{R}, t)\rangle$, where $\mathbf{r}$ and $\mathbf{R}$ are the electronic and nuclear coordinates, and t is time. The wavepacket is the solution to the time-dependent Schrödinger equation,

$$i\hbar \frac{d}{dt}|\Psi(\mathbf{r}, \mathbf{R}, t)\rangle = \hat{H}|\Psi(\mathbf{r}, \mathbf{R}, t)\rangle, \tag{1}$$

where the Hamiltonian $\hat{H}$ can be decomposed into a field-free molecular Hamiltonian $\hat{H}_0$ and an interaction term $\hat{H}_{\text{int}}$ that describes the interaction with the pump pulse (and, later, the probe pulse). Most commonly, the pump will be an optical laser pulse, so that the light-matter interaction is given in the dipole approximation by $\hat{H}_{\text{int}}(t) \approx -d\epsilon(t)$, where $\epsilon(t)$ is the time-dependent electromagnetic field and d the electric dipole operator. The wavepacket excited by the pump pulse will have certain qualitites imprinted by the pump laser, including a bandwidth, phase-properties, and overall energy distribution.

There are many strategies for solving the time-dependent Schrödinger equation in Equation (1) above. One possible ansatz expands the wavepacket in the complete basis of orthonormal eigenstates $|\Psi_j\rangle$ of the molecular Hamiltonian $\hat{H}_0$ (see e.g., Ref. [36]),

$$|\Psi(t)\rangle = \sum_j c_j(t) e^{-iE_j t/\hbar}|\Psi_j\rangle, \tag{2}$$

where $|\Psi(t)\rangle$ is the excited wavepacket, $|\Psi_j\rangle$ are orthonormal rovibronic wave functions and E_j are the corresponding energies. According to first-order perturbation theory, the time-dependent coefficients $c_j(t)$ become,

$$c_j(t) = iD_{js} \left[\frac{e}{\hbar} \int_{-\infty}^{t} dt e^{i\omega_{js}t} \epsilon(t') \right], \tag{3}$$

where D_{js} are the combined dipole transition moments and the Franck–Condon factors for excitation from the initial state s to the final state j. The expression in the square brackets corresponds to the energy- and the time-dependent complex excitation function with e the charge of an electron, the angular frequency $\omega_{js} = (E_j - E_s)/\hbar$ and $\epsilon(t')$ the excitation field [28,37,38]. The trouble with this solution is that it requires that the eigenstates of the system are calculated first, which renders the approach practical only in molecules with a small number of degrees of freedom, such as the diatomic molecule D_2 in the present article.

In the general case, one must propagate the wavepacket numerically while circumventing as much as possible the effect of the exponential scaling of the required grid (basis) with the number of degrees of freedom. One approach is to expand the molecular wavepacket in a basis of non-stationary Gaussian wavefunctions (coherent states). Methods that take this path include the ab-initio multiconfigurational Ehrenfest method (AI-MCE) [39,40] used in this article, the closely related ab-initio multiple spawning (AIMS) [41], the hybrid ab-initio multiple cloning (AIMC) method [42], the variational multiconfigurational Gaussians (v-MCG) [43] method, and the coupled coherent-states (CCS) method [44–46]. All of these methods trace their roots to seminal semi-classical work by Eric Heller [47].

In AI-MCE, the molecular wavepacket, $|\Psi(t)\rangle$, is expanded in terms of Ehrenfest functions with dynamically coupled expansion coefficients $D_k(t)$,

$$|\Psi(t)\rangle = \sum_{k=1}^{N} D_k(t)|\psi_k(t)\rangle, \tag{4}$$

where each Ehrenfest function, $|\psi_k(t)\rangle$, consists of a Gaussian nuclear coherent state, $|\bar{z}_k(t)\rangle$, distributed across N_s electronic states, $|\phi_k^i\rangle$, with the amplitudes $a_k^i(t)$,

$$|\psi_k(t)\rangle = \left[\sum_{i=1}^{N_s} a_k^i(t)|\phi_k^i\rangle\right]|\bar{z}_k(t)\rangle. \tag{5}$$

The coherent state, $|\bar{z}_k(t)\rangle$, is a product of 3×(number of atoms) one-dimensional Gaussian coherent states, shown here for x_α, the x-component of the coordinate of atom α,

$$\langle x_\alpha|\bar{z}_k(t)\rangle = \left(\frac{\gamma_\alpha}{\pi}\right)^{\frac{1}{4}} \exp\left(-\frac{\gamma_\alpha}{2}\left(x_\alpha - Q_{\alpha x}^k\right)^2 + \frac{i}{\hbar}P_{\alpha x}^k\left(x_\alpha - Q_{\alpha x}^k\right) + \frac{iP_{\alpha x}^k Q_{\alpha x}^k}{2\hbar}\right), \tag{6}$$

where the position is $Q_{\alpha x}^k$, the momentum $P_{\alpha x}^k$, and the width parameter γ_α. The Ehrenfest ansatz can be slow to converge, but, in the context of this article, the difference is mostly technical, as sampling methods have been developed to speed up the convergence of Equation (4) [42,48]. Each time-step of the propagation requires multiple calls to an ab-initio electronic structure software package to calculate energies for the ground and excited states, the potential gradients, and non-adiabatic couplings. For further details, see, e.g., Refs. [39,40,42,49]. In this article, we use the AI-MCE method to simulate the photoexcited dynamics of the molecules ethylene and CHD.

2.2. Computational Details for Dynamics

2.2.1. Deuterium (D_2)

In D_2, we consider a vibrational wavepacket in the $B^1\Sigma_u^+$ electronic state, excited by a single photon from the $X^1\Sigma_g^+$ ($v = 0$) ground state. The pump pulse energy is 14.3 eV, with the full width at half maximum (FWHM) duration 20 fs, such that the excited vibrational wavepacket is centered at the $v = 38$ vibrational state [27]. The potential energy curves are shown in Figure 1a. Since D_2 is a diatomic molecule, the wavepacket can be expanded in a basis of vibrational eigenfunctions using the ansatz in Equation (2). We use potential energy curves for the initial ground X-state [50] and the excited B-state [51], and dipole transition moments [52], calculated ab-initio by Wolniewicz et al. The molecular mass is taken from CODATA 2010 [53]. The vibrational eigenfunctions are calculated using an accurate 5th order Runge–Kutte algorithm, with the orthonormality of the vibrational eigenstates ensured via a Cholesky factorisation, and the transition moments D_{js} calculated for $B^1\Sigma_u^+(v) \leftarrow X^1\Sigma_g^+(v = 0)$.

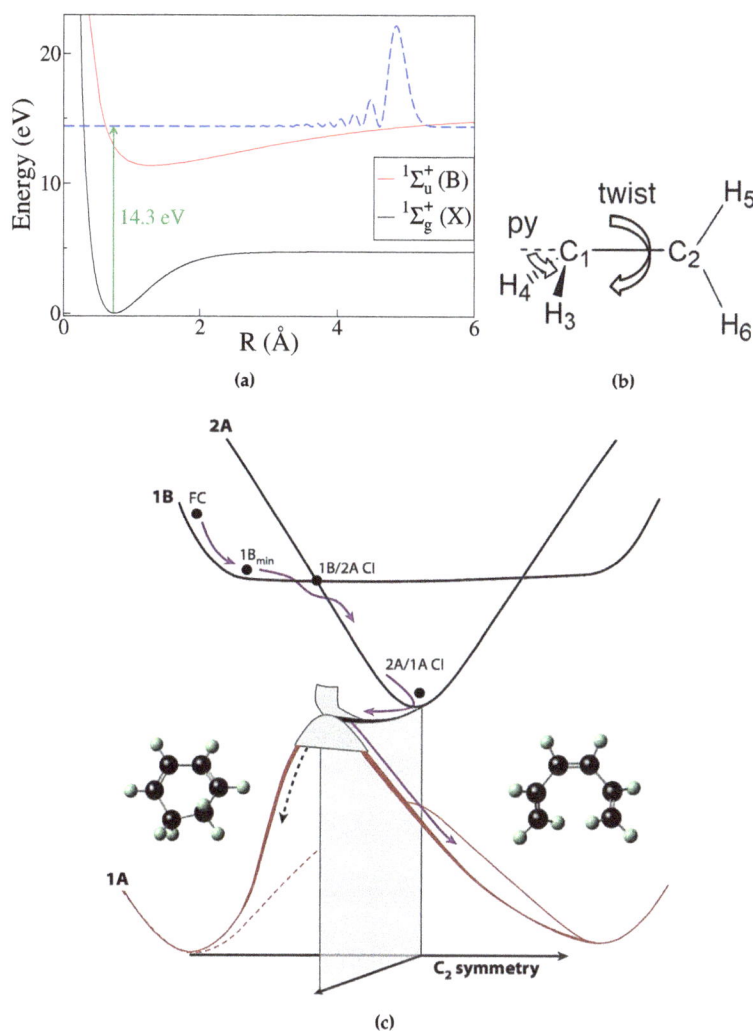

Figure 1. Overview of the molecules and dynamics discussed in this article. (**a**) D_2: Potential energy curves for the ground $X^1\Sigma_g^+$ and excited $B^1\Sigma_u^+$ state potential energy curves. The vibrational wavepacket is excited onto the $B^1\Sigma_u^+$ state centered on vibrational state $\nu = 38$ by a 14.3 eV pump pulse. The wavepacket (dashed line) is shown at the outer turning point; (**b**) Ethylene: After excitation to the S_1 state, the dynamics is dominated by C–C bond twist, pyramidalization, and changes in C–C bond length; (**c**) 1,3-cyclohexadiene (CHD): Upon excitation from the 1A ground state onto the 1B bright state in the Franck–Condon region, the molecule decays through a sequence of conical intersections back to the ground state in either the ring-open or ring-closed form (based on illustration in Ref. [54]).

2.2.2. Ethylene

In ethylene ($H_2C=CH_2$), we consider vertical excitation into the S_1 ($\pi\pi^*$) state. The excited molecule undergoes cis-trans isomerisation around the C=C bond, and decays via nonradiative decay through a twisted or pyramidalised conical intersection, or via H-atom migration to form ethylidene (CH_3CH), which then decays through a different conical intersection. The basic dynamics is sketched out in Figure 1b. The population on S_1 remains fairly constant for the first 30 fs, at which point the molecule begins to decay exponentially to S_0 with an approximate lifetime of $\approx$112 fs.

The simulations employ the AI-MCE method [55], with electronic potential energies, gradients, and nonadiabatic couplings calculated *on-the-fly* using the MOLPRO electronic structure package [56]. A total of 1000 Ehrenfest trajectories are initiated in the Franck–Condon region using a Wigner distribution [57], and are propagated for 150 fs. The electronic structure is calculated at the state-averaged SA3-CAS(2,2)-SCF/cc-pVDZ level (two electrons in π and π^* orbitals) including the S_0, the S_1 π-π^*, and the S_2 π^{*2} states. The calculated lifetime from the simulations (112 fs) agrees well with the 110 fs obtained with AIMS dynamics using the same CAS(2,2) active space [58]. Including more dynamic correlation via CASPT2 drops the lifetime to 89 fs [58] while including Rydberg states (predominantly the π-3s state) lowers the lifetime further to approximately 60 fs [59,60]. However, the influence of Rydberg states on the dynamics simulations is relatively small and experimental evidence for the role of Rydberg states remains under debate [61,62]. Importantly, the small active space in the present study allows us to converge the simulations [49], which is the key aspect for the present discussion of X-ray scattering signals.

2.2.3. 1,3-Cyclohexadiene (CHD)

The electrocyclic ring-opening reaction of CHD has been studied extensively in experiments [54], including ultrafast X-ray scattering [6] and time-resolved photoelectron spectroscopy [15]. Following vertical excitation into the $1B$ bright state, the molecule decays rapidly via two sets of conical intersections back to the $1A$ ground state, as outlined in Figure 1c. The reaction is very rapid and occurs in a ballistic manner. When the molecule reaches the ground state, there is an approximately even split between molecules that return to the ring-closed form and those that undergo ring-opening [6,54].

The simulations follow the procedure outlined in Ref. [6]. The same AI-MCE method as in the ethylene case is used, with adiabatic electronic potential energies, gradients, and nonadiabatic couplings calculated *on-the-fly* with MOLPRO [56] at the SA3-CAS(6,4)-SCF/cc-pVDZ level of theory, which has been shown to be suitable for the CHD ring-opening reaction [63]. Wigner-sampling is employed for the initial state, with 100 AI-MCE trajectories propagated for 200 fs. This is not sufficient to fully converge the simulations, but provides a reasonable and effectively semi-classical representation of the reaction dynamics.

2.3. X-ray Scattering

We now review briefly the theory of X-ray scattering. To a good approximation, X-ray photons scatter in a two-photon process via the squared field vector potential, A^2, taken in the first order of perturbation theory (see e.g., Refs. [64,65]). The resulting double-differential scattering cross-section is given by [66,67],

$$\frac{d^2S}{d\Omega d\omega_{k_1}} = \alpha \int_0^\infty dt \int_{-\infty}^\infty d\delta \, I_P(t) C_P(\delta) e^{-\iota \omega_{k_1} \delta} \mathcal{W}(t,\delta), \tag{7}$$

where $I_P(t) = |E_{k_0}|^2 e^{-(t-t_P)^2/\gamma_d^2}$ is the X-ray pulse intensity-profile, and $C_P(\delta) = \sqrt{\epsilon(\delta)} e^{\iota \omega_{k_0} \delta}$ the normalized X-ray pulse coherence function. The probe pulse duration, in terms of FWHM intensity, is $\tau_d = 2\gamma_d \sqrt{\ln 2}$, and $\epsilon(\delta) = e^{-(t-t_P)^2/2\gamma_d^2}$ is the Gaussian electric field envelope, with t_P the time-delay between pump and probe. The scattering from the material system is given by $\mathcal{W}(t,\delta)$, which is a function of the wavepacket $|\Psi(t)\rangle$ and the scattering operator $\hat{L}$,

$$\mathcal{W}(t,\delta) = \langle \Psi(t)|e^{\iota \hat{H}_0 \delta/2\hbar} \, \hat{L}^\dagger \, e^{-\iota \hat{H}_0 \delta/\hbar} \, \hat{L} \, e^{\iota \hat{H}_0 \delta/2\hbar}|\Psi(t)\rangle, \tag{8}$$

where the scattering operator $\hat{L}$ appears twice as appropriate for a two-photon process, and is given by

$$\hat{L} = \sum_j e^{\iota \mathbf{q} \mathbf{r}_j}, \tag{9}$$

where $\mathbf{r}_j$ is the position of electron j, and $\mathbf{q} = \mathbf{k}_0 - \mathbf{k}_1$ the momentum transfer vector with $\mathbf{k}_0$ and $\mathbf{k}_1$ the incoming and outgoing wavevectors for the X-rays. Matrix elements of the operator $\hat{L}$ correspond to elastic and inelastic scattering [68]. Assuming that the time-scale for nuclear motion is significantly slower than for electrons, and that electronic states are well separated, one arrives at the so-called elastic approximation [67,69],

$$W(t, \delta) \propto \sum_i \int |\chi_i(\mathbf{R}, t)|^2 |L_{ii}|^2 \, d\mathbf{R}, \tag{10}$$

in which we have used the Born–Huang ansatz for the wavepacket $|\Psi(\mathbf{R}, t)\rangle = \sum_i \chi_i(\mathbf{R}, t)|i(\mathbf{r}; \mathbf{R})\rangle$, with $\chi_i(\mathbf{R}, t)$ the nuclear wavepacket on each electronic state $|i(\mathbf{r}; \mathbf{R})\rangle$, and the elastic X-ray scattering matrix elements $L_{ii} = \langle i(\mathbf{r}; \mathbf{R})|\hat{L}|i(\mathbf{r}; \mathbf{R})\rangle_{\mathbf{r}}$ [70–72]. The elastic scattering can be simplified further if we adopt the independent atom model (IAM), whereby we ignore the specifics of each electronic state and instead assume that the bulk of the scattering corresponds to a coherent sum of the scattering from the individual constituent atoms. In this approximation, $L_{ii} \approx f_{IAM}$, irrespective of electronic state i. The f_{IAM} is given by,

$$f_{IAM}(\mathbf{q}) = \sum_{\alpha=1}^{N_{at}} f_\alpha^0(q) e^{i\mathbf{R}_\alpha \mathbf{q}}, \tag{11}$$

where $f_\alpha^0(q)$ are the tabulated atomic form factors [73], $\mathbf{R}_\alpha$ the atomic positions, and $q=|\mathbf{q}|$ the amplitude of the momentum transfer, which is a function of the scattering angle θ via $q = 2|\mathbf{k}_0| \sin \theta/2$. Finally, $|L_{ii}|^2 \approx |f_{IAM}|^2$ becomes,

$$|f_{IAM}(\mathbf{q})|^2 = \sum_{\alpha=1}^{N_{at}} \left| f_\alpha^0(q) \right|^2 + \sum_{\alpha \neq \beta}^{N_{at}} f_\alpha^0(q) f_\beta^0(q) e^{i\mathbf{R}_{\alpha\beta} \mathbf{q}}, \tag{12}$$

with $\mathbf{R}_{\alpha\beta} = \mathbf{R}_\alpha - \mathbf{R}_\beta$ the vector between each pair of atoms. The first sum on the right is the atomic term, which forms a constant background, and the second sum is the molecular term, which contains all structural interference relating to molecular geometry. Equation (12), when rotationally averaged, becomes the standard IAM formula shown in Equation (15). It is perhaps worthwhile to point out that elastic electron scattering results in an expression almost identical to Equation (12), with the main difference being that the form factors for electron scattering must account for the additional scattering from the nuclei [2,74,75].

3. Results

3.1. Simple Model

We begin by examining the effect of the delocalization of the nuclear wavepacket on the diffraction signal via a simple analytical model. For this, we use frozen Gaussian width parameters, γ_α, fitted to calculated ground state harmonic vibrational wavefunctions in a large set of organic molecules by Thompson et al. [76]. Consider a Gaussian nuclear wavepacket, $v(\mathbf{R})$, expressed as a product of three-dimensional Gaussian wavefunctions, one for each atom. The probability density for such a wavepacket is,

$$|v(\mathbf{R}_1, \ldots, \mathbf{R}_{N_{at}})|^2 = \prod_{\alpha=1}^{N_{at}} \left(\frac{\gamma_\alpha}{\pi} \right)^{\frac{3}{2}} e^{-\gamma_\alpha (\mathbf{R}_\alpha - \mathbf{Q}_\alpha)^2}, \tag{13}$$

where $\mathbf{R}_\alpha = (R_{\alpha x}, R_{\alpha y}, R_{\alpha z})$ are the Cartesian nuclear coordinates, $\mathbf{Q}_\alpha$ is the coordinate that determines the center of the wavepacket, N_{at} is the number of atoms, and γ_α the width parameter of the Gaussian associated with each atom. There is a large body of theoretical work that shows that such Gaussian wavepackets constitute a sensible basis for nuclear dynamics (see e.g., Refs. [40,43,44,47,77]). Time-dependent propagation of the wavepacket generally involves propagation (classical, semi-classical, or quantum) of the phase-space coordinates and sometimes the width parameter γ_α.

The X-ray scattering signal will be proportional to $\langle v(\mathbf{R})| \, |f_{\text{IAM}}(q)|^2 \, |v(\mathbf{R})\rangle_{\mathbf{R}}$, according to the elastic approximation in Equation (10). Solving this integral analytically, including rotational averaging and thus assuming no preferred orientation of the molecule, results in,

$$I_{\text{IAM}}^{\text{wavepacket}}(q) \propto \sum_{\alpha=1}^{N_{\text{at}}} \left|f_\alpha^0(q)\right|^2 + \sum_{\beta \neq \alpha}^{N_{\text{at}}} f_\alpha^0(q) f_\beta^0(q) \frac{\sin\left(qQ_{\alpha\beta}\right)}{qQ_{\alpha\beta}} e^{-q^2/2\gamma_{\alpha\beta}}, \tag{14}$$

with $Q_{\alpha\beta} = |\mathbf{Q}_\alpha - \mathbf{Q}_\beta|$ the distance between centers of the Gaussian wavepackets for each pair of atoms, and with the damping factor $\exp\left(-q^2/2\gamma_{\alpha\beta}\right)$ proportional to the combined Gaussian width of the two atoms given by $\gamma_{\alpha\beta} = 2\gamma_\alpha\gamma_\beta/\left(\gamma_\alpha + \gamma_\beta\right)$. The consequence of introducing the wavepacket in Equation (13) is thus an exponential damping of the molecular term in the scattering, which leaves the atomic term unaffected. It is instructive to contrast Equation (14) with the standard rotationally averaged IAM expression for scattering, originally derived by Debye [78],

$$I_{\text{IAM}}(q) \propto \sum_{\alpha=1}^{N_{\text{at}}} \left|f_\alpha^0(q)\right|^2 + \sum_{\beta \neq \alpha}^{N_{\text{at}}} f_\alpha^0(q) f_\beta^0(q) \frac{\sin\left(qR_{\alpha\beta}\right)}{qR_{\alpha\beta}}, \tag{15}$$

where $R_{\alpha\beta}$ are the distances between atoms (i.e., $\mathbf{R}_\alpha$ are the fixed nuclear positions). This expression is recovered from the wavepacket-damped expression in Equation (14) for strong localization ($\gamma_\alpha \to \infty$). Conversely, strong *delocalization* ($\gamma_\alpha \to 0$) extinguishes the structural interference in the molecular term, which is responsible for structural information. It is interesting to note that the damping in Equation (14) takes the same form as the temperature damping derived in electron diffraction [2], but has nothing to do with temperature; it is a consequence of the inherently delocalized nature of the wavepacket [21].

Numerically, we can examine the damping factor $\exp -q^2/2\gamma_{\alpha\beta}$ using the Gaussian widths determined by Thompson et al. [76]. The results, calculated in the range $0 < q < 20$ Å^{-1} are shown in Figure 2 for pairs of atoms. For each pair, the damping factor is calculated for $\gamma_\alpha \pm \sigma_\alpha$, where σ_α is the standard deviation obtained in the fitting procedure for the width parameters γ_α in Ref. [76]. Numerical values for the damping factors at $q = 10$ Å^{-1} are shown in Table 1. Invariably, the damping at $q = 10$ Å^{-1} is on the order of 0.7 or less (down to ≈ 0.4 for H, but then H-atoms contribute little to the scattering signal [75]). A factor of 0.7 corresponds to 30% of the signal irrevocably lost due to the delocalized nature of the target. It is important to point out that these damping factors constitute *lower* limits, since the width parameters used here are essentially minimum width parameters fitted to ground state harmonic oscillators. As such they are representative of the initially pumped wavepacket in the Franck–Condon region above the ground state geometry. It is therefore clear that, generally, we can expect significant degradation of the diffraction signal for $q > 10$ Å^{-1}. The subsequent evolution of the wavepacket then leads to further dispersion, and we will investigate the effects of the propagation of the wavepacket for the three molecules D_2, ethylene, and CHD, in Section 3.2.

Table 1. The damping factor (see Equation (14)) at $q = 10$ Å^{-1} for pairs of atoms, including: H, C, N, O, F, S, and Cl. A damping factor with value 1 corresponds to no damping (100% of the signal remains), and 0 to complete damping (the signal vanishes). In parenthesis, the value of the damping factor for $\gamma_\alpha - \sigma_\alpha$, with σ_α the standard deviation, is given to provide an approximate lower bound. The numerical values for γ_α and σ_α are taken from Ref. [76].

Atom	H	C	N	O	F	S	Cl
H	0.22 (0.17)	0.40 (0.33)	0.39 (0.31)	0.35 (0.27)	0.31 (0.21)	0.38 (0.28)	0.29 (0.17)
C		0.73 (0.65)	0.71 (0.61)	0.64 (0.54)	0.57 (0.40)	0.69 (0.54)	0.53 (0.33)
N			0.69 (0.58)	0.62 (0.50)	0.55 (0.38)	0.67 (0.51)	0.52 (0.32)
O				0.56 (0.44)	0.50 (0.33)	0.61 (0.45)	0.47 (0.28)
F					0.44 (0.25)	0.54 (0.34)	0.41 (0.21)
S						0.66 (0.45)	0.50 (0.28)
Cl							0.39 (0.17)

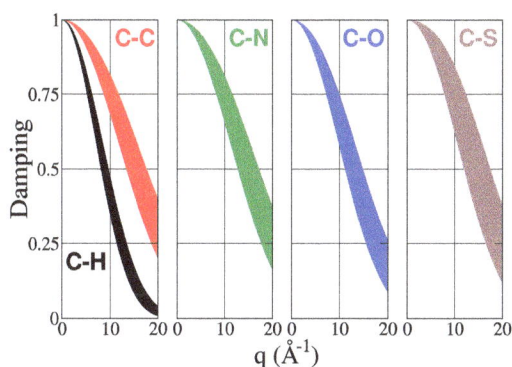

Figure 2. Damping factors (see Equation (14)) as a function of momentum transfer q (Å^{-1}) calculated using frozen Gaussian widths γ_α from Thompson et al. [76] for scattering from the following pairs of atoms: C−H, C−C, C−N, C−O, and C−S. The damping is given for $\gamma_\alpha \pm \sigma_\alpha$, where σ_α are the standard deviations in the fitted widths [76].

3.2. Simulations

Having considered a simple analytic model of the effect of delocalization on the diffraction pattern in the previous section, we now turn to simulations to examine the effect of the propagation of the molecular wavefunction on the scattering signal. We consider three different molecules to emphasize the generality of the discussed effects. In D$_2$ (Section 3.2.1), we examine the effect in a prototypical diatomic molecule, compare the results to a classical molecule, and evaluate the effect of nuclear mass on the wavepacket and thus the scattering signal. In ethylene (Section 3.2.2), we use the fact that quantum molecular dynamics simulations can be fully converged in such a comparatively small molecule to examine the sensitivity of the scattering signal to the number of trajectories used as a nuclear basis in the simulations. Finally, in CHD (Section 3.2.3), we examine the case of a larger polyatomic molecule, and in particular, evaluate the effect of the width parameters associated with the nuclear basis.

3.2.1. Deuterium (D$_2$)

We begin by considering the vibrational wavepacket in the *B*-state of the diatomic molecule D$_2$. The probability distribution of the wavepacket is shown in a contour plot in Figure 3a as a function of time and internuclear distance *R*. Upon excitation at the inner turning point by the 20 fs pump pulse, the wavepacket oscillates between the inner and outer turning points of the *B*-state potential with a period of 82 fs. Due to the anharmonic nature of the potential, the wavepacket gradually spreads across the entire *B*-state potential. It is worth emphasizing that this wavepacket has been observed in a recent experiment using a strong-field probe [27].

The elastic X-ray scattering corresponding to this wavepacket is shown in Figure 3b as a function of time and momentum transfer q. In order to focus on the structural component of the diffraction, and to remove the atom-specific effects of the form factors, we plot the *modified molecular intensity* [74],

$$M(q,t) = qI_{\text{mol}}(q,t) / \left|f_m^0(q)\right| \left|f_n^0(q)\right|, \tag{16}$$

with $f_m^0 = f_n^0 = f_H^0$ in the present case. The molecular intensity, $I_{\text{mol}}(q,t)$, is calculated as $I_{\text{mol}}(q,t) = \langle \Psi(t)| \sum_{i \neq j} f_i^0(q) f_j^0(q) \sin qR_{ij}/qR_{ij} |\Psi(t)\rangle$, which thus combines the elastic scattering approximation in Equation (10) and the IAM in Equation (12). The modified molecular intensity makes it easier to visualize the elastic scattering intensity at larger values of q, which otherwise decays rapidly due to the exponential decay of the atomic form factors $f^0(q)$ [74]. We assume an instant X-ray probe pulse, $\delta(t - t_P)$, to avoid convolution over the temporal X-ray pulse envelope, which would otherwise act to further '*soften*' the diffraction patterns and thus obscure the effect of the wavepacket on the scattering.

Since D_2 has a small reduced mass the wavepacket is strongly delocalized from the beginning, even when excited by a bandwidth limited pulse as in the present example. The resulting diffraction signal is thus strongly damped from the outset, as can be seen at high q values in Figure 3b. It is enlightening to compare the diffraction pattern in Figure 3b to that of a classical D_2 molecule, shown in Figure 3c, in which the nuclei are perfectly localized at all times. The diffraction from the classical molecule is shown for three full oscillations, at the same frequency as the quantum D_2 wavepacket ($T = 82$ fs). Comparing the diffraction patterns from the classical and quantum D_2 molecule, it is apparent that the signal in the classical case is stronger and better defined across all values of q, and remains identical for each oscillation. In contrast, the quantum wavepacket disperses with time, see $t/T > 5$ in Figure 3a, an effect which increases the decay of the diffraction signal over time, as seen in Figure 3b.

One possible objection to using D_2 as an example, despite the fact that the modified molecular intensity minimizes the effect of the atomic form factors on the scattering, is that D_2 is not an obvious candidate for a scattering experiment and that its small reduced mass exaggerates delocalization. We have therefore performed the same calculation but with a higher reduced mass corresponding to the potassium dimer, K_2. In order to aid comparison, we have not changed the potential, i.e., the potential energy curve remains that of the D_2 *B*-state, and all parameters except the reduced mass are left unchanged. We refer to this wavepacket as $\tilde{K}_2$ to avoid confusion with actual K_2. We first consider the wavepacket, which is plotted in Figure 4a. The greater reduced mass leads to a significant increase in the density of vibrational states, and thus in the number of eigenstates contained in the wavepacket in Equation (2). As a consequence, initial localization is stronger than in D_2, but conversely, the dispersion at longer times (on the normalized t/T timescale) is also greater. The scattering from the $\tilde{K}_2$ wavepacket, shown in Figure 4b, is thus more defined at earlier time, but suffers more from the dispersion of the wavepacket at larger multiples of the characteristic period, which leads to a more marked degradation of the diffraction signal at large values of q.

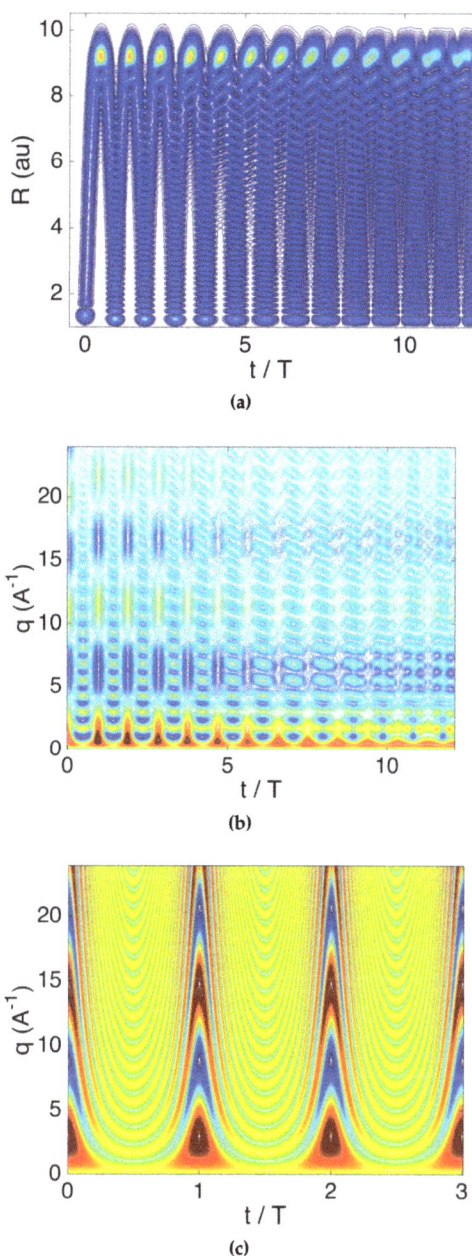

Figure 3. Results for D_2 excited to the electronic B-state by a 14.3 eV pulse with 20 fs duration. (a) Contour plot of the D_2 vibrational wavepacket probability density, $|v(R,t)|^2$, with R the internuclear distance in Bohr and t the time; (b) Contour plot of the corresponding modified (elastic) scattering intensity $M(q,t)$ for the wavepacket, with q the momentum transfer in Å^{-1} and t the time; (c) Contour plot of the modified (elastic) scattering intensity $M(q,t)$ for the *classical* D_2 molecule, shown over three oscillation periods, $t \in [0, 3T]$. Note that in all three plots the time t is given in units of the classic oscillation period $T = 82.5$ fs. The modified scattering intensity $M(q,t)$ is defined in Equation (16), and we use the form factor for H in the denominator.

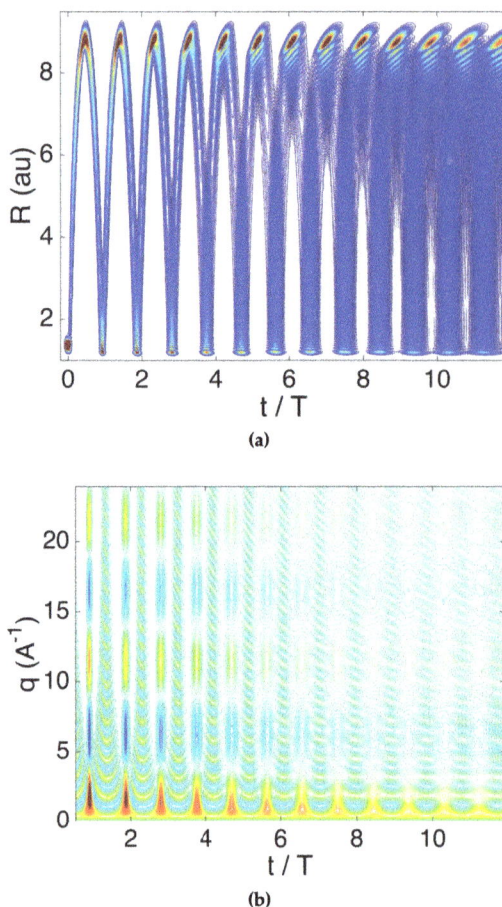

Figure 4. Results for $\tilde{K}_2$, calculated using the reduced mass of K_2 and the B-state potential of D_2, excited to the electronic B-state by a 14.3 eV pulse with 20 fs duration. (**a**) The $\tilde{K}_2$ vibrational wavepacket probability density, $|v(R,t)|^2$, with R the internuclear distance in Bohr and t the time; (**b**) Contour plot of the corresponding modified (elastic) scattering intensity, $M(q,t)$, with q the momentum transfer in Å^{-1} and t the time. In both plots, the time t is given in units of the classic oscillation period $T = 335.8$ fs. The modified scattering intensity $M(q,t)$ is defined in Equation (16), and we use the form factor for K in the denominator.

3.2.2. Ethylene

We now consider ethylene, a small molecule where we can be confident that the simulations are converged, to investigate the effect of the number of trajectories included in the simulations. In the case of ethylene, full convergence requires >500 trajectories [49]. In Figure 5, we compare the modified molecular intensity for elastic scattering from ethylene for a small set of 20 trajectories, shown in Figure 5a, with a large set of 1000 trajectories shown in Figure 5b. Comparison of the two signals indicates that the subset of 20 trajectories underestimates the dispersion at longer times, with the full set of 1000 trajectories showing a distinct deterioration of the signal over time. Interestingly, the limited subset of 20 trajectories *is* sufficient to capture the evolution of the dynamics qualitatively. In part, this reflects that the diffraction signal is dominated by the comparatively simple dynamics of the

C–C bond [75], which is well reproduced by the smaller subset of trajectories, but the observation is more general since we know from earlier studies that a small number of trajectories can do a good job of capturing the essence of a reaction path [79]. This observation is also commensurate with recent work on the ring-opening reaction of CHD, which found that the time-dependent diffraction pattern for the reaction could be reproduced accurately by a comparatively small number of trajectories representative of the dynamics [6]. Overall, this is good news for the analysis of ultrafast diffraction experiments because it suggests that a sensible analysis of experimental results does not require fully converged quantum molecular dynamics simulations (which are simply out of reach in many molecules of interest). However, it does raise another issue, which is the influence of the width parameters associated with trajectories, as discussed in the example of CHD next.

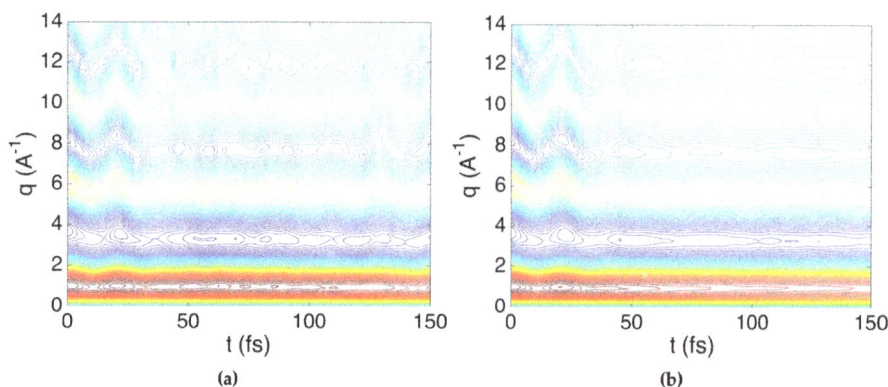

Figure 5. Contour plots of modified (elastic) scattering intensity, $M(q, t)$, for the photoexcited dynamics of ethylene. Time in fs and momentum transfer q in Å^{-1}, with $M(q, t)$ defined in Equation (16) (we use form factors for the C-atom in the denominator). (**a**) A small set of 20 trajectories; (**b**) A large set of 1000 trajectories, giving a better representation of dispersion of the wavepacket at long times.

3.2.3. 1,3-Cyclohexadiene (CHD)

We consider the elastic scattering from the polyatomic molecule CHD during the electrocyclic ring-opening reaction triggered by an optical pump pulse [6,15], as outlined in Section 2.2.3. A reduced-dimensionality representation of the dominant dynamics is shown in Figure 6, in terms of the length of the C–C bond that breaks during the ring-opening reaction. Here, we will focus on the effect of the wavepacket width parameters γ_α defined in Equation (6) on the diffraction signal. We show the scattering in terms of the modified molecular intensity, Equation (16), with $f_m^0 = f_n^0 = f_C^0$. Note that the simulations are identical in all three scenarios discussed below, and only the γ_α parameters are varied. Note that fully converged simulations with an oversampled nuclear basis would not display a dependence on γ_α in the scattering.

First, we examine the diffraction in the fully localized limit, i.e., for $\gamma \to \infty$, shown in Figure 8a. This is also known as the bracket-averaged Taylor expansion (BAT) approximation [67], and is equivalent to perfectly localized trajectories not dissimilar to the trajectories obtained in surface hopping simulations. In this case, the damping of the molecular (interference) term is absent, and as expected, there is little deterioration of the diffraction signal even at large values of q. In contrast, for the standard values of γ_α (taken from Thompson et al. [76]), there is a significant degradation of the diffraction signal at large values of q which persists at all times, as can be seen in Figure 8b. Given that the default width parameters are fitted to ground state molecules, it is not far-fetched that the most appropriate values will deviate from the standard γ_α values. We therefore consider, in the third scenario shown in Figure 8c, the diffraction when the width factors have been halved, i.e., $\gamma_\alpha/2$. This leads to an even more delocalized wavepacket and has a dramatic effect on the diffraction signal, with the signal becoming very weak for $q > 15$ Å^{-1} (intensity below the height of the lowest contour in Figure 8c). Again, considering that the standard values of γ_α constitute a conservative estimate of the width of the wavepacket, this emphasizes the strong effect of the damping due to the delocalization of wavepackets.

In addition to the effect of the width parameters discussed above, we also observe in Figure 8 that the dispersion plays an important role in degrading the signal over time, in particular for $t > 50$ fs, similarly to what was observed in the calculations for D_2 and ethylene. This 50 fs time-scale is consistent with Figure 6, which indicates that the dynamics is comparatively uniform at early times (the so-called 'ballistic' nature of this particular reaction [35]), but diverges at later times ($t > 50$ fs in Figure 6).

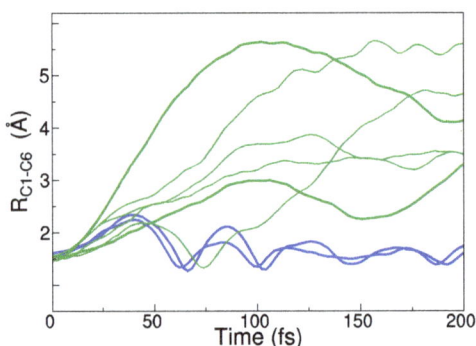

Figure 6. Representative C−C bondlength distances for the bond that breaks during the 1,3-cyclohexadiene (CHD) ring-opening reaction, shown as a function of time. Reproduced from Ref. [6]. Trajectories in blue lead back to the ring-closed form, while trajectories in green lead to the ring-open form of the molecule.

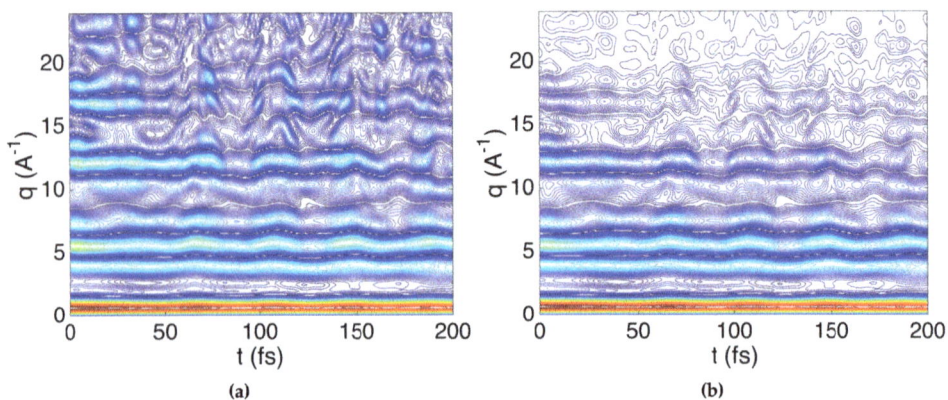

(a)

(b)

Figure 7. *Cont.*

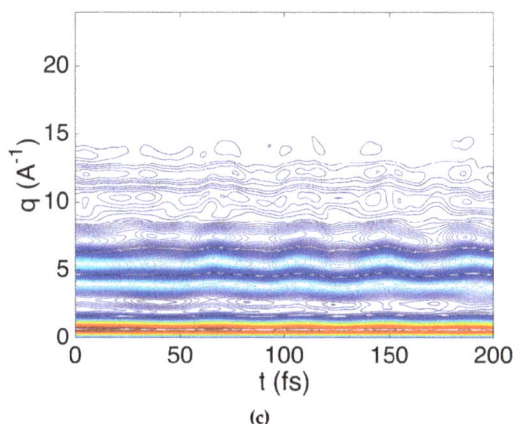

(c)

Figure 8. Contour plots of modified (elastic) scattering intensity, $M(q, t)$, for the ring-opening reaction of 1,3-cyclohexadiene (CHD) [6]. Time in fs and momentum transfer q in Å^{-1}, with $M(q, t)$ defined in Equation (16) (we use C-atom form factors in the denominator). (**a**) Perfectly localized trajectories with $\gamma_\alpha \to \infty$; (**b**) Standard values of the width parameters γ_α; (**c**) Increased width, i.e., more delocalized wavepackets, obtained via half-value $0.5\gamma_\alpha$ width parameters.

4. Discussion

Our analytical model in Section 3.1 indicates that degradation of the diffraction signal beyond q values of about 10–15 Å^{-1} is significant, with at least 50% of the signal lost by $q = 15$ Å^{-1}. This is a conservative estimate, since it is based on width factors fitted to ground state harmonic vibrational wavefunctions. These widths reflect the shape of the initially excited wavepacket in the Franck–Condon region above the ground state equilibrium geometry, which is why the Wigner distribution is used for sampling initial conditions in semi-classical propagation schemes [57]. However, subsequent propagation of the excited wavepacket leads to further dispersion. This is a general phenomenon, as emphasized by our simulations in three different molecules: D_2, 1,3-cyclohexadiene, and ethylene. In all three examples, the scattering signal degrades further with time as the wavepacket evolves.

In D_2 (Section 3.2.1), a diatomic molecule, we compare to a classically oscillating molecule and find that the diffraction signal is better defined and persists for all q when the nuclei are perfectly localized. Furthermore, we examine the effect of nuclear mass on the wavepacket, and find that the greater density of states associated with heavier atoms (or conversely a pump pulse with greater bandwidth) leads to greater initial localization. However, this sharper initial localization, which leads to better defined scattering, also leads to faster and more complete dispersion as the wavepacket evolves, degrading the scattering at longer times. In the ethylene molecule (Section 3.2.2), we examine the sensitivity of the scattering signal to the number of trajectories used in the simulations. We find that the number of trajectories correlates with the dispersion recorded in the scattering signal, such that a larger number of trajectories lead to an increased degradation of the signal. Strikingly, a small number of trajectories can qualitatively reproduce the scattering signal, indicating that a comparatively small number of trajectories can capture the essence of the reaction path and thus the diffraction signal, albeit while underestimating the dispersion at longer times. This finding is congruent with the analysis of recent ultrafast X-ray scattering experiments in 1,3-cyclohexadiene [6]. However, using not fully converged quantum molecular dynamics simulations, or equivalently just a small number of trajectories, introduces a sensitivity in the predicted diffraction patterns on the width parameters associated with each trajectory. In our simulations of 1,3-cyclohexadiene (Section 3.2.3), we examine this effect for the example of a larger, polyatomic molecule. Already with the standard and rather conservative width parameters, the scattering degrades notably at large values of q, and an additional

50% decrease in the width factors is sufficient to effectively quench the molecular diffraction signal for $q > 14$ Å^{-1}. This suggests that width parameters must be considered whenever ultrafast diffraction data for the time-evolution of molecular geometry is analysed.

Clearly, the concept of molecular geometry, in a classical sense, must be used with caution in the context of ultrafast processes and ultrafast diffraction in particular. The upside is that ultrafast diffraction experiments provide a sensitive probe of the delocalization of wavepackets, and one could consider using such experiments to determine, possibly time-dependent, width parameters that can be compared to quantum molecular simulations. Although molecular geometry is 'fuzzy' in time-evolving molecules, the wavepacket does have a distinct shape at all times. It is interesting to envision the use of ultrafast scattering to image the wavepacket in full detail, including for instance so-called quantum ripples [80]. This may only be possible in very small molecules, essentially diatomics, where direct inversion of the scattering signal is feasible. A similar resolution in a polyatomic molecule, although in principle possible, would certainly require advanced inversion algorithms and experiments that use alignment and/or tomographic techniques.

It is natural in this context to reflect on whether the dispersion of the wavepacket can be overcome via clever experimental setups. For instance, the shorter the duration of the pump pulse, the greater the initial localization, but this initial localization results in greater dispersion at later times. This is directly analogous to the calculations in D_2 and $\tilde{K}_2$ in this article, where the effect of an increased density of states for $\tilde{K}_2$ corresponds to the effect of shorter-duration, greater bandwidth pump pulses in an experiment. In practice, one often wishes to balance a degree of selectivity in the optical excitation against the achieved time resolution, since a pulse with very large bandwidth may simultaneously excite dynamics on several electronic states, leading to complex dynamics that are challenging to interpret [81]. Another possibility might be focusing. Most pump-probe experiments are performed with bandwidth-limited (i.e., transform-limited) pulses, which yield an initially localized wavepacket. As a consequence, the wavepacket naturally disperses and becomes more delocalized as it propagates for times $t > 0$. One could attempt to circumvent this by focusing the wavepacket at specific times using phase-shaped pump-pulses, enabling more accurate structure determinations at specific time points. However, first of all, such focusing would be unlikely to overcome the lower bound estimates of signal degradation at high q presented in our simple analytic model. Secondly, implementation of the focusing would add technical challenges to the already commanding experiments, essentially requiring a readjustment of the pulse shape for every time point on which to focus the wavepacket. While such an experiment might not necessarily help to elucidate reaction paths resulting from ordinary light exposure, it could add significant information about the potential energy surfaces involved and might help to optimize desired product yields.

5. Conclusions

The delocalization of nuclear wavepackets is a natural effect in molecular systems, and constitutes a fundamental property of wavepackets that cannot be easily circumvented. As the signal measured in scattering experiments is the transform of the molecular structure, the natural delocalization of structures during chemical reactions dampens the diffraction signal, particularly at large values of q, and implies a limit to the range of scattering vectors that yield useful information on the molecular dynamics. For the X-ray scattering experiments considered here, our simulations and analytical model suggest that there is little information beyond 10 to 15 Å^{-1}, at least for typical organic molecules. The sensitivity of scattering experiments to the shape of the wavepackets may offer an opportunity to experimentally measure them. Such measurements could be valuable for the development of advanced computational codes.

Ultrafast X-ray scattering experiments benefit from the exciting development of ultrashort pulsed free electron lasers. A challenge to date has been that the X-ray photon energy is limited, so that only rather small ranges of q can be probed. However, advances in accelerator technologies, in particular the build-out of LCLS-II at the SLAC National Accelerator Laboratory, promise harder X-rays and a concommittant expanded q range. With a typical detector geometry covering scattering up to 60 degrees, 30 keV X-ray photons will reach up to $q = 15$ Å^{-1}. It therefore seems likely that the advances offered by LCLS-II will enable experimental determinations of wavepacket motions up to the useful maximum suggested by our simulations.

Acknowledgments: A.K. acknowledges funding from the Leverhulme Trust (RPG-2013-365), the European Union (FP7-PEOPLE-2013-CIG-NEWLIGHT), and the hospitality of Roland Lindh (Uppsala University) including sabbatical support from the Wenner–Gren Foundations. P.M.W. acknowledges support from the National Science Foundation, Grant No. CBET- 1336105, and by Defence Threat Reduction Agency, Grant Number HDTRA1-14-1-0008. This study benefitted from stimulating conversations with Jeremy Hastings and Michael Minitti from the SLAC National Accelerator Laboratory.

Author Contributions: A.K. and P.M.W. conceived and designed the calculations; A.K. performed the calculations; A.K. and P.M.W. analyzed the data; A.K. and P.M.W. wrote the article.

Conflicts of Interest: The authors declare no conflict of interest.

Abbreviations

The following abbreviations are used in this manuscript:

XFEL: X-ray Free-Electron Laser
LCLS: Linac Coherent Light Source
IAM: Independent Atom Model
CHD: 1,3-cyclohexadiene
FWHM: Full Width at Half Maximum

References

1. Mark, H.; Wierl, R. Über Elektronenbeugung am einzelnen Molekül. *Naturwissenschaften* **1930**, *18*, 205.
2. Hargittai, I.; Hargittai, M. *Stereochemical Applications of Gas-Phase Electron Diffraction: Part A the Electron Diffraction Technique*, 1st ed.; VCH: New York, NY, USA, 1988.
3. Warren, B.E. *X-ray Diffraction*; Courier Corporation: North Chelmsford, MA, USA, 1969.
4. Rentzepis, P.M.; Helliwell, J., Eds. *Time Resolved Electron and X-ray Diffraction*, 1st ed.; Oxford University Press: New York, NY, USA, 1997.
5. Ischenko, A.A.; Weber, P.M.; Miller, R.J.D. Capturing Chemistry in Action with Electrons: Realization of Atomically Resolved Reaction Dynamics. *Chem. Rev.* **2017**, submitted.
6. Minitti, M.P.; Budarz, J.M.; Kirrander, A.; Robinson, J.S.; Ratner, D.; Lane, T.J.; Zhu, D.; Glownia, J.M.; Kozina, M.; Lemke, H.T.; et al. Imaging molecular motion: Femtosecond X-ray scattering of an electrocyclic chemical reaction. *Phys. Rev. Lett.* **2015**, *114*, 255501.
7. Yang, J.; Guehr, M.; Shen, X.; Li, R.; Vecchione, T.; Coffee, R.; Corbett, J.; Fry, A.; Hartmann, N.; Hast, C.; et al. Diffractive Imaging of Coherent Nuclear Motion in Isolated Molecules. *Phys. Rev. Lett.* **2016**, *117*, 153002.
8. Glownia, J.M.; Natan, A.; Cryan, J.P.; Hartsock, R.; Kozina, M.; Minitti, M.P.; Nelson, S.; Robinson, J.; Sato, T.; van Driel, T.; et al. Self-Referenced Coherent Diffraction X-ray Movie of Ångström- and Femtosecond-Scale Atomic Motion. *Phys. Rev. Lett.* **2016**, *117*, 153003.
9. Stankus, B.; Budarz, J.M.; Kirrander, A.; Rogers, D.; Robinson, J.; Lane, T.J.; Ratner, D.; Hastings, J.; Minitti, M.P.; Weber, P.M. Femtosecond photodissociation dynamics of 1,4-diiodobenzene by gas-phase X-ray scattering and photoelectron spectroscopy. *Faraday Discuss.* **2016**, *194*, 525–536.
10. Budarz, J.M.; Minitti, M.P.; Cofer-Shabica, D.V.; Stankus, B.; Kirrander, A.; Hastings, J.B.; Weber, P.M. Observation of Femtosecond Molecular Dynamics via Pump-probe Gas Phase X-ray Scattering. *J. Phys. B* **2016**, *49*, 034001.
11. Minitti, M.P.; Budarz, J.M.; Kirrander, A.; Robinson, J.; Lane, T.J.; Ratner, D.; Saita, K.; Northey, T.; Stankus, B.; Cofer-Shabica, V.; et al. Toward structural femtosecond chemical dynamics: Imaging chemistry in space and time. *Faraday Discuss.* **2014**, *171*, 81–91.
12. Levantino, M.; Schiro, G.; Lemke, H.T.; Cottone, G.; Glownia, J.M.; Zhu, D.; Chollet, M.; Ihee, H.; Cupane, A.; Cammarata, M. Ultrafast myoglobin structural dynamics observed with an X-ray free-electron laser. *Nat. Commun.* **2015**, *6*, 6772.

13. Kim, K.H.; Kim, J.G.; Nozawa, S.; Sato, T.; Oang, K.Y.; Kim, T.W.; Ki, H.; Jo, J.; Park, S.; Song, C.; et al. Direct observation of bond formation in solution with femtosecond X-ray scattering. *Nature* **2015**, *518*, 385–389.

14. Bostedt, C.; Bozek, J.D.; Bucksbaum, P.H.; Coffee, R.N.; Hastings, J.B.; Huang, Z.; Lee, R.W.; Schorb, S.; Corlett, J.N.; Denes, P.; et al. Ultra-fast and ultra-intense X-ray sciences: First results from the Linac Coherent Light Source free-electron laser. *J. Phys. B* **2013**, *46*, 164003.

15. Pemberton, C.C.; Zhang, Y.; Saita, K.; Kirrander, A.; Weber, P.M. From the (1B) Spectroscopic State to the Photochemical Product of the Ultrafast Ring-Opening of 1,3-Cyclohexadiene: A Spectral Observation of the Complete Reaction Path. *J. Phys. Chem. A* **2015**, *119*, 8832–8845.

16. Fleming, G. *Chemical Applications of Ultrafast Spectroscopy*; Oxford University Press: New York, NY, USA, 1986.

17. Lorincz, A.; Novak, F.A.; Rice, S.A. Relaxation of Large Molecules Following Ultrafast Excitation. In *Ultrafast Phenomena IV*; Auston, D., Eisenthal, K., Eds.; Springer: Berlin/Heidelberg, Germany, 1984; Volume 38, pp. 387–389.

18. Andor, L.; Lörincz, A.; Siemion, J.; Smith, D.D.; Rice, S.A. Shot-noise-limited detection scheme for two-beam laser spectroscopies. *Rev. Sci. Instrum.* **1984**, *55*, 64–67.

19. Rosker, M.J.; Dantus, M.; Zewail, A.H. Femtosecond real-time probing of reactions. I. The technique. *J. Chem. Phys.* **1988**, *89*, 6113–6127.

20. Thompson, J.; Weber, P.M.; Estrup, P.J. Pump-Probe Low Energy Electron Diffraction. In Proceedings of the SPIE Conference on Time Resolved Electron and X-ray Diffraction, San Diego, CA, USA, 9 July 1995; Volume 2521, pp. 113–122.

21. Geiser, J.D.; Weber, P.M. Pump-probe diffraction imaging of vibrational wave functions. *J. Chem. Phys.* **1998**, *108*, 8004–8011.

22. Dudek, R.C.; Weber, P.M. Ultrafast Diffraction Imaging of the Electrocyclic Ring-Opening Reaction of 1,3-Cyclohexadiene. *J. Phys. Chem. A* **2001**, *105*, 4167–4171.

23. Gosselin, J.L.; Minitti, M.P.; Rudakov, F.M.; Solling, T.I.; Weber, P.M. Energy Flow and Fragmentation Dynamics of *N,N*-Dimethylisopropylamine. *J. Phys. Chem. A* **2006**, *110*, 4251–4255.

24. Cardoza, J.D.; Rudakov, F.M.; Weber, P.M. Electronic Spectroscopy and Ultrafast Energy Relaxation Pathways in the Lowest Rydberg States of Trimethylamine. *J Phys. Chem. A* **2008**, *112*, 10736–10743.

25. Kirrander, A.; Fielding, H.H. Coherent control in the continuum: Autoionisation of Xe. *J. Phys. B* **2007**, *40*, 897.

26. Suominen, H.J.; Kirrander, A. How to observe coherent electron dynamics directly. *Phys. Rev. Lett.* **2014**, *112*, 043002.

27. Bainbridge, A.R.; Harrington, J.; Kirrander, A.; Cacho, C.; Springate, E.; Bryan, W.A.; Minns, R.S. VUV Excitation of a Vibrational Wavepacket in D_2 Measured through Strong-Field Dissociative Ionization. *New J. Phys.* **2015**, *17*, 103013.

28. Kirrander, A.; Jungen, C.; Fielding, H.H. Control of ionization and dissociation with optical pulse trains. *Phys. Chem. Chem. Phys.* **2010**, *12*, 8948–8952.

29. Weber, P.M.; Thantu, N. Photoionization via transient states: A coherent probe of molecular eigenstates. *Chem. Phys. Lett.* **1992**, *197*, 556–561.

30. Thantu, N.; Weber, P.M. Dependence of two photon ionization photoelectron spectra on laser coherence bandwidth. *Chem. Phys. Lett.* **1993**, *214*, 276–280.

31. Thantu, N.; Weber, P.M. Resonant two photon ionization of phenanthrene via its transient S_2 state. *Z. Phys. D* **1993**, *28*, 191–194.

32. Freed, K.F.; Nitzan, A. Intramolecular vibrational energy redistribution and the time evolution of molecular fluorescence. *J. Chem. Phys.* **1980**, *73*, 4765–4778.

33. Lorincz, A.; Smith, D.D.; Novak, F.; Kosloff, R.; Tannor, D.J.; Rice, S.A. Rotational state dependence of pyrazine fluorescence: Initial decays for the vibrationless $^1B_{3u}$ state. *J. Chem. Phys.* **1985**, *82*, 1067–1072.

34. Novak, F.; Kosloff, R.; Tannor, D.J.; Lorincz, A.; Smith, D.D.; Rice, S.A. Wave packet evolution in isolated pyrazine molecules: Coherence triumphs over chaos. *J. Chem. Phys.* **1985**, *82*, 1073–1078.

35. Garavelli, M.; Page, C.S.; Celani, P.; Olivucci, M.; Schmid, W.E.; Trushin, S.A.; Fuss, W. Reaction Path of a sub-200 fs Photochemical Electrocyclic Reaction. *J. Phys. Chem. A* **2001**, *105*, 4458–4469.

36. Kirrander, A.; Fielding, H.H.; Jungen, C. Excitation, dynamics and control of rotationally autoionizing Rydberg states of H_2. *J. Chem. Phys.* **2007**, *127*, 164301.

37. Kirrander, A.; Jungen, C.; Fielding, H.H. Localization of electronic wave packets in H_2. *J. Phys. B* **2008**, *41*, 074022.
38. Kirrander, A.; Fielding, H.H.; Jungen, C. Optical phase and the ionization-dissociation dynamics of excited H_2. *J. Chem. Phys.* **2010**, *132*, 024313.
39. Shalashilin, D.V. Nonadiabatic dynamics with the help of multiconfigurational Ehrenfest method: Improved theory and fully quantum 24D simulation of pyrazine. *J. Chem. Phys.* **2010**, *132*, 244111.
40. Shalashilin, D.V. Multiconfigurational Ehrenfest approach to quantum coherent dynamics in large molecular systems. *Faraday Discuss.* **2011**, *153*, 105–116.
41. Levine, B.G.; Coe, J.D.; Virshup, A.M.; Martinez, T.J. Implementation of *ab initio* multiple spawning in the MOLPRO quantum chemistry package. *Chem. Phys.* **2008**, *347*, 3–16.
42. Makhov, D.V.; Glover, W.J.; Martinez, T.J.; Shalashilin, D.V. *Ab initio* multiple cloning algorithm for quantum nonadiabatic molecular dynamics. *J. Chem. Phys.* **2014**, *141*, 054110.
43. Richings, G.; Polyak, I.; Spinlove, K.; Worth, G.; Burghardt, I.; Lasorne, B. Quantum dynamics simulations using Gaussian wavepackets: The vMCG method. *Int. Rev. Phys. Chem.* **2015**, *34*, 269–308.
44. Shalashilin, D.V.; Child, M.S. The phase space CCS approach to quantum and semiclassical molecular dynamics for high-dimensional systems. *Chem. Phys.* **2004**, *304*, 103–120.
45. Shalashilin, D.V.; Child, M.S.; Kirrander, A. Mechanisms of double ionization in strong laser field from simulation with Coupled Coherent States. *Chem. Phys.* **2008**, *347*, 257–262.
46. Kirrander, A.; Shalashilin, D.V. Quantum dynamics with fermion coupled coherent states. *Phys. Rev. A* **2011**, *84*, 033406.
47. Heller, E.J. The semiclassical way to molecular spectroscopy. *Acc. Chem. Res.* **1981**, *14*, 368–375.
48. Makhov, D.V.; Saita, K.; Martinez, T.J.; Shalashilin, D.V. *Ab initio* multiple cloning simulations of pyrrole photodissociation: TKER spectra and velocity map imaging. *Phys. Chem. Chem. Phys.* **2015**, *17*, 3316–3325.
49. Saita, K.; Shalashilin, D.V. On-the-fly *ab initio* molecular dynamics with multiconfigurational Ehrenfest method. *J. Chem. Phys.* **2012**, *137*, 22A506.
50. Wolniewicz, L. Nonadiabatic energies of the ground state of the hydrogen molecule. *J. Chem. Phys.* **1995**, *103*, 1792–1799.
51. Staszewska, G.; Wolniewicz, L. Adiabatic Energies of Excited $^1\Sigma_u$ States of the Hydrogen Molecule. *J. Mol. Spectrosc.* **2002**, *212*, 208–212.
52. Wolniewicz, L.; Staszewska, G. $^1\Sigma_u^+ \rightarrow X^1\Sigma_g^+$ transition moments for the hydrogen molecule. *J. Mol. Spectrosc.* **2003**, *217*, 181–185.
53. Mohr, P.J.; Taylor, B.N.; Newell, D.B. CODATA Recommended Values of the Fundamental Physical Constants: 2010. *Rev. Mod. Phys.* **2012**, *84*, 1527.
54. Deb, S.; Weber, P.M. The Ultrafast Pathway of Photon-Induced Electrocyclic Ring-Opening Reactions: The Case of 1,3-Cyclohexadiene. *Ann. Rev. Phys. Chem.* **2011**, *62*, 19–39.
55. Makhov, D.V.; Martinez, T.J.; Shalashilin, D.V. Toward fully quantum modelling of ultrafast photodissociation imaging experiments. Treating tunnelling in the ab initio multiple cloning approach. *Faraday Discuss.* **2016**, *194*, 81–94.
56. Werner, H.J.; Knowles, P.J.; Knizia, G.; Manby, F.R.; Schütz, M.; others. *MOLPRO, Version 2012.1, a Package of Ab Initio Programs.*
57. Brown, R.C.; Heller, E.J. Classical trajectory approach to photodissociation: The Wigner method. *J. Chem. Phys.* **1981**, *75*, 186–188.
58. Tao, H.; Levine, B.G.; Martinez, T.J. *Ab Initio* Multiple Spawning Dynamics Using Multi-State Second-Order Perturbation Theory. *J. Phys. Chem. A* **2009**, *113*, 13656–13662.
59. Mori, T.; Glover, W.J.; Schuurman, M.S.; Martinez, T.J. Role of Rydberg States in the Photochemical Dynamics of Ethylene. *J. Phys. Chem. A* **2012**, *116*, 2808–2818.
60. Sellner, B.; Barbatti, M.; Müller, T.; Domcke, W.; Lischka, H. Ultrafast non-adiabatic dynamics of ethylene including Rydberg states. *Mol. Phys.* **2013**, *111*, 2439–2450.
61. Champenois, E.G.; Shivaram, N.H.; Wright, T.W.; Yang, C.S.; Belkacem, A.; Cryan, J.P. Involvement of a low-lying Rydberg state in the ultrafast relaxation dynamics of ethylene. *J. Chem. Phys.* **2016**, *144*, 014303.
62. Kobayashi, T.; Horio, T.; Suzuki, T. Ultrafast Deactivation of the $\pi\pi^*$(V) State of Ethylene Studied Using Sub-20 fs Time-Resolved Photoelectron Imaging. *J. Phys. Chem. A* **2015**, *119*, 9518–9523.

63. Tao, H. First Principles Molecular Dynamics and Control of Photochemical Reactions. Ph.D. Thesis, Stanford University, Stanford, CA, USA, 2011.
64. Eisenberger, P.; Platzman, P.M. Compton Scattering of X-rays from Bound Electrons. *Phys. Rev. A* **1970**, *2*, 415–423.
65. De Groot, F.; Kotani, A. *Core Level Spectroscopy of Solids*, 1st ed.; CRC Press: Boca Raton, FL, USA, 2008.
66. Henriksen, N.E.; Møller, K.B. On the Theory of Time-Resolved X-ray Diffraction. *J. Phys. Chem. B* **2008**, *112*, 558.
67. Kirrander, A.; Saita, K.; Shalashilin, D.V. Ultrafast X-ray Scattering from Molecules. *J. Chem. Theory Comput.* **2016**, *12*, 957–967.
68. Carrascosa, A.M.; Kirrander, A. *Ab initio* calculation of inelastic scattering. *Phys. Chem. Chem. Phys.* **2017**, doi:10.1039/C7CP02054F.
69. Møller, K.B.; Henriksen, N.E. Time-Resolved X-ray Diffraction: The Dynamics of the Chemical Bond. *Struct. Bond.* **2012**, *142*, 185.
70. Northey, T.; Zotev, N.; Kirrander, A. *Ab Initio* Calculation of Molecular Diffraction. *J. Chem. Theory Comput.* **2014**, *10*, 4911.
71. Northey, T.; Carrascosa, A.M.; Schäfer, S.; Kirrander, A. Elastic X-ray scattering from state-selected molecules. *J. Chem. Phys.* **2016**, *145*, 154304.
72. Carrascosa, A.M.; Northey, T.; Kirrander, A. Imaging rotations and vibrations in polyatomic molecules with X-ray scattering. *Phys. Chem. Chem. Phys.* **2017**, *19*, 7853–7863.
73. Prince, E. (Ed.) *International Tables for Crystallography Volume C: Mathematical, Physical and Chemical Tables*; Springer International Publishing: New York, NY, USA, 2006; ISBN 978-1-4020-1900-5.
74. Shorokhov, D.; Park, S.T.; Zewail, A.H. Ultrafast Electron Diffraction: Dynamical Structures on Complex Energy Landscapes. *ChemPhysChem* **2005**, *6*, 2228–2250.
75. Stefanou, M.; Saita, K.; Shalashilin, D.; Kirrander, A. Comparison of ultrafast X-ray and electron scattering—A computational study. *Chem. Phys. Lett.* **2017**, doi:10.1016/j.cplett.2017.03.007.
76. Thompson, A.L.; Punwong, C.; Martinez, T.J. Optimization of width parameters for quantum dynamics with frozen Gaussian basis sets. *Chem. Phys.* **2010**, *370*, 70–77.
77. Heller, E.J. Frozen Gaussians: A very simple semiclassical approximation. *J. Chem. Phys.* **1981**, *75*, 2923–2931.
78. Debye, P. Zerstreuung von Röntgenstrahlen. *Ann. Phys.* **1915**, *46*, 809–823.
79. Ben-Nun, M.; Martinez, T.J. Direct Observation of Disrotatory Ring-Opening in Photoexcited Cyclobutene Using ab Initio Molecular Dynamics. *J. Am. Chem. Soc.* **2000**, *122*, 6299–6300.
80. Katsuki, H.; Chiba, H.; Girard, B.; Meier, C.; Ohmori, K. Visualizing Picometric Quantum Ripples of Ultrafast Wave-Packet Interference. *Science* **2006**, *311*, 1589–1592.
81. Bellshaw, D.; Horke, D.A.; Smith, A.D.; Watts, H.M.; Jager, E.; Springate, E.; Alexander, O.; Cacho, C.; Chapman, R.T.; Kirrander, A.; et al. *Ab-initio* Surface Hopping and Multiphoton Ionisation Study of the Photodissociation Dynamics of CS_2. *Chem. Phys. Lett.* **2017**, doi:10.1016/j.cplett.2017.02.058.

Article

Algorithm for Reconstruction of 3D Images of Nanorice Particles from Diffraction Patterns of Two Particles in Independent Random Orientations with an X-ray Laser

Sung Soon Kim, Sandi Wibowo and Dilano Kerzaman Saldin *

Department of Physics, University of Wisconsin-Milwaukee, P.O. Box 413,
Milwaukee, WI 53201, USA; sungskim@uwm.edu (S.S.K.); indosandi@gmail.com (S.W.)
* Correspondence: dksaldin@uwm.edu; Tel.: +1-(414)-229-6423

Academic Editor: Kiyoshi Ueda
Received: 31 March 2017; Accepted: 8 June 2017; Published: 23 June 2017

Abstract: The method of angular correlations recovers quantities from diffraction patterns of randomly oriented particles, as expected to be measured with an X-ray free electron laser (XFEL), proportional to quadratic functions of the spherical harmonic expansion coefficients of the diffraction volume of a single particle. We have previously shown that it is possible to reconstruct a randomly oriented icosahedral or helical virus from the average over all measured diffraction patterns of such correlations. We point out in this paper that a structure of even simpler particles of 50 Å or so in diameter and consisting of heavier atomic elements (to enhance scattering) that has been used as a test case for reconstructions from XFEL diffraction patterns can also be solved by this technique. Even though there has been earlier work on similar objects (prolate spheroids), one advantage of the present technique is its potential to also work with diffraction patterns not only due to single particles as has been suggested on the basis on nonoverlapping delta functions of angular scattering. Accordingly, we calculated from the diffraction patterns the angular momentum expansions of the pair correlations and triple correlations for general particle images and reconstructed those images in the standard way. Although the images looked pretty much the same, it is not totally clear to us that the angular correlations are exactly the same as different numbers of particles due to the possibility of constructive or destructive interference between the scattered waves from different particles. It is of course known that, for a large number of particles contributing to a diffraction parttern, the correlations converge to that of a single particle. It could be that the lack of perfect agreement between the images reconstructed with one and two particles is due to uncancelling constructive and destructive conditions that are not found in the case of solution scattering.

Keywords: XFEL; nanorice; angular correlations

---

1. Introduction

An X-ray free electron laser (XFEL) produces X-rays of unprecedented brilliance of about 10 billion times what was previously possible. As such, it has given rise to the speculation that it may be possible to determine the structures of uncrystallized individual biomolecules [1]. Although the ultimate aim is to determine the structures of biomolecules, it would be helpful to demonstrate the feasibility of the approach to simpler objects initially. In this vein, there has been some work already on reconstructing prolate spheroids [2] of metallic particles. What such experiments demonstrate is the feasibility of reconstructing the structure of particles of random unknown orientations. The aim of the present paper is to show that reconstruction of the structure of such particles is possible even with two particles in independent random orientations contributing to a single diffraction pattern. If it can be demonstrated for two randomly oriented nanoparticles, and we think that since the angular correlations seem to be identical, the number of identical particles may be increased without limit. In this respect, the method has some similarities with one proposed in the 1970s by Kam [3] to generalize the methods then current for small angle X-ray scattering (SAXS). In SAXS, an ensemble of many particles is suspended in solution in random orientations. A major problem with SAXS is that one attempts to recover the structure of a particle from a set of experimental data consisting of a single line plot $I(q)$. From the spherical harmonic expansion of a crude model of the particle, its shape is normally obtained by SAXS techniques. Kam's innovation was to point out that if an experiment could be performed that measures the diffraction patterns on a time scale shorter than the rotational diffusion time of the molecules, extra 3D information on the molecular structure could be obtained from the correlations in the angular variations of the measured intensities. There has also been work on determining the structure of Au nanoparticles to atomic resolution [4] by this technique. A major advantage of this method, if feasible, is the ability to determine the detailed structure of an ensemble of randomly oriented molecules. One would thereby be able to use the intensity enhancement caused by working with an ensemble of molecules rather than a single one, and yet not have to crystallize the molecules so that they are all in exactly the same orientation. Kirian et al. [5] have shown that even though there is nothing to be gained by having more particles per diffraction pattern since the noise goes up in the same proportion as the signal, one of the other results of their own analysis is that the noise can be reduced compared to the signal by averaging over many diffraction patterns, It should be pointed out that the need to avoid crystallization was the original aim of XFEL studies of biomolecules. If the experiment we propose is possible, it may be possible to avoid crystallization in structural studies and still avoid the very low scattered intensities that are inevitable in single particle studies even with an X-ray free electron laser (XFEL).

An experiment has recently been reported in which the structure of the mimivirus has been determined experimentally by a variant of the single particle methods described earlier [6]. At least for an icosahedral particle, as the mimivirus largely is, it has been previously shown by us [7] that it is possible to determine the structure of the particle by an analogous method to that described here from simulated diffraction patterns. In an experiment to recover time-resolved structural variations, starting from a knowledge of a nearby structure, using many of the same quantities, namely, the pair correlations, have also been shown to be capable of recovering time-resolved changes in a structure from the knowledge of a closely related structure of a single molecule in realistic simulations, including shot noise [8]. This possibility is unprecedented in structural work with XFEL diffraction patterns.

It is with the aim of further developing this idea that the work here is undertaken. Initially, following the basic theory presented here, the aim is only to reconstruct from an ensemble of randomly oriented particles of simple form, its low resolution 3D structure. Indeed, there has already been experimental work on "nanorice" particles of the type we describe here. While it may be true that such a simple structure may be obtained by SAXS alone, the demonstration of such a method even for a simple structure paves the way to its application to more complex structures that may not be accessible to SAXS. Furthermore, "machine learning" algorithms based on manifold embedding [9] have been used to perform an approximate sorting of the experimental diffraction patterns. Following this paper, we have begun to work with the authors of that paper on going all the way from measured diffraction patterns to structure. We describe here how the structure of "nanorice" particles on which experimental data is already available at the public web site cxidb.org can be reconstructed in principle. There has also been some work reported on the reconstruction of prolate and oblate spheroids by the method of cryptotomography [2]. However, this does require diffraction patterns from single particles to be identified beforehand from an experimental data set [10]. In the present paper, we describe an algorithm that works on either prolate or oblate spheroids and yet allows the possibility of working with diffraction patterns of multiple randomly oriented particles. To do this, we employ a method [11] of reconstructing the structure of a particle from XFEL diffraction patterns of random unknown orientations of the object.

This method has close similarities to a method proposed much earlier [3] for recovering the structure of particles randomly oriented, as in the method of small angle X-ray scattering (SAXS). Although, of course, in the 1970s, it was not possible to focus an X-ray beam to hit just a single particle, this has now become possible with the recent advent of the X-ray free electron laser (XFEL). It should be stressed that the method we describe in this paper is flexible enough to reconstruct a structure from "single-particle" experiments such as in the recent Single Particle Initiative (SPI) at the Linac Coherent Light Source (LCLS) in Stanford, CA, USA. What is used is the average over measured diffraction patterns of the angular correlations on each diffraction pattern. Since this uses diffraction data of perhaps millions of diffraction patterns, its expected shot noise would be significantly reduced by this averaging process [12]. Such methods are very different from those of traditional crystallography on crystals, as the orientation of each particle in generally unknown a priori and there are not even Bragg spots that enable their orientation to be determined via indexing [13].

We describe our method next. One of its advantages is that it has been shown to be equally applicable to ensembles of molecules of random orientations [8]. This could be an advantage experimentally, as it will allow a fuller use of available experimental data, and obviates the need for methods [10] that eliminate diffraction patterns from multiple particle hits from the analysis.

2. Method of Angular Correlations

The first step in using this method is to calculate angular cross correlations on each diffraction pattern in polar coordinates. Polar coordinates are natural for this problem since the particles differ mainly in their orientations (they may also differ in position, and this does not affect the diffraction pattern intensities that are insensitive to the phases of the scattered amplitudes). This is relevant so long as it is a single particle in the beam at one time. Otherwise, the intensities are sensitive for the relative displacements of the particles in the same diffraction pattens. Even in that case, one might hope that, due to the random nature of these displacements, such relative phases are unimportant [8].

Angular pair correlations are related to measured quantities from

$$C_2(q, q', \Delta\phi) = < \int I_p(q, \phi) I_p(q', \phi + \Delta\phi) d\Delta\phi >_p, \tag{1}$$

where $I_p(q, \phi)$ is the measured intensity at resolution ring q and azimuthal angle ϕ on diffraction pattern p, and $I_p(q', \phi + \Delta\phi)$ the corresponding intensity at resolution ring q' azimuthal angle $\phi + \Delta\phi$. Similar to the pair correlations, two-point angular triple correlations may be defined [14]:

$$C_3(q, q', \Delta\phi) = < \int I_p^2(q, \phi) I_p(q', \phi + \Delta\phi) d\Delta\phi >_p. \tag{2}$$

The next step is to calculate these quantities to form other quantities related to them in the following way:

$$B_l(q, q') = \frac{2l+1}{2} \int C_2(q, q'; \Delta\phi) P_l(x) dx \tag{3}$$

and

$$T_l(q, q') = \frac{2l+1}{2} \int C_3(q, q'; \Delta\phi) P_l(x) dx, \tag{4}$$

where $x = \cos \Delta\phi$.

The sequence of operations is summarized next.

This method is well suited to the problem of a curved Ewald sphere, as pointed out in, e.g., [11]. A curved Ewald sphere is inevitable at high resolution. The point is that the quantities actually used to construct the 3D diffraction volume, namely, the $B_l(q, q')$ and the $T_l(q, q')$, are extracted from the quantities measurable in an experiment, namely, the $C_2(q, q'; \Delta\phi)$ and $C_3(, q, q'; \Delta\phi)$, by removing the effect of the details of the experiment, such as the X-ray energy, which are contained in the arguments of the Legendre polynomials P_l. This allows for a nice separation of the quantities that depend on the experimental details, and the quantities that appear to contain only information about the 3D diffraction volume, which, in turn, gives information about the 3D structure of the particle. These arguments are equally applicable to pair correlation as to the two-point triple correlations defined by Kam [3]. Although Pedrini [15] and Kurta et al. [16] have proposed methods involving a three-point triple correlation function, the two-point triple correlation functions introduced by Kam [14] are easier to measure when photon counts are weak, as pixels with just two non-zero pixels make contributions. If the probability of a high resolution pixel is non-zero is 10^{-2} [12], the probability that two of them are non-zero is 10^{-4}. However, reasonable statistics are expected if one has the number of diffraction patterns expected to be measured in a typical experimental shift at even a present-day XFEL. In other words, in Kam's method [3], one calculated correlations from samples of many randomly oriented molecules. Nevertheless, and this is our crucial point, the correlations are similar to those from a single particle [17] We admit that this demonstration is only for rotations about a single axis, and for delta function like intensities. We investigate here whether it is still likely to hold for a random particle orientation in 3D, and for realistic amplitudes scattered from each particle. The angular pair correlations can be related to their angular momentum depompostions B_l by

$$C_2(q, q', \Delta\phi) = \sum_l F_l(q, q', \Delta\phi) B_l(q, q'),$$
(5)

where

$$F_l(q, q', \Delta\phi) = \frac{1}{4\pi} P_l[\cos\theta(q)\cos\theta(q') + \sin\theta(q)\sin\theta(q')\cos(\Delta\phi)],$$
(6)

where

$$\theta(q) = \pi/2 - \sin^{-1}[q/(2\kappa)]$$
(7)

and P_l is a Legendre polynomial.

Likewise, it can be shown that the triple correlations defined by (2) can be written as [14]

$$C_3(q, q'\Delta\phi) = \sum_l F_l(q, q', \Delta\phi) T_l(q, q'),$$
(8)

where

$$T_l(q,q) = \sum_{l_1,l_2,m_1,m_2,m}^{l_{max}} (-1)^m \left[\frac{(2l+1)(2l_2+1)(2l_1+1)}{4\pi} \right]^{1/2} \times$$
$$\begin{pmatrix} l_1 & l_2 & l \\ 0 & 0 & 0 \end{pmatrix} \begin{pmatrix} l_1 & l_2 & l \\ m_1 & m_2 & m \end{pmatrix} I_{l_2 m_2}(q) I_{l_1 m_1}(q) I_{lm}^*(q) =$$
$$\sum_{l_1,l_2,m_1,m_2,m}^{l_{max}} (-1)^m G(l_1 m_1; l_2 m_2; lm) I_{l_2 m_2}(q) I_{l_1 m_1}(q) I_{lm}^*(q),$$
(9)

where G is a Gaunt coeffcient [18,19] and

$$I_{lm}(q) = \int I(q, \theta, \phi) Y_{lm}^*(\theta, \phi) d\Omega.$$
(10)

Although, in general, the $I_{lm}(q)$'s are complex quantities, the quantities we are concerned with this paper, namely, the $I_{l0}(q)$'s, are real, and all m values are zeros for the azimuthal symmetry.

For a flat Ewald sphere, the quantities $B_l(q, q')$ and $T_l(q, q')$, which contain the structural information, may be found assuming that $\cos(\theta(q))$ and $\cos(\theta(q'))$ tend towards zero and $\sin(\theta(q))$ and $\sin(\theta(q'))$ tend toward unity. In this case, the quantities $B_l(q, q')$ and $T_l(q, q')$ that contain true structural information about the molecule can be found, for instance, (to within an arbitrary scaling factor) from the integrals

$$B_l(q,q) = \frac{2l+1}{2} \int C_2(q,q,\Delta\phi) P_l(\cos\Delta\phi) \sin\Delta\phi \, d\Delta\phi \tag{11}$$

and

$$T_l(q,q) = \frac{2l+1}{2} \int C_3(q,q,\Delta\phi) P_l(\cos\Delta\phi) \sin\Delta\phi \, d\Delta\phi \tag{12}$$

over the measured quantities C_2, and C_3. It should be stressed that, even in the case of a curved Ewald sphere the quantities containing the structural information about the particle, the $B_l(q,q)$ and $T_l(q,q)$ may be found by matrix inversion of Equations (5) and (8). Due to its expected azimuthal symmetry, nanorice particles have diffraction volumes characterized by spherical harmonic expansion coefficients of only azimuthal quantum numbers $m = 0$. In this case, the only relevant spherical harmonic expansion coefficients, I_{l0}, are real. Thus, the computations in this case ($q' = q$) may be performed entirely with real coefficients and only the diagonal parts of $B_l(q,q')$ and $T_l(q,q')$ are adopted.

It has been shown that these equations may be solved easily for particular common symmetries. It was pointed out by Caspar and Klug [20] that viruses tend to possess mainly icosahedral and helical symmetry. We have shown how to solve these equations for both icosahedral [7] and helical [21] cases, and are working on extending the method to particles that deviate a little from exact symmetry, using a form of perturbation theory [22]. In the case of a helical virus, we exploited the fact that if the virus is oriented with a helix axis parallel to the z-axis, the diffraction volume has exact cylindrical symmetry up to a certain resolution and may therefore be characterized by azimuthal quantum number $m = 0$.

Another case includes where the diffraction volume could be characterized by an $m = 0$ quantum number that is a diffraction volume of a nanorice particle. Experiments have been done on such particles at an XFEL and diffraction data from nanorice has been deposited in the cxidb.org web site. We investigate here whether this data may also be used to reconstruct the structure of nanorice by making use of the $m = 0$ condition. Because of the azimuthal symmetry of a nanorice particle about the major axis, it would be expected that such particles would also be characterized by $m = 0$. A question is if both a helical virus and a nanorice particle may be characterized by $m = 0$, how is it possible to distinguish them? The reason is that the intensities in the case of a helical virus must reflect the periodicity along the azimuthal axis and consequently split up into layer lines, but it does not do so in the case of a nanorice particle.

3. Simulation of Diffraction Patterns of Nanorice in Random Orientations

We tried to make the simulation as realistic as possible by assuming an X-ray beam of width 1000 Å, the design specification of the minimum focus of the Linac Coherence Light Source, the worlds' first X-ray free electron laser (XFEL). As for the size of a nanorice particle, we assumed each particle to be of a width about 50 Å to simulate a small protein. This is a realistic circumstance in which one expects a number of proteins to be illuminated. For the purposes of our test, we assumed only two nanorice particles illuminated in random relative orientations. We note this is a circumstance that is beyond the capabilities of all other algorithms that have been proposed for the XFEL problem, but is a realistic circumstance which is well suited to our method. There is an advantage to illuminating multiple particles as the total scattered signal goes up—with single particles, one is always struggling with few scattered photons.

For an initial model of nanorice, we assumed a ellipsoidal mask on a 3D Cartesian grid in real space, that is, we took the electron density of the nanorice particle to be uniform inside the particle and zero outside. As for the size of the particle in our solutions, the particle was entirely enclosed in a volume of $19 \times 19 \times 19$ central voxels. This model is shown on Figure 1. We then simulated diffraction patterns due to random orientations of nanorice particles as follows.

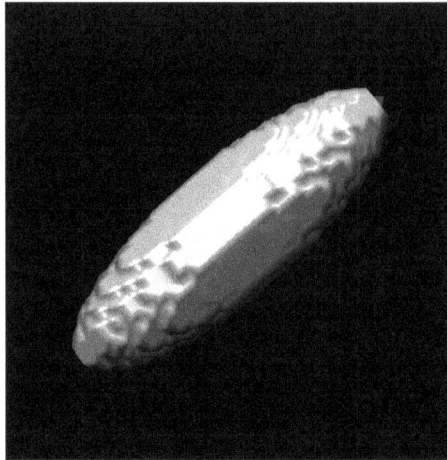

Figure 1. Model of a nanorice particle.

Since the angular correlations are quantities that are independent from all orientations of the particles, they are essentially independent from the number of particles contributing to each diffraction pattern, provided each takes all orientations. This has great importance for XFEL work, as often great pains are taken to restrict the number of particles in each of the measured diffraction patterns to just one, as required by all other methods of reconstructing particles from their diffraction patterns by the use of "hit-finder" software [10], for example. However, it might be argued that if not too much accuracy is required, this is unnecessary as angular correlations are independent from the number of particles illuminated and are the same for the same type of particle. This gives rise to the possibility that one might gain the same advantage. There seems to be no reason to not use multiple particles incident from a beam and actually get rid of the "hit-finder" [10] software. This paper demonstrates at least in theory that one may avoid "hit-finder" software altogether by this means. The only caveat is if the particles inevitably cluster, then the particles will differ in structure as well as orientation and it may be better to ensure that the same particle is hit always from different orientations by admitting only one particle at a time. This is a question that can only be answered by an experiment, but it is useful to know that there is an another, and possibly preferable option if the particles do not cluster. We reconstructed the object with about 2× oversampling in a region consisting of 41 × 41 × 41 voxels.

1. First, we calculated the 3D amplitude distribution in reciprocal space from their correlations, by the following steps:

$$A(\mathbf{q}) \quad = \quad \int \rho(\mathbf{r}) \exp(2\pi i \mathbf{q.r}) d\mathbf{r}, \tag{13}$$

$$I(\mathbf{q}) \quad = \quad |A(\mathbf{q})|^2 = I(q, \theta, \phi). \tag{14}$$

2. We then calculated the spherical harmonics expansion of the 3D diffraction volume via

$$I_{lm}(q) = \int I(q, \theta, \phi) Y_{lm}^*(\theta, \phi) d\Omega. \tag{15}$$

3. We then found the values of the expansion coefficients I_{lm} for random orientations of the diffraction volume using a Wigner D-matrix for random Euler angles α, β and γ

$$I_{lm}(q) \quad = \sum_{m'} D_{m,m'}^l(\alpha, \beta, \gamma) I_{lm'}(\theta, \phi). \tag{16}$$

4. We then calculated the diffraction pattern expected by slicing through the diffraction volume in each case through the plane $q_z = 0$

$$I(q, \theta = \frac{\pi}{2}, \phi) \quad = \quad \sum_{lm} I_{lm}(q) Y_{lm}(\theta = \frac{\pi}{2}, \phi). \tag{17}$$

This represents the intensity distribution expected of diffraction patterns from random orientations of a nanorice particle. For the two-particle case, we repeated these steps and either added amplitudes of scattering from each of the particles to simulate coherent illumination (assuming random interparticle displacements) or added scattered intensities to simulate incoherent illumination of multiple particles. Such typical diffraction patterns are shown in Figure 2.

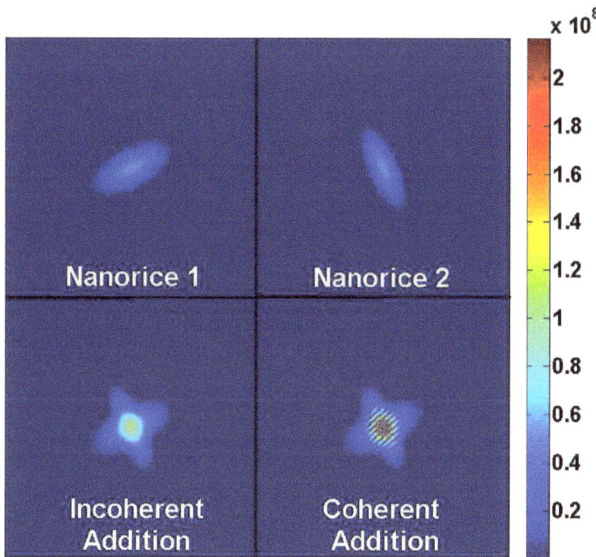

Figure 2. Simulated X-ray free electron laser (XFEL) diffraction patterns of one and two nanorice particles in random orientations.

4. Reconstruction of Diffraction Pattern of Nanorice

Having thus simulated diffraction patterns expected of random orientations of nanorice, our next step was to demonstrate that it is possible to reconstruct our model of nanorice from those patterns using the method outlined above. To this end, we calculated $B_l(q, q)$ and $T_l(q, q)$ as outlined above from the data in the simulated diffraction patterns.

The coefficients I_{lm} of a spherical harmonic expansion of the diffraction volume clearly depend on the orientation of the diffraction volume relative to the chosen z-axis. Two such 3D intensity distributions are displayed on Figures 3 and 4. By choosing a z-axis at the center of azimuthal symmetry, we eliminate the other components of I_{lm} except $m = 0$. Note that this is no loss of generality since the correlations do not determine the particle's orientation. As a matter of fact, the correlations are the same independent of the particles' orientation. On reconstruction of a real-space image of the particle made from the orentation-independent correlations, one is free to choose the particle's orientation for one's convenience .

It has been claimed that the recently proposed multi-tiered itrerative phasing (MTIP) algorithm [23] would not need to assume azimuthal symmetry. Since it is supposed to determine the the magnitudes of the quantities characterized by different magnetic quantum numbers, in principle, it is capable of deducing that $m = 0$. However, in their recent paper, Donatelli et al. [23] definitely found much better results on the assumption of the degree of angular symmetry. Following Starodub et al. [24], we obtain much the same information by performing an singular value decomposition (SVD) matrix decomposition of the pair correlations.

The quantities B_l and T_l depend on the angular momentum quantum number l but not on the azimuthal quantum number m. However, in general, the spherical harmonic expansion coefficients $I_{lm}(q)$ depend on both sets of quantum numbers. However, there is one orientation when m is fixed and the the the I_{lm} coefficients depend additionally only on l, and that is when a major axis of the ellipsoid representing the nanorice is coincident with the z-axis. Under these conditions, the particle, and also the diffraction volume, has azimuthal symmetry about the z-axis, and then can be characterized exactly by $m = 0$ for all l.

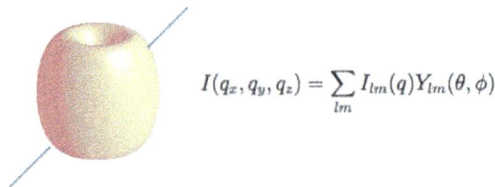

$$I(q_x, q_y, q_z) = \sum_{lm} I_{lm}(q) Y_{lm}(\theta, \phi)$$

Figure 3. Expansion in spherical harmonics with respect to an arbitrary axis.

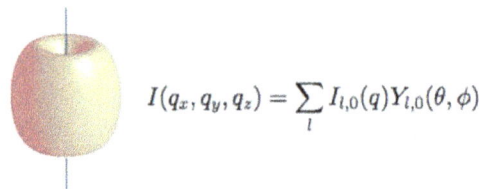

$$I(q_x, q_y, q_z) = \sum_{l} I_{l,0}(q) Y_{l,0}(\theta, \phi)$$

Figure 4. Expansion in spherical harmonics with respect to the z-axis.

We are not trying to reconstruct the particle in any particular orientation. Due to the random orientation of the particles contributing to the data, any orientation of reconstruction is as good as any other. We choose the orientation with the major axis of the ellipsoid along the z-axis. We can choose this orientation by assuming that only the $m = 0$ components of the $I_{lm}(q)$s exist. At this point, these coefficients depend only on l. The magnitudes of these spherical harmonic coefficients determined from

$$|I_{l0}(q)| = \sqrt{B_l(q,q)}. \tag{18}$$

Since I_{l0} is real, its only uncertainty is one of sign. These signs are found by an exhaustive search through

$$T_l(q,q) = \sum_{l_1,l_2}^{l_{max}} G(l_1 0, l_2 0, l 0) I_{l_1 0}(q) I_{l_2 0}(q) I_{l_0}(q), \tag{19}$$

where G is a Gaunt coefficient [18,19], and basically we search through all possible signs of the $I_{l0}(q)$ on the RHS (right hand side) (the magnitudes of the I_{l0} are known from the pair correlations as mentioned above) to get best agreement with the experimentally-determined LHS (left hand side). Again, ms are zeros for azimuthal symmetry.

Since an ellipsoid has azimuthal symmetry about a particular axis, we can choose that particular axis as the z-axis, thus eliminating any other components of the magnetic quantum number except $m = 0$. $|I_{l0}(q)|$ can be obtained directly from $B_l(q,q)$ via (18).

The only unknown here is sign of $I_{l0}(q)$. The sign can be determined by fitting all possible signs of $I_{l0}(q)$ to the values $T_l(q)$ of the triple correlations calculated directly from the diffraction patterns of random particle orientations. It should be stressed that the number of equations is equal to the number of distinct $T_l(q,q)$ values, namely, the numbers of q and l values, as is the number of unknowns $I_{l0}(q)$, so there is no information deficit.

After obtaining the signs of the $I_{l0}(q)$, the diffraction volume can be calculated from

$$I(\mathbf{q}) = \sum_l I_{l0}(q) Y_{l0}(\hat{\mathbf{q}}). \tag{20}$$

An iterative phasing algorithm [25,26] applied to these diffraction volumes can then recover the electron density of the particle, the real space object giving rise to the diffraction volume. The calculated intensities are displayed on Figures 5 and 6 and reconstructed electron density after phasing is displayed on Figures 7 and 8.

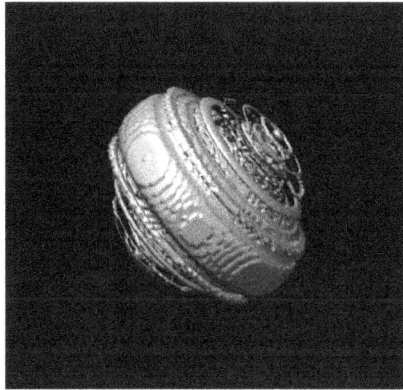

Figure 5. Reconstructed diffraction volume from one particle per shot (randomly oriented).

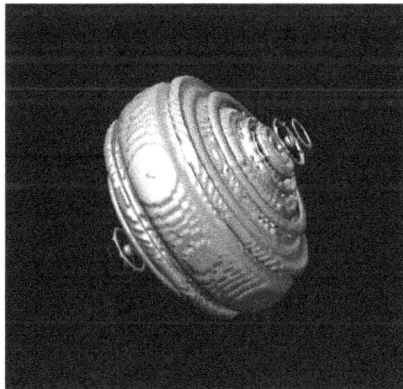

Figure 6. Reconstructed diffraction volume from two particles per shot (randomly oriented).

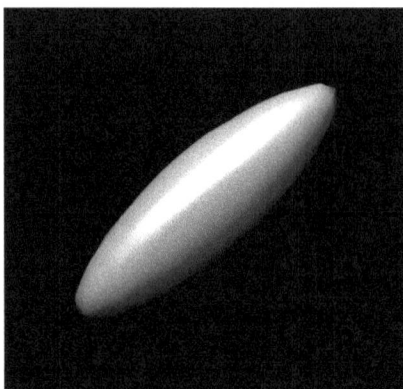

Figure 7. Image reconstructed from one particle per diffraction patten.

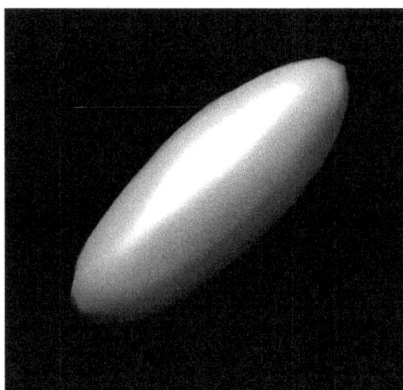

Figure 8. Image reconstructed from two particles per diffraction patten.

5. Angular Dependence of Scattering

We also investigated the effects of the angular dependence of the scattering. Our nanorice particles are about are about 50 Å in diameter. They are constituted of the chemical molecules Fe_2O_3 and thus on average are much stronger scatterers than a typical bioparticle. In fact, we estimate $\log |F(q)|^2$ to be about 10^8 at low resolutions, much higher than for a bioparticle. Of course, we see the precipitous fall off with angle, and Fung et al. estimated the number of high-resolution photons per Shannon pixel to be about 10^{-2} for a 100 kD protein. Since the nanoparticle contains many more electrons, we expect the scattering power of our nanoparticles to not need Poisson statistics. Consequently, we have plotted only $\log |F(q)|^2$, which will give the angular dependence of the measured signal (Figure 9).

6. Conclusions

It should be noted that the images computed from the diffraction patterns of single particles do not agree perfectly with those from two particles. This to be expected, as there will be extra interparticle interfences in the two-particle case. It is only if the number of particles contributing to each diffraction pattern becomes large that the arguments made by [27], for suggesting that the correlations from multiple particles, approach those of a single particle, due to the fact that the phases are random, allowing interparticle inteferences to be ignored.

The method of angular correlations [3,11] is shown to be very capable of recovering accurate 3D images of any particle of azimuthal symmetry as injected into an X-ray free electron laser (XFEL), even if there are multiple particles in random orientations. We tested this by means of a simulations in the case of nanorice, one of the early particles used as a test case with an XFEL. To this end, we simulated diffraction patterns expected of random orientations of the nanorice. From this data, one is able to find quantities that enable the reconstruction of an accurate image of nanorice particles. Note that this method is also applicable for reconstruction of artificial dumbbells, as demonstrated recently in a paper by Starodub et al. [24].

It should be noted that $m = 0$ is also assumed for a helical virus. The difference with that case is that the diffraction volume tends to be concentrated on "layer planes" implying a periodicity of the particle structure along the azimuthal axis. It is remarkable that the combination of "layer plane" intensities and azimuthal symmetry yield a helical structure in real space on phasing [21]. In the present case, the intensities are not concentrated on "layer planes" and yield a structure that is genuinely azimuthally symmetric, with no hint of helicity. Both are indications of the power of modern iterative phasing algorithms to find the correct structure from diffraction volumes.

Our next step will be to apply this method to nanorice experimental data from an XFEL, as deposited on [28], for example.

A problem with experimental data is that a huge number of diffraction patterns are measured, only a subset of which ("good" patterns) correspond to actual particle hits. In addition to all other methods that have been proposed for structure determination from XFEL diffraction patterns, one needs to filter out diffraction patterns corresponding to multiple particle hits. This is not necessary with our method. Only an elimination of patterns corresponding to no particle hits at all. Due to the huge number of diffraction patterns measured in an experiments, this needs to be done by computer. Fortunately, there has already been at least one machine learning algorithm [9] applied to discriminate between measured [9] "good" and "bad" patterns. We are collaborating with the first author of that paper [9] with an aim to apply our method to experimental data.

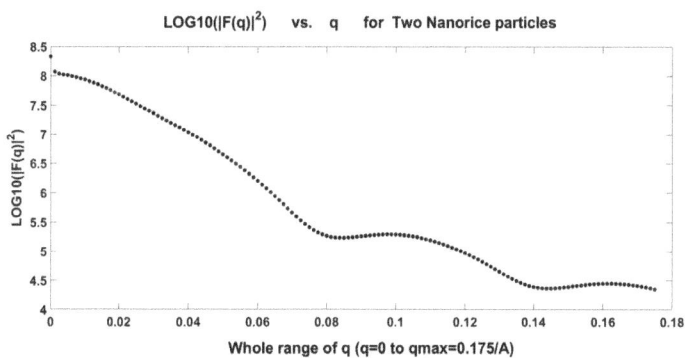

Figure 9. Logarithm of the intensities reconstructed from the angular correlations of two randomly spaced and oriented nanorice particles with widths of about 50 Å.

Acknowledgments: We acknowledge a grant from the U.S. National Science Foundation to fund the bioXFEL Science and Technology Center under Grant No. STC 1231306.

Author Contributions: Sung Soon Kim developed the 2 nanorice model, applied correlation method to this model with azimuthal symmetry, generated all figures in the manuscript though rekted caculateds and corrected minor mistakes in the manuscript. Sandi Wibowo conceived of the original theory of calculating nanorice from the angular correlations and developed the theory thereof. Diano Kerzaman Saldin was the supervisor of the project and wrote the first draft of the manuscript.

Conflicts of Interest: The authors declare no conflict of interest.

References

1. Neutze, R.; Wouts, R.; van de Spoel, D.; Weckert, E.; Hadju, J. Potential for bimolecular imaging with femtosecond X-ray pulses. *Nature* **2000**, *406*, 752–757.
2. Loh, N.D.; Bogan, M.J.; Elser, V.; Barty, A.; Boutet, S.; Bajt, S.; Hajdu, J.; Ekeberg, T.; Maia, F.R.N.C.; Schulz, J.; et al. Cryptotomography: Reconstructing 3D Fourier intensities from randomly oriented single shot diffraction patterns. *Phys. Rev. Lett.* **2010**, *104*, 239902.
3. Kam, Z. Determination of macromolecule structure in solution by spatial correlation of scattering fluctuations. *Macromolecules* **1977**, *10*, 927–934.
4. Mendez, D.; Lane, T.J.; Sung, J.; Seilberg, J.; Levard, C.; Watkins, H.; Cohen, A.E.; Soltis, M.; Sutton, S.; Spudich, J.; et al. Observation of correlated X-ray scattering at atomic resolution. *Philos. Trans. R. Soc. B* **2014**, *369*, 20130315.
5. Kirian, R.; Scmhidt, K.E.; Wang, X.; Doak, R.B.; Spence, J.C.H. Signal, noise, ans resolution in correlated fluctuations from snapshot small angle X-ray scattering. *Phys. Rev. E* **2011**, *84*, 011921.
6. Svenda, T.E.M.; Abergel, C.; Maia, F.R.N.C.; Selzer, V.; Claverie, J.-M.; Hantke, M.; Jönsson, O.; Nettelblad, C.; van de Schot, G.; Liang, M.; et al. Three-dimensional reconstruction of the giant mimivirus particle with an X-ray free-electron laser. *Phys. Rev. Lett.* **2015**, *114*, 098102.
7. Saldin, D.K.; Poon, H.C.; Schwander, P.; Uddin, M.; Schmidt, M. Reconstruction of an icosahedral virus from single particle diffraction experiments. *Opt. Express* **2011**, *19*, 17318–17335.
8. Pande, K.; Schmidt, M.; Schwander, P.; Saldin, D.K. Simulations on time-resolved structure determination of uncrystallized biomolecules in the presence of shot noise. *Struct. Dyn.* **2015**, *2*, 024104.
9. Yoon, C.H.; Schwander, P.; Abergel, C.; Andersson, I.; Andreasson, J.; Andrea, A.; Bogan, M.J.; Bajt, S.; Barthelmess, M.; Bart, A.; et al. Unsupervised classification of single particle X-ray diffraction snapshots by spectral clustering. *Opt. Express* **2011**, *19*, 16542–16549.
10. CHEETAH Software. Available online: www.desy.de/~barty/cheetah/Cheetah/Welcome.html (accessed on 21 July 2017).
11. Saldin, D.K.; Shneerson, V.L.; Fung, R.; Ourmazd, A. Structure of isolated biomolecules from ultra-short X-ray pulses: Exploiting the symmetry of random orientations. *J. Phys. Condens. Matter* **2009**, *21*, 134014.
12. Fung, R.; Shneerson, V.L.; Saldin, D.K.; Ourmazd, A. Structure from fleeting illumination of faint spinning objects in flight. *Nat. Phys.* **2009**, *5*, 64–67.
13. Tenboer, J.; Basu, S.; Zatsepin, N.; Pande, K.; Milathlianaki, D.; Frank, M.; Hunter, M.; Boutet, S.; Williams, G.J.; Koglin, J.E.; et al. Time-resolved serial crystallography captures high-resolution intermediates of photoactive yellow protein. *Science* **2014**, *346*, 1242–1246.
14. Kam, Z. The reconstruction of structure from electron micrographs of randomly oriented particles. *J. Theory Biol.* **1980**, *82*, 15–39.
15. Pedrini, B.; Menzel, A.; Guizar-Siciaros, M.; Guzenko, V.A.; Goreilik, S.; David, C.; Petterson, B.D.; Abela, R. Two-dimensional structure from random multiparticle X-ray scattering images using cross-correlations. *Nat. Commun.* **2013**, *4*, 1647.
16. Kurta, R.; Dronyak, R.; Altarelli, M.; Weckert, E.; Vartanyants, I.A. Solution of the phase problem for coherent scattering from a disordered system of identical particles. *New J. Phys.* **2013**, *15*, 013049.
17. Poon, H.C.; Saldin, D.K. Beyond the crystallization paradigm: Structure determination from diffraction patterns from ensembles of randomly oriented particles. *Ultramicroscopy* **2011**, *111*, 798–806.
18. Pendry, J.B. *Low Energy Electron Diffraction*; Academic Press: London, UK, 1974.
19. Messiah, A. *Quamum Mechanoics*; Wiley: New York, NY, USA, 1981.
20. Caspar, D.L.D.; Klug, A. Physical principles of the construction of regular viruses. In *Cold Spring Harbor Symposia on Quantitative Biology*; Cold Spring Harbor Laboratory Press: New York, NY, USA, 1962; Volume 27, pp. 1–24.
21. Poon, H.C.; Schwander, P.; Uddin, M.; Saldin, D.K. Fiber diffraction without fibers. *Phys. Rev. Lett.* **2013**, *110*, 265505.
22. Pande, K.; Schwander, P.; Schmidt, M.; Saldin, D.K. Deducing fast electron density changes in randomly oriented uncrstallized biomolecules in a pump-probe experiment. *Philos. Trans. R. Soc. B* **2014**, *2013*, 2013332.
23. Donatelli, J.J.; Zwart, P.H.; Sethian, J.A. Iterative phasing for fluctuation X-ray scattering. *Proc. Natl. Acad. Sci. USA* **2015**, *112*, 10286–10291.

24. Starodub, D.; Aquila, A.; Bajt, S.; Barthelmess, M.; Barty, A.; Bostedt, C.; Bozek, J.D.; Coppola, N.; Doak, R.B.; Epp, S.W.; et al. Single-particle structure determination by correlations of snapshot X-ray diffraction patterns. *Nat. Commun.* **2012**, *3*, 1276.

25. Oszlányi, G.; Süto, A. *Ab initio* structure solution by charge flipping. *Acta Crystallogr. A* **2004**, *60*, 134–141.

26. Oszlányi, G.; Süto, A. *Ab initio* structure by charge flipping. II. Use of weak reflections. *Acta Crystallogr. A* **2005**, *61*, 147–152.

27. Saldin, D.K.; Poon, H.C.; Shneerson, V.L.; Howells, M.; Chapman, H.N.; Kirian, R.; Schmidt, K.E.; Spence, J.C.H. Beyond small-angle X-ray scattering: Exploiting angular correlations. *Phys. Rev. B* **2010**, *81*, 174105.

28. Coherent X-ray Imaging Date Bank. Available online: http://www.cxidb.org/ (accessed on 21 July 2017).

applied sciences

MDPI

Meeting Report

Nobel Symposium on Free Electron Laser Research

Mats Larsson

Department of Physics, AlbaNova University Center, Stockholm University, Stockholm SE-106 91, Sweden; mats.larsson@fysik.su.se; Tel.: +46-8-5537-8647

Academic Editor: Kiyoshi Ueda
Received: 5 April 2017; Accepted: 17 April 2017; Published: 18 April 2017

Abstract: This meeting report describes the Nobel Symposium on Free Electron Laser Research, which was organized in Sigtuna, Sweden, 14–18 June 2015.

Keywords: free electron laser; X-rays; Nobel Symposia

1. Introduction

The Nobel Foundation initiated Nobel Symposia in 1965. According to the Nobel Symposia Instructions, "The symposium should be devoted to scientific or scholarly disciplines related to the Nobel Prize areas." Before a symposium is approved, the Nobel Committees are consulted. Two characteristics of a Nobel Symposium is a limited number of participants (30–40) and a chance for a limited number of younger researchers to follow the symposium as observers.

The local symposium committee, in the present case consisting of Mats Larsson, Ingolf Lindau, and Joseph Nordgren, all members of the Royal Swedish Academy of Sciences, is requested to have an international committee to assist in composing the scientific programme. The international committee consisted of Massimo Altarelli (European XFEL), Nora Berrah (UConn), John Galayda (SLAC), and Jon Marangos (Imperial College London).

The Nobel Symposium took place at Sigtuna outside Stockholm between 14 and 18 June 2015.

2. Aim and Scope and Presentations

The programme was divided into seven areas: atomic and molecular optical physics (AMO), life science, matter at extreme conditions, time-resolved phenomena, condensed matter physics, nonlinear processes, and accelerator physics. Each speaker was allotted an hour including questions, and the discussions were lively!

The main focus was of course on the impressive scientific development of X-ray free electron lasers (FELs), but there were also a few talks devoted to long wavelength FELs. The full programme including links to the talks as PDF files is available here: http://agenda.albanova.se/conferenceDisplay.py?confId=4905.

3. Conclusions

The invited speakers provide a very good overview of the cutting edge research on free electron lasers, related theory work, and both their history and their future.

As a curiosity, it can be mentioned that the conference photo displayed on the symposium webpage was initiated by the guest editor of this volume, Kiyoshi Ueda, who was also invited to speak at the symposium.

For several participants, this was the first time they had the opportunity to meet and hear the inventor of the free electron laser, John Madey, who delivered the keynote address Sadly, this was also the last time most of the participants met him; John Madey passed away just over a year after the symposium, on 5 July 2016.

Appl. Sci. **2017**, *7*, 408

Acknowledgments: The Nobel Symposium on Free Electron Laser Research was Number 158 of all Symposia since the start in 1965. It was financially supported by the Nobel Foundation and the Knut and Alice Wallenberg Foundation. The FEL consortium between Stockholm University, Uppsala University, and the Royal Institute of Technology supported the participation of about 30 younger researchers, and observers from Belarus and Ukraine.

Conflicts of Interest: The author declares no conflict of interest.

MDPI

St. Alban-Anlage 66

4052 Basel

Switzerland

Tel. +41 61 683 77 34

Fax +41 61 302 89 18

www.mdpi.com

Applied Sciences Editorial Office

E-mail: appliedsciences@mdpi.com

www.mdpi.com/journal/appliedsciences

www.ingramcontent.com/pod-product-compliance
Lightning Source LLC
Chambersburg PA
CBHW051703210326
41597CB00032B/5357